Lecture Notes in Mathematics 2078

Editors:
J.-M. Morel, Cachan
B. Teissier, Paris

For further volumes:
http://www.springer.com/series/304

Catherine Donati-Martin • Antoine Lejay
Alain Rouault
Editors

Séminaire de Probabilités XLV

Editors
Catherine Donati-Martin
Université de Versailles-St Quentin
Versailles, France

Antoine Lejay
Nancy-Université, INRIA
Vandoeuvre-lès-Nancy, France

Alain Rouault
Université de Versailles-St Quentin
Versailles, France

ISBN 978-3-319-00320-7   ISBN 978-3-319-00321-4 (eBook)
DOI 10.1007/978-3-319-00321-4
Springer Cham Heidelberg New York Dordrecht London

Lecture Notes in Mathematics ISSN print edition: 0075-8434
                              ISSN electronic edition: 1617-9692

Library of Congress Control Number: 2013941134

Mathematics Subject Classification (2010): 60-XX, 60JXX, 60J60, 60J10, 60J65, 60J55, 46L54

© Springer International Publishing Switzerland 2013
This work is subject to copyright. All rights are reserved by the Publisher, whether the whole or part of the material is concerned, specifically the rights of translation, reprinting, reuse of illustrations, recitation, broadcasting, reproduction on microfilms or in any other physical way, and transmission or information storage and retrieval, electronic adaptation, computer software, or by similar or dissimilar methodology now known or hereafter developed. Exempted from this legal reservation are brief excerpts in connection with reviews or scholarly analysis or material supplied specifically for the purpose of being entered and executed on a computer system, for exclusive use by the purchaser of the work. Duplication of this publication or parts thereof is permitted only under the provisions of the Copyright Law of the Publisher's location, in its current version, and permission for use must always be obtained from Springer. Permissions for use may be obtained through RightsLink at the Copyright Clearance Center. Violations are liable to prosecution under the respective Copyright Law.
The use of general descriptive names, registered names, trademarks, service marks, etc. in this publication does not imply, even in the absence of a specific statement, that such names are exempt from the relevant protective laws and regulations and therefore free for general use.
While the advice and information in this book are believed to be true and accurate at the date of publication, neither the authors nor the editors nor the publisher can accept any legal responsibility for any errors or omissions that may be made. The publisher makes no warranty, express or implied, with respect to the material contained herein.

Printed on acid-free paper

Springer is part of Springer Science+Business Media (www.springer.com)

Preface

The series of advanced courses, initiated in Séminaire de Probabilités XXXIII, continues with a course of Ivan Nourdin on Gaussian approximations by Malliavin calculus. The *Séminaire* also occasionally publishes a series of contributions on some given theme; in this spirit, some participants from September 2011 Conference on Stochastic Filtrations, held in Strasbourg and organized by Michel Émery, have contributed to this volume. The rest of the volume covers a wide range of themes, such as stochastic calculus and Markov processes, random matrices and free probability, and combinatorial optimization. These contributions come from the spontaneous submissions or were solicited by the editors.

We remind that the web site of the Séminaire is

http://portail.mathdoc.fr/SemProba/

and that all the articles of the Séminaire from Volume I in 1967 to Volume XXXVI in 2002 are freely accessible from the web site

http://www.numdam.org/numdam-bin/feuilleter?j=SPS

We thank the Cellule Math Doc for hosting all these articles within the NUMDAM project.

Versailles, France	C. Donati-Martin
Vandoeuvre-lès-Nancy, France	A. Lejay
Versailles, France	A. Rouault

Contents

Part I Specialized Course

Lectures on Gaussian Approximations with Malliavin Calculus 3
Ivan Nourdin

Part II Other Contributions

Some Sufficient Conditions for the Ergodicity of the Lévy Transformation 93
Vilmos Prokaj

Vershik's Intermediate Level Standardness Criterion and the Scale of an Automorphism 123
Stéphane Laurent

Filtrations Indexed by Ordinals; Application to a Conjecture of S. Laurent 141
Claude Dellacherie and Michel Émery

A Planar Borel Set Which Divides Every Non-negligible Borel Product 159
Michel Émery

Characterising Ocone Local Martingales with Reflections 167
Jean Brossard and Christophe Leuridan

Approximation and Stability of Solutions of SDEs Driven by a Symmetric α Stable Process with Non-Lipschitz Coefficients 181
Hiroya Hashimoto

Path Properties and Regularity of Affine Processes on General State Spaces 201
Christa Cuchiero and Josef Teichmann

Langevin Process Reflected on a Partially Elastic Boundary II............ 245
Emmanuel Jacob

Windings of Planar Stable Processes... 277
R.A. Doney and S. Vakeroudis

**An Elementary Proof that the First Hitting Time of an Open
Set by a Jump Process is a Stopping Time** 301
Alexander Sokol

**Catalytic Branching Processes via Spine Techniques
and Renewal Theory** 305
Leif Döring and Matthew I. Roberts

Malliavin Calculus and Self Normalized Sums 323
Solesne Bourguin and Ciprian A. Tudor

A Note on Stochastic Calculus in Vector Bundles 353
Pedro J. Catuogno, Diego S. Ledesma, and Paulo R. Ruffino

**Functional Co-monotony of Processes with Applications
to Peacocks and Barrier Options** 365
Gilles Pagès

Fluctuations of the Traces of Complex-Valued Random Matrices 401
Salim Noreddine

Functionals of the Brownian Bridge 433
Janosch Ortmann

Étude spectrale minutieuse de processus moins indécis que les autres..... 459
Laurent Miclo et Pierre Monmarché

Combinatorial Optimization Over Two Random Point Sets 483
Franck Barthe and Charles Bordenave

**A Simple Proof of Duquesne's Theorem on Contour Processes
of Conditioned Galton–Watson Trees**.. 537
Igor Kortchemski

Part I
Specialized Course

Lectures on Gaussian Approximations with Malliavin Calculus

Ivan Nourdin

Overview. In a seminal paper of 2005, Nualart and Peccati [40] discovered a surprising central limit theorem (called the "Fourth Moment Theorem" in the sequel) for sequences of multiple stochastic integrals of a fixed order: in this context, convergence in distribution to the standard normal law is equivalent to convergence of just the fourth moment. Shortly afterwards, Peccati and Tudor [46] gave a multidimensional version of this characterization.

Since the publication of these two beautiful papers, many improvements and developments on this theme have been considered. Among them is the work by Nualart and Ortiz-Latorre [39], giving a new proof only based on Malliavin calculus and the use of integration by parts on Wiener space. A second step is my joint paper [27] (written in collaboration with Peccati) in which, by bringing together Stein's method with Malliavin calculus, we were able (among other things) to associate quantitative bounds to the Fourth Moment Theorem. It turns out that Stein's method and Malliavin calculus fit together admirably well. Their interaction has led to some remarkable new results involving central and non-central limit theorems for functionals of infinite-dimensional Gaussian fields.

The current survey aims to introduce the main features of this recent theory. It originates from a series of lectures I delivered[1] at the Collège de France between January and March 2012, within the framework of the annual prize of the Fondation des Sciences Mathématiques de Paris. It may be seen as a teaser for the book [32], in which the interested reader will find much more than in this short survey.

---

[1] You may watch the videos of the lectures at http://www.sciencesmaths-paris.fr/index.php?page=175.

I. Nourdin (✉)
Université de Lorraine, Institut de Mathématiques Élie Cartan, B.P. 70239, 54506 Vandoeuvre-lès-Nancy Cedex, France
e-mail: inourdin@gmail.com

1 Breuer–Major Theorem

The aim of this first section is to illustrate, through a guiding example, the power of the approach we will develop in this survey.

Let $\{X_k\}_{k \geq 1}$ be a centered stationary Gaussian family. In this context, stationary just means that there exists $\rho : \mathbb{Z} \to \mathbb{R}$ such that $E[X_k X_l] = \rho(k-l)$, $k, l \geq 1$. Assume further that $\rho(0) = 1$, that is, each X_k is $\mathcal{N}(0, 1)$ distributed.

Let $\varphi : \mathbb{R} \to \mathbb{R}$ be a measurable function satisfying

$$E[\varphi^2(X_1)] = \frac{1}{\sqrt{2\pi}} \int_{\mathbb{R}} \varphi^2(x) e^{-x^2/2} dx < \infty. \tag{1}$$

Let $H_0, H_1, \ldots$ denote the sequence of Hermite polynomials. The first few Hermite polynomials are $H_0 = 1$, $H_1 = X$, $H_2 = X^2 - 1$ and $H_3 = X^3 - 3X$. More generally, the qth Hermite polynomial H_q is defined through the relation $XH_q = H_{q+1} + qH_{q-1}$. It is a well-known fact that, when it verifies (1), the function φ may be expanded in $L^2(\mathbb{R}, e^{-x^2/2} dx)$ (in a unique way) in terms of Hermite polynomials as follows:

$$\varphi(x) = \sum_{q=0}^{\infty} a_q H_q(x). \tag{2}$$

Let $d \geq 0$ be the first integer $q \geq 0$ such that $a_q \neq 0$ in (2). It is called the *Hermite rank* of φ; it will play a key role in our study. Also, let us mention the following crucial property of Hermite polynomials with respect to Gaussian elements. For any integer $p, q \geq 0$ and any jointly Gaussian random variables $U, V \sim \mathcal{N}(0, 1)$, we have

$$E[H_p(U) H_q(V)] = \begin{cases} 0 & \text{if } p \neq q \\ q! E[UV]^q & \text{if } p = q. \end{cases} \tag{3}$$

In particular (choosing $p = 0$) we have that $E[H_q(X_1)] = 0$ for all $q \geq 1$, meaning that $a_0 = E[\varphi(X_1)]$ in (2). Also, combining the decomposition (2) with (3), it is straightforward to check that

$$E[\varphi^2(X_1)] = \sum_{q=0}^{\infty} q! a_q^2. \tag{4}$$

We are now in position to state the celebrated Breuer–Major theorem.

Theorem 1 (Breuer and Major (1983); see [7]). *Let $\{X_k\}_{k \geq 1}$ and $\varphi : \mathbb{R} \to \mathbb{R}$ be as above. Assume further that $a_0 = E[\varphi(X_1)] = 0$ and that $\sum_{k \in \mathbb{Z}} |\rho(k)|^d < \infty$,* where ρ is the covariance function of $\{X_k\}_{k\geqslant 1}$ and d is the Hermite rank of φ (observe that $d \geqslant 1$). Then, as $n \to \infty$,

$$V_n = \frac{1}{\sqrt{n}} \sum_{k=1}^{n} \varphi(X_k) \overset{\text{law}}{\to} \mathcal{N}(0, \sigma^2), \tag{5}$$

with σ^2 given by

$$\sigma^2 = \sum_{q=d}^{\infty} q! a_q^2 \sum_{k \in \mathbb{Z}} \rho(k)^q \in [0, \infty). \tag{6}$$

(The fact that $\sigma^2 \in [0, \infty)$ is part of the conclusion.)

The proof of Theorem 1 is far from being obvious. The original proof consisted to show that *all* the moments of V_n converge to those of the Gaussian law $\mathcal{N}(0, \sigma^2)$. As anyone might guess, this required a high ability and a lot of combinatorics. In the proof we will offer, the complexity is the same as checking that the variance and the fourth moment of V_n converges to σ^2 and $3\sigma^4$ respectively, which is a drastic simplification with respect to the original proof. Before doing so, let us make some other comments.

Remark 1. 1. First, it is worthwhile noticing that Theorem 1 (strictly) contains the classical central limit theorem (CLT), which is not an evident claim at first glance. Indeed, let $\{Y_k\}_{k\geqslant 1}$ be a sequence of i.i.d. centered random variables with common variance $\sigma^2 > 0$, and let F_Y denote the common cumulative distribution function. Consider the pseudo-inverse F_Y^{-1} of F_Y, defined as

$$F_Y^{-1}(u) = \inf\{y \in \mathbb{R} : u \leqslant F_Y(y)\}, \quad u \in (0, 1).$$

When $U \sim \mathcal{U}_{[0,1]}$ is uniformly distributed, it is well-known that $F_Y^{-1}(U) \overset{\text{law}}{=} Y_1$. Observe also that $\frac{1}{\sqrt{2\pi}} \int_{-\infty}^{X_1} e^{-t^2/2} dt$ is $\mathcal{U}_{[0,1]}$ distributed. By combining these two facts, we get that $\varphi(X_1) \overset{\text{law}}{=} Y_1$ with

$$\varphi(x) = F_Y^{-1}\left(\frac{1}{\sqrt{2\pi}} \int_{-\infty}^{x} e^{-t^2/2} dt\right), \quad x \in \mathbb{R}.$$

Assume now that $\rho(0) = 1$ and $\rho(k) = 0$ for $k \neq 0$, that is, assume that the sequence $\{X_k\}_{k\geqslant 1}$ is composed of i.i.d. $\mathcal{N}(0, 1)$ random variables. Theorem 1 yields that

$$\frac{1}{\sqrt{n}} \sum_{k=1}^{n} Y_k \overset{\text{law}}{=} \frac{1}{\sqrt{n}} \sum_{k=1}^{n} \varphi(X_k) \overset{\text{law}}{\to} \mathcal{N}\left(0, \sum_{q=d}^{\infty} q! a_q^2\right),$$

thereby concluding the proof of the CLT since $\sigma^2 = E[\varphi^2(X_1)] = \sum_{q=d}^{\infty} q! a_q^2$, see (4). Of course, such a proof of the CLT is like to crack a walnut with a sledgehammer. This approach has nevertheless its merits: it shows that the independence assumption in the CLT is not crucial to allow a Gaussian limit. Indeed, this is rather the summability of a series which is responsible of this fact, see also the second point of this remark.

2. Assume that $d \geq 2$ and that $\rho(k) \sim |k|^{-D}$ as $|k| \to \infty$ for some $D \in (0, \frac{1}{d})$. In this case, it may be shown that $n^{dD/2-1} \sum_{k=1}^{n} \varphi(X_k)$ converges in law to a <u>non-Gaussian</u> (non degenerated) random variable. This shows in particular that, in the case where $\sum_{k \in \mathbb{Z}} |\rho(k)|^d = \infty$, we can get a non-Gaussian limit. In other words, the summability assumption in Theorem 1 is, roughly speaking, equivalent (when $d \geq 2$) to the asymptotic normality.

3. There exists a functional version of Theorem 1, in which the sum $\sum_{k=1}^{n}$ is replaced by $\sum_{k=1}^{[nt]}$ for $t \geq 0$. It is actually not that much harder to prove and, unsurprisingly, the limiting process is then the standard Brownian motion multiplied by σ.

Let us now prove Theorem 1. We first compute the limiting variance, which will justify the formula (6) we claim for σ^2. Thanks to (2) and (3), we can write

$$E[V_n^2] = \frac{1}{n} E\left[\left(\sum_{q=d}^{\infty} a_q \sum_{k=1}^{n} H_q(X_k)\right)^2\right] = \frac{1}{n} \sum_{p,q=d}^{\infty} a_p a_q \sum_{k,l=1}^{n} E[H_p(X_k) H_q(X_l)]$$

$$= \frac{1}{n} \sum_{q=d}^{\infty} q! a_q^2 \sum_{k,l=1}^{n} \rho(k-l)^q = \sum_{q=d}^{\infty} q! a_q^2 \sum_{r \in \mathbb{Z}} \rho(r)^q \left(1 - \frac{|r|}{n}\right) \mathbf{1}_{\{|r|<n\}}.$$

When $q \geq d$ and $r \in \mathbb{Z}$ are fixed, we have that

$$q! a_q^2 \rho(r)^q \left(1 - \frac{|r|}{n}\right) \mathbf{1}_{\{|r|<n\}} \to q! a_q^2 \rho(r)^q \quad \text{as } n \to \infty.$$

On the other hand, using that $|\rho(k)| = |E[X_1 X_{k+1}]| \leq \sqrt{E[X_1^2] E[X_{1+k}^2]} = 1$, we have

$$q! a_q^2 |\rho(r)|^q \left(1 - \frac{|r|}{n}\right) \mathbf{1}_{\{|r|<n\}} \leq q! a_q^2 |\rho(r)|^q \leq q! a_q^2 |\rho(r)|^d,$$

with $\sum_{q=d}^{\infty} \sum_{r \in \mathbb{Z}} q! a_q^2 |\rho(r)|^d = E[\varphi^2(X_1)] \times \sum_{r \in \mathbb{Z}} |\rho(r)|^d < \infty$, see (4). By applying the dominated convergence theorem, we deduce that $E[V_n^2] \to \sigma^2$ as $n \to \infty$, with $\sigma^2 \in [0, \infty)$ given by (6).

Let us next concentrate on the proof of (5). We shall do it in three steps of increasing generality (but of decreasing complexity!):

(i) When $\varphi = H_q$ has the form of a Hermite polynomial (for some $q \geq 1$).
(ii) When $\varphi = P \in \mathbb{R}[X]$ is a real polynomial.
(iii) In the general case when $\varphi \in L^2(\mathbb{R}, e^{-x^2/2}dx)$.

We first show that (ii) implies (iii). That is, let us assume that Theorem 1 is shown for polynomial functions φ, and let us show that it holds true for any function $\varphi \in L^2(\mathbb{R}, e^{-x^2/2}dx)$. We proceed by approximation. Let $N \geq 1$ be a (large) integer (to be chosen later) and write

$$V_n = \frac{1}{\sqrt{n}} \sum_{q=d}^{N} a_q \sum_{k=1}^{n} H_q(X_k) + \frac{1}{\sqrt{n}} \sum_{q=N+1}^{\infty} a_q \sum_{k=1}^{n} H_q(X_k) =: V_{n,N} + R_{n,N}.$$

Similar computations as above lead to

$$\sup_{n \geq 1} E[R_{n,N}^2] \leq \sum_{q=N+1}^{\infty} q! a_q^2 \times \sum_{r \in \mathbb{Z}} |\rho(r)|^d \to 0 \text{ as } N \to \infty. \tag{7}$$

(Recall from (4) that $E[\varphi^2(X_1)] = \sum_{q=d}^{\infty} q! a_q^2 < \infty$.) On the other hand, using (ii) we have that, for fixed N and as $n \to \infty$,

$$V_{n,N} \xrightarrow{\text{law}} \mathcal{N}\left(0, \sum_{q=d}^{N} q! a_q^2 \sum_{k \in \mathbb{Z}} \rho(k)^q\right). \tag{8}$$

It is then a routine exercise (details are left to the reader) to deduce from (7)–(8) that $V_n = V_{n,N} + R_{n,N} \xrightarrow{\text{law}} \mathcal{N}(0, \sigma^2)$ as $n \to \infty$, that is, that (iii) holds true.

Next, let us prove (i), that is, (5) when $\varphi = H_q$ is the qth Hermite polynomial. We actually need to work with a specific realization of the sequence $\{X_k\}_{k \geq 1}$. The space

$$\mathscr{H} := \overline{\text{span}\{X_1, X_2, \ldots\}}^{L^2(\Omega)}$$

being a real separable Hilbert space, it is isometrically isomorphic to either $\mathbb{R}^N$ (with $N \geq 1$) or $L^2(\mathbb{R}_+)$. Let us assume that $\mathscr{H} \simeq L^2(\mathbb{R}_+)$, the case where $\mathscr{H} \simeq \mathbb{R}^N$ being easier to handle. Let $\Phi : \mathscr{H} \to L^2(\mathbb{R}_+)$ be an isometry. Set $e_k = \Phi(X_k)$ for each $k \geq 1$. We have

$$\rho(k-l) = E[X_k X_l] = \int_0^{\infty} e_k(x) e_l(x) dx, \quad k, l \geq 1 \tag{9}$$

If $B = (B_t)_{t \geq 0}$ denotes a standard Brownian motion, we deduce that

$$\{X_k\}_{k \geq 1} \stackrel{\text{law}}{=} \left\{ \int_0^\infty e_k(t) dB_t \right\}_{k \geq 1},$$

these two families being indeed centered, Gaussian and having the same covariance structure (by construction of the e_k's). On the other hand, it is a well-known result of stochastic analysis (which follows from an induction argument through the Itô formula) that, for any function $e \in L^2(\mathbb{R}_+)$ such that $\|e\|_{L^2(\mathbb{R}_+)} = 1$, we have

$$H_q\left(\int_0^\infty e(t) dB_t\right) = q! \int_0^\infty dB_{t_1} e(t_1) \int_0^{t_1} dB_{t_2} e(t_2) \ldots \int_0^{t_{q-1}} dB_{t_q} e(t_q). \quad (10)$$

(For instance, by Itô's formula we can write

$$\left(\int_0^\infty e(t) dB_t\right)^2 = 2 \int_0^\infty dB_{t_1} e(t_1) \int_0^{t_1} dB_{t_2} e(t_2) + \int_0^\infty e(t)^2 dt$$

$$= 2 \int_0^\infty dB_{t_1} e(t_1) \int_0^{t_1} dB_{t_2} e(t_2) + 1,$$

which is nothing but (10) for $q = 2$, since $H_2 = X^2 - 1$.) At this stage, let us adopt the two following notational conventions:

(a) If φ (resp. ψ) is a function of r (resp. s) arguments, then the tensor product $\varphi \otimes \psi$ is the function of $r + s$ arguments given by $\varphi \otimes \psi(x_1, \ldots, x_{r+s}) = \varphi(x_1, \ldots, x_r)\psi(x_{r+1}, \ldots, x_{r+s})$. Also, if $q \geq 1$ is an integer and e is a function, the tensor product function $e^{\otimes q}$ is the function $e \otimes \ldots \otimes e$ where e appears q times.

(b) If $f \in L^2(\mathbb{R}_+^q)$ is symmetric (meaning that $f(x_1, \ldots, x_q) = f(x_{\sigma(1)}, \ldots, x_{\sigma(q)})$ for all permutation $\sigma \in \mathfrak{S}_q$ and almost all $x_1, \ldots, x_q \in \mathbb{R}_+$) then

$$I_q^B(f) = \int_{\mathbb{R}_+^q} f(t_1, \ldots, t_q) dB_{t_1} \ldots dB_{t_q}$$

$$:= q! \int_0^\infty dB_{t_1} \int_0^{t_1} dB_{t_2} \ldots \int_0^{t_{q-1}} dB_{t_q} f(t_1, \ldots, t_q).$$

With these new notations at hand, observe that we can rephrase (10) in a simple way as

$$H_q\left(\int_0^\infty e(t) dB_t\right) = I_q^B(e^{\otimes q}). \quad (11)$$

It is now time to introduce a very powerful tool, the so-called *Fourth Moment Theorem* of Nualart and Peccati. This wonderful result lies at the heart of the approach we shall develop in these lecture notes. We will prove it in Sect. 5.

Theorem 2 (Nualart and Peccati (2005); see [40]). *Fix an integer $q \geq 2$, and let $\{f_n\}_{n \geq 1}$ be a sequence of symmetric functions of $L^2(\mathbb{R}_+^q)$. Assume that $E[I_q^B(f_n)^2] = q! \|f_n\|_{L^2(\mathbb{R}_+^q)}^2 \to \sigma^2$ as $n \to \infty$ for some $\sigma > 0$. Then, the following three assertions are equivalent as $n \to \infty$:*

(1) $I_q^B(f_n) \overset{\text{law}}{\to} \mathcal{N}(0, \sigma^2)$;

(2) $E[I_q^B(f_n)^4] \overset{\text{law}}{\to} 3\sigma^4$;

(3) $\|f_n \otimes_r f_n\|_{L^2(\mathbb{R}_+^{2q-2r})} \to 0$ *for each* $r = 1, \ldots, q-1$, *where $f_n \otimes_r f_n$ is the function of $L^2(\mathbb{R}_+^{2q-2r})$ defined by*

$$f_n \otimes_r f_n(x_1, \ldots, x_{2q-2r})$$
$$= \int_{\mathbb{R}_+^r} f_n(x_1, \ldots, x_{q-r}, y_1, \ldots, y_r) f_n(x_{q-r+1}, \ldots, x_{2q-2r}, y_1, \ldots, y_r) dy_1 \ldots dy_r.$$

Remark 2. In other words, Theorem 2 states that the convergence in law of a normalized sequence of multiple Wiener–Itô integrals $I_q^B(f_n)$ towards the Gaussian law $\mathcal{N}(0, \sigma^2)$ is equivalent to convergence of just the fourth moment to $3\sigma^4$. This surprising result has been the starting point of a new line of research, and has quickly led to several applications, extensions and improvements. One of these improvements is the following quantitative bound associated to Theorem 2 that we shall prove in Sect. 5 by combining Stein's method with the Malliavin calculus.

Theorem 3 (Nourdin and Peccati (2009); see [27]). *If $q \geq 2$ is an integer and f is a symmetric element of $L^2(\mathbb{R}_+^q)$ satisfying $E[I_q^B(f)^2] = q! \|f\|_{L^2(\mathbb{R}_+^q)}^2 = 1$, then*

$$\sup_{A \in \mathscr{B}(\mathbb{R})} \left| P[I_q^B(f) \in A] - \frac{1}{\sqrt{2\pi}} \int_A e^{-x^2/2} dx \right| \leq 2\sqrt{\frac{q-1}{3q}} \sqrt{|E[I_q^B(f)^4] - 3|}.$$

Let us go back to the proof of (i), that is, to the proof of (5) for $\varphi = H_q$. Recall that the sequence $\{e_k\}$ has be chosen for (9) to hold. Using (10) (see also (11)), we can write $V_n = I_q^B(f_n)$, with

$$f_n = \frac{1}{\sqrt{n}} \sum_{k=1}^n e_k^{\otimes q}.$$

We already showed that $E[V_n^2] \to \sigma^2$ as $n \to \infty$. So, according to Theorem 2, to get (i) it remains to check that $\|f_n \otimes_r f_n\|_{L^2(\mathbb{R}_+^{2q-2r})} \to 0$ for any $r = 1, \ldots, q-1$. We have

$$f_n \otimes_r f_n = \frac{1}{n} \sum_{k,l=1}^{n} e_k^{\otimes q} \otimes_r e_l^{\otimes q} = \frac{1}{n} \sum_{k,l=1}^{n} \langle e_k, e_l \rangle_{L^2(\mathbb{R}_+)}^{r} e_k^{\otimes q-r} \otimes e_l^{\otimes q-r}$$

$$= \frac{1}{n} \sum_{k,l=1}^{n} \rho(k-l)^r e_k^{\otimes q-r} \otimes e_l^{\otimes q-r},$$

implying in turn

$$\|f_n \otimes_r f_n\|_{L^2(\mathbb{R}_+^{2q-2r})}^2$$

$$= \frac{1}{n^2} \sum_{i,j,k,l=1}^{n} \rho(i-j)^r \rho(k-l)^r \langle e_i^{\otimes q-r} \otimes e_j^{\otimes q-r}, e_k^{\otimes q-r} \otimes e_l^{\otimes q-r} \rangle_{L^2(\mathbb{R}_+^{2q-2r})}$$

$$= \frac{1}{n^2} \sum_{i,j,k,l=1}^{n} \rho(i-j)^r \rho(k-l)^r \rho(i-k)^{q-r} \rho(j-l)^{q-r}.$$

Observe that $|\rho(k-l)|^r |\rho(i-k)|^{q-r} \leqslant |\rho(k-l)|^q + |\rho(i-k)|^q$. This, together with other obvious manipulations, leads to the bound

$$\|f_n \otimes_r f_n\|_{L^2(\mathbb{R}_+^{2q-2r})}^2 \leqslant \frac{2}{n} \sum_{k \in \mathbb{Z}} |\rho(k)|^q \sum_{|i|<n} |\rho(i)|^r \sum_{|j|<n} |\rho(j)|^{q-r}$$

$$\leqslant \frac{2}{n} \sum_{k \in \mathbb{Z}} |\rho(k)|^d \sum_{|i|<n} |\rho(i)|^r \sum_{|j|<n} |\rho(j)|^{q-r}$$

$$= 2 \sum_{k \in \mathbb{Z}} |\rho(k)|^d \times n^{-\frac{q-r}{q}} \sum_{|i|<n} |\rho(i)|^r \times n^{-\frac{r}{q}} \sum_{|j|<n} |\rho(j)|^{q-r}.$$

Thus, to get that $\|f_n \otimes_r f_n\|_{L^2(\mathbb{R}_+^{2q-2r})} \to 0$ for any $r = 1, \ldots, q-1$, it suffices to show that

$$s_n(r) := n^{-\frac{q-r}{q}} \sum_{|i|<n} |\rho(i)|^r \to 0 \text{ for any } r = 1, \ldots, q-1.$$

Let $r = 1, \ldots, q-1$. Fix $\delta \in (0, 1)$ (to be chosen later) and let us decompose $s_n(r)$ into

$$s_n(r) = n^{-\frac{q-r}{q}} \sum_{|i|<[n\delta]} |\rho(i)|^r + n^{-\frac{q-r}{q}} \sum_{[n\delta] \leqslant |i|<n} |\rho(i)|^r =: s_{1,n}(\delta, r) + s_{2,n}(\delta, r).$$

Using Hölder inequality, we get that

$$s_{1,n}(\delta, r) \leqslant n^{-\frac{q-r}{r}} \left(\sum_{|i|<[n\delta]} |\rho(i)|^q \right)^{r/q} (1 + 2[n\delta])^{\frac{q-r}{q}} \leqslant \text{cst} \times \delta^{1-r/q},$$

as well as

$$s_{2,n}(\delta,r) \leqslant n^{-\frac{q-r}{r}} \left(\sum_{[n\delta] \leqslant |i| < n} |\rho(i)|^q \right)^{r/q} (2n)^{\frac{q-r}{q}} \leqslant \text{cst} \times \left(\sum_{|i| \geqslant [n\delta]} |\rho(i)|^q \right)^{r/q}.$$

Since $1 - r/q > 0$, it is a routine exercise (details are left to the reader) to deduce that $s_n(r) \to 0$ as $n \to \infty$. Since this is true for any $r = 1, \ldots, q-1$, this concludes the proof of (i).

It remains to show (ii), that is, convergence in law (5) whenever φ is a real polynomial. We shall use the multivariate counterpart of Theorem 2, which was obtained shortly afterwards by Peccati and Tudor. Since only a weak version (where all the involved multiple Wiener–Itô integrals have different orders) is needed here, we state the result of Peccati and Tudor only in this situation. We refer to Sect. 6 for a more general version and its proof.

Theorem 4 (Peccati and Tudor (2005); see [46]). *Consider l integers $q_1, \ldots, q_l \geqslant 1$, with $l \geqslant 2$. Assume that all the q_i's are pairwise different. For each $i = 1, \ldots, l$, let $\{f_n^i\}_{n \geqslant 1}$ be a sequence of symmetric functions of $L^2(\mathbb{R}_+^{q_i})$ satisfying $E[I_{q_i}^B(f_n^i)^2] = q_i! \|f_n^i\|_{L^2(\mathbb{R}_+^{q_i})}^2 \to \sigma_i^2$ as $n \to \infty$ for some $\sigma_i > 0$. Then, the following two assertions are equivalent as $n \to \infty$:*

(1) $I_{q_i}^B(f_n^i) \stackrel{\text{law}}{\to} \mathcal{N}(0, \sigma_i^2)$ for all $i = 1, \ldots, l$;

(2) $\left(I_{q_1}^B(f_n^1), \ldots, I_{q_l}^B(f_n^l) \right) \stackrel{\text{law}}{\to} \mathcal{N}(0, \text{diag}(\sigma_1^2, \ldots, \sigma_l^2))$.

In other words, Theorem 4 proves the surprising fact that, for such a sequence of vectors of multiple Wiener–Itô integrals, componentwise convergence to Gaussian always implies joint convergence. We shall combine Theorem 4 with (i) to prove (ii). Let φ have the form of a real polynomial. In particular, it admits a decomposition of the type $\varphi = \sum_{q=d}^N a_q H_q$ for some *finite* integer $N \geqslant d$. Together with (i), Theorem 4 yields that

$$\left(\frac{1}{\sqrt{n}} \sum_{k=1}^n H_d(X_k), \ldots, \frac{1}{\sqrt{n}} \sum_{k=1}^n H_N(X_k) \right) \stackrel{\text{law}}{\to} \mathcal{N}(0, \text{diag}(\sigma_d^2, \ldots, \sigma_N^2)),$$

where $\sigma_q^2 = q! \sum_{k \in \mathbb{Z}} \rho(k)^q$, $q = d, \ldots, N$. We deduce that

$$V_n = \frac{1}{\sqrt{n}} \sum_{q=d}^N a_q \sum_{k=1}^n H_q(X_k) \stackrel{\text{law}}{\to} \mathcal{N}\left(0, \sum_{q=d}^N a_q^2 q! \sum_{k \in \mathbb{Z}} \rho(k)^q \right),$$

which is the desired conclusion in (ii) and conclude the proof of Theorem 1. □

To Go Further. In [33], one associates quantitative bounds to Theorem 1 by using a similar approach.

2 Universality of Wiener Chaos

Before developing the material which will be necessary for the proof of the Fourth Moment Theorem 2 (as well as other related results), to motivate the reader let us study yet another consequence of this beautiful result.

For *any* sequence $X_1, X_2, \ldots$ of i.i.d. random variables with mean 0 and variance 1, the central limit theorem asserts that $V_n = (X_1 + \ldots + X_n)/\sqrt{n} \overset{\text{law}}{\to} \mathcal{N}(0, 1)$ as $n \to \infty$. It is a particular instance of what is commonly referred to as a "universality phenomenon" in probability. Indeed, we observe that the limit of the sequence V_n does not rely on the specific law of the X_i's, but only of the fact that its first two moments are 0 and 1 respectively.

Another example that exhibits a universality phenomenon is given by Wigner's theorem in the random matrix theory. More precisely, let $\{X_{ij}\}_{j>i\geq 1}$ and $\{X_{ii}/\sqrt{2}\}_{i\geq 1}$ be two independent families composed of i.i.d. random variables with mean 0, variance 1, and all the moments. Set $X_{ji} = X_{ij}$ and consider the $n \times n$ random matrix $M_n = (\frac{X_{ij}}{\sqrt{n}})_{1 \leq i,j \leq n}$. The matrix M_n being symmetric, its eigenvalues $\lambda_{1,n}, \ldots, \lambda_{n,n}$ (possibly repeated with multiplicity) belong to $\mathbb{R}$. Wigner's theorem then asserts that the spectral measure of M_n, that is, the random probability measure defined as $\frac{1}{n} \sum_{k=1}^{n} \delta_{\lambda_{k,n}}$, converges almost surely to the semicircular law $\frac{1}{2\pi} \sqrt{4 - x^2} \mathbf{1}_{[-2,2]}(x) dx$, whatever the exact distribution of the entries of M_n are.

In this section, our aim is to prove yet another universality phenomenon, which is in the spirit of the two afore-mentioned results. To do so, we need to introduce the following two blocks of basic ingredients:

(i) Three sequences $\mathbf{X} = (X_1, X_2, \ldots)$, $\mathbf{G} = (G_1, G_2, \ldots)$ and $\mathbf{E} = (\varepsilon_1, \varepsilon_2, \ldots)$ of i.i.d. random variables, all with mean 0, variance 1 and finite fourth moment. We are more specific with $\mathbf{G}$ and $\mathbf{E}$, by assuming further that $G_1 \sim \mathcal{N}(0, 1)$ and $P(\varepsilon_1 = 1) = P(\varepsilon_1 = -1) = 1/2$. (As we will see, $\mathbf{E}$ will actually play no role in the statement of Theorem 5; we will however use it to build a interesting counterexample, see Remark 3(1).)

(ii) A fixed integer $d \geq 1$ as well as a sequence $g_n : \{1, \ldots, n\}^d \to \mathbb{R}, n \geq 1$ of real functions, each g_n satisfying in addition that, for all $i_1, \ldots, i_d = 1, \ldots, n$,

(a) $g_n(i_1, \ldots, i_d) = g_n(i_{\sigma(1)}, \ldots, i_{\sigma(d)})$ for all permutation $\sigma \in \mathfrak{S}_d$.
(b) $g_n(i_1, \ldots, i_d) = 0$ whenever $i_k = i_l$ for some $k \neq l$.
(c) $d! \sum_{i_1, \ldots, i_d=1}^{n} g_n(i_1, \ldots, i_d)^2 = 1$.

(Of course, conditions (a) and (b) are becoming immaterial when $d = 1$.) If $\mathbf{x} = (x_1, x_2, \ldots)$ is a given real sequence, we also set

$$Q_d(g_n, \mathbf{x}) = \sum_{i_1, \ldots, i_d=1}^{n} g_n(i_1, \ldots, i_d) x_{i_1} \ldots x_{i_d}.$$

Using (b) and (c), it is straightforward to check that, for any $n \geq 1$, we have $E[Q_d(g_n, \mathbf{X})] = 0$ and $E[Q_d(g_n, \mathbf{X})^2] = 1$.

We are now in position to state our new universality phenomenon.

Theorem 5 (Nourdin, Peccati and Reinert (2010); see [34]). *Assume that $d \geq 2$. Then, as $n \to \infty$, the following two assertions are equivalent:*

(α) $Q_d(g_n, \mathbf{G}) \overset{law}{\to} \mathcal{N}(0, 1)$;

(β) $Q_d(g_n, \mathbf{X}) \overset{law}{\to} \mathcal{N}(0, 1)$ *for any sequence $\mathbf{X}$ as given in (i).*

Before proving Theorem 5, let us address some comments.

Remark 3. 1. In reality, the universality phenomenon in Theorem 5 is a bit more subtle than in the CLT or in Wigner's theorem. To illustrate what we have in mind, let us consider an explicit situation (in the case $d = 2$). Let $g_n : \{1, \ldots, n\}^2 \to \mathbb{R}$ be the function given by

$$g_n(i, j) = \frac{1}{2\sqrt{n-1}} \mathbf{1}_{\{i=1, j \geq 2 \text{ or } j=1, i \geq 2\}}.$$

It is easy to check that g_n satisfies the three assumptions (a)-(b)-(c) and also that

$$Q_2(g_n, \mathbf{x}) = x_1 \times \frac{1}{\sqrt{n-1}} \sum_{k=2}^{n} x_k.$$

The classical CLT then implies that $Q_2(g_n, \mathbf{G}) \overset{law}{\to} G_1 G_2$ and $Q_2(g_n, \mathbf{E}) \overset{law}{\to} \varepsilon_1 G_2$. Moreover, it is a classical and easy exercise to check that $\varepsilon_1 G_2$ is $\mathcal{N}(0, 1)$ distributed. Thus, what we just showed is that, although $Q_2(g_n, \mathbf{E}) \overset{law}{\to} \mathcal{N}(0, 1)$ as $n \to \infty$, the assertion (β) in Theorem 5 fails when choosing $\mathbf{X} = \mathbf{G}$ (indeed, the product of two independent $\mathcal{N}(0, 1)$ random variables is not gaussian). This means that, in Theorem 5, we cannot replace the sequence $\mathbf{G}$ in (α) by any other sequence (at least, not by $\mathbf{E}$!).

2. Theorem 5 is completely false when $d = 1$. For an explicit counterexample, consider for instance $g_n(i) = \mathbf{1}_{\{i=1\}}$, $i = 1, \ldots, n$. We then have $Q_1(g_n, \mathbf{x}) = x_1$. Consequently, the assertion (α) is trivially verified (it is even an equality in law!) but the assertion (β) is never true unless $X_1 \sim \mathcal{N}(0, 1)$.

Proof of Theorem 5. Of course, only the implication (α)$\to$(β) must be shown. Let us divide its proof into three steps.

Step 1. Set $e_i = \mathbf{1}_{[i-1, i]}$, $i \geq 1$, and let $f_n \in L^2(\mathbb{R}_+^d)$ be the symmetric function defined as

$$f_n = \sum_{i_1, \ldots, i_d = 1}^{n} g_n(i_1, \ldots, i_d) e_{i_1} \otimes \ldots \otimes e_{i_d}.$$

By the very definition of $I_d^B(f_n)$, we have

$$I_d^B(f_n) = d! \sum_{i_1,\ldots,i_d=1}^{n} g_n(i_1,\ldots,i_d) \int_0^\infty dB_{t_1} e_{i_1}(t_1) \int_0^{t_1} dB_{t_2} e_{i_2}(t_2) \ldots \int_0^{t_{d-1}} dB_{t_d} e_{i_d}(t_d).$$

Observe that

$$\int_0^\infty dB_{t_1} e_{i_1}(t_1) \int_0^{t_1} dB_{t_2} e_{i_2}(t_2) \ldots \int_0^{t_{d-1}} dB_{t_d} e_{i_d}(t_d)$$

is *not* almost surely zero (if and) only if $i_d \leq i_{d-1} \leq \ldots \leq i_1$. By combining this fact with assumption (b), we deduce that

$$I_d^B(f_n) = d! \sum_{1 \leq i_d < \ldots < i_1 \leq n} g_n(i_1,\ldots,i_d)$$

$$\times \int_0^\infty dB_{t_1} e_{i_1}(t_1) \int_0^{t_1} dB_{t_2} e_{i_2}(t_2) \ldots \int_0^{t_{d-1}} dB_{t_d} e_{i_d}(t_d)$$

$$= d! \sum_{1 \leq i_d < \ldots < i_1 \leq n} g_n(i_1,\ldots,i_d)(B_{i_1} - B_{i_1-1})\ldots(B_{i_d} - B_{i_d-1})$$

$$= \sum_{i_1,\ldots,i_d=1}^{n} g_n(i_1,\ldots,i_d)(B_{i_1} - B_{i_1-1})\ldots(B_{i_d} - B_{i_d-1}) \stackrel{\text{law}}{=} Q_d(g_n, \mathbf{G}).$$

That is, the sequence $Q_d(g_n, \mathbf{G})$ in (α) has actually the form of a multiple Wiener–Itô integral. On the other hand, going back to the definition of $f_n \otimes_{d-1} f_n$ and using that $\langle e_i, e_j \rangle_{L^2(\mathbb{R}_+)} = \delta_{ij}$ (Kronecker symbol), we get

$$f_n \otimes_{d-1} f_n = \sum_{i,j=1}^{n} \left(\sum_{k_2,\ldots,k_d=1}^{n} g_n(i,k_2,\ldots,k_d) g_n(j,k_2,\ldots,k_d) \right) e_i \otimes e_j,$$

so that

$$\|f_n \otimes_{d-1} f_n\|_{L^2(\mathbb{R}_+^2)}^2$$

$$= \sum_{i,j=1}^{n} \left(\sum_{k_2,\ldots,k_d=1}^{n} g_n(i,k_2,\ldots,k_d) g_n(j,k_2,\ldots,k_d) \right)^2$$

$$\geq \sum_{i=1}^{n} \left(\sum_{k_2,\ldots,k_d=1}^{n} g_n(i,k_2,\ldots,k_d)^2 \right)^2 \quad \text{(by summing only over } i=j\text{)}$$

$$\geq \max_{1 \leq i \leq n} \left(\sum_{k_2,\ldots,k_d=1}^{n} g_n(i,k_2,\ldots,k_d)^2 \right)^2$$

$$= \tau_n^2, \tag{12}$$

where

$$\tau_n := \max_{1 \leq i \leq n} \sum_{k_2,\ldots,k_d=1}^{n} g_n(i,k_2,\ldots,k_d)^2. \tag{13}$$

Now, assume that (α) holds. By Theorem 2 and because $Q_d(g_n, \mathbf{G}) \stackrel{\text{law}}{=} I_d^B(f_n)$, we have in particular that $\|f_n \otimes_{d-1} f_n\|_{L^2(\mathbb{R}_+^2)} \to 0$ as $n \to \infty$. Using the inequality (12), we deduce that $\tau_n \to 0$ as $n \to \infty$.

Step 2. We claim that the following result (whose proof is given in Step 3) allows to conclude the proof of (α) $\to$ (β).

Theorem 6 (Mossel, O'Donnel and Oleszkiewicz (2010); see [20]). *Let $\mathbf{X}$ and $\mathbf{G}$ be given as in (i) and let $g_n : \{1,\ldots,n\}^d \to \mathbb{R}$ be a function satisfying the three conditions (a)-(b)-(c). Set $\gamma = \max\{3, E[X_1^4]\} \geq 1$ and let τ_n be the quantity given by (13). Then, for all function $\varphi : \mathbb{R} \to \mathbb{R}$ of class $\mathscr{C}^3$ with $\|\varphi'''\|_\infty < \infty$, we have*

$$\left| E[\varphi(Q_d(g_n, \mathbf{X}))] - E[\varphi(Q_d(g_n, \mathbf{G}))] \right| \leq \frac{\gamma}{3}(3+2\gamma)^{\frac{3}{2}(d-1)} d^{3/2} \sqrt{d!} \, \|\varphi'''\|_\infty \sqrt{\tau_n}.$$

Indeed, assume that (α) holds. By Step 1, we have that $\tau_n \to 0$ as $n \to \infty$. Next, Theorem 6 together with (α), lead to (β) and therefore conclude the proof of Theorem 5.

Step 3: Proof of Theorem 6. During the proof, we will need the following auxiliary lemma, which is of independent interest.

Lemma 1 (Hypercontractivity). *Let $n \geq d \geq 1$, and consider a multilinear polynomial $P \in \mathbb{R}[x_1,\ldots,x_n]$ of degree d, that is, P is of the form*

$$P(x_1,\ldots,x_n) = \sum_{\substack{S \subset \{1,\ldots,n\} \\ |S|=d}} a_S \prod_{i \in S} x_i.$$

Let $\mathbf{X}$ be as in (i). Then,

$$E\left[P(X_1,\ldots,X_n)^4\right] \leq \left(3 + 2E[X_1^4]\right)^{2d} E\left[P(X_1,\ldots,X_n)^2\right]^2. \tag{14}$$

Proof. The proof follows ideas from [20] and is by induction on n. The case $n = 1$ is trivial. Indeed, in this case we have $d = 1$ so that $P(x_1) = ax_1$; the conclusion therefore asserts that (recall that $E[X_1^2] = 1$, implying in turn that $E[X_1^4] \geq E[X_1^2]^2 = 1$)

$$a^4 E[X_1^4] \leq a^4 \left(3 + 2E[X_1^4]\right)^2,$$

which is evident. Assume now that $n \geq 2$. We can write

$$P(x_1,\ldots,x_n) = R(x_1,\ldots,x_{n-1}) + x_n S(x_1,\ldots,x_{n-1}),$$

where $R, S \in \mathbb{R}[x_1, \ldots, x_{n-1}]$ are multilinear polynomials of $n - 1$ variables. Observe that R has degree d, while S has degree $d - 1$. Now write $\mathbf{P} = P(X_1, \ldots, X_n)$, $\mathbf{R} = R(X_1, \ldots, X_{n-1})$, $\mathbf{S} = S(X_1, \ldots, X_{n-1})$ and $\alpha = E[X_1^4]$. Clearly, $\mathbf{R}$ and $\mathbf{S}$ are independent of X_n. We have, using $E[X_n] = 0$ and $E[X_n^2] = 1$:

$$E[\mathbf{P}^2] = E[(\mathbf{R} + \mathbf{S}X_n)^2] = E[\mathbf{R}^2] + E[\mathbf{S}^2]$$
$$E[\mathbf{P}^4] = E[(\mathbf{R} + \mathbf{S}X_n)^4] = E[\mathbf{R}^4] + 6E[\mathbf{R}^2\mathbf{S}^2] + 4E[X_n^3]E[\mathbf{R}\mathbf{S}^3] + E[X_n^4]E[\mathbf{S}^4].$$

Observe that $E[\mathbf{R}^2\mathbf{S}^2] \leq \sqrt{E[\mathbf{R}^4]}\sqrt{E[\mathbf{S}^4]}$ and

$$E[X_n^3]E[\mathbf{R}\mathbf{S}^3] \leq \alpha^{\frac{3}{4}}\left(E[\mathbf{R}^4]\right)^{\frac{1}{4}}\left(E[\mathbf{S}^4]\right)^{\frac{3}{4}} \leq \alpha\sqrt{E[\mathbf{R}^4]}\sqrt{E[\mathbf{S}^4]} + \alpha E[\mathbf{S}^4],$$

where the last inequality used both $x^{\frac{1}{4}}y^{\frac{3}{4}} \leq \sqrt{xy} + y$ (by considering $x < y$ and $x > y$) and $\alpha^{\frac{3}{4}} \leq \alpha$ (because $\alpha \geq E[X_n^4] \geq E[X_n^2]^2 = 1$). Hence

$$E[\mathbf{P}^4] \leq E[\mathbf{R}^4] + 2(3 + 2\alpha)\sqrt{E[\mathbf{R}^4]}\sqrt{E[\mathbf{S}^4]} + 5\alpha E[\mathbf{S}^4]$$
$$\leq E[\mathbf{R}^4] + 2(3 + 2\alpha)\sqrt{E[\mathbf{R}^4]}\sqrt{E[\mathbf{S}^4]} + (3 + 2\alpha)^2 E[\mathbf{S}^4]$$
$$= \left(\sqrt{E[\mathbf{R}^4]} + (3 + 2\alpha)\sqrt{E[\mathbf{S}^4]}\right)^2.$$

By induction, we have $\sqrt{E[\mathbf{R}^4]} \leq (3 + 2\alpha)^d E[\mathbf{R}^2]$ and $\sqrt{E[\mathbf{S}^4]} \leq (3 + 2\alpha)^{d-1} E[\mathbf{S}^2]$. Therefore

$$E[\mathbf{P}^4] \leq (3 + 2\alpha)^{2d}\left(E[\mathbf{R}^2] + E[\mathbf{S}^2]\right)^2 = (3 + 2\alpha)^{2d} E[\mathbf{P}^2]^2,$$

and the proof of the lemma is concluded. □

We are now in position to prove Theorem 6. Following [20], we use the Lindeberg replacement trick. Without loss of generality, we assume that $\mathbf{X}$ and $\mathbf{G}$ are stochastically independent. For $i = 0, \ldots, n$, let $\mathbf{W}^{(i)} = (G_1, \ldots, G_i, X_{i+1}, \ldots, X_n)$. Fix a particular $i = 1, \ldots, n$ and write

$$U_i = \sum_{\substack{1 \leq i_1, \ldots, i_d \leq n \\ i_1 \neq i, \ldots, i_d \neq i}} g_n(i_1, \ldots, i_d) W_{i_1}^{(i)} \ldots W_{i_d}^{(i)},$$

$$V_i = \sum_{\substack{1 \leq i_1, \ldots, i_d \leq n \\ \exists j : i_j = i}} g_n(i_1, \ldots, i_d) W_{i_1}^{(i)} \ldots \widehat{W_i^{(i)}} \ldots W_{i_d}^{(i)}$$

$$= d \sum_{i_2, \ldots, i_d = 1}^{n} g_n(i, i_2, \ldots, i_d) W_{i_2}^{(i)} \ldots W_{i_d}^{(i)},$$

where $\widehat{W_i^{(i)}}$ means that this particular term is dropped (observe that this notation bears no ambiguity: indeed, since g_n vanishes on diagonals, each string $i_1,\ldots,i_d$ contributing to the definition of V_i contains the symbol i exactly once). For each i, note that U_i and V_i are independent of the variables X_i and G_i, and that

$$Q_d(g_n, \mathbf{W}^{(i-1)}) = U_i + X_i V_i \quad \text{and} \quad Q_d(g_n, \mathbf{W}^{(i)}) = U_i + G_i V_i.$$

By Taylor's theorem, using the independence of X_i from U_i and V_i, we have

$$\left| E[\varphi(U_i + X_i V_i)] - E[\varphi(U_i)] - E[\varphi'(U_i)V_i]E[X_i] - \frac{1}{2}E[\varphi''(U_i)V_i^2]E[X_i^2] \right|$$
$$\leq \frac{1}{6}\|\varphi'''\|_\infty E[|X_i|^3]E[|V_i|^3].$$

Similarly,

$$\left| E[\varphi(U_i + G_i V_i)] - E[\varphi(U_i)] - E[\varphi'(U_i)V_i]E[G_i] - \frac{1}{2}E[\varphi''(U_i)V_i^2]E[G_i^2] \right|$$
$$\leq \frac{1}{6}\|\varphi'''\|_\infty E[|G_i|^3]E[|V_i|^3].$$

Due to the matching moments up to second order on one hand, and using that $E[|X_i|^3] \leq \gamma$ and $E[|G_i|^3] \leq \gamma$ on the other hand, we obtain that

$$\left| E[\varphi(Q_d(g_n, \mathbf{W}^{(i-1)}))] - E[\varphi(Q_d(g_n, \mathbf{W}^{(i)}))] \right|$$
$$= \left| E[\varphi(U_i + G_i V_i)] - E[\varphi(U_i + X_i V_i)] \right|$$
$$\leq \frac{\gamma}{3}\|\varphi'''\|_\infty E[|V_i|^3].$$

By Lemma 1, we have

$$E[|V_i|^3] \leq E[V_i^4]^{\frac{3}{4}} \leq (3+2\gamma)^{\frac{3}{2}(d-1)} E[V_i^2]^{\frac{3}{2}}.$$

Using the independence between $\mathbf{X}$ and $\mathbf{G}$, the properties of g_n (which is symmetric and vanishes on diagonals) as well as $E[X_i] = E[G_i] = 0$ and $E[X_i^2] = E[G_i^2] = 1$, we get

$$E[V_i^2]^{3/2} = \left(dd! \sum_{i_2,\ldots,i_d=1}^n g_n(i, i_2,\ldots,i_d)^2 \right)^{3/2}$$
$$\leq (dd!)^{3/2} \sqrt{\max_{1 \leq j \leq n} \sum_{j_2,\ldots,j_d=1}^n g_n(j, j_2,\ldots,j_d)^2 \times \sum_{i_2,\ldots,i_d=1}^n g_n(i, i_2,\ldots,i_d)^2},$$

implying in turn that

$$\sum_{i=1}^n E[V_i^2]^{3/2} \leq (dd!)^{3/2} \sqrt{\max_{1 \leq j \leq n} \sum_{j_2,\ldots,j_d=1}^n g_n(j,j_2,\ldots,j_{d_k})^2}$$

$$\times \sum_{i_1,\ldots,i_d=1}^n g_n(i_1,i_2,\ldots,i_d)^2,$$

$$= d^{3/2} \sqrt{d!} \sqrt{\tau_n}.$$

By collecting the previous bounds, we get

$$|E[\varphi(Q_d(g_n,\mathbf{X}))] - E[\varphi(Q_d(g_n,\mathbf{G}))]|$$

$$\leq \sum_{i=1}^n |E[\varphi(Q_d(g_n,\mathbf{W}^{(i-1)}))] - E[\varphi(Q_d(g_n,\mathbf{W}^{(i)}))]|$$

$$\leq \frac{\gamma}{3}\|\varphi'''\|_\infty \sum_{i=1}^n E[|V_i|^3] \leq \frac{\gamma}{3}(3+2\gamma)^{\frac{3}{2}(d-1)}\|\varphi'''\|_\infty \sum_{i=1}^n E[V_i^2]^{\frac{3}{2}}$$

$$\leq \frac{\gamma}{3}(3+2\gamma)^{\frac{3}{2}(d-1)} d^{3/2} \sqrt{d!} \|\varphi'''\|_\infty \sqrt{\tau_n}, \qquad \square$$

which is exactly what was claimed in Theorem 6.

As a final remark, let us observe that Theorem 6 contains the CLT as a special case. Indeed, fix $d=1$ and let $g_n : \{1,\ldots,n\} \to \mathbb{R}$ be the function given by $g_n(i) = \frac{1}{\sqrt{n}}$. We then have $\tau_n = 1/n$. It is moreover clear that $Q_1(g_n,\mathbf{G}) \sim \mathcal{N}(0,1)$. Then, for any function $\varphi : \mathbb{R} \to \mathbb{R}$ of class $\mathscr{C}^3$ with $\|\varphi'''\|_\infty < \infty$ and any sequence $\mathbf{X}$ as in (i), Theorem 6 implies that

$$\left| E\left[\varphi\left(\frac{X_1+\ldots+X_n}{\sqrt{n}}\right)\right] - \frac{1}{\sqrt{2\pi}} \int_\mathbb{R} \varphi(y) e^{-y^2/2} dy \right|$$

$$\leq \max\{E[X_1^4]/3, 1\} \|\varphi'''\|_\infty \times \frac{1}{\sqrt{n}},$$

from which it is straightforward to deduce the CLT.

To Go Further. In [34], Theorem 5 is extended to the case where the target law is the centered Gamma law. In [48], there is a version of Theorem 5 in which the sequence $\mathbf{G}$ is replaced by $\mathbf{P}$, a sequence of i.i.d. Poisson random variables. Finally, let us mention that both Theorems 5 and 6 have been extended to the free probability framework (see Sect. 11) in [13].

3 Stein's Method

In this section, we shall introduce some basic features of the so-called Stein method, which is the first step toward the proof of the Fourth Moment Theorem 2. Actually, we will not need the full force of this method, only a basic estimate.

A random variable X is $\mathcal{N}(0, 1)$ distributed if and only if $E[e^{itX}] = e^{-t^2/2}$ for all $t \in \mathbb{R}$. This simple fact leads to the idea that a random variable X has a law which is *close* to $\mathcal{N}(0, 1)$ if and only if $E[e^{itX}]$ is *approximately* $e^{-t^2/2}$ for all $t \in \mathbb{R}$. This last claim is nothing but the usual criterion for the convergence in law through the use of characteristic functions.

Stein's seminal idea is somehow similar. He noticed in [52] that X is $\mathcal{N}(0, 1)$ distributed if and only if $E[f'(X) - Xf(X)] = 0$ for all function f belonging to a sufficiently rich class of functions (for instance, the functions which are $\mathscr{C}^1$ and whose derivative grows at most polynomially). He then wondered whether a suitable quantitative version of this identity may have fruitful consequences. This is actually the case and, even for specialists (at least for me!), the reason why it works so well remains a bit mysterious. Surprisingly, the simple following statement (due to Stein [52]) happens to contain all the elements of Stein's method that are needed for our discussion. (For more details or extensions of the method, one can consult the recent books [9, 32] and the references therein.)

Lemma 2 (Stein (1972); see [52]). *Let* $N \sim \mathcal{N}(0, 1)$ *be a standard Gaussian random variable. Let* $h : \mathbb{R} \to [0, 1]$ *be any continuous function. Define* $f : \mathbb{R} \to \mathbb{R}$ *by*

$$f(x) = e^{\frac{x^2}{2}} \int_{-\infty}^{x} (h(a) - E[h(N)]) e^{-\frac{a^2}{2}} da \tag{15}$$

$$= -e^{\frac{x^2}{2}} \int_{x}^{\infty} (h(a) - E[h(N)]) e^{-\frac{a^2}{2}} da. \tag{16}$$

Then f *is of class* $\mathscr{C}^1$, *satisfies* $|f(x)| \leq \sqrt{\pi/2}$, $|f'(x)| \leq 2$ *and*

$$f'(x) = xf(x) + h(x) - E[h(N)] \tag{17}$$

for all $x \in \mathbb{R}$.

Proof. The equality between (15) and (16) comes from

$$0 = E[h(N) - E[h(N)]] = \frac{1}{\sqrt{2\pi}} \int_{-\infty}^{+\infty} (h(a) - E[h(N)]) e^{-\frac{a^2}{2}} da.$$

Using (16) we have, for $x \geq 0$:

$$|xf(x)| = \left| xe^{\frac{x^2}{2}} \int_x^{+\infty} (h(a) - E[h(N)]) e^{-\frac{a^2}{2}} da \right|$$

$$\leq xe^{\frac{x^2}{2}} \int_x^{+\infty} e^{-\frac{a^2}{2}} da \leq e^{\frac{x^2}{2}} \int_x^{+\infty} ae^{-\frac{a^2}{2}} da = 1.$$

Using (15) we have, for $x \leq 0$:

$$|xf(x)| = \left| xe^{\frac{x^2}{2}} \int_{-\infty}^x (h(a) - E[h(N)]) e^{-\frac{a^2}{2}} da \right|$$

$$\leq |x| e^{\frac{x^2}{2}} \int_{|x|}^{+\infty} e^{-\frac{a^2}{2}} da \leq e^{\frac{x^2}{2}} \int_{|x|}^{+\infty} ae^{-\frac{a^2}{2}} da = 1.$$

The identity (17) is readily checked. We deduce, in particular, that

$$|f'(x)| \leq |xf(x)| + |h(x) - E[h(N)]| \leq 2$$

for all $x \in \mathbb{R}$. On the other hand, by (15)–(16), we have, for every $x \in \mathbb{R}$,

$$|f(x)| \leq e^{x^2/2} \min\left(\int_{-\infty}^x e^{-y^2/2} dy, \int_x^\infty e^{-y^2/2} dy \right) = e^{x^2/2} \int_{|x|}^\infty e^{-y^2/2} dy \leq \sqrt{\frac{\pi}{2}},$$

where the last inequality is obtained by observing that the function $s : \mathbb{R}_+ \to \mathbb{R}$ given by $s(x) = e^{x^2/2} \int_x^\infty e^{-y^2/2} dy$ attains its maximum at $x = 0$ (indeed, we have

$$s'(x) = xe^{x^2/2} \int_x^\infty e^{-y^2/2} dy - 1 \leq e^{x^2/2} \int_x^\infty ye^{-y^2/2} dy - 1 = 0$$

so that s is decreasing on $\mathbb{R}_+$) and that $s(0) = \sqrt{\pi/2}$.

The proof of the lemma is complete. □

To illustrate how Stein's method is a powerful approach, we shall use it to prove the celebrated Berry–Esseen theorem. (Our proof is based on an idea introduced by Ho and Chen in [16], see also Bolthausen [5].)

Theorem 7 (Berry and Esseen (1956); see [15]). Let $\mathbf{X} = (X_1, X_2, \ldots)$ be a sequence of i.i.d. random variables with $E[X_1] = 0$, $E[X_1^2] = 1$ and $E[|X_1|^3] < \infty$, and define

$$V_n = \frac{1}{\sqrt{n}} \sum_{k=1}^n X_k, \quad n \geq 1,$$

to be the associated sequence of normalized partial sums. Then, for any $n \geq 1$, one has

$$\sup_{x \in \mathbb{R}} \left| P(V_n \leq x) - \frac{1}{\sqrt{2\pi}} \int_{-\infty}^{x} e^{-u^2/2} du \right| \leq \frac{33 \, E[|X_1|^3]}{\sqrt{n}}. \tag{18}$$

Remark 4. One may actually show that (18) holds with the constant 0.4784 instead of 33. This has been proved by Korolev and Shevtsova [18] in 2010. (They do not use Stein's method.) On the other hand, according to Esseen [15] himself, it is impossible to expect a universal constant smaller than 0.4097.

Proof of (18). For each $n \geq 2$, let $C_n > 0$ be the best possible constant satisfying, for all i.i.d. random variables $X_1, \ldots, X_n$ with $E[|X_1|^3] < \infty$, $E[X_1^2] = 1$ and $E[X_1] = 0$, that

$$\sup_{x \in \mathbb{R}} \left| P(V_n \leq x) - \frac{1}{\sqrt{2\pi}} \int_{-\infty}^{x} e^{-u^2/2} du \right| \leq \frac{C_n \, E[|X_1|^3]}{\sqrt{n}}. \tag{19}$$

As a first (rough) estimation, we first observe that, since X_1 is centered with $E[X_1^2] = 1$, one has $E[|X_1|^3] \geq E[X_1^2]^{\frac{3}{2}} = 1$, so that $C_n \leq \sqrt{n}$. This is of course not enough to conclude, since we need to show that $C_n \leq 33$.

For any $x \in \mathbb{R}$ and $\varepsilon > 0$, introduce the function

$$h_{x,\varepsilon}(u) = \begin{cases} 1 & \text{if } u \leq x - \varepsilon \\ \text{linear} & \text{if } x - \varepsilon < u < x + \varepsilon \\ 0 & \text{if } u \geq x + \varepsilon \end{cases}.$$

It is immediately checked that, for all $n \geq 2$, $\varepsilon > 0$ and $x \in \mathbb{R}$, we have

$$E[h_{x-\varepsilon,\varepsilon}(V_n)] \leq P(V_n \leq x) \leq E[h_{x+\varepsilon,\varepsilon}(V_n)].$$

Moreover, for $N \sim \mathcal{N}(0,1)$, $\varepsilon > 0$ and $x \in \mathbb{R}$, we have, using that the density of N is bounded by $\frac{1}{\sqrt{2\pi}}$,

$$E[h_{x+\varepsilon,\varepsilon}(N)] - \frac{4\varepsilon}{\sqrt{2\pi}} \leq E[h_{x-\varepsilon,\varepsilon}(N)] \leq P(N \leq x)$$

$$\leq E[h_{x+\varepsilon,\varepsilon}(N)] \leq E[h_{x-\varepsilon,\varepsilon}(N)] + \frac{4\varepsilon}{\sqrt{2\pi}}.$$

Therefore, for all $n \geq 2$ and $\varepsilon > 0$, we have

$$\sup_{x \in \mathbb{R}} \left| P(V_n \leq x) - \frac{1}{\sqrt{2\pi}} \int_{-\infty}^{x} e^{-u^2/2} du \right| \leq \sup_{x \in \mathbb{R}} \left| E[h_{x,\varepsilon}(V_n)] - E[h_{x,\varepsilon}(N)] \right| + \frac{4\varepsilon}{\sqrt{2\pi}}.$$

Assume for the time being that, for all $\varepsilon > 0$,

$$\sup_{x \in \mathbb{R}} |E[h_{x,\varepsilon}(V_n)] - E[h_{x,\varepsilon}(N)]| \leq \frac{6 E[|X_1|^3]}{\sqrt{n}} + \frac{3 C_{n-1} E[|X_1|^3]^2}{\varepsilon n}. \quad (20)$$

We deduce that, for all $\varepsilon > 0$,

$$\sup_{x \in \mathbb{R}} \left| P(V_n \leq x) - \frac{1}{\sqrt{2\pi}} \int_{-\infty}^{x} e^{-u^2/2} du \right| \leq \frac{6 E[|X_1|^3]}{\sqrt{n}} + \frac{3 C_{n-1} E[|X_1|^3]^2}{\varepsilon n} + \frac{4\varepsilon}{\sqrt{2\pi}}.$$

By choosing $\varepsilon = \sqrt{\frac{C_{n-1}}{n}} E[|X_1|^3]$, we get that

$$\sup_{x \in \mathbb{R}} \left| P(V_n \leq x) - \frac{1}{\sqrt{2\pi}} \int_{-\infty}^{x} e^{-u^2/2} du \right| \leq \frac{E[|X_1|^3]}{\sqrt{n}} \left[6 + \left(3 + \frac{4}{\sqrt{2\pi}} \right) \sqrt{C_{n-1}} \right],$$

so that $C_n \leq 6 + \left(3 + \frac{4}{\sqrt{2\pi}} \right) \sqrt{C_{n-1}}$. It follows by induction that $C_n \leq 33$ (recall that $C_n \leq \sqrt{n}$ so that $C_2 \leq 33$ in particular), which is the desired conclusion.

We shall now use Stein's Lemma 2 to prove that (20) holds. Fix $x \in \mathbb{R}$ and $\varepsilon > 0$, and let f denote the Stein solution associated with $h = h_{x,\varepsilon}$, that is, f satisfies (15). Observe that h is continuous, and therefore f is $\mathscr{C}^1$. Recall from Lemma 2 that $\|f\|_\infty \leq \sqrt{\frac{\pi}{2}}$ and $\|f'\|_\infty \leq 2$. Set also $\tilde{f}(x) = xf(x)$, $x \in \mathbb{R}$. We then have

$$|\tilde{f}(x) - \tilde{f}(y)| = |f(x)(x-y) + (f(x) - f(y))y| \leq \left(\sqrt{\frac{\pi}{2}} + 2|y| \right) |x-y|. \quad (21)$$

On the other hand, set

$$V_n^i = V_n - \frac{X_i}{\sqrt{n}}, \quad i = 1, \ldots, n.$$

Observe that V_n^i and X_i are independent by construction. One can thus write

$$E[h(V_n)] - E[h(N)] = E[f'(V_n) - V_n f(V_n)]$$

$$= \sum_{i=1}^{n} E\left[f'(V_n) \frac{1}{n} - f(V_n) \frac{X_i}{\sqrt{n}} \right]$$

$$= \sum_{i=1}^{n} E\left[f'(V_n) \frac{1}{n} - (f(V_n) - f(V_n^i)) \frac{X_i}{\sqrt{n}} \right] \quad \text{because } E[f(V_n^i) X_i] = E[f(V_n^i)] E[X_i] = 0$$

$$= \sum_{i=1}^{n} E\left[f'(V_n) \frac{1}{n} - f'\left(V_n^i + \theta \frac{X_i}{\sqrt{n}} \right) \frac{X_i^2}{n} \right] \quad \text{with } \theta \sim \mathscr{U}_{[0,1]} \text{ independent of } X_1, \ldots, X_n.$$

We have $f'(x) = \tilde{f}(x) + h(x) - E[h(N)]$, so that

$$E[h(V_n)] - E[h(N)] = \sum_{i=1}^{n} \left(a_i(\tilde{f}) - b_i(\tilde{f}) + a_i(h) - b_i(h) \right), \qquad (22)$$

where

$$a_i(g) = E[g(V_n) - g(V_n^i)] \frac{1}{n} \quad \text{and} \quad b_i(g) = E\left[\left(g\left(V_n^i + \theta \frac{X_i}{\sqrt{n}} \right) - g(V_n^i) \right) X_i^2 \right] \frac{1}{n}.$$

(Here again, we have used that V_n^i and X_i are independent.) Hence, to prove that (20) holds true, we must bound four terms.

1st term. One has, using (21) as well as $E[|X_1|] \leq E[X_1^2]^{\frac{1}{2}} = 1$ and $E[|V_n^i|] \leq E[(V_n^i)^2]^{\frac{1}{2}} \leq 1$,

$$|a_i(\tilde{f})| \leq \frac{1}{n\sqrt{n}} \left(E[|X_1|] \sqrt{\frac{\pi}{2}} + 2E[|X_1|]E[|V_n^i|] \right) \leq \left(\sqrt{\frac{\pi}{2}} + 2 \right) \frac{1}{n\sqrt{n}}.$$

2nd term. Similarly and because $E[\theta] = \frac{1}{2}$, one has

$$|b_i(\tilde{f})| \leq \frac{1}{n\sqrt{n}} \left(E[\theta]E[|X_1|^3] \sqrt{\frac{\pi}{2}} + 2E[\theta]E[|X_1|^3]E[|V_n^i|] \right)$$

$$\leq \left(\frac{1}{2}\sqrt{\frac{\pi}{2}} + 1 \right) \frac{E[|X_1|^3]}{n\sqrt{n}}.$$

3rd term. By definition of h, we have

$$h(v) - h(u) = (v-u) \int_0^1 h'(u + s(v-u)) ds = -\frac{v-u}{2\varepsilon} E\left[\mathbf{1}_{[x-\varepsilon, x+\varepsilon]}(u + \hat{\theta}(v-u)) \right],$$

with $\hat{\theta} \sim \mathcal{U}_{[0,1]}$ independent of θ and $X_1, \ldots, X_n$, so that

$$|a_i(h)| \leq \frac{1}{2\varepsilon n\sqrt{n}} E\left[|X_i| \mathbf{1}_{[x-\varepsilon, x+\varepsilon]}\left(V_n^i + \hat{\theta} \frac{X_i}{\sqrt{n}} \right) \right]$$

$$= \frac{1}{2\varepsilon n\sqrt{n}} E\left[|X_i| P\left(x - \frac{y}{\sqrt{n}} - \varepsilon \leq V_n^i \leq x - \frac{y}{\sqrt{n}} + \varepsilon \right) \Big|_{y=\hat{\theta} X_i} \right]$$

$$\leq \frac{1}{2\varepsilon n\sqrt{n}} \sup_{y \in \mathbb{R}} P\left(x - \frac{y}{\sqrt{n}} - \varepsilon \leq V_n^i \leq x - \frac{y}{\sqrt{n}} + \varepsilon \right).$$

We are thus left to bound $P(a \leq V_n^i \leq b)$ for all $a, b \in \mathbb{R}$ with $a \leq b$. For that, set $\tilde{V}_n^i = \frac{1}{\sqrt{n-1}} \sum_{j \neq i} X_j$, so that $V_n^i = \sqrt{1 - \frac{1}{n}} \tilde{V}_n^i$. We then have, using in particular (19) (with $n - 1$ instead of n) and the fact that the standard Gaussian density is bounded by $\frac{1}{\sqrt{2\pi}}$,

$$P(a \leq V_n^i \leq b) = P\left(\frac{a}{\sqrt{1-\frac{1}{n}}} \leq \tilde{V}_n^i \leq \frac{b}{\sqrt{1-\frac{1}{n}}}\right)$$

$$= P\left(\frac{a}{\sqrt{1-\frac{1}{n}}} \leq N \leq \frac{b}{\sqrt{1-\frac{1}{n}}}\right)$$

$$+ P\left(\frac{a}{\sqrt{1-\frac{1}{n}}} \leq \tilde{V}_n^i \leq \frac{b}{\sqrt{1-\frac{1}{n}}}\right)$$

$$- P\left(\frac{a}{\sqrt{1-\frac{1}{n}}} \leq N \leq \frac{b}{\sqrt{1-\frac{1}{n}}}\right)$$

$$\leq \frac{b-a}{\sqrt{2\pi}\sqrt{1-\frac{1}{n}}} + \frac{2 C_{n-1} E[|X_1|^3]}{\sqrt{n-1}}.$$

We deduce that

$$|a_i(h)| \leq \frac{1}{\sqrt{2\pi n}\sqrt{n-1}} + \frac{C_{n-1} E[|X_1|^3]}{n\sqrt{n}\sqrt{n-1}\varepsilon}.$$

4th term. Similarly, we have

$$|b_i(h)| = \frac{1}{2n\sqrt{n\varepsilon}} \left| E\left[X_i^3 \theta \mathbf{1}_{[x-\varepsilon,x+\varepsilon]}\left(V_n^i + \hat{\theta}\theta\frac{X_i}{\sqrt{n}}\right)\right]\right|$$

$$\leq \frac{E[|X_1|^3]}{4n\sqrt{n\varepsilon}} \sup_{y \in \mathbb{R}} P\left(x - \frac{y}{\sqrt{n}} - \varepsilon \leq V_n^i \leq x - \frac{y}{\sqrt{n}} + \varepsilon\right)$$

$$\leq \frac{E[|X_1|^3]}{2\sqrt{2\pi n}\sqrt{n-1}} + \frac{C_{n-1} E[|X_1|^3]^2}{2n\sqrt{n}\sqrt{n-1}\varepsilon}.$$

Plugging these four estimates into (22) and by using the fact that $n \geq 2$ (and therefore $n - 1 \geq \frac{n}{2}$) and $E[|X_1|^3] \geq 1$, we deduce the desired conclusion. □

To Go Further. Stein's method has developed considerably since its first appearance in 1972. A comprehensive and very nice reference to go further is the book [9] by Chen, Goldstein and Shao, in which several applications of Stein's method are carefully developed.

4 Malliavin Calculus in a Nutshell

The second ingredient for the proof of the Fourth Moment Theorem 2 is the Malliavin calculus (the first one being Stein's method, as developed in the previous section). So, let us introduce the reader to the basic operators of Malliavin calculus. For the sake of simplicity and to avoid technicalities that would be useless in this survey, we will only consider the case where the underlying Gaussian process (fixed once for all throughout the sequel) is a classical Brownian motion $B = (B_t)_{t \geq 0}$ defined on some probability space $(\Omega, \mathscr{F}, P)$; we further assume that the σ-field $\mathscr{F}$ is generated by B.

For a detailed exposition of Malliavin calculus (in a more general context) and for missing proofs, we refer the reader to the textbooks [32, 38].

Dimension One. In this first section, we would like to introduce the basic operators of Malliavin calculus in the simplest situation (where only *one* Gaussian random variable is involved). While easy, it is a sufficiently rich context to encapsulate all the essence of this theory. We first need to recall some useful properties of Hermite polynomials.

Proposition 1. *The family* $(H_q)_{q \in \mathbb{N}} \subset \mathbb{R}[X]$ *of Hermite polynomials has the following properties.*

(a) $H_q' = qH_{q-1}$ *and* $H_{q+1} = XH_q - qH_{q-1}$ *for all* $q \in \mathbb{N}$.
(b) The family $\left(\frac{1}{\sqrt{q!}} H_q\right)_{q \in \mathbb{N}}$ *is an orthonormal basis of* $L^2(\mathbb{R}, \frac{1}{\sqrt{2\pi}} e^{-x^2/2} dx)$.
(c) Let (U, V) *be a Gaussian vector with* $U, V \sim \mathcal{N}(0, 1)$. *Then, for all* $p, q \in \mathbb{N}$,

$$E[H_p(U)H_q(V)] = \begin{cases} q! E[UV]^q & \text{if } p = q \\ 0 & \text{otherwise.} \end{cases}$$

Proof. This is well-known. For a proof, see, e.g., [32, Proposition 1.4.2]. □

Let $\varphi : \mathbb{R} \to \mathbb{R}$ be an element of $L^2(\mathbb{R}, \frac{1}{\sqrt{2\pi}} e^{-x^2/2} dx)$. Proposition 1(b) implies that φ may be expanded (in a unique way) in terms of Hermite polynomials as follows:

$$\varphi = \sum_{q=0}^{\infty} a_q H_q. \tag{23}$$

When φ is such that $\sum q q! a_q^2 < \infty$, let us define

$$D\varphi = \sum_{q=0}^{\infty} q a_q H_{q-1}. \tag{24}$$

Since the Hermite polynomials satisfy $H'_q = qH_{q-1}$ (Proposition 1(a)), observe that

$$D\varphi = \varphi'$$

(in the sense of distributions). Let us now define the *Ornstein–Uhlenbeck semigroup* $(P_t)_{t \geq 0}$ by

$$P_t \varphi = \sum_{q=0}^{\infty} e^{-qt} a_q H_q. \tag{25}$$

Plainly, $P_0 = \mathrm{Id}$, $P_t P_s = P_{t+s}$ ($s, t \geq 0$) and

$$DP_t = e^{-t} P_t D. \tag{26}$$

Since $(P_t)_{t \geq 0}$ is a semigroup, it admits a generator L defined as

$$L = \frac{d}{dt}\Big|_{t=0} P_t.$$

Of course, for any $t \geq 0$ one has that

$$\frac{d}{dt} P_t = \lim_{h \to 0} \frac{P_{t+h} - P_t}{h} = \lim_{h \to 0} P_t \frac{P_h - \mathrm{Id}}{h} = P_t \lim_{h \to 0} \frac{P_h - \mathrm{Id}}{h} = P_t \frac{d}{dh}\Big|_{h=0} P_h = P_t L,$$

and, similarly, $\frac{d}{dt} P_t = L P_t$. Moreover, going back to the definition of $(P_t)_{t \geq 0}$, it is clear that the domain of L is the set of functions $\varphi \in L^2(\mathbb{R}, \frac{1}{\sqrt{2\pi}} e^{-x^2/2} dx)$ such that $\sum q^2 q! a_q^2 < \infty$ and that, in this case,

$$L\varphi = -\sum_{q=0}^{\infty} q a_q H_q.$$

We have the following integration by parts formula, whose proof is straightforward (start with the case $\varphi = H_p$ and $\psi = H_q$, and then use bilinearity and approximation to conclude in the general case) and left to the reader.

Proposition 2. *Let φ be in the domain of L and ψ be in the domain of D. Then*

$$\int_{\mathbb{R}} L\varphi(x)\psi(x) \frac{e^{-x^2/2}}{\sqrt{2\pi}} dx = -\int_{\mathbb{R}} D\varphi(x) D\psi(x) \frac{e^{-x^2/2}}{\sqrt{2\pi}} dx. \tag{27}$$

We shall now extend all the previous operators in a situation where, instead of dealing with a random variable of the form $F = \varphi(N)$ (that involves only *one* Gaussian random variable N), we deal more generally with a random variable F that is measurable with respect to the Brownian motion $(B_t)_{t \geq 0}$.

Wiener Integral. For any adapted[2] and square integrable stochastic process $u = (u_t)_{t \geq 0}$, let us denote by $\int_0^\infty u_t dB_t$ its Itô integral. Recall from any standard textbook of stochastic analysis that the Itô integral is a linear functional that takes its values on $L^2(\Omega)$ and has the following basic features, coming mainly from the independence property of the increments of B:

$$E\left[\int_0^\infty u_s dB_s\right] = 0 \qquad (28)$$

$$E\left[\int_0^\infty u_s dB_s \times \int_0^\infty v_s dB_s\right] = E\left[\int_0^\infty u_s v_s ds\right]. \qquad (29)$$

In the particular case where $u = f \in L^2(\mathbb{R}_+)$ is *deterministic*, we say that $\int_0^\infty f(s)dB_s$ is the *Wiener integral of f*; it is then easy to show that

$$\int_0^\infty f(s)dB_s \sim \mathcal{N}\left(0, \int_0^\infty f^2(s)ds\right). \qquad (30)$$

Multiple Wiener–Itô Integrals and Wiener Chaoses. Let $f \in L^2(\mathbb{R}_+^q)$. Let us see how one could give a "natural" meaning to the q-fold multiple integral

$$I_q^B(f) = \int_{\mathbb{R}_+^q} f(s_1, \ldots, s_q) dB_{s_1} \ldots dB_{s_q}.$$

To achieve this goal, we shall use an iterated Itô integral; the following heuristic "calculations" are thus natural within this framework:

$$\int_{\mathbb{R}_+^q} f(s_1, \ldots, s_q) dB_{s_1} \ldots dB_{s_q}$$

$$= \sum_{\sigma \in \mathfrak{S}_q} \int_{\mathbb{R}_+^q} f(s_1, \ldots, s_q) \mathbf{1}_{\{s_{\sigma(1)} > \ldots > s_{\sigma(q)}\}} dB_{s_1} \ldots dB_{s_q}$$

$$= \sum_{\sigma \in \mathfrak{S}_q} \int_0^\infty dB_{s_{\sigma(1)}} \int_0^{s_{\sigma(1)}} dB_{s_{\sigma(2)}} \cdots \int_0^{s_{\sigma(q-1)}} dB_{s_{\sigma(q)}} f(s_1, \ldots, s_q)$$

$$= \sum_{\sigma \in \mathfrak{S}_q} \int_0^\infty dB_{t_1} \int_0^{t_1} dB_{t_2} \cdots \int_0^{t_{q-1}} dB_{t_q} f(t_{\sigma^{-1}(1)}, \ldots, t_{\sigma^{-1}(q)})$$

$$= \sum_{\sigma \in \mathfrak{S}_q} \int_0^\infty dB_{t_1} \int_0^{t_1} dB_{t_2} \cdots \int_0^{t_{q-1}} dB_{t_q} f(t_{\sigma(1)}, \ldots, t_{\sigma(q)}). \qquad (31)$$

Now, we can use (31) as a natural candidate for being $I_q^B(f)$.

---

[2] Any adapted process u that is either càdlàg or càglàd admits a progressively measurable version. We will always assume that we are dealing with it.

Definition 1. Let $q \geq 1$ be an integer.

1. When $f \in L^2(\mathbb{R}_+^q)$, we set

$$I_q^B(f) = \sum_{\sigma \in \mathfrak{S}_q} \int_0^\infty dB_{t_1} \int_0^{t_1} dB_{t_2} \cdots \int_0^{t_{q-1}} dB_{t_q}\, f(t_{\sigma(1)}, \ldots, t_{\sigma(q)}). \tag{32}$$

The random variable $I_q^B(f)$ is called the qth multiple Wiener–Itô integral of f.

2. The set $\mathcal{H}_q^B$ of random variables of the form $I_q^B(f)$, $f \in L^2(\mathbb{R}_+^q)$, is called the qth Wiener chaos of B. We also use the convention $\mathcal{H}_0^B = \mathbb{R}$.

The following properties are readily checked.

Proposition 3. *Let $q \geq 1$ be an integer and let $f \in L^2(\mathbb{R}_+^q)$.*

1. *If f is symmetric (meaning that $f(t_1, \ldots, t_q) = f(t_{\sigma(1)}, \ldots, t_{\sigma(q)})$ for any $t \in \mathbb{R}_+^q$ and any permutation $\sigma \in \mathfrak{S}_q$), then*

$$I_q^B(f) = q! \int_0^\infty dB_{t_1} \int_0^{t_1} dB_{t_2} \cdots \int_0^{t_{q-1}} dB_{t_q}\, f(t_1, \ldots, t_q). \tag{33}$$

2. *We have*

$$I_q^B(f) = I_q^B(\tilde{f}), \tag{34}$$

where $\tilde{f}$ stands for the symmetrization of f given by

$$\tilde{f}(t_1, \ldots, t_q) = \frac{1}{q!} \sum_{\sigma \in \mathfrak{S}_q} f(t_{\sigma(1)}, \ldots, t_{\sigma(q)}). \tag{35}$$

3. *For any $p, q \geq 1$, $f \in L^2(\mathbb{R}_+^p)$ and $g \in L^2(\mathbb{R}_+^q)$,*

$$E[I_q^B(f)] = 0 \tag{36}$$

$$E[I_p^B(f) I_q^B(g)] = p! \langle \tilde{f}, \tilde{g} \rangle_{L^2(\mathbb{R}_+^p)} \quad \text{if } p = q \tag{37}$$

$$E[I_p^B(f) I_q^B(g)] = 0 \quad \text{if } p \neq q. \tag{38}$$

The space $L^2(\Omega)$ can be decomposed into the infinite orthogonal sum of the spaces $\mathcal{H}_q^B$. (It is a statement which is analogous to the content of Proposition 1(b), and it is precisely here that we need to assume that the σ-field $\mathscr{F}$ is generated by B.) It follows that any square-integrable random variable $F \in L^2(\Omega)$ admits the following chaotic expansion:

$$F = E[F] + \sum_{q=1}^\infty I_q^B(f_q), \tag{39}$$

where the functions $f_q \in L^2(\mathbb{R}_+^q)$ are symmetric and uniquely determined by F. In practice and when F is "smooth" enough, one may rely on Stroock's formula (see [53] or [38, Exercise 1.2.6]) to compute the functions f_q explicitly.

The following result contains a very useful property of multiple Wiener–Itô integrals. It is in the same spirit as Lemma 1.

Theorem 8 (Nelson (1973); see [21]). *Let $f \in L^2(\mathbb{R}_+^q)$ with $q \geq 1$. Then, for all $r \geq 2$,*

$$E\big[|I_q^B(f)|^r\big] \leq [(r-1)^q q!]^{r/2} \|f\|_{L^2(\mathbb{R}_+^q)}^r < \infty. \tag{40}$$

Proof. See, e.g., [32, Corollary 2.8.14]. (The proof uses the hypercontractivity property of $(P_t)_{t \geq 0}$ defined as (48).) □

Multiple Wiener–Itô integrals are linear by construction. Let us see how they behave with respect to multiplication. To this aim, we need to introduce the concept of *contractions*.

Definition 2. When $r \in \{1, \ldots, p \wedge q\}$, $f \in L^2(\mathbb{R}_+^p)$ and $g \in L^2(\mathbb{R}_+^q)$, we write $f \otimes_r g$ to indicate the rth contraction of f and g, defined as being the element of $L^2(\mathbb{R}_+^{p+q-2r})$ given by

$$(f \otimes_r g)(t_1, \ldots, t_{p+q-2r}) \tag{41}$$
$$= \int_{\mathbb{R}_+^r} f(t_1, \ldots, t_{p-r}, x_1, \ldots, x_r) g(t_{p-r+1}, \ldots, t_{p+q-2r}, x_1, \ldots, x_r) dx_1 \ldots dx_r.$$

By convention, we set $f \otimes_0 g = f \otimes g$ as being the tensor product of f and g, that is,

$$(f \otimes_0 g)(t_1, \ldots, t_{p+q}) = f(t_1, \ldots, t_p) g(t_{p+1}, \ldots, t_{p+q}).$$

Observe that

$$\|f \otimes_r g\|_{L^2(\mathbb{R}_+^{p+q-2r})} \leq \|f\|_{L^2(\mathbb{R}_+^p)} \|g\|_{L^2(\mathbb{R}_+^q)}, \quad r = 0, \ldots, p \wedge q \tag{42}$$

by Cauchy–Schwarz, and that $f \otimes_p g = \langle f, g \rangle_{L^2(\mathbb{R}_+^p)}$ when $p = q$. The next result is the fundamental *product formula* between two multiple Wiener–Itô integrals.

Theorem 9. *Let $p, q \geq 1$ and let $f \in L^2(\mathbb{R}_+^p)$ and $g \in L^2(\mathbb{R}_+^q)$ be two symmetric functions. Then*

$$I_p^B(f) I_q^B(g) = \sum_{r=0}^{p \wedge q} r! \binom{p}{r} \binom{q}{r} I_{p+q-2r}^B(f \widetilde{\otimes}_r g), \tag{43}$$

where $f \otimes_r g$ stands for the contraction (41).

Proof. Theorem 9 can be established by at least two routes, namely by induction (see, e.g., [38, page 12]) or by using the concept of diagonal measure in the context of the Engel–Rota–Wallstrom theory (see [45]). Let us proceed with a heuristic proof following this latter strategy. Going back to the very definition of $I_p^B(f)$, we see that the diagonals are avoided. That is, $I_p^B(f)$ can be seen as

$$I_p^B(f) = \int_{\mathbb{R}_+^p} f(s_1, \ldots, s_p) \mathbf{1}_{\{s_i \neq s_j, i \neq j\}} dB_{s_1} \ldots dB_{s_p}$$

The same holds for $I_q^B(g)$. Then we have (just as through Fubini)

$$I_p^B(f) I_q^B(g)$$
$$= \int_{\mathbb{R}_+^{p+q}} f(s_1, \ldots, s_p) \mathbf{1}_{\{s_i \neq s_j, i \neq j\}} g(t_1, \ldots, t_q) \mathbf{1}_{\{t_i \neq t_j, i \neq j\}} dB_{s_1} \ldots dB_{s_p} dB_{t_1} \ldots dB_{t_q}.$$

While there is no diagonals in the first and second blocks, there are all possible mixed diagonals in the joint writing. Hence we need to take into account all these diagonals (whence the combinatorial coefficients in the statement, which count all possible diagonal sets of size r) and then integrate out (using the rule $(dB_t)^2 = dt$). We thus obtain

$$I_p^B(f) I_q^B(g) = \sum_{r=0}^{p \wedge q} r! \binom{p}{r} \binom{q}{r} \int_{\mathbb{R}_+^{p+q-2r}} (f \otimes_r g)(x_1, \ldots, x_{p+q-2r}) dB_{x_1} \ldots dB_{x_{p+q-2r}}$$

which is exactly the claim (43). □

Malliavin Derivatives. We shall extend the operator D introduced in (24). Let $F \in L^2(\Omega)$ and consider its chaotic expansion (39).

Definition 3. 1. When $m \geq 1$ is an integer, we say that F belongs to the Sobolev–Watanabe space $\mathbb{D}^{m,2}$ if

$$\sum_{q=1}^{\infty} q^m q! \|f_q\|_{L^2(\mathbb{R}_+^q)}^2 < \infty. \tag{44}$$

2. When (44) holds with $m = 1$, the Malliavin derivative $DF = (D_t F)_{t \geq 0}$ of F is the element of $L^2(\Omega \times \mathbb{R}_+)$ given by

$$D_t F = \sum_{q=1}^{\infty} q I_{q-1}^B \left(f_q(\cdot, t) \right). \tag{45}$$

3. More generally, when (44) holds with an m bigger than or equal to 2 we define the mth Malliavin derivative $D^m F = (D_{t_1, \ldots, t_m} F)_{t_1, \ldots, t_m \geq 0}$ of F as the element of $L^2(\Omega \times \mathbb{R}_+^m)$ given by

$$D_{t_1,\ldots,t_m}F = \sum_{q=m}^{\infty} q(q-1)\ldots(q-m+1)I_{q-m}^B\left(f_q(\cdot,t_1,\ldots,t_m)\right). \quad (46)$$

The exponent 2 in the notation $\mathbb{D}^{m,2}$ is because it is related to the space $L^2(\Omega)$. (There exists a space $\mathbb{D}^{m,p}$ related to $L^p(\Omega)$ but we will not use it in this survey.) On the other hand, it is clear by construction that D is a linear operator. Also, using (37)–(38) it is easy to compute the L^2-norm of DF in terms of the kernels f_q appearing in the chaotic expansion (39) of F:

Proposition 4. *Let $F \in \mathbb{D}^{1,2}$. We have*

$$E\left[\|DF\|_{L^2(\mathbb{R}_+)}^2\right] = \sum_{q=1}^{\infty} qq! \|f_q\|_{L^2(\mathbb{R}_+^q)}^2.$$

Proof. By (45), we can write

$$E\left[\|DF\|_{L^2(\mathbb{R}_+)}^2\right] = \int_{\mathbb{R}_+} E\left[\left(\sum_{q=1}^{\infty} q I_{q-1}^B\left(f_q(\cdot,t)\right)\right)^2\right]dt$$

$$= \sum_{p,q=1}^{\infty} pq \int_{\mathbb{R}_+} E\left[I_{p-1}^B\left(f_p(\cdot,t)\right) I_{q-1}^B\left(f_q(\cdot,t)\right)\right]dt.$$

Using (38), we deduce that

$$E\left[\|DF\|_{L^2(\mathbb{R}_+)}^2\right] = \sum_{q=1}^{\infty} q^2 \int_{\mathbb{R}_+} E\left[I_{q-1}^B\left(f_q(\cdot,t)\right)^2\right]dt.$$

Finally, using (37), we get that

$$E\left[\|DF\|_{L^2(\mathbb{R}_+)}^2\right] = \sum_{q=1}^{\infty} q^2(q-1)! \int_{\mathbb{R}_+} \|f_q(\cdot,t)\|_{L^2(\mathbb{R}_+^{q-1})}^2 dt = \sum_{q=1}^{\infty} qq! \|f_q\|_{L^2(\mathbb{R}_+^q)}^2.$$

$\square$

Let H_q be the qth Hermite polynomial (for some $q \geq 1$) and let $e \in L^2(\mathbb{R}_+)$ have norm 1. Recall (10) and Proposition 1(a). We deduce that, for any $t \geq 0$,

$$D_t\left(H_q\left(\int_0^{\infty} e(s)dW_s\right)\right) = D_t(I_q^B(e^{\otimes q})) = qI_{q-1}^B(e^{\otimes q-1})e(t)$$

$$= qH_{q-1}\left(\int_0^{\infty} e(s)dB_s\right)e(t) = H_q'\left(\int_0^{\infty} e(s)dB_s\right)D_t\left(\int_0^{\infty} e(s)dB_s\right).$$

More generally, the Malliavin derivative D verifies the *chain rule*:

Theorem 10. *Let $\varphi : \mathbb{R} \to \mathbb{R}$ be both of class $\mathscr{C}^1$ and Lipschitz, and let $F \in \mathbb{D}^{1,2}$. Then, $\varphi(F) \in \mathbb{D}^{1,2}$ and*

$$D_t \varphi(F) = \varphi'(F) D_t F, \quad t \geq 0. \tag{47}$$

Proof. See, e.g., [38, Proposition 1.2.3]. □

Ornstein–Uhlenbeck Semigroup. We now introduce the extension of (25) in our infinite-dimensional setting.

Definition 4. The Ornstein–Uhlenbeck semigroup is the family of linear operators $(P_t)_{t \geq 0}$ defined on $L^2(\Omega)$ by

$$P_t F = \sum_{q=0}^{\infty} e^{-qt} I_q^B(f_q), \tag{48}$$

where the symmetric kernels f_q are given by (39).

A crucial property of $(P_t)_{t \geq 0}$ is the Mehler formula, that gives an alternative and often useful representation formula for P_t. To be able to state it, we need to introduce a further notation. Let (B, B') be a two-dimensional Brownian motion defined on the product probability space $(\boldsymbol{\Omega}, \boldsymbol{\mathscr{F}}, \mathbf{P}) = (\Omega \times \Omega', \mathscr{F} \otimes \mathscr{F}', P \times P')$. Let $F \in L^2(\Omega)$. Since F is measurable with respect to the Brownian motion B, we can write $F = \Psi_F(B)$ with Ψ_F a measurable mapping determined $P \circ B^{-1}$ a.s.. As a consequence, for any $t \geq 0$ the random variable $\Psi_F(e^{-t}B + \sqrt{1-e^{-2t}}B')$ is well-defined $P \times P'$ a.s. (note indeed that $e^{-t}B + \sqrt{1-e^{-2t}}B'$ is again a Brownian motion for any $t > 0$). We then have the following formula.

Theorem 11 (Mehler's formula). *For every $F = F(B) \in L^2(\Omega)$ and every $t \geq 0$, we have*

$$P_t(F) = E'\bigl[\Psi_F(e^{-t}B + \sqrt{1-e^{-2t}}B')\bigr], \tag{49}$$

where E' denotes the expectation with respect to P'.

Proof. By using standard arguments, one may show that the linear span of random variables F having the form $F = \exp\{\int_0^\infty h(s) dB_s\}$ with $h \in L^2(\mathbb{R}_+)$ is dense in $L^2(\Omega)$. Therefore, it suffices to consider the case where F has this particular form. On the other hand, we have the following identity, see, e.g., [32, Proposition 1.4.2(vi)]: for all $c, x \in \mathbb{R}$,

$$e^{cx - c^2/2} = \sum_{q=0}^{\infty} \frac{c^q}{q!} H_q(x),$$

with H_q the qth Hermite polynomial. By setting $c = \|h\|_{L^2(\mathbb{R}_+)} = \|h\|$ and $x = \int_0^\infty \frac{h(s)}{\|h\|} dB_s$, we deduce that

$$\exp\left\{\int_0^\infty h(s)dB_s\right\} = e^{\frac{1}{2}\|h\|^2} \sum_{q=0}^\infty \frac{\|h\|^q}{q!} H_q\left(\int_0^\infty \frac{h(s)}{\|h\|} dB_s\right),$$

implying in turn, using (10), that

$$\exp\left\{\int_0^\infty h(s)dB_s\right\} = e^{\frac{1}{2}\|h\|^2} \sum_{q=0}^\infty \frac{1}{q!} I_q^B(h^{\otimes q}). \tag{50}$$

Thus, for $F = \exp\{\int_0^\infty h(s)dB_s\}$,

$$P_t F = e^{\frac{1}{2}\|h\|^2} \sum_{q=0}^\infty \frac{e^{-qt}}{q!} I_q^B(h^{\otimes q}).$$

On the other hand,

$$E'[\Psi_F(e^{-t} B + \sqrt{1-e^{-2t}} B')] = E'\left[\exp\int_0^\infty h(s)(e^{-t} dB_s + \sqrt{1-e^{-2t}} dB'_s)\right]$$

$$= \exp\left(e^{-t}\int_0^\infty h(s)dB_s\right) \exp\left(\frac{1-e^{-2t}}{2}\|h\|^2\right)$$

$$= \exp\left(\frac{1-e^{-2t}}{2}\|h\|^2\right) e^{\frac{e^{-2t}}{2}\|h\|^2} \sum_{q=0}^\infty \frac{e^{-qt}}{q!} I_q^B(h^{\otimes q}) \quad \text{by (50)}$$

$$= P_t F.$$

The desired conclusion follows. □

Generator of the Ornstein–Uhlenbeck Semigroup. Recall the definition (44) of the Sobolev–Watanabe spaces $\mathbb{D}^{m,2}$, $m \geq 1$, and that the symmetric kernels $f_q \in L^2(\mathbb{R}_+^q)$ are uniquely defined through (39).

Definition 5. 1. The generator of the Ornstein–Uhlenbeck semigroup is the linear operator L defined on $\mathbb{D}^{2,2}$ by

$$LF = -\sum_{q=0}^\infty q I_q^B(f_q).$$

2. The pseudo-inverse of L is the linear operator L^{-1} defined on $L^2(\Omega)$ by

$$L^{-1} F = -\sum_{q=1}^\infty \frac{1}{q} I_q^B(f_q).$$

It is obvious that, for any $F \in L^2(\Omega)$, we have that $L^{-1}F \in \mathbb{D}^{2,2}$ and

$$LL^{-1}F = F - E[F]. \tag{51}$$

Our terminology for L^{-1} is explained by the identity (51). Another crucial property of L is contained in the following result, which is the exact generalization of Proposition 2.

Proposition 5. *Let $F \in \mathbb{D}^{2,2}$ and $G \in \mathbb{D}^{1,2}$. Then*

$$E[LF \times G] = -E[\langle DF, DG \rangle_{L^2(\mathbb{R}_+)}]. \tag{52}$$

Proof. By bilinearity and approximation, it is enough to show (52) for $F = I_p^B(f)$ and $G = I_q^B(g)$ with $p, q \geq 1$ and $f \in L^2(\mathbb{R}_+^p)$, $g \in L^2(\mathbb{R}_+^q)$ symmetric. When $p \neq q$, we have

$$E[LF \times G] = -pE[I_p^B(f)I_q^B(g)] = 0$$

and

$$E[\langle DF, DG \rangle_{L^2(\mathbb{R}_+)}] = pq \int_0^\infty E[I_{p-1}^B(f(\cdot, t))I_{q-1}^B(g(\cdot, t))]dt = 0$$

by (38), so the desired conclusion holds true in this case. When $p = q$, we have

$$E[LF \times G] = -pE[I_p^B(f)I_p^B(g)] = -pp!\langle f, g \rangle_{L^2(\mathbb{R}_+^p)}$$

and

$$E[\langle DF, DG \rangle_{L^2(\mathbb{R}_+)}] = p^2 \int_0^\infty E[I_{p-1}^B(f(\cdot, t))I_{p-1}^B(g(\cdot, t))]dt$$

$$= p^2(p-1)! \int_0^\infty \langle f(\cdot, t), g(\cdot, t) \rangle_{L^2(\mathbb{R}_+^{p-1})} dt = pp!\langle f, g \rangle_{L^2(\mathbb{R}_+^p)}$$

by (37), so the desired conclusion holds true as well in this case. □

We are now in position to state and prove an integration by parts formula which will play a crucial role in the sequel.

Theorem 12. *Let $\varphi : \mathbb{R} \to \mathbb{R}$ be both of class $\mathscr{C}^1$ and Lipschitz, and let $F \in \mathbb{D}^{1,2}$ and $G \in L^2(\Omega)$. Then*

$$\mathrm{Cov}(G, \varphi(F)) = E[\varphi'(F)\langle DF, -DL^{-1}G \rangle_{L^2(\mathbb{R}_+)}]. \tag{53}$$

Proof. Using the assumptions made on F and φ, we can write:

$$\text{Cov}(G, \varphi(F)) = E[L(L^{-1}G) \times \varphi(F)] \quad \text{(by (51))}$$
$$= E[\langle D\varphi(F), -DL^{-1}G \rangle_{L^2(\mathbb{R}_+)}] \quad \text{(by (52))}$$
$$= E[\varphi'(F)\langle D\varphi(F), -DL^{-1}G \rangle_{L^2(\mathbb{R}_+)}] \quad \text{(by (47))},$$

which is the announced formula. □

Theorem 12 admits a useful extension to indicator functions. Before stating and proving it, we recall the following classical result from measure theory.

Proposition 6. *Let C be a Borel set in $\mathbb{R}$, assume that $C \subset [-A, A]$ for some $A > 0$, and let μ be a finite measure on $[-A, A]$. Then, there exists a sequence (h_n) of continuous functions with support included in $[-A, A]$ and such that $h_n(x) \in [0, 1]$ and $\mathbf{1}_C(x) = \lim_{n \to \infty} h_n(x)$ μ-a.e.*

Proof. This is an immediate corollary of Lusin's theorem, see, e.g., [50, page 56]. □

Corollary 1. *Let C be a Borel set in $\mathbb{R}$, assume that $C \subset [-A, A]$ for some $A > 0$, and let $F \in \mathbb{D}^{1,2}$ be such that $E[F] = 0$. Then*

$$E\left[F \int_{-\infty}^{F} \mathbf{1}_C(x) dx\right] = E[\mathbf{1}_C(F) \langle DF, -DL^{-1}F \rangle_{L^2(\mathbb{R}_+)}].$$

Proof. Let λ denote the Lebesgue measure and let P_F denote the law of F. By Proposition 6 with $\mu = (\lambda + P_F)|_{[-A,A]}$ (that is, μ is the restriction of $\lambda + P_F$ to $[-A, A]$), there is a sequence (h_n) of continuous functions with support included in $[-A, A]$ and such that $h_n(x) \in [0, 1]$ and $\mathbf{1}_C(x) = \lim_{n \to \infty} h_n(x)$ μ-a.e. In particular, $\mathbf{1}_C(x) = \lim_{n \to \infty} h_n(x)$ λ-a.e. and P_F-a.e. By Theorem 12, we have moreover that

$$E\left[F \int_{-\infty}^{F} h_n(x) dx\right] = E[h_n(F) \langle DF, -DL^{-1}F \rangle_{L^2(\mathbb{R}_+)}].$$

The dominated convergence applies and yields the desired conclusion. □

As a corollary of both Theorem 12 and Corollary 1, we shall prove that the law of any multiple Wiener–Itô integral is always absolutely continuous with respect to the Lebesgue measure except, of course, when its kernel is identically zero.

Corollary 2 (Shigekawa; see [51]). *Let $q \geq 1$ be an integer and let f be a non zero element of $L^2(\mathbb{R}_+^q)$. Then the law of $F = I_q^B(f)$ is absolutely continuous with respect to the Lebesgue measure.*

Proof. Without loss of generality, we further assume that f is symmetric. The proof is by induction on q. When $q = 1$, the desired property is readily checked because $I_1^B(f) \sim \mathcal{N}(0, \|f\|_{L^2(\mathbb{R}_+)}^2)$, see (30). Now, let $q \geq 2$ and assume that the statement of Corollary 2 holds true for $q - 1$, that is, assume that the law of $I_{q-1}^B(g)$ is absolutely continuous for any symmetric element g of $L^2(\mathbb{R}_+^{q-1})$ such that $\|g\|_{L^2(\mathbb{R}_+^{q-1})} > 0$. Let f be a symmetric element of $L^2(\mathbb{R}_+^q)$ with $\|f\|_{L^2(\mathbb{R}_+^q)} > 0$. Let $h \in L^2(\mathbb{R})$ be such that $\left\|\int_0^\infty f(\cdot, s)h(s)ds\right\|_{L^2(\mathbb{R}_+^{q-1})} \neq 0$. (Such an h necessarily exists because, otherwise, we would have that $f(\cdot, s) = 0$ for almost all $s \geq 0$ which, by symmetry, would imply that $f \equiv 0$; this would be in contradiction with our assumption.) Using the induction assumption, we have that the law of

$$\langle DF, h\rangle_{L^2(\mathbb{R}_+)} = \int_0^\infty D_s F\, h(s)ds = qI_{q-1}^B\left(\int_0^\infty f(\cdot, s)h(s)ds\right)$$

is absolutely continuous with respect to the Lebesgue measure. In particular,

$$P(\langle DF, h\rangle_{L^2(\mathbb{R}_+)} = 0) = 0,$$

implying in turn, because $\{\|DF\|_{L^2(\mathbb{R}_+)} = 0\} \subset \{\langle DF, h\rangle_{L^2(\mathbb{R}_+)} = 0\}$, that

$$P(\|DF\|_{L^2(\mathbb{R}_+)} > 0) = 1. \tag{54}$$

Now, let C be a Borel set in $\mathbb{R}$. Using Corollary 1, we can write, for every $n \geq 1$,

$$E\left[\mathbf{1}_{C\cap[-n,n]}(F)\frac{1}{q}\|DF\|_{L^2(\mathbb{R}_+)}^2\right] = E\left[\mathbf{1}_{C\cap[-n,n]}(F)\langle DF, -DL^{-1}F\rangle_{L^2(\mathbb{R}_+)}\right]$$

$$= E\left[F\int_{-\infty}^F \mathbf{1}_{C\cap[-n,n]}(y)dy\right].$$

Assume that the Lebesgue measure of C is zero. The previous equality implies that

$$E\left[\mathbf{1}_{C\cap[-n,n]}(F)\frac{1}{q}\|DF\|_{L^2(\mathbb{R}_+)}^2\right] = 0, \quad n \geq 1.$$

But (54) holds as well, so $P(F \in C \cap [-n, n]) = 0$ for all $n \geq 1$. By monotone convergence, we actually get $P(F \in C) = 0$. This shows that the law of F is absolutely continuous with respect to the Lebesgue measure. The proof of Corollary 2 is concluded. □

To Go Further. In the literature, the most quoted reference on Malliavin calculus is the excellent book [38] by Nualart. It contains many applications of this theory (such as the study of the smoothness of probability laws or the anticipating stochastic calculus) and constitutes, as such, an unavoidable reference to go further.

5 Stein Meets Malliavin

We are now in a position to prove the Fourth Moment Theorem 2. As we will see, to do so we will combine the results of Sect. 3 (Stein's method) with those of Sect. 4 (Malliavin calculus), thus explaining the title of the current section! It is a different strategy with respect to the original proof, which was based on the use of the Dambis–Dubins–Schwarz theorem.

We start by introducing the distance we shall use to measure the closeness of the laws of random variables.

Definition 6. The total variation distance between the laws of two real-valued random variables Y and Z is defined by

$$d_{TV}(Y, Z) = \sup_{C \in \mathscr{B}(\mathbb{R})} |P(Y \in C) - P(Z \in C)|, \qquad (55)$$

where $\mathscr{B}(\mathbb{R})$ stands for the set of Borel sets in $\mathbb{R}$.

When $C \in \mathscr{B}(\mathbb{R})$, we have that $P(Y \in C \cap [-n, n]) \to P(Y \in C)$ and $P(Z \in C \cap [-n, n]) \to P(Z \in C)$ as $n \to \infty$ by the monotone convergence theorem. So, without loss we may restrict the supremum in (55) to be taken over *bounded* Borel sets, that is,

$$d_{TV}(Y, Z) = \sup_{\substack{C \in \mathscr{B}(\mathbb{R}) \\ C \text{ bounded}}} |P(Y \in C) - P(Z \in C)|. \qquad (56)$$

We are now ready to derive a bound for the Gaussian approximation of any centered element F belonging to $\mathbb{D}^{1,2}$.

Theorem 13 (Nourdin and Peccati (2009); see [27]). *Consider $F \in \mathbb{D}^{1,2}$ with $E[F] = 0$. Then, with $N \sim \mathcal{N}(0, 1)$,*

$$d_{TV}(F, N) \leq 2 E\left[|1 - \langle DF, -DL^{-1}F \rangle_{L^2(\mathbb{R}_+)}|\right]. \qquad (57)$$

Proof. Let C be a bounded Borel set in $\mathbb{R}$. Let $A > 0$ be such that $C \subset [-A, A]$. Let λ denote the Lebesgue measure and let P_F denote the law of F. By Proposition 6 with $\mu = (\lambda + P_F)|_{[-A, A]}$ (the restriction of $\lambda + P_F$ to $[-A, A]$), there is a sequence (h_n) of continuous functions such that $h_n(x) \in [0, 1]$ and $\mathbf{1}_C(x) = \lim_{n \to \infty} h_n(x)$ μ-a.e. By the dominated convergence theorem, $E[h_n(F)] \to P(F \in C)$ and $E[h_n(N)] \to P(N \in C)$ as $n \to \infty$. On the other hand, using Lemma 2 (and denoting by f_n the function associated with h_n) as well as (53) we can write, for each n,

$$\begin{aligned}
|E[h_n(F)] - E[h_n(N)]| &= |E[f_n'(F)] - E[Ff_n(F)]| \\
&= |E[f_n'(F)(1 - \langle DF, -DL^{-1}F \rangle_{L^2(\mathbb{R}_+)}]| \\
&\leq 2 E[|1 - \langle DF, -DL^{-1}F \rangle_{L^2(\mathbb{R}_+)}|].
\end{aligned}$$

Letting n go to infinity yields

$$|P(F \in C) - P(N \in C)| \leq 2\, E\big[|1 - \langle DF, -DL^{-1}F\rangle_{L^2(\mathbb{R}_+)}|\big],$$

which, together with (56), implies the desired conclusion. □

Wiener Chaos and the Fourth Moment Theorem. In this section, we apply Theorem 13 to a chaotic random variable F, that is, to a random variable having the specific form of a multiple Wiener–Itô integral. We begin with a technical lemma which, among other, shows that the fourth moment of F is necessarily greater than $3E[F^2]^2$. We recall from Definition 2 the meaning of $f \widetilde{\otimes}_r f$.

Lemma 3. *Let $q \geq 1$ be an integer and consider a symmetric function $f \in L^2(\mathbb{R}_+^q)$. Set $F = I_q^B(f)$ and $\sigma^2 = E[F^2] = q!\|f\|_{L^2(\mathbb{R}_+^q)}^2$. The following two identities hold:*

$$E\left[\left(\sigma^2 - \frac{1}{q}\|DF\|_{L^2(\mathbb{R}_+)}^2\right)^2\right] = \sum_{r=1}^{q-1} \frac{r^2}{q^2} r!^2 \binom{q}{r}^4 (2q-2r)! \|f \widetilde{\otimes}_r f\|_{L^2(\mathbb{R}_+^{2q-2r})}^2 \tag{58}$$

and

$$E[F^4] - 3\sigma^4 = \frac{3}{q}\sum_{r=1}^{q-1} r\, r!^2 \binom{q}{r}^4 (2q-2r)! \|f \widetilde{\otimes}_r f\|_{L^2(\mathbb{R}_+^{2q-2r})}^2 \tag{59}$$

$$= \sum_{r=1}^{q-1} q!^2 \binom{q}{r}^2 \left\{\|f \otimes_r f\|_{L^2(\mathbb{R}_+^{2q-2r})}^2 + \binom{2q-2r}{q-r}\|f \widetilde{\otimes}_r f\|_{L^2(\mathbb{R}_+^{2q-2r})}^2\right\}. \tag{60}$$

In particular,

$$E\left[\left(\sigma^2 - \frac{1}{q}\|DF\|_{L^2(\mathbb{R}_+)}^2\right)^2\right] \leq \frac{q-1}{3q}\left(E[F^4] - 3\sigma^4\right). \tag{61}$$

Proof. We follow [28] for (58)–(59) and [40] for (60). For any $t \geq 0$, we have $D_t F = q I_{q-1}^B(f(\cdot, t))$ so that, using (43),

$$\frac{1}{q}\|DF\|_{L^2(\mathbb{R}_+)}^2 = q\int_0^\infty I_{q-1}^B(f(\cdot,t))^2 dt$$

$$= q\int_0^\infty \sum_{r=0}^{q-1} r!\binom{q-1}{r}^2 I_{2q-2-2r}^B\big(f(\cdot,t) \widetilde{\otimes}_r f(\cdot,t)\big) dt$$

$$= q \int_0^\infty \sum_{r=0}^{q-1} r! \binom{q-1}{r}^2 I_{2q-2-2r}^B(f(\cdot,t) \otimes_r f(\cdot,t))dt$$

$$= q \sum_{r=0}^{q-1} r! \binom{q-1}{r}^2 I_{2q-2-2r}^B \left(\int_0^\infty f(\cdot,t) \otimes_r f(\cdot,t)dt \right)$$

$$= q \sum_{r=0}^{q-1} r! \binom{q-1}{r}^2 I_{2q-2-2r}^B(f \otimes_{r+1} f)$$

$$= q \sum_{r=1}^{q} (r-1)! \binom{q-1}{r-1}^2 I_{2q-2r}^B(f \otimes_r f)$$

$$= q! \|f\|_{L^2(\mathbb{R}_+^q)}^2 + q \sum_{r=1}^{q-1} (r-1)! \binom{q-1}{r-1}^2 I_{2q-2r}^B(f \otimes_r f). \tag{62}$$

Since $E[F^2] = q!\|f\|_{L^2(\mathbb{R}_+^q)}^2 = \sigma^2$, the identity (58) follows now from (62) and the orthogonality properties of multiple Wiener–Itô integrals. Recall the hypercontractivity property (40) of multiple Wiener–Itô integrals, and observe the relations $-L^{-1}F = \frac{1}{q}F$ and $D(F^3) = 3F^2 DF$. By combining formula (53) with an approximation argument (the derivative of $\varphi(x) = x^3$ being not bounded), we infer that

$$E[F^4] = E[F \times F^3] = \frac{3}{q} E[F^2 \|DF\|_{L^2(\mathbb{R}_+)}^2]. \tag{63}$$

Moreover, the multiplication formula (43) yields

$$F^2 = I_q^B(f)^2 = \sum_{s=0}^{q} s! \binom{q}{s}^2 I_{2q-2s}^B(f \tilde{\otimes}_s f). \tag{64}$$

By combining this last identity with (62) and (63), we obtain (59) and finally (61). It remains to prove (60). Let σ be a permutation of $\{1,\ldots,2q\}$ (this fact is written in symbols as $\sigma \in \mathfrak{S}_{2q}$). If $r \in \{0,\ldots,q\}$ denotes the cardinality of $\{\sigma(1),\ldots,\sigma(q)\} \cap \{1,\ldots,q\}$ then it is readily checked that r is also the cardinality of $\{\sigma(q+1),\ldots,\sigma(2q)\} \cap \{q+1,\ldots,2q\}$ and that

$$\int_{\mathbb{R}_+^{2q}} f(t_1,\ldots,t_q) f(t_{\sigma(1)},\ldots,t_{\sigma(q)}) f(t_{q+1},\ldots,t_{2q})$$

$$\times f(t_{\sigma(q+1)},\ldots,t_{\sigma(2q)}) dt_1 \ldots dt_{2q}$$

$$= \int_{\mathbb{R}_+^{2q-2r}} (f \otimes_r f)(x_1,\ldots,x_{2q-2r})^2 dx_1 \ldots dx_{2q-2r}$$

$$= \|f \otimes_r f\|_{L^2(\mathbb{R}_+^{2q-2r})}^2. \tag{65}$$

Moreover, for any fixed $r \in \{0, \ldots, q\}$, there are $\binom{q}{r}^2 (q!)^2$ permutations $\sigma \in \mathfrak{S}_{2q}$ such that $\#\{\sigma(1), \ldots, \sigma(q)\} \cap \{1, \ldots, q\} = r$. (Indeed, such a permutation is completely determined by the choice of: (a) r distinct elements $y_1, \ldots, y_r$ of $\{1, \ldots, q\}$; (b) $q-r$ distinct elements $y_{r+1}, \ldots, y_q$ of $\{q+1, \ldots, 2q\}$; (c) a bijection between $\{1, \ldots, q\}$ and $\{y_1, \ldots, y_q\}$; (d) a bijection between $\{q+1, \ldots, 2q\}$ and $\{1, \ldots, 2q\} \setminus \{y_1, \ldots, y_q\}$.) Now, observe that the symmetrization of $f \otimes f$ is given by

$$f \widetilde{\otimes} f(t_1, \ldots, t_{2q}) = \frac{1}{(2q)!} \sum_{\sigma \in \mathfrak{S}_{2q}} f(t_{\sigma(1)}, \ldots, t_{\sigma(q)}) f(t_{\sigma(q+1)}, \ldots, t_{\sigma(2q)}).$$

Therefore,

$$\|f \widetilde{\otimes} f\|^2_{L^2(\mathbb{R}_+^{2q})}$$

$$= \frac{1}{(2q)!^2} \sum_{\sigma, \sigma' \in \mathfrak{S}_{2q}} \int_{\mathbb{R}_+^{2q}} f(t_{\sigma(1)}, \ldots, t_{\sigma(q)}) f(t_{\sigma(q+1)}, \ldots, t_{\sigma(2q)})$$
$$\times f(t_{\sigma'(1)}, \ldots, t_{\sigma'(q)}) f(t_{\sigma'(q+1)}, \ldots, t_{\sigma'(2q)}) dt_1 \ldots dt_{2q}$$

$$= \frac{1}{(2q)!} \sum_{\sigma \in \mathfrak{S}_{2q}} \int_{\mathbb{R}_+^{2q}} f(t_1, \ldots, t_q) f(t_{q+1}, \ldots, t_{2q})$$
$$\times f(t_{\sigma(1)}, \ldots, t_{\sigma(q)}) f(t_{\sigma(q+1)}, \ldots, t_{\sigma(2q)}) dt_1 \ldots dt_{2q}$$

$$= \frac{1}{(2q)!} \sum_{r=0}^{q} \sum_{\substack{\sigma \in \mathfrak{S}_{2q} \\ \{\sigma(1), \ldots, \sigma(q)\} \cap \{1, \ldots, q\} = r}} \int_{\mathbb{R}_+^{2q}} f(t_1, \ldots, t_q) f(t_{q+1}, \ldots, t_{2q})$$
$$\times f(t_{\sigma(1)}, \ldots, t_{\sigma(q)}) f(t_{\sigma(q+1)}, \ldots, t_{\sigma(2q)}) dt_1 \ldots dt_{2q}.$$

Using (65), we deduce that

$$(2q)! \|f \widetilde{\otimes} f\|^2_{L^2(\mathbb{R}_+^{2q})} = 2(q!)^2 \|f\|^4_{L^2(\mathbb{R}_+^q)} + (q!)^2 \sum_{r=1}^{q-1} \binom{q}{r}^2 \|f \otimes_r f\|^2_{L^2(\mathbb{R}_+^{2q-2r})}. \tag{66}$$

Using the orthogonality and isometry properties of multiple Wiener–Itô integrals, the identity (64) yields

$$E[F^4] = \sum_{r=0}^{q}(r!)^2\binom{q}{r}^4(2q-2r)!\|f\tilde{\otimes}_r f\|^2_{L^2(\mathbb{R}_+^{2q-2r})}$$

$$= (2q)!\|f\tilde{\otimes} f\|^2_{L^2(\mathbb{R}_+^{2q})} + (q!)^2\|f\|^4_{L^2(\mathbb{R}_+^q)}$$

$$+ \sum_{r=1}^{q-1}(r!)^2\binom{q}{r}^4(2q-2r)!\|f\tilde{\otimes}_r f\|^2_{L^2(\mathbb{R}_+^{2q-2r})}.$$

By inserting (66) in the previous identity (and because $(q!)^2\|f\|^4_{L^2(\mathbb{R}_+^q)} = E[F^2]^2 = \sigma^4$), we get (60). □

As a consequence of Lemma 3, we deduce the following bound on the total variation distance for the Gaussian approximation of a normalized multiple Wiener–Itô integral. This is nothing but Theorem 3 but we restate it for convenience.

Theorem 14 (Nourdin and Peccati (2009); see [27]). *Let $q \geq 1$ be an integer and consider a symmetric function $f \in L^2(\mathbb{R}_+^q)$. Set $F = I_q^B(f)$, assume that $E[F^2] = 1$, and let $N \sim \mathcal{N}(0,1)$. Then*

$$d_{TV}(F,N) \leq 2\sqrt{\frac{q-1}{3q}|E[F^4]-3|}. \qquad (67)$$

Proof. Since $L^{-1}F = -\frac{1}{q}F$, we have $\langle DF, -DL^{-1}F\rangle_{L^2(\mathbb{R}_+)} = \frac{1}{q}\|DF\|^2_{L^2(\mathbb{R}_+)}$. So, we only need to apply Theorem 13 and then formula (61) to conclude. □

The estimate (67) allows to deduce an easy proof of the following characterization of CLTs on Wiener chaos. (This is the Fourth Moment Theorem 2 of Nualart and Peccati!). We note that our proof differs from the original one, which is based on the use of the Dambis–Dubins–Schwarz theorem.

Corollary 3 (Nualart and Peccati (2005); see [40]). *Let $q \geq 1$ be an integer and consider a sequence (f_n) of symmetric functions of $L^2(\mathbb{R}_+^q)$. Set $F_n = I_q^B(f_n)$ and assume that $E[F_n^2] \to \sigma^2 > 0$ as $n \to \infty$. Then, as $n \to \infty$, the following three assertions are equivalent:*

(i) $F_n \xrightarrow{Law} N \sim \mathcal{N}(0,\sigma^2)$;
(ii) $E[F_n^4] \to E[N^4] = 3\sigma^4$;
(iii) $\|f_n\tilde{\otimes}_r f_n\|_{L^2(\mathbb{R}_+^{2q-2r})} \to 0$ for all $r = 1,\ldots,q-1$.
(iv) $\|f_n \otimes_r f_n\|_{L^2(\mathbb{R}_+^{2q-2r})} \to 0$ for all $r = 1,\ldots,q-1$.

Proof. Without loss of generality, we may and do assume that $\sigma^2 = 1$ and $E[F_n^2] = 1$ for all n. The implication (ii) → (i) is a direct application of Theorem 14. The implication (i) → (ii) comes from the Continuous Mapping

Theorem together with an approximation argument (observe that $\sup_{n \geq 1} E[F_n^4] < \infty$ by the hypercontractivity relation (40)). The equivalence between (ii) and (iii) is an immediate consequence of (59). The implication (iv) → (iii) is obvious (as $\|f_n \tilde{\otimes}_r f_n\| \leq \|f_n \otimes_r f_n\|$) whereas the implication (ii) → (iv) follows from (60).
□

Quadratic Variation of the Fractional Brownian Motion. In this section, we aim to illustrate Theorem 13 in a concrete situation. More precisely, we shall use Theorem 13 in order to derive an explicit bound for the second-order approximation of the quadratic variation of a fractional Brownian motion on [0, 1].

Let $B^H = (B_t^H)_{t \geq 0}$ be a fractional Brownian motion with Hurst index $H \in (0, 1)$. This means that B^H is a centered Gaussian process with covariance function given by

$$E[B_t^H B_s^H] = \frac{1}{2}(t^{2H} + s^{2H} - |t-s|^{2H}), \quad s, t \geq 0.$$

It is easily checked that B^H is selfsimilar of index H and has stationary increments.

Fractional Brownian motion has been successfully used in order to model a variety of natural phenomena coming from different fields, including hydrology, biology, medicine, economics or traffic networks. A natural question is thus the identification of the Hurst parameter from real data. To do so, it is popular and classical to use the quadratic variation (on, say, [0, 1]), which is observable and given by

$$S_n = \sum_{k=0}^{n-1} (B_{(k+1)/n}^H - B_{k/n}^H)^2, \quad n \geq 1.$$

One may prove (see, e.g., [25, (2.12)]) that

$$n^{2H-1} S_n \xrightarrow{\text{proba}} 1 \quad \text{as } n \to \infty. \tag{68}$$

We deduce that the estimator $\hat{H}_n$, defined as

$$\hat{H}_n = \frac{1}{2} - \frac{\log S_n}{2 \log n},$$

satisfies $\hat{H}_n \xrightarrow{\text{proba}} 1$ as $n \to \infty$. To study the asymptotic normality, consider

$$F_n = \frac{n^{2H}}{\sigma_n} \sum_{k=0}^{n-1} \left[(B_{(k+1)/n}^H - B_{k/n}^H)^2 - n^{-2H} \right] \stackrel{\text{(law)}}{=} \frac{1}{\sigma_n} \sum_{k=0}^{n-1} \left[(B_{k+1}^H - B_k^H)^2 - 1 \right],$$

where $\sigma_n > 0$ is so that $E[F_n^2] = 1$. We then have the following result.

Theorem 15. Let $N \sim \mathcal{N}(0,1)$ and assume that $H \leq 3/4$. Then, $\lim_{n\to\infty} \sigma_n^2/n = 2\sum_{r\in\mathbb{Z}} \rho^2(r)$ if $H \in (0, \frac{3}{4})$, with $\rho : \mathbb{Z} \to \mathbb{R}$ given by

$$\rho(r) = \frac{1}{2}(|r+1|^{2H} + |r-1|^{2H} - 2|r|^{2H}), \qquad (69)$$

and $\lim_{n\to\infty} \sigma_n^2/(n\log n) = \frac{9}{16}$ if $H = \frac{3}{4}$. Moreover, there exists a constant $c_H > 0$ (depending only on H) such that, for every $n \geq 1$,

$$d_{TV}(F_n, N) \leq c_H \times \begin{cases} \frac{1}{\sqrt{n}} & \text{if } H \in (0, \frac{5}{8}) \\ \frac{(\log n)^{3/2}}{\sqrt{n}} & \text{if } H = \frac{5}{8} \\ n^{4H-3} & \text{if } H \in (\frac{5}{8}, \frac{3}{4}) \\ \frac{1}{\log n} & \text{if } H = \frac{3}{4} \end{cases} \qquad (70)$$

As an immediate consequence of Theorem 15, provided $H < 3/4$ we obtain that

$$\sqrt{n}(n^{2H-1}S_n - 1) \xrightarrow{\text{law}} \mathcal{N}\left(0, 2\sum_{r\in\mathbb{Z}} \rho^2(r)\right) \quad \text{as } n \to \infty, \qquad (71)$$

implying in turn

$$\sqrt{n}\log n(\hat{H}_n - H) \xrightarrow{\text{law}} \mathcal{N}\left(0, \frac{1}{2}\sum_{r\in\mathbb{Z}} \rho^2(r)\right) \quad \text{as } n \to \infty. \qquad (72)$$

Indeed, we can write

$$\log x = x - 1 - \int_1^x du \int_1^u \frac{dv}{v^2} \quad \text{for all } x > 0,$$

so that (by considering $x \geq 1$ and $0 < x < 1$)

$$|\log x + 1 - x| \leq \frac{(x-1)^2}{2}\left\{1 + \frac{1}{x^2}\right\} \quad \text{for all } x > 0.$$

As a result,

$$\sqrt{n}\log n(\hat{H}_n - H) = -\frac{\sqrt{n}}{2}\log(n^{2H-1}S_n) = -\frac{\sqrt{n}}{2}(n^{2H-1}S_n - 1) + R_n$$

with

$$|R_n| \leq \frac{(\sqrt{n}(n^{2H-1}S_n - 1))^2}{4\sqrt{n}}\left\{1 + \frac{1}{(n^{2H-1}S_n)^2}\right\}.$$

Using (68) and (71), it is clear that $R_n \overset{proba}{\to} 0$ as $n \to \infty$ and then that (72) holds true.

Now we have motivated it, let us go back to the proof of Theorem 15. To perform our calculations, we will mainly follow ideas taken from [3]. We first need the following ancillary result.

Lemma 4. 1. *For any $r \in \mathbb{Z}$, let $\rho(r)$ be defined by (69). If $H \neq \frac{1}{2}$, one has $\rho(r) \sim H(2H-1)|r|^{2H-2}$ as $|r| \to \infty$. If $H = \frac{1}{2}$ and $|r| \geq 1$, one has $\rho(r) = 0$. Consequently, $\sum_{r \in \mathbb{Z}} \rho^2(r) < \infty$ if and only if $H < 3/4$.*
2. *For all $\alpha > -1$, we have $\sum_{r=1}^{n-1} r^\alpha \sim \frac{n^{\alpha+1}}{\alpha+1}$ as $n \to \infty$.*

Proof. 1. The sequence ρ is symmetric, that is, one has $\rho(n) = \rho(-n)$. When $r \to \infty$,

$$\rho(r) = H(2H-1)r^{2H-2} + o(r^{2H-2}).$$

Using the usual criterion for convergence of Riemann sums, we deduce that $\sum_{r \in \mathbb{Z}} \rho^2(r) < \infty$ if and only if $4H - 4 < -1$ if and only if $H < \frac{3}{4}$.
2. For $\alpha > -1$, we have:

$$\frac{1}{n} \sum_{r=1}^{n} \left(\frac{r}{n}\right)^\alpha \to \int_0^1 x^\alpha dx = \frac{1}{\alpha+1} \quad \text{as } n \to \infty.$$

We deduce that $\sum_{r=1}^{n} r^\alpha \sim \frac{n^{\alpha+1}}{\alpha+1}$ as $n \to \infty$. □

We are now in position to prove Theorem 15.

Proof of Theorem 15. Without loss of generality, we will rather use the second expression of F_n:

$$F_n = \frac{1}{\sigma_n} \sum_{k=0}^{n-1} \left[(B_{k+1}^H - B_k^H)^2 - 1 \right].$$

Consider the linear span $\mathscr{H}$ of $(B_k^H)_{k \in \mathbb{N}}$, that is, $\mathscr{H}$ is the closed linear subspace of $L^2(\Omega)$ generated by $(B_k^H)_{k \in \mathbb{N}}$. It is a real separable Hilbert space and, consequently, there exists an isometry $\Phi : \mathscr{H} \to L^2(\mathbb{R}_+)$. For any $k \in \mathbb{N}$, set $e_k = \Phi(B_{k+1}^H - B_k^H)$; we then have, for all $k, l \in \mathbb{N}$,

$$\int_0^\infty e_k(s) e_l(s) ds = E[(B_{k+1}^H - B_k^H)(B_{l+1}^H - B_l^H)] = \rho(k-l) \quad (73)$$

with ρ given by (69). Therefore,

$$\{B_{k+1}^H - B_k^H : k \in \mathbb{N}\} \overset{law}{=} \left\{ \int_0^\infty e_k(s) dB_s : k \in \mathbb{N} \right\} = \{I_1^B(e_k) : k \in \mathbb{N}\},$$

where B is a Brownian motion and $I_p^B(\cdot)$, $p \geq 1$, stands for the pth multiple Wiener–Itô integral associated to B. As a consequence we can, without loss of generality, replace F_n by

$$F_n = \frac{1}{\sigma_n} \sum_{k=0}^{n-1} \left[\left(I_1^B(e_k)\right)^2 - 1 \right].$$

Now, using the multiplication formula (43), we deduce that

$$F_n = I_2^B(f_n), \quad \text{with } f_n = \frac{1}{\sigma_n} \sum_{k=0}^{n-1} e_k \otimes e_k.$$

By using the same arguments as in the proof of Theorem 1, we obtain the exact value of σ_n:

$$\sigma_n^2 = 2 \sum_{k,l=0}^{n-1} \rho^2(k-l) = 2 \sum_{|r|<n} (n-|r|)\rho^2(r).$$

Assume that $H < \frac{3}{4}$ and write

$$\frac{\sigma_n^2}{n} = 2 \sum_{r \in \mathbb{Z}} \rho^2(r) \left(1 - \frac{|r|}{n}\right) 1_{\{|r|<n\}}.$$

Since $\sum_{r \in \mathbb{Z}} \rho^2(r) < \infty$ by Lemma 4, we obtain by dominated convergence that, when $H < \frac{3}{4}$,

$$\lim_{n \to \infty} \frac{\sigma_n^2}{n} = 2 \sum_{r \in \mathbb{Z}} \rho^2(r). \tag{74}$$

Assume now that $H = \frac{3}{4}$. We then have $\rho^2(r) \sim \frac{9}{64|r|}$ as $|r| \to \infty$, implying in turn

$$n \sum_{|r|<n} \rho^2(r) \sim \frac{9n}{64} \sum_{0<|r|<n} \frac{1}{|r|} \sim \frac{9n \log n}{32}$$

and

$$\sum_{|r|<n} |r|\rho^2(r) \sim \frac{9}{64} \sum_{|r|<n} 1 \sim \frac{9n}{32}$$

as $n \to \infty$. Hence, when $H = \frac{3}{4}$,

$$\lim_{n \to \infty} \frac{\sigma_n^2}{n \log n} = \frac{9}{16}. \tag{75}$$

On the other hand, recall that the convolution of two sequences $\{u(n)\}_{n\in\mathbb{Z}}$ and $\{v(n)\}_{n\in\mathbb{Z}}$ is the sequence $u * v$ defined as $(u * v)(j) = \sum_{n\in\mathbb{Z}} u(n)v(j-n)$, and observe that $(u * v)(l-i) = \sum_{k\in\mathbb{Z}} u(k-l)v(k-i)$ whenever $u(n) = u(-n)$ and $v(n) = v(-n)$ for all $n \in \mathbb{Z}$. Set

$$\rho_n(k) = |\rho(k)| \mathbf{1}_{\{|k| \leq n-1\}}, \quad k \in \mathbb{Z}, n \geq 1.$$

We then have (using (58) for the first equality, and noticing that $f_n \otimes_1 f_n = f_n \widetilde{\otimes}_1 f_n$),

$$E\left[\left(1 - \frac{1}{2}\|D[I_2^B(f_n)]\|_{L^2(\mathbb{R}_+)}^2\right)^2\right]$$

$$= 8\|f_n \otimes_1 f_n\|_{L^2(\mathbb{R}_+^2)}^2 = \frac{8}{\sigma_n^4} \sum_{i,j,k,l=0}^{n-1} \rho(k-l)\rho(i-j)\rho(k-i)\rho(l-j)$$

$$\leq \frac{8}{\sigma_n^4} \sum_{i,l=0}^{n-1} \sum_{j,k\in\mathbb{Z}} \rho_n(k-l)\rho_n(i-j)\rho_n(k-i)\rho_n(l-j)$$

$$= \frac{8}{\sigma_n^4} \sum_{i,l=0}^{n-1} (\rho_n * \rho_n)(l-i)^2 \leq \frac{8n}{\sigma_n^4} \sum_{k\in\mathbb{Z}} (\rho_n * \rho_n)(k)^2 = \frac{8n}{\sigma_n^4} \|\rho_n * \rho_n\|_{\ell^2(\mathbb{Z})}^2.$$

Recall Young's inequality: if $s, p, q \geq 1$ are such that $\frac{1}{p} + \frac{1}{q} = 1 + \frac{1}{s}$, then

$$\|u * v\|_{\ell^s(\mathbb{Z})} \leq \|u\|_{\ell^p(\mathbb{Z})} \|v\|_{\ell^q(\mathbb{Z})}. \tag{76}$$

Let us apply (76) with $u = v = \rho_n$, $s = 2$ and $p = \frac{4}{3}$. We get $\|\rho_n * \rho_n\|_{\ell^2(\mathbb{Z})}^2 \leq \|\rho_n\|_{\ell^{4/3}(\mathbb{Z})}^4$, so that

$$E\left[\left(1 - \frac{1}{2}\|D[I_2^B(f_n)]\|_{L^2(\mathbb{R}_+)}^2\right)^2\right] \leq \frac{8n}{\sigma_n^4}\left(\sum_{|k|<n} |\rho(k)|^{\frac{4}{3}}\right)^3. \tag{77}$$

Recall the asymptotic behavior of $\rho(k)$ as $|k| \to \infty$ from Lemma 4(1). Hence

$$\sum_{|k|<n} |\rho(k)|^{\frac{4}{3}} = \begin{cases} O(1) & \text{if } H \in (0, \frac{5}{8}) \\ O(\log n) & \text{if } H = \frac{5}{8} \\ O(n^{(8H-5)/3}) & \text{if } H \in (\frac{5}{8}, 1). \end{cases} \tag{78}$$

Assume first that $H < \frac{3}{4}$ and recall (74). This, together with (77) and (78), imply that

$$E\left[\left|1-\frac{1}{2}\|D[I_2^B(f_n)]\|^2_{L^2(\mathbb{R}_+)}\right|\right] \leq \sqrt{E\left[\left(1-\frac{1}{2}\|D[I_2^B(f_n)]\|^2_{L^2(\mathbb{R}_+)}\right)^2\right]}$$

$$\leq c_H \times \begin{cases} \frac{1}{\sqrt{n}} & \text{if } H \in (0, \frac{5}{8}) \\ \frac{(\log n)^{3/2}}{\sqrt{n}} & \text{if } H = \frac{5}{8} \\ n^{4H-3} & \text{if } H \in (\frac{5}{8}, \frac{3}{4}) \end{cases}.$$

Therefore, the desired conclusion holds for $H \in (0, \frac{3}{4})$ by applying Theorem 13. Assume now that $H = \frac{3}{4}$ and recall (75). This, together with (77) and (78), imply that

$$E\left[\left|1-\frac{1}{2}\|D[I_2^B(f_n)]\|^2_{L^2(\mathbb{R}_+)}\right|\right] \leq \sqrt{E\left[\left(1-\frac{1}{2}\|D[I_2^B(f_n)]\|^2_{L^2(\mathbb{R}_+)}\right)^2\right]}$$

$$= O(1/\log n),$$

and leads to the desired conclusion for $H = \frac{3}{4}$ as well. □

To Go Further. In [27], one may find a version of Theorem 13 where N is replaced by a centered Gamma law (see also [56]). In [1], one associate to Corollary 3 an almost sure central limit theorem. In [6], the case where H is bigger than 3/4 in Theorem 15 is analyzed.

6 The Smart Path Method

The aim of this section is to prove Theorem 4 (that is, the multidimensional counterpart of the Fourth Moment Theorem), and even a more general version of it. Following the approach developed in the previous section for the one-dimensional case, a possible way for achieving this goal would have consisted in extending Stein's method to the multivariate setting, so to combine them with the tools of Malliavin calculus. This is indeed the approach developed in [35] and it works well. In this survey, we will actually proceed differently (we follow [28]), by using the so-called "smart path method" (which is a popular method in spin glasses theory, see, e.g., Talagrand [54]).

Let us first illustrate this approach in dimension one. Let $F \in \mathbb{D}^{1,2}$ with $E[F] = 0$, let $N \sim \mathcal{N}(0, 1)$ and let $h : \mathbb{R} \to \mathbb{R}$ be a $\mathscr{C}^2$ function satisfying $\|\varphi''\|_\infty < \infty$. Imagine we want to estimate $E[h(F)] - E[h(N)]$. Without loss of generality, we may assume that N and F are stochastically independent. We further have:

$$E[h(F)] - E[h(N)] = \int_0^1 \frac{d}{dt} E[h(\sqrt{t}F + \sqrt{1-t}N)]dt$$
$$= \int_0^1 \left(\frac{1}{2\sqrt{t}} E[h'(\sqrt{t}F + \sqrt{1-t}N)F] - \frac{1}{2\sqrt{1-t}} E[h'(\sqrt{t}F + \sqrt{1-t}N)N]\right) dt.$$

For any $x \in \mathbb{R}$ and $t \in [0,1]$, Theorem 12 implies that

$$E[h'(\sqrt{t}F + \sqrt{1-t}x)F] = \sqrt{t}\, E[h''(\sqrt{t}F + \sqrt{1-t}x)\langle DF, -DL^{-1}F\rangle_{L^2(\mathbb{R}_+)}],$$

whereas a classical integration by parts yields

$$E[h'(\sqrt{t}x + \sqrt{1-t}N)N] = \sqrt{1-t}\, E[h''(\sqrt{t}x + \sqrt{1-t}N)].$$

We deduce, since N and F are independent, that

$$E[h(F)] - E[h(N)] = \frac{1}{2}\int_0^1 E[h''(\sqrt{t}x + \sqrt{1-t}N)(\langle DF, -DL^{-1}F\rangle_{L^2(\mathbb{R}_+)} - 1)]dt, \tag{79}$$

implying in turn

$$\left|E[h(F)] - E[h(N)]\right| \le \frac{1}{2}\|h''\|_\infty E\left[\left|1 - \langle DF, -DL^{-1}F\rangle_{L^2(\mathbb{R}_+)}\right|\right], \tag{80}$$

compare with (57).

It happens that this approach extends easily to the multivariate setting. To see why, we will adopt the following short-hand notation: for every $h : \mathbb{R}^d \to \mathbb{R}$ of class $\mathscr{C}^2$, we set

$$\|h''\|_\infty = \max_{i,j=1,\ldots,d} \sup_{x \in \mathbb{R}^d} \left|\frac{\partial^2 h}{\partial x_i \partial x_j}(x)\right|.$$

Theorem 16 below is a first step towards Theorem 4, and is nothing but the multivariate counterpart of (79)–(80).

Theorem 16. *Fix $d \ge 2$ and let $F = (F_1, \ldots, F_d)$ be such that $F_i \in \mathbb{D}^{1,2}$ with $E[F_i] = 0$ for any i. Let $C \in \mathscr{M}_d(\mathbb{R})$ be a symmetric and positive matrix, and let N be a centered Gaussian vector with covariance C. Then, for any $h : \mathbb{R}^d \to \mathbb{R}$ belonging to $\mathscr{C}^2$ and such that $\|h''\|_\infty < \infty$, we have*

$$\left|E[h(F)] - E[h(N)]\right| \le \frac{1}{2}\|h''\|_\infty \sum_{i,j=1}^d E\left[\left|C(i,j) - \langle DF_j, -DL^{-1}F_i\rangle_{L^2(\mathbb{R}_+)}\right|\right]. \tag{81}$$

Proof. Without loss of generality, we assume that N is independent of the underlying Brownian motion B. Let h be as in the statement of the theorem. For any

$t \in [0, 1]$, set $\Psi(t) = E[h(\sqrt{1-t}F + \sqrt{t}N)]$, so that

$$E[h(N)] - E[h(F)] = \Psi(1) - \Psi(0) = \int_0^1 \Psi'(t)dt.$$

We easily see that Ψ is differentiable on $(0, 1)$ with

$$\Psi'(t) = \sum_{i=1}^d E\left[\frac{\partial h}{\partial x_i}(\sqrt{1-t}F + \sqrt{t}N)\left(\frac{1}{2\sqrt{t}}N_i - \frac{1}{2\sqrt{1-t}}F_i\right)\right].$$

By integrating by parts, we can write

$$E\left[\frac{\partial h}{\partial x_i}(\sqrt{1-t}F + \sqrt{t}N)N_i\right] = E\left\{E\left[\frac{\partial h}{\partial x_i}(\sqrt{1-t}x + \sqrt{t}N)N_i\right]_{|x=F}\right\}$$

$$= \sqrt{t}\sum_{j=1}^d C(i,j)\, E\left\{E\left[\frac{\partial^2 h}{\partial x_i \partial x_j}(\sqrt{1-t}x + \sqrt{t}N)\right]_{|x=F}\right\}$$

$$= \sqrt{t}\sum_{j=1}^d C(i,j)\, E\left[\frac{\partial^2 h}{\partial x_i \partial x_j}(\sqrt{1-t}F + \sqrt{t}N)\right].$$

By using Theorem 12 in order to perform the integration by parts, we can also write

$$E\left[\frac{\partial h}{\partial x_i}(\sqrt{1-t}F + \sqrt{t}N)F_i\right] = E\left\{E\left[\frac{\partial h}{\partial x_i}(\sqrt{1-t}F + \sqrt{t}x)F_i\right]_{|x=N}\right\}$$

$$= \sqrt{1-t}\sum_{j=1}^d E\left\{E\left[\frac{\partial^2 h}{\partial x_i \partial x_j}(\sqrt{1-t}F + \sqrt{t}x)\langle DF_j, -DL^{-1}F_i\rangle_{L^2(\mathbb{R}_+)}\right]_{|x=N}\right\}$$

$$= \sqrt{1-t}\sum_{j=1}^d E\left[\frac{\partial^2 h}{\partial x_i \partial x_j}(\sqrt{1-t}F + \sqrt{t}N)\langle DF_j, -DL^{-1}F_i\rangle_{L^2(\mathbb{R}_+)}\right].$$

Hence

$$\Psi'(t) = \frac{1}{2}\sum_{i,j=1}^d E\left[\frac{\partial^2 h}{\partial x_i \partial x_j}(\sqrt{1-t}F + \sqrt{t}N)\left(C(i,j) - \langle DF_j, -DL^{-1}F_j\rangle_{L^2(\mathbb{R}_+)}\right)\right],$$

and the desired conclusion follows. □

We are now in position to prove Theorem 2 (using a different approach compared to the original proof; here, we rather follow [39]). We will actually even show the following more general version.

Theorem 17 (Peccati and Tudor (2005); see [46]). *Let $d \geq 2$ and $q_d, \ldots, q_1 \geq 1$ be some fixed integers. Consider vectors*

$$F_n = (F_{1,n}, \ldots, F_{d,n}) = (I^B_{q_1}(f_{1,n}), \ldots, I^B_{q_d}(f_{d,n})), \quad n \geq 1,$$

with $f_{i,n} \in L^2(\mathbb{R}_+^{q_i})$ symmetric. Let $C \in \mathcal{M}_d(\mathbb{R})$ be a symmetric and positive matrix, and let N be a centered Gaussian vector with covariance C. Assume that

$$\lim_{n \to \infty} E[F_{i,n} F_{j,n}] = C(i,j), \quad 1 \leq i, j \leq d. \tag{82}$$

Then, as $n \to \infty$, the following two conditions are equivalent:

(a) F_n converges in law to N;
(b) for every $1 \leq i \leq d$, $F_{i,n}$ converges in law to $\mathcal{N}(0, C(i,i))$.

Proof. By symmetry, we assume without loss of generality that $q_1 \leq \ldots \leq q_d$. The implication $(a) \Rightarrow (b)$ being trivial, we only concentrate on $(b) \Rightarrow (a)$. So, assume (b) and let us show that (a) holds true. Thanks to (81), we are left to show that, for each $i, j = 1, \ldots, d$,

$$\langle DF_{j,n}, -DL^{-1} F_{i,n} \rangle_{L^2(\mathbb{R}_+)} = \frac{1}{q_i} \langle DF_{j,n}, DF_{i,n} \rangle_{L^2(\mathbb{R}_+)} \xrightarrow{L^2(\Omega)} C(i,j) \quad \text{as } n \to \infty. \tag{83}$$

Observe first that, using the product formula (43),

$$\frac{1}{q_i} \langle DF_{j,n}, DF_{i,n} \rangle_{L^2(\mathbb{R}_+)} = q_j \int_0^\infty I^B_{q_i-1}(f_{i,n}(\cdot,t)) I^B_{q_j-1}(f_{j,n}(\cdot,t)) dt$$

$$= q_j \sum_{r=0}^{q_i \wedge q_j - 1} r! \binom{q_i-1}{r} \binom{q_j-1}{r} I^B_{q_i+q_j-2-2r}\left(\int_0^\infty f_{i,n}(\cdot,t) \otimes_r f_{j,n}(\cdot,t) dt\right)$$

$$= q_j \sum_{r=0}^{q_i \wedge q_j - 1} r! \binom{q_i-1}{r} \binom{q_j-1}{r} I^B_{q_i+q_j-2-2r}(f_{i,n} \otimes_{r+1} f_{j,n})$$

$$= q_j \sum_{r=1}^{q_i \wedge q_j} (r-1)! \binom{q_i-1}{r-1} \binom{q_j-1}{r-1} I^B_{q_i+q_j-2r}(f_{i,n} \otimes_r f_{j,n}). \tag{84}$$

Now, let us consider all the possible cases for q_i and q_j with $j \geq i$.

First case: $q_i = q_j = 1$. We have $\langle DF_{j,n}, DF_{i,n} \rangle_{L^2(\mathbb{R}_+)} = \langle f_{i,n}, f_{j,n} \rangle_{L^2(\mathbb{R}_+)} = E[F_{i,n} F_{j,n}]$. But it is our assumption that $E[F_{i,n} F_{j,n}] \to C(i,j)$ so (83) holds true in this case.

Second case: $q_i = 1$ and $q_j \geq 2$. We have $\langle DF_{j,n}, DF_{i,n} \rangle_{L^2(\mathbb{R}_+)} = \langle f_{i,n}, DF_{j,n} \rangle_{L^2(\mathbb{R}_+)} = I^B_{q_j-1}(f_{i,n} \otimes_1 f_{j,n})$. We deduce that

$$E[\langle DF_{j,n}, DF_{i,n}\rangle^2_{L^2(\mathbb{R}_+)}] = (q_j-1)!\|f_{i,n}\tilde\otimes_1 f_{j,n}\|^2_{L^2(\mathbb{R}_+^{q_j-1})}$$

$$\leq (q_j-1)!\|f_{i,n}\otimes_1 f_{j,n}\|^2_{L^2(\mathbb{R}_+^{q_j-1})}$$

$$= (q_j-1)!\langle f_{i,n}\otimes f_{i,n}, f_{j,n}\otimes_{q_j-1} f_{j,n}\rangle_{L^2(\mathbb{R}_+^2)}$$

$$\leq (q_j-1)!\|f_{i,n}\|^2_{L^2(\mathbb{R}_+)}\|f_{j,n}\otimes_{q_j-1} f_{j,n}\|_{L^2(\mathbb{R}_+^2)}$$

$$= (q_j-1)!E[F_{i,n}^2]\|f_{j,n}\otimes_{q_j-1} f_{j,n}\|_{L^2(\mathbb{R}_+^2)}.$$

At this stage, observe the following two facts. First, because $q_i\neq q_j$, we have $C(i,j)=0$ necessarily. Second, since $E[F_{j,n}^2]\to C(j,j)$ and $F_{j,n}\overset{\text{Law}}{\to}\mathcal{N}(0,C(j,j))$, we have by Theorem 3 that $\|f_{j,n}\otimes_{q_j-1} f_{j,n}\|_{L^2(\mathbb{R}_+^2)}\to 0$. Hence, (83) holds true in this case as well.

Third case: $q_i=q_j\geq 2$. By (84), we can write

$$\frac{1}{q_i}\langle DF_{j,n}, DF_{i,n}\rangle_{L^2(\mathbb{R}_+)} = E[F_{i,n}F_{j,n}] + q_i\sum_{r=1}^{q_i-1}(r-1)!\binom{q_i-1}{r-1}^2 I_{2q_i-2r}^B(f_{i,n}\otimes_r f_{j,n}).$$

We deduce that

$$E\left[\left(\frac{1}{q_i}\langle DF_{j,n}, DF_{i,n}\rangle_{L^2(\mathbb{R}_+)} - C(i,j)\right)^2\right]$$

$$= (E[F_{i,n}F_{j,n}] - C(i,j))^2$$

$$+ q_i^2\sum_{r=1}^{q_i-1}(r-1)!^2\binom{q_i-1}{r-1}^4(2q_i-2r)!\|f_{i,n}\tilde\otimes_r f_{j,n}\|^2_{L^2(\mathbb{R}_+^{2q_i-2r})}.$$

The first term of the right-hand side tends to zero by assumption. For the second term, we can write, whenever $r\in\{1,\ldots,q_i-1\}$,

$$\|f_{i,n}\tilde\otimes_r f_{j,n}\|^2_{L^2(\mathbb{R}_+^{2q_i-2r})} \leq \|f_{i,n}\otimes_r f_{j,n}\|^2_{L^2(\mathbb{R}_+^{2q_i-2r})}$$

$$= \langle f_{i,n}\otimes_{q_i-r} f_{i,n}, f_{j,n}\otimes_{q_i-r} f_{j,n}\rangle_{L^2(\mathbb{R}_+^{2r})}$$

$$\leq \|f_{i,n}\otimes_{q_i-r} f_{i,n}\|_{L^2(\mathbb{R}_+^{2r})}\|f_{j,n}\otimes_{q_i-r} f_{j,n}\|_{L^2(\mathbb{R}_+^{2r})}.$$

Since $F_{i,n}\overset{\text{Law}}{\to}\mathcal{N}(0,C(i,i))$ and $F_{j,n}\overset{\text{Law}}{\to}\mathcal{N}(0,C(j,j))$, by Theorem 3 we have that $\|f_{i,n}\otimes_{q_i-r} f_{i,n}\|_{L^2(\mathbb{R}_+^{2r})}\|f_{j,n}\otimes_{q_i-r} f_{j,n}\|_{L^2(\mathbb{R}_+^{2r})}\to 0$, thereby showing that (83) holds true in our third case.

Fourth case: $q_j > q_i \geqslant 2$. By (84), we have

$$\frac{1}{q_i}\langle DF_{j,n}, DF_{i,n}\rangle_{L^2(\mathbb{R}_+)} = q_j \sum_{r=1}^{q_i}(r-1)!\binom{q_i-1}{r-1}\binom{q_j-1}{r-1}I^B_{q_i+q_j-2r}(f_{i,n}\otimes_r f_{j,n}).$$

We deduce that

$$E\left[\frac{1}{q_i}\langle DF_{j,n}, DF_{i,n}\rangle^2_{L^2(\mathbb{R}_+)}\right]$$
$$= q_j^2 \sum_{r=1}^{q_i}(r-1)!^2\binom{q_i-1}{r-1}^2\binom{q_j-1}{r-1}^2(q_i+q_j-2r)!\|f_{i,n}\widetilde{\otimes}_r f_{j,n}\|^2_{L^2(\mathbb{R}_+^{q_i+q_j-2r})}.$$

For any $r \in \{1,\ldots,q_i\}$, we have

$$\|f_{i,n}\widetilde{\otimes}_r f_{j,n}\|^2_{L^2(\mathbb{R}_+^{q_i+q_j-2r})} \leqslant \|f_{i,n}\otimes_r f_{j,n}\|^2_{L^2(\mathbb{R}_+^{q_i+q_j-2r})}$$
$$= \langle f_{i,n}\otimes_{q_i-r} f_{i,n}, f_{j,n}\otimes_{q_j-r} f_{j,n}\rangle_{L^2(\mathbb{R}_+^{2r})}$$
$$\leqslant \|f_{i,n}\otimes_{q_i-r} f_{i,n}\|_{L^2(\mathbb{R}_+^{2r})}\|f_{j,n}\otimes_{q_j-r} f_{j,n}\|_{L^2(\mathbb{R}_+^{2r})}$$
$$\leqslant \|f_{i,n}\|^2_{L^2(\mathbb{R}_+^{q_i})}\|f_{j,n}\otimes_{q_j-r} f_{j,n}\|_{L^2(\mathbb{R}_+^{2r})}.$$

Since $F_{j,n} \overset{\text{Law}}{\to} \mathcal{N}(0, C(j,j))$ and $q_j - r \in \{1,\ldots,q_j-1\}$, by Theorem 3 we have that $\|f_{j,n}\otimes_{q_j-r} f_{j,n}\|_{L^2(\mathbb{R}_+^{2r})} \to 0$. We deduce that (83) holds true in our fourth case.

Summarizing, we have that (83) is true for any i and j, and the proof of the theorem is done. □

When the integers $q_d,\ldots,q_1$ are pairwise disjoint in Theorem 17, notice that (82) is automatically verified with $C(i,j) = 0$ for all $i \neq j$, see indeed (38). As such, we recover the version of Theorem 17 (that is, Theorem 4) which was stated and used in Lecture 1 to prove Breuer–Major theorem.

To Go Further. In [35], Stein's method is combined with Malliavin calculus in a multivariate setting to provide bounds for the Wasserstein distance between the laws of $N \sim \mathcal{N}_d(0, C)$ and $F = (F_1,\ldots,F_d)$ where each $F_i \in \mathbb{D}^{1,2}$ verifies $E[F_i] = 0$. Compare with Theorem 16.

7 Cumulants on the Wiener Space

In this section, following [29] our aim is to analyze the cumulants of a given element F of $\mathbb{D}^{1,2}$ and to show how the formula we shall obtain allows one to give yet another proof of the Fourth Moment Theorem 2.

Let F be a random variable with, say, all the moments (to simplify the exposition). Let ϕ_F denote its characteristic function, that is, $\phi_F(t) = E[e^{itF}]$, $t \in \mathbb{R}$. Then, it is well-known that we may recover the moments of F from ϕ_F through the identity

$$E[F^j] = (-i)^j \frac{d^j}{dt^j}|_{t=0} \phi_F(t).$$

The cumulants of F, denoted by $\{\kappa_j(F)\}_{j \geq 1}$, are defined in a similar way, just by replacing ϕ_F by $\log \phi_F$ in the previous expression:

$$\kappa_j(F) = (-i)^j \frac{d^j}{dt^j}|_{t=0} \log \phi_F(t).$$

The first few cumulants are

$$\kappa_1(F) = E[F],$$
$$\kappa_2(F) = E[F^2] - E[F]^2 = \text{Var}(F),$$
$$\kappa_3(F) = E[F^3] - 3E[F^2]E[F] + 2E[F]^3.$$

It is immediate that

$$\kappa_j(F+G) = \kappa_j(F) + \kappa_j(G) \quad \text{and} \quad \kappa_j(\lambda F) = \lambda^j \kappa_j(F) \tag{85}$$

for all $j \geq 1$, when $\lambda \in \mathbb{R}$ and F and G are *independent* random variables (with all the moments). Also, it is easy to express moments in terms of cumulants and vice-versa. Finally, let us observe that the cumulants of $F \sim \mathcal{N}(0, \sigma^2)$ are all zero, except for the second one which is σ^2. This fact, together with the two properties (85), gives a quick proof of the classical CLT and illustrates that cumulants are often relevant when wanting to decide whether a given random variable is approximately normally distributed.

The following simple lemma is a useful link between moments and cumulants.

Lemma 5. *Let F be a random variable (in a given probability space $(\Omega, \mathscr{F}, P)$) having all the moments. Then, for all $m \in \mathbb{N}$,*

$$E[F^{m+1}] = \sum_{s=0}^{m} \binom{m}{s} \kappa_{s+1}(F) E[F^{m-s}].$$

Proof. We can write

$E[F^{m+1}]$

$= (-i)^{m+1} \frac{d^{m+1}}{dt^{m+1}}|_{t=0} \phi_F(t) = (-i)^{m+1} \frac{d^m}{dt^m}|_{t=0} \left(\phi_F(t) \frac{d}{dt} \log \phi_F(t) \right)$

$$= (-i)^{m+1} \sum_{s=0}^{m} \binom{m}{s} \left(\frac{d^{s+1}}{dt^{s+1}} \big|_{t=0} \log \phi_F(t) \right) \left(\frac{d^{m-s}}{dt^{m-s}} \big|_{t=0} \phi_F(t) \right) \quad \text{by Leibniz rule}$$

$$= \sum_{s=0}^{m} \binom{m}{s} \kappa_{s+1}(F) E[F^{m-s}]. \qquad \square$$

From now on, we will deal with a random variable F with all moments that is further measurable with respect to the Brownian motion $(B_t)_{t \geq 0}$. We let the notation of Sect. 4 prevail and we consider the chaotic expansion (39) of F. We further assume (only to avoid technical issues) that F belongs to $\mathbb{D}^\infty$, meaning that $F \in \mathbb{D}^{m,2}$ for all $m \geq 1$ and that $E[\|D^m F\|_{L^p(\mathbb{R}_+^m)}^p] < \infty$ for all $m \geq 1$ and all $p \geq 2$. This assumption allows us to introduce recursively the following (well-defined) sequence of random variables related to F. Namely, set $\Gamma_0(F) = F$ and

$$\Gamma_{j+1}(F) = \langle DF, -DL^{-1}\Gamma_j(F) \rangle_{L^2(\mathbb{R}_+)}.$$

The following result contains a neat expression of the cumulants of F in terms of the family $\{\Gamma_s(F)\}_{s \in \mathbb{N}}$.

Theorem 18 (Nourdin and Peccati (2010); see [29]). *Let $F \in \mathbb{D}^\infty$. Then, for any $s \in \mathbb{N}$,*

$$\kappa_{s+1}(F) = s! E[\Gamma_s(F)].$$

Proof. The proof is by induction. It consists in computing $\kappa_{s+1}(F)$ using the induction hypothesis, together with Lemma 5 and (53). First, the result holds true for $s = 0$, as it only says that $\kappa_1(F) = E[\Gamma_0(F)] = E[F]$. Assume now that $m \geq 1$ is given and that $\kappa_{s+1}(F) = s! E[\Gamma_s(F)]$ for all $s \leq m - 1$. We can then write

$$\kappa_{m+1}(F) = E[F^{m+1}] - \sum_{s=0}^{m-1} \binom{m}{s} \kappa_{s+1}(F) E[F^{m-s}] \quad \text{by Lemma 5}$$

$$= E[F^{m+1}] - \sum_{s=0}^{m-1} s! \binom{m}{s} E[\Gamma_s(F)] E[F^{m-s}] \quad \text{by the induction hypothesis.}$$

On the other hand, by applying (53) repeatedly, we get

$$E[F^{m+1}] = E[F^m] E[\Gamma_0(F)] + \text{Cov}(F^m, \Gamma_0(F)) = E[F^m] E[\Gamma_0(F)] + m E[F^{m-1} \Gamma_1(F)]$$

$$= E[F^m] E[\Gamma_0(F)] + m E[F^{m-1}] E[\Gamma_1(F)] + m \text{Cov}(F^{m-1}, \Gamma_1(F))$$

$$= E[F^m] E[\Gamma_0(F)] + m E[F^{m-1}] E[\Gamma_1(F)] + m(m-1) E[F^{m-2} \Gamma_2(F)]$$

$$= \ldots$$

$$= \sum_{s=0}^{m} s! \binom{m}{s} E[F^{m-s}] E[\Gamma_s(F)].$$

Thus

$$\kappa_{m+1}(F) = E[F^{m+1}] - \sum_{s=0}^{m-1} s! \binom{m}{s} E[\Gamma_s(F)] E[F^{m-s}] = m! E[\Gamma_m(F)],$$

and the desired conclusion follows. □

Let us now focus on the computation of cumulants associated to random variables having the form of a multiple Wiener–Itô integral. The following statement provides a compact representation for the cumulants of such random variables.

Theorem 19. *Let $q \geq 2$ and assume that $F = I_q^B(f)$, where $f \in L^2(\mathbb{R}_+^q)$. We have $\kappa_1(F) = 0$, $\kappa_2(F) = q! \|f\|_{L^2(\mathbb{R}_+^q)}^2$ and, for every $s \geq 3$,*

$$\kappa_s(F) = q!(s-1)! \sum c_q(r_1, \ldots, r_{s-2}) \langle (\ldots ((f \tilde{\otimes}_{r_1} f) \tilde{\otimes}_{r_2} f) \ldots \tilde{\otimes}_{r_{s-3}} f) \tilde{\otimes}_{r_{s-2}} f, f \rangle_{L^2(\mathbb{R}_+^q)}, \quad (86)$$

where the sum $\sum$ runs over all collections of integers $r_1, \ldots, r_{s-2}$ such that:

(i) $1 \leq r_1, \ldots, r_{s-2} \leq q$;
(ii) $r_1 + \ldots + r_{s-2} = \frac{(s-2)q}{2}$;
(iii) $r_1 < q$, $r_1 + r_2 < \frac{3q}{2}$, $\ldots$, $r_1 + \ldots + r_{s-3} < \frac{(s-2)q}{2}$;
(iv) $r_2 \leq 2q - 2r_1$, $\ldots$, $r_{s-2} \leq (s-2)q - 2r_1 - \ldots - 2r_{s-3}$;

and where the combinatorial constants $c_q(r_1, \ldots, r_{s-2})$ are recursively defined by the relations

$$c_q(r) = q(r-1)! \binom{q-1}{r-1}^2,$$

and, for $a \geq 2$,

$$c_q(r_1, \ldots, r_a) = q(r_a - 1)! \binom{aq - 2r_1 - \ldots - 2r_{a-1} - 1}{r_a - 1} \binom{q-1}{r_a - 1} c_q(r_1, \ldots, r_{a-1}).$$

Remark 5. 1. If sq is odd, then $\kappa_s(F) = 0$, see indeed condition (ii). This fact is easy to see in any case: use that $\kappa_s(-F) = (-1)^s \kappa_s(F)$ and observe that, when q is odd, then $F \stackrel{\text{(law)}}{=} -F$ (since $B \stackrel{\text{(law)}}{=} -B$).
2. If $q = 2$ and $F = I_2^B(f)$ with $f \in L^2(\mathbb{R}_+^2)$, then the only possible integers $r_1, \ldots, r_{s-2}$ verifying (i)–(iv) in the previous statement are $r_1 = \ldots = r_{s-2} = 1$. On the other hand, we immediately compute that $c_2(1) = 2$, $c_2(1, 1) = 4$, $c_2(1, 1, 1) = 8$, and so on. Therefore,

$$\kappa_s(I_2^B(f)) = 2^{s-1}(s-1)! \langle (\ldots (f \otimes_1 f) \ldots f) \otimes_1 f, f \rangle_{L^2(\mathbb{R}_+^2)}, \quad (87)$$

and we recover the classical expression of the cumulants of a double integral.

3. If $q \geq 2$ and $F = I_q^B(f)$, $f \in L^2(\mathbb{R}_+^q)$, then (86) for $s = 4$ reads

$$\kappa_4(I_q^B(f)) = 6q! \sum_{r=1}^{q-1} c_q(r, q-r) \langle (f \tilde{\otimes}_r f) \tilde{\otimes}_{q-r} f, f \rangle_{L^2(\mathbb{R}_+^q)}$$

$$= \frac{3}{q} \sum_{r=1}^{q-1} rr!^2 \binom{q}{r}^4 (2q - 2r)! \langle (f \tilde{\otimes}_r f) \otimes_{q-r} f, f \rangle_{L^2(\mathbb{R}_+^q)}$$

$$= \frac{3}{q} \sum_{r=1}^{q-1} rr!^2 \binom{q}{r}^4 (2q - 2r)! \langle f \tilde{\otimes}_r f, f \otimes_r f \rangle_{L^2(\mathbb{R}_+^{2q-2r})}$$

$$= \frac{3}{q} \sum_{r=1}^{q-1} rr!^2 \binom{q}{r}^4 (2q - 2r)! \| f \tilde{\otimes}_r f \|_{L^2(\mathbb{R}_+^{2q-2r})}^2, \tag{88}$$

and we recover the expression for $\kappa_4(F)$ given in (59) by a different route.

Proof of Theorem 19. Let us first show the following formula: for any $s \geq 2$, we claim that

$$\Gamma_{s-1}(F)$$

$$= \sum_{r_1=1}^{q} \cdots \sum_{r_{s-1}=1}^{[(s-1)q-2r_1-\ldots-2r_{s-2}] \wedge q} c_q(r_1, \ldots, r_{s-1}) \mathbf{1}_{\{r_1 < q\}} \cdots \mathbf{1}_{\{r_1 + \ldots + r_{s-2} < \frac{(s-1)q}{2}\}}$$

$$\times I_{sq-2r_1-\ldots-2r_{s-1}}^B \big((\ldots (f \tilde{\otimes}_{r_1} f) \tilde{\otimes}_{r_2} f) \ldots f) \tilde{\otimes}_{r_{s-1}} f \big). \tag{89}$$

We shall prove (89) by induction. When $s = 2$, identity (89) simply reads $\Gamma_1(F) = \sum_{r=1}^{q} c_q(r) I_{2q-2r}^B(f \tilde{\otimes}_r f)$ and is nothing but (62). Assume now that (89) holds for $\Gamma_{s-1}(F)$, and let us prove that it continues to hold for $\Gamma_s(F)$. We have, using the product formula (43) and following the same line of reasoning as in the proof of (62),

$$\Gamma_s(F) = \langle DF, -DL^{-1}\Gamma_{s-1}F \rangle_{L^2(\mathbb{R}_+)}$$

$$= \sum_{r_1=1}^{q} \cdots \sum_{r_{s-1}=1}^{[(s-1)q-2r_1-\ldots-2r_{s-2}] \wedge q} qc_q(r_1, \ldots, r_{s-1}) \mathbf{1}_{\{r_1 < q\}} \cdots \mathbf{1}_{\{r_1 + \ldots + r_{s-2} < \frac{(s-1)q}{2}\}}$$

$$\times \mathbf{1}_{\{r_1 + \ldots + r_{s-1} < \frac{sq}{2}\}}$$

$$\times \langle I_{q-1}^B(f), I_{sq-2r_1-\ldots-2r_{s-1}-1}^B \big((\ldots (f \tilde{\otimes}_{r_1} f) \tilde{\otimes}_{r_2} f) \ldots f) \tilde{\otimes}_{r_{s-1}} f \big) \rangle_{L^2(\mathbb{R}_+)}$$

$$= \sum_{r_1=1}^{q} \cdots \sum_{r_{s-1}=1}^{[(s-1)q-2r_1-\ldots-2r_{s-2}] \wedge q} \sum_{r_s=1}^{[sq-2r_1-\ldots-2r_{s-1}] \wedge q} c_q(r_1, \ldots, r_{s-1}) \times q(r_s - 1)!$$

$$\times \binom{sq - 2r_1 - \ldots - 2r_{s-1} - 1}{r_s - 1}\binom{q-1}{r_s-1}\mathbf{1}_{\{r_1<q\}}\cdots\mathbf{1}_{\{r_1+\ldots+r_{s-2}<\frac{(s-1)q}{2}\}}$$

$$\times \mathbf{1}_{\{r_1+\ldots+r_{s-1}<\frac{sq}{2}\}}I^B_{(s+1)q-2r_1-\ldots-2r_s}((\ldots(f\widetilde{\otimes}_{r_1}f)\widetilde{\otimes}_{r_2}f)\ldots f)\widetilde{\otimes}_{r_s}f),$$

which is the desired formula for $\Gamma_s(F)$. The proof of (89) for all $s \geq 1$ is thus finished. Now, let us take the expectation on both sides of (89). We get

$$\kappa_s(F) = (s-1)! E[\Gamma_{s-1}(F)]$$

$$= (s-1)!$$

$$\times \sum_{r_1=1}^{q}\cdots\sum_{r_{s-1}=1}^{[(s-1)q-2r_1-\ldots-2r_{s-2}]\wedge q} c_q(r_1,\ldots,r_{s-1})\mathbf{1}_{\{r_1<q\}}\cdots\mathbf{1}_{\{r_1+\ldots+r_{s-2}<\frac{(s-1)q}{2}\}}$$

$$\times \mathbf{1}_{\{r_1+\ldots+r_{s-1}=\frac{sq}{2}\}} \times (\ldots(f\widetilde{\otimes}_{r_1}f)\widetilde{\otimes}_{r_2}f)\ldots f)\widetilde{\otimes}_{r_{s-1}}f.$$

Observe that, if $2r_1 + \ldots + 2r_{s-1} = sq$ and $r_{s-1} \leq (s-1)q - 2r_1 - \ldots - 2r_{s-2}$ then $2r_{s-1} = q + (s-1)q - 2r_1 - \ldots - 2r_{s-2} \geq q + r_{s-1}$, so that $r_{s-1} \geq q$. Therefore,

$$\kappa_s(F) = (s-1)!$$

$$\times \sum_{r_1=1}^{q}\cdots\sum_{r_{s-2}=1}^{[(s-2)q-2r_1-\ldots-2r_{s-3}]\wedge q} c_q(r_1,\ldots,r_{s-2},q)\mathbf{1}_{\{r_1<q\}}\cdots\mathbf{1}_{\{r_1+\ldots+r_{s-3}<\frac{(s-2)q}{2}\}}$$

$$\times \mathbf{1}_{\{r_1+\ldots+r_{s-2}=\frac{(s-2)q}{2}\}}\langle(\ldots(f\widetilde{\otimes}_{r_1}f)\widetilde{\otimes}_{r_2}f)\ldots f)\widetilde{\otimes}_{r_{s-2}}f, f\rangle_{L^2(\mathbb{R}^q_+)},$$

which is the announced result, since $c_q(r_1,\ldots,r_{s-2},q) = q! c_q(r_1,\ldots,r_{s-2})$. $\square$

We conclude this section by providing yet another proof (based on our new formula (86)) of the Fourth Moment Theorem 2. More precisely, let us show by another route that, if $q \geq 2$ is fixed and if $(F_n)_{n\geq 1}$ is a sequence of the form $F_n = I^B_q(f_n)$ with $f_n \in L^2(\mathbb{R}^q_+)$ such that $E[F_n^2] = q! \|f_n\|^2_{L^2(\mathbb{R}^q_+)} = 1$ for all $n \geq 1$ and $E[F_n^4] \to 3$ as $n \to \infty$, then $F_n \to \mathcal{N}(0,1)$ in law as $n \to \infty$.

To this end, observe that $\kappa_1(F_n) = 0$ and $\kappa_2(F_n) = 1$. To estimate $\kappa_s(F_n)$, $s \geq 3$, we consider the expression (86). Let $r_1,\ldots,r_{s-2}$ be some integers such that (i)–(iv) in Theorem 19 are satisfied. Using Cauchy–Schwarz and then successively

$$\|g\widetilde{\otimes}_r h\|_{L^2(\mathbb{R}^{p+q-2r}_+)} \leq \|g\otimes_r h\|_{L^2(\mathbb{R}^{p+q-2r}_+)} \leq \|g\|_{L^2(\mathbb{R}^p_+)}\|h\|_{L^2(\mathbb{R}^q_+)}$$

whenever $g \in L^2(\mathbb{R}_+^p)$, $h \in L^2(\mathbb{R}_+^q)$ and $r = 1, \ldots, p \wedge q$, we get that

$$|\langle (\ldots (f_n \widetilde{\otimes}_{r_1} f_n) \widetilde{\otimes}_{r_2} f_n) \ldots f_n) \widetilde{\otimes}_{r_{s-2}} f_n, f_n \rangle_{L^2(\mathbb{R}_+^q)}|$$

$$\leq \|(\ldots (f_n \widetilde{\otimes}_{r_1} f_n) \widetilde{\otimes}_{r_2} f_n) \ldots f_n) \widetilde{\otimes}_{r_{s-2}} f_n\|_{L^2(\mathbb{R}_+^q)} \|f_n\|_{L^2(\mathbb{R}_+^q)}$$

$$\leq \|f_n \widetilde{\otimes}_{r_1} f_n\|_{L^2(\mathbb{R}_+^{2q-2r_1})} \|f_n\|_{L^2(\mathbb{R}_+^q)}^{s-2}$$

$$= (q!)^{1-\frac{s}{2}} \|f_n \widetilde{\otimes}_{r_1} f_n\|_{L^2(\mathbb{R}_+^{2q-2r_1})}. \tag{90}$$

Since $E[F_n^4] - 3 = \kappa_4(F_n) \to 0$, we deduce from (88) that $\|f_n \widetilde{\otimes}_r f_n\|_{L^2(\mathbb{R}_+^{2q-2r})} \to 0$ for all $r = 1, \ldots, q-1$. Consequently, by combining (86) with (90), we get that $\kappa_s(F_n) \to 0$ as $n \to \infty$ for all $s \geq 3$, implying in turn that $F_n \to \mathcal{N}(0,1)$ in law. □

To Go Further. The multivariate version of Theorem 18 may be found in [23].

8 A New Density Formula

In this section, following [37] we shall explain how the quantity $\langle DF, -DL^{-1}F \rangle_{L^2(\mathbb{R}_+)}$ is related to the density of $F \in \mathbb{D}^{1,2}$ (provided it exists). More specifically, when $F \in \mathbb{D}^{1,2}$ is such that $E[F] = 0$, let us introduce the function $g_F : \mathbb{R} \to \mathbb{R}$, defined by means of the following identity:

$$g_F(F) = E[\langle DF, -DL^{-1}F \rangle_{L^2(\mathbb{R}_+)} | F]. \tag{91}$$

A key property of the random variable $g_F(F)$ is as follows.

Proposition 7. *If $F \in \mathbb{D}^{1,2}$ satisfies $E[F] = 0$, then $P(g_F(F) \geq 0) = 1$.*

Proof. Let C be a Borel set of $\mathbb{R}$ and set $\phi_n(x) = \int_0^x \mathbf{1}_{C \cap [-n,n]}(t) dt$, $n \geq 1$ (with the usual convention $\int_0^x = -\int_x^0$ for $x < 0$). Since ϕ_n is increasing and vanishing at zero, we have $x\phi_n(x) \geq 0$ for all $x \in \mathbb{R}$. In particular,

$$0 \leq E[F\phi_n(F)] = E\left[F \int_0^F \mathbf{1}_{C \cap [-n,n]}(t) dt\right] = E\left[F \int_{-\infty}^F \mathbf{1}_{C \cap [-n,n]}(t) dt\right].$$

Therefore, we deduce from Corollary 1 that $E\left[g_F(F) \mathbf{1}_{C \cap [-n,n]}(F)\right] \geq 0$. By dominated convergence, this yields $E[g_F(F) \mathbf{1}_C(F)] \geq 0$, implying in turn that $P(g_F(F) \geq 0) = 1$. □

The following theorem gives a new density formula for F in terms of the function g_F. We will then study some of its consequences.

Theorem 20 (Nourdin and Viens (2009); see [37]). *Let $F \in \mathbb{D}^{1,2}$ with $E[F] = 0$. Then, the law of F admits a density with respect to Lebesgue measure (say, $\rho : \mathbb{R} \to \mathbb{R}$) if and only if $P(g_F(F) > 0) = 1$. In this case, the support of ρ, denoted by $\operatorname{supp} \rho$, is a closed interval of $\mathbb{R}$ containing zero and we have, for (almost) all $x \in \operatorname{supp} \rho$:*

$$\rho(x) = \frac{E[|F|]}{2g_F(x)} \exp\left(-\int_0^x \frac{y\,dy}{g_F(y)}\right). \tag{92}$$

Proof. Assume that $P(g_F(F) > 0) = 1$ and let C be a Borel set. Let $n \geq 1$. Corollary 1 yields

$$E\left[F \int_{-\infty}^{F} \mathbf{1}_{C \cap [-n,n]}(t)dt\right] = E\left[\mathbf{1}_{C \cap [-n,n]}(F) g_F(F)\right]. \tag{93}$$

Suppose that the Lebesgue measure of C is zero. Then $\int_{-\infty}^{F} \mathbf{1}_{C \cap [-n,n]}(t)dt = 0$, so that $E\left[\mathbf{1}_{C \cap [-n,n]}(F) g_F(F)\right] = 0$ by (93). But, since $P(g_F(F) > 0) = 1$, we get that $P(F \in C \cap [-n,n]) = 0$ and, by letting $n \to \infty$, that $P(F \in C) = 0$. Therefore, the Radon–Nikodym criterion is verified, hence implying that the law of F has a density.

Conversely, assume that the law of F has a density, say ρ. Let $\phi : \mathbb{R} \to \mathbb{R}$ be a continuous function with compact support, and let Φ denote any antiderivative of ϕ. Note that Φ is necessarily bounded. We can write:

$$E[\phi(F) g_F(F)] = E[\Phi(F) F] \quad \text{by (53)}$$

$$= \int_{\mathbb{R}} \Phi(x) \, x \, \rho(x) dx \underset{(*)}{=} \int_{\mathbb{R}} \phi(x) \left(\int_x^\infty y\rho(y)dy\right) dx = E\left[\phi(F) \frac{\int_F^\infty y\rho(y)dy}{\rho(F)}\right].$$

Equation $(*)$ was obtained by integrating by parts, after observing that

$$\int_x^\infty y\rho(y)dy \to 0 \text{ as } |x| \to \infty$$

(for $x \to +\infty$, this is because $F \in L^1(\Omega)$; for $x \to -\infty$, this is because F has mean zero). Therefore, we have shown that, P-a.s.,

$$g_F(F) = \frac{\int_F^\infty y\rho(y)dy}{\rho(F)}. \tag{94}$$

(Notice that $P(\rho(F) > 0) = \int_{\mathbb{R}} \mathbf{1}_{\{\rho(x)>0\}} \rho(x)dx = \int_{\mathbb{R}} \rho(x)dx = 1$, so that identity (94) always makes sense.) Since $F \in \mathbb{D}^{1,2}$, one has (see, e.g., [38, Proposition 2.1.7]) that $\operatorname{supp} \rho = [\alpha, \beta]$ with $-\infty \leq \alpha < \beta \leq +\infty$. Since F

has zero mean, note that $\alpha < 0$ and $\beta > 0$ necessarily. For every $x \in (\alpha, \beta)$, define

$$\varphi(x) = \int_x^\infty y\rho(y)\,dy. \tag{95}$$

The function φ is differentiable almost everywhere on (α, β), and its derivative is $-x\rho(x)$. In particular, since $\varphi(\alpha) = \varphi(\beta) = 0$ and φ is strictly increasing before 0 and strictly decreasing afterwards, we have $\varphi(x) > 0$ for all $x \in (\alpha, \beta)$. Hence, (94) implies that $P(g_F(F) > 0) = 1$.

Finally, let us prove (92). Let φ still be defined by (95). On the one hand, we have $\varphi'(x) = -x\rho(x)$ for almost all $x \in \operatorname{supp}\rho$. On the other hand, by (94), we have, for almost all $x \in \operatorname{supp}\rho$,

$$\varphi(x) = \rho(x)g_F(x). \tag{96}$$

By putting these two facts together, we get the following ordinary differential equation satisfied by φ:

$$\frac{\varphi'(x)}{\varphi(x)} = -\frac{x}{g_F(x)} \quad \text{for almost all } x \in \operatorname{supp}\rho.$$

Integrating this relation over the interval $[0, x]$ yields

$$\log\varphi(x) = \log\varphi(0) - \int_0^x \frac{y\,dy}{g_F(y)}.$$

Taking the exponential and using $0 = E(F) = E(F_+) - E(F_-)$ so that $E|F| = E(F_+) + E(F_-) = 2E(F_+) = 2\varphi(0)$, we get

$$\varphi(x) = \frac{1}{2}E[|F|]\exp\left(-\int_0^x \frac{y\,dy}{g_F(y)}\right).$$

Finally, the desired conclusion comes from (96). □

As a consequence of Theorem 20, we have the following statement, yielding sufficient conditions in order for the law of F to have a support equal to the real line.

Corollary 4. *Let $F \in \mathbb{D}^{1,2}$ with $E[F] = 0$. Assume that there exists $\sigma_{\min} > 0$ such that*

$$g_F(F) \geq \sigma_{\min}^2, \quad P\text{-a.s.} \tag{97}$$

Then the law of F, which has a density ρ by Theorem 20, has $\mathbb{R}$ for support and (92) holds almost everywhere in $\mathbb{R}$.

Proof. It is an immediate consequence of Theorem 20, except for the fact that $\operatorname{supp}\rho = \mathbb{R}$. For the moment, we just know that $\operatorname{supp}\rho = [\alpha, \beta]$ with $-\infty \leq$

$\alpha < 0 < \beta \leqslant +\infty$. Identity (94) yields

$$\int_x^\infty y\rho(y)\,dy \geqslant \sigma_{\min}^2 \rho(x) \quad \text{for almost all } x \in (\alpha, \beta). \tag{98}$$

Let φ be defined by (95), and recall that $\varphi(x) > 0$ for all $x \in (\alpha, \beta)$. When multiplied by $x \in [0, \beta)$, the inequality (98) gives $\frac{\varphi'(x)}{\varphi(x)} \geqslant -\frac{x}{\sigma_{\min}^2}$. Integrating this relation over the interval $[0, x]$ yields $\log \varphi(x) - \log \varphi(0) \geqslant -\frac{x^2}{2\sigma_{\min}^2}$, i.e., since $\varphi(0) = \frac{1}{2}E|F|$,

$$\varphi(x) = \int_x^\infty y\rho(y)\,dy \geqslant \frac{1}{2}E|F|e^{-\frac{x^2}{2\sigma_{\min}^2}}. \tag{99}$$

Similarly, when multiplied by $x \in (\alpha, 0]$, inequality (98) gives $\frac{\varphi'(x)}{\varphi(x)} \leqslant -\frac{x}{\sigma_{\min}^2}$. Integrating this relation over the interval $[x, 0]$ yields $\log \varphi(0) - \log \varphi(x) \leqslant \frac{x^2}{2\sigma_{\min}^2}$, i.e. (99) still holds for $x \in (\alpha, 0]$. Now, let us prove that $\beta = +\infty$. If this were not the case, by definition, we would have $\varphi(\beta) = 0$; on the other hand, by letting x tend to β in the above inequality, because φ is continuous, we would have $\varphi(\beta) \geqslant \frac{1}{2}E|F|e^{-\frac{\beta^2}{2\sigma_{\min}^2}} > 0$, which contradicts $\beta < +\infty$. The proof of $\alpha = -\infty$ is similar. In conclusion, we have shown that $\operatorname{supp}\rho = \mathbb{R}$. □

Using Corollary 4, we deduce a neat criterion for normality.

Corollary 5. *Let $F \in \mathbb{D}^{1,2}$ with $E[F] = 0$ and assume that F is not identically zero. Then F is Gaussian if and only if $\operatorname{Var}(g_F(F)) = 0$.*

Proof. By (53) (choose $\varphi(x) = x$, $G = F$ and recall that $E[F] = 0$), we have

$$E[\langle DF, -DL^{-1}F\rangle_{\mathfrak{H}}] = E[F^2] = \operatorname{Var} F. \tag{100}$$

Therefore, the condition $\operatorname{Var}(g_F(F)) = 0$ is equivalent to $P(g_F(F) = \operatorname{Var} F) = 1$. Let $F \sim \mathcal{N}(0, \sigma^2)$ with $\sigma > 0$. Using (94), we immediately check that $g_F(F) = \sigma^2$, P-a.s. Conversely, if $g_F(F) = \sigma^2 > 0$ P-a.s., then Corollary 4 implies that the law of F has a density ρ, given by $\rho(x) = \frac{E|F|}{2\sigma^2}e^{-\frac{x^2}{2\sigma^2}}$ for almost all $x \in \mathbb{R}$, from which we immediately deduce that $F \sim \mathcal{N}(0, \sigma^2)$. □

Observe that if $F \sim \mathcal{N}(0, \sigma^2)$ with $\sigma > 0$, then $E|F| = \sqrt{2/\pi}\,\sigma$, so that the formula (92) for ρ agrees, of course, with the usual one in this case.

As a "concrete" application of (92), let us consider the following situation. Let $K : [0, 1]^2 \to \mathbb{R}$ be a square-integrable kernel such that $K(t, s) = 0$ for $s > t$, and consider the centered Gaussian process $X = (X_t)_{t \in [0,1]}$ defined as

$$X_t = \int_0^1 K(t, s)\,dB_s = \int_0^t K(t, s)\,dB_s, \quad t \in [0, 1]. \tag{101}$$

Fractional Brownian motion is an instance of such a process, see, e.g., [25, Sect. 2.3]. Consider the maximum

$$Z = \sup_{t \in [0,1]} X_t. \tag{102}$$

Assume further that the kernel K satisfies

$$\exists c, \alpha > 0, \quad \forall s, t \in [0,1]^2, s \neq t, \quad 0 < \int_0^1 (K(t,u) - K(s,u))^2 du \leq c|t-s|^\alpha. \tag{103}$$

This latter assumption ensures (see, e.g., [11]) that: (i) $Z \in \mathbb{D}^{1,2}$; (ii) the law of Z has a density with respect to Lebesgue measure; (iii) there exists a (a.s.) unique random point $\tau \in [0,1]$ where the supremum is attained, that is, such that $Z = X_\tau = \int_0^1 K(\tau, s) dB_s$; and (iv) $D_t Z = K(\tau, t)$, $t \in [0,1]$. We claim the following formula.

Proposition 8. *Let Z be given by (102), X be defined as (101) and $K \in L^2([0,1]^2)$ be satisfying (103). Then, the law of Z has a density ρ whose support is $\mathbb{R}_+$, given by*

$$\rho(x) = \frac{E|Z - E[Z]|}{2h_Z(x)} \exp\left(-\int_{E[Z]}^x \frac{(y - E[Z]) dy}{h_Z(y)}\right), \quad x \geq 0.$$

Here,

$$h_Z(x) = \int_0^\infty e^{-u} E\left[R(\tau_0, \tau_u) | Z = x\right] du,$$

where $R(s,t) = E[X_s X_t]$, $s, t \in [0,1]$, and τ_u is the (almost surely) unique random point where

$$X_t^{(u)} = \int_0^1 K(t,s)(e^{-u} dB_s + \sqrt{1 - e^{-2u}} dB_s')$$

attains its maximum on $[0,1]$, with (B, B') a two-dimensional Brownian motion defined on the product probability space $(\boldsymbol{\Omega}, \boldsymbol{\mathscr{F}}, \mathbf{P}) = (\Omega \times \Omega', \mathscr{F} \otimes \mathscr{F}', P \times P')$.

Proof. Set $F = Z - E[Z]$. We have $-D_t L^{-1} F = \sum_{q=1}^\infty I_{q-1}^B(f_q(\cdot, t))$ and $D_t F = \sum_{q=1}^\infty q I_{q-1}^B(f_q(\cdot, t))$. Thus

$$\int_0^\infty e^{-u} P_u(D_t F) du = \sum_{q=1}^\infty I_{q-1}^B(f_q(\cdot, t)) \int_0^\infty e^{-u} q e^{-(q-1)u} du = \sum_{q=1}^\infty I_{q-1}^B(f_q(\cdot, t)).$$

Consequently,

$$-D_t L^{-1} F = \int_0^\infty e^{-u} P_u(D_t F) du, \quad t \in [0,1].$$

By Mehler's formula (49), and since $DF = DZ = K(\tau, \cdot)$ with $\tau = \mathrm{argmax}_{t \in [0,1]} \int_0^1 K(t,s) dB_s$, we deduce that

$$-D_t L^{-1} F = \int_0^\infty e^{-u} E'[K(\tau_u, t)] du,$$

implying in turn

$$g_F(F) = E[\langle DF, -DL^{-1}F \rangle_{L^2([0,1])} | F] = \int_0^1 dt \int_0^\infty du\, e^{-u} K(\tau_0, t) E[E'[K(\tau_u, t) | F]]$$

$$= \int_0^\infty e^{-u} E\left[E'\left[\int_0^1 K(\tau_0, t) K(\tau_u, t) dt \Big| F\right]\right] du$$

$$= \int_0^\infty e^{-u} E\left[E'\left[R(\tau_0, \tau_u) | F\right]\right] du$$

$$= \int_0^\infty e^{-u} \mathbf{E}[R(\tau_0, \tau_u) | F] du.$$

The desired conclusion follows now from Theorem 20 and the fact that $F = Z - E[Z]$. □

To Go Further. Reference [37] contains concentration inequalities for centered random variables $F \in \mathbb{D}^{1,2}$ satisfying $g_F(F) \leq \alpha F + \beta$. The paper [41] shows how Theorem 20 can lead to optimal Gaussian density estimates for a class of stochastic equations with additive noise.

9 Exact Rates of Convergence

In this section, we follow [30]. Let $\{F_n\}_{n \geq 1}$ be a sequence of random variables in $\mathbb{D}^{1,2}$ such that $E[F_n] = 0$, $\mathrm{Var}(F_n) = 1$ and $F_n \overset{\text{law}}{\to} N \sim \mathcal{N}(0,1)$ as $n \to \infty$. Our aim is to develop tools for computing the exact asymptotic expression of the (suitably normalized) sequence

$$P(F_n \leq x) - P(N \leq x), \quad n \geq 1,$$

when $x \in \mathbb{R}$ is fixed. This will complement the content of Theorem 13.

A Technical Computation. For every fixed x, we denote by $f_x : \mathbb{R} \to \mathbb{R}$ the function

$$f_x(u) = e^{u^2/2} \int_{-\infty}^{u} \left(\mathbf{1}_{(-\infty,x]}(a) - \Phi(x)\right) e^{-a^2/2} da$$

$$= \sqrt{2\pi} e^{u^2/2} \times \begin{cases} \Phi(u)(1 - \Phi(x)) & \text{if } u \leq x \\ \Phi(x)(1 - \Phi(u)) & \text{if } u \geq x \end{cases}, \quad (104)$$

where $\Phi(x) = \frac{1}{\sqrt{2\pi}} \int_{-\infty}^{x} e^{-a^2/2} da$. We have the following result.

Proposition 9. *Let $N \sim \mathcal{N}(0,1)$. We have, for every $x \in \mathbb{R}$,*

$$E[f_x'(N)N] = \frac{1}{3}(x^2 - 1)\frac{e^{-x^2/2}}{\sqrt{2\pi}}. \quad (105)$$

Proof. Integrating by parts (the bracket term is easily shown to vanish), we first obtain that

$$E[f_x'(N)N] = \int_{-\infty}^{+\infty} f_x'(u) u \frac{e^{-u^2/2}}{\sqrt{2\pi}} du = \int_{-\infty}^{+\infty} f_x(u)(u^2 - 1)\frac{e^{-u^2/2}}{\sqrt{2\pi}} du$$

$$= \frac{1}{\sqrt{2\pi}} \int_{-\infty}^{+\infty} (u^2 - 1) \left(\int_{-\infty}^{u} [\mathbf{1}_{(-\infty,x]}(a) - \Phi(x)] e^{-a^2/2} da \right) du.$$

Integrating by parts once again, this time using the relation $u^2 - 1 = \frac{1}{3}(u^3 - 3u)'$, we deduce that

$$\int_{-\infty}^{+\infty} (u^2 - 1) \left(\int_{-\infty}^{u} [\mathbf{1}_{(-\infty,x]}(a) - \Phi(x)] e^{-a^2/2} da \right) du$$

$$= -\frac{1}{3} \int_{-\infty}^{+\infty} (u^3 - 3u) [\mathbf{1}_{(-\infty,x]}(u) - \Phi(x)] e^{-u^2/2} du$$

$$= -\frac{1}{3} \left(\int_{-\infty}^{x} (u^3 - 3u) e^{-u^2/2} du - \Phi(x) \int_{-\infty}^{+\infty} (u^3 - 3u) e^{-u^2/2} du \right)$$

$$= \frac{1}{3}(x^2 - 1) e^{-x^2/2}, \quad \text{since } [(u^2 - 1)e^{-u^2/2}]' = -(u^3 - 3u)e^{-u^2/2}. \quad \square$$

A General Result. Assume that $\{F_n\}_{n \geq 1}$ is a sequence of (sufficiently regular) centered random variables with unitary variance such that the sequence

$$\varphi(n) := \sqrt{E[(1 - \langle DF_n, -DL^{-1}F_n \rangle_{L^2(\mathbb{R}_+)})^2]}, \quad n \geq 1, \quad (106)$$

converges to zero as $n \to \infty$. According to Theorem 13 one has that, for any $x \in \mathbb{R}$ and as $n \to \infty$,

$$P(F_n \leq x) - P(N \leq x) \leq d_{TV}(F_n, N) \leq 2\varphi(n) \to 0, \tag{107}$$

where $N \sim \mathcal{N}(0, 1)$. The forthcoming result provides a useful criterion in order to compute an exact asymptotic expression (as $n \to \infty$) for the quantity

$$\frac{P(F_n \leq x) - P(N \leq x)}{\varphi(n)}, \quad n \geq 1.$$

Theorem 21 (Nourdin and Peccati (2010); see [30]). *Let $\{F_n\}_{n \geq 1}$ be a sequence of random variables belonging to $\mathbb{D}^{1,2}$, and such that $E[F_n] = 0$, $\mathrm{Var}[F_n] = 1$. Suppose moreover that the following three conditions hold:*

(i) *we have $0 < \varphi(n) < \infty$ for every n and $\varphi(n) \to 0$ as $n \to \infty$.*
(ii) *the law of F_n has a density with respect to Lebesgue measure for every n.*
(iii) *as $n \to \infty$, the two-dimensional vector $\left(F_n, \frac{\langle DF_n, -DL^{-1}F_n \rangle_{L^2(\mathbb{R}_+)} - 1}{\varphi(n)} \right)$ converges in distribution to a centered two-dimensional Gaussian vector (N_1, N_2), such that $E[N_1^2] = E[N_2^2] = 1$ and $E[N_1 N_2] = \rho$.*

Then, as $n \to \infty$, one has for every $x \in \mathbb{R}$,

$$\frac{P(F_n \leq x) - P(N \leq x)}{\varphi(n)} \to \frac{\rho}{3}(1 - x^2)\frac{e^{-x^2/2}}{\sqrt{2\pi}}. \tag{108}$$

Proof. For any integer n and any $\mathscr{C}^1$-function f with a bounded derivative, we know by Theorem 12 that

$$E[F_n f(F_n)] = E[f'(F_n)\langle DF_n, -DL^{-1}F_n \rangle_{L^2(\mathbb{R}_+)}].$$

Fix $x \in \mathbb{R}$ and observe that the function f_x defined by (104) is not $\mathscr{C}^1$ due to the singularity in x. However, by using a regularization argument given assumption (ii), one can show that the identity

$$E[F_n f_x(F_n)] = E[f_x'(F_n)\langle DF_n, -DL^{-1}F_n \rangle_{L^2(\mathbb{R}_+)}]$$

is true for any n. Therefore, since $P(F_n \leq x) - P(N \leq x) = E[f_x'(F_n)] - E[F_n f_x(F_n)]$, we get

$$\frac{P(F_n \leq x) - P(N \leq x)}{\varphi(n)} = E\left[f_x'(F_n) \times \frac{1 - \langle DF_n, -DL^{-1}F_n \rangle_{L^2(\mathbb{R}_+)}}{\varphi(n)} \right].$$

Reasoning as in Lemma 2, one may show that f_x is Lipschitz with constant 2. Since $\varphi(n)^{-1}(1 - \langle DF_n, -DL^{-1}F_n \rangle_{L^2(\mathbb{R}_+)})$ has variance 1 by definition of $\varphi(n)$, we deduce that the sequence

$$f'_x(F_n) \times \frac{1 - \langle DF_n, -DL^{-1}F_n \rangle_{L^2(\mathbb{R}_+)}}{\varphi(n)}, \quad n \geq 1,$$

is uniformly integrable. Definition (104) shows that $u \to f'_x(u)$ is continuous at every $u \neq x$. This yields that, as $n \to \infty$ and due to assumption (iii),

$$E\left[f'_x(F_n) \times \frac{1 - \langle DF_n, -DL^{-1}F_n \rangle_{L^2(\mathbb{R}_+)}}{\varphi(n)} \right] \to -E[f'_x(N_1)N_2] = -\rho\, E[f'_x(N_1)N_1].$$

Consequently, relation (108) now follows from formula (105). □

The Double Integrals Case and a Concrete Application. When applying Theorem 21 in concrete situations, the main issue is often to check that condition (ii) therein holds true. In the particular case of sequences belonging to the second Wiener chaos, we can go further in the analysis, leading to the following result.

Proposition 10. *Let $N \sim \mathcal{N}(0,1)$ and let $F_n = I_2^B(f_n)$ be such that $f_n \in L^2(\mathbb{R}_+^2)$ is symmetric for all $n \geq 1$. Write $\kappa_p(F_n)$, $p \geq 1$, to indicate the sequence of the cumulants of F_n. Assume that $\kappa_2(F_n) = E[F_n^2] = 1$ for all $n \geq 1$ and that $\kappa_4(F_n) = E[F_n^4] - 3 \to 0$ as $n \to \infty$. If we have in addition that*

$$\frac{\kappa_3(F_n)}{\sqrt{\kappa_4(F_n)}} \to \alpha \quad \text{and} \quad \frac{\kappa_8(F_n)}{(\kappa_4(F_n))^2} \to 0, \tag{109}$$

then, for all $x \in \mathbb{R}$,

$$\frac{P(F_n \leq x) - P(N \leq x)}{\sqrt{\kappa_4(F_n)}} \to \frac{\alpha}{6\sqrt{2\pi}} (1 - x^2) e^{-\frac{x^2}{2}} \quad \text{as } n \to \infty. \tag{110}$$

Remark 6. Due to (109), we see that (110) is equivalent to

$$\frac{P(F_n \leq x) - P(N \leq x)}{\kappa_3(F_n)} \to \frac{1}{6\sqrt{2\pi}} (1 - x^2) e^{-\frac{x^2}{2}} \quad \text{as } n \to \infty.$$

Since each F_n is centered, one also has that $\kappa_3(F_n) = E[F_n^3]$.

Proof. We shall apply Theorem 21. Thanks to (60), we get that

$$\frac{\kappa_4(F_n)}{6} = \frac{E[F_n^4] - 3}{6} = 8 \|f_n \otimes_1 f_n\|_{L^2(\mathbb{R}_+^2)}^2.$$

By combining this identity with (58) (here, it is worth observing that $f_n \otimes_1 f_n$ is symmetric, so that the symmetrization $f_n \tilde{\otimes}_1 f_n$ is immaterial), we see that the quantity $\varphi(n)$ appearing in (106) is given by $\sqrt{\kappa_4(F_n)/6}$. In particular, condition (i) in Theorem 21 is met (here, let us stress that one may show that $\kappa_4(F_n) > 0$ for all n by means of (60)). On the other hand, since F_n is a non-zero double integral, its law has a density with respect to Lebesgue measure, according to Theorem 2. This means that condition (ii) in Theorem 21 is also in order. Hence, it remains to check condition (iii). Assume that (109) holds. Using (87) in the cases $p = 3$ and $p = 8$, we deduce that

$$\frac{\kappa_3(F_n)}{\sqrt{\kappa_4(F_n)}} = \frac{8 \langle f_n, f_n \otimes_1 f_n \rangle_{L^2(\mathbb{R}_+^2)}}{\sqrt{6}\varphi(n)}$$

and

$$\frac{\kappa_8(F_n)}{(\kappa_4(F_n))^2} = \frac{17920 \|(f_n \otimes_1 f_n) \otimes_1 (f_n \otimes_1 f_n)\|^2_{L^2(\mathbb{R}_+^2)}}{\varphi(n)^4}.$$

On the other hand, set

$$Y_n = \frac{\frac{1}{2}\|DF_n\|^2_{L^2(\mathbb{R}_+)} - 1}{\varphi(n)}.$$

By (62), we have $\frac{1}{2}\|DY_n\|^2_{L^2(\mathbb{R}_+)} - 1 = 2 I_2^B(f_n \otimes_1 f_n)$. Therefore, by (58), we get that

$$E\left[\left(\frac{1}{2}\|DY_n\|^2_{L^2(\mathbb{R}_+)} - 1\right)^2\right] = \frac{128}{\varphi(n)^4}\|(f_n \otimes_1 f_n) \otimes_1 (f_n \otimes_1 f_n)\|_{L^2(\mathbb{R}_+^2)}$$

$$= \frac{\kappa_8(F_n)}{140 (\kappa_4(F_n))^2} \to 0 \quad \text{as } n \to \infty.$$

Hence, by Theorem 3, we deduce that $Y_n \overset{\text{Law}}{\to} \mathcal{N}(0, 1)$. We also have

$$E[Y_n F_n] = \frac{4}{\varphi(n)} \langle f_n \otimes_1 f_n, f_n \rangle_{L^2(\mathbb{R}_+^2)} = \frac{\sqrt{6}\kappa_3(F_n)}{2\sqrt{\kappa_4(F_n)}} \to \frac{\alpha\sqrt{6}}{2} =: \rho \quad \text{as } n \to \infty.$$

Therefore, to conclude that condition (iii) in Theorem 21 holds true, it suffices to apply Theorem 17. □

To give a concrete application of Proposition 10, let us go back to the quadratic variation of fractional Brownian motion. Let $B^H = (B_t^H)_{t \geq 0}$ be a fractional Brownian motion with Hurst index $H \in (0, \frac{1}{2})$ and let

$$F_n = \frac{1}{\sigma_n} \sum_{k=0}^{n-1} [(B_{k+1}^H - B_k^H)^2 - 1],$$

where $\sigma_n > 0$ is so that $E[F_n^2] = 1$. Recall from Theorem 15 that $\lim_{n\to\infty} \sigma_n^2/n = 2\sum_{r\in\mathbb{Z}} \rho^2(r) < \infty$, with $\rho : \mathbb{Z} \to \mathbb{R}_+$ given by (69); moreover, there exists a constant $c_H > 0$ (depending only on H) such that, with $N \sim \mathcal{N}(0,1)$,

$$d_{TV}(F_n, N) \leq \frac{c_H}{\sqrt{n}}, \quad n \geq 1. \tag{111}$$

The next two results aim to show that one can associate a *lower* bound to (111). We start by the following auxiliary result.

Proposition 11. *Fix an integer $s \geq 2$, let F_n be as above and let ρ be given by (69). Recall that $H < \frac{1}{2}$, so that $\rho \in \ell^1(\mathbb{Z})$. Then, the sth cumulant of F_n behaves asymptotically as*

$$\kappa_s(F_n) \sim n^{1-s/2} \, 2^{s/2-1}(s-1)! \, \frac{\langle \rho^{*(s-1)}, \rho \rangle_{\ell^2(\mathbb{Z})}}{\|\rho\|_{\ell^2(\mathbb{Z})}^s} \quad \text{as } n \to \infty. \tag{112}$$

Proof. As in the proof of Theorem 15, we have that $F_n \stackrel{\text{law}}{=} I_2^B(f_n)$ with $f_n = \frac{1}{\sigma_n} \sum_{k=0}^{n-1} e_k^{\otimes 2}$. Now, let us proceed with the proof. It is divided into several steps.

First step. Using the formula (87) giving the cumulants of $F_n = I_2^B(f_n)$ as well as the very definition of the contraction $\otimes_1$, we immediately check that

$$\kappa_s(F_n) = \frac{2^{s-1}(s-1)!}{\sigma_n^s} \sum_{k_1,\ldots,k_s=0}^{n-1} \rho(k_s - k_{s-1}) \ldots \rho(k_2 - k_1)\rho(k_1 - k_s).$$

Second step. Since $H < \frac{1}{2}$, we have that $\rho \in \ell^1(\mathbb{Z})$. Therefore, by applying Young inequality repeatedly, we have

$$\| |\rho|^{*(s-1)} \|_{\ell^\infty(\mathbb{Z})} \leq \|\rho\|_{\ell^1(\mathbb{Z})} \| |\rho|^{*(s-2)} \|_{\ell^\infty(\mathbb{Z})} \leq \ldots \leq \|\rho\|_{\ell^1(\mathbb{Z})}^{s-1} < \infty.$$

In particular, we have that $\langle |\rho|^{*(s-1)}, |\rho| \rangle_{\ell^2(\mathbb{Z})} \leq \|\rho\|_{\ell^1(\mathbb{Z})}^s < \infty$.

Third step. Thanks to the result shown in the previous step, observe first that

$$\sum_{k_2,\ldots,k_s\in\mathbb{Z}} |\rho(k_2)\rho(k_2-k_3)\rho(k_3-k_4)\ldots\rho(k_{s-1}-k_s)\rho(k_s)| = \langle |\rho|^{*(s-1)}, |\rho| \rangle_{\ell^2(\mathbb{Z})}$$

$$< \infty.$$

Hence, one can apply dominated convergence to get, as $n \to \infty$, that

$$\frac{\sigma_n^s}{2^{s-1}(s-1)!\,n} \kappa_s(F_n)$$

$$= \frac{1}{n} \sum_{k_1=0}^{n-1} \sum_{k_2,\ldots,k_s=-k_1}^{n-1-k_1} \rho(k_2)\rho(k_2-k_3)\rho(k_3-k_4)\ldots\rho(k_{s-1}-k_s)\rho(k_s)$$

$$
= \sum_{k_2,\ldots,k_s \in \mathbb{Z}} \rho(k_2)\rho(k_2 - k_3)\rho(k_3 - k_4)\ldots\rho(k_{s-1} - k_s)\rho(k_s)
$$

$$
\times \left[1 \wedge \left(1 - \frac{\max\{k_2,\ldots,k_s\}}{n}\right) - 0 \vee \left(\frac{\min\{k_2,\ldots,k_s\}}{n}\right)\right] \mathbf{1}_{\{|k_2|<n,\ldots,|k_s|<n\}}
$$

$$
\to \sum_{k_2,\ldots,k_s \in \mathbb{Z}} \rho(k_2)\rho(k_2 - k_3)\rho(k_3 - k_4)\ldots\rho(k_{s-1} - k_s)\rho(k_s) = \langle \rho^{*(s-1)}, \rho \rangle_{\ell^2(\mathbb{Z})}.
$$

(113)

Since $\sigma_n \sim \sqrt{2n}\,\|\rho\|_{\ell^2(\mathbb{Z})}$ as $n \to \infty$, the desired conclusion follows. □

Corollary 6. *Let F_n be as above (with $H < \frac{1}{2}$), let $N \sim \mathcal{N}(0,1)$, and let ρ be given by (69). Then, for all $x \in \mathbb{R}$, we have*

$$
\sqrt{n}\bigl(P(F_n \leq x) - P(N \leq x)\bigr) \to \frac{\langle \rho^{*2}, \rho \rangle_{\ell^2(\mathbb{Z})}}{3\|\rho\|_{\ell^2(\mathbb{Z})}^2}\,(1 - x^2)\,e^{-\frac{x^2}{2}} \quad \text{as } n \to \infty.
$$

In particular, we deduce that there exists $d_H > 0$ such that

$$
\frac{d_H}{\sqrt{n}} \leq \bigl|P(F_n \leq 0) - P(N \leq 0)\bigr| \leq d_{TV}(F_n, N), \quad n \geq 1. \tag{114}
$$

Proof. The desired conclusion follows immediately by combining Propositions 10 and 11. □

By paying closer attention to the used estimates, one may actually show that (114) holds true for any $H < \frac{5}{8}$ (not only $H < \frac{1}{2}$). See [32, Theorem 9.5.1] for the details.

To Go Further. The paper [30] contains several other examples of application of Theorem 21 and Proposition 10. In [4], one shows that the deterministic sequence

$$
\max\{E[F_n^3], E[F_n^4] - 3\}, \quad n \geq 1,
$$

completely characterizes the rate of convergence (with respect to smooth distances) in CLTs involving chaotic random variables.

10 An Extension to the Poisson Space (Following the Invited Talk by Giovanni Peccati)

Let $B = (B_t)_{t \geq 0}$ be a Brownian motion, let F be any centered element of $\mathbb{D}^{1,2}$ and let $N \sim \mathcal{N}(0,1)$. We know from Theorem 13 that

$$
d_{TV}(F, N) \leq 2\,E[|1 - \langle DF, -DL^{-1}F \rangle_{L^2(\mathbb{R}_+)}|]. \tag{115}
$$

The aim of this section, which follows [43, 44], is to explain how to deduce inequalities of the type (115), when F is a regular functional of a *Poisson measure* η and when the target law N is either Gaussian or Poisson.

We first need to introduce the basic concepts in this framework.

Poisson Measure. In what follows, we shall use the symbol $Po(\lambda)$ to indicate the Poisson distribution of parameter $\lambda > 0$ (that is, $\mathscr{P}_\lambda \sim Po(\lambda)$ if and only if $P(\mathscr{P}_\lambda = k) = e^{-\lambda}\frac{\lambda^k}{k!}$ for all $k \in \mathbb{N}$), with the convention that $Po(0) = \delta_0$ (Dirac mass at 0). Set $A = \mathbb{R}^d$ with $d \geq 1$, let $\mathscr{A}$ be the Borel σ-field on A, and let μ be a positive, σ-finite and atomless measure over $(A, \mathscr{A})$. We set $\mathscr{A}_\mu = \{B \in \mathscr{A} : \mu(B) < \infty\}$.

Definition 7. A Poisson measure η with control μ is an object of the form $\{\eta(B)\}_{B \in \mathscr{A}_\mu}$ with the following features:

(1) for all $B \in \mathscr{A}_\mu$, we have $\eta(B) \sim Po(\mu(B))$.
(2) for all $B, C \in \mathscr{A}_\mu$ with $B \cap C \neq \emptyset$, the random variables $\eta(B)$ and $\eta(C)$ are independent.

Also, we note $\hat{\eta}(B) = \eta(B) - \mu(B)$.

Remark 7. 1. As a simple example, note that for $d = 1$ and $\mu = \lambda \times$ Leb (with 'Leb' the Lebesgue measure) the process $\{\eta([0, t])\}_{t \geq 0}$ is nothing but a Poisson process with intensity λ.
2. Let μ be a σ-finite atomless measure over $(A, \mathscr{A})$, and observe that this implies that there exists a sequence of disjoint sets $\{A_j : j \geq 1\} \subset \mathscr{A}_\mu$ such that $\cup_j A_j = A$. For every $j = 1, 2, \ldots$ belonging to the set J_0 of those indices such that $\mu(A_j) > 0$ consider the following objects: $X^{(j)} = \{X_k^{(j)} : k \geq 1\}$ is a sequence of i.i.d. random variables with values in A_j and with common distribution $\frac{\mu_{|A_j}}{\mu(A_j)}$; P_j is a Poisson random variable with parameter $\mu(A_j)$. Assume moreover that: (i) $X^{(j)}$ is independent of $X^{(k)}$ for every $k \neq j$, (ii) P_j is independent of P_k for every $k \neq j$, and (iii) the classes $\{X^{(j)}\}$ and $\{P_j\}$ are independent. Then, it is a straightforward computation to verify that the random point measure

$$\eta(\cdot) = \sum_{j \in J_0} \sum_{k=1}^{P_j} \delta_{X_k^{(j)}}(\cdot),$$

where δ_x indicates the Dirac mass at x and $\sum_{k=1}^0 = 0$ by convention, is a Poisson random measure with control μ. See e.g. [49, Sect. 1.7].

Multiple Integrals and Chaotic Expansion. As a preliminary remark, let us observe that $E[\hat{\eta}(B)] = 0$ and $E[\hat{\eta}(B)^2] = \mu(B)$ for all $B \in \mathscr{A}_\mu$. For any $q \geq 1$, set $L^2(\mu^q) = L^2(A^q, \mathscr{A}^q, \mu^q)$. We want to appropriately define

$$I_q^{\hat{\eta}}(f) = \int_{A^q} f(x_1,\ldots,x_q)\hat{\eta}(dx_1)\ldots\hat{\eta}(dx_q)$$

when $f \in L^2(\mu^q)$. To reach our goal, we proceed in a classical way. We first consider the subset $\mathscr{E}(\mu^q)$ of simple functions, which is defined as

$\mathscr{E}(\mu^q)$
$= \mathrm{span}\left\{\mathbf{1}_{B_1} \otimes \ldots \otimes \mathbf{1}_{B_q}, \text{ with } B_1,\ldots,B_q \in \mathscr{A}_\mu \text{ such that } B_i \cap B_j = \emptyset \text{ for all } i \neq j\right\}.$

When $f = \mathbf{1}_{B_1} \otimes \ldots \otimes \mathbf{1}_{B_q}$ with $B_1,\ldots,B_q \in \mathscr{A}_\mu$ such that $B_i \cap B_j = \emptyset$ for all $i \neq j$, we naturally set

$$I_q^{\hat{\eta}}(f) := \hat{\eta}(B_1)\ldots\hat{\eta}(B_q) = \int_{A^q} f(x_1,\ldots,x_q)\hat{\eta}(dx_1)\ldots\hat{\eta}(dx_q).$$

(For such a simple function f, note that the right-hand side in the previous formula makes perfectly sense by considering $\hat{\eta}$ as a signed measure.) We can extend by linearity the definition of $I_q^{\hat{\eta}}(f)$ to any $f \in \mathscr{E}(\mu^q)$. It is then a simple computation to check that

$$E[I_p^{\hat{\eta}}(f)I_q^{\hat{\eta}}(g)] = p!\delta_{p,q}\,\langle \tilde{f},\tilde{g}\rangle_{L^2(\mu^p)}$$

for all $f \in \mathscr{E}(\mu^p)$ and $g \in \mathscr{E}(\mu^q)$, with $\tilde{f}$ (resp. $\tilde{g}$) the symmetrization of f (resp. g) and $\delta_{p,q}$ the Kronecker symbol. Since $\mathscr{E}(\mu^q)$ is dense in $L^2(\mu^q)$ (it is precisely here that the fact that μ has no atom is crucial!), we can define $I_q^{\hat{\eta}}(f)$ by isometry to any $f \in L^2(\mu^q)$. Relevant properties of $I_q^{\hat{\eta}}(f)$ include $E[I_q^{\hat{\eta}}(f)] = 0$, $I_q^{\hat{\eta}}(f) = I_q^{\hat{\eta}}(\tilde{f})$ and (importantly!) the fact that $I_q^{\hat{\eta}}(f)$ is a *true* multiple integral when $f \in \mathscr{E}(\mu^q)$.

Definition 8. Fix $q \geq 1$. The set of random variables of the form $I_q^{\hat{\eta}}(f)$ is called the qth Poisson–Wiener chaos.

In this framework, we have an analogue of the chaotic decomposition (39)—see e.g. [45, Corollary 10.0.5] for a proof.

Theorem 22. *For all $F \in L^2(\sigma\{\eta\})$ (that is, for all random variable F which is square integrable and measurable with respect to η), we have*

$$F = E[F] + \sum_{q=1}^\infty I_q^{\hat{\eta}}(f_q), \tag{116}$$

where the kernels f_q are (μ^q-a.e.) symmetric elements of $L^2(\mu^q)$ and are uniquely determined by F.

Multiplication Formula and Contractions. When $f \in \mathscr{E}(\mu^p)$ and $g \in \mathscr{E}(\mu^q)$ are symmetric, we define, for all $r = 0,\ldots,p \wedge q$ and $l = 0,\ldots,r$:

$$f \star_r^l g(x_1, \ldots, x_{p+q-r-l}) = \int_{A^l} f(y_1, \ldots, y_l, x_1, \ldots, x_{r-l}, x_{r-l+1}, \ldots, x_{p-l})$$
$$\times g(y_1, \ldots, y_l, x_1, \ldots, x_{r-l}, x_{p-l+1}, \ldots, x_{p+q-r-l})$$
$$\times \mu(dy_1) \ldots \mu(dy_l).$$

We then have the following product formula, compare with (43).

Theorem 23 (Product formula). *Let $p, q \geq 1$ and let $f \in \mathcal{E}(\mu^p)$ and $g \in \mathcal{E}(\mu^q)$ be symmetric. Then*

$$I_p^{\hat{\eta}}(f) I_q^{\hat{\eta}}(g) = \sum_{r=0}^{p \wedge q} r! \binom{p}{r} \binom{q}{r} \sum_{l=0}^{r} \binom{r}{l} I_{p+q-r-l}^{\hat{\eta}}\left(\widetilde{f \star_r^l g}\right).$$

Proof. Recall that, when dealing with functions in $\mathcal{E}(\mu^p)$, $I_p^{\hat{\eta}}(f)$ is a *true* multiple integral (by seeing $\hat{\eta}$ as a signed measure). We deduce

$$I_p^{\hat{\eta}}(f) I_q^{\hat{\eta}}(g) = \int_{A^{p+q}} f(x_1, \ldots, x_p) g(y_1, \ldots, y_q) \hat{\eta}(dx_1) \ldots \hat{\eta}(dx_p) \hat{\eta}(dy_1) \ldots \hat{\eta}(dy_q).$$

By definition of f (the same applies for g), we have that $f(x_1, \ldots, x_p) = 0$ when $x_i = x_j$ for some $i \neq j$. Consider $r = 0, \ldots, p \wedge q$, as well as pairwise disjoint indices $i_1, \ldots, i_r \in \{1, \ldots, p\}$ and pairwise disjoint indices $j_1, \ldots, j_r \in \{1, \ldots, q\}$. Set $\{k_1, \ldots, k_{p-r}\} = \{1, \ldots, p\} \setminus \{i_1, \ldots, i_r\}$ and $\{l_1, \ldots, l_{q-r}\} = \{1, \ldots, q\} \setminus \{j_1, \ldots, j_r\}$. We have, since μ is atomless and using $\hat{\eta}(dx) = \eta(dx) - \mu(dx)$,

$$\int_{A^{p+q}} f(x_1, \ldots, x_p) g(y_1, \ldots, y_q) \mathbf{1}_{\{x_{i_1} = y_{j_1}, \ldots, x_{i_r} = y_{j_r}\}}$$
$$\times \hat{\eta}(dx_1) \ldots \hat{\eta}(dx_p) \hat{\eta}(dy_1) \ldots \hat{\eta}(dy_q)$$
$$= \int_{A^{p+q-2r}} f(x_{k_1}, \ldots, x_{k_{p-r}}, x_{i_1}, \ldots, x_{i_r}) g(y_{l_1}, \ldots, y_{l_{q-r}}, x_{i_1}, \ldots, x_{i_r})$$
$$\times \hat{\eta}(dx_{k_1}) \ldots \hat{\eta}(dx_{k_{p-r}}) \hat{\eta}(dy_{l_1}) \ldots \hat{\eta}(dy_{l_{q-r}}) \eta(dx_{i_1}) \ldots \eta(dx_{i_r})$$
$$= \int_{A^{p+q-2r}} f(x_1, \ldots, x_{p-r}, a_1, \ldots, a_r) g(y_1, \ldots, y_{q-r}, a_1, \ldots, a_r)$$
$$\times \hat{\eta}(dx_1) \ldots \hat{\eta}(dx_{p-r}) \hat{\eta}(dy_1) \ldots \hat{\eta}(dy_{q-r}) \eta(da_1) \ldots \eta(da_r).$$

By decomposing over the hyperdiagonals $\{x_i = y_j\}$, we deduce that

$$I_p^{\hat{\eta}}(f) I_q^{\hat{\eta}}(g) = \sum_{r=0}^{p \wedge q} r! \binom{p}{r} \binom{q}{r} \int_{A^{p+q-2r}} f(x_1, \ldots, x_{p-r}, a_1, \ldots, a_r)$$
$$\times g(y_1, \ldots, y_{q-r}, a_1, \ldots, a_r)$$
$$\times \hat{\eta}(dx_1) \ldots \hat{\eta}(dx_{p-r}) \hat{\eta}(dy_1) \ldots \hat{\eta}(dy_{q-r}) \eta(da_1) \ldots \eta(da_r),$$

and we get the desired conclusion by using the relationship

$$\eta(da_1)\ldots\eta(da_r) = \big(\hat{\eta}(da_1) + \mu(da_1)\big)\ldots\big(\hat{\eta}(da_r) + \mu(da_r)\big). \qquad \square$$

Malliavin Operators. Each time we deal with a random element F of $L^2(\{\sigma(\eta)\})$, in what follows we always consider its chaotic expansion (116).

Definition 9. 1. Set $\mathrm{Dom}\, D = \{F \in L^2(\sigma\{\eta\}) : \sum q q! \|f_q\|^2_{L^2(\mu^q)} < \infty\}$. If $F \in \mathrm{Dom}\, D$, we set

$$D_t F = \sum_{q=1}^{\infty} q I^{\hat{\eta}}_{q-1}(f_q(\cdot, t)), \quad t \in A.$$

The operator D is called the Malliavin derivative.

2. Set $\mathrm{Dom}\, L = \{F \in L^2(\sigma\{\eta\}) : \sum q^2 q! \|f_q\|^2_{L^2(\mu^q)} < \infty\}$. If $F \in \mathrm{Dom}\, L$, we set

$$LF = -\sum_{q=1}^{\infty} q I^{\hat{\eta}}_q(f_q).$$

The operator L is called the generator of the Ornstein–Uhlenbeck semigroup.

3. If $F \in L^2(\sigma\{\eta\})$, we set

$$L^{-1} F = -\sum_{q=1}^{\infty} \frac{1}{q} I^{\hat{\eta}}_q(f_q).$$

The operator L^{-1} is called the pseudo-inverse of L.

It is readily checked that $LL^{-1} F = F - E[F]$ for $F \in L^2(\sigma\{\eta\})$. Moreover, proceeding *mutatis mutandis* as in the proof of Theorem 12, we get the following result.

Proposition 12. *Let $F \in L^2(\sigma\{\eta\})$ and let $G \in \mathrm{Dom}\, D$. Then*

$$\mathrm{Cov}(F, G) = E[\langle DG, -DL^{-1}F \rangle_{L^2(\mu)}]. \qquad (117)$$

The operator D does not satisfy the chain rule. Instead, it admits an "add-one cost" representation which plays an identical role.

Theorem 24 (Nualart and Vives (1990); see [42]). *Let $F \in \mathrm{Dom}\, D$. Since F is measurable with respect to η, we can view it as $F = F(\eta)$ with a slight abuse of notation. Then*

$$D_t F = F(\eta + \delta_t) - F(\eta), \quad t \in A, \qquad (118)$$

where δ_t stands for the Dirac mass at t.

Proof. By linearity and approximation, it suffices to prove the claim for $F = I_q^{\hat{\eta}}(f)$, with $q \geqslant 1$ and $f \in \mathscr{E}(\mu^q)$ symmetric. In this case, we have

$$F(\eta + \delta_t) = \int_{A^q} f(x_1, \ldots, x_q)(\hat{\eta}(dx_1) + \delta_t(dx_1)) \ldots (\hat{\eta}(dx_q) + \delta_t(dx_q)).$$

Let us expand the integrator. Each member of such an expansion such that there is strictly more than one Dirac mass in the resulting expression gives a contribution equal to zero, since f vanishes on diagonals. We therefore deduce that

$$F(\eta + \delta_t) = F(\eta)$$
$$+ \sum_{l=1}^{q} \int_{A^q} f(x_1, \ldots, x_{l-1}, t, x_{l+1}, \ldots, x_q)$$
$$\times \hat{\eta}(dx_1) \ldots \hat{\eta}(dx_{l-1})\hat{\eta}(dx_{l+1}) \ldots \hat{\eta}(dx_q)$$
$$= F(\eta) + q I_{q-1}^{\hat{\eta}}(f(t, \cdot)) \quad \text{by symmetry of } f$$
$$= F(\eta) + D_t F. \qquad \square$$

As an immediate corollary of the previous theorem, we get the formula

$$D_t(F^2) = (F + D_t F)^2 - F^2 = 2F\, D_t F + (D_t F)^2, \quad t \in A,$$

which shows how D is far from satisfying the chain rule (47).

Gaussian Approximation. It happens that it is the following distance which is appropriate in our framework.

Definition 10. The Wasserstein distance between the laws of two real-valued random variables Y and Z is defined by

$$d_W(Y, Z) = \sup_{h \in \text{Lip}(1)} |E[h(Y)] - E[h(Z)]|, \qquad (119)$$

where $\text{Lip}(1)$ stands for the set of Lipschitz functions $h : \mathbb{R} \to \mathbb{R}$ with constant 1.

Since we are here dealing with Lipschitz functions h, we need a suitable version of Stein's lemma. Compare with Lemma 2.

Lemma 6 (Stein (1972); see [52]). *Suppose $h : \mathbb{R} \to \mathbb{R}$ is a Lipschitz constant with constant 1. Let $N \sim \mathcal{N}(0, 1)$. Then, there exists a solution to the equation*

$$f'(x) - x f(x) = h(x) - E[h(N)], \quad x \in \mathbb{R},$$

that satisfies $\|f'\|_\infty \leqslant \sqrt{\frac{2}{\pi}}$ *and* $\|f''\|_\infty \leqslant 2$.

Proof. Let us recall that, according to Rademacher's theorem, a function which is Lipschitz continuous on $\mathbb{R}$ is almost everywhere differentiable. Let $f : \mathbb{R} \to \mathbb{R}$ be the (well-defined!) function given by

$$f(x) = -\int_0^\infty \frac{e^{-t}}{\sqrt{1-e^{-2t}}} E[h(e^{-t}x + \sqrt{1-e^{-2t}}N)N]dt. \quad (120)$$

By dominated convergence we have that $f_h \in \mathscr{C}^1$ with

$$f'(x) = -\int_0^\infty \frac{e^{-2t}}{\sqrt{1-e^{-2t}}} E[h'(e^{-t}x + \sqrt{1-e^{-2t}}N)N]dt.$$

We deduce, for any $x \in \mathbb{R}$,

$$|f'(x)| \leq E|N| \int_0^\infty \frac{e^{-2t}}{\sqrt{1-e^{-2t}}} dt = \sqrt{\frac{2}{\pi}}. \quad (121)$$

Now, let $F : \mathbb{R} \to \mathbb{R}$ be the function given by

$$F(x) = \int_0^\infty E[h(N) - h(e^{-t}x + \sqrt{1-e^{-2t}}N)]dt, \quad x \in \mathbb{R}.$$

Observe that F is well-defined since $h(N) - h(e^{-t}x + \sqrt{1-e^{-2t}}N)$ is integrable due to

$$\left| h(N) - h(e^{-t}x + \sqrt{1-e^{-2t}}N) \right| \leq e^{-t}|x| + (1 - \sqrt{1-e^{-2t}})|N|$$
$$\leq e^{-t}|x| + e^{-2t}|N|,$$

where the last inequality follows from $1 - \sqrt{1-u} = u/(\sqrt{1-u} + 1) \leq u$ if $u \in [0, 1]$. By dominated convergence, we immediately see that F is differentiable with

$$F'(x) = -\int_0^\infty e^{-t} E[h'(e^{-t}x + \sqrt{1-e^{-2t}}N)]dt.$$

By integrating by parts, we see that $F'(x) = f(x)$. Moreover, by using the notation introduced in Sect. 4, we can write

$f'(x) - xf(x)$

$= LF(x)$, by decomposing in Hermite polynomials, since $LH_q = -qH_q = H_q'' - XH_q'$

$= -\int_0^\infty LP_t h(x)dt$, since $F(x) = \int_0^\infty (E[h(N)] - P_t h(x))dt$

$= -\int_0^\infty \frac{d}{dt} P_t h(x)dt$

$= P_0 h(x) - P_\infty h(x) = h(x) - E[h(N)].$

This proves the claim for $\|f'\|_\infty$. The claim for $\|f''\|_\infty$ is a bit more difficult to achieve; we refer to Stein [52, pp. 25–28] to keep the length of this survey within bounds. □

We can now derive a bound for the Gaussian approximation of any centered element F belonging to Dom$D$, compare with (115).

Theorem 25 (Peccati, Solé, Taqqu and Utzet (2010); see [44]). *Consider $F \in$ Dom$D$ with $E[F] = 0$. Then, with $N \sim \mathcal{N}(0,1)$,*

$$d_W(F, N) \leq \sqrt{\frac{2}{\pi}} E\left[|1 - \langle DF, -DL^{-1}F\rangle_{L^2(\mu)}|\right] + E\left[\int_A (D_t F)^2 |D_t L^{-1} F| \mu(dt)\right].$$

Proof. Let $h \in \text{Lip}(1)$ and let f be the function of Lemma 6. Using (118) and a Taylor formula, we can write

$$D_t f(F) = f(F + D_t F) - f(F) = f'(F) D_t F + R(t),$$

with $|R(t)| \leq \frac{1}{2} \|f''\|_\infty (D_t F)^2 \leq (D_t F)^2$. We deduce, using (117) as well,

$$E[h(F)] - E[h(N)] = E[f'(F)] - E[Ff(F)]$$
$$= E[f'(F)] - E[\langle Df(F), -DL^{-1}F\rangle_{L^2(\mu)}]$$
$$= E[f'(F)(1 - \langle DF, -DL^{-1}F\rangle_{L^2(\mu)})]$$
$$+ \int_A (-D_t L^{-1} F) R(t) \mu(dt).$$

Consequently, since $\|f'\|_\infty \leq \sqrt{\frac{2}{\pi}}$,

$$d_W(F, N) = \sup_{h \in \text{Lip}(1)} |E[h(F)] - E[h(N)]|$$

$$\leq \sqrt{\frac{2}{\pi}} E\left[|1 - \langle DF, -DL^{-1}F\rangle_{L^2(\mu)}|\right] + E\left[\int_A (D_t F)^2 |D_t L^{-1} F| \mu(dt)\right].$$

□

Poisson Approximation. To conclude this section, we will prove a very interesting result, which may be seen as a Poisson counterpart of Theorem 25.

Theorem 26 (Peccati (2012); see [43]). *Let $F \in \text{Dom}D$ with $E[F] = \lambda > 0$ and F taking its values in $\mathbb{N}$. Let $\mathscr{P}_\lambda \sim Po(\lambda)$. Then,*

$$\sup_{C \subset \mathbb{N}} |P(F \in C) - P(\mathscr{P}_\lambda \in C)| \leq \frac{1 - e^{-\lambda}}{\lambda} E|\lambda - \langle DF, -DL^{-1}F\rangle_{L^2(\mu)}| \quad (122)$$

$$+ \frac{1 - e^{-\lambda}}{\lambda^2} E \int |D_t F(D_t F - 1) D_t L^{-1} F| \mu(dt).$$

Just as a mere illustration, consider the case where $F = \eta(B) = I_1^{\hat{\eta}}(\mathbf{1}_B)$ with $B \in \mathscr{A}_\mu$. We then have $DF = -DL^{-1}F = \mathbf{1}_B$, so that $\langle DF, -DL^{-1}F\rangle_{L^2(\mu)} = \int \mathbf{1}_B d\mu = \mu(B)$ and $DF(DF-1) = 0$ a.e. The right-hand side of (122) is therefore zero, as it was expected since $F \sim Po(\lambda)$.

During the proof of Theorem 26, we shall use an analogue of Lemma 2 in the Poisson context, which reads as follows.

Lemma 7 (Chen (1975); see [8]). *Let $C \subset \mathbb{N}$, let $\lambda > 0$ and let $\mathscr{P}_\lambda \sim Po(\lambda)$. The equation with unknown $f : \mathbb{N} \to \mathbb{R}$,*

$$\lambda f(k+1) - kf(k) = \mathbf{1}_C(k) - P(\mathscr{P}_\lambda \in C), \quad k \in \mathbb{N}, \tag{123}$$

admits a unique solution such that $f(0) = 0$, denoted by f_C. Moreover, by setting $\Delta f(k) = f(k+1) - f(k)$, we have $\|\Delta f_C\|_\infty \leq \frac{1-e^{-\lambda}}{\lambda}$ and $\|\Delta^2 f_C\|_\infty \leq \frac{2}{\lambda}\|\Delta f_C\|_\infty$.

Proof. We only provide a proof for the bound on Δf_C; the estimate on $\Delta^2 f_C$ is proved e.g. by Daly in [10]. Multiplying both sides of (123) by $\lambda^k/k!$ and summing up yields that, for every $k \geq 1$,

$$f_C(k) = \frac{(k-1)!}{\lambda^k} \sum_{r=0}^{k-1} \frac{\lambda^r}{r!}[\mathbf{1}_C(r) - P(\mathscr{P}_\lambda \in C)] \tag{124}$$

$$= \sum_{j \in C} f_{\{j\}}(k) \tag{125}$$

$$= -f_{C^c}(k) \tag{126}$$

$$= -\frac{(k-1)!}{\lambda^k} \sum_{r=k}^{\infty} \frac{\lambda^r}{r!}[\mathbf{1}_C(r) - P(\mathscr{P}_\lambda \in C)], \tag{127}$$

where C^c denotes the complement of C in $\mathbb{N}$. (Identity (125) comes from the additivity property of $C \mapsto f_C$, identity (126) is because $f_{\mathbb{N}} \equiv 0$ and identity (126) is due to

$$\sum_{r=0}^{\infty} \frac{\lambda^r}{r!}[\mathbf{1}_C(r) - P(\mathscr{P}_\lambda \in C)] = E[\mathbf{1}_C(\mathscr{P}_\lambda) - E[\mathbf{1}_C(\mathscr{P}_\lambda)]] = 0.)$$

Since $f_C(k) - f_C(k+1) = f_{C^c}(k+1) - f_{C^c}(k)$ (due to (126)), it is sufficient to prove that, for every $k \geq 1$ and every $C \subset \mathbb{N}$, $f_C(k+1) - f_C(k) \leq (1 - e^{-\lambda})/\lambda$. One has the following fact: for every $j \geq 1$ the mapping $k \mapsto f_{\{j\}}(k)$ is negative and decreasing for $k = 1, \ldots, j$ and positive and decreasing for $k \geq j+1$. Indeed, we use (124) to deduce that, if $1 \leq k \leq j$,

$$f_{\{j\}}(k) = -e^{-\lambda} \frac{\lambda^j}{j!} \sum_{r=1}^{k} \lambda^{-r} \frac{(k-1)!}{(k-r)!} \quad \text{(which is negative and decreasing in } k\text{)},$$

whereas (127) implies that, if $k \geqslant j+1$,

$$f_{\{j\}}(k) = e^{-\lambda} \frac{\lambda^j}{j!} \sum_{r=0}^{\infty} \lambda^r \frac{(k-1)!}{(k+r)!} \quad \text{(which is positive and decreasing in } k\text{)}.$$

Using (125), one therefore infers that $f_C(k+1) - f_C(k) \leqslant f_{\{k\}}(k+1) - f_{\{k\}}(k)$, for every $k \geqslant 0$. Since

$$f_{\{k\}}(k+1) - f_{\{k\}}(k) = e^{-\lambda} \left[\sum_{r=0}^{k-1} \frac{\lambda^r}{r!k} + \sum_{r=k+1}^{\infty} \frac{\lambda^{r-1}}{r!} \right]$$

$$= \frac{e^{-\lambda}}{\lambda} \left[\sum_{r=1}^{k} \frac{\lambda^r}{r!} \times \frac{r}{k} + \sum_{r=k+1}^{\infty} \frac{\lambda^r}{r!} \right]$$

$$\leqslant \frac{1 - e^{-\lambda}}{\lambda},$$

the proof is concluded. □

We are now in a position to prove Theorem 26.

Proof of Theorem 26. The main ingredient is the following simple inequality, which is a kind of Taylor formula: for all $k, a \in \mathbb{N}$,

$$\left| f(k) - f(a) - \Delta f(a)(k-a) \right| \leqslant \frac{1}{2} \|\Delta^2 f\|_{\infty} |(k-a)(k-a-1)|. \quad (128)$$

Assume for the time being that (128) holds true and fix $C \subset \mathbb{N}$. We have, using Lemma 7 and then (117)

$$\left| P(F \in C) - P(\mathscr{P}_\lambda \in C) \right| = \left| E[\lambda f_C(F+1)] - E[F f_C(F)] \right|$$

$$= \left| \lambda E[\Delta f_C(F)] - E[(F - \lambda) f_C(F)] \right|$$

$$= \left| \lambda E[\Delta f_C(F)] - E[\langle D f_C(F), -DL^{-1} F \rangle_{L^2(\mu)}] \right|.$$

Now, combining (118) with (128), we can write

$$D_t f_C(F) = \Delta f_C(F) D_t F + S(t),$$

with $S(t) \leqslant \frac{1}{2} \|\Delta^2 f_C\|_{\infty} |D_t F(D_t F - 1)| \leqslant \frac{1-e^{-\lambda}}{\lambda^2} |D_t F(D_t F - 1)|$, see indeed Lemma 7 for the last inequality. Putting all these bounds together and since $\|\Delta f_C\|_{\infty} \leqslant \frac{1-e^{-\lambda}}{\lambda}$ by Lemma 7, we get the desired conclusion.

So, to conclude the proof, it remains to show that (128) holds true. Let us first assume that $k \geqslant a + 2$. We then have

$$f(k) = f(a) + \sum_{j=a}^{k-1} \Delta f(j) = f(a) + \Delta f(a)(k-a) + \sum_{j=a}^{k-1}(\Delta f(j) - \Delta f(a))$$

$$= f(a) + \Delta f(a)(k-a) + \sum_{j=a}^{k-1}\sum_{l=a}^{j-1} \Delta^2 f(l)$$

$$= f(a) + \Delta f(a)(k-a) + \sum_{l=a}^{k-2} \Delta^2 f(l)(k-l-1),$$

so that

$$|f(k) - f(a) - \Delta f(a)(k-a)| \leq \|\Delta^2 f\|_\infty \sum_{l=a}^{k-2}(k-l-1)$$

$$= \frac{1}{2}\|\Delta^2 f\|_\infty (k-a)(k-a-1),$$

that is, (128) holds true in this case. When $k = a$ or $k = a+1$, (128) is obviously true. Finally, consider the case $k \leq a-1$. We have

$$f(k) = f(a) - \sum_{j=k}^{a-1} \Delta f(j) = f(a) + \Delta f(a)(k-a) + \sum_{j=k}^{a-1}(\Delta f(a) - \Delta f(j))$$

$$= f(a) + \Delta f(a)(k-a) + \sum_{j=k}^{a-1}\sum_{l=j}^{a-1} \Delta^2 f(l)$$

$$= f(a) + \Delta f(a)(k-a) + \sum_{l=k}^{a-1} \Delta^2 f(l)(l-k+1),$$

so that

$$|f(k) - f(a) - \Delta f(a)(k-a)| \leq \|\Delta^2 f\|_\infty \sum_{l=k}^{a-1}(l-k+1)$$

$$= \frac{1}{2}\|\Delta^2 f\|_\infty (a-k)(a-k+1),$$

that is, (128) holds true in this case as well. The proof of Theorem 26 is done. □

To Go Further. A multivariate extension of Theorem 25 can be found in [47]. Reference [19] contains several explicit applications of the tools developed in this section.

11 Fourth Moment Theorem and Free Probability

To conclude this survey, we shall explain how the Fourth Moment Theorem 2 extends in the theory of free probability, which provides a convenient framework for investigating limits of random matrices. We start with a short introduction to free probability. We refer to [22] for a systematic presentation and to [2] for specific results on Wigner multiple integrals.

Free Tracial Probability Space. A *non-commutative probability space* is a von Neumann algebra $\mathscr{A}$ (that is, an algebra of operators on a complex separable Hilbert space, closed under adjoint and convergence in the weak operator topology) equipped with a *trace* φ, that is, a unital linear functional (meaning preserving the identity) which is weakly continuous, positive (meaning $\varphi(X) \geq 0$ whenever X is a non-negative element of $\mathscr{A}$; i.e. whenever $X = YY^*$ for some $Y \in \mathscr{A}$), faithful (meaning that if $\varphi(YY^*) = 0$ then $Y = 0$), and tracial (meaning that $\varphi(XY) = \varphi(YX)$ for all $X, Y \in \mathscr{A}$, even though in general $XY \neq YX$).

Random Variables. In a non-commutative probability space, we refer to the self-adjoint elements of the algebra as *random variables*. Any random variable X has a *law*: this is the unique probability measure μ on $\mathbb{R}$ with the same moments as X; in other words, μ is such that

$$\int_{\mathbb{R}} x^k d\mu(x) = \varphi(X^k), \quad k \geq 1. \tag{129}$$

(The existence and uniqueness of μ follow from the positivity of φ, see [22, Proposition 3.13].)

Convergence in Law. We say that a sequence $(X_{1,n}, \ldots, X_{k,n})$, $n \geq 1$, of random vectors *converges in law* to a random vector $(X_{1,\infty}, \ldots, X_{k,\infty})$, and we write

$$(X_{1,n}, \ldots, X_{k,n}) \xrightarrow{\text{law}} (X_{1,\infty}, \ldots, X_{k,\infty}),$$

to indicate the convergence in the sense of (joint) moments, that is,

$$\lim_{n \to \infty} \varphi\left(Q(X_{1,n}, \ldots, X_{k,n})\right) = \varphi\left(Q(X_{1,\infty}, \ldots, X_{k,\infty})\right), \tag{130}$$

for any polynomial Q in k non-commuting variables.

We say that a sequence (F_n) of *non-commutative stochastic processes* (that is, each F_n is a one-parameter family of self-adjoint operators $F_n(t)$ in $(\mathscr{A}, \varphi)$) *converges in the sense of finite-dimensional distributions* to a non-commutative stochastic process F_∞, and we write

$$F_n \xrightarrow{\text{f.d.d.}} F_\infty,$$

to indicate that, for any $k \geq 1$ and any $t_1, \ldots, t_k \geq 0$,

$$(F_n(t_1), \ldots, F_n(t_k)) \overset{\text{law}}{\to} (F_\infty(t_1), \ldots, F_\infty(t_k)).$$

Free Independence. In the free probability setting, the notion of *independence* (introduced by Voiculescu in [55]) goes as follows. Let $\mathscr{A}_1, \ldots, \mathscr{A}_p$ be unital subalgebras of $\mathscr{A}$. Let $X_1, \ldots, X_m$ be elements chosen from among the $\mathscr{A}_i$'s such that, for $1 \leq j < m$, two consecutive elements X_j and X_{j+1} do not come from the same $\mathscr{A}_i$ and are such that $\varphi(X_j) = 0$ for each j. The subalgebras $\mathscr{A}_1, \ldots, \mathscr{A}_p$ are said to be *free* or *freely independent* if, in this circumstance,

$$\varphi(X_1 X_2 \cdots X_m) = 0. \tag{131}$$

Random variables are called freely independent if the unital algebras they generate are freely independent. Freeness is in general much more complicated than classical independence. For example, if X, Y are free and $m, n \geq 1$, then by (131),

$$\varphi\big((X^m - \varphi(X^m)1)(Y^n - \varphi(Y^n)1)\big) = 0.$$

By expanding (and using the linear property of φ), we get

$$\varphi(X^m Y^n) = \varphi(X^m)\varphi(Y^n), \tag{132}$$

which is what we would expect under classical independence. But, by setting $X_1 = X_3 = X - \varphi(X)1$ and $X_2 = X_4 = Y - \varphi(Y)$ in (131), we also have

$$\varphi\big((X - \varphi(X)1)(Y - \varphi(Y)1)(X - \varphi(X)1)(Y - \varphi(Y)1)\big) = 0.$$

By expanding, using (132) and the tracial property of φ (for instance $\varphi(XYX) = \varphi(X^2 Y)$) we get

$$\varphi(XYXY) = \varphi(Y)^2 \varphi(X^2) + \varphi(X)^2 \varphi(Y^2) - \varphi(X)^2 \varphi(Y)^2,$$

which is different from $\varphi(X^2)\varphi(Y^2)$, which is what one would have obtained if X and Y were classical independent random variables. Nevertheless, if X, Y are freely independent, then their joint moments are determined by the moments of X and Y separately, exactly as in the classical case.

Semicircular Distribution. The *semicircular distribution* $\mathscr{S}(m, \sigma^2)$ with mean $m \in \mathbb{R}$ and variance $\sigma^2 > 0$ is the probability distribution

$$\mathscr{S}(m, \sigma^2)(dx) = \frac{1}{2\pi\sigma^2} \sqrt{4\sigma^2 - (x-m)^2}\, \mathbf{1}_{\{|x-m| \leq 2\sigma\}}\, dx. \tag{133}$$

If $m = 0$, this distribution is symmetric around 0, and therefore its odd moments are all 0. A simple calculation shows that the even centered moments are given by (scaled) *Catalan numbers*: for non-negative integers k,

$$\int_{m-2\sigma}^{m+2\sigma} (x-m)^{2k} \mathscr{S}(m,\sigma^2)(dx) = C_k \sigma^{2k},$$

where $C_k = \frac{1}{k+1}\binom{2k}{k}$ (see, e.g., [22, Lecture 2]). In particular, the variance is σ^2 while the centered fourth moment is $2\sigma^4$. The semicircular distribution plays here the role of the Gaussian distribution. It has the following similar properties:

1. If $S \sim \mathscr{S}(m,\sigma^2)$ and $a,b \in \mathbb{R}$, then $aS + b \sim \mathscr{S}(am+b, a^2\sigma^2)$.
2. If $S_1 \sim \mathscr{S}(m_1,\sigma_1^2)$ and $S_2 \sim \mathscr{S}(m_2,\sigma_2^2)$ are freely independent, then $S_1 + S_2 \sim \mathscr{S}(m_1+m_2, \sigma_1^2 + \sigma_2^2)$.

Free Brownian Motion. A *free Brownian motion* $S = \{S(t)\}_{t \geq 0}$ is a non-commutative stochastic process with the following defining characteristics:

(1) $S(0) = 0$.
(2) For $t_2 > t_1 \geq 0$, the law of $S(t_2) - S(t_1)$ is the semicircular distribution of mean 0 and variance $t_2 - t_1$.
(3) For all n and $t_n > \cdots > t_2 > t_1 > 0$, the increments $S(t_1), S(t_2) - S(t_1), \ldots, S(t_n) - S(t_{n-1})$ are freely independent.

We may think of free Brownian motion as "infinite-dimensional matrix-valued Brownian motion".

Wigner Integral. Let $S = \{S(t)\}_{t \geq 0}$ be a free Brownian motion. Let us quickly sketch out the construction of the *Wigner integral* of f with respect to S. For an indicator function $f = \mathbf{1}_{[u,v]}$, the Wigner integral of f is defined by

$$\int_0^\infty \mathbf{1}_{[u,v]}(x) dS(x) = S(v) - S(u).$$

We then extend this definition by linearity to simple functions of the form $f = \sum_{i=1}^k \alpha_i \mathbf{1}_{[u_i,v_i]}$, where $[u_i, v_i]$ are disjoint intervals of $\mathbb{R}_+$. Simple computations show that

$$\varphi\left(\int_0^\infty f(x) dS(x)\right) = 0 \tag{134}$$

$$\varphi\left(\int_0^\infty f(x) dS(x) \times \int_0^\infty g(x) dS(x)\right) = \langle f, g \rangle_{L^2(\mathbb{R}_+)}. \tag{135}$$

By isometry, the definition of $\int_0^\infty f(x) dS(x)$ is extended to all $f \in L^2(\mathbb{R}_+)$, and (134)–(135) continue to hold in this more general setting.

Multiple Wigner Integral. Let $S = \{S(t)\}_{t \geq 0}$ be a free Brownian motion, and let $q \geq 1$ be an integer. When $f \in L^2(\mathbb{R}_+^q)$ is real-valued, we write f^* to indicate the function of $L^2(\mathbb{R}_+^q)$ given by $f^*(t_1, \ldots, t_q) = f(t_q, \ldots, t_1)$.

Following [2], let us quickly sketch out the construction of the *multiple Wigner integral* of f with respect to S. Let $D^q \subset \mathbb{R}_+^q$ be the collection of all diagonals, i.e.

$$D^q = \{(t_1, \ldots, t_q) \in \mathbb{R}_+^q : t_i = t_j \text{ for some } i \neq j\}. \tag{136}$$

For an indicator function $f = \mathbf{1}_A$, where $A \subset \mathbb{R}_+^q$ has the form $A = [u_1, v_1] \times \ldots \times [u_q, v_q]$ with $A \cap D^q = \emptyset$, the qth multiple Wigner integral of f is defined by

$$I_q^S(f) = (S(v_1) - S(u_1)) \ldots (S(v_q) - S(u_q)).$$

We then extend this definition by linearity to simple functions of the form $f = \sum_{i=1}^k \alpha_i \mathbf{1}_{A_i}$, where $A_i = [u_1^i, v_1^i] \times \ldots \times [u_q^i, v_q^i]$ are disjoint q-dimensional rectangles as above which do not meet the diagonals. Simple computations show that

$$\varphi(I_q^S(f)) = 0 \tag{137}$$

$$\varphi(I_q^S(f) I_q^S(g)) = \langle f, g^* \rangle_{L^2(\mathbb{R}_+^q)}. \tag{138}$$

By isometry, the definition of $I_q^S(f)$ is extended to all $f \in L^2(\mathbb{R}_+^q)$, and (137)–(138) continue to hold in this more general setting. If one wants $I_q^S(f)$ to be a random variable, it is necessary for f to be *mirror symmetric*, that is, $f = f^*$ (see [17]). Observe that $I_1^S(f) = \int_0^\infty f(x) dS(x)$ when $q = 1$. We have moreover

$$\varphi(I_p^S(f) I_q^S(g)) = 0 \text{ when } p \neq q, \ f \in L^2(\mathbb{R}_+^p) \text{ and } g \in L^2(\mathbb{R}_+^q). \tag{139}$$

When $r \in \{1, \ldots, p \wedge q\}$, $f \in L^2(\mathbb{R}_+^p)$ and $g \in L^2(\mathbb{R}_+^q)$, let us write $f \overset{r}{\frown} g$ to indicate the rth *contraction* of f and g, defined as being the element of $L^2(\mathbb{R}_+^{p+q-2r})$ given by

$$f \overset{r}{\frown} g(t_1, \ldots, t_{p+q-2r}) \tag{140}$$

$$= \int_{\mathbb{R}_+^r} f(t_1, \ldots, t_{p-r}, x_1, \ldots, x_r) g(x_r, \ldots, x_1, t_{p-r+1}, \ldots, t_{p+q-2r}) dx_1 \ldots dx_r.$$

By convention, set $f \overset{0}{\frown} g = f \otimes g$ as being the tensor product of f and g. Since f and g are not necessarily symmetric functions, the position of the identified variables $x_1, \ldots, x_r$ in (140) is important, in contrast to what happens in classical probability. Observe moreover that

$$\|f \overset{r}{\frown} g\|_{L^2(\mathbb{R}_+^{p+q-2r})} \leq \|f\|_{L^2(\mathbb{R}_+^p)} \|g\|_{L^2(\mathbb{R}_+^q)} \tag{141}$$

by Cauchy–Schwarz, and also that $f \overset{q}{\frown} g = \langle f, g^* \rangle_{L^2(\mathbb{R}_+^q)}$ when $p = q$.

We have the following *product formula* (see [2, Proposition 5.3.3]), valid for any $f \in L^2(\mathbb{R}_+^p)$ and $g \in L^2(\mathbb{R}_+^q)$:

$$I_p^S(f) I_q^S(g) = \sum_{r=0}^{p \wedge q} I_{p+q-2r}^S(f \stackrel{r}{\frown} g). \tag{142}$$

We deduce (by a straightforward induction) that, for any $e \in L^2(\mathbb{R}_+)$ and any $q \geq 1$,

$$U_q\left(\int_0^\infty e(x) dS_x\right) = I_q^S(e^{\otimes q}), \tag{143}$$

where $U_0 = 1$, $U_1 = X$, $U_2 = X^2 - 1$, $U_3 = X^3 - 2X$, ..., is the sequence of Tchebycheff polynomials of second kind (determined by the recursion $XU_k = U_{k+1} + U_{k-1}$), $\int_0^\infty e(x) dS(x)$ is understood as a Wigner integral, and $e^{\otimes q}$ is the qth tensor product of e. This is the exact analogue of (10) in our context.

We are now in a position to offer a free analogue of the Fourth Moment Theorem 3, which reads as follows.

Theorem 27 (Kemp, Nourdin, Peccati and Speicher (2011); see [17]). *Fix an integer $q \geq 2$ and let $\{S_t\}_{t \geq 0}$ be a free Brownian motion. Whenever $f \in L^2(\mathbb{R}_+^q)$, set $I_q^S(f)$ to denote the qth multiple Wigner integrals of f with respect to S. Let $\{F_n\}_{n \geq 1}$ be a sequence of Wigner multiple integrals of the form*

$$F_n = I_q^S(f_n),$$

where each $f_n \in L^2(\mathbb{R}_+)$ is mirror-symmetric, that is, is such that $f_n = f_n^*$. Suppose moreover that $\varphi(F_n^2) \to 1$ as $n \to \infty$. Then, as $n \to \infty$, the following two assertions are equivalent:

(i) $F_n \stackrel{Law}{\to} S_1 \sim \mathscr{S}(0, 1)$;
(ii) $\varphi(F_n^4) \to 2 = \varphi(S_1^4)$.

Proof (following [24]). Without loss of generality and for sake of simplicity, we suppose that $\varphi(F_n^2) = 1$ for all n (instead of $\varphi(F_n^2) \to 1$ as $n \to \infty$). The proof of the implication $(i) \Rightarrow (ii)$ being trivial by the very definition of the convergence in law in a free tracial probability space, we only concentrate on the proof of $(ii) \Rightarrow (i)$.

Fix $k \geq 3$. By iterating the product formula (142), we can write

$$F_n^k = I_q^S(f_n)^k = \sum_{(r_1,\ldots,r_{k-1}) \in A_{k,q}} I_{kq-2r_1-\ldots-2r_{k-1}}^S((\ldots((f_n \stackrel{r_1}{\frown} f_n) \stackrel{r_2}{\frown} f_n) \ldots) \stackrel{r_{k-1}}{\frown} f_n),$$

where

$$A_{k,q} = \{(r_1,\ldots,r_{k-1}) \in \{0,1,\ldots,q\}^{k-1} : r_2 \leq 2q - 2r_1,\ r_3 \leq 3q - 2r_1 - 2r_2, \ldots,$$
$$r_{k-1} \leq (k-1)q - 2r_1 - \ldots - 2r_{k-2}\}.$$

By taking the φ-trace in the previous expression and taking into account that (137) holds, we deduce that

$$\varphi(F_n^k) = \varphi(I_q^S(f_n)^k) = \sum_{(r_1,\ldots,r_{k-1}) \in B_{k,q}} (\ldots((f_n \stackrel{r_1}{\frown} f_n) \stackrel{r_2}{\frown} f_n)\ldots) \stackrel{r_{k-1}}{\frown} f_n, \quad (144)$$

with

$$B_{k,q} = \{(r_1,\ldots,r_{k-1}) \in A_{k,q} : 2r_1 + \ldots + 2r_{k-1} = kq\}.$$

Let us decompose $B_{k,q}$ into $C_{k,q} \cup E_{k,q}$, with $C_{k,q} = B_{k,q} \cap \{0,q\}^{k-1}$ and $E_{k,q} = B_{k,q} \setminus C_{k,q}$. We then have

$$\varphi(F_n^k) = \sum_{(r_1,\ldots,r_{k-1}) \in C_{k,q}} \left((\ldots((f_n \stackrel{r_1}{\frown} f_n) \stackrel{r_2}{\frown} f_n)\ldots) \stackrel{r_{k-1}}{\frown} f_n\right)$$

$$+ \sum_{(r_1,\ldots,r_{k-1}) \in E_{k,q}} \left((\ldots((f_n \stackrel{r_1}{\frown} f_n) \stackrel{r_2}{\frown} f_n)\ldots) \stackrel{r_{k-1}}{\frown} f_n\right).$$

Using the two relationships $f_n \stackrel{0}{\frown} f_n = f_n \otimes f_n$ and

$$f_n \stackrel{q}{\frown} f_n = \int_{\mathbb{R}_+^q} f_n(t_1,\ldots,t_q) f_n(t_q,\ldots,t_1) dt_1 \ldots dt_q = \|f_n\|_{L^2(\mathbb{R}_+^q)}^2 = 1,$$

it is evident that $(\ldots((f_n \stackrel{r_1}{\frown} f_n) \stackrel{r_2}{\frown} f_n)\ldots) \stackrel{r_{k-1}}{\frown} f_n = 1$ for all $(r_1,\ldots,r_{k-1}) \in C_{k,q}$. We deduce that

$$\varphi(F_n^k) = \#C_{k,q} + \sum_{(r_1,\ldots,r_{k-1}) \in E_{k,q}} \left((\ldots((f_n \stackrel{r_1}{\frown} f_n) \stackrel{r_2}{\frown} f_n)\ldots) \stackrel{r_{k-1}}{\frown} f_n\right).$$

On the other hand, by applying (144) with $q = 1$, we get that

$$\varphi(S_1^k) = \varphi(I_1^S(\mathbf{1}_{[0,1]})^k) = \sum_{(r_1,\ldots,r_{k-1}) \in B_{k,1}} (\ldots((\mathbf{1}_{[0,1]} \stackrel{r_1}{\frown} \mathbf{1}_{[0,1]}) \stackrel{r_2}{\frown} \mathbf{1}_{[0,1]})\ldots) \stackrel{r_{k-1}}{\frown} \mathbf{1}_{[0,1]}$$

$$= \sum_{(r_1,\ldots,r_{k-1}) \in B_{k,1}} 1 = \#B_{k,1}.$$

But it is clear that $C_{k,q}$ is in bijection with $B_{k,1}$ (by dividing all the r_i's in $C_{k,q}$ by q). Consequently,

$$\varphi(F_n^k) = \varphi(S_1^k) + \sum_{(r_1,\ldots,r_{k-1}) \in E_{k,q}} \left((\ldots((f_n \stackrel{r_1}{\frown} f_n) \stackrel{r_2}{\frown} f_n)\ldots) \stackrel{r_{k-1}}{\frown} f_n\right). \quad (145)$$

Now, assume that $\varphi(F_n^4) \to \varphi(S_1^4) = 2$ and let us show that $\varphi(F_n^k) \to \varphi(S_1^k)$ for all $k \geq 3$. Using that $f_n = f_n^*$, observe that

$$f_n \stackrel{r}{\frown} f_n(t_1,\ldots,t_{2q-2r})$$

$$= \int_{\mathbb{R}_+^r} f_n(t_1,\ldots,t_{q-r},s_1,\ldots,s_r) f_n(s_r,\ldots,s_1,t_{q-r+1},\ldots,t_{2q-2r}) ds_1 \ldots ds_r$$

$$= \int_{\mathbb{R}_+^r} f_n(s_r,\ldots,s_1,t_{q-r},\ldots,t_1) f_n(t_{2q-2r},\ldots,t_{q-r+1},s_1,\ldots,s_r) ds_1 \ldots ds_r$$

$$= f_n \stackrel{r}{\frown} f_n(t_{2q-2r},\ldots,t_1) = (f_n \stackrel{r}{\frown} f_n)^*(t_1,\ldots,t_{2q-2r}),$$

that is, $f_n \stackrel{r}{\frown} f_n = (f_n \stackrel{r}{\frown} f_n)^*$. On the other hand, the product formula (142) leads to $F_n^2 = \sum_{r=0}^{q} I_{2q-2r}^S(f_n \stackrel{r}{\frown} f_n)$. Since two multiple integrals of different orders are orthogonal (see (139)), we deduce that

$$\varphi(F_n^4) = \|f_n \otimes f_n\|_{L^2(\mathbb{R}_+^{2q})}^2 + (\|f_n\|_{L^2(\mathbb{R}_+^q)}^2)^2 + \sum_{r=1}^{q-1} \langle f_n \stackrel{r}{\frown} f_n, (f_n \stackrel{r}{\frown} f_n)^*\rangle_{L^2(\mathbb{R}_+^{2q-2r})}$$

$$= 2\|f_n\|_{L^2([0,1]^q)}^4 + \sum_{r=1}^{q-1} \|f_n \stackrel{r}{\frown} f_n\|_{L^2(\mathbb{R}_+^{2q-2r})}^2$$

$$= 2 + \sum_{r=1}^{q-1} \|f_n \stackrel{r}{\frown} f_n\|_{L^2(\mathbb{R}_+^{2q-2r})}^2. \tag{146}$$

Using that $\varphi(F_n^4) \to 2$, we deduce that

$$\|f_n \stackrel{r}{\frown} f_n\|_{L^2(\mathbb{R}_+^{2q-2r})}^2 \to 0 \quad \text{for all } r = 1,\ldots,q-1. \tag{147}$$

Fix $(r_1,\ldots,r_{k-1}) \in E_{k,q}$ and let $j \in \{1,\ldots,k-1\}$ be the smallest integer such that $r_j \in \{1,\ldots,q-1\}$. Then:

$$\left|(\ldots((f_n \stackrel{r_1}{\frown} f_n) \stackrel{r_2}{\frown} f_n)\ldots) \stackrel{r_{k-1}}{\frown} f_n\right|$$

$$= \left|(\ldots((f_n \stackrel{r_1}{\frown} f_n) \stackrel{r_2}{\frown} f_n)\ldots \stackrel{r_{j-1}}{\frown} f_n) \stackrel{r_j}{\frown} f_n) \stackrel{r_{j+1}}{\frown} f_n)\ldots) \stackrel{r_{k-1}}{\frown} f_n\right|$$

$$= \left|(\ldots((f_n \otimes \ldots \otimes f_n) \stackrel{r_j}{\frown} f_n) \stackrel{r_{j+1}}{\frown} f_n)\ldots) \stackrel{r_{k-1}}{\frown} f_n\right| \quad (\text{since } f_n \stackrel{q}{\frown} f_n = 1)$$

$$= \left|(\ldots((f_n \otimes \ldots \otimes f_n) \otimes (f_n \stackrel{r_j}{\frown} f_n)) \stackrel{r_{j+1}}{\frown} f_n)\ldots) \stackrel{r_{k-1}}{\frown} f_n\right|$$

$$\leq \|(f_n \otimes \ldots \otimes f_n) \otimes (f_n \stackrel{r_j}{\frown} f_n)\| \|f_n\|^{k-j-1} \quad (\text{Cauchy-Schwarz})$$

$$= \|f_n \stackrel{r_j}{\frown} f_n\| \quad (\text{since } \|f_n\|^2 = 1)$$

$$\to 0 \quad \text{as } n \to \infty \quad \text{by (147)}.$$

Therefore, we deduce from (145) that $\varphi(F_n^k) \to \varphi(S_1^k)$, which is the desired conclusion and concludes the proof of the theorem. □

During the proof of Theorem 27, we actually showed (see indeed (146)) that the two assertions (*i*)–(*ii*) are both equivalent to a third one, namely

(*iii*): $\|f_n \overset{r}{\frown} f_n\|^2_{L^2(\mathbb{R}_+^{2q-2r})} \to 0$ for all $r = 1, \ldots, q-1$.

Combining (*iii*) with Corollary 3, we immediately deduce an interesting transfer principle for translating results between the classical and free chaoses.

Corollary 7. *Fix an integer $q \geq 2$, let $\{B_t\}_{t \geq 0}$ be a standard Brownian motion and let $\{S_t\}_{t \geq 0}$ be a free Brownian motion. Whenever $f \in L^2(\mathbb{R}_+^q)$, we write $I_q^B(f)$ (resp. $I_q^S(f)$) to indicate the qth multiple Wiener integrals of f with respect to B (resp. S). Let $\{f_n\}_{n \geq 1} \subset L^2(\mathbb{R}_+^q)$ be a sequence of symmetric functions and let $\sigma > 0$ be a finite constant. Then, as $n \to \infty$, the following two assertions hold true.*

(i) *$E[I_q^B(f_n)] \to q!\sigma^2$ if and only if $\varphi(I_q^S(f_n)^2) \to \sigma^2$.*

(ii) *If the asymptotic relations in (i) are verified, then $I_q^B(f_n) \overset{law}{\to} \mathcal{N}(0, q!\sigma^2)$ if and only if $I_q^S(f_n) \overset{law}{\to} \mathcal{S}(0, \sigma^2)$.*

To Go Further. A multivariate version of Theorem 27 (free counterpart of Theorem 17) can be found in [36]. In [31] (resp. [14]), one exhibits a version of Theorem 27 in which the semicircular law in the limit is replaced by the free Poisson law (resp. the so-called tetilla law). An extension of Theorem 27 in the context of the q-Brownian motion (which is an interpolation between the standard Brownian motion corresponding to $q = 1$ and the free Brownian motion corresponding to $q = 0$) is given in [12].

Acknowledgements It is a pleasure to thank the Fondation des Sciences Mathématiques de Paris for its generous support during the academic year 2011–2012 and for giving me the opportunity to speak about my recent research in the prestigious Collège de France. I am grateful to all the participants of these lectures for their assiduity. Also, I would like to warmly thank two anonymous referees for their very careful reading and for their valuable suggestions and remarks. Finally, my last thank goes to Giovanni Peccati, not only for accepting to give a lecture (resulting to the material developed in Sect. 10) but also (and especially!) for all the nice theorems we recently discovered together. I do hope it will continue this way as long as possible!

References

1. B. Bercu, I. Nourdin, M.S. Taqqu, Almost sure central limit theorems on the Wiener space. Stoch. Proc. Appl. **120**(9), 1607–1628 (2010)
2. P. Biane, R. Speicher, Stochastic analysis with respect to free Brownian motion and analysis on Wigner space. Probab. Theory Rel. Fields **112**, 373–409 (1998)

3. H. Biermé, A. Bonami, J. Léon, Central limit theorems and quadratic variations in terms of spectral density. Electron. J. Probab. **16**, 362–395 (2011)
4. H. Biermé, A. Bonami, I. Nourdin, G. Peccati, Optimal Berry-Esseen rates on the Wiener space: the barrier of third and fourth cumulants. ALEA Lat. Am. J. Probab. Math. Stat. **9**(2), 473–500 (2012)
5. E. Bolthausen, An estimate of the remainder in a combinatorial central limit theorem. Z. Wahrscheinlichkeitstheorie verw. Gebiete **66**, 379–386 (1984)
6. J.-C. Breton, I. Nourdin, Error bounds on the non-normal approximation of Hermite power variations of fractional Brownian motion. Electron. Comm. Probab. **13**, 482–493 (2008) (electronic)
7. P. Breuer, P. Major, Central limit theorems for non-linear functionals of Gaussian fields. J. Mult. Anal. **13**, 425–441 (1983)
8. L.H.Y. Chen, Poisson approximation for dependent trials. Ann. Probab. **3**(3), 534–545 (1975)
9. L.H.Y. Chen, L. Goldstein, Q.-M. Shao, *Normal Approximation by Stein's Method*. Probability and Its Applications. Springer, New York (2010)
10. F. Daly, Upper bounds for Stein-type operators. Electron. J. Probab. **13**(20), 566–587 (2008) (electronic)
11. L. Decreusefond, D. Nualart, Hitting times for Gaussian processes. Ann. Probab. **36**(1), 319–330 (2008)
12. A. Deya, S. Noreddine, I. Nourdin, Fourth moment theorem and q-Brownian chaos. Commun. Math. Phys. 1–22 (2012)
13. A. Deya, I. Nourdin, Invariance principles for homogeneous sums of free random variables. arXiv preprint arXiv:1201.1753 (2012)
14. A. Deya, I. Nourdin, Convergence of Wigner integrals to the tetilla law. ALEA **9**, 101–127 (2012)
15. C.G. Esseen, A moment inequality with an application to the central limit theorem. Skand. Aktuarietidskr. **39**, 160–170 (1956)
16. S.-T. Ho, L.H.Y. Chen, An L_p bound for the remainder in a combinatorial central limit theorem. Ann. Probab. **6**(2), 231–249 (1978)
17. T. Kemp, I. Nourdin, G. Peccati, R. Speicher, Wigner chaos and the fourth moment. Ann. Probab. **40**(4), 1577–1635 (2012)
18. V.Yu. Korolev, I.G. Shevtsova, On the upper bound for the absolute constant in the Berry-Esseen inequality. Theory Probab. Appl. **54**(4), 638–658 (2010)
19. R. Lachièze-Rey, G. Peccati, Fine Gaussian fluctuations on the Poisson space, I: contractions, cumulants and geometric random graphs. arXiv preprint arXiv:1111.7312 (2011)
20. E. Mossel, R. O'Donnell, K. Oleszkiewicz, Noise stability of functions with low influences: invariance and optimality. Ann. Math. **171**(1), 295–341 (2010)
21. E. Nelson, The free Markoff field. J. Funct. Anal. **12**, 211–227 (1973)
22. A. Nica, R. Speicher, *Lectures on the Combinatorics of Free Probability* (Cambridge University Press, Cambridge, 2006)
23. S. Noreddine, I. Nourdin, On the Gaussian approximation of vector-valued multiple integrals. J. Multiv. Anal. **102**(6), 1008–1017 (2011)
24. I. Nourdin, Yet another proof of the Nualart-Peccati criterion. Electron. Comm. Probab. **16**, 467–481 (2011)
25. I. Nourdin, *Selected Aspects of Fractional Brownian Motion* (Springer, New York, 2012)
26. I. Nourdin, G. Peccati, Non-central convergence of multiple integrals. Ann. Probab. **37**(4), 1412–1426 (2007)
27. I. Nourdin, G. Peccati, Stein's method on Wiener chaos. Probab. Theory Rel. Fields **145**(1), 75–118 (2009)
28. I. Nourdin, G. Peccati, Stein's method meets Malliavin calculus: a short survey with new estimates. *Recent Advances in Stochastic Dynamics and Stochastic Analysis* (World Scientific, Singapore, 2010), pp. 207–236
29. I. Nourdin, G. Peccati, Cumulants on the Wiener space. J. Funct. Anal. **258**, 3775–3791 (2010)
30. I. Nourdin, G. Peccati, Stein's method and exact Berry-Esseen asymptotics for functionals of Gaussian fields. Ann. Probab. **37**(6), 2231–2261 (2010)

31. I. Nourdin, G. Peccati, Poisson approximations on the free Wigner chaos. arXiv preprint arXiv:1103.3925 (2011)
32. I. Nourdin, G. Peccati, *Normal Approximations Using Malliavin Calculus: from Stein's Method to Universality*. Cambridge Tracts in Mathematics (Cambridge University Press, Cambridge, 2012)
33. I. Nourdin, G. Peccati, M. Podolskij, Quantitative Breuer–Major Theorems. Stoch. Proc. App. **121**(4), 793–812 (2011)
34. I. Nourdin, G. Peccati, G. Reinert, Invariance principles for homogeneous sums: universality of Gaussian Wiener chaos. Ann. Probab. **38**(5), 1947–1985 (2010)
35. I. Nourdin, G. Peccati, A. Réveillac, Multivariate normal approximation using Stein's method and Malliavin calculus. Ann. Inst. H. Poincaré (B) Probab. Stat. **46**(1), 45–58 (2010)
36. I. Nourdin, G. Peccati, R. Speicher, Multidimensional semicircular limits on the free Wigner chaos, in *Ascona Proceedings*, Birkhäuser Verlag (2013)
37. I. Nourdin, F.G. Viens, Density estimates and concentration inequalities with Malliavin calculus. Electron. J. Probab. **14**, 2287–2309 (2009) (electronic)
38. D. Nualart, *The Malliavin Calculus and Related Topics*, 2nd edn. (Springer, Berlin, 2006)
39. D. Nualart, S. Ortiz-Latorre, Central limit theorems for multiple stochastic integrals and Malliavin calculus. Stoch. Proc. Appl. **118**(4), 614–628 (2008)
40. D. Nualart, G. Peccati, Central limit theorems for sequences of multiple stochastic integrals. Ann. Probab. **33**(1), 177–193 (2005)
41. D. Nualart, L. Quer-Sardanyons, Optimal Gaussian density estimates for a class of stochastic equations with additive noise. Infinite Dimensional Anal. Quant. Probab. Relat. Top. **14**(1), 25–34 (2011)
42. D. Nualart, J. Vives, Anticipative calculus for the Poisson process based on the Fock space. Séminaire de Probabilités, vol. XXIV, LNM 1426 (Springer, New York, 1990), pp. 154–165
43. G. Peccati, The Chen-Stein method for Poisson functionals. arXiv:1112.5051v3 (2012)
44. G. Peccati, J.-L. Solé, M.S. Taqqu, F. Utzet, Stein's method and normal approximation of Poisson functionals. Ann. Probab. **38**(2), 443–478 (2010)
45. G. Peccati, M.S. Taqqu, *Wiener Chaos: Moments, Cumulants and Diagrams* (Springer, New York, 2010)
46. G. Peccati, C.A. Tudor, Gaussian limits for vector-valued multiple stochastic integrals. Séminaire de Probabilités, vol. XXXVIII, LNM 1857 (Springer, New York, 2005), pp. 247–262
47. G. Peccati, C. Zheng, Multidimensional Gaussian fluctuations on the Poisson space. Electron. J. Probab. **15**, paper 48, 1487–1527 (2010) (electronic)
48. G. Peccati, C. Zheng, Universal Gaussian fluctuations on the discrete Poisson chaos. arXiv preprint arXiv:1110.5723v1
49. M. Penrose, *Random Geometric Graphs* (Oxford University Press, Oxford, 2003)
50. W. Rudin, *Real and Complex Analysis*, 3rd edn. (McGraw-Hill, New York, 1987)
51. I. Shigekawa, Derivatives of Wiener functionals and absolute continuity of induced measures. J. Math. Kyoto Univ. **20**(2), 263–289 (1980)
52. Ch. Stein, A bound for the error in the normal approximation to the distribution of a sum of dependent random variables. In: *Proceedings of the Sixth Berkeley Symposium on Mathematical Statistics and Probability, Vol. II: Probability Theory*, 583–602. University of California Press, Berkeley, California (1972)
53. D.W. Stroock, Homogeneous chaos revisited. In: Séminaire de Probabilités, vol. XXI. Lecture Notes in Math. vol. 1247 (Springer, Berlin, 1987), pp. 1–8
54. M. Talagrand, *Spin Glasses, a Challenge for Mathematicians* (Springer, New York, 2003)
55. D.V. Voiculescu, Symmetries of some reduced free product C^*-algebras. *Operator algebras and their connection with topology and ergodic theory*, Springer Lecture Notes in Mathematics, vol. 1132, 556–588 (1985)
56. R. Zintout, Total variation distance between two double Wiener-Itô integrals. Statist. Probab. Letter, to appear (2013)

Part II
Other Contributions

Some Sufficient Conditions for the Ergodicity of the Lévy Transformation

Vilmos Prokaj*

Abstract We propose a possible way of attacking the question posed originally by Daniel Revuz and Marc Yor in their book published in 1991. They asked whether the Lévy transformation of the Wiener-space is ergodic. Our main results are formulated in terms of a strongly stationary sequence of random variables obtained by evaluating the iterated paths at time one. Roughly speaking, this sequence has to approach zero "sufficiently fast". For example, one of our results states that if the expected hitting time of small neighborhoods of the origin do not grow faster than the inverse of the size of these sets then the Lévy transformation is strongly mixing, hence ergodic.

1 Introduction

We work on the canonical space for continuous processes, that is, on the set of continuous functions $\mathcal{C}[0, \infty)$ equipped with the Borel σ-field $\mathcal{B}(\mathcal{C}[0, \infty))$ and the Wiener measure **P**. On this space the canonical process $\beta_t(\omega) = \omega(t)$ is a Brownian motion and the Lévy transformation **T**, given by the formula

$$(\mathbf{T}\beta)_t = \int_0^t \mathrm{sign}(\beta_s)\mathrm{d}\beta_s,$$

---

*The European Union and the European Social Fund have provided financial support to the project under the grant agreement no. TÁMOP 4.2.1./B-09/1/KMR-2010-0003.

V. Prokaj (✉)
Department of Probability and Statistics, Eötvös Loránd University,
Pázmány P. sétány 1/C, Budapest, H-1117 Hungary
e-mail: prokaj@cs.elte.hu

is almost everywhere defined and preserves the measure **P**. A long standing open question is the ergodicity of this transformation. It was probably first mentioned in written form in Revuz and Yor [11] (pp. 257). Since then there were some work on the question, see Dubins and Smorodinsky [3]; Dubins et al. [4]; Fujita [5]; Malric [7, 8]. One of the recent deep result of Marc Malric, see [9], is the topological recurrence of the transformation, that is, the orbit of a typical Brownian path meets any non empty open set almost surely. Brossard and Leuridan [2] provide an alternative presentation of the proof.

In this paper we consider mainly the strong mixing property of the Lévy transformation. Our main results are formulated in terms of a strongly stationary sequence of random variables defined by evaluating the iterated paths at time one. Put $Z_n = \min_{0 \leq k < n} |(\mathbf{T}^k \beta)_1|$. We show in Theorem 8 that if

$$\liminf_{n \to \infty} \frac{Z_{n+1}}{Z_n} < 1, \quad \text{almost surely,} \qquad (*)$$

then **T** is strongly mixing, hence ergodic.

We will say that a family of real valued variables $\{\xi_i : i \in I\}$ is tight if the family of the probability measures $\{\mathbf{P} \circ \xi_i^{-1} : i \in I\}$ is tight, that is if $\sup_{i \in I} \mathbf{P}(|\xi_i| > K) \to 0$ as $K \to \infty$.

In Theorem 11 below, we will see that the tightness of the family $\{nZ_n : n \geq 1\}$ implies $(*)$, in particular if $\mathbf{E}(Z_n) = O(1/n)$ then the Lévy transformation is strongly mixing, hence ergodic. Another way of expressing the same idea, uses the hitting time $\nu(x) = \inf\{n \geq 0 : Z_n < x\}$ of the x-neighborhood of zero by the sequence $((\mathbf{T}^k \beta)_1)_{k \geq 0}$ for $x > 0$. In the same Theorem we will see that the tightness of $\{x\nu(x) : x \in (0, 1)\}$ is also sufficient for $(*)$. In particular, if $\mathbf{E}(\nu(x)) = O(1/x)$ as $x \to 0$, that is, the expected hitting time of small neighborhoods of the origin do not grow faster than the inverse of the size of these sets, then the Lévy transformation is strongly mixing, hence ergodic.

It is natural to compare our result with the density theorem of Marc Malric. We obtain that to settle the question of ergodicity one should focus on specific open sets only, but for those sets deeper understanding of the hitting time is required.

In the next section we sketch our argument, formulating the intermediate steps. Most of the proofs are given in Sect. 3. Note, that we do not use the topological recurrence theorem of Marc Malric, instead all of our argument is based on his density result of the zeros of the iterated paths, see [8]. This theorem states that the set

$$\{t \geq 0 : \exists n, (\mathbf{T}^n \beta)_t = 0\} \quad \text{is dense in } [0, \infty) \text{ almost surely.} \qquad (1)$$

Hence the argument given below may eventually lead to an alternative proof of the topological recurrence theorem as well.

2 Results and Tools

2.1 Integral-Type Transformations

Recall, that a measure preserving transformation T of a probability space $(\Omega, \mathcal{B}, \mathbf{P})$ is ergodic, if

$$\lim_{n \to \infty} \frac{1}{n} \sum_{k=0}^{n-1} \mathbf{P}(A \cap T^{-k}B) = \mathbf{P}(A)\mathbf{P}(B), \quad \text{for } A, B \in \mathcal{B},$$

and strongly mixing provided that

$$\lim_{n \to \infty} \mathbf{P}(A \cap T^{-n}B) = \mathbf{P}(A)\mathbf{P}(B), \quad \text{for } A, B \in \mathcal{B}.$$

The next theorem, whose proof is given in Sect. 3.2, uses that ergodicity and strong mixing can be interpreted as asymptotic independence when the base set Ω is a Polish space. Here the special form of the Lévy transformation and the one-dimensional setting are not essential, hence we will use the phrase *integral-type* for the transformation of the d-dimensional Wiener space in the form

$$T\beta = \int_0^{\cdot} h(s, \beta) \mathrm{d}\beta_s \qquad (2)$$

where h is a progressive $d \times d$-matrix valued function. It is measure-preserving, that is, $T\beta$ is a d-dimensional Brownian motion, if and only if $h(t, \omega)$ is an orthogonal matrix $dt \times d\mathbf{P}$ almost everywhere, that is, $h^T h = I_d$, where h^T denotes the transpose of h and I_d is the identity matrix of size $d \times d$. Recall that $\|a\|_{HS} = \operatorname{Tr}\left(aa^T\right)^{1/2}$ is the Hilbert–Schmidt norm of the matrix a.

Theorem 1. *Let T be an integral-type measure-preserving transformation of the d-dimensional Wiener-space as in (2) and denote by $X_n(t)$ the process*

$$X_n(t) = \int_0^t h_s^{(n)} \mathrm{d}s \quad \text{with} \quad h_s^{(n)} = h(s, T^{n-1}\beta) \cdots h(s, T\beta)h(s, \beta). \qquad (3)$$

Then

(i) T is strongly mixing if and only if $X_n(t) \xrightarrow{p} 0$ for all $t \geq 0$.

(ii) T is ergodic if and only if $\dfrac{1}{N}\sum_{n=1}^{N} \|X_n(t)\|_{HS}^2 \xrightarrow{p} 0$ for all $t \geq 0$.

The two parts of Theorem 1 can be proved along similar lines, see Sect. 3.2. Note, that the Hilbert–Schmidt norm of an orthogonal transformation in dimension d is $\sqrt{d}$ hence by (3) we have the trivial bound: $\|X_n(t)\|_{HS} \leq t\sqrt{d}$. By this boundedness the convergence in probability is equivalent to the convergence in L^1 in both parts of Theorem 1.

2.2 Lévy Transformation

Throughout this section $\beta^{(n)} = \beta \circ \mathbf{T}^n$ denotes the nth iterated path under the Lévy transformation $\mathbf{T}$. Then $h_t^{(n)} = \prod_{k=0}^{n-1} \text{sign}(\beta_t^{(k)})$.

By boundedness, the convergence of $X_n(t)$ in probability is the same as the convergence in L^2. Writing out $X_n^2(t)$ we obtain that:

$$X_n^2(t) = 2 \int_{0<u<v<t} h_u^{(n)} h_v^{(n)} du dv. \tag{4}$$

Combining (4) and (i) of Theorem 1 we obtain that $\mathbf{T}$ is strongly mixing provided that

$$\mathbf{E}\left(h_s^{(n)} h_t^{(n)}\right) \to 0, \quad \text{for almost all } 0 < s < t. \tag{5}$$

By scaling, $\mathbf{E}\left(h_s^{(n)} h_t^{(n)}\right)$ depends only on the ratio s/t, and the sufficient condition (5) is even simplifies to

$$\mathbf{E}\left(h_s^{(n)} h_1^{(n)}\right) \to 0, \quad \text{for almost every } s \in (0, 1).$$

Since $h_s^{(n)} h_1^{(n)}$ takes values in $\{-1, +1\}$ we actually have to show that $\mathbf{P}\left(h_s^{(n)} h_1^{(n)} = 1\right) - \mathbf{P}\left(h_s^{(n)} h_1^{(n)} = -1\right) \to 0$. It is quite natural to prove this limiting relation by a kind of coupling. In the present setting this means a transformation S of the state space $\mathcal{C}[0, \infty)$ preserving the Wiener measure and mapping most of the event $\{h_s^{(n)} h_1^{(n)} = 1\}$ to $\{h_s^{(n)} h_1^{(n)} = -1\}$ for n large.

The transformation S will be the reflection of the path after a suitably chosen stopping time τ, i.e.,

$$(S\beta)_t = 2\beta_{t \wedge \tau} - \beta_t.$$

Proposition 2. *Let $C > 0$ and $s \in (0, 1)$. If there exists a stopping time τ such that*

(a) $s < \tau < 1$ almost surely,
(b) $\nu = \inf\left\{n \geq 0 : \beta_\tau^{(n)} = 0\right\}$ is finite almost surely,
(c) $|\beta_\tau^{(k)}| > C\sqrt{1-\tau}$ for $0 \leq k < \nu$ almost surely.

then

$$\limsup_{n \to \infty} \left|\mathbf{E}\left(h_s^{(n)} h_1^{(n)}\right)\right| \leq \mathbf{P}\left(\sup_{t \in [0,1]} |\beta| > C\right)$$

One can relax the requirement that τ is a stopping time in Proposition 2.

Proposition 3. *Assume that for any $s < 1$ and $C > 0$ time there exists a random time τ with properties (a), (b) and (c) in Proposition 2.*

Then there are also a stopping times with these properties for any $s < 1$, $C > 0$.

For a given $s \in (0, 1)$ and $C > 0$, to prove the existence of the random time τ with the prescribed properties it is natural to consider all time points not only time one. That is, for a given path $\beta^{(0)}$ how large is the random set of "good time points", which will be denoted by $A(C, s)$:

$$A(C, s) = \{t > 0 : \text{exist } n, \gamma, \text{ such that } st < \gamma < t,$$

$$\beta_\gamma^{(n)} = 0 \text{ and } \inf_{0 \le k < n} |\beta_\gamma^{(k)}| > C\sqrt{t - \gamma}\right\}. \quad (6)$$

Note that it may happen that $n = 0$ and then the infimum $\inf_{0 \le k < n} |\beta_\gamma^{(k)}|$ is infinite.

Some basic properties of $A(C, s)$ for easier reference:

(a) Invariance under scaling. For $x \ne 0$, let Θ_x denote the scaling of the path, $(\Theta_x \omega)(t) = x^{-1} \omega(x^2 t)$. Then, since $\mathbf{T}\Theta_x = \Theta_x \mathbf{T}$ clearly holds for the Lévy transformation $\mathbf{T}$, we have

$$t \in A(C, s)(\omega) \quad \Leftrightarrow \quad x^{-2} t \in A(C, s)(\Theta_x \omega) \quad (7)$$

(b) Since the scaling Θ_x preserves the Wiener-measure, the previous point implies that $\mathbf{P}(t \in A(C, s))$ does not depend on $t > 0$.

Observe that $A(C, s)$ contains an open interval on the right of every zero of $\beta^{(n)}$ for all $n \ge 0$. Indeed, if γ is a zero of $\beta^{(n)}$ for some $n \ge 0$, then by choosing the smallest n such that $\beta_\gamma^{(n)} = 0$, one gets that $t \in A(C, s)$ for all $t > \gamma$ such that $t - \gamma$ is small enough. Since the union of the set of zeros of the iterated paths is dense, see [8], we have that the set of good time points is a dense open set. Unfortunately this is not enough for our purposes; a dense open set might be of small Lebesgue measure. To prove that the set of good time points is of full Lebesgue measure, we borrow a notion from real analysis.

Definition 4. Let $H \subset \mathbb{R}$ and denote by $f(x, \varepsilon)$ the supremum of the lengths of the intervals contained in $(x - \varepsilon, x + \varepsilon) \setminus H$. Then H is *porous* at x if $\limsup_{\varepsilon \to 0+} f(x, \varepsilon)/\varepsilon > 0$.

A set H is called porous when it is porous at each point $x \in H$.

Observe that if H is porous at x then its lower density

$$\liminf_{\varepsilon \to 0+} \frac{\lambda([x - \varepsilon, x + \varepsilon] \cap H)}{2\varepsilon} \le 1 - \limsup_{\varepsilon \to 0+} \frac{f(x, \varepsilon)}{2\varepsilon} < 1,$$

where λ denotes the Lebesgue measure. By Lebesgue's density theorem, see [12, pp. 13], the density of a measurable set exists and equals to 1 at almost every point of the set. Since the closure of a porous set is also porous we obtain the well known fact that a porous set is of zero Lebesgue measure.

Lemma 5. *Let H be a random closed subset of $[0, \infty)$. If H is scaling invariant, that is cH has the same law as H for all $c > 0$, then*

$$\{1 \notin H\} \subset \{H \text{ is porous at } 1\} \quad \text{and} \quad \mathbf{P}((\{H \text{ is porous at } 1\} \setminus \{1 \notin H\})) = 0.$$

That is, the events $\{1 \notin H\}$ and $\{H \text{ is porous at } 1\}$ are equal up to a null sets.
In particular, if H is porous at 1 almost surely, then $\mathbf{P}(1 \notin H) = 1$.

Proof. Recall that a random closed set H is a random element in the space of the closed subset of $[0, \infty)$—we denote it by $\mathcal{F}$—, endowed with the smallest σ-algebra containing the sets $C_G = \{F \in \mathcal{F} : F \cap G \neq \emptyset\}$, for all open $G \subset [0, \infty)$. Then it is easy to see, that $\{\omega : H(\omega) \text{ is porous at } 1\}$ is an event and

$$\mathbf{H} = \{(t, \omega) \in [0, \infty) \times \Omega : t \in H(\omega)\},$$
$$\mathbf{H}_p = \{(t, \omega) \in [0, \infty) \times \Omega : H(\omega) \text{ is porous at } t\}$$

are measurable subsets of $[0, \infty) \times \Omega$. We will also use the notation

$$H_p(\omega) = \{t \in [0, \infty) : (t, \omega) \in \mathbf{H}_p\} = \{t \in [0, \infty) : H(\omega) \text{ is porous at } t\}.$$

Then for each $\omega \in \Omega$ the set $H(\omega) \cap H_p(\omega)$ is a porous set, hence of Lebesgue measure zero; see the remark before Lemma 5. Whence Fubini theorem yields that

$$(\lambda \otimes \mathbf{P})(\mathbf{H} \cap \mathbf{H}_p) = \mathbf{E}(\lambda(H \cap H_p)) = 0.$$

Using Fubini theorem again we get

$$0 = (\lambda \otimes \mathbf{P})(\mathbf{H} \cap \mathbf{H}_p) = \int_0^\infty \mathbf{P}(t \in H \cap H_p) dt.$$

Since $\mathbf{P}(t \in H \cap H_p)$ does not depend on t by the scaling invariance of H we have that $\mathbf{P}(1 \in H \cap H_p) = 0$. Now $\{1 \in H \cap H_p\} = \{1 \in H_p\} \setminus \{1 \notin H\}$, so we have shown that

$$\mathbf{P}(\{H \text{ is porous at } 1\} \setminus \{1 \notin H\}) = 0.$$

The first part of the claim $\{1 \notin H\} \subset \{H \text{ is porous at } 1\}$ is obvious, since $H(\omega)$ is closed and if $1 \notin H(\omega)$ then there is an open interval containing 1 and disjoint from H. □

We want to apply this lemma to $[0, \infty) \setminus A(C, s)$, the random set of bad time points. We have seen in (7) that the law of $[0, \infty) \setminus A(C, s)$ has the scaling property. For easier reference we state explicitly the corollary of the above argument, that is the combination of (i) in Theorem 1, Propositions 2–3 and Lemma 5:

Corollary 6. *If $[0, \infty) \setminus A(C, s)$ is almost surely porous at 1 for any $C > 0$ and $s \in (0, 1)$ then the Lévy transformation is strongly mixing.*

The condition formulated in terms $A(C, s)$ requires that small neighborhoods of time 1 contain sufficiently large subintervals of $A(C, s)$. Looking at only the left and only the right neighborhoods we can obtain Theorems 7 and 8 below, respectively.

To state these results we introduce the following notations, for $t > 0$

- $$\gamma_n(t) = \max \left\{ s \leq t : \beta_s^{(n)} = 0 \right\}$$

  is the last zero before t,

- $$\gamma_n^*(t) = \max_{0 \leq k \leq n} \gamma_k(t),$$

  the last time s before t such that $\beta^{(0)}, \ldots, \beta^{(n)}$ has no zero in $(s, t]$,

- $$Z_n(t) = \min_{0 \leq k < n} |\beta_t^{(k)}|.$$

When $t = 1$ we omit it from the notation, that is, $\gamma_n = \gamma_n(1)$, $\gamma_n^* = \gamma_n^*(1)$ and $Z_n = Z_n(1)$.

Theorem 7. *Let*

$$Y = \limsup_{n \to \infty} \frac{Z_n(\gamma_n^*)}{\sqrt{1 - \gamma_n^*}}. \tag{8}$$

Then Y is a $\mathbf{T}$ invariant, $\{0, \infty\}$ valued random variable and

(i) *either $\mathbf{P}(Y = 0) = 1$;*
(ii) *or $0 < \mathbf{P}(Y = 0) < 1$, and then $\mathbf{T}$ is not ergodic;*
(iii) *or $\mathbf{P}(Y = 0) = 0$, that is $Y = \infty$ almost surely, and $\mathbf{T}$ is strongly mixing.*

Theorem 8. *Let*

$$X = \liminf_{n \to \infty} \frac{Z_{n+1}}{Z_n}. \tag{9}$$

Then X is a $\mathbf{T}$ invariant, $\{0, 1\}$ valued random variable and

(i) *either $\mathbf{P}(X = 1) = 1$;*
(ii) *or $0 < \mathbf{P}(X = 1) < 1$, and then $\mathbf{T}$ is not ergodic;*
(iii) *or $\mathbf{P}(X = 1) = 0$, that is $X = 0$ almost surely, and $\mathbf{T}$ is strongly mixing.*

Remark. In Theorem 8, the first possibility $X = 1$ looks very unlikely. If one is able to exclude it, then the Lévy $\mathbf{T}$ transformation is either strongly mixing or not ergodic and the invariant random variable X witnesses it.

The statements in Theorems 7 and 8 have similar structure, and the easy parts, the invariance of X and Y are proved in Sect. 3.4, while the more difficult parts are proved in Sects. 3.5 and 3.6, respectively.

We can complement Theorems 7 and 8 with the next statement, which shows that X, Y and the goodness of time 1 for all $C > 0$ and $s \in (0, 1)$ are strongly connected. Its proof is deferred to Sect. 3.7 since it uses the side results of the proofs of Theorems 7 and 8.

Theorem 9. *Set*

$$A = \bigcap_{s \in (0,1)} \bigcap_{C>0} A(C, s).$$

Then the events $\{1 \in A\}$, $\{Y = \infty\}$ and $\{X = 0\}$ are equal up to null events. In particular, $X = 1/(1 + Y)$ almost surely.

We close this section with a sufficient condition for $X < 1$ almost surely. For $x > 0$, let $\nu(x) = \inf\{n \geq 0 : |\beta_1^{(n)}| < x\}$. By the next Corollary of the density theorem of Malric [8], recalled in (1), $\nu(x)$ is finite almost surely for all $x > 0$.

Corollary 10. $\inf_n |\beta^{(n)}|$ *is identically zero almost surely, that is*

$$\mathbf{P}\left(\inf_{n \geq 0} |\beta_t^{(n)}| = 0, \forall t \geq 0\right) = 1$$

Recall that a family of real valued variables $\{\xi_i : i \in I\}$ is tight if $\sup_{i \in I} \mathbf{P}(|\xi_i| > K) \to 0$ as $K \to \infty$.

Theorem 11. *The tightness of the families $\{x\nu(x) : x \in (0, 1)\}$ and $\{nZ_n : n \geq 1\}$ are equivalent and both imply $X < 1$ almost surely, hence also the strong mixing property of the Lévy transformation.*

For the sake of completeness we state the next corollary, which is just an easy application of the Markov inequality.

Corollary 12. *If there exists an unbounded, increasing function $f : [0, \infty) \to [0, \infty)$ such that $\sup_{x \in (0,1)} \mathbf{E}(f(x\nu(x))) < \infty$ or $\sup_n \mathbf{E}(f(nZ_n)) < \infty$ then the Lévy transformation is strongly mixing.*

In particular, if $\sup_{x \in (0,1)} \mathbf{E}(x\nu(x)) < \infty$ or $\sup_n \mathbf{E}(nZ_n) < \infty$ then the Lévy transformation is strongly mixing.

3 Proofs

3.1 General Results

First, we characterize strong mixing and ergodicity of measure-preserving transformation over a Polish space. This will be the key to prove Theorem 1. Although it seems to be natural, the author was not able to locate it in the literature.

Proposition 13. *Let $(\Omega, \mathcal{B}, \mathbf{P}, T)$ be a measure-preserving system, where Ω is a Polish space and $\mathcal{B}$ is its Borel σ-field. Then*

(i) T is strongly mixing if and only if $\mathbf{P} \circ (T^0, T^n)^{-1} \xrightarrow{w} \mathbf{P} \otimes \mathbf{P}$.
(ii) T is ergodic if and only if $\frac{1}{n} \sum_{k=0}^{n-1} \mathbf{P} \circ (T^0, T^k)^{-1} \xrightarrow{w} \mathbf{P} \otimes \mathbf{P}$.

Both part of the statement follows obviously from the following common generalization.

Proposition 14. *Let Ω be a Polish space and μ_n, μ be probability measures on the product $(\Omega \times \Omega, \mathcal{B} \times \mathcal{B})$, where $\mathcal{B}$ is a Borel σ-field of Ω.*
Assume that for all n the marginals of μ_n and μ are the same, that is for $A \in \mathcal{B}$ we have $\mu_n(A \times \Omega) = \mu(A \times \Omega)$ and $\mu_n(\Omega \times A) = \mu(\Omega \times A)$.
Then $\mu_n \xrightarrow{w} \mu$ if and only if $\mu_n(A \times B) \to \mu(A \times B)$ for all $A, B \in \mathcal{B}$.

Proof. Assume first that $\mu_n(A \times B) \to \mu(A \times B)$ for $A, B \in \mathcal{B}$. By portmanteau theorem, see Billingsley [1, Theorem 2.1], it is enough to show that for closed sets $F \subset \Omega \times \Omega$ the limiting relation

$$\limsup_{n \to \infty} \mu_n(F) \leq \mu(F) \tag{10}$$

holds. To see this, consider first a compact subset F of $\Omega \times \Omega$ and an open set G such that $F \subset G$. We can take a finite covering of F with open rectangles $F \subset \cup_{i=1}^r A_i \times B_i \subset G$, where $A_i, B_i \subset \Omega$ are open. Since the difference of rectangles can be written as finite disjoint union of rectangles we can write

$$(A_i \times B_i) \setminus \bigcup_{k<i}(A_k \times B_k) = \bigcup_j (A'_{i,j} \times B'_{i,j}),$$

where $\left\{A'_{i,j} \times B'_{i,j} : i, j\right\}$ is a finite collection of disjoint rectangles. By assumption

$$\lim_{n \to \infty} \mu_n\left(A'_{i,j} \times B'_{i,j}\right) = \mu\left(A'_{i,j} \times B'_{i,j}\right),$$

which yields

$$\limsup_{n \to \infty} \mu_n(F) \leq \lim_{n \to \infty} \mu_n\left(\bigcup_i (A_i \times B_i)\right) = \mu\left(\bigcup_i (A_i \times B_i)\right) \leq \mu(G).$$

Taking infimum over $G \supset F$, (10) follows for compact sets.

For a general closed F, let $\varepsilon > 0$ and denote by $\mu^1(A) = \mu(A \times \Omega)$, $\mu^2(A) = \mu(\Omega \times A)$ the marginals of μ. By the tightness of $\{\mu^1, \mu^2\}$, one can find a compact set C such that $\mu^1(C^c) = \mu(C^c \times \Omega) \leq \varepsilon$ and $\mu^2(C^c) = \mu(\Omega \times C^c) \leq \varepsilon$. Then

$$\mu_n(F) \leq \mu_n(F \cap (C \times C)) + 2\varepsilon.$$

Since $F' = F \cap (C \times C)$ is compact, we have that

$$\limsup_{n\to\infty} \mu_n(F) \leq \limsup_{n\to\infty} \mu_n(F') + 2\varepsilon \leq \mu(F') + 2\varepsilon \leq \mu(F) + 2\varepsilon.$$

Letting $\varepsilon \to 0$ finishes this part of the proof.

For the converse, note that μ^1 and μ^2 are regular since Ω is a Polish space and μ^1, μ^2 are probability measures on its Borel σ-field.

Fix $\varepsilon > 0$. For $A_i \in \mathcal{B}$ one can find, using the regularity of μ^i, closed sets F_i and open sets G_i such that $F_i \subset A_i \subset G_i$ and $\mu^i(G_i \setminus F_i) \leq \varepsilon$. Then

$$(G_1 \times G_2) \setminus (F_1 \times F_2) \subset ((G_1 \setminus F_1) \times \Omega) \cup (\Omega \times (G_2 \setminus F_2))$$

yields that

$$\mu_n(A_1 \times A_2) \leq \mu_n(G_1 \times G_2) \leq \mu_n(F_1 \times F_2) + 2\varepsilon,$$
$$\mu_n(A_1 \times A_2) \geq \mu_n(F_1 \times F_2) \geq \mu_n(G_1 \times G_2) - 2\varepsilon,$$

hence by portmanteau theorem $\mu_n \xrightarrow{w} \mu$ gives

$$\limsup_{n\to\infty} \mu_n(A_1 \times A_2) \leq \mu(F_1 \times F_2) + 2\varepsilon \leq \mu(A_1 \times A_2) + 2\varepsilon$$

$$\liminf_{n\to\infty} \mu_n(A_1 \times A_2) \geq \mu(G_1 \times G_2) - 2\varepsilon \geq \mu(A_1 \times A_2) - 2\varepsilon.$$

Letting $\varepsilon \to 0$ we get $\lim_{n\to\infty} \mu_n(A_1 \times A_2) = \mu(A_1 \times A_2)$. □

3.2 Proof of Theorem 1

Proof of the sufficiency of the conditions in Theorem 1. We start with the strong mixing case. We want to show that

$$X_n(t) = \int_0^t h_s^{(n)} ds \xrightarrow{P} 0, \quad \text{for all } t \geq 0, \tag{11}$$

where $h_s^{(n)}$ is given by (3), implies the strong mixing of the integral-type measure-preserving transformation T.

Actually, we show by characteristic function method that (11) implies that the finite dimensional marginals of $(\beta, \beta^{(n)})$ converge in distribution to the appropriate marginals of a $2d$-dimensional Brownian motion. Then, since the sequence $(\beta, \beta^{(n)})_{n\geq 0}$ is tight, not only the finite dimensional marginals but the sequence of processes $(\beta, \beta^{(n)})$ converges in distribution to a $2d$-dimensional Brownian motion. By Proposition 13 this is equivalent with the strong mixing property of T.

Let $\underline{t} = (t_1, \ldots, t_k)$ be a finite subset of $[0, \infty)$. Then the characteristic function of $(\beta_{t_1}, \ldots, \beta_{t_k}, \beta_{t_1}^{(n)}, \ldots, \beta_{t_k}^{(n)})$ can be written as

$$\phi_n(\alpha) = \mathbf{E}\left(\exp\left\{i \int_0^\infty f \, d\beta + i \int_0^\infty g \, d\beta^{(n)}\right\}\right)$$
$$= \mathbf{E}\left(\exp\left\{i \int_0^\infty (f + g h^{(n)}) \, d\beta\right\}\right), \qquad (12)$$

where f, g are deterministic step function obtained from the time vector $\underline{t}$ and $\alpha = (a_1, \ldots, a_k, b_1, \ldots, b_k)$; here a_i, b_j are d-dimensional row vectors and

$$f = \sum_{j=1}^k a_j \mathbf{1}_{[0,t_j]}, \quad \text{and} \quad g = \sum_{j=1}^k b_j \mathbf{1}_{[0,t_j]}.$$

We have to show that

$$\phi_n(\alpha) \to \phi(\alpha) = \exp\left\{-\frac{1}{2}\int_0^\infty (|f|^2 + |g|^2)\right\} \quad \text{as } n \to \infty.$$

Using that $\beta^{(n)} = \int h^{(n)} d\beta$ and

$$M_t = \exp\left\{i \int_0^t (f(s) + g(s)h_s^{(n)}) \, d\beta_s + \frac{1}{2}\int_0^t |f(s) + g(s)h_s^{(n)}|^2 \, ds\right\}$$

is a uniformly integrable martingale starting from 1, we obtain that $\mathbf{E}(M_\infty) = 1$ and

$$\phi(\alpha) = \phi(\alpha)\mathbf{E}(M_\infty) =$$
$$\mathbf{E}\left(\exp\left\{i \int_0^\infty (f(s) + g(s)h_s^{(n)}) \, d\beta_s + \int_0^\infty g(s)h_s^{(n)} f^T(s) \, ds\right\}\right) \qquad (13)$$

As $\exp\{i \int_{[0\infty)} (f + g h^{(n)}) d\beta\}$ is of modulus one, we get from (12) and (13) that

$$|\phi(\alpha) - \phi_n(\alpha)| \le \mathbf{E}\left(\left|\exp\left\{\int_0^\infty g(s)h_s^{(n)} f^T(s) \, ds\right\} - 1\right|\right). \qquad (14)$$

Note that $f^T g$ is a matrix valued function of the form $f^T g = \sum_{j=1}^k c_j \mathbf{1}_{[0,t_j]}$, hence

$$\int_0^\infty g(s) h_s^{(n)} f^T(s) \, ds = \int_0^\infty \text{Tr}\left(f^T(s) g(s) h_s^{(n)}\right) ds = \sum_{j=1}^k \text{Tr}\left(c_j X_n(t_j)\right),$$

and $|\int_0^\infty g(s) h_s^{(n)} f^T(s) ds| \leq M = \int_0^\infty |f(s)| |g(s)| ds < \infty$. With this notation, using $|e^x - 1| \leq |x| e^{|x|}$ for $x \in \mathbb{R}$ and $|\mathrm{Tr}(ab)| \leq \|a\|_{HS} \|b\|_{HS}$, we can continue (14) to get

$$|\phi_n(\alpha) - \phi(\alpha)| \leq \mathbf{E}\left(\left|\exp\left\{\int_0^\infty g(s) h_s^{(n)} f^T(s) ds\right\} - 1\right|\right)$$

$$\leq e^M \mathbf{E}\left(\left|\sum_{j=1}^k \mathrm{Tr}\left(c_j X_n(t_j)\right)\right|\right) \quad (15)$$

$$\leq e^M \sum_{j=1}^k \|c_j\|_{HS} \mathbf{E}(\|X_n(t_j)\|_{HS}).$$

Since $\|X_n(t_j)\|_{HS} \leq t_j \sqrt{d}$ and $X_n(t_j) \xrightarrow{p} 0$ by assumption, we obtained that $\phi_n(\alpha) \to \phi(\alpha)$ and the statement follows.

To prove (ii) we use the same method. We introduce κ_n which is a random variable independent of the sequence $(\beta^{(n)})_{n \in \mathbb{Z}}$ and uniformly distributed on $\{0, 1, \ldots, n-1\}$. Ergodicity can be formulated as $(\beta, \beta^{(\kappa_n)})$ converges in distribution to a $2d$-dimensional Brownian motion. The joint characteristic function ψ_n of $(\beta_{t_1}, \ldots, \beta_{t_k}, \beta_{t_1}^{(\kappa_n)}, \ldots, \beta_{t_k}^{(\kappa_n)})$ can be expressed, similarly as above,

$$\psi_n = \frac{1}{n} \sum_{\ell=0}^{n-1} \phi_\ell$$

where ϕ_ℓ is as in the first part of the proof. Using the estimation (15) obtained in the first part

$$|\phi(\alpha) - \psi_n(\alpha)| \leq \frac{1}{n} \sum_{\ell=0}^{n-1} |\phi(\alpha) - \phi_\ell(\alpha)|$$

$$\leq \frac{e^M}{n} \sum_{\ell=0}^{n-1} \sum_{j=0}^k \|c_j\|_{HS} \mathbf{E}(\|X_\ell(t_j)\|_{HS})$$

$$= e^M \sum_{j=1}^k \|c_j\|_{HS} \mathbf{E}\left(\frac{1}{n} \sum_{\ell=0}^{n-1} \|X_\ell(t_j)\|_{HS}\right).$$

Now $|\phi(\alpha) - \psi_n(\alpha)| \to 0$ follows from our condition in part (ii) by the Cauchy–Schwartz inequality, since

$$\left(\frac{1}{n}\sum_{\ell=0}^{n-1}\|X_\ell(t_j)\|_{HS}\right)^2 \le \frac{1}{n}\sum_{\ell=0}^{n-1}\|X_\ell(t_j)\|_{HS}^2 \overset{p}{\to} 0.$$

and $\frac{1}{n}\sum_{\ell=0}^{n-1}\|X_\ell(t_j)\|_{HS}^2 \le t_j^2 d$. □

Proof of the necessity of the conditions in Theorem 1. Recall that the quadratic variation of an m-dimensional martingale $M = (M_1, \ldots, M_m)$ is a matrix valued process whose (j,k) entry is $\langle M_j, M_k \rangle$. The proof of the following fact can be found in [6], see Corollary 6.6 of Chap. VI.

Let $(M^{(n)})$ be a sequence of m-dimensional, continuous local martingales. If $M^{(n)} \overset{d}{\to} M$ then $(M^{(n)}, \langle M^{(n)} \rangle) \overset{d}{\to} (M, \langle M^{(n)} \rangle)$.

By enlarging the probability space, we may assume that there is a random variable U, which is uniform on $(0,1)$ and independent of β. Denote by $\kappa_n = [nU]$ the integer part of nU. Let $\mathcal{G}$ be the smallest filtration satisfying the usual hypotheses, making U $\mathcal{G}_0$ measurable and β adapted to $\mathcal{G}$. Then β is a Brownian motion in $\mathcal{G}$; $(\beta, \beta^{(n)})$ and $(\beta, \beta^{(\kappa_n)})$ are continuous martingales in $\mathcal{G}$. The quadratic covariations are

$$\langle \beta^{(n)}, \beta \rangle_t = \int_0^t h_s^{(n)} ds = X_n(t), \quad \text{and} \quad \langle \beta^{(\kappa_n)}, \beta \rangle_t = \sum_{k=0}^{n-1} \mathbb{1}_{(\kappa_n = k)} X_k(t).$$

By Proposition 3, the strong mixing property and the ergodicity of T are respectively equivalent to the convergence in distribution of $(\beta, \beta^{(n)})$ and $(\beta, \beta^{(\kappa_n)})$ to a $2d$-dimensional Brownian motion.

By the fact just recalled, the strong mixing property of T implies that $\langle \beta^{(n)}, \beta \rangle_t \overset{d}{\to} 0$, while its ergodicity ensures that $\langle \beta^{(\kappa_n)}, \beta \rangle_t \overset{d}{\to} 0$ for every $t \ge 0$. Since the limit is deterministic, the convergence also holds in probability. The "only if" part of (i) follows immediately.

For the "only if" part of (ii) we add that

$$\|\langle \beta^{(\kappa_n)}, \beta \rangle_t\|_{HS}^2 = \left\|\sum_{k=0}^{n-1} \mathbb{1}_{(\kappa_n=k)} X_k(t)\right\|_{HS}^2 = \sum_{k=0}^{n-1} \mathbb{1}_{(\kappa_n=k)} \|X_k(t)\|_{HS}^2$$

Since $\|X_k(t)\|_{HS}^2 \le t^2 d$ the convergence in probability of $\langle \beta^{(\kappa_n)}, \beta \rangle_t$ to zero is also a convergence of $\|\langle \beta^{(\kappa_n)}, \beta \rangle_t\|_{HS}^2$ to zero in $L^1(\mathbf{P})$, which implies the convergence in $L^1(\mathbf{P})$ to zero of the conditional expectation

$$\mathbf{E}(\|\langle \beta^{(\kappa_n)}, \beta \rangle_t\|_{HS}^2 | \sigma(\beta)) = \frac{1}{n}\sum_{k=0}^{n-1} \|X_k(t)\|_{HS}^2.$$

The "only if" part of (ii) follows. □

3.3 First Results for the Lévy Transformation

We will use the following property of the Lévy transformation many times. Recall that $\mathbf{T}^n \beta = \beta \circ \mathbf{T}^n$ is also denoted by $\beta^{(n)}$. We will also use the notation $h_t^{(n)} = \prod_{k=0}^{n-1} \text{sign}(\beta_t^{(k)})$ for $n \geq 1$ and $h^{(0)} = 1$.

Lemma 15. *On an almost sure event the following property holds: For any interval $I \subset [0, \infty)$, point $a \in I$ and integer $n > 0$, if*

$$\sup_{t \in I} |\beta_t - \beta_a| < \min_{0 \leq k < n} |(\mathbf{T}^k \beta)_a| \tag{16}$$

then

(i) $\mathbf{T}^k \beta$ has no zero in I, for $0 \leq k \leq n-1$,
(ii) $(\mathbf{T}^k \beta)_t - (\mathbf{T}^k \beta)_a = h_a^{(k)} (\beta_t - \beta_a)$ for $t \in I$ and $0 \leq k \leq n$.
  In particular, $|(\mathbf{T}^k \beta)_t - (\mathbf{T}^k \beta)_a| = |\beta_t - \beta_a|$ for $t \in I$ and $0 \leq k \leq n$.

Proof. In the next argument we only use that if β is a Brownian motion and L is its local time at level zero then the points of increase for L is exactly the zero set of β and $\mathbf{T}\beta = |\beta| - L$ almost surely. Then there is Ω' of full probability such that on Ω' both properties hold for $\mathbf{T}^n \beta$ for all $n \geq 0$ simultaneously.

Let $N = N(I) = \inf\{n \geq 0 : \mathbf{T}^n \beta \text{ has a zero in } I\}$. Since $\mathbf{T}$ acts as $\mathbf{T}\beta = |\beta| - L$, if β has no zero in I we have

$$\mathbf{T}\beta_t = \text{sign}(\beta_a)\beta_t - L_a, \quad \text{for } t \in I.$$

But, then $\mathbf{T}\beta_t - \mathbf{T}\beta_a = \text{sign}(\beta_a)(\beta_t - \beta_a)$ and $|\mathbf{T}\beta_t - \mathbf{T}\beta_a| = |\beta_t - \beta_a|$ for $t \in I$. Iterating it we obtain that

$$\begin{aligned} (\mathbf{T}^k \beta)_t - (\mathbf{T}^k \beta)_a &= h_a^{(k)} (\beta_t - \beta_a), \\ |(\mathbf{T}^k \beta)_t - (\mathbf{T}^k \beta)_a| &= |\beta_t - \beta_a|, \end{aligned} \quad \text{on } \{k \leq N\} \text{ and for } t \in I. \tag{17}$$

Now assume that (16) holds. Then, necessarily $n \leq N$ as the other possibility would lead to a contradiction. Indeed, if $N < n$ then N is finite, $\mathbf{T}^N \beta$ has a zero t_0 in I and

$$0 = |\mathbf{T}^N \beta_{t_0}| = |\mathbf{T}^N \beta_a| - |\mathbf{T}^N \beta_{t_0} - \mathbf{T}^N \beta_a| \geq \min_{0 \leq k < n} |\mathbf{T}^k \beta_a| - \sup_{t \in I} |\beta_t - \beta_a| > 0.$$

So (16) implies that $n \leq N$, which proves (i) by the definition of N and also (ii) by (17). □

Combined with the densities of zeros, Lemma 15 implies Corollary 10 stated above.

Proof of Corollary 10. The statement here is that $\inf_{n\geq 0}|(\mathbf{T}^n\beta)_t| = 0$ for all $t \geq 0$.

Assume that for $\omega \in \Omega$ there is some $t > 0$, such that $\inf_{n\geq 0}|(\mathbf{T}^n\beta)_t|$ is not zero at ω. Then there is a neighborhood I of t such that

$$\sup_{s\in I}|\beta_s - \beta_t| < \inf_k|(\mathbf{T}^k\beta)_t|.$$

Using Lemma 15, we would get that for this ω the iterated paths $\mathbf{T}^k\beta(\omega)$, $k \geq 0$ has no zero in I. However, since

$$\{t \geq 0 : \exists k, (\mathbf{T}^k\beta)_t = 0\}$$

is dense in $[0,\infty)$ almost surely by the result of Malric [8], ω belongs to the exceptional negligible set. □

Proof of Proposition 2. Let $C > 0$ and $s \in (0, 1)$ as in the statement and assume that τ is a stopping time satisfying (a)–(c), that is, $s < \tau < 1$, and for the almost surely finite random index ν we have $\beta_\tau^{(\nu)} = 0$ and $\min_{0\leq k<\nu}|\beta_\tau^{(k)}| > C\sqrt{1-\tau}$. Recall that S denotes the reflection of the trajectories after τ.

Set $\varepsilon_n = h_s^{(n)}h_1^{(n)}$ for $n > 0$ and

$$A_C = \left\{\sup_{t\in[\tau,1]}|\beta_t^{(0)} - \beta_\tau^{(0)}| \leq C\sqrt{1-\tau}\right\}.$$

We show below that on the event $A_C \cap \{n > \nu\}$, we have $\varepsilon_n = -\varepsilon_n \circ S$. Since S preserves the Wiener measure $\mathbf{P}$, this implies that

$$|\mathbf{E}(\varepsilon_n)| = \frac{1}{2}|\mathbf{E}(\varepsilon_n + \varepsilon_n \circ S)| \leq \frac{1}{2}\mathbf{E}(|\varepsilon_n + \varepsilon_n \circ S|)$$

$$= \mathbf{P}(\varepsilon_n = \varepsilon_n \circ S)$$

$$\leq \mathbf{P}(A_C^c \cup \{n \leq \nu\}) \leq \mathbf{P}(A_C^c) + \mathbf{P}(n \leq \nu).$$

When $n \to \infty$, this yields

$$\limsup_{n\to\infty}\left|\mathbf{E}\left(h_s^{(n)}h_1^{(n)}\right)\right| \leq \mathbf{P}(A_C^c) = \mathbf{P}\left(\sup_{s\in[0,1]}|\beta_s| > C\right),$$

by the Markov property and the scaling property of the Brownian motion.

It remains to show that on $A_C \cap \{n > \nu\}$ the identity $\varepsilon_n = -\varepsilon_n \circ S$ holds. By definition of S, the trajectory of β and $\beta \circ S$ coincide on $[0, \tau]$, hence $h^{(k)}$ and $h^{(k)} \circ S$ coincide on $[0, \tau]$ for $k > 0$. In particular, $h_\tau^{(k)} = h_\tau^{(k)} \circ S$ and $h_s^{(k)} = h_s^{(k)} \circ S$ for all k since $\tau > s$.

On the event A_C we can apply Lemma 15 with $I = [\tau, 1]$, $a = \tau$ and $n = \nu$ to both β and $S \circ \beta$ to get that

$$\beta_t^{(k)} - \beta_\tau^{(k)} = h_\tau^{(k)}(\beta_t - \beta_\tau),$$
$$\beta_t^{(k)} \circ S - \beta_\tau^{(k)} \circ S = -h_\tau^{(k)}(\beta_t - \beta_\tau), \qquad k \le \nu, \, t \in [\tau, 1]. \qquad (18)$$

We have used that $h_\tau^{(k)} = h_\tau^{(k)} \circ S$ and $\beta_t \circ S_t - \beta_\tau \circ S = -(\beta_t - \beta_\tau)$ for $t \ge \tau$ by the definition of S.

Using that on A_C

$$|\beta_\tau^{(k)}| > C\sqrt{1-\tau} \ge |\beta_1 - \beta_\tau|, \quad \text{for } k < \nu$$

we get immediately from (18) that $\text{sign}(\beta_1^{(k)}) = \text{sign}(\beta_1^{(k)}) \circ S$ for $k < \nu$.

Since $\beta_\tau^{(\nu)} = (\beta_\tau^{(\nu)}) \circ S = 0$, for $k = \nu$ (18) gives that $\beta^{(\nu)}$ and $\beta^{(\nu)} \circ S$ coincide on $[0, \tau]$ and are opposite of each other on $[\tau, 1]$. Hence, $\beta^{(k)}$ and $\beta^{(k)} \circ S$ coincide on $[0, 1]$ for every $k > \nu$.

As a result on the event A_C,

$$\text{sign}(\beta_1^{(k)}) \circ S = \begin{cases} \text{sign}(\beta_1^{(k)}), & \text{if } k \ne \nu, \\ -\text{sign}(\beta_1^{(k)}), & \text{if } k = \nu \end{cases}$$

hence $h_1^{(n)} \circ S = -h_1^{(n)}$ on $A_C \cap \{n > \nu\}$. Since $h_s^{(n)} \circ S = h_s^{(n)}$ for all n we are done. □

Proof of Proposition 3. Let $C > 0$ and $s \in (0, 1)$. Call τ the infimum of those time points that satisfy (b) and (c) of Proposition 2 with C replaced by $2C$, namely $\tau = \inf_n \tau_n$, where

$$\tau_n = \inf\left\{t > s : \beta_t^{(n)} = 0, \forall k < n, |\beta_t^{(k)}| > 2C\sqrt{(1-t) \vee 0}\right\}.$$

By assumption $\tau_n < 1$ for some $n \ge 0$. Furthermore, there exists some finite index ν such that $\tau = \tau_\nu$. Otherwise, there would exist a subsequence $(\tau_n)_{n \in D}$ bounded by 1 and converging to τ. For every k one has $k < n$ for infinitely many $n \in D$, hence $|\beta_{\tau_n}^{(k)}| \ge 2C\sqrt{1 - \tau_n}$ by the choice of D. Letting $n \to \infty$ yields $\left|\beta_\tau^{(k)}\right| \ge 2C\sqrt{1-\tau} > 0$ for every k. This can happen only with probability zero by Corollary 10.

As ν is almost surely finite and $\tau = \tau_\nu$ we get that $\beta_\tau^{(\nu)} = 0$ and

$$\inf\{|\beta_\tau^{(k)}| : k < \nu\} \ge 2C\sqrt{1-\tau} > C\sqrt{1-\tau}.$$

We have that $\tau > s$ holds almost surely, since s is not a zero of any $\beta^{(n)}$ almost surely, so τ satisfies (a)–(c) of Lemma 2. □

3.4 Easy Steps of the Proof of Theorems 7 and 8

The main step of the proof of these theorems, that will be given in Sects. 3.6 and 3.7, is that if $Y > 0$ almost surely (or $X < 1$ almost surely), then for any $C > 0$, $s \in (0, 1)$ the set of the bad time points $[0, \infty) \setminus A(C, s)$ is almost surely porous at 1. Then Corollary 6 applies and the Lévy transformation $\mathbf{T}$ is strongly mixing.

If $Y > 0$ does not hold almost surely, then either $Y = 0$ or Y is a non-constant variable invariant for $\mathbf{T}$, hence in latter case the Lévy transformation $\mathbf{T}$ is not ergodic. These are the first two cases in Theorem 7. Similar analysis applies to X and Theorem 8.

To show the invariance of Y recall that $\gamma_n^* \to 1$ by the density theorem of the zeros due to Malric [8] and $\gamma_0 < 1$, both property holding almost surely. Hence, for every large enough n, $\gamma_{n+1}^* > \gamma_0$, therefore $\gamma_{n+1}^* = \gamma_n^* \circ \mathbf{T}$,

$$Z_n(\gamma_n^*) \circ \mathbf{T} = \min_{0 \leq k < n} |\beta_{\gamma_n^* \circ \mathbf{T}}^{(k+1)}| = \min_{1 \leq k < n+1} |\beta_{\gamma_{n+1}^*}^{(k)}| \geq Z_{n+1}(\gamma_{n+1}^*),$$

and

$$\frac{Z_n(\gamma_n^*)}{\sqrt{1 - \gamma_n^*}} \circ \mathbf{T} \geq \frac{Z_{n+1}(\gamma_{n+1}^*)}{\sqrt{1 - \gamma_{n+1}^*}}.$$

Taking limit superior we obtain that $Y \circ \mathbf{T} \geq Y$. Using that $\mathbf{T}$ is measure-preserving we conclude $Y \circ \mathbf{T} = Y$ almost surely, that is, Y is $\mathbf{T}$ invariant.

To show the invariance of X directly, without referring to Theorem 9, we use Corollary 10, which says that almost surely $\inf_{n \geq 0} |\beta_t^{(n)}| = 0$ for all $t \geq 0$. Thus $Z_n \to 0$ and since $|\beta_1^{(0)}| > 0$ almost surely, for every large enough n, $Z_n < |\beta_1^{(0)}|$, therefore $(Z_{n+1}/Z_n) \circ \mathbf{T} = (Z_{n+2}/Z_{n+1})$. Hence $X \circ \mathbf{T} = X$.

3.5 Proof of Theorem 7

Fix $C > 0$ and $s \in (0, 1)$ and consider the random set

$$\tilde{A}(C, s) = \{t > 0 : \text{exist } n \geq 1 \text{ such that } st < \gamma_n(t) = \gamma_n^*(t) \text{ and}$$

$$\min_{0 \leq k < n} |\beta_{\gamma_n(t)}^{(k)}| > C\sqrt{t - \gamma_n(t)}\bigg\} \subset A(C, s). \quad (19)$$

The difference between $A(C, s)$ and $\tilde{A}(C, s)$ is that in the latter case we only consider last zeros satisfying $\gamma_n(t) > \gamma_k(t)$ for $k = 0, \ldots, n-1$, whereas in the case of $A(C, s)$ we consider any zero of the iterated paths. Note also, that here $n > 0$, so the zeros of β itself are not used, while n can be zero in the definition of $A(C, s)$.

We prove below the next proposition.

Proposition 16. *Almost surely on the event* $\{Y > 0\}$, *the closed set* $[0, \infty) \setminus \tilde{A}(C, s)$ *is porous at* 1 *for any* $C > 0$ *and* $s \in (0, 1)$.

This result readily implies that if $Y > 0$ almost surely, then $[0, \infty) \setminus \tilde{A}(C, s)$ and the smaller random closed set $[0, \infty) \setminus A(C, s)$ are both almost surely porous at 1 for any $C > 0$ and $s \in (0, 1)$. Then the strong mixing property of **T** follows by Corollary 6.

It remains to show that $Y = \infty$ almost surely on the event $\{Y > 0\}$, which proves that $Y \in \{0, \infty\}$ almost surely. This is the content of the next proposition.

Proposition 17. *Set*

$$\tilde{A}(s) = \bigcap_{C>0} \tilde{A}(C, s), \quad \text{for } s \in (0, 1) \text{ and } \quad \tilde{A} = \bigcap_{s \in (0,1)} \tilde{A}(s).$$

Then the events $\{Y > 0\}$, $\{Y = \infty\}$, $\{1 \in \tilde{A}(s)\}$, $s \in (0, 1)$ *and* $\{1 \in \tilde{A}\}$ *are equal up to null sets.*

Proof of Proposition 17. Recall that $Y = \limsup_{n \to \infty} Y_n$ with

$$Y_n = \frac{\min_{0 \leq k < n} |\beta^{(k)}_{\gamma^*_n}|}{\sqrt{1 - \gamma^*_n}}.$$

With this notation, on $\{1 \in \tilde{A}(C, s)\}$ there is a random $n \geq 1$ such that $Y_n > C$. Here, the restriction $n \geq 1$ in the definition of $\tilde{A}(C, s)$ is useful. This way, we get that $\sup_{n \geq 1} Y_n \geq C$ on $\{1 \in \tilde{A}(C, s)\}$ and $\sup_{n \geq 1} Y_n = \infty$ on $\{1 \in \tilde{A}(s)\}$. Since $Y_n < \infty$ almost surely for all $n \geq 1$, we also have that $Y = \infty$ almost surely on $\{1 \in \tilde{A}(s)\}$.

Next, the law of the random closed set $[0, \infty) \setminus \tilde{A}(C, s)$ is invariant by scaling, hence by Proposition 16 and Lemma 5,

$$\{Y > 0\} \subset \{[0, \infty) \setminus \tilde{A}(C, s) \text{ is porous at } 1\} \subset \{1 \in \tilde{A}(C, s)\}, \quad \text{almost surely.}$$

The inclusions $\tilde{A}(C, s) \subset \tilde{A}(C', s)$ for $C > C'$ and $\tilde{A}(C, s) \subset \tilde{A}(C, s')$ for $1 > s' > s > 0$ yield

$$\tilde{A} = \bigcap_{k=1}^{\infty} \tilde{A}(k, 1 - 1/k).$$

Thus, $\{Y > 0\} \subset \{1 \in \tilde{A}\}$ almost surely.

Hence, up to null events,

$$\{Y > 0\} \subset \{1 \in \tilde{A}\} \subset \{1 \in \tilde{A}(s)\} \subset \{Y = \infty\} \subset \{Y > 0\}$$

for any $s \in (0, 1)$, which completes the proof. □

Proof of Proposition 16. By Malric's density theorem of zeros, recalled in (1), $\gamma_n^* \to 1^-$ almost surely. Hence it is enough to show that on the event $\{Y > 0\} \cap \{\gamma_n^* \to 1^-\}$ the set $\tilde{H} = [0, \infty) \setminus \tilde{A}(C, s)$ is porous at 1.

Let $\xi = Y/2$ and

$$I_n = (\gamma_n^*, \gamma_n^* + r_n), \quad \text{where} \quad r_n = \left(\frac{\xi \wedge C}{C}\right)^2 (1 - \gamma_n^*).$$

We claim that if

$$\xi > 0, \quad \gamma_n = \gamma_n^* > s, \quad \text{and} \quad |\beta_{\gamma_n}^{(k)}| > \xi\sqrt{1 - \gamma_n}, \quad \text{for } 0 \leq k < n. \tag{20}$$

then $I_n \subset \tilde{A}(C, s) \cap (\gamma_n^*, 1)$ with $r_n/(1 - \gamma_n^*) > 0$ not depending on n. Since on $\{Y > 0\} \cap \{\gamma_n^* \to 1^-\}$ the condition (20) holds for infinitely many n, we obtain the porosity at 1.

So assume that (20) holds for n at a given ω. As $I_n \subset (\gamma_n^*, 1)$, for $t \in I_n$ we have that $s < t < 1$ and $st < s < \gamma_n(t) = \gamma_n^*(t) = \gamma_n = \gamma_n^*$, that is, the first requirement in (19): $st < \gamma_n(t) = \gamma_n^*(t)$ holds for any $t \in I_n$. For the other requirement, note that $t - \gamma_n(t) < r_n \leq (1 - \gamma_n^*)\xi^2/C^2$ yields

$$\min_{0 \leq k < n} |\beta_{\gamma_n}^{(k)}| > \xi\sqrt{1 - \gamma_n^*} > C\sqrt{t - \gamma_n(t)}, \quad \text{for } t \in I_n. \qquad \square$$

3.6 Proof of Theorem 8

Compared to Theorem 7 in the proof of Theorem 8 we consider an even larger set $[0, \infty) \setminus \check{A}(C, s)$, where

$$\check{A}(C, s) = \{t > 0 : \exists n \geq 1, \, st < \gamma_n(t) = \gamma_n^*(t),$$

$$\min_{0 \leq k < n} |\beta_{\gamma_n(t)}^{(k)}| > C\sqrt{t - \gamma_n(t)},$$

$$\max_{u \in [\gamma_n(t), t]} |\beta_u - \beta_{\gamma_n(t)}| < \sqrt{t - \gamma_n(t)}\} \subset \tilde{A}(C, s) \subset A(C, s).$$

Here we also require that the fluctuation of β between $\gamma_n(t)$ and t is not too big.

We will prove the next proposition below.

Proposition 18. *Let $C > 1$, and $s \in (0, 1)$. Then almost surely on the event $\{X < 1\}$, the closed set $[0, \infty) \setminus \check{A}(C, s)$ is porous at 1.*

This result implies that if $X < 1$ almost surely, then for any $C > 0$, $s \in (0, 1)$ the random closed set $[0, \infty) \setminus \check{A}(C, s)$ is porous at 1 almost surely, and so is the smaller set $[0, \infty) \setminus A(C, s)$. Then the strong mixing of **T** follows from Corollary 6.

To complete the proof of Theorem 8, it remains to show that $X = 0$ almost surely on the event $\{X < 1\}$. This is the content of next proposition. In order to prove Theorem 9 we introduce a new parameter $L > 0$.

$$\check{A}_L(C, s) = \{t > 0 : \exists n \geq 1, \ st < \gamma_n(t) = \gamma_n^*(t),$$

$$\min_{0 \leq k < n} |\beta_{\gamma_n(t)}^{(k)}| > C\sqrt{t - \gamma_n(t)}, \ \max_{u \in [\gamma_n(t), t]} |\beta_u - \beta_{\gamma_n(t)}| < L\sqrt{t - \gamma_n(t)}\}$$

Then $\check{A}(C, s) = \check{A}_1(C, s)$.

Proposition 19. *Fix $L \geq 1$ and set*

$$\check{A}_L(s) = \bigcap_{C > 0} \check{A}_L(C, s), \quad \text{for } s \in (0, 1) \text{ and } \quad \check{A}_L = \bigcap_{s \in (0,1)} \check{A}_L(s).$$

Then the events $\{X = 0\}$, $\{X < 1\}$, $\{1 \in \check{A}_L\}$ and $\{1 \in \check{A}_L(s)\}$, $s \in (0, 1)$ are equal up to null sets.

Proof of Proposition 19. Fix $s \in (0, 1)$ $L \geq 1$ and let $C > L$. Assume that $1 \in \check{A}_L(C, s)$. Let $n > 0$ be an index which witnesses the containment. Then, as $C > L$ we can apply Lemma 15 to see that the absolute increments of $\beta^{(0)}, \ldots, \beta^{(n)}$ between γ_n and 1 are the same. This implies that

$$|\beta_1^{(k)}| \geq |\beta_{\gamma_n}^{(k)}| - |\beta_1^{(k)} - \beta_{\gamma_n}^{(k)}| = |\beta_{\gamma_n}^{(k)}| - |\beta_1 - \beta_{\gamma_n}|, \quad \text{for } 0 \leq k \leq n,$$

hence

$$Z_n \geq \min_{0 \leq k < n} |\beta_{\gamma_n}^{(k)}| - |\beta_1 - \beta_{\gamma_n}| > C\sqrt{1 - \gamma_n} - L\sqrt{1 - \gamma_n}$$

whereas

$$Z_{n+1} \leq |\beta_1^{(n)}| = |\beta_1^{(n)} - \beta_{\gamma_n}^{(n)}| = |\beta_1 - \beta_{\gamma_n}| < L\sqrt{1 - \gamma_n}.$$

Thus

$$\inf_{n \geq 0} \frac{Z_{n+1}}{Z_n} \leq \frac{L}{C - L}, \quad \text{on } \{1 \in \check{A}_L(C, s)\} \text{ almost surely,}$$

and

$$\inf_{n\geq 0} \frac{Z_{n+1}}{Z_n} = 0, \quad \text{on } \{1 \in \check{A}_L(s)\} \text{ almost surely.} \tag{21}$$

But, $Z_{n+1}/Z_n > 0$ almost surely for all n, hence $X = \liminf_{n\to\infty} Z_{n+1}/Z_n = 0$ almost surely on $\{1 \in \check{A}_L(s)\}$. This proves $\{1 \in \check{A}_L(s)\} \subset \{X = 1\}$.

Next, the law of the random closed set $[0, \infty) \setminus \check{A}_L(C, s)$ is clearly invariant under scaling, hence by Proposition 18 and Lemma 5

$$\{X < 1\} \subset \left\{[0, \infty) \setminus \check{A}_L(C, s) \text{ is porous at } 1\right\} = \left\{1 \in \check{A}_L(C, s)\right\}, \tag{22}$$

each relation holding up to a null set.

The inclusion $\check{A}_L(C', s') \subset \check{A}_L(C, s)$ for $C' \geq C > 0$ and $0 < s \leq s' < 1$ yields

$$\check{A}_L = \bigcap_{k=1}^{\infty} \check{A}_L(k, 1 - 1/k).$$

Hence, $\{X < 1\} \subset \{1 \in \check{A}_L\} \subset \{1 \in \check{A}_L(s)\}$ almost surely, which together with $\{1 \in \check{A}_L(s)\} \subset \{X = 0\}$ completes the proof. □

To prove Proposition 18 we need a corollary of the Blumenthal $0 - 1$ law.

Corollary 20. *Let (x_n) be a sequence of non-zero numbers tending to zero, $\mathbf{P}$ the Wiener measure on $\mathcal{C}[0, \infty)$ and $D \subset \mathcal{C}[0, \infty)$ be a Borel set such that $\mathbf{P}(D) > 0$. Then $\mathbf{P}(\Theta_{x_n}^{-1}(D) \text{ i.o.}) = 1$.*

Proof. Recall that the canonical process on $\mathcal{C}[0, \infty)$ was denoted by β. We also use the notation $\mathcal{B}_t = \sigma\{\beta_s : s \leq t\}$.

We approximate D with $D_n \in \mathcal{B}_{t_n}$ such that $\sum \mathbf{P}(D \triangle D_n) < \infty$, where $\triangle$ denotes the symmetric difference operator. Passing to a subsequence if necessary, we may assume that $t_n x_n^2 \to 0$. Then, since $\Theta_{x_n}^{-1}(D_n) \in \mathcal{B}_{t_n x_n^2}$, we have that $\{\Theta_{x_n}^{-1}(D_n), \text{i.o.}\} \in \cap_{s>0} \mathcal{B}_s$, and the Blumenthal $0 - 1$ law ensures that $\mathbf{P}(\Theta_{x_n}^{-1}(D_n), \text{i.o.}) \in \{0, 1\}$.

But $\sum \mathbf{P}(\Theta_{x_n}^{-1}(D) \triangle \Theta_{x_n}^{-1}(D_n)) < \infty$ since Θ_{x_n} preserves $\mathbf{P}$. Borel–Cantelli lemma shows that, almost surely, $\Theta_{x_n}^{-1}(D) \triangle \Theta_{x_n}^{-1}(D_n)$ occurs for finitely many n. Hence $\mathbf{P}(\Theta_{x_n}^{-1}(D), \text{i.o.}) \in \{0, 1\}$.

Fatou lemma applied to the indicator functions of $\Theta_{x_n}^{-1}(D)^c$ yields

$$\mathbf{P}(\Theta_{x_n}^{-1}(D), \text{i.o.}) \geq \limsup_{n\to\infty} \mathbf{P}(\Theta_{x_n}^{-1}(D)) = \mathbf{P}(D) > 0.$$

Hence $\mathbf{P}(\Theta_{x_n}^{-1}(D), \text{i.o.}) = 1$. □

Proof of Proposition 18. We work on the event $\{X < 1\}$. Set $\xi = (1/X - 1)/2$. Then $1 < \xi + 1 < 1/X$ and

$$1 < 1 + \xi < \limsup_{n\to\infty} \frac{Z_n}{Z_{n+1}} = \limsup_{n\to\infty} \frac{Z_n}{|\beta_1^{(n)}|}.$$

Hence

$$\min_{0 \le k < n} |\beta_1^{(k)}| = Z_n > (1+\xi)|\beta_1^{(n)}|, \quad \text{for infinitely many } n.$$

Let $n_1 < n_2 < \ldots$ the enumeration of those indices, and set $x_k = h_1^{(n_k)} \beta_1^{(n_k)}$ for $k \ge 1$. The inequality $|\beta_1^{(n_k)}| < (1+\xi)^{-1}|\beta_1^{(n_k - 1)}|$ shows that $x_k \to 0$.

Call B the Brownian motion defined by $B_t = \beta_{t+1} - \beta_1$ and for real numbers $\delta, C > 0$ set

$$D(\delta, C) = \left\{ w \in \mathcal{C}[0, \infty) : \sup_{t \le 2} |w(t)| < 1 + \delta; \right.$$

$w + 1$ has a zero in $[0, 1]$, but no zero in $(1, 2]$;

$$\left. \max_{t \in [\gamma, 2]} |w(t) + 1| \le \frac{\delta \wedge C}{2C}, \text{ where } \gamma \text{ is the last zero of } w + 1 \text{ in } [0, 2] \right\}.$$

For each $\delta, C > 0$ the Wiener measure puts positive, although possibly very small, probability on $D(\delta, C)$. Then Corollary 20 yields that the Brownian motion B takes values in the random sets $\Theta_{x_k}^{-1} D(\xi, C)$ for infinitely many k on $\{\xi > 0\} = \{X < 1\}$ almost surely; since the random variables x_k, ξ are $\mathcal{B}_1$-measurable, and B is independent of $\mathcal{B}_1$.

For $k \ge 1$ let $\tilde{\gamma}_k = \gamma_{n_k}(1 + x_k^2)$, that is, the last zero of $\beta^{(n_k)}$ before $1 + x_k^2$ and set

$$I_k = (\tilde{\gamma}_k + \tfrac{1}{2} r_k, \tilde{\gamma}_k + r_k), \quad \text{where} \quad r_k = \left(\frac{\xi \wedge C}{C} \right)^2 x_k^2.$$

This interval is similar to the one used in the proof of Proposition 16, but now we use only the right half of the interval $(\tilde{\gamma}_k, \tilde{\gamma}_k + r_k)$.

Next we show that

$$B \in \Theta_{x_k}^{-1} D(\xi, C), \quad \text{and} \quad s \le (1 + x_k^2)^{-1} \tag{23}$$

implies

$$I_k \subset \check{A}(C, s) \cap (1, 1 + 2x_k^2). \tag{24}$$

By definition $r_k/(4x_k^2)$, the ratio of the lengths of I_k and $(1, 1 + 2x_k^2)$, does not depend on k. Then the porosity of $[0, \infty) \setminus \check{A}(C, s)$ at 1 follows for almost every point of $\{X < 1\}$, as we have seen that (23) holds for infinitely many k almost surely on $\{X < 1\}$.

So assume that (23) holds for k at a given ω. The key observations are that then

$$\beta_{1+t}^{(\ell)} - \beta_1^{(\ell)} = h_1^{(\ell)} B_t, \qquad \text{for } 0 \le \ell \le n_k, 0 \le t \le 2x_k^2, \tag{25}$$

$$\gamma_\ell(t) < 1, \qquad \text{for } 0 \le \ell < n_k \text{ and } 1 \le t \le 1 + 2x_k^2, \tag{26}$$

$$\gamma_{n_k}(t) = \tilde{\gamma}_k > 1, \qquad \text{for } t \in [\tilde{\gamma}_k, 1 + 2x_k^2]. \tag{27}$$

First, we prove (25)–(27) and then with their help we derive $I_k \subset \check{A}(C, s)$.

To get (25) and (26) we apply Lemma 15 to $I = [1, 1+2x_k^2]$, $n = n_k$ and $a = 1$. This can be done since we have

$$\min_{0 \le \ell < n_k} |\beta_1^{(\ell)}| > (1 + \xi)|x_k|, \quad \text{by the choice of } n_k, \tag{28}$$

$$\max_{t \in [1, 1+2x_k^2]} |\beta_t - \beta_1| < (1 + \xi)|x_k|, \quad \text{since } \Theta_{x_k} B \in D(\xi, C) \text{ by (23)}. \tag{29}$$

(i) of Lemma 15 is exactly (26), while (ii) of the same lemma gives (25) if we note that $B_t = \beta_{1+t} - \beta_1$ by definition.

Equation (27) claims two things: $\beta^{(n_k)}$ has a zero in $(1, 1 + x_k^2]$, but has no zero in $(1 + x_k^2, 1 + 2x_k^2]$. Write (25) with $\ell = n_k$:

$$\beta_{1+t}^{(n_k)} = \beta_1^{(n_k)} + h_1^{(n_k)} B_t = h_1^{(n_k)}(x_k + B_t), \quad \text{for } 0 \le t \le 2x_k^2.$$

Next, we use that $\Theta_{x_k} B \in D(\xi, C)$, whence $1 + \Theta_{x_k} B$ has a zero in $[0, 1]$ but no zero in $(1, 2]$. Then the relation

$$x_k \left[1 + (\Theta_{x_k} B)_v\right] = x_k + B_{x_k^2 v} = h_1^{(n_k)} \beta_{1+x_k^2 v}^{(n_k)} \tag{30}$$

justifies (27).

To finish the proof, it remains to show that $I_k \subset \check{A}(C, s)$, since by (27) $\tilde{\gamma}_k$ the last zero of $\beta^{(n_k)}$ before $1 + x_k^2$ is greater than 1, so $I_k \subset (1, 1 + 2x_k^2)$ holds.

Fix $t \in I_k$. We need to check the next three properties.

(1) $st < \gamma_{n_k}(t) = \gamma_{n_k}^*(t)$.

By (27) $\gamma_{n_k}(t) = \tilde{\gamma}_k > 1$ and by the definition of I_k we have $1 < \tilde{\gamma}_k < t < \tilde{\gamma}_k + r_k \le \tilde{\gamma}_k + x_k^2$. Hence,

$$\gamma_{n_k}(t) = \tilde{\gamma}_k > \frac{\tilde{\gamma}_k}{\tilde{\gamma}_k + x_k^2} t > \frac{1}{1 + x_k^2} t \ge st,$$

where we used $s \le (1 + x_k^2)^{-1}$, the second part of (23).
By (26), $\gamma_{n_k}(t) = \gamma_{n_k}^*(t)$, as $t \in I_k \subset [1, 1 + 2x_k^2]$.

(2) $\min_{0\le\ell<n_k} |\beta_{\tilde{\gamma}_k}^{(\ell)}| > C\sqrt{t-\tilde{\gamma}_k}$.

Since $x_k = h_1^{(n_k)}\beta_1^{(n_k)}$, $\beta_{\tilde{\gamma}_k}^{(n_k)} = 0$ and $\tilde{\gamma}_k \in [1, 1+x_k^2]$, (25) yields

$$\max_{0\le\ell<n_k} |\beta_{\tilde{\gamma}_k}^{(\ell)} - \beta_1^{(\ell)}| = |\beta_{\tilde{\gamma}_k}^{(n_k)} - \beta_1^{(n_k)}| = |\beta_1^{(n_k)}| = |x_k|.$$

Then, by the triangle inequality and (28)

$$\min_{0\le\ell<n_k} |\beta_{\tilde{\gamma}_k}^{(\ell)}| \ge \min_{0\le\ell<n_k} |\beta_1^{(\ell)}| - \max_{0\le\ell<n_k} |\beta_{\tilde{\gamma}_k}^{(\ell)} - \beta_1^{(\ell)}|$$
$$> (1+\xi)|x_k| - |x_k| = \xi|x_k|.$$

On the other hand $\sqrt{t-\tilde{\gamma}_k} < \sqrt{r_k} \le |x_k|\xi/C$, hence

$$\min_{0\le\ell<n_k} |\beta_{\tilde{\gamma}_k}^{(\ell)}| > \xi|x_k| \ge C\sqrt{t-\tilde{\gamma}_k}.$$

(3) $\max_{u\in[\tilde{\gamma}_k,t]} |\beta_u - \beta_{\tilde{\gamma}_k}| < \sqrt{t-\tilde{\gamma}_k}$.

$1 + \Theta_{x_k} B$ has a zero in $[0, 1]$ but no zero in $(1, 2]$, since $\Theta_{x_k} B \in D(\xi, C)$. Denote as above by γ its last zero in $[0, 1]$. Then by relation (30) we have that $\tilde{\gamma}_k = 1 + x_k^2 \gamma$ and

$$\max_{u\in[\tilde{\gamma}_k,1+2x_k^2]} |\beta_u^{(n_k)}| = |x_k| \max_{v\in[\gamma,2]} |1+(\Theta_{x_k}B)_v| \le |x_k|\frac{\xi\wedge C}{2C} = \frac{\sqrt{r_k}}{2}.$$

Writing (25) with $\ell = n_k$ and using that $\beta_{\tilde{\gamma}_k}^{(n_k)} = 0$ and $t < 1+2x_k^2$ we obtain

$$\max_{u\in[\tilde{\gamma}_k,t]} |\beta_u - \beta_{\tilde{\gamma}_k}| = \max_{u\in[\tilde{\gamma}_k,t]} |\beta_u^{(n_k)}| \le \max_{u\in[\tilde{\gamma}_k,1+2x_k^2]} |\beta_u^{(n_k)}| \le \frac{\sqrt{r_k}}{2}.$$

By the definition of I_k we have $t - \tilde{\gamma}_k > \frac{1}{2}r_k$. Hence

$$\max_{u\in[\tilde{\gamma}_k,t]} |\beta_u - \beta_{\tilde{\gamma}_k}| \le \frac{\sqrt{r_k}}{2} < \frac{\sqrt{r_k}}{2}\sqrt{\frac{t-\tilde{\gamma}_k}{\frac{1}{2}r_k}} < \sqrt{t-\tilde{\gamma}_k}. \quad \square$$

3.7 Proof of Theorem 9

In this subsection we prove the equality of the events $\{X = 0\}$, $\{Y = \infty\}$ and $\{1 \in A\}$ up to null sets, where

$$A = \bigcap_{s \in (0,1)} A(s), \quad \text{with} \quad A(s) = \bigcap_{C>0} A(C,s).$$

We keep the notation introduced in Propositions 17 and 19 for $\check{A}_L(s)$, $\check{A}_L$ and $\tilde{A}$.

Recall that $\check{A}_L \subset \tilde{A} \subset A$ by definition for any $L \geq 1$. Then by Propositions 17 and 19 we have

$$\{X = 0\} = \{1 \in \check{A}_L\} \subset \{1 \in \tilde{A}\} = \{Y = \infty\} \subset \{1 \in A\}. \tag{31}$$

For $C > 0$ let

$$\tau_C = \inf\left\{ t \geq \tfrac{1}{2} : \exists n \geq 0, \, \beta_t^{(n)} = 0 \min_{0 \leq k < n} |\beta_t^{(k)}| \geq C\sqrt{(1-t) \vee 0} \right\}.$$

We show below that

$$\{1 \in A\} \subset \bigcap_{C>0} \{\tau_C < 1\}, \quad \text{up to null a set,} \tag{32}$$

and

$$\mathbf{P}\left(\bigcap_{C>0} \{\tau_C < 1\} \right) \leq \mathbf{P}(X = 0). \tag{33}$$

Then the claim follows by concatenating (31) and (32), and observing that the largest and the smallest events in the obtained chain of almost inclusions has the same probability by (33).

We start with (32). If $1 \in A$ then $1 \in A(C, s)$ for every $s \in (0, 1)$, especially for $s_0 = \gamma_0 \vee 1/2$, where γ_0 is the last zero of β before 1, we have $1 \in A(C, s_0)$. Then, by the definition of $A(C, s_0)$ there is an integer $n \geq 0$ and a real number $\gamma \in (s_0, 1)$ such that $\beta_\gamma^{(n)} = 0$ and $\min_{0 \leq k < n} |\beta_1^{(k)}| > C\sqrt{1-\gamma}$. The integer n cannot be zero since $\beta^{(0)} = \beta$ has no zero in $(s_0, 1)$. Thus $\tau_C \leq \gamma < 1$, which shows the inclusion.

Next, we turn to (32). Fix $C > L \geq 1$ and let

$$\gamma = \sup\{s \in [\tau_C, 1] : \beta_s = \beta_{\tau_C}\}.$$

Let us show that

$$\left\{ \tau_C < 1 \text{ and } \max_{\tau_C \leq t \leq 1} |\beta_t - \beta_{\tau_C}| < L\sqrt{1-\gamma} \right\} \subset \left\{ 1 \in \check{A}_L(C, \tfrac{1}{2}) \right\}. \tag{34}$$

Indeed, on the event on the left hand side of (34) there exists a random index n such that $\beta_{\tau_C}^{(n)} = 0$ and

$$\min_{0 \leq k \leq n-1} |\beta_{\tau_C}^{(k)}| \geq C\sqrt{1-\tau_C} > L\sqrt{1-\gamma} > \max_{\tau_C \leq t \leq 1} |\beta_t - \beta_{\tau_C}|.$$

Then we can apply Lemma 15 with $I = [\tau_C, 1]$, $a = \tau_C$ and $n = n$. We obtain that $\beta^{(k)}$ has no zero in $[\tau_C, 1]$ for $k = 0, \ldots, n-1$, and the absolute increments $|\beta_t^{(k)} - \beta_{\tau_C}^{(k)}|$, are the same for $k = 0, \ldots, n$ and $t \in [\tau_C, 1]$. In particular, $\beta_\gamma^{(k)} = \beta_{\tau_C}^{(k)}$ for every $0 \leq k \leq n$, γ is the last zero of $\beta^{(n)}$ in $[\tau_C, 1]$ and $\gamma = \gamma_n = \gamma_n^*$. Moreover,

$$\min_{0 \leq k < n} |\beta_{\gamma_n^*}^{(k)}| = \min_{0 \leq k < n} |\beta_{\tau_C}^{(k)}| \geq C\sqrt{1 - \tau_C} > C\sqrt{1 - \gamma_n^*}.$$

So n and γ_n^* witnesses that $1 \in \check{A}_L(C, \frac{1}{2})$, since we also have that

$$\max_{t \in [\gamma_n^*, 1]} |\beta_t - \beta_{\gamma_n^*}| \leq \max_{t \in [\tau_C, 1]} |\beta_t - \beta_{\tau_C}| < L\sqrt{1 - \gamma_n^*}.$$

From (34), by the strong Markov property and the scaling invariance of β, we obtain

$$\mathbf{P}(\tau_C < 1) \times \mathbf{P}\left(\max_{t \in [0,1]} |\beta_t| \leq L\sqrt{1 - \gamma_0}\right) \leq \mathbf{P}\left(1 \in \check{A}_L(C, \tfrac{1}{2})\right).$$

Letting C go to infinity and using Proposition 19, this yields

$$\mathbf{P}\left(\bigcap_{C>0} \{\tau_C < 1\}\right) \times \mathbf{P}\left(\max_{t \in [0,1]} |\beta_t| \leq L\sqrt{1 - \gamma_0}\right) \leq \mathbf{P}\left(1 \in \check{A}_L(\tfrac{1}{2})\right)$$

$$= \mathbf{P}(X = 0).$$

This is true for all $L \geq 1$. Thus (33) is obtained by letting L go to infinity.

3.8 Proof of Theorem 11

In this subsection we prove that the tightness of $\{xv(x) : x \in (0, 1)\}$ and $\{nZ_n : n \geq 1\}$ are equivalent and both implies $X < 1$ almost surely.

Fix $K > 0$. By definition $\{(K/n)v(K/n) > K\} = \{nZ_n \geq K\}$ for any $n \geq 1$. For small $x > 0$ values there is n such that $K/n < x < 2K/n$ and $xv(x) \leq (2K/n)v(K/n)$ by the monotonicity of v. But, then $\{xv(x) > 2K\} \subset \{nZ_n > K\}$. Hence

$$\limsup_{x \to 0^+} \mathbf{P}(xv(x) > 2K) \leq \limsup_{n \to \infty} \mathbf{P}(nZ_n \geq K) \leq \limsup_{x \to 0^+} \mathbf{P}(xv(x) > K).$$

So the tightness of the two families are equivalent and it is enough to prove that when $\{xv(x) : x \in (0, 1)\}$ is tight then $X < 1$ almost surely.

We have the next easy lemma, whose proof is sketched at the end of this subsection.

Lemma 21.

$$X = \liminf_{n\to\infty} \frac{Z_{n+1}}{Z_n} = \liminf_{x\to 0+} \frac{|\beta_1^{(v(x))}|}{x}.$$

Then we have that

$$\mathbb{1}_{(X>1-\delta)} \leq \liminf_{x\to 0+} \mathbb{1}_{(|\beta_1^{(v(x))}|/x>1-\delta)}.$$

Hence, by Fatou lemma

$$\mathbf{P}(X > 1-\delta) \leq \liminf_{x\to 0+} \mathbf{P}\Big(|\beta_1^{(v(x))}| > x(1-\delta)\Big).$$

Let $x \in (0,1)$ and $K > 0$. Since on the event

$$\left\{v(x) \leq \frac{K}{x}\right\} \cap \left\{|\beta_1^{(v(x))}| > x(1-\delta)\right\}$$

at least one of the standard normal variables $\beta_1^{(k)}$, $0 \leq k \leq K/x$ takes values in a set of size $2x\delta$, namely in $(-x, -x(1-\delta)) \cup (x(1-\delta), x)$,

$$\mathbf{P}\left(\frac{|\beta_1^{(v(x))}|}{x} > 1-\delta\right)$$

$$\leq \mathbf{P}\left(v(x) > \frac{K}{x}\right) + \left(\frac{K}{x} + 1\right)\mathbf{P}\left(1-\delta < \frac{|\beta_1|}{x} < 1\right)$$

$$\leq \mathbf{P}(xv(x) > K) + (K+1)\delta.$$

In the last step we used that the standard normal density is bounded by $1/\sqrt{2\pi}$, whence $\mathbf{P}\left(1-\delta < \frac{|\beta_1|}{x} < 1\right) \leq \delta x$.

By the tightness assumption for any $\varepsilon > 0$ there exists K_ε such that $\sup_{x\in(0,1)} \mathbf{P}(xv(x) > K_\varepsilon) \leq \varepsilon$. Hence,

$$\mathbf{P}(X = 1) = \lim_{\delta\to 0+} \mathbf{P}(X > 1-\delta) \leq \lim_{\delta\to 0+} \varepsilon + (K_\varepsilon + 1)\delta = \varepsilon.$$

Since, this is true for all $\varepsilon > 0$, we get that $\mathbf{P}(X = 1) = 0$ and the proof of Theorem 11 is complete. □

Proof of Lemma 21. Since $Z_{v(x)} = |\beta_1^{(v(x))}|$ Lemma 21 is a particular case of the following claim: if (a_n) is a decreasing sequence of positive numbers tending to zero then

$$\liminf_{k\to\infty} \frac{a_{k+1}}{a_k} = \liminf_{x\to 0+} \frac{a_{n(x)}}{x},$$

where $n(x) = \inf\{k \geq 1 : a_k < x\}$. First, for $x < a_1$ the relation $a_{n(x)-1} \geq x > a_{n(x)}$ gives

$$\frac{a_{n(x)}}{a_{n(x)-1}} \leq \frac{a_{n(x)}}{x}$$

and

$$\liminf_{k \to \infty} \frac{a_{k+1}}{a_k} \leq \liminf_{x \to 0^+} \frac{a_{n(x)}}{x}.$$

For the opposite direction, for every $k \geq 0$, $a_{n(a_k)} < a_k$, therefore $a_{n(a_k)} \leq a_{k+1}$ as (a_n) is non-increasing. Since $a_k \to 0$ as $k \to \infty$, one gets

$$\liminf_{x \to 0^+} \frac{a_{n(x)}}{x} \leq \liminf_{k \to \infty} \frac{a_{n(a_k)}}{a_k} \leq \liminf_{k \to \infty} \frac{a_{k+1}}{a_k}. \qquad \square$$

Acknowledgements Crucial part of this work was done while visiting the University of Strasbourg, in February of 2011. I am very grateful to IRMA and especially to professor Michel Émery for their invitation and for their hospitality.

Conversations with Christophe Leuridan and Jean Brossard, and a few days later with Marc Yor and Marc Malric inspired the first formulation of Theorems 7 and 8.

The author is grateful to the anonymous referee for his detailed reports and suggestions that improved the presentation of the results significantly. Especially, the referee proposed a simpler argument for stronger result in Theorem 11, pointed out a sloppiness in the proof of Theorem 8, suggested a better formulation of Theorems 7 and 8 and one of his/her remarks motivated Theorem 9.

The original proof of Corollary 20 was based on the Erdős–Rényi generalization of the Borel–Cantelli lemma, see [10]. The somewhat shorter proof in the text was proposed by the referee.

References

1. P. Billingsley, *Convergence of Probability Measures* (Wiley, New York, 1968)
2. J. Brossard, C. Leuridan, Densité des orbites des trajectoires browniennes sous l'action de la transformation de Lévy. Ann. Inst. H. Poincaré Probab. Stat. **48**(2), 477–517 (2012). doi:10.1214/11-AIHP463
3. L.E. Dubins, M. Smorodinsky, The modified, discrete, Lévy-transformation is Bernoulli. In *Séminaire de Probabilités, XXVI, Lecture Notes in Math.*, vol. 1526 (Springer, Berlin, 1992), pp. 157–161
4. L.E. Dubins, M. Émery, M. Yor, On the Lévy transformation of Brownian motions and continuous martingales. In *Séminaire de Probabilités, XXVII, Lecture Notes in Math.*, vol. 1557 (Springer, Berlin, 1993), pp. 122–132
5. T. Fujita, A random walk analogue of Lévy's theorem. Studia Sci. Math. Hungar. **45**(2), 223–233 (2008). doi:10.1556/SScMath.45.2008.2.50
6. J. Jacod, A.N. Shiryaev, Limit theorems for stochastic processes, in *Grundlehren der Mathematischen Wissenschaften [Fundamental Principles of Mathematical Sciences]*, vol. 288 (Springer, Berlin, 1987)

7. M. Malric, Transformation de Lévy et zéros du mouvement brownien. Probab. Theory Relat. Fields **101**(2), 227–236 (1995). doi:10.1007/BF01375826
8. M. Malric, Densité des zéros des transformés de Lévy itérés d'un mouvement brownien. C. R. Math. Acad. Sci. Paris **336**(6), 499–504 (2003)
9. M. Malric, Density of paths of iterated Lévy transforms of Brownian motion. ESAIM Probab. Stat. **16**, 399–424 (2012). doi:10.1051/ps/2010020
10. T.F. Móri, G.J. Székely, On the Erdős-Rényi generalization of the Borel-Cantelli lemma. Studia Sci. Math. Hungar. **18**(2–4), 173–182 (1983)
11. D. Revuz, M. Yor, Continuous martingales and Brownian motion, *Grundlehren der Mathematischen Wissenschaften [Fundamental Principles of Mathematical Sciences]*, vol. 293. (Springer, Berlin, 1991)
12. F. Riesz, B. Sz.-Nagy, *Functional Analysis* (Frederick Ungar Publishing, New York, 1955). Translated by Leo F. Boron

Vershik's Intermediate Level Standardness Criterion and the Scale of an Automorphism

Stéphane Laurent

Abstract In the case of r_n-adic filtrations, Vershik's standardness criterion takes a particular form, hereafter called Vershik's intermediate level criterion. This criterion, whose nature is combinatorial, has been intensively used in the ergodic-theoretic literature, but it is not easily applicable by probabilists because it is stated in a language specific to the theory of measurable partitions and has not been translated into probabilistic terms. We aim to provide an easily applicable probabilistic statement of this criterion. Finally, Vershik's intermediate level criterion is illustrated by revisiting Vershik's definition of the scale of an invertible measure-preserving transformation.

1 Introduction

Although many efforts have been devoted to translating Vershik's theory of decreasing sequence of measurable partitions into a theory of filtrations written in the language of stochastic processes [4, 8–10], in the ergodic-theoretic literature many papers dealing with standard filtrations still remain difficult to read by probabilists outside the class of experts in this topic. Difficulties do not lie in the basic concepts of ergodic theory such as the ones presented in introductory books on measure-preserving systems, but rather in the language of the theory of measurable partitions initiated by Rokhlin (see [15]). Rokhlin's correspondence (see [3]) between measurable partitions and complete σ- fields is not a complicated thing, but the approach to filtrations is somewhat geometrical in the language of

S. Laurent (✉)
Université de Liège
QuantOM, HEC-Management School, Paris, France
e-mail: laurent_step@yahoo.fr

partitions, whereas probabilists are more comfortable with a filtration considered as the history of some stochastic process whose dynamics is clearly described.

Particularly, standardness is studied by many ergodic-theoretic papers (such as [5–7]) in the context of r_n-*adic filtrations*: those filtrations $\mathcal{F} = (\mathcal{F}_n)_{n \leq 0}$ for which $\mathcal{F}_n = \mathcal{F}_{n-1} \vee \sigma(\varepsilon_n)$ for each $n \leq 0$, where the *innovation* ε_n is a random variable independent of $\mathcal{F}_{n-1}$ and uniformly distributed on r_n possible values, for some sequence $(r_n)_{n \leq 0}$ of positive integers. For such filtrations, Vershik's general standardness criterion, which has received some attention in the probabilistic literature [4, 10], and which we call *Vershik's second level (standardness) criterion* for more clarity, takes a particular form, which is in fact the original form of Vershik's standardness criterion (he focused on r_n-adic filtrations), and which we call *Vershik's intermediate level (standardness) criterion*. Its statement involves tree automorphisms and characterizes standardness in terms of a problem of combinatorial nature. Ergodicians directly apply Vershik's intermediate level criterion in their works, but this criterion has not been translated in the probabilistic literature (an attempt was done in the author's PhD. thesis [8]), thereby causing difficulties for a probabilist reader. The present paper provides a probabilistic statement of Vershik's intermediate level criterion. This statement is not as brief as Vershik's analogous statement in the language of measurable partitions, but it is directly applicable to investigate standardness of an r_n-adic filtration without resorting to notions unfamiliar to probabilists, except possibly the notion of a tree automorphism. Roughly speaking, tree automorphisms lie in the heart of Vershik's intermediate level criterion because any two local innovations $(\varepsilon_n, \ldots, \varepsilon_0)$ of an r_n-adic filtration $\mathcal{F}$ differ from each other by the action of some $\mathcal{F}_{n-1}$-measurable random tree automorphism.

We will show that in the context of r_n-adic filtrations, Vershik's intermediate level criterion is equivalent to Vershik's second level criterion. All results in the present paper are self-contained, except the proofs of rather elementary statements for which we will refer to [10]. Section 2 aims to provide the non-specialist reader with some motivations for the development of Vershik's two criteria by recalling their relations with the notions of productness and standardness. In Sect. 3 we state Vershik's second level criterion, similarly to [4] and [10], and its elementary properties. Vershik's intermediate level criterion is then the purpose of Sect. 4. In Sect. 5, we will illustrate Vershik's intermediate level criterion by formulating his definition of the *scale of an automorphism* [13] in terms of this criterion, thereby shedding a new light on the scale. Vershik used another definition which involves the orbits of the automorphism. Equivalence of both definitions was announced by the author in [10], without proof, and [10] further shows how to derive the scale of Bernoulli automorphisms from the theorem on productness of the split-word process presented in [9]. With our definition, many properties of the scale of an automorphism stated by Vershik in [13] appear to be direct consequences of elementary properties of Vershik's intermediate level criterion or more general results of the theory of filtrations. For instance we will see that the scale of a completely ergodic automorphism is nonempty as a consequence of Vershik's theorem on lacunary isomorphism, whereas Vershik proved this proposition by a direct construction.

2 Standardness and Productness

We briefly present the meaningful notions of productness and standardness and their relations with Vershik's intermediate level criterion and Vershik's second level criterion. These notions manifestly motivate the development of these criteria.

Vershik's pioneering work mainly deals with r_n-adic filtrations. For a given sequence $(r_n)_{n \leq 0}$ of integers $r_n \geq 2$, a filtration $\mathcal{F}$ is said to be r_n-adic if $\mathcal{F}_n = \mathcal{F}_{n-1} \vee \sigma(\varepsilon_n)$ for every $n \leq 0$ where ε_n is a random variable independent of $\mathcal{F}_{n-1}$ and uniformly distributed on a finite set consisting of r_n elements. Such random variables ε_n are called *innovations* of $\mathcal{F}$. The process $(\varepsilon_n)_{n \leq 0}$ is then a sequence of independent random variables and it is itself called an innovation of $\mathcal{F}$, and we also say that $(\varepsilon_n, \ldots, \varepsilon_0)$ is a *local innovation* of $\mathcal{F}$. In other words, the innovation ε_n is a random variable generating an independent complement of $\mathcal{F}_{n-1}$ in $\mathcal{F}_n$. (We say that a σ-field $\mathcal{C}$ is an *independent complement* of a σ-field $\mathcal{B}$ in a σ-field $\mathcal{A} \supset \mathcal{B}$ if $\mathcal{C}$ is independent of $\mathcal{B}$ and $\mathcal{A} = \mathcal{B} \vee \mathcal{C}$.) Independent complements are not unique in general, as testified by the following lemma whose proof is left as an easy exercise.

Lemma 2.1. *Let $(\mathcal{B}, \mathcal{A})$ be an increasing pair of σ-fields and V be a random variable generating an independent complement of $\mathcal{B}$ in $\mathcal{A}$. If V is uniformly distributed on some finite set F, and if Φ is any $\mathcal{B}$-measurable random permutation of F, then $\Phi(V)$ also generates an independent complement of $\mathcal{B}$ in $\mathcal{A}$.*

Actually one can conversely prove that every independent complement of $\mathcal{B}$ in $\mathcal{A}$ is generated by $\Phi(V)$ for some $\mathcal{B}$-measurable random permutation Φ of F; but we will not need this result.

Both following theorems are presented to motivate our work. They will be admitted and not used in the present paper. We say that a filtration $\mathcal{F} = (\mathcal{F}_n)_{n \leq 0}$ is *essentially separable* when its final σ-field $\mathcal{F}_0$ is essentially separable (i.e., separable modulo negligible events).

One of Vershik's main achievements in his pioneering work [11, 12] was a criterion characterizing *productness* of r_n-adic filtrations. The first theorem below is our rephrasement of this result.

Theorem 2.2. *An essentially separable r_n-adic filtration satisfies Vershik's intermediate level criterion if and only if it is of product type.*

A filtration is said to be of *product type* if it is generated by some sequence of independent random variables. In particular, an r_n-adic filtration is of product type if it is the filtration generated by uniform, independent r_n-ary random drawings, that is, a sequence (ε_n) of independent random variables with ε_n uniformly distributed on r_n distinct values for each n. This criterion stated by Vershik in the language of measurable partitions is in general simply called the standardness criterion, or Vershik's standardness criterion, in ergodic-theoretic research articles dealing with r_n-adic filtrations. We call our proposed probabilistic translation of this criterion *Vershik's intermediate level (standardness) criterion*.

For r_n-adic filtrations the equivalence between Vershik's intermediate level criterion and productness (Theorem 2.2) as well as its equivalence with the I-cosiness criterion (not introduced in the present paper) can be deduced from the results in the literature and from the equivalence between Vershik's intermediate and second level criteria (Theorem 4.9), but both these equivalences are not very difficult to prove directly. In fact, Laurent provides the self-joining criterion corresponding to Vershik's intermediate level criterion in [9], called *Vershik's self-joining criterion*, and both equivalences mentioned above can be directly proved by means of the same ideas used in [9] to prove the analogous statements for Vershik's self-joining criterion. This is even easier with Vershik's intermediate level criterion.

For arbitrary filtrations the statement of Vershik's intermediate level criterion does not make sense. But Vershik also provided in his pioneering work an equivalent statement of this criterion which makes sense for arbitrary filtrations, and the probabilistic analogue of this criterion was provided by Émery and Schachermayer in [4]. In the present paper, agreeing with the terminology of [10], we call it *Vershik's second level (standardness) criterion*, or, shortly, the *second Vershik property*, and we also say that a filtration is *Vershikian* when it satisfies this property. The equivalence between Vershik's intermediate and second level criteria will be proved in the present paper (Theorem 4.9). The following theorem, proved in [4], was stated by Vershik in [14] in the language of measurable partitions.

Theorem 2.3. *An essentially separable filtration satisfies Vershik's second level standardness criterion if and only if it is standard.*

Standardness for an arbitrary filtration is defined with the help of the notion of immersion. A filtration $\mathcal{F}$ is said to be *immersed* in a filtration $\mathcal{G}$ if every $\mathcal{F}$-martingale is a $\mathcal{G}$-martingale (this implies $\mathcal{F} \subset \mathcal{G}$). We refer to [4] and [9] for more details on the immersion property. Then a filtration $\mathcal{F}$ is said to be *standard* if, up to isomorphism, it is immersed in the filtration generated by some sequence of independent random variables each having a diffuse law, or, equivalently (see [9]), in the filtration generated by some sequence of independent random variables.

3 Vershik's Second Level Standardness Criterion

We will state Vershik's second level criterion in Sect. 3.2 after having introduced some preliminary notions in Sect. 3.1.

3.1 The Kantorovich Metric and Vershik's Progressive Predictions

The *Kantorovich* distance plays a major role in the statement of the second level Vershik property. Given a separable metric space (E, ρ), the Kantorovich distance ρ' on the set E' of probabilities on E is defined by

$$\rho'(\mu, \nu) = \inf_{\Lambda \in \mathcal{J}(\mu,\nu)} \iint \rho(x, y) \, d\Lambda(x, y),$$

where $\mathcal{J}(\mu, \nu)$ is the set of all joinings of μ and ν, that is, the set of all probabilities on $E \times E$ whose first and second marginal measures are respectively μ and ν.

In general, the topology induced by ρ' on the set E' of probability on E is finer than the topology of weak convergence. These two topologies coincide when (E, ρ) is compact, hence in particular (E', ρ') is itself compact in this case. The metric space (E', ρ') is complete and separable whenever (E, ρ) is (see e.g. [1]).

The following lemma will be used to prove the equivalence between Vershik's intermediate and second level properties.

Lemma 3.1. *Let $r \geq 2$ be an integer and let f and g be functions from $\{1, \ldots, r\}$ to a Polish metric space (E, ρ). Denote by ν the uniform probability on $\{1, \ldots, r\}$. Then the infimum in the Kantorovich distance $\rho'\bigl(f(\nu), g(\nu)\bigr)$ is attained for the joint law of a random pair $\bigl(f(\varepsilon), g(\varepsilon')\bigr)$ where ε is a random variable distributed according to ν and $\varepsilon' = \sigma(\varepsilon)$ for some permutation σ of $\{1, \ldots, r\}$.*

Proof. Any joining of $f(\nu)$ and $g(\nu)$ is the law of a random pair $\bigl(f(\varepsilon), g(\varepsilon')\bigr)$ where $\varepsilon \sim \nu$ and $\varepsilon' \sim \nu$; and the expectation $\mathbb{E}\bigl[\rho\bigl(f(\varepsilon), g(\varepsilon')\bigr)\bigr]$ is a linear form of the joint law of ε and ε'. Therefore, in the set of all joinings, there exists at least an extremal point where the minimal possible value of this expectation is attained; but Birkhoff's theorem says that the law of $(\varepsilon, \varepsilon')$ is an extremal joining whenever $\varepsilon' = \sigma(\varepsilon)$ for some permutation σ. □

Now, let $\mathcal{F}$ be a filtration, E a Polish metric space and $X \in L^1(\mathcal{F}_0; E)$. The Vershik second level property of X involves *Vershik's progressive predictions* $\pi_n X$ of X, which correspond to the so-called *universal projectors* in [11] and [14]. They are recursively defined as follows: we put $\pi_0 X = X$, and $\pi_{n-1} X = \mathcal{L}(\pi_n X \mid \mathcal{F}_{n-1})$ (the conditional law of $\pi_n X$ given $\mathcal{F}_{n-1}$); thus, the n-th progressive prediction $\pi_n X$ of X with respect to $\mathcal{F}$ is a random variable taking its values in the Polish space $E^{(n)}$, which is recursively defined by $E^{(0)} = E$ and $E^{(n-1)} = (E^{(n)})'$, denoting as before by E' the space of all probability measures on any separable metric space E. The state space $E^{(n)}$ of $\pi_n X$ is Polish when endowed with the distance ρ_n obtained by iterating $|n|$ times the construction of the Kantorovich distance starting with ρ: we recursively define ρ_n by putting $\rho_0 = \rho$ and by defining $\rho_{n-1} = (\rho_n)'$ as the Kantorovich distance issued from ρ_n.

Finally, in order to state Vershik's second level criterion, we introduce the *dispersion* disp X of (the law of) an integrable random variable X in a Polish metric space. It is defined as the expectation of $\rho(X', X'')$ where X' and X'' are two independent copies of X, that is, two independent random variables having the same law as X.

3.2 Vershik's Second Level Criterion

Let $\mathcal{F}$ be a filtration, E a Polish metric space and $X \in L^1(\mathcal{F}_0; E)$. We say that the random variable X satisfies *Vershik's second level (standardness) criterion*, or the *second Vershik property*, or, for short, that X is *Vershikian* (with respect to $\mathcal{F}$) if $\operatorname{disp} \pi_n X \longrightarrow 0$ as n goes to $-\infty$. Then we extend this definition to σ-fields $\mathcal{E}_0 \subset \mathcal{F}_0$ and to the whole filtration as follows: we say that a σ-field $\mathcal{E}_0 \subset \mathcal{F}_0$ is *Vershikian* if each random variable $X \in L^1(\mathcal{E}_0; [0, 1])$ is Vershikian with respect to $\mathcal{F}$, and we say that the filtration $\mathcal{F}$ is *Vershikian* if the final σ-field $\mathcal{F}_0$ is Vershikian.

The following proposition is proved in [10] when (E, ρ) is a compact metric space, but it is easy to check that the proof remains valid for a Polish metric space.

Proposition 3.2. *For any Polish metric space (E, ρ), a random variable $X \in L^1(\mathcal{F}_0, E)$ is Vershikian if and only if the σ-field $\sigma(X)$ is Vershikian.*

Below we state Vershik's theorem on lacunary isomorphism which will be used in Sect. 5 to prove nonemptiness of the scale of a completely ergodic automorphism. A filtration is said to be *Kolmogorovian* if $\mathcal{F}_{-\infty} := \cap_{n \leq 0} \mathcal{F}_n$ is the degenerate σ-field.

Theorem 3.3. *Let $\mathcal{F} = (\mathcal{F}_n)_{n \leq 0}$ be an essentially separable filtration. If $\mathcal{F}$ is Kolmogorovian, there exists a strictly increasing map $\phi : -\mathbb{N} \to -\mathbb{N}$ such that the extracted filtration $(\mathcal{F}_{\phi(n)})_{n \leq 0}$ is Vershikian.*

The first version of this theorem was stated and proved by Vershik in the context of r_n-adic filtrations, but was proved without using any standardness criterion: Vershik directly showed that it is possible to extract a filtration of product type from every Kolmogorovian r_n-adic filtration. The analogous proof for conditionally non-atomic filtrations (that is, filtrations admitting innovations with diffuse law) was provided by Émery and Schachermayer [4]. This proof is somewhat constructive but quite technical, whereas a short proof of the general version of the theorem on lacunary isomorphism stated above, based on Vershik's second level criterion, is given in [10].

It is not straightforward to see from the definition of the Vershik property that every filtration extracted from a Vershikian filtration is itself Vershikian, whereas this is very easy to see from the definition of the I-cosiness criterion which is known to be equivalent to the Vershik property (see [4, 10]). It is also easy to see that this property holds for Vershik's intermediate level criterion, but this one only concerns r_n-adic filtrations. Note that it is not restrictive to take $\phi(0) = 0$ in Theorem 3.3 since the Vershik property is an asymptotic one (see [10]).

We will also use the following lemma in Sect. 5. It is proven in [10].

Lemma 3.4. *For any Polish metric space E, a random variable $X \in L^1(\mathcal{F}_0; E)$ is Vershikian if and only if the filtration generated by the stochastic process $(\pi_n X)_{n \leq 0}$ is Vershikian.*

The next lemma says that the second level Vershik property is hereditary for immersions; we refer to [10] for its proof.

Lemma 3.5. *Let $\mathcal{F}$ be a filtration, $\mathcal{E}$ a filtration immersed in $\mathcal{F}$, and E a Polish metric space. A random variable $X \in L^1(\mathcal{E}_0; E)$ is Vershikian with respect to $\mathcal{F}$ if and only it is Vershikian with respect to $\mathcal{E}$. Consequently, if the filtration $\mathcal{F}$ is Vershikian, then so is also $\mathcal{E}$.*

4 Vershik's Intermediate Level Criterion

Vershik's intermediate level criterion is the object of Sect. 4.3. In Sect. 4.1 we mainly fix some notations about tree automorphisms which will be needed to prove the equivalence between Vershik's two criteria, and in Sect. 4.2 we introduce the split-word processes which will be needed to state Vershik's intermediate level criterion.

Throughout this section we will speak of *words* on a set A called the *alphabet*. A *word* w on A is an element of A^ℓ, or equivalently an application from $\{1, \ldots, \ell\}$ to A, for some integer $\ell \geq 1$ called the *length* of w. A word of length ℓ is shortly termed as an ℓ-word. The *letters* of an ℓ-word w are $w(1), \ldots, w(\ell)$. When A is treated as a Polish space it is understood that the set A^ℓ of ℓ-words on A is treated as the corresponding product Polish space.

4.1 Tree Automorphisms

All notions defined below are relative to a given sequence $(r_n)_{n \leq 0}$ consisting of integers $r_n \geq 2$, from which we define the sequence $(\ell_n)_{n \leq 0}$ by $\ell_n = \prod_{k=n+1}^{0} r_k$ for all $n \leq 0$.

Define the sets $B_n = \prod_{k=n+1}^{0} \{1, \ldots, r_k\}$ for $n \leq -1$. The group G_n of tree automorphisms of B_n is a subgroup of the group of permutations of B_n recursively defined as follows. The group G_{-1} is the whole group of permutations of $\{1, \ldots, r_0\}$, and a permutation $\tau \in G_n$ maps an element $b_n = (c_n, b_{n+1}) \in B_n = \{1, \ldots, r_{n+1}\} \times B_{n+1}$ to $\tau(b_n) = \big(\sigma(c_n), \psi(\sigma(c_n))(b_{n+1})\big)$ where σ is a permutation of $\{1, \ldots, r_{n+1}\}$ and ψ is a map from $\{1, \ldots, r_{n+1}\}$ to G_{n+1}.

Lemma 4.1. *Let $\mathcal{F}$ be an r_n-adic filtration and $(\varepsilon_n)_{n \leq 0}$ an innovation of $\mathcal{F}$. If τ is a random $\mathcal{F}_n$-measurable tree automorphism then $\tau(\varepsilon_{n+1}, \ldots, \varepsilon_0)$ is a local innovation of $\mathcal{F}$.*

Proof. This is easily proved by recursion with the help of Lemma 2.1. □

Fig. 1 A labeled tree

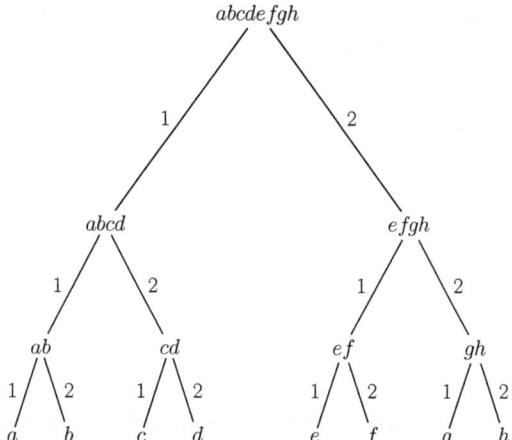

For any set A, there is a natural action of G_n on the set of ℓ_n-words on A. First introduce the lexicographic order on B_n, which is made visual by drawing a tree as in Fig. 1 and then by numbering from left to right the branches of this tree; for this order, the position $p(b)$ of $b = (b_{n+1}, \ldots, b_0) \in B_n$ is given by $p(b) = 1 + \sum_{k=n+1}^{0}(b_k - 1)\ell_k$. Now introduce the following notation.

Notation 4.2. *Given a word w of length ℓ_n, $n \leq -1$, and a branch $b \in B_n$, we denote by $t_n(w, b) = w(p(b))$ the letter of w whose index is the position of b for the lexicographic order on B_n.*

In other words, the i-th letter of w is $t_n(w, b)$ for the branch $b = p^{-1}(i)$. The application t_n is made visual on Fig. 1: the letter $t_n(w, b)$ of w is the label at the leaf of the branch b. Then the action of G_n on A^{ℓ_n} is defined as follows. For a tree automorphism $\tau \in G_n$ and a word $w \in A^{\ell_n}$ we define $\tau.w$ as the word satisfying $t_n(\tau.w, b) = t_n(w, \tau(b))$ for every branch $b \in B_n$, that is, the $p(b)$-th letter of $\tau.w$ is the $p(\tau(b))$-letter of w.

Now we introduce additional notations for later use.

Notation 4.3. *Given an underlying sequence $(r_n)_{n \leq 0}$, define as above the sequence $(\ell_n)_{n \leq 0}$. For any $n \leq 0$ and any word w of length $\ell_{n-1} = r_n \ell_n$ on an alphabet A, we denote by $\tilde{w}$ the word of length r_n on the alphabet A^{ℓ_n} obtained from w, that is, the j-th letter of $\tilde{w}$ is $\tilde{w}(j) = w_{(j-1)\ell_n + 1} \ldots w_{j\ell_n}$.*

Then note that the action of a tree automorphism in G_{n-1} on a word $w \in A^{\ell_{n-1}}$ consists of r_n tree automorphisms in G_n respectively acting on the subwords $\tilde{w}(1)$, ..., $\tilde{w}(r_n)$ together with a permutation of these subwords. This yields relation (1) below.

Notation 4.4. *For a fixed sequence $(r_n)_{n \leq 0}$ and given two words w, w' of length ℓ_n on a Polish metric space (A, θ), we define the distance δ_n^θ between w and w' by*

$$\delta_n^\theta(w, w') = \frac{1}{\ell_n} \sum_{i=1}^{\ell_n} \theta(w(i), w'(i))$$

and the associated distance between the orbits of w and w' under the action of G_n by

$$d_n^\theta(w, w') = \min_{\tau \in G_n} \delta_n^\theta(w, \tau.w').$$

When this causes no ambiguity we write d_n and δ_n instead of d_n^θ and δ_n^θ.

It can be easily checked that, with the notations above, the recurrence relation

$$d_{n-1}^\theta(w, w') = \min_{\sigma \in \mathcal{S}_{r_n}} \frac{1}{r_n} \sum_{i=1}^{r_n} d_n^\theta(\tilde{w}(i), \tilde{w}'(\sigma(i))) \tag{1}$$

holds for all words w, w' of length ℓ_{n-1} on the alphabet A; here $\mathcal{S}_{r_n}$ denotes the group of permutations of $\{1, \ldots, r_n\}$.

4.2 Split-Word Processes

Throughout this section, we consider a Polish metric space (A, θ). The set A is termed as *alphabet*.

Given a sequence $(r_n)_{n \leq 0}$ of integers $r_n \geq 2$, called the *splitting sequence*, we will soon define an r_n-adic split-word process on A. We firstly define the *length sequence* $(\ell_n)_{n \leq 0}$ by $\ell_n = \prod_{k=n+1}^{0} r_k$ for all $n \leq 0$. Next, according to Notation 4.3, every word w of length $\ell_{n-1} = r_n \ell_n$ on the alphabet A is naturally identified as a word $\tilde{w}$ of length r_n on the alphabet A^{ℓ_n}, and we define the *splitting map* $s_n : A^{\ell_{n-1}} \times \{1, 2, \ldots, r_n\} \longrightarrow A^{\ell_n}$ for each $n \leq 0$ by $s_n(w, j) = \tilde{w}(j)$. That is, to each word w of length ℓ_{n-1} on A and each integer $j \in [1, r_n]$, the splitting map s_n assigns the j-th letter of w treated as a r_n-word on A^{ℓ_n}.

With the help of the notations s_n and $\tilde{w}$ introduced above, we say that, with respect to some filtration $\mathcal{F}$, a process $(W_n, \varepsilon_n)_{n \leq 0}$ is a *split-word process on the alphabet A with splitting sequence* $(r_n)_{n \leq 0}$, or an r_n-*adic split-word process on A*, if for each $n \leq 0$:

- W_n is a random ℓ_n-word on A.
- ε_n is a random variable uniformly distributed on $\{1, 2, \ldots, r_n\}$ and is independent of $\mathcal{F}_{n-1}$, and $W_n = s_n(W_{n-1}, \varepsilon_n) = \tilde{W}_{n-1}(\varepsilon_n)$, that is, the word W_n is the ε_n-th letter of W_{n-1} treated as an r_n-word on A^{ℓ_n}.
- W_n and ε_n are $\mathcal{F}_n$-measurable.

An r_n-adic split-word process $(W_n, \varepsilon_n)_{n \leq 0}$ generates an r_n-adic filtration for which $(\varepsilon_n)_{n \leq 0}$ is an innovation. Note that one has $W_0 = t_n(W_n, \varepsilon_{n+1}, \ldots, \varepsilon_0)$ with Notation 4.2; and note also that the process $(W_n, \varepsilon_n)_{n \leq 0}$ is Markovian with respect to the filtration $\mathcal{F}$, hence the filtration it generates is immersed in $\mathcal{F}$.

Given a sequence $(\gamma_n)_{n \leq 0}$ of probability measures γ_n on A^{ℓ_n}, the existence of such a process $(W_n, \varepsilon_n)_{n \leq 0}$ with $W_n \sim \gamma_n$ occurs whenever each γ_n is the image under the splitting map s_n of the independent product of γ_{n-1} with the uniform probability on $\{1, 2, \ldots, r_n\}$. For example, γ_n can be taken as the projection on ℓ_n consecutive coordinates of a stationary probability measure on $A^{\mathbb{Z}}$.

Example 4.5 (The "ordinary" split-word processes). The ordinary split-word process with splitting sequence $(r_n)_{n \leq 0}$ is the process $(W_n, \varepsilon_n)_{n \leq 0}$ defined above when the probability γ_n on A^{ℓ_n} is the product probability of some probability measure μ on A. Standardness of the filtration $\mathcal{F}$ generated by an ordinary split-word process is known to be characterized by the following asymptotic condition on the splitting sequence:

$$\sum_{n=-\infty}^{0} \frac{\log r_n}{\ell_n} = +\infty. \qquad (2)$$

This result is presented by Laurent [9]. Maybe the most difficult part of it is Ceillier's proof [2] that $\mathcal{F}$ is standard under condition (2) when μ is the uniform probability on a finite alphabet A. In Sect. 5 we will deduce from this result that condition (2) defines the scale of Bernoulli automorphisms.

In the next two lemmas, we consider a Polish metric alphabet (A, θ). The equivalence between Vershik's intermediate level and second level criteria will be proved with the help of Lemma 4.7.

Lemma 4.6. *Let $\mathcal{F}$ be an r_n-adic filtration and $(\varepsilon_n)_{n \leq 0}$ an innovation of $\mathcal{F}$. For every $\mathcal{F}_0$-measurable random variable W_0 in A, there exists an r_n-adic $\mathcal{F}$-split-word process $(W_n, \varepsilon_n)_{n \leq 0}$ with final letter W_0.*

Proof. We firstly construct W_{-1}. Since W_0 is measurable with respect to $\mathcal{F}_{-1} \vee \sigma(\varepsilon_0)$ there exist a $\mathcal{F}_{-1}$-measurable random variable F_{-1} and a Borel function f such that $W_0 = f(F_{-1}, \varepsilon_0)$. Define W_{-1} as the r_0-word whose j-th letter is $f(F_{-1}, j)$. Now, assuming that $W_n, \ldots, W_0$ are constructed, we construct W_{n-1} in the same way: we write $W_n = g(F_{n-1}, \varepsilon_n)$ for some Borel function g and some $\mathcal{F}_{n-1}$-measurable random variable F_{n-1}, and we define W_{n-1} as the ℓ_{n-1}-word such that, with Notation 4.3, $g(F_{n-1}, j)$ is the j-th letter of $\tilde{W}_{n-1}(j)$ for every integer $j \in [1, r_n]$. □

It is easy to check that the split-word process $(W_n, \varepsilon_n)_{n \leq 0}$ in the lemma above is unique, but we will not need this fact.

Lemma 4.7. Let $(r_n)_{n \leq 0}$ be a splitting sequence and $(\ell_n)_{n \leq 0}$ the corresponding length sequence. There exist some maps $\iota_n \colon A^{\ell_n} \to A^{(n)}$, $n \leq 0$, satisfying the following properties:

- the map ι_n induces an isometry from the quotient space $\frac{A^{\ell_n}}{G_n}$ to $A^{(n)}$, when $\frac{A^{\ell_n}}{G_n}$ is equipped with the distance d_n^θ (Notation 4.4) and $A^{(n)}$ is equipped with the iterated Kantorovich distance θ_n (Sect. 3.1);
- for any r_n-adic split-word process $(W_n, \varepsilon_n)_{n \leq 0}$ on A, one has $\pi_n W_0 = \iota_n(W_n)$.

Proof. Firstly, it is not difficult to check that $\pi_n W_0 = \iota_n(W_n)$ where the maps $\iota_n \colon A^{\ell_n} \to A^{(n)}$ are recursively defined as follows. Given an integer $k \geq 2$ and a Polish space (E, ρ), denote by $D_k \colon E^k \to E'$ the map defined by $D_k(x_1, \ldots, x_k) = \frac{1}{k}(\delta_{x_1} + \cdots + \delta_{x_k})$. Then define $\iota_0(w) = w$ and, using Notation 4.3, define $\iota_{n-1}(w) = D_{r_n}(\iota_n(\tilde{w}(1)), \ldots, \iota_n(\tilde{w}(r_n)))$. From this construction it is easy to see that the map ι_n is invariant under the action of G_n, and then defines a map from $\frac{A^{\ell_n}}{G_n}$ to $A^{(n)}$. By Lemma 3.1, the Kantorovich distance between $D_k(x)$ and $D_k(x')$ for any $x, x' \in E^k$ is given by $\rho'(D_k(x), D_k(x')) = \min_{\sigma \in S_k} \frac{1}{k} \sum_{i=1}^{k} \rho(x_i, x'_{\sigma(i)})$. Using this fact and the recurrence relation (1) on d_n^θ, it is easy to check by recursion that $d_n^\theta(w, w')$ is the Kantorovich distance between $\iota_n(w)$ and $\iota_n(w')$ for all words $w, w' \in A^{\ell_n}$. □

4.3 Vershik's Intermediate Level Criterion

The statement of Vershik's intermediate level criterion, as well as its equivalence with Vershik's second level criterion, are based on the following lemma. Recall that the pseudo-distance d_n^θ is defined in Notation 4.4.

Lemma 4.8. With respect to an underlying (r_n)-adic filtration $\mathcal{F}$, let $(W_n, \varepsilon_n)_{n \leq 0}$ be a split-word process on a Polish metric alphabet (A, θ). Then $\operatorname{disp} \pi_n W_0 = \tilde{\mathbb{E}}[d_n^\theta(W_n^*, W_n^{**})]$ where (W_n^*) and (W_n^{**}) are independent copies of the process (W_n) on some probability space $(\tilde{\Omega}, \tilde{\mathcal{A}}, \tilde{\mathbb{P}})$. In other words, $\operatorname{disp} \pi_n W_0 = \operatorname{disp}(G_n \cdot W_n)$ denoting by $G_n \cdot W_n$ the orbit of W_n under the action of G_n (Sect. 4.1).

Proof. This straightforwardly results from Lemma 4.7. □

Then, given any Polish metric space (A, θ), we say that, with respect to an r_n-adic filtration $\mathcal{F}$, a random variable $W_0 \in L^1(\mathcal{F}_0; A)$ satisfies *Vershik's intermediate level criterion* if $\tilde{\mathbb{E}}[d_n^\theta(W_n^*, W_n^{**})]$ goes to 0 as n goes to $-\infty$ with the notations of the lemma above. We shortly say that W_0 satisfies the *intermediate Vershik property*. This definition makes sense in view of the lemma above, which shows that any property on the sequence of expectations $\tilde{\mathbb{E}}[d_n^\theta(W_n^*, W_n^{**})]$ only depends on W_0 (and of the underlying filtration), and in view of Lemma 4.6, which guarantees the existence of a split-word process with final letter W_0.

We also extend the definition of the intermediate Vershik property to σ-fields $\mathcal{E}_0 \subset \mathcal{F}_0$ and to the whole filtration $\mathcal{F}$ as for the second Vershik property. We then immediately get the following theorem from Lemma 4.8.

Theorem 4.9. *With respect to some r_n-adic filtration, the intermediate Vershik property and the second Vershik property are equivalent (for a random variable, a σ-field or the whole filtration).*

Corollary 4.10. *The analogue of Proposition 3.2 for the intermediate Vershik property holds true.*

The following lemma will be used in Sect. 5.

Lemma 4.11. *Let $\mathcal{F}$ be an ambient r_n-adic filtration and (A, θ) be a Polish metric space. A random variable $W_0 \in L^1(\mathcal{F}_0; A)$ satisfies the intermediate Vershik property if and only if there exists a sequence $(w^{(n)})_{n \leq 0}$ consisting of words $w^{(n)}$ of length ℓ_n such that $\mathbb{E}[d_n^\theta(W_n, w^{(n)})]$ goes to 0 for any split-word process $(W_n, \varepsilon_n)_{n \leq 0}$ with final letter W_0.*

Proof. With the notations of Lemma 4.8, this obviously results from the inequalities

$$\inf_w \mathbb{E}[d_n^\theta(W_n, w)] \leq \tilde{\mathbb{E}}[d_n^\theta(W_n^*, W_n^{**})] \leq 2 \inf_w \mathbb{E}[d_n^\theta(W_n, w)],$$

which are easy to prove. □

5 The Scale of an Automorphism

Vershik defined the scale of an automorphism in [13]. We will see that his definition can be rephrased in terms of Vershik's intermediate level criterion.

Let T be an invertible measure-preserving transformation (in other words, an automorphism) of a Lebesgue space (E, ν). Vershik's definition of the scale of T is the following one. With the same terminology used to describe the split-word processes, consider a splitting sequence $(r_n)_{n \leq 0}$ and the corresponding length sequence $(\ell_n)_{n \leq 0}$. Given $x \in E$, one defines a word $w_n(x)$ of length ℓ_n on E for each $n \leq 0$ by $w_n(x) = (x, Tx, \ldots, T^{\ell_n - 1}x)$. Thus w_n is a random word on E. The *scale* of T is the set of splitting sequences $(r_n)_{n \leq 0}$ (consisting of integers $r_n \geq 2$) satisfying the following property: for every $f \in L^1(\nu)$ there exists a sequence $(c^{(n)})_{n \leq 0}$ consisting of vectors $c^{(n)} \in \mathbb{R}^{\ell_n}$ such that the sequence of random variables $d_n(f(w_n), c^{(n)})$ goes to 0 in probability, where d_n is the pseudo-metric d_n^θ defined in Notation 4.4 with $(A, \theta) = (\mathbb{R}, |\cdot|)$. The scale of T is denoted by $\mathfrak{S}(T)$.

Denote by γ_n the law of the random word w_n on E, and consider the (unique up to isomorphism) filtration $\mathcal{F}$ of the (unique in law) r_n-adic split-word process $(W_n, \varepsilon_n)_{n \leq 0}$ whose law is given by $W_n \sim \gamma_n$ (the γ_n obviously satisfy the consistency condition required for this process to exist). Then, in view of Lemma 4.11 and

Corollary 4.10, the sequence $(r_n)_{n \leq 0}$ belongs to the scale of T if and only if Vershik's intermediate level criterion with respect to $\mathcal{F}$ holds for the final letter W_0.

It is clear, owing to Corollary 4.10, that this property defines an invariant of T: if $S = \phi \circ T \circ \phi^{-1}$ is an invertible measure-preserving transformation conjugate to T then the property is equivalent to Vershik's intermediate level criterion holding for the final letter $\phi(W_0)$ of the split-word process $\big(\phi(W_n), \varepsilon_n\big)_{n \leq 0}$. Some other basic properties of the scale given by Vershik in [13] can be derived by straightforward applications of basic properties of Vershik's intermediate level criterion, such as Corollary 4.10.

Remark 5.1. Actually, as pointed out by Vershik in [13], his definition above is useful only for completely ergodic transformations T (i.e. all powers of T are ergodic), but it may be extended to arbitrary transformations T by replacing each vector $c^{(n)}$ by a vector-valued function $c^{(n)}(x)$ constant on the ergodic components of T^{ℓ_n}. We will not investigate this extension, except for our concluding remarks at the end of this section; the reader has to be aware that some results in [13] are not valid with the definition given above when T is not completely ergodic.

In the sequel we denote by $(W_n, \varepsilon_n)_{n \leq 0}$ the (unique in law) split-word process associated to T for a given sequence $(r_n)_{n \leq 0}$, and by $\mathcal{F}$ the r_n-adic filtration it generates (unique up to isomorphism). In particular, W_0 is an E-valued random variable with law ν.

The following proposition shows in particular that the filtration of the ordinary split-word process (Example 4.5) is the filtration $\mathcal{F}$ when T is the Bernoulli shift on $(A, \mu)^{\mathbb{Z}}$. A *generator* of an automorphism T is a measurable function f from (E, ν) to a Lebesgue space (A, μ) such that $\sigma(X) = \vee_{i=-\infty}^{\infty} \sigma\big(f(T^i X)\big)$ for some ($\iff$ for every) random variable X distributed on E according to ν. For a (possibly non Bernoulli) shift on $A^{\mathbb{Z}}$, a natural generator is the function $f : A^{\mathbb{Z}} \to A$ which sends a sequence in $A^{\mathbb{Z}}$ to its coordinate at index 0.

Proposition 5.2. *If f is a generator of T, then $\mathcal{F}$ is the filtration generated by the split-word process $\big(f(W_n), \varepsilon_n\big)_{n \leq 0}$.*

Proof. It suffices to show that the first letter $W_n(1)$ of W_n is measurable with respect to $\vee_{m=-\infty}^{n} \sigma\big(f(W_n), \varepsilon_n\big)$ for every $n \leq 0$. For notational convenience, we only treat the case $n = 0$, and it will be clear how to similarly treat the case of any $n \leq 0$. One has $f(W_n) = \big(f(T^{P_n} W_0), \ldots, f(T^{Q_n} W_0)\big)$ where $P_n \leq 0$ and $Q_n \geq 0$ are random integers measurable with respect to $\sigma(\varepsilon_{n+1}, \ldots, \varepsilon_0)$, and satisfy $P_n \to -\infty$ and $Q_n \to +\infty$ since they obviously satisfy $P_n \leq -\sum_{i=n+1}^{0} \ell_i \mathbb{1}_{\varepsilon_i \neq 1}$ and $Q_n \geq \sum_{i=n+1}^{0} \ell_i \mathbb{1}_{\varepsilon_i \neq r_i}$, thereby showing that W_0 is measurable with respect to $\vee_{m=-\infty}^{0} \sigma\big(f(W_n), \varepsilon_n\big)$. □

A famous theorem by Rokhlin says that any aperiodic transformation T has a countable generator f, that is, f takes its values in a countable space. Recall that T is said to be *aperiodic* when $\mathbb{P}(W_0 = T^i W_0 \text{ for some } i \geq 1) = 0$. In particular, if T is ergodic, it is aperiodic.

In the sequel we denote by $D_n = W_n(1)$ the first letter of W_n for every $n \leq 0$. Obviously, $\mathcal{F}$ is also generated by the process $(D_n, \varepsilon_n)_{n \leq 0}$ and $D_n = (T^{\ell_n})^{\varepsilon_n - 1}(D_{n-1})$. We state the proposition below only by way of remark.

Proposition 5.3. *If T is aperiodic then $\mathcal{F}$ is generated by the process $(W_n)_{n \leq 0}$.*

Proof. It suffices to show that $\sigma(\varepsilon_n) \subset \sigma(D_{n-1}, D_n)$ when assuming aperiodicity of T. Let $K_n \leq J_n := (\varepsilon_n - 1)\ell_n$ be the smallest integer such that $D_n = T^{K_n}(D_{n-1})$, hence the equality $T^{K_n}(D_{n-1}) = T^{J_n}(D_{n-1})$ almost surely holds. Therefore $\mathbb{P}(K_n \neq J_n) \leq \sum_{k \neq j} \mathbb{P}\left(T^k(D_{n-1}) = T^j(D_{n-1})\right) = 0$. □

Now, for an aperiodic T, we will prove that $(r_n)_{n \leq 0} \in \mathfrak{S}(T)$ means that the whole filtration $\mathcal{F}$ is Vershikian (Theorem 5.5).

Lemma 5.4. *If T is aperiodic then $\sigma(\pi_n D_0) = \sigma(D_n)$.*

Proof. It suffices to show that $\mathcal{L}(D_n \mid \mathcal{F}_{n-1})$ generates the same σ-field as D_{n-1} for every $n \leq 0$ since the σ-field generated by the conditional law of a random variable X given any σ-field depends on X only through the σ-field $\sigma(X)$. To do so, put $S = T^{\ell_n}$. Conditionally on $\mathcal{F}_{n-1}$, the random variable D_n is uniformly chosen among $D_{n-1}, S(D_{n-1}), \ldots, S^{r_n - 1}(D_{n-1})$. Hence, the conditional law $\mathcal{L}(D_n \mid \mathcal{F}_{n-1})$ determines the set $\{D_{n-1}, S(D_{n-1}), \ldots, S^{r_n - 1}(D_{n-1})\}$. Let σ be a random permutation of $I := \{0, 1, \ldots, r_n - 1\}$ such that $S^{\sigma(j)}(D_{n-1}) = S^j\left(S^{\sigma(0)}(D_{n-1})\right)$ for every $j \in I$. We will show that σ almost surely equals the identity map of I; the lemma will obviously follow. Let $K = 0$ if σ is the identity permutation and K be a strictly positive integer such that $\{S^K(D_{n-1}) = D_{n-1}\}$ otherwise. Then $\mathbb{P}(K \neq 0) \leq \sum_{k > 0} \mathbb{P}\left(S^k(D_{n-1}) = D_{n-1}\right) = 0$. □

Theorem 5.5. *For an aperiodic T, the sequence $(r_n)_{n \leq 0}$ belongs to the scale of T if and only if the r_n-adic filtration $\mathcal{F}$ is Vershikian.*

Proof. This stems from Lemmas 5.4 and 3.4. □

This theorem along with Proposition 5.2 yield our last claim in Example 4.5: the scale of a Bernoulli shift coincides with the set of all sequences $(r_n)_{n \leq 0}$ for which the corresponding ordinary split-word process generates a standard filtration, and therefore it consists in all sequences $(r_n)_{n \leq 0}$ satisfying condition (2) given in Example 4.5.

Now we will give a proof of the following result based on the theorem on lacunary isomorphism (Theorem 3.3). Vershik proved it in [13] by a direct construction.

Proposition 5.6. *The scale of a completely ergodic invertible measure-preserving transformation is not empty.*

Our proof is an application of the theorem on lacunary isomorphism based on Corollary 5.8 which is derived from Proposition 5.7 below. The following notations are used in this proposition. We consider a splitting sequence $(r_n)_{n \leq 0}$ and for each

$n \leq 0$, we put $\mathcal{J}_n = D_n^{-1}(\mathfrak{I}_n)$ where $\mathfrak{I}_n$ is the σ-field of T^{ℓ_n}-invariant events. A random variable is $\mathcal{J}_n$-measurable if and only if it is of the form $f(D_n)$ where f is a T^{ℓ_n}-invariant Borel function. Therefore $(\mathcal{J}_n)_{n \leq 0}$ is a decreasing sequence of σ-fields. Note also that for any $\mathcal{J}_n$-measurable random variable J_n and any integer i which is multiple of ℓ_n, the random pair (J_n, D_n) has the same distribution as $(J_n, T^i D_n)$. Recall that the ergodic theorem says that

$$\frac{1}{k} \sum_{i=0}^{k-1} f(T^i D_n) \xrightarrow{L^1} \mathbb{E}[f(D_n) \mid \mathcal{J}_n] \quad \text{as } k \to +\infty$$

for all $f \in L^1(\nu)$.

Proposition 5.7. *Let T be an invertible measure-preserving transformation and $\mathcal{F}$ the r_n-adic filtration associated to T as above. Then $\mathcal{F}_{-\infty} = \lim \nearrow \mathcal{J}_n$.*

Proof. Recall that we denote by $D_n = W_n(1)$ the first letter of W_n. It is clear that $\{D_n \in A\} \in \mathcal{F}_{-\infty}$ for every Borel set A which is invariant by T^{ℓ_n}, thereby yielding the inclusion $\lim \nearrow \mathcal{J}_n \subset \mathcal{F}_{-\infty}$.

Conversely, putting $\mathcal{J}_{-\infty} = \lim \nearrow \mathcal{J}_n$, it suffices to show that $\mathbb{E}[Z \mid \mathcal{F}_n]$ tends in L^1 to a $\mathcal{J}_{-\infty}$-measurable random variable for each $Z \in L^1(\mathcal{F}_0)$. It is not difficult to check that this property holds whenever it holds for all random variables $Z = f(D_{n_0})$ where $f \in L^1(\nu)$ and $n_0 \geq 0$. Then, given $f \in L^1(\nu)$, we will show that $\mathbb{E}[f(D_{n_0}) \mid \mathcal{F}_n] \xrightarrow{L^1} \mathbb{E}[f(D_{n_0}) \mid \mathcal{J}_{n_0}]$ for every $n_0 \leq 0$. Conditionally on $\mathcal{F}_n$ (with $n \leq n_0$), D_{n_0} is uniformly chosen among $D_n, T^{\ell_{n_0}}(D_n), \ldots, T^{(\ell_n/\ell_{n_0}-1)\ell_{n_0}}(D_n)$, therefore

$$\mathbb{E}[f(D_{n_0}) \mid \mathcal{F}_n] = \frac{1}{\ell_n/\ell_{n_0}} \sum_{j=0}^{\ell_n/\ell_{n_0}-1} f\big((T^{\ell_{n_0}})^j (D_n)\big)$$

$$= \frac{1}{\ell_n/\ell_{n_0}} \sum_{j=0}^{\ell_n/\ell_{n_0}-1} f\big((T^{\ell_{n_0}})^j (T^X D_{n_0})\big)$$

where X is a random integer which is multiple of ℓ_{n_0} and is independent of D_{n_0}, and then the result follows from the ergodic theorem. □

Corollary 5.8. *The filtration $\mathcal{F}$ is Kolmogorovian if and only if T^{ℓ_n} is ergodic for all $n \leq 0$.*

Proposition 5.6 is then straightforwardly shown by applying the theorem on lacunary isomorphism (Theorem 3.3), by noting that any filtration $(\mathcal{F}_{\phi(n)})_{n \leq 0}$ extracted from $\mathcal{F}$ with $\phi(0) = 0$ is the filtration of the split-word process associated to T with another splitting sequence.

Now we prove the proposition below as another illustration of our definition of the scale. We do not know whether the converse inclusion holds.

Proposition 5.9. *Let S and T be invertible measure-preserving transformations. Then $\mathfrak{S}(S \times T) \subset \mathfrak{S}(S) \cap \mathfrak{S}(T)$.*

Proof. Let $(W_n, \varepsilon_n)_{n \leq 0}$ be the r_n-adic split-word process associated to $S \times T$ and $\mathcal{F}$ the filtration it generates. For each $n \leq 0$, one has $W_n = (Y_n, Z_n)$ where $(Y_n, \varepsilon_n)_{n \leq 0}$ and $(Z_n, \varepsilon_n)_{n \leq 0}$ are the r_n-adic split-word processes associated to S and T respectively, each of them generating a filtration immersed in $\mathcal{F}$. Therefore the result follows from the fact that $\sigma(W_0)$ is Vershikian if $(r_n)_{n \leq 0} \in \mathfrak{S}(S \times T)$ (Proposition 3.2) and from the hereditariness of the Vershik property for immersion (Lemma 3.5). □

We close this section by translating Vershik's general definition of the scale (Remark 5.1) into a probabilistic statement generalizing the preceding definition. Using the notations of the first definition given at the beginning of this section, Vershik's general definition says that (r_n) belongs to $\mathfrak{S}(T)$ if $d_n(W_n) \to 0$ in probability where $d_n(W_n) = \inf_w \mathbb{E}[d_n(W_n, w) \mid \mathcal{J}_n]$ (where d_n is the pseudo-metric introduced in Notation 4.4 and the $\mathcal{J}_n$ are the σ-fields introduced before Proposition 5.7). Thus this definition obviously coincides with the first one when the T^{ℓ_n} are ergodic. In addition one also has $d_n(W_n) = \inf_w \mathbb{E}[d_n(W_n, w) \mid \mathcal{F}_{-\infty}]$, owing to the following lemma.

Lemma 5.10. *For every $n \leq 0$, the random variable D_n is conditionally independent of $\mathcal{F}_{-\infty}$ given $\mathcal{J}_n$.*

Proof. Recall that $\mathcal{F}_{-\infty} = \lim \nearrow \mathcal{J}_n$ (Proposition 5.7). The statement of the lemma amounts to saying that $\mathbb{E}[g(D_n) \mid \mathcal{J}_n] = \mathbb{E}[g(D_n) \mid \mathcal{F}_{-\infty}]$ for $g \in L^1(\nu)$, which is equivalent to $\mathbb{E}[g(D_n) \mid \mathcal{J}_n] = \mathbb{E}[g(D_n) \mid \mathcal{J}_m]$ for every $m \leq n$. To prove this equality, we take a random variable $J_m \in L^\infty(\mathcal{J}_m)$, hence $J_m = f(D_m)$ where the function $f \in L^\infty(\nu)$ is T^{ℓ_m}-invariant. Since $D_m = (T^{\ell_n})^X D_n$ where X is a random integer independent of D_0 and uniformly distributed on $\{0, -1, \ldots, -\ell_m/\ell_n + 1\}$,

$$\mathbb{E}[J_m g(D_n)] = \frac{1}{\ell_m/\ell_n} \sum_{i=-\ell_m/\ell_n+1}^{0} \mathbb{E}\left[f((T^{\ell_n})^i D_n) g(D_n) \right] = \mathbb{E}[J_n g(D_n)]$$

where $J_n = \frac{1}{\ell_m/\ell_n} \sum_{i=-\ell_m/\ell_n+1}^{0} f((T^{\ell_n})^i D_n)$. But J_n is measurable with respect to $\mathcal{J}_n$ since

$$\sum_{i=-\ell_m/\ell_n+1}^{0} f((T^{\ell_n})^i T^{\ell_n} D_n) = f(T^{\ell_n} D_n) + \sum_{i=-\ell_m/\ell_n+2}^{0} f((T^{\ell_n})^i D_n)$$

and $f(T^{\ell_n} D_n) = f(T^{-\ell_m + \ell_n} D_n)$. Therefore we get

$$\mathbb{E}[J_m g(D_n)] = \mathbb{E}[J_n \mathbb{E}[g(D_n) \mid \mathcal{J}_n]]$$

but we similarly prove that $\mathbb{E}[J_m \mathbb{E}[g(D_n) \mid \mathcal{J}_n]] = \mathbb{E}[J_n \mathbb{E}[g(D_n) \mid \mathcal{J}_n]]$. □

Thus, the general definition of the statement "$r_n \in \mathfrak{S}(T)$" becomes $\inf_w \mathbb{E}[d_n(W_n, w) \mid \mathcal{F}_{-\infty}] \to 0$, and it sounds like the intermediate Vershik property of W_0 conditionally on $\mathcal{F}_{-\infty}$. We expect that criteria for filtrations (such as Vershik's criteria and the I-cosiness criterion) can be more generally stated conditionally on $\mathcal{F}_{-\infty}$ in such a way that most results (such as the theorem on lacunary isomorphism) remain to be true. But, currently, we do not feel the motivation to develop this generalization.

Acknowledgements I thank Michel Émery and Anatoly Vershik for the interest they have expressed in this work, and I also thank Michel Émery for helpful comments on a previous version of the paper.

References

1. F. Bolley, Separability and completeness for the Wasserstein distance. Séminaire de Probabilités XLI. Springer Lect. Notes Math. **1934**, 371–377 (2008)
2. G. Ceillier, The filtration of the split-words process. Probab. Theory Relat. Fields **153**, 269–292 (2012)
3. Y. Coudène, Une version mesurable du théorème de Stone-Weierstrass. Gazette des mathématiciens **91**, 10–17 (2002)
4. M. Émery, W. Schachermayer, On Vershik's standardness criterion and Tsirelson's notion of cosiness. Séminaire de Probabilités XXXV, Springer Lect. Notes Math. **1755**, 265–305 (2001)
5. D. Heicklen, C. Hoffman, $[T, T^{-1}]$ is not standard. Ergod. Theory Dyn. Syst. **18**, 875–878 (1998)
6. C. Hoffman, A zero entropy T such that the $[T, \mathrm{Id}]$ endomorphism is nonstandard. Proc. A.M.S. **128**(1), 183–188 (1999)
7. C. Hoffman, D. Rudolph, If the $[T, \mathrm{Id}]$ automorphism is Bernoulli then the $[T, \mathrm{Id}]$ endomorphism is standard. Proc. A.M.S. **128**(1), 183–188 (1999)
8. S. Laurent, Filtrations à temps discret négatif. PhD Thesis, Université de Strasbourg, Strasbourg, 2004
9. S. Laurent, On standardness and I-cosiness. Séminaire de Probabilités XLIII. Springer Lect. Notes Math. **2006**, 127–186 (2010)
10. S. Laurent, On Vershikian and I-cosy random variables and filtrations. Teoriya Veroyatnostei i ee Primeneniya **55**, 104–132 (2010). Also published in: Theory Probab. Appl. **55**, 54–76 (2011)
11. A.M. Vershik, Decreasing sequences of measurable partitions, and their applications. Dokl. Akad. Nauk SSSR **193**, 748–751 (1970). English translation: Soviet Math. Dokl. **11**, 1007–1011 (1970)
12. A.M. Vershik, Approximation in measure theory (in Russian). PhD Dissertation, Leningrad University, 1973. Expanded and updated version: [14]
13. A.M. Vershik, Four definitions of the scale of an automorphism. Funktsional'nyi Analiz i Ego Prilozheniya **7**(3), 1–17 (1973). English translation: Funct. Anal. Appl. **7**(3), 169–181 (1973)
14. A.M. Vershik, The theory of decreasing sequences of measurable partitions (in Russian). Algebra i Analiz **6**(4), 1–68 (1994). English translation: St. Petersburg Math. J. **6**(4), 705–761 (1995)
15. A.M. Vershik, V. A. Rokhlin and the modern theory of measurable partitions. Am. Math. Soc. Transl. **202**(2), 11–20 (2001)

Filtrations Indexed by Ordinals; Application to a Conjecture of S. Laurent

Claude Dellacherie and Michel Émery*

> [...] attention, car, à partir de ce moment, nous voilà aux prises avec le cardinal.
>
> A. DUMAS, *Les trois mousquetaires.*

Abstract The following fact has been conjectured by Stéphane Laurent (Conjecture 3.18, page 160 of Séminaire de Probabilités XLIII): *Let $\mathscr{F} = (\mathscr{F}_t)$ and $\mathscr{G} = (\mathscr{G}_t)$ be two filtrations on some probability space, and suppose that every $\mathscr{F}$-martingale is also a $\mathscr{G}$-martingale. For $s < t$, if $\mathscr{G}_t$ is generated by $\mathscr{G}_s$ and by countably[2] many events, then $\mathscr{F}_t$ is generated by $\mathscr{F}_s$ and by countably many events.* In this statement, "and by countably many events" can equivalently be replaced with "and by some separable σ-algebra", or with "and by some random variable valued in some Polish space".

We propose a rather intuitive proof of this conjecture, based on the following necessary and sufficient condition: *Given a probability space, let $\mathscr{D}$ be a σ-algebra*

---

*The support of the ANR programme ProbaGeo is gratefully acknowledged.

[2]We use the terms "countable" and "countably" in their broad sense: "countable" will mean "finite or countably infinite".

C. Dellacherie
Laboratoire de Mathématiques Raphaël Salem, CNRS et Université de Rouen, Avenue de l'Université, 76801 Saint-Étienne-du-Rouvray, France
e-mail: claude.dellacherie@univ-rouen.fr

M. Émery (✉)
IRMA, CNRS et Université Unique de Strasbourg, 7 rue René Descartes, 67084 Strasbourg Cedex, France
e-mail: michel.emery@math.unistra.fr

of measurable sets and $\mathscr{C}$ a sub-σ-algebra of $\mathscr{D}$. Then $\mathscr{D}$ is generated by $\mathscr{C}$ and by countably many events if and only if there exists no strictly increasing filtration $\mathscr{F} = (\mathscr{F}_\alpha)_{\alpha < \aleph_1}$, indexed by the set $[0, \aleph_1[$ of all countable ordinals, and satisfying $\mathscr{C} \subseteq \mathscr{F}_\alpha \subseteq \mathscr{D}$ for each α.

Another question then arises: can the martingale hypothesis on $\mathscr{F}$ and $\mathscr{G}$ be replaced by a more general condition involving the null events but not the values of the probability? We propose such a weaker hypothesis, but we are no longer able to derive the conclusion from it; so the question is left open.

1 Introduction

To solve Stéphane Laurent's conjecture (see the abstract), we measure the relative size of a σ-algebra $\mathscr{D}$ with respect to a sub-σ-algebra $\mathscr{C}$, by successively adding to $\mathscr{C}$ new $\mathscr{D}$-events in such a way that the σ-algebra generated by $\mathscr{C}$ and these events increases at each step; when the relative size is infinite, it is characterized by the transfinite number of steps. This idea is made rigorous in Corollary 2, which says in particular that $\mathscr{D}$ is generated by $\mathscr{C}$ and by some r.v. iff no strictly increasing filtration indexed by the set of all countable ordinals can be inserted between $\mathscr{C}$ and $\mathscr{D}$.

When $\mathscr{C}$ is trivial, this characterizes the absolute size of $\mathscr{D}$, and we find that $\mathscr{D}$ is essentially separable iff it contains no strictly increasing filtration indexed by all countable ordinals (Corollary 3).

Laurent's conjecture can be stated as a question involving four σ-algebras; using Corollary 2, this question is solved by Proposition 1, whose translation into the language of filtrations (Laurent's formulation) is Corollary 5.

In Proposition 1, some conditional independence is assumed, but the conclusion is "qualitative": it does not explicitly feature the probability, only the null events. Hence, one is naturally led to ask whether Proposition 1 remains true when its conditional independence hypothesis is replaced by some kind of "qualitative conditional independence", which should imply usual conditional independence and be stable under replacement of $\mathbb{P}$ by an equivalent probability. We propose a possible definition of qualitative conditional independence, but we are not able to assert whether or not the conclusion of Proposition 1 is still implied by this weaker hypothesis; so we leave the question unanswered.

2 Notation and Elementary Facts from Set Theory

Recall that two sets have the same *cardinal* (or cardinality) if they are related by some bijection; we shall denote by $|S|$ the cardinality of a set S. And two well-ordered sets define the same *ordinal* if they are related by some order-preserving bijection. Any ordinal α can (and will) be identified with the well-ordered set $[0, \alpha[$ of all ordinals β such that $\beta < \alpha$; any cardinal c can (and will) be identified with the smallest ordinal having cardinality c. There exists a (unique) order-preserving, one-to-one correspondence $\alpha \mapsto \aleph_\alpha$ between ordinals and infinite cardinals. The smallest infinite cardinal is $\omega = \aleph_0$, which corresponds to the well-ordered set $\mathbb{N}$ of natural integers; the next one is $\aleph_1$, the first uncountable ordinal, so $[0, \aleph_1[$ is the well-ordered set of all countable ordinals.

Every ordinal α has a successor, namely $\alpha + 1$; every cardinal c has *two* successors: $c + 1$, the smallest ordinal $> c$, is its successor as an ordinal; while c', the smallest cardinal $> c$, is its successor as a cardinal. If c is finite, $c' = c + 1$ (ordinary addition of natural numbers). But if c is infinite, $c = \aleph_\alpha$ for some ordinal α, and $c' = \aleph_{\alpha+1}$; in that case, $c + 1$ has the same cardinality as c, so $c' > c + 1$.

We shall use only one non-elementary fact from the theory of cardinals:

Theorem 1. *If S is any infinite set, $|S \times S| = |S|$.*

A proof can be found in any book expounding cardinal arithmetic; see for instance Theorem 3.5 in [4].

Corollary 1. *a) If c is an infinite cardinal, and if $(S_i)_{i \in I}$ is a family of sets such that $|I| \leq c$ and $|S_i| \leq c$ for each i, then $\left|\bigcup_{i \in I} S_i\right| \leq c$.*
b) If S is an infinite set, the set of all finite subsets of S has the same cardinality as S.

Proof. a) Choose a set S with cardinality c (for instance, $S = c$). If $|S_i| = c$ for each i and if the S_i are pairwise disjoint, the union $\bigcup_{i \in I} S_i$ is equipotent to the product $I \times S$, for this product is the union of its fibers $\{i\} \times S$. Hence $\left|\bigcup_{i \in I} S_i\right| = |I \times S| \leq c^2 = c$ (by Theorem 1). This majoration subsists a fortiori when $|S_i| \leq c$ or when the S_i are not disjoint.
b) Put $c = |S|$; by Theorem 1, $c^2 = c$, and, by iteration, $c^n = c$ for each integer $n \geq 1$. Now, for $n \geq 1$, the set of all subsets of S with n elements has cardinality c_n satisfying $c \leq c_n \leq c^n = c$; so $c_n = c$. Last, the set of all finite subsets of S is a countable union of sets with cardinal c, plus one singleton; by a), its cardinal is c. □

Theorem 1 and Corollary 1 will be used to prove a generalized form of Laurent's conjecture. The reader only interested in the conjecture as stated by Laurent, just needs the familiar, countable case: a product of two countable sets, a union of countably many countable sets, or the set of all finite parts of a countable set, are always countable. (S)he simply has to replace the arbitrary infinite cardinal c featuring throughout the paper by the countable cardinal ω, and its successor c' by $\aleph_1$.

3 Extensions of σ-Algebras

From now on, an ambient probability space $(\Omega, \mathscr{A}, \mathbb{P})$ is fixed. We assume that it is complete (every subset of any negligible event is in $\mathscr{A}$), and by a σ-algebra, we always mean *an $(\mathscr{A}, \mathbb{P})$-complete sub-$\sigma$-algebra of $\mathscr{A}$ (i.e., a sub-σ-algebra containing all events with probability 0 or 1).*

On the contrary, by a *Boolean algebra*, we mean *any* Boolean sub-algebra of $\mathscr{A}$, not necessarily completed.

If $\mathscr{C}$ is a σ-algebra, the operator of conditional expectation w.r.t. $\mathscr{C}$ will be denoted by $\mathbb{E}^{\mathscr{C}}$; and if A is an event, $\mathbb{P}^{\mathscr{C}}(A)$ will stand for $\mathbb{E}^{\mathscr{C}} 1_A$.

If $\mathscr{C}$ and $\mathscr{D}$ are two σ-algebras such that $\mathscr{D} \supseteq \mathscr{C}$, we say that $\mathscr{D}$ is an *extension* of $\mathscr{C}$. The extension is said to be *finite* if $\mathscr{D}$ is generated by $\mathscr{C}$ and by finitely many events; else it is *infinite*.

When the extension is finite, one can define its *multiplicity;* we borrow this notion from M. T. Barlow, J. Pitman and M. Yor, pp. 284–285 of [2].

Definition 1. If $\mathscr{D}$ is a finite extension of $\mathscr{C}$, the *multiplicity* of the extension is the smallest integer m such that $\mathscr{D}$ is generated by $\mathscr{C}$ and by some m-partition $\{D_1, \ldots, D_m\}$ of Ω. We shall denote it by $\mathrm{mult}(\mathscr{D} : \mathscr{C})$.

Clearly, $\mathrm{mult}(\mathscr{D} : \mathscr{C}) \geq 1$, with equality if and only if $\mathscr{D} = \mathscr{C}$.

Remark that in Definition 1, as the D_i form a partition, $\mathscr{D}$ is also generated by $\mathscr{C}$ and $D_1, \ldots, D_{m-1}$; so m is not [3] *the smallest number of events needed to generate $\mathscr{D}$ from $\mathscr{C}$*. But multiplicity is nonetheless a natural measure of the size of the extension; M. T. Barlow, J. Pitman and M. Yor give in [2] four equivalent definitions of m, and more can be found in Proposition 12 of [1] (where a slightly different notion of multiplicity is used: it is no longer a constant, but a $\mathscr{C}$-measurable, integer-valued r.v.). Here are two of these characterizations:

Lemma 1. *Given an extension $\mathscr{D}$ of $\mathscr{C}$ and an integer $m \geq 1$, the following three statements are equivalent:*

(i) $\mathrm{mult}(\mathscr{D} : \mathscr{C}) \leq m$;

(ii) for each $\mathscr{D}$-measurable r.v. X, one can find m $\mathscr{C}$-measurable r.v. $Y_1, \ldots, Y_m$ such that

$$\prod_{i=1}^{m} (X - Y_i) = 0 ;$$

(iii) for any $m+1$ pairwise disjoint events $D_1, \ldots, D_{m+1}$ in $\mathscr{D}$, one has

$$\prod_{i=1}^{m+1} \mathbb{P}^{\mathscr{C}}(D_i) = 0 .$$

---

[3] Far from it! It is an easy exercise to verify that the smallest number is $\lceil \log_2 m \rceil$; but we just need to know that it is strictly smaller than m.

As mentioned on p. 285 of [2], "elementary but tedious arguments" show this equivalence. Such arguments are explicitly written in the proof of Proposition 12 in [1]; we refer to them for a proof of Lemma 1. The intuitive meaning of property (ii) is obvious: "when $\mathscr{C}$ is known", any $\mathscr{D}$-measurable r.v. may assume at most m different values.

How does this notion of multiplicity extend to infinite extensions? In general, partitions can no longer be used (for instance, the σ-algebra of Lebesgue-measurable subsets of $\Omega = \mathbb{R}$ is not generated by the negligible sets and any partition[4]); but observe that, in the finite case, three different numbers can be defined:

- m_1, the smallest number such that $\mathscr{D}$ is generated by $\mathscr{C}$ and by m_1 events,
- m, the smallest number such that $\mathscr{D}$ is generated by $\mathscr{C}$ and by an m-partition,
- m_2, the smallest number such that $\mathscr{D}$ is generated by $\mathscr{C}$ and by a Boolean algebra with m_2 elements.

There is a one-to-one correspondence between all finite Boolean algebras of subsets of Ω and all finite partitions of Ω: the Boolean algebra is generated by the partition, and in turn, the partition consists of all atoms of the Boolean algebra. If the partition has m elements, the Boolean algebra has 2^m elements; and one has

$$(\lceil \log_2 m \rceil =) \, m_1 < m < m_2 \, (= 2^m) \, .$$

In the infinite case, when numbers are replaced by cardinals, we have seen that m does not make sense; but both m_1 and m_2 do, and they turn out to be equal, as is easily seen (this will be shown in Lemma 2). This makes it reasonable to define $\mathrm{mult}(\mathscr{D} : \mathscr{C})$ as the common value of m_1 and m_2:

Definition 2. If $\mathscr{D}$ is an infinite extension of $\mathscr{C}$, the *multiplicity* of the extension is the smallest cardinal c such that $\mathscr{D} = \sigma(\mathscr{C} \cup \mathscr{E})$ for some set $\mathscr{E} \subseteq \mathscr{D}$ with cardinality $|\mathscr{E}| = c$. As in the finite case, it will be denoted by $\mathrm{mult}(\mathscr{D} : \mathscr{C})$.

In that case the multiplicity is necessarily an infinite cardinal, by definition of an infinite extension.

Lemma 2. a) If $\mathscr{E}$ is an infinite set of events, the Boolean algebra it generates has the same cardinal as $\mathscr{E}$.
b) Let $\mathscr{D}$ be an extension of $\mathscr{C}$. If $\mathrm{mult}(\mathscr{D} : \mathscr{C})$ is infinite, it is also the smallest cardinal c such that $\mathscr{D} = \sigma(\mathscr{C} \cup \mathscr{B})$ for some Boolean algebra $\mathscr{B} \subseteq \mathscr{D}$ with cardinality $|\mathscr{B}| = c$.

Proof. The Boolean algebra $\mathscr{B}$ generated by a class $\mathscr{E}$ of events can be obtained from $\mathscr{E}$ in three steps: first, stabilize $\mathscr{E}$ under complementation; second, stabilize

---

[4]Let $\mathscr{L}$ denote the Lebesgue σ-algebra of $\mathbb{R}$, $\mathscr{N}$ the sub-σ-algebra consisting of all negligible or co-negligible sets, and $\mathscr{P} \subseteq \mathscr{L}$ a partition of $\mathbb{R}$. If each element of $\mathscr{P}$ is negligible, then $\sigma(\mathscr{N} \cup \mathscr{P}) = \mathscr{N} \subsetneq \mathscr{L}$; on the contrary, if some $P \in \mathscr{P}$ is not negligible, then P is an a.s. atom of $\sigma(\mathscr{N} \cup \mathscr{P})$, whence $\sigma(\mathscr{N} \cup \mathscr{P}) \subsetneq \mathscr{L}$ again.

the so-obtained class under finite intersections; last, stabilize under finite unions (see for instance Sect. I-2 in Neveu [6]). If $\mathscr{E}$ is infinite, at each of these three steps, the cardinality remains the same owing to Corollary 1 b); so $|\mathscr{B}| = |\mathscr{E}|$. This establishes a), and b) immediately follows. □

Stéphane Laurent's Conjecture 3.18 in [5] concerns separable extensions, that is, extensions with countable multiplicity: $\mathrm{mult}(\mathscr{D} : \mathscr{C}) \leqslant \omega$. Countable multiplicity is equivalent to demanding that $\mathscr{D}$ is generated by $\mathscr{C}$ and some random variable, or by $\mathscr{C}$ and some countable Boolean algebra.

Lemma 3. *If $\mathscr{C}$, $\mathscr{C}'$, $\mathscr{D}'$ and $\mathscr{D}$ are four σ-algebras such that $\mathscr{C} \subseteq \mathscr{C}' \subseteq \mathscr{D}' \subseteq \mathscr{D}$, then*

$$\mathrm{mult}(\mathscr{D}' : \mathscr{C}') \leqslant \mathrm{mult}(\mathscr{D} : \mathscr{C}) .$$

Proof. It suffices to separately establish that $\mathrm{mult}(\mathscr{D} : \mathscr{C})$ decreases when $\mathscr{C}$ is replaced by a bigger σ-algebra, and decreases when $\mathscr{D}$ is replaced by a smaller one.

The dependence in $\mathscr{C}$ is trivial, since if $\mathscr{C} \subseteq \mathscr{C}' \subseteq \mathscr{D}$, every $\mathscr{E}$ (which may be a partition or not) such that $\mathscr{D} = \sigma(\mathscr{C} \cup \mathscr{E})$ also satisfies $\mathscr{D} = \sigma(\mathscr{C}' \cup \mathscr{E})$.

It remains to show that $\mathrm{mult}(\mathscr{D}' : \mathscr{C}) \leqslant \mathrm{mult}(\mathscr{D} : \mathscr{C})$ whenever $\mathscr{C} \subseteq \mathscr{D}' \subseteq \mathscr{D}$. We shall consider two cases.

First case: $\mathrm{mult}(\mathscr{D} : \mathscr{C})$ is finite. The result is immediate using for instance property (ii) from Lemma 1.

Second case: $\mathrm{mult}(\mathscr{D} : \mathscr{C})$ is infinite; call it c. By Lemma 2 b), there exists a Boolean algebra $\mathscr{B}$ with cardinality c, such that $\mathscr{D} = \sigma(\mathscr{C} \cup \mathscr{B})$; our aim is to find some set $\mathscr{E}$, with cardinality at most c, such that $\mathscr{D}' = \sigma(\mathscr{C} \cup \mathscr{E})$.

For each event $B \in \mathscr{B}$, choose and fix a version of $\mathbb{P}^{\mathscr{D}'}(B)$ (with values in $[0, 1]$) and, for each rational number r, call E_r^B the event $\{\mathbb{P}^{\mathscr{D}'}(B) > r\}$. Notice that $\mathbb{P}^{\mathscr{D}'}(B)$ can be uniformly approximated to any given accuracy by linear combinations of finitely many $\mathbf{1}_{E_r^B}$.

By Corollary 1 a), the set $\mathscr{E} = \{E_r^B, B \in \mathscr{B}, r \in \mathbb{Q}\} \subseteq \mathscr{D}'$ has cardinality $\leqslant c$; so it suffices to verify that $\mathscr{D}' = \sigma(\mathscr{C} \cup \mathscr{E})$.

The Boolean algebra generated by $\mathscr{C} \cup \mathscr{B}$ consists of all sets of the form $\sum_{i=1}^n (C_i \cap B_i)$, with n finite, $C_i \in \mathscr{C}$, $B_i \in \mathscr{B}$, and $\sum$ meaning a disjoint union. As a Boolean algebra, it is dense (for the "distance" $d(A', A'') = \mathbb{P}(A' \triangle A'') = \|\mathbf{1}_{A'} - \mathbf{1}_{A''}\|_{L^1}$) in the σ-algebra $\mathscr{D}$ it generates (see Sect. I-5 from Neveu [6]). In particular, given any $D' \in \mathscr{D}'$ and $\varepsilon > 0$, one can find n, C_i and B_i so that $\sum_{i=1}^n \mathbf{1}_{C_i} \mathbf{1}_{B_i}$ is ε-close to $\mathbf{1}_{D'}$ in L^1. Taking conditional expectations w.r.t. $\mathscr{D}'$ gives

$$\left\|\sum_{i=1}^n \mathbf{1}_{C_i}\, \mathbb{P}^{\mathscr{D}'}(B_i) - \mathbf{1}_{D'}\right\|_{L^1} < \varepsilon . \qquad (*)$$

We have seen earlier that with arbitrarily small error, $\mathbb{P}^{\mathscr{D}'}(B_i)$ in $(*)$ can be replaced by a combination of finitely many $\mathbf{1}_{E_r^{B_i}}$. So, finally, $(*)$ shows that $\mathbf{1}_{D'}$ is an L^1-limit of nice functions of C_i and $E_r^{B_i}$, and the claim $D' \in \sigma(\mathscr{C} \cup \mathscr{E})$ is established. □

Remark that, in the infinite case, the above proof uses the probability and the completeness of $\mathscr{C}$. The analogue of Lemma 3 for an ambient measurable space with no measure is not always true. For a simple counter-example, take $\Omega = \mathbb{R}$, $\mathscr{C} = \mathscr{C}' = \{\emptyset, \Omega\}$, $\mathscr{D} =$ the Borel σ-field, and $\mathscr{D}' =$ the σ-field of all countable or co-countable subsets of Ω. The σ-field $\mathscr{D}$ is separable, but $\mathscr{D}'$ is not.

4 Filtrations with Well-Ordered Time-Axis

Filtrations indexed by well-ordered sets will play a central rôle in the proof of Laurent's conjecture.

Definition 3. Given a totally ordered set T, a *filtration* indexed by T is a family $(\mathscr{F}_t)_{t \in T}$ of σ-algebras such that $\mathscr{F}_s \subseteq \mathscr{F}_t$ for $s \leq t$.

If furthermore $\mathscr{F}_s \subsetneq \mathscr{F}_t$ whenever $s < t$, we say that the filtration is *strict*.

If there exists an $s \in T$ such that $\mathscr{F}_t = \mathscr{F}_s$ for all $t \geq s$, we say that the filtration is *eventually stationary*. (If T has a maximal element, every filtration indexed by T is eventually stationary; so this notion is interesting only when T has no maximal element!)

A filtration $(\mathscr{F}_t)_{t \in T}$ is said to be *inserted between two σ-algebras $\mathscr{C}$ and $\mathscr{D}$* if $\mathscr{C} \subseteq \mathscr{F}_t \subseteq \mathscr{D}$ for each t.

Lemma 4. *Let c be a cardinal, and call c' the smallest cardinal $> c$ (so $[0, c'[$ is the well-ordered set of all ordinals α such that $|\alpha| \leq c$).*

If $(\mathscr{F}_\alpha)_{\alpha \in [0,c'[}$ is any filtration, the set $\bigcup_{\alpha \in [0,c'[} \mathscr{F}_\alpha$ is a σ-algebra (we shall denote this σ-algebra by $\mathscr{F}_{c'-}$).

Proof. If c is finite, $c' = c + 1$ and $\bigcup_{\alpha < c'} \mathscr{F}_\alpha = \bigcup_{\alpha \leq c} \mathscr{F}_\alpha = \mathscr{F}_c$ is a σ-algebra.

If c is infinite, the union $\mathscr{U} = \bigcup_{\alpha < c'} \mathscr{F}_\alpha$ is a Boolean algebra, simply because the index set $[0, c'[$ is totally ordered. To show that $\mathscr{U}$ is a σ-algebra, take any sequence (E_n) in $\mathscr{U}$; for each n, E_n belongs to $\mathscr{F}_{\alpha_n}$ for some α_n in $[0, c'[$. As $|\alpha_n| \leq c$, the union $\beta = \bigcup_{n \in \mathbb{N}} \alpha_n$ (which is an ordinal because it is well-ordered) also has cardinality $|\beta| \leq c$ by Corollary 1 a). Consequently, $\beta \in [0, c'[$ and $E_n \in \mathscr{F}_\beta$; so both sets $\bigcup_n E_n$ and $\bigcap_n E_n$ belong to $\mathscr{F}_\beta$ and a fortiori to $\mathscr{U}$. □

If c is a cardinal which is not the successor of any other cardinal, there may exist filtrations $(\mathscr{F}_\alpha)_{\alpha \in [0,c[}$ such that the set $\bigcup_{\alpha \in [0,c[} \mathscr{F}_\alpha$ is not a σ-algebra. An example is the familiar case when $c = \aleph_0$: if $(\mathscr{F}_n)_{n \in \mathbb{N}}$ is a filtration, in general the Boolean algebra $\bigcup_n \mathscr{F}_n$ is not a σ-algebra. Another example is the case when $c = \aleph_\omega$. Using the fact that $\aleph_\omega = \bigcup_{n \in \mathbb{N}} \aleph_n$, the reader will easily exhibit situations where $\bigcup_{\alpha < \aleph_\omega} \mathscr{F}_\alpha$ is not a σ-algebra.

Lemma 5. *Let c be an infinite cardinal, c' its successor, and $\mathscr{F} = (\mathscr{F}_\alpha)_{\alpha \in [0,c'[}$ a filtration indexed by all ordinals with cardinality $\leq c$.*

If $\mathrm{mult}(\mathscr{F}_{c'-} : \mathscr{F}_0) \leq c$, the filtration $\mathscr{F}$ is eventually stationary.

Proof. If $\mathrm{mult}(\mathscr{F}_{c'-} : \mathscr{F}_0) \leqslant c$, by Definitions 1 and 2 there exists some $\mathscr{E} \subseteq \mathscr{F}_{c'-}$ with cardinality $|\mathscr{E}| \leqslant c$ and such that $\mathscr{F}_{c'-} = \sigma(\mathscr{F}_0 \cup \mathscr{E})$. By Lemma 4, for every $E \in \mathscr{E}$, there exists an α_E in $[0, c'[$ such that $E \in \mathscr{F}_{\alpha_E}$. Since $|\alpha_E| \leqslant c$ and $|\mathscr{E}| \leqslant c$, according to Corollary 1 a) the ordinal $\beta = \bigcup_{E \in \mathscr{E}} \alpha_E$ also satisfies $|\beta| \leqslant c$, so it belongs to $[0, c'[$. From $E \in \mathscr{F}_{\alpha_E} \subseteq \mathscr{F}_\beta$, one gets $\mathscr{F}_0 \cup \mathscr{E} \subseteq \mathscr{F}_\beta$, and

$$\mathscr{F}_{c'-} = \sigma(\mathscr{F}_0 \cup \mathscr{E}) \subseteq \mathscr{F}_\beta \subseteq \mathscr{F}_{c'-} ;$$

this implies $\mathscr{F}_\beta = \mathscr{F}_{c'-}$, showing that the filtration is stationary from β on. □

Lemma 6. *Let $\mathscr{C}$ be a σ-algebra, $\mathscr{D}$ an extension of $\mathscr{C}$, and c an infinite cardinal with successor c'. If $\mathrm{mult}(\mathscr{D} : \mathscr{C}) > c$, there exists a strict filtration $(\mathscr{F}_\alpha)_{\alpha \in [0,c'[}$ inserted between $\mathscr{C}$ and $\mathscr{D}$.*

Proof. Fixing $\mathscr{C}$ and $\mathscr{D}$ such that $\mathrm{mult}(\mathscr{D} : \mathscr{C}) > c$, we shall show that there exists a transfinite sequence $(E_\alpha)_{\alpha \in [0,c'[}$ of elements of $\mathscr{D}$ such that, for each $\alpha \in [0, c'[$, E_α does not belong to the σ-algebra $\mathscr{F}_\alpha = \sigma(\mathscr{C} \cup \{E_\beta, \ \beta < \alpha\})$. As E_α obviously belongs to $\mathscr{F}_{\alpha+1} = \sigma(\mathscr{C} \cup \{E_\beta, \ \beta \leqslant \alpha\})$, the filtration $(\mathscr{F}_\alpha)_{\alpha \in [0,c'[}$ will then be strict; and this will prove the lemma, because $\mathscr{C} \subseteq \mathscr{F}_\alpha \subseteq \mathscr{D}$.

The argument goes by transfinite induction: for an $\alpha \in [0, c'[$, suppose that all E_β with $\beta < \alpha$ have already been constructed, and call $\mathscr{E}$ the set $\{E_\beta, \ \beta < \alpha\} \subseteq \mathscr{D}$. Its cardinality satisfies $|\mathscr{E}| \leqslant |[0, \alpha[| = |\alpha| \leqslant c$. From $\mathrm{mult}(\mathscr{D} : \mathscr{C}) > c$ one can then deduce $\mathscr{F}_\alpha = \sigma(\mathscr{C} \cup \mathscr{E}) \subsetneq \mathscr{D}$; and it suffices to pick any E_α in $\mathscr{D} \setminus \mathscr{F}_\alpha$. □

Corollary 2. *Let c be an infinite cardinal, c' its successor, $\mathscr{C}$ a σ-algebra and $\mathscr{D}$ an extension of $\mathscr{C}$. The following three statements are equivalent:*

(i) $\mathrm{mult}(\mathscr{D} : \mathscr{C}) \leqslant c$;

(ii) every filtration indexed by $[0, c'[$ and inserted between $\mathscr{C}$ and $\mathscr{D}$ is eventually stationary;

(iii) no strict filtration indexed by $[0, c'[$ can be inserted between $\mathscr{C}$ and $\mathscr{D}$.

Proof. (i) $\Rightarrow$ (ii). If $\mathrm{mult}(\mathscr{D} : \mathscr{C}) \leqslant c$ and if $(\mathscr{F}_\alpha)_{\alpha \in [0,c'[}$ is inserted between $\mathscr{C}$ and $\mathscr{D}$, one has $\mathscr{C} \subseteq \mathscr{F}_0 \subseteq \mathscr{F}_{c'-} \subseteq \mathscr{D}$; so Lemma 3 gives $\mathrm{mult}(\mathscr{F}_{c'-} : \mathscr{F}_0) \leqslant c$ and Lemma 5 says that $\mathscr{F}$ is eventually stationary.

(ii) $\Rightarrow$ (iii). An eventually stationary filtration indexed by $[0, c'[$ cannot be strict.

(iii) $\Rightarrow$ (i). Lemma 6 says that $\neg$ (i) $\Rightarrow$ $\neg$ (iii). □

Taking in particular $c = \omega = \aleph_0$, an extension is separable if and only if no strict filtration indexed by $[0, \aleph_1[$ can be inserted in it.

In statement (iii) of Corollary 2, the index set $[0, c'[$ can equivalently be replaced by the closed interval $[0, c']$, simply because if a filtration $\mathscr{F}$ indexed by $[0, c'[$ is strict, it remains strict when extended to $[0, c']$ (by defining for instance $\mathscr{F}_{c'}$ as the σ-algebra $\mathscr{F}_{c'-}$).

5 Essential Multiplicity

This short section (not needed in the sequel) is devoted to the special case of Corollary 2 when $\mathscr{C}$ is degenerate and $\mathscr{D} = \mathscr{A}$.

Recall that a complete probability space $(\Omega, \mathscr{A}, \mathbb{P})$ has been fixed, and that all σ-algebras we consider are $(\mathscr{A}, \mathbb{P})$-complete sub-$\sigma$-algebras of $\mathscr{A}$. The smallest possible σ-algebra is the *degenerate σ-algebra* $\mathscr{N}$, consisting of all events $A \in \mathscr{A}$ such that $\mathbb{P}(A) \in \{0, 1\}$; so we can define the *essential multiplicity* of $\mathscr{A}$ as

$$\text{ess mult}\, \mathscr{A} = \text{mult}(\mathscr{A} : \mathscr{N}).$$

It is a child's play to verify that ess mult $\mathscr{A}$ is finite if and only if the probability space $(\Omega, \mathscr{A}, \mathbb{P})$ is essentially finite (that is, the Hilbert space $L^2(\Omega, \mathscr{A}, \mathbb{P})$ is finite-dimensional[5]); more precisely, in that case one has $\dim L^2(\Omega, \mathscr{A}, \mathbb{P}) = $ ess mult $\mathscr{A}$.

It is just as easy to see that ess mult $\mathscr{A}$ is countable if and only if $\mathscr{A}$ is essentially separable, that is, generated by $\mathscr{N}$ and by some r.v. (equivalently, by $\mathscr{N}$ and by countably many r.v. with values in Polish spaces; this is tantamount to $L^2(\Omega, \mathscr{A}, \mathbb{P})$ being a separable Hilbert space).

(When ess mult $\mathscr{A}$ is an uncountable cardinal, readers familiar with the Kolmogorov–Maharam decomposition of an arbitrary $(\Omega, \mathscr{A}, \mathbb{P})$ can check as an exercise that ess mult $\mathscr{A}$ is precisely the biggest cardinal featuring in that decomposition.)

Taking $\mathscr{C} = \mathscr{N}$ and $\mathscr{D} = \mathscr{A}$ in Corollary 2 immediately gives the following characterization:

Corollary 3. *Let c be an infinite cardinal, c' its successor, and $(\Omega, \mathscr{A}, \mathbb{P})$ a complete probability space. The following three statements are equivalent:*

(i) ess mult $\mathscr{A} \leqslant c$;
(ii) on $(\Omega, \mathscr{A}, \mathbb{P})$, every filtration indexed by $[0, c'[$ is eventually stationary;
(iii) on $(\Omega, \mathscr{A}, \mathbb{P})$ there exists no strict filtration indexed by $[0, c'[$.

In particular, when $c = \omega$, the corollary says that *a complete probability space $(\Omega, \mathscr{A}, \mathbb{P})$ is essentially separable if and only if it admits no strict filtration indexed by the interval $[0, \aleph_1[$ of all countable ordinals.*

(But, on essentially separable probability spaces, strict filtrations indexed by other uncountable totally ordered sets are commonplace; think for instance of Brownian or Poisson filtrations.)

---

[5]The exponent 2 plays no particular rôle; here and in the next paragraph, the Banach space L^1, the metrizable vector space L^0 or any L^p could be used just as well.

6 Conditional Independence

This section shows that conditional independence implies inequalities between multiplicities of extensions. For infinite extensions, a crucial rôle will be played by Corollary 2.

If $\mathscr{B}$ is a σ-algebra, recall that $\mathbb{E}^{\mathscr{B}}$ stands for the operator of conditional expectation w.r.t. $\mathscr{B}$. Given three σ-algebras $\mathscr{B}$, $\mathscr{C}'$ and $\mathscr{C}''$, recall that $\mathscr{C}'$ and $\mathscr{C}''$ are conditionally independent w.r.t. $\mathscr{B}$ (and one writes $\mathscr{C}' \underset{\mathscr{B}}{\perp\!\!\!\perp} \mathscr{C}''$) if $\mathbb{E}^{\mathscr{B}} C' \, \mathbb{E}^{\mathscr{B}} C'' = \mathbb{E}^{\mathscr{B}}[C'C'']$ for all $C' \in L^\infty(\mathscr{C}')$ and $C'' \in L^\infty(\mathscr{C}'')$. The next two lemmas state properties of conditional independence. They are classical (and elementary) exercises from the first chapter of any course on Markov processes; we recall them without proof.

Lemma 7. *Let $\mathscr{B}$, $\mathscr{C}'$ and $\mathscr{C}''$ be three σ-algebras. If $\mathscr{C}' \underset{\mathscr{B}}{\perp\!\!\!\perp} \mathscr{C}''$, then $\mathscr{C}' \cap \mathscr{C}'' \subseteq \mathscr{B}$.*

Lemma 8. *Let $\mathscr{B}$, $\mathscr{C}'$ and $\mathscr{C}''$ be three σ-algebras. The following are equivalent:*

(i) $\mathscr{C}' \underset{\mathscr{B}}{\perp\!\!\!\perp} \mathscr{C}''$;
(ii) $\sigma(\mathscr{B} \cup \mathscr{C}') \underset{\mathscr{B}}{\perp\!\!\!\perp} \sigma(\mathscr{B} \cup \mathscr{C}'')$.

If furthermore $\mathscr{B} \subseteq \mathscr{C}'$, the next property is also equivalent to (i) and (ii):
(iii) *for every $C'' \in L^1(\mathscr{C}'')$, $\mathbb{E}^{\mathscr{C}'} C''$ is $\mathscr{B}$-measurable (and therefore a.s. equal to $\mathbb{E}^{\mathscr{B}} C''$).*

The following property is classical too. (It says for instance that any Markov process is also second-order Markovian.)

Corollary 4. *Let $\mathscr{B}$, $\mathscr{B}'$, $\mathscr{C}'$ and $\mathscr{C}''$ be four σ-algebras such that $\mathscr{B} \subseteq \mathscr{B}' \subseteq \mathscr{C}'$. If $\mathscr{C}' \underset{\mathscr{B}}{\perp\!\!\!\perp} \mathscr{C}''$, then $\mathscr{C}' \underset{\mathscr{B}'}{\perp\!\!\!\perp} \mathscr{C}''$.*

Proof. Immediate by Lemma 8 (iii): if $\mathbb{E}^{\mathscr{C}'} C''$ is $\mathscr{B}$-measurable, it is a fortiori $\mathscr{B}'$-measurable. □

Lemma 9. *Let $\mathscr{B}$, $\mathscr{C}'$, $\mathscr{C}''$ and $\mathscr{D}$ be four σ-algebras such that $\mathscr{B} \subseteq \mathscr{C}' \subseteq \mathscr{D}$ and $\mathscr{B} \subseteq \mathscr{C}'' \subseteq \mathscr{D}$; suppose that $\mathscr{C}' \underset{\mathscr{B}}{\perp\!\!\!\perp} \mathscr{C}''$. If $\mathrm{mult}(\mathscr{D} : \mathscr{C}'')$ is finite, then*

$$\mathrm{mult}(\mathscr{C}' : \mathscr{B}) \leq \mathrm{mult}(\mathscr{D} : \mathscr{C}'') \, .$$

Proof. Call m the integer $\mathrm{mult}(\mathscr{D} : \mathscr{C}'')$, and take any $m+1$ pairwise disjoint events $C'_1, \ldots, C'_{m+1}$ in $\mathscr{C}'$. Since $\mathscr{C}' \subseteq \mathscr{D}$, by Lemma 1 applied to the extension $\mathscr{D}$ of $\mathscr{C}''$ one has $\prod_{i=1}^{m+1} \mathbb{P}^{\mathscr{C}''}(C'_i) = 0$. Now, $\mathbb{P}^{\mathscr{C}''}(C'_i) = \mathbb{P}^{\mathscr{B}}(C'_i)$ owing to Lemma 8 (iii) (with $\mathscr{C}'$ and $\mathscr{C}''$ exchanged). Consequently, $\prod_{i=1}^{m+1} \mathbb{P}^{\mathscr{B}}(C'_i) = 0$, and, resorting to Lemma 1 again, one has $\mathrm{mult}(\mathscr{C}' : \mathscr{B}) \leq m$. □

Lemma 9 is quite simple and its proof is elementary. The next proposition is deeper; it extends the scope of this lemma to infinite multiplicity. Our proof uses filtrations indexed by ordinals and relies on the results from previous sections; we do not know any simpler argument.

Proposition 1. *Let $\mathcal{B}$, $\mathcal{C}'$, $\mathcal{C}''$ and $\mathcal{D}$ be four σ-algebras such that $\mathcal{B} \subseteq \mathcal{C}' \subseteq \mathcal{D}$ and $\mathcal{B} \subseteq \mathcal{C}'' \subseteq \mathcal{D}$. If $\mathcal{C}' \underset{\mathcal{B}}{\perp\!\!\!\perp} \mathcal{C}''$, then*

$$\mathrm{mult}(\mathcal{C}' : \mathcal{B}) \leq \mathrm{mult}(\mathcal{D} : \mathcal{C}'') \quad \text{and} \quad \mathrm{mult}(\mathcal{C}'' : \mathcal{B}) \leq \mathrm{mult}(\mathcal{D} : \mathcal{C}') \,.$$

Proof. It suffices to prove the first inequality; the other one will then follow by exchanging $\mathcal{C}'$ and $\mathcal{C}''$. We have to show that, given any cardinal c,

$$\mathrm{mult}(\mathcal{D} : \mathcal{C}'') \leq c \quad \Longrightarrow \quad \mathrm{mult}(\mathcal{C}' : \mathcal{B}) \leq c \,.$$

If c is finite, this holds by Lemma 9; hence we will suppose c to be infinite, and *the proposition will be proved by establishing that, for c an infinite cardinal,*

$$\mathrm{mult}(\mathcal{C}' : \mathcal{B}) > c \quad \Longrightarrow \quad \mathrm{mult}(\mathcal{D} : \mathcal{C}'') > c \,.$$

So we suppose $\mathrm{mult}(\mathcal{C}' : \mathcal{B}) > c$; call c' the successor of c. By Corollary 2, some strict filtration $\mathcal{F}$ indexed by $[0, c'[$ is inserted between $\mathcal{B}$ and $\mathcal{C}'$.

Define a new filtration $\mathcal{G} = (\mathcal{G}_\alpha)_{\alpha \in [0, c'[}$ by the formula $\mathcal{G}_\alpha = \sigma(\mathcal{C}'' \cup \mathcal{F}_\alpha)$. This filtration $\mathcal{G}$ is inserted between $\mathcal{C}''$ and $\mathcal{D}$ because $\mathcal{F}_\alpha \subseteq \mathcal{C}' \subseteq \mathcal{D}$. We shall show that $\mathcal{G}$ is strict; using Corollary 2 again, this will imply $\mathrm{mult}(\mathcal{D} : \mathcal{C}'') > c$, thus proving the proposition.

To see that $\mathcal{G}$ is strict, take α and β in $[0, c'[$ such that $\alpha < \beta$. By hypothesis,

$$\mathcal{C}' \underset{\mathcal{B}}{\perp\!\!\!\perp} \mathcal{C}'' \,.$$

In this formula, we may replace $\mathcal{C}'$ by $\mathcal{F}_\beta$ (which is smaller since $\mathcal{F}$ is inserted between $\mathcal{B}$ and $\mathcal{C}'$), and $\mathcal{B}$ by $\mathcal{F}_\alpha$ (owing to $\mathcal{B} \subseteq \mathcal{F}_\alpha \subseteq \mathcal{F}_\beta$ and to Corollary 4); so

$$\mathcal{F}_\beta \underset{\mathcal{F}_\alpha}{\perp\!\!\!\perp} \mathcal{C}'' \,.$$

Next, by (i) $\Rightarrow$ (ii) in Lemma 8, $\mathcal{C}''$ may be replaced with $\sigma(\mathcal{C}'' \cup \mathcal{F}_\alpha)$, that is, $\mathcal{G}_\alpha$; and one has

$$\mathcal{F}_\beta \underset{\mathcal{F}_\alpha}{\perp\!\!\!\perp} \mathcal{G}_\alpha \,. \tag{$**$}$$

Lemma 7 now yields $\mathcal{F}_\beta \cap \mathcal{G}_\alpha \subseteq \mathcal{F}_\alpha$, whence $\mathcal{F}_\beta \setminus \mathcal{F}_\alpha \subseteq \mathcal{G}_\alpha^c$. Using $\mathcal{F}_\beta \subseteq \mathcal{G}_\beta$, one finally obtains $\mathcal{F}_\beta \setminus \mathcal{F}_\alpha \subseteq \mathcal{G}_\beta \setminus \mathcal{G}_\alpha$; so strictness propagates from $\mathcal{F}$ to $\mathcal{G}$. □

Remark that the standing assumption that all σ-algebras are $(\mathcal{A}, \mathbb{P})$-complete is needed in Lemma 9 and Proposition 1. Without this assumption, counter-examples are easily produced: take $\mathcal{C}' = \mathcal{C}'' = \mathcal{D} = \mathcal{N}$, the σ-algebra of all events with probability 0 or 1, and $\mathcal{B}$ any non $(\mathcal{A}, \mathbb{P})$-complete sub-$\sigma$-field of $\mathcal{N}$. One always has $\mathcal{C}' \underset{\mathcal{B}}{\perp\!\!\!\perp} \mathcal{C}''$ and $\mathrm{mult}(\mathcal{D} : \mathcal{C}'') = 1$, but $\mathcal{C}' \not\supseteq \mathcal{B}$, thus violating both statements.

7 Immersion

Definition 4. Given a totally ordered set T and two filtrations $\mathscr{F} = (\mathscr{F}_t)_{t \in T}$ and $\mathscr{G} = (\mathscr{G}_t)_{t \in T}$, one says that $\mathscr{F}$ is *immersed in* $\mathscr{G}$ if every $\mathscr{F}$-martingale is a $\mathscr{G}$-martingale.

Immersion is much stronger than requiring only $\mathscr{F}_t \subseteq \mathscr{G}_t$ for every t. More precisely, a well-known characterization involves conditional independence:

Lemma 10. *Let $\mathscr{F} = (\mathscr{F}_t)_{t \in T}$ and $\mathscr{G} = (\mathscr{G}_t)_{t \in T}$ be two filtrations such that $\mathscr{F}_t \subseteq \mathscr{G}_t$ for all t. Then $\mathscr{F}$ is immersed in $\mathscr{G}$ if and only if for all s and t such that $s < t$,*

$$\mathscr{F}_t \underset{\mathscr{F}_s}{\perp\!\!\!\perp} \mathscr{G}_s \ .$$

Proof. Immersion holds if and only if, for each $t \in T$ and each $F_t \in L^1(\mathscr{F}_t)$, the $\mathscr{F}$-martingale $M_s = \mathbb{E}^{\mathscr{F}_s} F_t$ is also a $\mathscr{G}$-martingale. This means that $M_s = \mathbb{E}^{\mathscr{G}_s} F_t$ for all $s < t$; so $\mathscr{F}$ is immersed in $\mathscr{G}$ if and only if for all $F_t \in L^1(\mathscr{F}_t)$ and all $s < t$, $\mathbb{E}^{\mathscr{G}_s} F_t = \mathbb{E}^{\mathscr{F}_s} F_t$. By (i) $\Leftrightarrow$ (iii) in Lemma 8, this is equivalent to $\mathscr{F}_t \underset{\mathscr{F}_s}{\perp\!\!\!\perp} \mathscr{G}_s$. □

As for an example, in the proof of Proposition 1, formula (∗∗) says that $\mathscr{F}$ is immersed in $\mathscr{G}$.

Corollary 5. *Let T be a totally ordered set, $\mathscr{F} = (\mathscr{F}_t)_{t \in T}$ and $\mathscr{G} = (\mathscr{G}_t)_{t \in T}$ two filtrations with $\mathscr{F}$ immersed in $\mathscr{G}$.*

(i) If s and t are two instants in T such that $s < t$, then

$$\mathrm{mult}(\mathscr{F}_t : \mathscr{F}_s) \leqslant \mathrm{mult}(\mathscr{G}_t : \mathscr{G}_s) \ .$$

(ii) The (cardinal-valued) map $t \mapsto \mathrm{mult}(\mathscr{G}_t : \mathscr{F}_t)$ is increasing.

Proof. For $s < t$, one has $\mathscr{F}_s \subseteq \mathscr{F}_t \subseteq \mathscr{G}_t$ and $\mathscr{F}_s \subseteq \mathscr{G}_s \subseteq \mathscr{G}_t$; Lemma 10 translates the immersion hypothesis into $\mathscr{F}_t \underset{\mathscr{F}_s}{\perp\!\!\!\perp} \mathscr{G}_s$. So Proposition 1 applies, and yields

$$\mathrm{mult}(\mathscr{F}_t : \mathscr{F}_s) \leqslant \mathrm{mult}(\mathscr{G}_t : \mathscr{G}_s) \qquad \text{and} \qquad \mathrm{mult}(\mathscr{G}_s : \mathscr{F}_s) \leqslant \mathrm{mult}(\mathscr{G}_t : \mathscr{F}_t) \ .$$

The first inequality is (i); the second one gives (ii). □

Laurent's Conjecture 3.18 of [5] is a particular case of Corollary 5 (i): if $\mathscr{F}$ is immersed in $\mathscr{G}$ and if $\mathrm{mult}(\mathscr{G}_t : \mathscr{G}_s) \leqslant \omega$ (i.e., $\mathscr{G}_t$ is a separable extension of $\mathscr{G}_s$), then $\mathrm{mult}(\mathscr{F}_t : \mathscr{F}_s) \leqslant \omega$ too (so $\mathscr{F}_t$ is a separable extension of $\mathscr{F}_s$).

8 Qualitative Conditional Independence

The inequalities between multiplicities of extensions, established in Proposition 1 and Corollary 5, are not fully satisfactory. On the one hand, the definition of multiplicity does not really involve the ambient probability $\mathbb{P}$, but only its equivalence class: $\mathrm{mult}(\mathscr{D} : \mathscr{C})$ remains invariant when $\mathbb{P}$ is replaced by an equivalent probability. But on the other hand, the hypotheses (conditional independence in Proposition 1 and immersion in Corollary 5) used to establish these inequalities are not invariant under equivalent changes of measure.

The present section deals with hypotheses ("qualitative conditional independence" and "qualitative immersion"), which are weaker than conditional independence and immersion, and depend upon $\mathbb{P}$ through its negligible events only. One may hope that the multiplicity inequalities from Proposition 1 and Corollary 5 subsist under these weakened hypotheses; this is an open question, which we have not been able to settle.

A non-probabilistic substitute to independence, called qualitative independence, has been known and studied for a long time (see for instance Sect. 3.3 in Rényi [7]). No probability is needed; two σ-fields $\mathscr{C}'$ and $\mathscr{C}''$ are classically said to be qualitatively independent if, for all events $C' \in \mathscr{C}'$ and $C'' \in \mathscr{C}''$, one has

$$C' \neq \varnothing \quad \text{and} \quad C'' \neq \varnothing \quad \Longrightarrow \quad C' \cap C'' \neq \varnothing$$

(the reverse implication is trivial).

This notion poorly suits our needs, because, to work with a $\mathscr{B}$-conditional version of qualitative independence, we must have a definition which neglects null events. The analogous property which lends itself to $\mathscr{B}$-conditioning is the following:

Definition 5. Let $\mathscr{C}'$ and $\mathscr{C}''$ be two complement-stable sets of events. We say that $\mathscr{C}'$ and $\mathscr{C}''$ are *qualitatively independent*, and we write $\mathscr{C}' \perp\!\!\!\perp \mathscr{C}''$, if one has

$$\mathbb{P}(C') > 0 \quad \text{and} \quad \mathbb{P}(C'') > 0 \quad \Longrightarrow \quad \mathbb{P}(C' \cap C'') > 0 \qquad \text{(qi)}$$

for all $C' \in \mathscr{C}'$ and $C'' \in \mathscr{C}''$ (the reverse implication is trivial).

The $\mathscr{B}$-conditional version is now easily defined:

Definition 6. Let $\mathscr{C}'$ and $\mathscr{C}''$ be two complement-stable sets of events. Given a σ-algebra $\mathscr{B}$, we say that $\mathscr{C}'$ and $\mathscr{C}''$ are *qualitatively conditionally independent* w.r.t. $\mathscr{B}$, and we write $\mathscr{C}' \underset{\mathscr{B}}{\perp\!\!\!\perp} \mathscr{C}''$, if the almost sure inclusion

$$\{\mathbb{P}^{\mathscr{B}}(C') > 0 \text{ and } \mathbb{P}^{\mathscr{B}}(C'') > 0\} \subseteq \{\mathbb{P}^{\mathscr{B}}(C' \cap C'') > 0\} \qquad \text{(qci)}$$

holds for all events $C' \in \mathscr{C}'$ and $C'' \in \mathscr{C}''$.

Notice that the reverse a.s. inclusion always holds; so $\subseteq$ can be replaced with $=$ in this definition.

Remark also that these definitions are sensible because $\mathscr{C}'$ and $\mathscr{C}''$ are stable under complementation. If one wishes to define qualitative (conditional) independence of two events C' and C'', it is reasonable to take complements and to demand that each of the four pairs (C', C''), (C', C''^c), (C'^c, C'') and (C'^c, C''^c) satisfies the implication (qi) (the a.s. relation (qci)).

Definition 6 can be translated into another language, that of $\mathscr{B}$-halos. Fix the σ-algebra $\mathscr{B}$. For any event $C \in \mathscr{A}$, define the $\mathscr{B}$-halo $\overline{C}^{\mathscr{B}}$ of C (simply written $\overline{C}$ if no ambiguity is possible) as the event $\{\mathbb{P}^{\mathscr{B}}(C) > 0\}$. This event is $\mathscr{B}$-measurable, and defined up to negligibility only; it is (up to negligibility) the smallest $\mathscr{B}$-measurable event a.s. containing C; so it does not really depend upon $\mathbb{P}$, but only upon the class of negligible sets. (One could similarly introduce $\overset{\circ}{C} = \{\mathbb{P}^{\mathscr{B}}(C) = 1\}$, the biggest $\mathscr{B}$-measurable event included in C; it satisfies $(\overset{\circ}{C})^c = \overline{C^c}$, etc.)

Here are some immediate properties of $\mathscr{B}$-halos, valid for all events C, C' and C'' in $\mathscr{A}$ and B in $\mathscr{B}$:

$$B \supseteq C \Leftrightarrow B \supseteq \overline{C} ; \tag{h1}$$

$$C = \overline{C} \Leftrightarrow C \in \mathscr{B} ; \tag{h2}$$

$$C' \subseteq C'' \Rightarrow \overline{C'} \subseteq \overline{C''} ; \tag{h3}$$

$$C \text{ is negligible} \Leftrightarrow \overline{C} \text{ is negligible} ; \tag{h4}$$

$$\overline{C' \cup C''} = \overline{C'} \cup \overline{C''} ; \tag{h5}$$

$$\overline{B \cap C} = B \cap \overline{C} ; \tag{h6}$$

$$\overline{C' \cap C''} \subseteq \overline{C'} \cap \overline{C''} . \tag{h7}$$

The reverse of (h7), namely, $\overline{C' \cap C''} \supseteq \overline{C'} \cap \overline{C''}$, is nothing but (qci) from Definition 6; so, taking (h7) into account, (qci) is also equivalent to $\overline{C' \cap C''} = \overline{C'} \cap \overline{C''}$.

It is clear from the definitions that conditional independence implies qualitative conditional independence. But these properties are not equivalent, for the latter is invariant under equivalent changes of probability.

Lemma 7 has a qualitative analogue:

Lemma 7a. *Let $\mathscr{B}$ be a σ-algebra and $\mathscr{C}'$ and $\mathscr{C}''$ two sets of events stable under complementation. If $\mathscr{C}' \underset{\mathscr{B}}{\perp\!\!\!\perp} \mathscr{C}''$, then $\mathscr{C}' \cap \mathscr{C}'' \subseteq \mathscr{B}$.*

Proof. For simplicity, we write $\overline{C}$ for $\overline{C}^{\mathscr{B}}$.

Taking any $A \in \mathscr{C}' \cap \mathscr{C}''$, apply (qci) to A (which is in $\mathscr{C}'$) and A^c (which is in $\mathscr{C}''$); this gives $\overline{A} \cap \overline{A^c} \subseteq \overline{A \cap A^c} = \overline{\varnothing} = \varnothing$. A fortiori, $\overline{A} \cap A^c = \varnothing$, that is, $\overline{A} \subseteq A$. So $\overline{A} = A$, and $A \in \mathscr{B}$ by (h2). □

Equivalence (i) ⇔ (iii) from Lemma 8 also has a qualitative analogue:

Lemma 8a. *Let $\mathscr{B}$ and $\mathscr{C}$ be two σ-algebras such that $\mathscr{B} \subseteq \mathscr{C}$; let $\mathscr{D}$ be a set of events stable under complementation. The following are equivalent:*

(i) $\mathscr{C} \underset{\mathscr{B}}{\perp\!\!\!\perp} \mathscr{D}$;

(ii) *for every $D \in \mathscr{D}$, one has $\overline{D}^{\mathscr{B}} \subseteq \overline{D}^{\mathscr{C}}$;*

(iii) *for every $D \in \mathscr{D}$, one has $\overline{D}^{\mathscr{B}} = \overline{D}^{\mathscr{C}}$;*

(iv) *for every $D \in \mathscr{D}$, $\overline{D}^{\mathscr{C}}$ belongs to $\mathscr{B}$.*

Proof. If A is an event, its $\mathscr{C}$-halo is $\overline{A}^{\mathscr{C}}$, but its $\mathscr{B}$-halo will simply be written $\overline{A}$ to lighten the notation.

(i) ⇒ (ii). Assuming qualitative conditional independence, take $D \in \mathscr{D}$ and call C the complement of $\overline{D}^{\mathscr{C}}$. As $\overline{D}^{\mathscr{C}} \supseteq D$, $C \cap D$ is negligible; by (h4), so is also $\overline{C \cap D}$, which equals $\overline{C} \cap \overline{D}$ by (i) and because $C \in \mathscr{C}$ and $D \in \mathscr{D}$. A fortiori, $C \cap \overline{D}$ is negligible too, that is, $\overline{D} \subseteq C^c = \overline{D}^{\mathscr{C}}$.

(ii) ⇒ (iii). One always has $\overline{D} \supseteq \overline{D}^{\mathscr{C}}$, because $\overline{D}$ contains D and is $\mathscr{C}$-mesurable (remember that $\mathscr{B} \subseteq \mathscr{C}$).

(iii) ⇒ (i). Given $C \in \mathscr{C}$ and $D \in \mathscr{D}$, the complement B of $\overline{C \cap D}$ does not meet $C \cap D$, so $B \cap C \cap D$ is negligible. By (h4), $\overline{(B \cap C) \cap D}$ is negligible; as B and C are in $\mathscr{C}$, (h6) says that $(B \cap C) \cap \overline{D}^{\mathscr{C}}$ is negligible. Now, $\overline{D}^{\mathscr{C}} = \overline{D}$ by (iii); hence $B \cap C \cap \overline{D}$ is negligible, and also $\overline{B \cap C \cap \overline{D}}$ by (h4). Since $B \cap \overline{D} \in \mathscr{B}$, (h6) says that $(B \cap \overline{D}) \cap \overline{C}$ is negligible; so $\overline{C} \cap \overline{D} \subseteq B^c = \overline{C \cap D}$, and C and D satisfy (qci).

(iii) ⇒ (iv) is trivial.

(iv) ⇒ (ii). If $\overline{D}^{\mathscr{C}} \in \mathscr{B}$, then $\overline{D}^{\mathscr{C}} \supseteq \overline{D}$ by (h1). □

Lemma 8a allows us to extend Corollary 4 and Lemma 9 to qualitative conditional independence, with exactly the same proofs:

Corollary 4a. *Let $\mathscr{B}$, $\mathscr{B}'$, $\mathscr{C}'$ and $\mathscr{C}''$ be four σ-algebras such that $\mathscr{B} \subseteq \mathscr{B}' \subseteq \mathscr{C}'$. If $\mathscr{C}' \underset{\mathscr{B}}{\perp\!\!\!\perp} \mathscr{C}''$, then $\mathscr{C}' \underset{\mathscr{B}'}{\perp\!\!\!\perp} \mathscr{C}''$.*

Proof. By (i) ⇔ (iv) in Lemma 8a, the corollary boils down to the following trivial implication:

$$\forall C'' \in \mathscr{C}'' \quad \overline{C''}^{\mathscr{C}'} \in \mathscr{B} \quad \Longrightarrow \quad \forall C'' \in \mathscr{C}'' \quad \overline{C''}^{\mathscr{C}'} \in \mathscr{B}'.$$ □

Lemma 9a. *Let $\mathscr{B}$, $\mathscr{C}'$, $\mathscr{C}''$ and $\mathscr{D}$ be four σ-algebras such that $\mathscr{B} \subseteq \mathscr{C}' \subseteq \mathscr{D}$ and $\mathscr{B} \subseteq \mathscr{C}'' \subseteq \mathscr{D}$; suppose that $\mathscr{C}' \underset{\mathscr{B}}{\perp\!\!\!\perp} \mathscr{C}''$. If $\mathrm{mult}(\mathscr{D} : \mathscr{C}'')$ is finite, then*

$$\mathrm{mult}(\mathscr{C}' : \mathscr{B}) \leq \mathrm{mult}(\mathscr{D} : \mathscr{C}'') .$$

Proof. Call m the integer mult($\mathscr{D} : \mathscr{C}''$), and take any $m+1$ pairwise disjoint events $C'_1, \ldots, C'_{m+1}$ in $\mathscr{C}'$. Since $\mathscr{C}' \subseteq \mathscr{D}$, by Lemma 1 applied to the extension $\mathscr{D}$ of $\mathscr{C}''$ one has $\prod_{i=1}^{m+1} \mathbb{P}^{\mathscr{C}''}(C'_i) = 0$; equivalently, $\bigcap_{i=1}^{m+1} \overline{C'_i}^{\mathscr{C}''}$ is negligible. Now, $\overline{C'_i}^{\mathscr{C}''} = \overline{C'_i}^{\mathscr{B}}$ by Lemma 8a (iii). Consequently, $\bigcap_{i=1}^{m+1} \overline{C'_i}^{\mathscr{B}}$ is negligible, and $\prod_{i=1}^{m+1} \mathbb{P}^{\mathscr{B}}(C'_i) = 0$. Resorting to Lemma 1 again, one has mult($\mathscr{C}' : \mathscr{B}$) $\leq m$. □

Putting aside for a minute the question whether the finiteness hypothesis can be lifted, we remain in the realm of finite extensions and rephrase Lemma 9a in terms of immersions. The next definition is a qualitative analogue of Lemma 10.

Definition 4a. Given a totally ordered set T and two filtrations $\mathscr{F} = (\mathscr{F}_t)_{t \in T}$ and $\mathscr{G} = (\mathscr{G}_t)_{t \in T}$, we say that $\mathscr{F}$ is *qualitatively immersed in* $\mathscr{G}$ if $\mathscr{F}$ is included in $\mathscr{G}$ (that is, $\mathscr{F}_t \subseteq \mathscr{G}_t$ for each t) and if $\mathscr{F}_t \perp\!\!\!\perp_{\mathscr{F}_s} \mathscr{G}_s$ for all s and t such that $s < t$.

Qualitative immersion is weaker than immersion, and invariant under equivalent changes of probability.

The next statement is a trivial reformulation of Lemma 9a in terms of qualitative immersion:

Corollary 5a. *Let T be a totally ordered set, $\mathscr{F} = (\mathscr{F}_t)_{t \in T}$ and $\mathscr{G} = (\mathscr{G}_t)_{t \in T}$ two filtrations with $\mathscr{F}$ qualitatively immersed in $\mathscr{G}$.*

(i) If s and t are two instants in T such that $s < t$ and mult($\mathscr{G}_t : \mathscr{G}_s$) is finite, then

$$\mathrm{mult}(\mathscr{F}_t : \mathscr{F}_s) \leq \mathrm{mult}(\mathscr{G}_t : \mathscr{G}_s) \ .$$

(ii) If the map $t \mapsto \mathrm{mult}(\mathscr{G}_t : \mathscr{F}_t)$ takes only finite values, it is increasing.

An open problem is to get rid of the finiteness hypothesis in Lemma 9a and Corollary 5a: does Proposition 1 remain true when its conditional independence hypothesis is replaced with qualitative conditional independence? We have not been able to conclude; the difficulty comes from the qualitative analogue of the implication (i) $\Rightarrow$ (ii) in Lemma 8. It is easy to see that

$$\mathscr{C}' \perp\!\!\!\perp_{\mathscr{B}} \mathscr{C}'' \quad \Longrightarrow \quad \mathrm{Boole}(\mathscr{B} \cup \mathscr{C}') \perp\!\!\!\perp_{\mathscr{B}} \mathrm{Boole}(\mathscr{B} \cup \mathscr{C}'') \ ,$$

where Boole $\mathscr{E}$ stands for the Boolean algebra generated by the set $\mathscr{E} \subseteq \mathscr{A}$; but Boole($\mathscr{B} \cup \mathscr{C}'$) and Boole($\mathscr{B} \cup \mathscr{C}''$) in this formula can not always be replaced by $\sigma(\mathscr{B} \cup \mathscr{C}')$ and $\sigma(\mathscr{B} \cup \mathscr{C}'')$.

A counter-example is easily shown to exist, using the construction described in [3]. The main result there is the existence of *a triple* $(\mathscr{U}, \mathscr{V}, A)$, where $\mathscr{U}$ and $\mathscr{V}$ are two independent σ-algebras, and A is an event in $\sigma(\mathscr{U} \cup \mathscr{V})$ such that, for all events $U \in \mathscr{U}$ and $V \in \mathscr{V}$, one has

$$\mathbb{P}(U \cap V) > 0 \quad \Longrightarrow \quad 0 < \mathbb{P}(A \mid U \cap V) < 1 \ .$$

Given such a triple $(\mathscr{U}, \mathscr{V}, A)$, one has $0 < \mathbb{P}(A) < 1$ (take $U = V = \Omega$), and $A \notin \mathscr{U}$ (else, take $U = A$ and $V = \Omega$). One also has $\mathscr{V} \perp\!\!\!\perp \sigma(\mathscr{U} \cup \{A\})$ because, for all $U \in \mathscr{U}$ and $V \in \mathscr{V}$,

$$\mathbb{P}(V) > 0 \quad \text{and} \quad \mathbb{P}(U \cap A) > 0$$
$$\implies \mathbb{P}[V \cap (U \cap A)] = \mathbb{P}(U)\,\mathbb{P}(V)\,\mathbb{P}(A \mid U \cap V) > 0,$$

and similarly with A replaced by its complement A^c.

By Corollary 4a (with $\mathscr{B}$ the degenerate σ-algebra), from $\mathscr{V} \perp\!\!\!\perp \sigma(\mathscr{U} \cup \{A\})$ we may deduce $\mathscr{V} \perp\!\!\!\perp_{\mathscr{U}} \sigma(\mathscr{U} \cup \{A\})$. Putting now $\mathscr{B} = \mathscr{U}$, $\mathscr{C}' = \mathscr{V}$ and $\mathscr{C}'' = \sigma(\mathscr{U} \cup \{A\})$, one has

$$\mathscr{C}' \perp\!\!\!\perp_{\mathscr{B}} \mathscr{C}'' \quad \text{but not} \quad \sigma(\mathscr{B} \cup \mathscr{C}') \perp\!\!\!\perp_{\mathscr{B}} \mathscr{C}'',$$

since this would violate Lemma 7a. Indeed, $\sigma(\mathscr{B} \cup \mathscr{C}') \cap \mathscr{C}'' = \mathscr{C}''$ is not included in $\mathscr{B}$, because A belongs to $\sigma(\mathscr{U} \cup \mathscr{V}) = \sigma(\mathscr{B} \cup \mathscr{C}')$ and also to $\sigma(\mathscr{U} \cup \{A\}) = \mathscr{C}''$, but it does not belong to $\mathscr{U} = \mathscr{B}$.

So, the existence of $(\mathscr{U}, \mathscr{V}, A)$ shows that in the formula $\mathscr{C}' \perp\!\!\!\perp_{\mathscr{B}} \mathscr{C}''$, one may not always replace $\mathscr{C}'$ by $\sigma(\mathscr{B} \cup \mathscr{C}')$; the qualitative analogue of (i) $\Rightarrow$ (ii) in Lemma 8 is false.

Acknowledgements We thank Vincent Vigon and Stéphane Laurent for their remarks on an earlier version of the manuscript.

Michel Émery is grateful to Stéphane Laurent for drawing his attention to this question, and for uncountable enjoyable conversations on this subject.

References

1. M.T. Barlow, M. Émery, F.B. Knight, S. Song, M. Yor, Autour d'un théorème de Tsirelson sur des filtrations browniennes et non browniennes. Séminaire de Probabilités XXXII, Springer Lecture Notes in Mathematics, vol. 1686 (Springer, Berlin, 1998), pp. 264–305
2. M.T. Barlow, J. Pitman, M. Yor, On Walsh's Brownian Motions. Séminaire de Probabilités XXIII, Springer Lecture Notes in Mathematics, vol. 1372 (Springer, Berlin, 1989), pp. 275–293
3. M. Émery, A planar Borel set which divides every Borel product. Séminaire de Probabilités XLV, Springer Lecture Notes in Mathematics, vol. 2078 (Springer, Berlin, 2013), pp. 159–165
4. T. Jech, *Set Theory. The Third Millenium Edition, Revised and Expanded*. Springer Monographs in Mathematics (Springer, Berlin, 2003)
5. S. Laurent, On standardness and I-cosiness. Séminaire de Probabilités XLIII, Springer Lecture Notes in Mathematics, vol. 2006 (Springer, Berlin, 2011), pp. 127–186
6. J. Neveu, *Bases Mathématiques du Calcul des Probabilités*. Deuxième édition (Masson, Paris, 1970)
7. A. Rényi, *Foundations of Probability* (Holden-Day, San Francisco, 1970)

A Planar Borel Set Which Divides Every Non-negligible Borel Product

Michel Émery*

> [...] tout événement est comme un moule d'une forme
> particulière [...].
>
> M. PROUST, *Albertine disparue.*

Abstract In the unit square $[0, 1] \times [0, 1]$ endowed with the Lebesgue measure λ, we construct a Borel subset A with the following property: if U and V are any two non-negligible Borel subsets of $[0, 1]$, then $0 < \lambda\bigl(A \cap (U \times V)\bigr) < \lambda(U \times V)$.

Given a probability space $(\Omega, \mathscr{A}, \mathbb{P})$, let $\mathscr{B} \subset \mathscr{A}$ be a Boolean algebra which generates the σ-field $\mathscr{A}$ (possibly up to negligible events). It is well known that $\mathscr{B}$ is dense in $\mathscr{A}$, in the sense that every event $A \in \mathscr{A}$ is the a.s. limit of some sequence from $\mathscr{B}$; and A is also a.s. equal to some event from $\mathscr{B}_{\sigma\delta}$ and to some event from $\mathscr{B}_{\delta\sigma}$. (See for instance Chapter I in Neveu [2].)

It is not difficult to exhibit examples where *some $A \in \mathscr{A}$ divides each non-negligible event from $\mathscr{B}$*, that is,

$$\forall B \in \mathscr{B} \qquad \mathbb{P}(B) > 0 \implies 0 < \mathbb{P}(A \cap B) < \mathbb{P}(B).$$

Consider for instance an i.i.d. sequence of signs $\varepsilon = (\varepsilon_n)_{n \geq 1}$ such that $\mathbb{P}[\varepsilon_n = 1] = \mathbb{P}[\varepsilon_n = -1] = \frac{1}{2}$; call $(\mathscr{F}_n)_{n \geq 1}$ its natural filtration. The σ-field $\mathscr{F}_\infty$ generated

---

*The support of the ANR programme ProbaGeo is gratefully acknowledged.

M. Émery (✉)
IRMA, CNRS et Université Unique de Strasbourg, 7 rue René Descartes, 67084 Strasbourg Cedex, France
e-mail: michel.emery@math.unistra.fr

by ε is also generated by the Boolean algebra $\mathscr{B} = \bigcup_n \mathscr{F}_n$, which consists of the events depending upon finitely many ε_n. The r.v. $S = \sum_n \varepsilon_n/n$ is a.s. finite, but unbounded from both sides; as a consequence, $\mathbb{P}[S > 0 \mid \mathscr{F}_n]$ and $\mathbb{P}[S < 0 \mid \mathscr{F}_n]$ are a.s. > 0, and the event $A = \{S > 0\}$ divides each non-negligible event from $\mathscr{B}$.

The usual realization of the random variables ε_n on the sample space $\Omega = [0, 1]$ via dyadic expansions transforms the preceding example into a Borel set $A \subset [0, 1]$ which divides every non-negligible dyadic interval. Note that, since every non-negligible interval contains a non-negligible dyadic interval, A also divides each non-negligible event from the Boolean algebra of finite unions of arbitrary intervals (that algebra is much bigger than $\mathscr{B}$, the Boolean algebra of finite unions of dyadic intervals).

Of course, there also exist situations where no event from $\mathscr{A}$ can divide each non-negligible element from $\mathscr{B}$; the most trivial case is when $\mathscr{B} = \mathscr{A}$.

The present note is devoted to the case when Ω is the unit square $[0, 1] \times [0, 1]$ (endowed with its Borel σ-field and the Lebesgue measure) and when $\mathscr{B}$ is the Boolean algebra of all finite unions of Borel products; by a Borel product, we mean a product $B = U \times V$ of two Borel subsets of $[0, 1]$. In this framework, we are going to exhibit a Borel set $A \subset [0, 1] \times [0, 1]$ dividing each non-negligible Borel product. We have been led to study this question when working on "qualitative conditional independence" in [1]; the existence of such an A stood as an obstruction to what we wanted to do (see [1] for more details).

Proposition 1. *Fix a sequence $(p_n)_{n \geq 1}$ such that $0 \leq p_n \leq 1$, $\sum_n p_n = \infty$, and $\sum_n p_n^2 < \infty$.*

Let $X = (X_n)_{n \geq 1}$ and $Y = (Y_n)_{n \geq 1}$ be two independent sequences of independent r.v. with values in $\{0, 1\}$, such that

$$\mathbb{P}[X_n = 1] = \mathbb{P}[Y_n = 1] = p_n \quad \text{and} \quad \mathbb{P}[X_n = 0] = \mathbb{P}[Y_n = 0] = 1 - p_n.$$

Call N the integer-valued r.v. $\sum_{n \geq 1} X_n Y_n$ (this series is a.s. convergent because $\mathbb{E}[N] = \sum_n p_n^2 < \infty$). For all events $U \in \sigma(X)$ and $V \in \sigma(Y)$, one has

$$\mathbb{P}(U \cap V) > 0 \quad \Longrightarrow \quad 0 < \mathbb{P}[N \text{ is even} \mid U \cap V] < 1.$$

As the (sequence-valued) r.v. X has a diffuse law, the sample space $(\Omega, \sigma(X), \mathbb{P})$ is isomorphic to $([0, 1], \text{Borel}, \text{Lebesgue})$; similarly with Y (an explicit isomorphism could easily be exhibited). Consequently, by taking a product, $(\Omega, \sigma(X, Y), \mathbb{P})$ is isomorphic to $([0, 1] \times [0, 1], \text{Borel}, \text{Lebesgue})$. On this square, X and Y are Borel functions of the coordinates, and, according to Proposition 1, $\{N \text{ is even}\}$ is a Borel subset of the square which divides each non-negligible Borel product.

The proof of Proposition 1 relies on an estimate in a finite probability space:

Lemma 1. *Let I be a finite set, and let $(p_i)_{i \in I}$ be any family of numbers such that $0 \leq p_i \leq 1$; let also Z be a r.v. taking values in $\{0, 1\}^I$, whose components Z_i are independent, with $\mathbb{P}[Z_i = 1] = p_i$ and $\mathbb{P}[Z_i = 0] = 1 - p_i$. If H and K are two subsets of $\{0, 1\}^I$ such that* *for all* $x \in H$ *and* $y \in K$, $\sum_{i \in I} x_i y_i$ *is even,*

then

$$\mathbb{P}[Z \in H]\,\mathbb{P}[Z \in K] \leq \prod_{i \in I} \max(p_i, 1-p_i) \,.$$

The proof of Lemma 1 will be given later; we first show how this lemma entails Proposition 1.

Proof of Proposition 1 (Lemma 1 is admitted). Given $U \in \sigma(X)$ and $V \in \sigma(Y)$ such that $\mathbb{P}(U \cap V) > 0$, we must establish two facts. Firstly, $\mathbb{P}[N \text{ is even} | U \cap V] < 1$; this means that $U \cap V$ is not a.s. included in the event $\{N \text{ is even}\}$. And secondly, $\mathbb{P}[N \text{ is even} | U \cap V] > 0$, which means that $U \cap V$ is not a.s. included in $\{N \text{ is odd}\}$.

The second fact is a consequence of the first one: it suffices to introduce the three new sequences $(p'_n)_{n \geq 0} = (1, p_1, p_2, \ldots)$, $X' = (1, X_1, X_2, \ldots)$, $Y' = (1, Y_1, Y_2, \ldots)$, and to observe that $\sigma(X') = \sigma(X)$, $\sigma(Y') = \sigma(Y)$ and that $N' = \sum_{n \geq 0} X'_n Y'_n = 1 + N$ is even if and only if N is odd.

Therefore, only the first fact needs to be proved. So, *fixing now* $U \in \sigma(X)$ *and* $V \in \sigma(Y)$ *such that the event* $U \cap V$ *is a.s. included in* $\{N \text{ is even}\}$, we have to show that $\mathbb{P}(U \cap V) = 0$.

Given $n \geq 1$, put $N_{>n} = \sum_{i>n} X_i Y_i$ and $\varepsilon_n = \mathbb{P}[N_{>n} \text{ is odd}]$. Later, n will tend to infinity, so $N_{>n}$ will go to 0 and ε_n will tend to 0 too; but for the time being, n and ε_n are fixed.

For $x = (x_1, \ldots, x_n)$ and $y = (y_1, \ldots, y_n)$ in $\{0, 1\}^n$, set

$$E_x = \{(X_1, \ldots, X_n) = x\} \quad \text{and} \quad F_y = \{(Y_1, \ldots, Y_n) = y\} \,.$$

Define

$$H = \{x \in \{0,1\}^n \,:\, \mathbb{P}(E_x) > 0 \quad \text{and} \quad \mathbb{P}[U \mid E_x] > \sqrt{\varepsilon_n}\} \,; \qquad E = \bigcup_{x \in H} E_x \,;$$

$$K = \{y \in \{0,1\}^n \,:\, \mathbb{P}(F_y) > 0 \quad \text{and} \quad \mathbb{P}[V \mid F_y] > \sqrt{\varepsilon_n}\} \,; \qquad F = \bigcup_{y \in K} F_y \,.$$

The event E is nothing but $\{(X_1, \ldots, X_n) \in H\}$. For $x \in \{0,1\}^n \setminus H$, either $\mathbb{P}(E_x) = 0$, or $\mathbb{P}(E_x) > 0$ and $\mathbb{P}[U \mid E_x] \leq \sqrt{\varepsilon_n}$; in both cases, one has $\mathbb{P}(U \cap E_x) \leq \sqrt{\varepsilon_n}\, \mathbb{P}(E_x)$. Consequently,

$$\mathbb{P}(U \setminus E) = \sum_{x \notin H} \mathbb{P}(U \cap E_x) \leq \sum_{x \notin H} \sqrt{\varepsilon_n}\, \mathbb{P}(E_x) \leq \sqrt{\varepsilon_n} \sum_{x \in \{0,1\}^n} \mathbb{P}(E_x) = \sqrt{\varepsilon_n} \,,$$

$$(*)$$

so U is approximately included in E. Similarly, one has $\mathbb{P}(V \setminus F) \leq \sqrt{\varepsilon_n}$, and V is approximately included in F.

For x and y such that $\mathbb{P}(E_x) > 0$, $\mathbb{P}(F_y) > 0$ and $\sum_{i=1}^{n} x_i y_i$ is odd, one has the estimate

$$\mathbb{P}[U \cap V \mid E_x \cap F_y] \leq \mathbb{P}[N \text{ is even} \mid E_x \cap F_y] \\ = \mathbb{P}[N_{>n} \text{ is odd} \mid E_x \cap F_y] = \mathbb{P}[N_{>n} \text{ is odd}] = \varepsilon_n \qquad (**)$$

(we have first used the fact that N is a.s. even on the event $U \cap V$, then the formula $N = \sum_{i=1}^{n} x_i y_i + N_{>n}$ valid on $E_x \cap F_y$, and last the independence of E_x, F_y and $N_{>n}$). Taking now x in H and y in K, one has $\mathbb{P}[U \mid E_x] > \sqrt{\varepsilon_n}$ and $\mathbb{P}[V \mid F_y] > \sqrt{\varepsilon_n}$, whence, by independence of $\sigma(X)$ and $\sigma(Y)$,

$$\mathbb{P}[U \cap V \mid E_x \cap F_y] = \mathbb{P}[U \mid E_x] \mathbb{P}[V \mid F_y] > \sqrt{\varepsilon_n} \sqrt{\varepsilon_n} = \varepsilon_n \;.$$

This violates the estimate $(**)$; consequently, $\sum_{i=1}^{n} x_i y_i$ must be even, and H and K satisfy the hypothesis of Lemma 1. Using that lemma, and calling Z a random vector with the same law as $(X_1, \ldots, X_n)$ and $(Y_1, \ldots, Y_n)$, one has

$$\mathbb{P}(E \cap F) = \mathbb{P}(E) \mathbb{P}(F) = \mathbb{P}[(X_1, \ldots, X_n) \in H] \mathbb{P}[(Y_1, \ldots, Y_n) \in K] \\ = \mathbb{P}[Z \in H] \mathbb{P}[Z \in K] \leq \prod_{i=1}^{n} \max(p_i, 1-p_i) \;. \qquad (***)$$

Last, writing $U \cap V \subset (E \cap F) \cup (U \setminus E) \cup (V \setminus F)$ and taking into account the estimates $(*)$ and $(***)$, one has

$$\mathbb{P}(U \cap V) \leq \mathbb{P}(E \cap F) + \mathbb{P}(U \setminus E) + \mathbb{P}(V \setminus F) \\ \leq \prod_{i=1}^{n} \max(p_i, 1-p_i) + \sqrt{\varepsilon_n} + \sqrt{\varepsilon_n}.$$

Time has come to unfix n and let it tend to infinity. As previously observed, ε_n goes to zero; and so does also $\prod_{i=1}^{n} \max(p_i, 1-p_i)$, because $\max(p_i, 1-p_i) = 1 - p_i$ for all but finitely many i (this is due to $\sum p_i^2 < \infty$), and because the series $\sum p_i$ is divergent. So $\mathbb{P}(U \cap V) = 0$; the proposition is proved. □

We still owe you the proof of Lemma 1. This combinatorial estimate will be established via simple linear-algebraic arguments, at the mild cost of some definitions.

From now on, a finite set I is fixed. The set $\{0,1\}^I$ will be identified with the power set $\mathscr{P}(I)$; for instance, the sum $\sum_{i \in I} x_i y_i$ in Lemma 1 is but $|x \cap y|$, the number of elements of $x \cap y$. We endow $\{0,1\}$ with its field structure (addition modulo 2 and usual multiplication), and call it $\mathbb{F}$. Then $\{0,1\}^I = \mathscr{P}(I)$ is further identified with $\mathbb{F}^I$, which is a vector space over the field $\mathbb{F}$. In this vector space, we define a symmetric bilinear form $(x, y) \mapsto x \cdot y \in \mathbb{F}$ by $x \cdot y = \sum_i x_i y_i$ (addition modulo 2).

For x and y in $\mathbb{F}^I$, we say that x *is orthogonal to* y, and we write $x \perp y$, whenever $x \cdot y = 0$; via the identification of $\mathbb{F}^I$ with $\mathscr{P}(I)$, this means that $|x \cap y|$ is even. Note that $x \perp x$ does not imply $x = 0$ (if x has an even number of elements, it is not necessarily empty). We also say that *two subsets H and K of $\mathbb{F}^I$ are orthogonal* if every element of H is orthogonal to every element of K; this property was precisely the hypothesis on H and K in Lemma 1. This hypothesis is preserved when H is replaced with its linear span in the vector space $\mathbb{F}^I$; so *without loss of generality, in the proof of Lemma 1 we may (and will) suppose that H is a linear subspace of $\mathbb{F}^I$*. Similarly, the orthogonality hypothesis is preserved if K is replaced with the bigger set

$$H^\perp = \{x \in \mathbb{F}^I : \forall h \in H \quad h \cdot x = 0\} \subset \mathbb{F}^I,$$

which is a linear subspace containing K and orthogonal to H. So in the proof of Lemma 1, *we will suppose that H and K are linear subspaces* of $\mathbb{F}^I$, and $K = H^\perp$.

Linear subspaces of $\mathbb{F}^I$ can easily be characterized: a subset of $\mathbb{F}^I$ is a linear subspace iff it is an additive subgroup, i.e., iff it contains the null vector 0 (that is, the empty set $\varnothing \in \mathscr{P}(I)$) and is stable under additions (i.e., symmetric set-differences in $\mathscr{P}(I)$). This characterization will not be used; we shall only need the necessary (but not sufficient) condition called Fact 1 below.

Our proof of Lemma 1 relies on just one more notion:

Definition 1. Let L be a subset of $\mathbb{F}^I$ (i.e., a set of subsets of I) and r an element of $\mathbb{F}^I$ (i.e., a subset of I). We say that L *is coded* (resp. *exactly coded*) by r if the map

$$\ell \in L \quad \mapsto \quad \ell \cap r \in \mathscr{P}(r)$$

is an injection (resp. a bijection) from L to $\mathscr{P}(r)$.

Using this definition, three (independent) facts will be established:

Fact 1. *Every linear subspace H of $\mathbb{F}^I$ is exactly coded by some subset of I.*

Fact 2. *If a set $H \subset \mathbb{F}^I$ is exactly coded by $r \subset I$, the set $H^\perp$ is coded by $I \setminus r$.*

Fact 3. *If $L \subset \mathbb{F}^I$ is coded by r, then (with the notation of Lemma 1)*

$$\mathbb{P}[Z \in L] \leq \prod_{i \in I \setminus r} \max(p_i, 1-p_i).$$

Lemma 1, with H and K assumed to be two linear subspaces of $\mathbb{F}^I$ such that $K = H^\perp$, immediately results from Facts 1, 2 and 3 and from the trivial identity

$$\Big(\prod_{j\in I\setminus r} m_j\Big)\Big(\prod_{k\in r} m_k\Big) = \prod_{i\in I} m_i \; .$$

Proof of Fact 1. *(Every linear subspace $H \subset \mathbb{F}^I$ is exactly coded by some $r \subset I$.)*

Let H be a linear subspace of $\mathbb{F}^I$, B a basis of H, and $d = |B|$ the dimension[2] of H. Each vector $b \in B$ has the form $b = (b_i)_{i\in I}$, with $b_i \in \mathbb{F}$. The matrix $M = (b_i)_{i\in I,\ b\in B}$ has size $|I| \times d$ and rank d, since its d columns (the elements of B) are linearly independent. Hence one can extract from M a square, $d \times d$ submatrix with rank d; in other words, there exists a subset r of I with $|r| = d$, such that the d vectors $b \cap r$ are linearly independent in $\mathbb{F}^r$. As $\dim(\mathbb{F}^r) = |r| = d$, these d vectors form a basis B' of $\mathbb{F}^r$; thus the linear map $h \mapsto h \cap r$, which bijects the basis B of H with the basis B' of $\mathbb{F}^r$, must be a bijection from H to $\mathbb{F}^r$, and H is exactly coded by r. □

Proof of Fact 2. *(If $H \subset \mathbb{F}^I$ is exactly coded[3] by r, then $H^\perp$ is coded by $I \setminus r$.)*

Let H be exactly coded by r. Putting $r^c = I \setminus r$, we must verify that the map $x \mapsto x \cap r^c$ injects the linear subspace $H^\perp$ into $\mathbb{F}^{r^c}$. This map is linear, so it suffices to check that its kernel is $\{0\}$, or, in terms of sets,

$$x \in H^\perp \quad \text{and} \quad x \cap r^c = \varnothing \quad \Longrightarrow \quad x = \varnothing \; .$$

Fixing an x in $H^\perp$ such that $x \cap r^c = \varnothing$, it suffices to show that $x \cap r = \varnothing$; so, for any given $i \in r$, we shall establish that $i \notin x$.

As H is exactly coded by r, and as $\{i\}$ is a subset of r, there exists an $h \in H$ such that $h \cap r = \{i\}$. A fortiori, $h \cap x$ is included in $\{i\}$, because x is included in r. Now, h is in H and x in $H^\perp$; so $h \perp x$, and the set $h \cap x$ has an even number of elements. Being included in the singleton $\{i\}$, this even set $h \cap x$ must be empty, and i, which belongs to h, cannot belong to x. □

Proof of Fact 3. *(If $L \subset \mathbb{F}^I$ is coded by r, then $\mathbb{P}[Z \in L] \leqslant \prod_{i\notin r} \max(p_i, 1-p_i)$.)*

We keep using the set-theoretic language: for s any subset of I, $Z \cap s$ stands for the random sub-vector $(Z_i)_{i\in s}$ of $Z = (Z_i)_{i\in I}$; and $Z \cap s$ and $Z \cap t$ are independent when s and t are two disjoint subsets of I. For $s \subset I$, put $\Pi(s) = \prod_{i\in s} \max(p_i, 1-p_i)$.

For each $i \in I$ and each $\lambda \in \{0, 1\}$, one has

$$\mathbb{P}[Z_i = \lambda] = \begin{cases} p_i & \text{if } \lambda = 1 \\ 1 - p_i & \text{if } \lambda = 0 \end{cases} \leqslant \max(p_i, 1 - p_i) \; ;$$

hence, for $z \subset s \subset I$, one has the estimate (which does not depend upon z)

---

[2] All linear notions (dimension, rank, independence) are understood with $\mathbb{F}$ as the field of scalars.
[3] A weaker hypothesis suffices, namely, for each $i \in r$ there is at least one $h \in H$ with $h \cap r = \{i\}$.

$$\mathbb{P}[Z \cap s = z] = \prod_{i \in s} \mathbb{P}[Z_i = z_i] \leq \prod_{i \in s} \max(p_i, 1 - p_i) = \Pi(s) \,.$$

Let now L be coded by r, put $r^c = I \setminus r$, and use the previous estimate to write

$$\mathbb{P}[Z \in L] = \sum_{\ell \in L} \mathbb{P}[Z = \ell] = \sum_{\ell \in L} \mathbb{P}[Z \cap r = \ell \cap r] \, \mathbb{P}[Z \cap r^c = \ell \cap r^c]$$
$$\leq \sum_{\ell \in L} \mathbb{P}[Z \cap r = \ell \cap r] \, \Pi(r^c) = \Pi(r^c) \sum_{\ell \in L} \mathbb{P}[Z \cap r = \ell \cap r] \,.$$

When ℓ ranges over L, the sets $\ell \cap r$ are different from each other (because L is coded by r), so the events $\{Z \cap r = \ell \cap r\}$ are pairwise disjoint. As a consequence, $\sum_{\ell \in L} \mathbb{P}[Z \cap r = \ell \cap r] \leq 1$, and finally $\mathbb{P}[Z \in L] \leq \Pi(r^c)$. □

Lemma 1 and Proposition 1 are now fully proved.

A reader who would like to play with the tools from this note can show as an exercise that in Fact 2, the coding of $H^\perp$ by $I \setminus r$ is exact iff H is a linear subspace.

Another possible exercise is to establish the combinatorial formula (featuring possibly negative numbers)

$$\mathbb{P}[Z \in H^\perp] = \frac{1}{|H|} \sum_{h \in H} \prod_{i \in h} (1 - 2p_i) \,,$$

which is valid for H any linear subspace of $\mathbb{F}^I$ and for Z as in Lemma 1. (Hint: for $x \in \mathbb{F}^I$, show that $\sum_{h \in H} (-1)^{h \cdot x} = |H| \mathbf{1}_{x \in H^\perp}$; replace x by Z, and take expectations.)

Acknowledgements I thank Claude Dellacherie for many useful comments, Gaël Ceillier, Jacques Franchi, Nicolas Juillet and Vincent Vigon for a sizeable simplification of the proof, and Wilfrid S. Kendall who suggested the term "coded".

References

1. C. Dellacherie, M. Émery, Filtrations indexed by ordinals; application to a conjecture of S. Laurent. Séminaire de Probabilités XLV, Springer Lecture Notes in Mathematics, vol. 2078 (Springer, Berlin, 2013), pp. 141–157
2. J. Neveu, *Mathematical Foundations of the Calculus of Probability* (Holden-Day, San Francisco, 1965)

Characterising Ocone Local Martingales with Reflections

Jean Brossard and Christophe Leuridan

Abstract Let $M = (M_t)_{t \geq 0}$ be any continuous real-valued stochastic process such that $M_0 = 0$. Chaumont and Vostrikova proved that if there exists a sequence $(a_n)_{n \geq 1}$ of positive real numbers converging to 0 such that M satisfies the reflection principle at levels 0, a_n and $2a_n$, for each $n \geq 1$, then M is an Ocone local martingale. They also asked whether the reflection principle at levels 0 and a_n only (for each $n \geq 1$) is sufficient to ensure that M is an Ocone local martingale.

We give a positive answer to this question, using a slightly different approach, which provides the following intermediate result. Let a and b be two positive real numbers such that $a/(a+b)$ is not dyadic. If M satisfies the reflection principle at the level 0 and at the first passage-time in $\{-a, b\}$, then M is close to a local martingale in the following sense: $|\mathbf{e}[M_{S \circ M}]| \leq a + b$ for every stopping time S in the canonical filtration of $\mathbf{w} = \{w \in \mathscr{C}(\mathbf{r}_+, \mathbf{r}) : w(0) = 0\}$ such that the stopped process $M_{\cdot \wedge (S \circ M)}$ is uniformly bounded.

Keywords Ocone martingales • Reflection principle

AMS 2000 subject classifications. 60G44, 60G42, 60J65.

1 Introduction

Let $(M_t)_{t \geq 0}$ denote a continuous local martingale, defined on some probability space $(\Omega, \mathscr{A}, P)$, such that $M_0 = 0$. Let $\mathscr{F}^M$ denote its natural filtration and $\mathscr{H}$ the set of all predictable processes with respect to $\mathscr{F}^M$ with values in $\{-1, 1\}$. Then for

---

J. Brossard · C. Leuridan (✉)
Institut Fourier, BP 74 38402 Saint Martin d'Hères CEDEX France
e-mail: jean.brossard@ujf-grenoble.fr; christophe.leuridan@ujf-grenoble.fr

every $H \in \mathcal{H}$, the local martingale

$$H \cdot M = \int_0^{\cdot} H_s dM_s$$

has the same quadratic variation as M. In particular, if M is a Brownian motion, then $H \cdot M$ is still a Brownian motion.

A natural problem is to determine when $H \cdot M$ has the same law as M for every $H \in \mathcal{H}$. Ocone proved in [4] that a necessary and sufficient condition is that M is a Gaussian martingale conditionally on its quadratic variation $\langle M \rangle$. Such processes are called *Ocone local martingales*. Various characterisations of these processes have been given, by Ocone himself, by Dubins, Émery and Yor in [3], by Vostrikova and Yor in [6]. We refer to [2] for a more complete presentation.

The following characterisation, given by Dubins, Émery and Yor, is particularly illuminating: M is an Ocone local martingale if and only if there exists a Brownian motion β (possibly defined on a larger probability space) which is independent of $\langle M \rangle$ and such that $M_t = \beta_{\langle M \rangle_t}$ for every t. Loosely speaking, Ocone local martingales are the processes obtained by the composition of a Brownian motion and an independent time-change.

Another characterisation of Ocone local martingales is based on their invariance with respect to reflections. For every positive real r, call h_r the map from $\mathbf{r}_+$ to $\{-1, 1\}$ defined by

$$h_r(t) = \mathbf{1}_{[t \le r]} - \mathbf{1}_{[t > r]}.$$

Then $h_r \cdot M = \rho_r \circ M$, where ρ_r is the *reflection at time r*. Let $\mathbf{w}$ denote the set of all continuous functions $w : \mathbf{r}_+ \to \mathbf{r}$ such that $w(0) = 0$. The transformation ρ_r maps $\mathbf{w}$ into itself and is defined by

$$\rho_r(w)(t) = \begin{cases} w(t) & \text{if } t \le r, \\ 2w(r) - w(t) & \text{if } t \ge r. \end{cases}$$

The functions h_r are sufficient to characterise Ocone local martingales: Theorem A of [4] states that if $h_r \cdot M$ has the same law as M for every positive r, then $H \cdot M$ has the same law as M for every $H \in \mathcal{H}$. In other words, if the law of M is invariant by the reflections at fixed times, then M is an Ocone local martingale. Note that it is not necessary to assume that M is a local martingale since the invariance by the reflections at fixed times implies that for every $t \ge s \ge 0$, the law of the increment $M_t - M_s$ is symmetric conditionally on $\mathcal{F}_s^M$.

The celebrated reflection principle due to André [1] shows that it may be worthwhile to consider reflections at first-passage times, which we now define. For every real a and $w \in \mathbf{w}$, note $T_a(w)$ the first-passage time of w at level a. The reflection at time T_a transforms w into $\rho_{T_a}(w)$ where

$$\rho_{T_a}(w)(t) = \begin{cases} w(t) & \text{if } t \le T_a(w), \\ 2a - w(t) & \text{if } t \ge T_a(w). \end{cases}$$

Note that $\rho_{T_a}(w) = w$ if $T_a(w)$ is infinite.

Chaumont and Vostrikova recently established in [2] that any continuous process whose law is invariant by the reflections at first-passage times is an Ocone local martingale. Actually, their result is even stronger.

Theorem 1 (Theorem 1 of [2]). *Let M be any continuous stochastic process such that $M_0 = 0$. If there exists a sequence $(a_n)_{n\geq 1}$ of positive real numbers converging to 0 such that the law of M is invariant by the reflections at times $T_0 = 0$, T_{a_n} and T_{2a_n}, then M is an Ocone local martingale. Moreover, if $T_{a_1} \circ M$ is almost surely finite, then M is almost surely divergent.*

We note that the assumption that the law of M is invariant by the reflection ρ_0 is missing in [2] and that it cannot be omitted: consider for example the deterministic process defined by $M_t = -t$. However, if $\inf\{t \geq 0 : M_t > 0\}$ is 0 almost surely, the invariance by ρ_0 is a consequence of the invariance by the reflections $\rho_{T_{a_n}}$.

To prove Theorem 1 above, Chaumont and Vostrikova establish a discrete version of the theorem and they apply it to some discrete approximations of M. The discrete version (Theorem 3 in [2]) states that if $(M_n)_{n\geq 0}$ is a discrete time skip-free process (this means that $M_0 = 0$ and $M_n - M_{n-1} \in \{-1, 0, 1\}$ for every $n \geq 1$) whose law is invariant by the reflections at times T_0, T_1 and T_2, then $(M_n)_{n\geq 0}$ is a discrete Ocone martingale (this means that $(M_n)_{n\geq 0}$ is obtained by the composition of a symmetric Bernoulli random walk with an independent skip-free time change).

The fact that the three invariances by the reflections at times T_0, T_1, and T_2 are actually useful (two of them would not be sufficient) explains the surprising requirement that the law of $(M_t)_{t\geq 0}$ is invariant by reflections at times T_{a_n} and T_{2a_n} in Theorem 1 of [2]. Chaumont and Vostrikova ask whether the assumption on T_{2a_n} can be removed. Their study of the discrete case could lead to believe that it cannot. Yet, we give in this paper a positive answer to this question. Here is our main result.

Theorem 2. *Let M be any continuous stochastic process such that $M_0 = 0$. If there exists a sequence $(a_n)_{n\geq 1}$ of positive real numbers converging to 0 such that the law of M is invariant by the reflections at times $T_0 = 0$ and T_{a_n}, then M is an Ocone local martingale. Moreover, if $T_{a_1} \circ M$ is almost surely finite, then M is almost surely divergent.*

We provide a simpler proof of this stronger statement (the final steps in the approximation method of [2] were rather technical). Let us now indicate the steps of the proof and the plan of the paper.

Our proof first uses some stability properties of the set of all stopping times T such that ρ_T preserves the law of M. These properties are established in Sect. 2.

In Sect. 3, we show that for any positive real numbers a and b such that $a/(a+b)$ is not dyadic, if the reflections ρ_0 and $\rho_{T_{-a}\wedge T_b}$ preserve the law of M, then M is close to a local martingale in the following sense: for every stopping time S in the canonical filtration of $\mathbf{w}$ such that the stopped process $M_{\cdot\wedge(S\circ M)}$ is uniformly bounded, $|\mathbf{e}[M_{S\circ M}]| \leq a + b$. To prove this, we build a nondecreasing sequence $(\tau_n)_{n\geq 0}$ of stopping times, increasing while finite ($\tau_n < \tau_{n+1}$ if $\tau_n < +\infty$), starting with $\tau_0 = 0$, such that the reflections ρ_{τ_n} preserve the law of M and such that the increments of M on each interval $[\tau_n, \tau_{n+1}]$ are bounded by $a + b$.

The proof that the reflections ρ_{τ_n} actually preserve the law of M is given in Sect. 4. The final step of the proof of Theorem 2 is in Sect. 5.

To prove these results, it is more convenient to work in the canonical space. From now on, $\mathscr{W}$ denotes the σ-field on **w** generated by the canonical projections, $X = (X_t)_{t \geq 0}$ the coordinate process on $(\mathbf{w}, \mathscr{W})$, and $\mathscr{F}^0$ its natural filtration of the space **w** (without any completion). Moreover, Q denotes the law of M and $\mathbf{e}_Q$ is the expectation with respect to Q.

2 Stability Properties

Call $\mathscr{T}_Q$ the set of all stopping times T of the filtration $\mathscr{F}^0$ such that the reflection ρ_T preserves Q. In this section, we establish some stability properties of $\mathscr{T}_Q$. Let us begin with a preliminary lemma.

Lemma 1. *Let S and T be $\mathscr{F}^0$-stopping times. Let w_1 and w_2 in **w**. If w_1 and w_2 coincide on $[0, T(w_1) \wedge T(w_2)]$, then*

- $T(w_1) = T(w_2)$;
- *either $S(w_1) = S(w_2)$ or $S(w_1) \wedge S(w_2) > T(w_1) = T(w_2)$.*

Thus, the random times S and T are in the same order on w_1 as on w_2.

Proof. The first point is an application of Galmarino's test (see [5], Chap. I, Exercise 4.21). The second follows by the same argument, since the inequality $S(w_1) \wedge S(w_2) \leq T(w_1) = T(w_2)$ would imply that w_1 and w_2 coincide on $[0, S(w_1) \wedge S(w_2)]$. □

Corollary 1. *Let T be an $\mathscr{F}^0$-stopping time. Then*

1. $T \circ \rho_T = T$
2. ρ_T *is an involution.*
3. *for every $A \in \mathscr{F}_T^0$, $\rho_T^{-1}(A) = A$. In particular, if S is another stopping time, the events $\{S < T\}$, $\{S = T\}$ and $\{S > T\}$ are invariant by ρ_T.*

Proof. The first point is a consequence of the application of the application of Lemma 1 to the paths w and $\rho_T(w)$. The secund point follows. The third point is another application of Galmarino's test (see [5], Chap. I, Exercise 4.21) since w and $\rho_T(w)$ coincide on $[0, T(w)]$. □

The next lemma states that $\mathscr{T}_Q$ is stable by the optional mixtures.

Lemma 2. *Let (S_n) be a (finite or infinite) sequence of $\mathscr{F}^0$-stopping times and (A_n) a measurable partition of $(\mathbf{w}, \mathscr{W})$ such that $A_n \in \mathscr{F}_{S_n}$ for every n. Then*

$$T := \sum_n S_n \mathbf{1}_{A_n}$$

is an $\mathscr{F}^0$-stopping time. If $S_n \in \mathscr{T}_Q$ for every n, then $T \in \mathscr{T}_Q$.

Proof. Note that T is an $\mathscr{F}^0$-stopping time since for every $t \in \mathbf{r}_+$,

$$\{T \leq t\} = \bigcup_n (A_n \cap \{S_n \leq t\}) \in \mathscr{F}_t.$$

Fix any bounded measurable function ϕ from $\mathbf{w}$ to $\mathbf{r}$. Since for each n, the event A_n and the probability Q are invariant by ρ_{S_n}, one has

$$\mathbf{e}_Q[\phi \circ \rho_T] = \sum_n \mathbf{e}_Q[(\phi \circ \rho_{S_n}) \mathbf{1}_{A_n}]$$

$$= \sum_n \mathbf{e}_Q[(\phi \mathbf{1}_{A_n}) \circ \rho_{S_n}]$$

$$= \sum_n \mathbf{e}_Q[\phi \mathbf{1}_{A_n}]$$

$$= \mathbf{e}_Q[\phi].$$

Hence ρ_T preserves Q. □

Corollary 2. *For every S and T in $\mathscr{T}_Q$, $S \wedge T$ and $S \vee T$ are in $\mathscr{T}_Q$.*

Proof. As the events $\{S < T\}$, $\{S = T\}$ and $\{S > T\}$ belong to $\mathscr{F}_S \cap \mathscr{F}_T$, the result is a direct application of Lemma 2. □

The following lemmas will be used to prove a subtler result: if S and T are in $\mathscr{T}_Q$, then $S \circ \rho_T$ is in $\mathscr{T}_Q$.

Lemma 3. *Let S and T be $\mathscr{F}^0$-stopping times. Then the following holds.*

- *For every $t \geq 0$, $\rho_T^{-1}(\mathscr{F}_t) = \mathscr{F}_t$.*
- *$S \circ \rho_T$ is an $\mathscr{F}^0$-stopping time.*

Proof. Fix $t \geq 0$. Then $\rho_T^{-1}(\mathscr{F}_t)$ is the σ-field generated by the random variables $X_s \circ \rho_T$ for $s \in [0, t]$, and the equality

$$X_s \circ \rho_T = (2X_T - X_s)\mathbf{1}_{[T \leq s]} + X_s \mathbf{1}_{[T > s]}$$

shows that these random variables are measurable for $\mathscr{F}_t$. Thus $\rho_T^{-1}(\mathscr{F}_t) \subset \mathscr{F}_t$. Since ρ_T is an involution, the reverse inclusion follows, which proves the first statement.

For each $t \geq 0$, $\{S \circ \rho_T \leq t\} = \rho_T^{-1}(\{S \leq t\}) \in \mathscr{F}_t$, which proves the second statement. □

Lemma 4. *Let S and T be $\mathscr{F}^0$-stopping times and $w \in \mathbf{w}$. Then the following holds.*

- *If $S(w) \leq T(w)$, then $S(\rho_T(w)) = S(w)$ and $\rho_{S \circ \rho_T}(w) = \rho_S(w)$.*
- *If $S(w) \geq T(w)$, then $T(\rho_S(\rho_T(w))) = T(w)$ and $\rho_{S \circ \rho_T}(w) = \rho_T(\rho_S(\rho_T(w)))$.*

Proof. If $S(w) \leq T(w)$, then w and $\rho_T(w)$ coincide on $[0, S(w)]$, thus $S(\rho_T(w)) = S(w)$ and $\rho_{S \circ \rho_T}(w) = \rho_S(w)$.

If $S(w) \geq T(w)$, then $S(\rho_T(w)) \geq T(\rho_T(w)) = T(w)$ by Corollary 1. The trajectories $\rho_S(\rho_T(w))$, $\rho_T(w)$ and w coincide on $[0, T(w)]$, thus $T(\rho_S(\rho_T(w))) = T(w)$. But, to get $\rho_T \circ \rho_S \circ \rho_T(w)$ from w, one must successively:

- multiply by -1 the increments after $T(w)$;
- multiply by -1 the increments after $S(\rho_T(w))$;
- multiply by -1 the increments after $T(\rho_S(\rho_T(w)))$.

Since $T(\rho_S(\rho_T(w))) = T(w)$, one gets $\rho_{S \circ \rho_T}(w) = \rho_T \circ \rho_S \circ \rho_T(w)$. □

Lemma 5. *For every S and T in $\mathcal{T}_Q$, $S \circ \rho_T$ belongs to $\mathcal{T}_Q$.*

Proof. By Lemma 4 and Corollary 1, one has, for every $B \in \mathcal{W}$,

$$Q[\rho_{S \circ \rho_T}^{-1}(B)] = Q[\rho_{S \circ \rho_T}^{-1}(B) \, ; \, S \leq T] + Q[\rho_{S \circ \rho_T}^{-1}(B) \, ; \, S > T]$$
$$= Q[\rho_S^{-1}(B \cap \{S \leq T\})] + Q[(\rho_T \circ \rho_S \circ \rho_T)^{-1}(B \cap \{S > T\})]$$
$$= Q[B \cap \{S \leq T\}] + Q[B \cap \{S > T\}] = Q[B].$$

Thus $S \circ \rho_T$ belongs to $\mathcal{T}_Q$. □

Here is a simple application of our last lemmas.

Corollary 3. *For every $a \in \mathbf{r}$, $T_{-a} = T_a \circ \rho_0$ and $\rho_{T_{-a}} = \rho_0 \circ \rho_{T_a} \circ \rho_0$. Thus, if $0 \in \mathcal{T}_Q$ and $T_a \in \mathcal{T}_Q$, then $T_{-a} \in \mathcal{T}_Q$.*

Proof. The first equality is obvious. The second equality follows from Lemma 4. One can deduce the last point either from the first equality by Lemma 5 or directly from the second equality. □

3 Reflections at 0 and at the Hitting Time of $\{-a, b\}$

We keep the notations of the previous section and we fix two positive real numbers a, b such that $a/(a+b)$ is not dyadic. Note that $T = T_{-a} \wedge T_b$ is the hitting time of $\{-a, b\}$. This section is devoted to the proof of the following result.

Proposition 1. *Let Q be a probability measure on $(\mathbf{w}, \mathcal{W})$. If $0 \in \mathcal{T}_Q$ and $T \in \mathcal{T}_Q$, then, for every finite stopping time S in the canonical filtration of $\mathbf{w}$ such that the stopped process $X_{. \wedge S}$ is uniformly bounded, one has*

$$|\mathbf{e}_Q[X_S]| \leq a + b.$$

Note that the process X may not be a local martingale. The law of any process which stops when its absolute value hits $\min(a, b)$ fulfills the assumptions provided it is invariant by T_0.

The requirement that $a/(a+b)$ is not dyadic may seem surprising, and one could think that it is just a technicality provided by the method used to prove the result. In fact, Proposition 1 becomes false if this assumption is removed. A simple counterexample is given by the continuous stochastic process $(M_t)_{t \geq 0}$ defined by

$$M_t = \begin{cases} t\xi & \text{if } t \leq 1, \\ \xi + (t-1)\eta & \text{if } t > 1, \end{cases}$$

where ξ and η are independent symmetric Bernoulli random variables. Indeed, the law Q of M is invariant by reflections at times 0 and $T_{-1} \wedge T_1$ since $T_{-1} \wedge T_1 = 1$ Q-almost surely. Yet, for every $c > 1$, the random variable $X_{T_{-2} \wedge T_c}$ is uniform on $\{-2, c\}$ and its expectation $(c-2)/2$ can be made as large as one wants.

The proof of Proposition 1 uses an increasing sequence of stopping times defined as follows. Call D the set of $c \in]-a, b[$ such that $(c+a)/(b+a)$ is not dyadic. For every $x \in D$, set

$$f(x) = \begin{cases} 2x + a & \text{if } x < (b-a)/2, \\ 2x - b & \text{if } x > (b-a)/2. \end{cases}$$

This defines a map f from D to D. Conjugating f by the affine map which sends $-a$ on 0 and b on 1 gives the classical map $x \mapsto 2x \bmod 1$ restricted to the non-dyadic elements of $]0, 1[$.

By hypothesis, $0 \in D$, so one can define an infinite sequence $(c_n)_{n \geq 0}$ of elements of D by $c_0 = 0$, and $c_n = f(c_{n-1})$ for $n \geq 1$. By definition, c_{n-1} is the middle point of the subinterval $[c_n, d_n]$ of $[-a, b]$, where

$$d_n = \begin{cases} -a & \text{if } c_{n-1} < (b-a)/2, \\ b & \text{if } c_{n-1} > (b-a)/2. \end{cases}$$

Note that $|c_n - c_{n-1}| = d(c_{n-1}, \{-a, b\})$.

We define a sequence $(\tau_n)_{n \geq 0}$ of stopping times on $\mathbf{w}$ by setting $\tau_0 = 0$, and for every $n \geq 1$,

$$\tau_n(w) = \inf\{t \geq \tau_{n-1}(w) : |w(t) - w(\tau_{n-1}(w))| = |c_n - c_{n-1}|\}.$$

Note that $c_n \neq c_{n-1}$ for every $n \geq 1$, hence the sequence $(\tau_n(w))_{n \geq 0}$ is increasing. Moreover, since $(|c_n - c_{n-1}|)_{n \geq 1}$ does not converge to 0, the continuity of w forces the sequence $(\tau_n(w))_{n \geq 0}$ to be unbounded. By convention, we set $\tau_\infty = +\infty$.

Note that if $a = -1$ and $b = 2$, then $c_n = 0$ for every even n and $c_n = 1$ for every odd n and the sequence $(\tau_n)_{n \geq 0}$ is similar to the sequences used in [2].

The proof of Proposition 1 relies on the following key statement.

Proposition 2. *If $0 \in \mathcal{T}_Q$ and $T \in \mathcal{T}_Q$, then $\tau_n \in \mathcal{T}_Q$ for every $n \geq 0$.*

This statement, that will be proved in the next section, has a remarkable consequence.

Corollary 4. *If $0 \in \mathcal{T}_Q$ and $T \in \mathcal{T}_Q$, then the sequence $(Y_n)_{n \geq 0}$ of random variables defined on the probability space $(\mathbf{w}, \mathcal{W}, Q)$ by*

$$Y_n(w) = X_{\tau_{D_n}}(w), \text{ where } D_n(w) = \max\{k \leq n : \tau_k(w) < +\infty\}.$$

is a martingale in the filtration $(\mathcal{F}^0_{\tau_n})_{n \geq 0}$.

Proof. Fix $n \geq 0$. The equality

$$Y_n(w) = \sum_{k=0}^{n-1} \mathbf{1}_{[\tau_k(w) < +\infty \,;\, \tau_{k+1}(w) = +\infty]} X_{\tau_k}(w) + \mathbf{1}_{[\tau_n(w) < +\infty]} X_{\tau_n}(w)$$

shows that Y_n is measurable for $\mathcal{F}^0_{\tau_n}$. Moreover, from the equality

$$Y_{n+1}(w) - Y_n(w) = (X_{\tau_{n+1}}(w) - X_{\tau_n}(w)) \mathbf{1}_{[\tau_{n+1}(w) < +\infty]},$$

we deduce that $(Y_{n+1} - Y_n) \circ \rho_{\tau_n} = -(Y_{n+1} - Y_n)$. Take $A \in \mathcal{F}^0_{\tau_n}$. By Corollary 1, $\rho_{\tau_n}^{-1}(A) = A$. Since the reflection ρ_{τ_n} preserves Q, we get

$$\mathbf{e}_Q[(Y_{n+1} - Y_n)\mathbf{1}_A] = \mathbf{e}_Q[((Y_{n+1} - Y_n)\mathbf{1}_A) \circ \rho_{\tau_n}] = -\mathbf{e}_Q[(Y_{n+1} - Y_n)\mathbf{1}_A],$$

which shows that $\mathbf{e}_Q[Y_{n+1} - Y_n | \mathcal{F}^0_{\tau_n}] = 0$. $\square$

We are now ready to prove Proposition 1.

Proof. Fix $C \in \mathbf{r}_+$ such that $|X_{t \wedge S}(w)| \leq C$ for every $t \in \mathbf{r}_+$ and $w \in \mathbf{w}$. For each $w \in \mathbf{w}$, set $N(w) = \inf\{n \geq 1 : \tau_n(w) \geq S(w)\}$. Since $S(w)$ is finite and $\tau_n(w)$ is unbounded as $n \to +\infty$, $N(w)$ is finite.

For every $n \geq 0$, $\{N \leq n\} = \{\tau_n \geq S\} \in \mathcal{F}^0_{\tau_n}$. Thus N is a stopping time and $(Y_{n \wedge N})_{n \geq 0}$ is a martingale in the filtration $(\mathcal{F}^0_{\tau_n})_{n \geq 0}$. Note that:

- for all $n < N(w)$, one has $\tau_{D_n}(w) < S(w)$ hence $|Y_n(w)| = |X_{D_n(w)}| \leq C$,
- and $|Y_N(w)| \leq |Y_{N-1}(w)| + |Y_N(w) - Y_{N-1}(w)| \leq C + (a+b)/2$.

This shows that the martingale $(Y_{n \wedge N})_{n \geq 0}$ is uniformly bounded, hence it converges in $L^1(Q)$ to Y_N and $\mathbf{e}_Q[Y_N] = \mathbf{e}_Q[Y_0] = 0$.

Note that $\tau_{N-1} < S < +\infty$, hence, by definition, $Y_N = X_{\tau_N}$ or $Y_N = X_{\tau_{N-1}}$. The inequalities $\tau_{N-1} < S \leq \tau_N$ and the fact that the increments of X are bounded by $a+b$ on each interval $[\tau_{n-1}, \tau_n[$ yield $|X_S - Y_N| \leq a+b$. Since $\mathbf{e}_Q[Y_N] = 0$, the proof is complete. $\square$

4 Proof of Proposition 2

We keep the notations of the previous section, and we introduce for every $n \geq 1$,

$$\varepsilon_n = \frac{Y_n - Y_{n-1}}{c_n - c_{n-1}} = \frac{X_{\tau_n} - X_{\tau_{n-1}}}{c_n - c_{n-1}} \mathbf{1}_{[\tau_n < +\infty]}.$$

For every $e = (e_n)_{n \geq 1} \in \{-1, 0, 1\}^\infty$, set

$$m_0(e) = \inf\{n \geq 1 : e_n = 0\}, \quad m(e) = \inf\{n \geq 1 : e_n = -1\}.$$

Call Σ the set of all sequences $e = (e_n)_{n \geq 1} \in \{-1, 0, 1\}^\infty$ such that $e_n = 0$ for all $n \geq m_0(e)$. Then $\varepsilon = (\varepsilon_n)_{n \geq 1}$ can be seen as a random variable with values in Σ.

The first key point is that T is always one of the times $(\tau_n)_{n \geq 1}$.

Lemma 6. *One has $T = \tau_{m \circ \varepsilon}$ (remind the convention $\tau_\infty = +\infty$). Thus, for every $n \geq 1$, $\{m \circ \varepsilon = n\} = \{T = \tau_n < +\infty\}$.*

Proof. Fix $w \in \mathbf{w}$ and set $m = m(\varepsilon(w))$ and $m_0 = m_0(\varepsilon(w))$.
For every $n \geq 1$, by definition of τ_n and ε_n, one has

$$\varepsilon_n(w) = \pm 1 \text{ if } \tau_n(w) < +\infty,$$
$$\varepsilon_n(w) = 0 \text{ if } \tau_n(w) = +\infty.$$

In particular, $\tau_n(w) = +\infty$ for every $n \geq m_0$ since the sequence $(\tau_n)_{n \geq 0}$ is non decreasing. Thus, whether $m \leq m_0$ or $m \geq m_0$, one has $\tau_{m \wedge m_0}(w) = \tau_m(w)$.

For every $k < m \wedge m_0$, $\varepsilon_k(w) = 1$ hence $w(\tau_k(w)) - w(\tau_{k-1}(w)) = c_k - c_{k-1}$. A recursion then gives $w(\tau_k(w)) = c_k \in]-a, b[$. Moreover, for $\tau_k(w) \leq t < \tau_{k+1}(w)$,

$$|w(t) - c_k| = |w(t) - w(\tau_k(w))| < |c_{k+1} - c_k| = d(c_k, \{-a, b\}).$$

Hence for every $t \in [0, \tau_m(w)[$, $w(t) \notin \{-a, b\}$. This proves that $T(w) \geq \tau_m(w)$.

If m is infinite, then $T(w)$ is infinite.

If m is finite, the equality

$$w(\tau_m(w)) - w(\tau_{m-1}(w)) = -(c_m - c_{m-1}) = d_m - c_{m-1}$$

implies $w(\tau_m(w)) = d_m \in \{-a, b\}$, hence $T(w) = \tau_m(w)$.

The proof of the first statement is complete. Since the sequence $(\tau_n(w))_{n \geq 0}$ is increasing and unbounded, the second statement follows. □

We can now describe the effect of the reflection ρ_T on the sequence $\varepsilon = (\varepsilon_n)_{n \geq 1}$. For every $e = (e_n)_{n \geq 1} \in \Sigma$, define $r(e) = (f_n)_{n \geq 1} \in \Sigma$ by

$$f_n = \begin{cases} e_n & \text{if } n \leq m(e), \\ -e_n & \text{if } n > m(e). \end{cases}$$

Set $g(e) = r(-e)$ and $\gamma = \rho_T \circ \rho_0$. Note that g and γ are bijective maps.

Corollary 5. *With the notation above, the following properties hold.*

1. *The reflections ρ_0, ρ_T and their composition $\gamma = \rho_T \circ \rho_0$ preserve the stopping times τ_n.*
2. *One has $\varepsilon \circ \rho_0 = -\varepsilon$, $\varepsilon \circ \rho_T = r \circ \varepsilon$ and $\varepsilon \circ \gamma = g \circ \varepsilon$.*

Proof. Let $w \in \mathbf{w}$. The trajectories w and $\rho_T(w)$ have the same increments on $[0, T(w)]$ and have opposite increments on $[T(w), +\infty[$. Since $T(w) = \tau_{m \circ \varepsilon}(w)$, the results on ρ_T follow immediately. The other statements are obvious. $\square$

For $n \in \mathbf{n}$, note $\mathbf{1}_n = (1, \ldots, 1) \in \{-1, 1\}^n$. For $(e_1, \ldots, e_n) \in \{-1, 1\}^n$ and $\sigma \in \Sigma$, note $(e_1, \ldots, e_n, \sigma) \in \Sigma$ the sequence obtained by concatenation. The next formula will play the same role as Lemma 1 of [2].

Lemma 7. *Let $N = a_0 + 2a_1 + \cdots + 2^{n-1}a_{n-1}$ be a natural integer written in base 2 with n digits (the digit a_{n-1} may be 0). Then for every $\sigma \in \Sigma$,*

$$g^N(\mathbf{1}_n, \sigma) = ((-1)^{a_0}, \ldots, (-1)^{a_{n-1}}, \sigma).$$

Moreover, if $n \geq 1$,

$$g^{N-2^{n-1}}(\mathbf{1}_{n-1}, -1, \sigma) = ((-1)^{a_0}, \ldots, (-1)^{a_{n-1}}, \sigma).$$

Proof. The first formula will be proved by induction on the number of digits. If $n = 0$, then $N = 0$ and the formula is obvious.

Assume the formula holds for all integers written with n digits. Let $N = a_0 + 2a_1 + \cdots + 2^n a_n$ be an integer written with $n + 1$ digits.

If $a_n = 0$, then it suffices to write N with n digits and to apply the induction hypothesis to the sequence $(1, \sigma)$.

If $a_n = 1$, let us apply the induction hypothesis to the integer $2^n - 1 = 1 + 2 + \cdots + 2^{n-1}$ and to the sequence $(1, \sigma)$. We get

$$g^{2^n-1}(\mathbf{1}_{n+1}, \sigma) = (-\mathbf{1}_n, 1, \sigma).$$

Applying g once more yields

$$g^{2^n}(\mathbf{1}_{n+1}, \sigma) = (\mathbf{1}_n, -1, \sigma).$$

Applying the induction hypothesis to the integer

$$N - 2^n = a_0 + 2a_1 + \cdots + 2^{n-1}a_{n-1}$$

and to the sequence $(-1, \sigma)$ yields

$$g^N(\mathbf{1}_{n+1}, \sigma) = ((-1)^{a_0}, \ldots, (-1)^{a_{n-1}}, -1, \sigma),$$

which achieves the proof of the first formula.

In particular, if $n \geq 1$, $g^{2^{n-1}}(\mathbf{1}_n, \sigma) = (\mathbf{1}_{n-1}, -1, \sigma)$, hence

$$g^{-2^{n-1}}(\mathbf{1}_{n-1}, -1, \sigma) = (\mathbf{1}_n, \sigma).$$

The second formula follows. □

Introduce $\Sigma_n \subset \{-1, 0, 1\}^n$ the subset of n-uples such that each component after a 0 is 0. Define the map g from Σ_n to itself as before.

Corollary 6. *For every $n \geq 1$ and $e = (e_1, \ldots, e_n) \in \Sigma_n$, there exists an integer $M(e)$ in $[1 - 2^{n-1}, 2^{n-1}]$ such that the event $A_e = \{(\varepsilon_1, \ldots, \varepsilon_n) = (e_1, \ldots, e_n)\}$ belongs to $\mathscr{F}^0_{T \circ \gamma^{M(e)}}$ and $\tau_n = T \circ \gamma^{M(e)}$ on A_e.*

Proof. Write $(e_1, \ldots, e_n) = ((-1)^{a_0}, \ldots, (-1)^{a_{d-1}}, 0, \ldots, 0)$ with $0 \leq d \leq n$ and $a_0, \ldots, a_{d-1}$ in $\{0, 1\}$.

If $d = n$, set $M(e) = 2^{n-1} - a_0 - \cdots - 2^{n-1} a_{n-1}$. Then by Lemmas 7 and 6,

$$A_e = \{g^{M(e)} \circ (\varepsilon_1, \ldots, \varepsilon_n) = (\mathbf{1}_{n-1}, -1)\}$$
$$= \{(\varepsilon_1, \ldots, \varepsilon_n) \circ \gamma^{M(e)} = (\mathbf{1}_{n-1}, -1)\}$$
$$= \{m \circ \varepsilon \circ \gamma^{M(e)} = n\}$$
$$= \{T \circ \gamma^{M(e)} = \tau_n \circ \gamma^{M(e)} < +\infty\}$$

Thus $A_e \in \mathscr{F}^0_{T \circ \gamma^{M(e)}}$, and $\tau_n = \tau_n \circ \gamma^{M(e)} = T \circ \gamma^{M(e)}$ on A_e.

If $d \leq n - 1$, set $M(e) = -a_0 - \cdots - 2^{d-1} a_{d-1}$. Then by Lemmas 7 and 6,

$$A_e = \{g^{M(e)} \circ (\varepsilon_1, \ldots, \varepsilon_n) = (\mathbf{1}_d, 0, \ldots, 0)\}$$
$$= \{(\varepsilon_1, \ldots, \varepsilon_n) \circ \gamma^{M(e)} = (\mathbf{1}_d, 0, \ldots, 0)\}$$
$$= \{m_0 \circ \varepsilon \circ \gamma^{M(e)} = d + 1 \,;\, m \circ \varepsilon \circ \gamma^{M(e)} = +\infty\}$$
$$= \{\tau_d \circ \gamma^{M(e)} < +\infty \,;\, T \circ \gamma^{M(e)} = \tau_{d+1} \circ \gamma^{M(e)} = +\infty\}$$

Thus $A_e \in \mathscr{F}^0_{T \circ \gamma^{M(e)}}$, and $\tau_n = \tau_n \circ \gamma^{M(e)} = +\infty = T \circ \gamma^{M(e)}$ on A_e. □

The last corollary and the stability properties given in Lemmas 5 and 2 show that if $0 \in \mathscr{T}_Q$ and $T \in \mathscr{T}_Q$, then $\tau_n \in \mathscr{T}_Q$ for all $n \geq 0$ (recall that $\tau_0 = 0$). This ends the proof of Proposition 2.

5 Proof of the Main Theorem

Let us now prove Theorem 2.

Proof. Call Q the law of M as before. The first step of the proof is the observation that for every positive integer n, $T_{-a_n} \in \mathcal{T}_Q$ by Corollary 3. Hence for all positive integers m and n, $T_{-a_n} \wedge T_{a_m} \in \mathcal{T}_Q$ by Corollary 2. Lemma 8, which will be stated and proved below, ensures that the ratios $a_n/(a_n + a_m)$ are not dyadic for arbitrarily large m and n. For such m and n, Proposition 1 applies and yields $\mathbf{e}_Q[X_S] \leq a_n + a_m$ for every finite stopping time S (in the canonical filtration $\mathcal{W}$) such that the stopped process $X_{\cdot \wedge S}$ is uniformly bounded. Since $(a_n)_{n \geq 1}$ converges to 0, this proves that $\mathbf{e}_Q[X_S] = 0$, hence X is a local martingale under Q.

The next arguments are the same as in [2] and we now summarize them.

Q-almost surely, the process X admits a quadratic variation $\langle X \rangle$ (defined as a limit in probability of sums of squared increments), which is preserved by the reflections ρ_0 and $\rho_{T_{a_n}}$. Consider a regular version of the conditional law of X with respect to $\langle X \rangle$. For any continuous non-decreasing function $f : \mathbf{r}_+ \to \mathbf{r}_+$ such that $f(0) = 0$, call Q_f the law of X conditionally on $\langle X \rangle = f$. Then for almost every f (for the law of $\langle X \rangle$ under Q), the probability Q_f is invariant by the reflections ρ_0 and $\rho_{T_{a_n}}$.

By the part of the theorem which is already proven, X is a local martingale under Q_f. But $\langle X \rangle = f$ almost surely under Q_f. Calling ϕ the right-continuous inverse of f, one gets that the process $B = (X_{\phi(s)})_{0 \leq s < f(+\infty)}$ is a Brownian motion with lifetime $f(+\infty)$.

Consider, in some suitable enlargement of the probability space $(\mathbf{w}, \mathcal{W}, Q)$, a Brownian motion W, independent of X. For almost every f, the Brownian motion W is still independent of X under Q_f. Since the local martingale X converges Q-almost surely to a random variable X_∞ on the event $\{\langle X \rangle_\infty < +\infty\}$, one gets a Brownian motion B defined on the whole interval $[0, +\infty[$ and independent of $\langle X \rangle$ by setting

$$B_s = X_\infty + W_{s - \langle X \rangle_\infty} \text{ on the event } \{\langle X \rangle_\infty \leq s\}.$$

Since $X_t = B_{\langle X \rangle_t}$ almost surely for all $t \geq 0$, this shows that X is an Ocone local martingale under Q.

Assume now that $\langle X \rangle_\infty$ is finite with positive probability. Then for some $s \in \mathbf{r}_+$, $\langle X \rangle_\infty \leq s$ with positive probability. But with positive probability, B does not visit a_1 before time s. By independence of B and $\langle X \rangle$,

$$Q[T_{a_1} = +\infty] \geq Q[T_{a_1} \circ B > s] \, Q[\langle X \rangle_\infty \leq s] > 0.$$

This shows that if T_{a_1} is finite Q-almost surely, then $\langle X \rangle_\infty$ is infinite Q-almost surely, hence X is almost surely divergent. □

Note that the proof of the last statement (if T_{a_1} is finite Q-almost surely, then X is almost surely divergent) given in the discrete case by Chaumont and Vostrikova (Lemma 2 of [2]) is not correct because they prove the implication

$$T_a(M) \vee T_{-a}(M) < +\infty \text{ a.s.} \implies T_{a+2}(M) \wedge T_{-a-2}(M) < +\infty \text{ a.s.},$$

which is not sufficient to perform an induction. Yet, the same arguments that Chaumont and Vostrikova used to prove their Lemma 1 are sufficient to prove their Lemma 2. Our Lemma 7 generalises these arguments, and the case in which some stopping time T_a is infinite is covered by the possibility for the sequence of signs $\sigma \in \Sigma$ to be eventually 0.

Lemma 8. *If $c > b > a > 0$, then at least one of the three following ratios $a/(a+b)$, $b/(b+c)$ and $a/(a+c)$ is not dyadic.*

Proof. The three ratios above belong to $]0, 1/2[$. Assume that they are dyadic. Then

$$\frac{a}{a+b} = \frac{i}{2^p}, \quad \frac{b}{b+c} = \frac{j}{2^q}, \quad \frac{a}{a+c} = \frac{k}{2^r},$$

where i, j and k are odd positive integers and p, q and r are integers greater or equal to 2. Thus

$$\frac{2^r - k}{k} = \frac{c}{a} = \frac{b}{a} \times \frac{c}{b} = \frac{2^p - i}{i} \times \frac{2^q - j}{j},$$

$$ij(2^r - k) = k(2^p - i)(2^q - j),$$

$$2^r ij + 2^q ik + 2^p jk - 2^{p+q} k = 2ijk.$$

This is a contradiction since the left-hand side is a multiple of 4 whereas the right-hand side is not. □

Acknowledgements The authors thank Loïc Chaumont who aroused our interest in this topic, and the referee for a careful reading and for a simplified proof of Lemma 2.

References

1. D. André, Solution directe du problème résolu par M. Bertrand. C. R. Acad. Sci. Paris **105**, 436–437 (1887)
2. L. Chaumont, L. Vostrikova, Reflection principle and Ocone martingales. Stoch. Process. Appl. **119**(10), 3816–3833 (2009)
3. L., Dubins, M., Émery, M. Yor, On the Lévy transformation of Brownian motions and continuous martingales. *Séminaire de Probabilités, XXVII*. Lecture Notes in Math. vol. 1557 (Springer, Berlin, 1993), pp. 122–132

4. D. Ocone, A symmetry characterization of conditionally independent increment martingales. *Barcelona Seminar on Stochastic Analysis* (St. Feliu de Guíxols, 1991), pp. 147–167. Progr. Probab. **32**, (Birkhäuser, Basel, 1993)
5. D. Revuz, M. Yor, *Continuous Martingales and Brownian Motion*, 3d edn. Grundlehren der Mathematischen Wissenschaften, [Fundamental Principles of Mathematical Sciences] vol. 293 (Springer, Berlin, 1999)
6. L. Vostrikova, M. Yor, Some invariance properties (of the laws) of Ocone's martingales. *Séminaire de Probabilités, XXXIV*. Lecture Notes in Math., vol. 1729 (Springer, Berlin, 2000), pp. 417–431

Approximation and Stability of Solutions of SDEs Driven by a Symmetric α Stable Process with Non-Lipschitz Coefficients

Hiroya Hashimoto

Abstract Firstly, we investigate Euler–Maruyama approximation for solutions of stochastic differential equations (SDEs) driven by a symmetric α stable process under Komatsu condition for coefficients. The approximation implies naturally the existence of strong solutions. Secondly, we study the stability of solutions under Komatsu condition, and also discuss it under Belfadli–Ouknine condition.

Keywords Euler–Maruyama approximation • Stability of solution • Symmetric α stable process

1 Introduction

Euler–Maruyama approximation is a key tool in the theory of stochastic differential equations (SDEs) as well as Picard approximation is. In this domain the theory on stability properties of solutions is considered as one of the cornerstones. This article is devoted firstly to study Euler–Maruyama approximation in the pathwise sense, and secondly to investigate stability problems also in the pathwise sense.

We consider these problems in the case where SDEs with non-Lipschitz coefficients are driven by a symmetric α stable process ($1 < \alpha < 2$). SDEs driven by a symmetric α stable process more generally those of pure jumps type arise naturally in connection with applications, for example, mathematical finance.

We briefly sketch some known results in this area. In the case where the driving process is a Brownian motion ($\alpha = 2$), it has been shown that Euler–Maruyama approximation is convergent in the pathwise sense under non-Lipschitz condition for coefficients (Yamada [18], see also Kaneko–Nakao [9]).

H. Hashimoto (✉)
Sanwa Kagaku Kenkyusho Co., Ltd., Nagoya, Japan
e-mail: hiro_hashimoto@skk-net.com

Stability problems for solutions have been developed very well by Émery [5] and Protter [15], in the framework of SDEs driven by semimartingales with Lipschitz coefficients.

Stability problems in law sense for martingale problem solutions of SDEs driven by jump processes have been discussed by many authors, for example, Kasahara–Yamada [10] and Janicki–Michna–Weron [8]. A number of papers devoted to these problems are seen in the references in [8]. Stability problems in the pathwise sense for solutions of Brownian SDEs with non-Lipschitz coefficients have been discussed in Kawabata–Yamada [11].

In the theory of SDEs driven by a symmetric α stable process, some non-Lipschitz conditions for coefficients which guarantee the pathwise uniqueness for solutions are known [2, 3, 7, 12, 17, 19]. Komatsu condition ([12], see also [2]) is an analogue of Yamada–Watanabe condition for one-dimensional Brownian SDEs. A Nagumo type modification of Komatsu condition is shown in [7]. In multi-dimensional case, Tsuchiya [17] considered rather recently the pathwise uniqueness of solutions of SDEs driven by a symmetric α process. Belfadli–Ouknine condition which was found very recently can be seen as the counterpart of Nakao–Le Gall condition in the Brownian motion case [3, 13, 14].

In this situation, it seems to be very natural to investigate Euler–Maruyama approximation as well as stability problems under some non-Lipschitz conditions in the case where SDEs are driven by a symmetric α stable process.

Our paper is organized as follows.

In Sect. 2, we show that Euler–Maruyama approximation is convergent in the pathwise sense under Komatsu condition for coefficients. Theorem 1 in the section corresponds to the main result stated in [18] for Brownian SDEs. Euler–Maruyama approximation in this section implies naturally the existence of strong solutions for SDEs driven by a symmetric α stable process.

In Sect. 3, the stability of solutions for SDEs in the pathwise sense under Komatsu condition is proved. The related result in Brownian motion case has been given in [11].

In Sect. 4, the stability of solutions also in the pathwise sense is discussed under Belfadli–Ouknine condition for coefficients.

2 Euler–Maruyama Approximation

Let $(\Omega, \mathscr{F}, \{\mathscr{F}_t\}, \mathbf{P})$ be a filtered probability space with usual conditions and $Z = \{Z(t); t \geq 0\}$ be a $\mathscr{F}_t$-symmetric α stable process such that $Z(0) = 0$, càdlàg (right continuous left limit) $\mathscr{F}_t$-adapted and

$$\mathbf{E}[\exp\{i\xi(Z(t) - Z(s))\}|\mathscr{F}_s] = \exp\{-(t-s)|\xi|^\alpha\} \text{ a.s. for any } s < t,\ \xi \in \mathbb{R}.$$

In the present section we consider the following SDE:

$$X(t) = X(0) + \int_0^t \sigma(s, X(s-))dZ(s), \tag{1}$$

where $Z(t)$ is a symmetric α stable process ($1 < \alpha < 2$), and $\sigma(t, x)$ is a Borel measurable real function with respect to (t, x).

We assume that the coefficient function $\sigma(t, x)$ satisfies the following condition:

Condition (A)

(i) there exists a positive constant M_1 such that $|\sigma(t, x)| \leq M_1$,
(ii) $\sigma(t, x)$ is uniformly continuous on $[0, \infty) \times \mathbb{R}$,
(iii) there exists a non-negative increasing function ρ defined on $[0, \infty)$ such that: $\rho(0) = 0$, $\int_{0+} \rho^{-1}(x)dx = \infty$

$$|\sigma(t, x) - \sigma(t, y)|^\alpha \leq \rho(|x - y|), \quad \forall x, \forall y \in \mathbb{R}.$$

Remark 1. The condition (A) is called Komatsu condition. Komatsu [12] has proved that under the condition (A) for given bounded initial value $X(0)$ the pathwise uniqueness holds for (1) (see also [2]).

Let $0 < T < \infty$ be a fixed constant. Let Δ be a partition of the interval $[0, T]$, such that $\Delta : 0 = t_0 < t_1 < \cdots < t_k < t_{k+1} < \cdots < t_n = T$. The norm of Δ, $\|\Delta\|$ is defined as $\|\Delta\| := \max_{1 \leq k \leq n}(t_k - t_{k-1})$, and we put $\eta_\Delta(t) := t_k$ for $t_k \leq t < t_{k+1}$.

Euler–Maruyama approximation for (1) is the following:

$X_\Delta(0) := X(0)$,

and

$X_\Delta(t_k) := X_\Delta(t_{k-1}) + \sigma(t_{k-1}, X_\Delta(t_{k-1}-))(Z(t_k) - Z(t_{k-1})), \quad k = 1, 2, \ldots, n.$

For $t_k \leq t < t_{k+1}$, $X_\Delta(t)$ is defined as

$$X_\Delta(t) := X_\Delta(t_k) + \sigma(t_k, X_\Delta(t_k-))(Z(t) - Z(t_k)).$$

Using the notation η_Δ, $X_\Delta(t)$ satisfies the equation:

$$X_\Delta(t) := X(0) + \int_0^t \sigma(\eta_\Delta(s), X_\Delta(\eta_\Delta(s)-))dZ(s).$$

Theorem 1. *We assume condition (A) for (1). Let $X(t)$ be a unique solution of (1) with bounded initial value $X(0)$. Then Euler–Maruyama approximation $X_\Delta(t)$ satisfies*

$$\lim_{\|\Delta\|\to 0} \mathbf{E}\left[\sup_{0\le t\le T} |X_\Delta(t) - X(t)|^\beta\right] = 0 \quad \text{for any } \beta \in (1,\alpha).$$

For the proof of Theorem 1, we prepare several lemmas.

Lemma 1. *Under the same assumption as in Theorem 1, we have*

$$\lim_{|u-v|\to 0} \mathbf{E}[|X_\Delta(u) - X_\Delta(v)|^\beta] = 0 \quad \text{for } \beta \in (1,\alpha), \text{ uniformly with respect to } \Delta.$$

Proof. Choose β' and $p > 0$ such that $1 < \beta < \beta' < \alpha$, and $1/p + 1/\beta' = 1/\beta$. For $v \le u$, we see

$$X_\Delta(u) - X_\Delta(v) = \int_v^u \sigma(\eta_\Delta(s), X_\Delta(\eta_\Delta(s)-))dZ(s).$$

Let $[Y, Y]$ be the quadratic variation of a semimartingale Y (see for examples [4, 15]). By Émery's inequality ([5], page 191 in [15]),

$$\mathbf{E}[([X_\Delta, X_\Delta](u) - [X_\Delta, X_\Delta](v))^{\beta/2}]^{1/\beta}$$

$$\le \mathbf{E}\left[\sup_{v\le s\le u} |\sigma(\eta_\Delta(s), X_\Delta(\eta_\Delta(s)-))|^p\right]^{1/p} \mathbf{E}[([Z, Z](u) - [Z, Z](v))^{\beta'/2}]^{1/\beta'}. \tag{2}$$

Also, by Burkholder–Davis–Gundy's inequality (see for examples [4, 15]), we have

$$c_\beta \mathbf{E}\left[\sup_{v\le s\le u} |X_\Delta(s) - X_\Delta(v)|^\beta\right]^{1/\beta}$$

$$\le \mathbf{E}[([X_\Delta, X_\Delta](u) - [X_\Delta, X_\Delta](v))^{\beta/2}]^{1/\beta} \tag{3}$$

and

$$\mathbf{E}[([Z, Z](u) - [Z, Z](v))^{\beta'/2}]^{1/\beta'}$$

$$\le C_{\beta'}\mathbf{E}\left[\sup_{v\le s\le u} |Z(s) - Z(v)|^{\beta'}\right]^{1/\beta'} \tag{4}$$

where c_β and $C_{\beta'}$ are positive constants which depend on β and β' with respectively. By (i) in (A), the right-hand side of (2) can be bounded by $M_1\mathbf{E}[([Z, Z](u) - [Z, Z](v))^{\beta'/2}]^{1/\beta'}$. Then by (2), (3) and (4), we have

$$\mathbf{E}\left[\sup_{v\leq s\leq u}|X_\Delta(s)-X_\Delta(v)|^\beta\right]^{1/\beta} \leq \frac{C_{\beta'}}{c_\beta}M_1\mathbf{E}\left[\sup_{v\leq s\leq u}|Z(s)-Z(v)|^{\beta'}\right]^{1/\beta'}.$$

Using Doob's inequality (see for example Theorem 1.7 of [16])

$$\mathbf{E}\left[\sup_{0\leq s\leq u-v}|Z(s)|^{\beta'}\right]^{1/\beta'} \leq \frac{\beta'}{\beta'-1}\mathbf{E}[|Z(u-v)|^{\beta'}]^{1/\beta'},$$

we have

$$\mathbf{E}\left[\sup_{v\leq s\leq u}|X_\Delta(s)-X_\Delta(v)|^\beta\right]^{1/\beta} \leq \frac{C_{\beta'}}{c_\beta}\frac{\beta'}{\beta'-1}M_1\mathbf{E}[|Z(u-v)|^{\beta'}]^{1/\beta'}.$$

The right-hand side in the above inequality does not depend on Δ. Thus, we can conclude

$$\lim_{|u-v|\to 0}\|X_\Delta(u)-X_\Delta(v)\|_{L^\beta} = 0, \quad \text{uniformly with respect to } \Delta. \qquad \square$$

Lemma 2. *Under the same assumption as in Theorem 1, we have*

$$\lim_{\|\Delta\|\to 0}\mathbf{E}[|X_\Delta(t)-X(t)|^{\alpha-1}] = 0, \quad \text{for } t\in[0,T].$$

Proof. As is well known (for example see [12]) that the generator $\mathscr{L}$ of $Z(t)$ is defined by

$$\mathscr{L}f(x) := \int [f(x+y)-f(x)-I_{\{|y|\leq 1\}}yf'(x)]|y|^{-1-\alpha}dy.$$

We choose a sequence $\{a_m\}$ such that $1 = a_0 > a_1 > \cdots$ and $\int_{a_m}^{a_{m-1}}\rho^{-1}(x)dx = m$. For this choice we can choose a sequence of sufficiently smooth even functions $\{\varphi_m\}$ such that,

$$\varphi_m(x) = \begin{cases} 0 & |x|\leq a_m \\ \text{between 0 and } 1/(m\rho(|x|)) & a_m < |x| < a_{m-1} \\ 0 & |x|\geq a_{m-1} \end{cases} \qquad (5)$$

and $\int_{-\infty}^{\infty}\varphi_m(x)dx = 1$. Following the argument employed by Komatsu [12], if we put $u(x) := |x|^{\alpha-1}$ and also $u_m := u*\varphi_m$ where $*$ stands for the convolution operator, then we have $\mathscr{L}u_m = K_\alpha\varphi_m$, where $K_\alpha = -2\pi\alpha^{-1}\cot(\alpha\pi/2)$. We note that K_α does not depend on m.

By the definition of u_m, we have

$$|x|^{\alpha-1} - a_{m-1}^{\alpha-1} \leq u_m(x) \leq |x|^{\alpha-1} + a_{m-1}^{\alpha-1}. \tag{6}$$

On the other hand, Itô formula implies

$$u_m(X_\Delta(t) - X(t))$$
$$= K_\alpha \int_0^t \varphi_m(X_\Delta(s) - X(s)) |\sigma(\eta_\Delta(s), X_\Delta(\eta_\Delta(s)-)) - \sigma(s, X(s-))|^\alpha ds$$
$$+ M_m(t)$$

where $M_m(t)$ is a martingale. So, we have

$$\mathbf{E}[u_m(X_\Delta(t) - X(t))]$$
$$= K_\alpha \mathbf{E}\bigg[\int_0^t \varphi_m(X_\Delta(s) - X(s))$$
$$|\sigma(\eta_\Delta(s), X_\Delta(\eta_\Delta(s)-)) - \sigma(s, X(s-))|^\alpha ds\bigg].$$

Using the left-hand side inequality of (6), we have

$$0 \leq \mathbf{E}[|X_\Delta(t) - X(t)|^{\alpha-1}]$$
$$\leq a_{m-1}^{\alpha-1} + K_\alpha \mathbf{E}\bigg[\int_0^t \varphi_m(X_\Delta(s) - X(s))$$
$$|\sigma(\eta_\Delta(s), X_\Delta(\eta_\Delta(s)-)) - \sigma(s, X(s-))|^\alpha ds\bigg]$$
$$\leq a_{m-1}^{\alpha-1} + 2K_\alpha \mathbf{E}\bigg[\int_0^t \|\varphi_m\| |\sigma(\eta_\Delta(s), X_\Delta(\eta_\Delta(s)-)) - \sigma(s, X_\Delta(s))|^\alpha ds\bigg]$$
$$+ 2K_\alpha \mathbf{E}\bigg[\int_0^t \varphi_m(X_\Delta(s) - X(s)) |\sigma(s, X_\Delta(s)) - \sigma(s, X(s-))|^\alpha ds\bigg]$$
$$= a_{m-1}^{\alpha-1} + J_\Delta^{(1)} + J_\Delta^{(2)} \quad \text{say}.$$

By (iii) in (A), (5) implies $J_\Delta^{(2)} \leq (2K_\alpha T)/m$. Let $\varepsilon > 0$ be fixed, we choose a fixed integer m such that $a_{m-1}^{\alpha-1} < \varepsilon/3$ and $J_\Delta^{(2)} < \varepsilon/3$. By (ii) in (A), we choose $\delta_1 > 0$ such that, for $|t - t'| < \delta_1$, $|x - x'| < \delta_1$

$$|\sigma(t, x) - \sigma(t', x')| < \frac{\varepsilon}{12 K_\alpha T \|\varphi_m\|} \tag{7}$$

holds. Lemma 1 implies immediately

$$\lim_{|u-v|\to 0} |X_\Delta(u) - X_\Delta(v)| = 0 \quad \text{(in probability), uniformly with respect to } \Delta.$$

So we can choose $\delta_2 > 0$ such that for $|u - v| < \delta_2$

$$\mathbf{P}(|X_\Delta(u) - X_\Delta(v)| > \delta_1) \leq \frac{\varepsilon}{12 K_\alpha T \|\varphi_m\| (2M_1)^\alpha} \tag{8}$$

holds.

Let $\|\Delta\| \leq \min(\delta_1, \delta_2)$. Then we have $|t - \eta_\Delta(t)| \leq \min(\delta_1, \delta_2)$. The inequalities (7) and (8) imply

$$J_\Delta^{(1)} = 2K_\alpha \|\varphi_m\| \mathbf{E}\left[\int_0^t I_{\{|X_\Delta(s)-X_\Delta(\eta_\Delta(s))|\leq\delta_1\}} |\sigma(s, X(s)) - \sigma(\eta_\Delta(s), X_\Delta(\eta_\Delta(s)-))|^\alpha ds\right]$$

$$+ 2K_\alpha \|\varphi_m\| \mathbf{E}\left[\int_0^t I_{\{|X_\Delta(s)-X_\Delta(\eta_\Delta(s))|>\delta_1\}} |\sigma(s, X(s)) - \sigma(\eta_\Delta(s), X_\Delta(\eta_\Delta(s)-))|^\alpha ds\right]$$

$$< 2K_\alpha \|\varphi_m\| T \frac{\varepsilon}{12 K_\alpha T \|\varphi_m\|} + 2K_\alpha \|\varphi_m\| T \frac{\varepsilon (2M_1)^\alpha}{12 K_\alpha T \|\varphi_m\| (2M_1)^\alpha}$$

$$= \frac{\varepsilon}{3}.$$

Thus, we can conclude that

$$\lim_{\|\Delta\|\to 0} \mathbf{E}[|X_\Delta(t) - X(t)|^{\alpha-1}] = 0. \qquad \square$$

Lemma 3. *Under the same assumption as in Theorem 1,*

[a] $\sup_{0 \leq t \leq T} |X_\Delta(t) - X(t)| \to 0$ *in probability* $(\|\Delta\| \to 0)$,
[b] *the class of random variables*

$$\left\{\sup_{0\leq t\leq T} |X_\Delta(t) - X(t)|^\beta, \Delta\right\}$$

is uniformly integrable.

Proof. We consider the probability $\mathbf{P}(\sup_{0 \leq t \leq T} |X_\Delta(t) - X(t)| > \lambda)$. By Giné–Marcus's inequality (page 213 in [1, 6]) there exists a constant $C > 0$ independent of $\lambda > 0$, such that

$$\mathbf{P}\left(\sup_{0\le t\le T}|X_\Delta(t)-X(t)|>\lambda\right)$$
$$=\mathbf{P}\left(\sup_{0\le t\le T}\left|\int_0^t\{\sigma(\eta_\Delta(s),X_\Delta(\eta_\Delta(s)-))-\sigma(s,X(s-))\}dZ(s)\right|>\lambda\right)$$
$$\le \lambda^{-\alpha}C\int_0^T \mathbf{E}[|\sigma(\eta_\Delta(s),X_\Delta(\eta_\Delta(s)-))-\sigma(s,X(s-))|^\alpha]ds. \tag{9}$$

Note that
$$|X_\Delta(\eta_\Delta(s)-)-X(s-)|$$
$$\le |X_\Delta(\eta_\Delta(s)-)-X_\Delta(s-)|+|X_\Delta(s-)-X(s-)|,$$

and also note that for fixed $u\in[0,T]$, $X_\Delta(u)=X_\Delta(u-)$ a.s. holds. By Lemmas 1 and 2, we can conclude that $|X_\Delta(\eta_\Delta(s)-)-X(s-)|$ converges to 0 in probability when $\|\Delta\|\to 0$. Since the function $\sigma(t,x)$ is bounded and uniformly continuous with respect to (t,x), the inequality (9) implies [a].

Choose $\beta', \tilde{\beta}, p$ such that $\beta<\beta'<\tilde{\beta}<\alpha$, $1/\tilde{\beta}+1/p=1/\beta'$. By an analogous argument as in the proof of Lemma 1, we have

$$\mathbf{E}\left[\sup_{0\le t\le T}|X_\Delta(t)-X(t)|^{\beta'}\right]^{1/\beta'}$$
$$\le \frac{1}{c_{\beta'}}\mathbf{E}[([X_\Delta-X,X_\Delta-X](T))^{\beta'/2}]^{1/\beta'}$$
$$\le \frac{1}{c_{\beta'}}\mathbf{E}\left[\sup_{0\le t\le T}|\sigma(\eta_\Delta(s),X_\Delta(\eta_\Delta(s)-))-\sigma(s,X(s-))|^p\right]^{1/p}$$
$$\times \mathbf{E}[([Z,Z](T))^{\tilde{\beta}/2}]^{1/\tilde{\beta}}$$
$$\le \frac{2M_1}{c_{\beta'}}\mathbf{E}[([Z,Z](T))^{\tilde{\beta}/2}]^{1/\tilde{\beta}}$$
$$\le 2M_1\frac{C_{\tilde{\beta}}}{c_{\beta'}}\mathbf{E}\left[\sup_{0\le t\le T}|Z(t)|^{\tilde{\beta}}\right]^{1/\tilde{\beta}}$$
$$\le 2M_1\frac{C_{\tilde{\beta}}}{c_{\beta'}}\frac{\tilde{\beta}}{\tilde{\beta}-1}\mathbf{E}[|Z(T)|^{\tilde{\beta}}]^{1/\tilde{\beta}}$$
$$<\infty.$$

Thus, we can conclude that the class $\{\sup_{0\le t\le T}|X_\Delta(t)-X(t)|^\beta, \Delta\}$ is uniformly integrable. □

Proof of Theorem 1. Lemma 3 implies immediately Theorem 1. □

Remark 2 (A construction of the strong solution of (1)). Let $X_\Delta(t)$ and $X_{\Delta'}(t)$ be two Euler–Maruyama approximations with the same bounded initial value $X(0)$. By an analogous method employed in the proof of Theorem 1, we can show

$$\lim_{\|\Delta\|\to 0, \|\Delta'\|\to 0,} \mathbf{E}\left[\sup_{0\le t\le T} |X_\Delta(t) - X_{\Delta'}(t)|^\beta\right] = 0, \qquad \forall \beta < \alpha \quad (10)$$

without using the existence of a solution of (1). From this fact we can construct a strong solution of (1) with initial value $X(0)$, in the following way.

Choose a sequence of positive numbers $\varepsilon_i > 0, i = 1, 2, \cdots$ such that

$$\sum_{i=1}^\infty 4^i \varepsilon_i < \infty. \quad (11)$$

By (10) we can choose a series of partitions $\Delta_i, i = 1, 2, \cdots$ such that

(i) $\|\Delta_i\| \to 0\ (i \to \infty)$,
(ii) $\mathbf{E}[\sup_{0\le t\le T} |X_{\Delta_i}(t) - X_{\Delta_{i+1}}(t)|^\beta] < \varepsilon_i, i = 1, 2, \cdots$.

Since

$$\mathbf{P}\left(\sup_{0\le t\le T} |X_{\Delta_i}(t) - X_{\Delta_{i+1}}(t)| > \frac{1}{2^i}\right)$$

$$= \mathbf{P}\left(\sup_{0\le t\le T} |X_{\Delta_i}(t) - X_{\Delta_{i+1}}(t)|^\beta > \left(\frac{1}{2^i}\right)^\beta\right)$$

$$\le \mathbf{P}\left(\sup_{0\le t\le T} |X_{\Delta_i}(t) - X_{\Delta_{i+1}}(t)|^\beta > \frac{1}{4^i}\right)$$

$$\le 4^i \mathbf{E}\left[\sup_{0\le t\le T} |X_{\Delta_i}(t) - X_{\Delta_{i+1}}(t)|^\beta\right]$$

$$< 4^i \varepsilon_i,$$

we have by (11),

$$\sum_{i=1}^\infty \mathbf{P}\left(\sup_{0\le t\le T} |X_{\Delta_i}(t) - X_{\Delta_{i+1}}(t)| > \frac{1}{2^i}\right) < \sum_{i=1}^\infty 4^i \varepsilon_i < \infty.$$

Then, by Borel–Cantelli lemma $X_{\Delta_i}(t)$ converges uniformly on $[0, T]$ a.s. ($i \to \infty$). Put $X(t) := \lim_{i\to\infty} X_{\Delta_i}(t), t \in [0, T]$. Then, we see that $X(t)$ is a càdlàg, and we have

$$\lim_{i\to\infty} \mathbf{E}\left[\sup_{0\le t\le T} |X_{\Delta_i}(t) - X(t)|^\beta\right] = 0.$$

We will show that $(X(t), Z(t))$ satisfies (1). For the purpose of the proof, we have

$$\mathbf{E}\left[\sup_{0\le t\le T}\left|X(t) - X(0) - \int_0^t \sigma(s, X(s-))dZ(s)\right|^\beta\right]$$

$$\le 2\mathbf{E}\left[\sup_{0\le t\le T} |X(t) - X_{\Delta_i}(t)|^\beta\right]$$

$$+ 2\mathbf{E}\left[\sup_{0\le t\le T}\left|\int_0^t \{\sigma(s, X(s-)) - \sigma(\eta_{\Delta_i}(s), X_{\Delta_i}(\eta_{\Delta_i}(s)-))\}dZ(s)\right|^\beta\right]$$

$$= \mathbf{E}[N_i^{(1)}] + \mathbf{E}[N_i^{(2)}] \quad \text{say}.$$

Obviously, we have $\lim_{i\to\infty} \mathbf{E}[N_i^{(1)}] = 0$. By the same discussion employed in the proof of Lemma 3, we can see that $\lim_{i\to\infty} \mathbf{E}[N_i^{(2)}] = 0$. Thus, we have

$$\mathbf{E}\left[\sup_{0\le t\le T}\left|X(t) - X(0) - \int_0^t \sigma(s, X(s-))dZ(s)\right|^\beta\right] = 0.$$

So, we can conclude

$$X(t) = X(0) + \int_0^t \sigma(s, X(s-))dZ(s).$$

By the definition of Euler–Maruyama approximation, $X_{\Delta_i}(t)$ is $\sigma(Z(s); 0 \le s \le t)$ measurable. This implies immediately $X(t)$ is $\sigma(Z(s); 0 \le s \le t)$ measurable. It means that $X(t)$ is a strong solution of (1).

3 Stability of Solutions Under Komatsu Condition

Consider the following sequence of SDEs driven by a same symmetric α stable process $Z(t)$:

$$X(t) = X(0) + \int_0^t \sigma(s, X(s-))dZ(s). \tag{12}$$

$$X_n(t) = X_n(0) + \int_0^t \sigma_n(s, X_n(s-))dZ(s), \quad n = 1, 2, \cdots \tag{13}$$

We assume that the coefficient functions $\sigma(t, x)$, $\sigma_n(t, x)$, $n = 1, 2, \cdots$ satisfy the following condition.

Condition (B)

(i) there exists a positive constant M_2 such that $|\sigma(t, x)| \leq M_2$, $|\sigma_n(t, x)| \leq M_2$, $n = 1, 2, \cdots$,
(ii) $\lim_{n \to \infty} \sup_{t,x} |\sigma_n(t, x) - \sigma(t, x)| = 0$,
(iii) there exists an increasing function ρ defined on $[0, \infty)$ such that $\rho(0) = 0$, $\int_{0+} \rho^{-1}(x) dx = \infty$

$$|\sigma_n(t, x) - \sigma_n(t, y)|^\alpha \leq \rho(|x - y|), \quad \forall x, \forall y \in \mathbb{R},\ t \in [0, \infty),\ n = 1, 2, \cdots$$

$$|\sigma(t, x) - \sigma(t, y)|^\alpha \leq \rho(|x - y|), \quad \forall x, \forall y \in \mathbb{R},\ t \in [0, \infty)$$

Remark 3. The pathwise uniqueness holds for solutions of (12) and (13) under the condition (B) [2, 12].

Remark 4. By (i) of (B), we can assume the function ρ is bounded.

The main result of this section is following:

Theorem 2. *Let $T > 0$ be fixed. Assume that there exists a positive number M_0 such that $|X(0)| \leq M_0$, a.s. and $|X_n(0)| \leq M_0$, a.s., $n = 1, 2, \cdots$. Assume also that*

$$\lim_{n \to \infty} \mathbf{E}[|X_n(0) - X(0)|^\alpha] = 0.$$

Then under the condition (B)

$$\lim_{n \to \infty} \mathbf{E}\left[\sup_{0 \leq t \leq T} |X_n(t) - X(t)|^\beta\right] = 0$$

holds, for $\beta < \alpha$.

For the proof of Theorem 2, we prepare some lemmas. In the following of the section we assume that ρ is bounded (see Remark 4).

Lemma 4. *Under the same assumption as in Theorem 2,*

$$\lim_{n \to \infty} \mathbf{E}[|X_n(t) - X(t)|^{\alpha-1}] = 0 \text{ holds, } t \in [0, T].$$

Proof. Put $u(x) := |x|^{\alpha-1}$. Choose a sequence $\{a_m\}$ such that $1 = a_0 > a_1 > \cdots$, $\lim_{m \to \infty} a_m = 0$ and $\int_{a_m}^{a_{m-1}} \rho^{-1}(x) dx = m$. For this choice, choose a sequence $\{\varphi_m\}$, $m = 1, 2, \cdots$ of sufficiently smooth even functions such that

$$\varphi_m(x) = \begin{cases} 0 & |x| \le a_m \\ \text{between 0 and } 1/(m\rho(|x|)) & a_m < |x| < a_{m-1} \\ 0 & |x| \ge a_{m-1} \end{cases} \quad (14)$$

and $\int_{-\infty}^{\infty} \varphi_m(x)dx = 1$. After Komatsu, put $u_m := u * \varphi_m$, then we have $\mathscr{L}u_m = K_\alpha \varphi_m$, where $K_\alpha = -2\pi\alpha^{-1}\cot(\alpha\pi/2)$. By (ii) of (B) we can choose a sequence $\{\varepsilon_n\}$ of decreasing positive numbers $\varepsilon_n \downarrow 0$ such that

$$\sup_{t,x} |\sigma_n(t,x) - \sigma(t,x)|^\alpha \le \varepsilon_n. \quad (15)$$

Corresponding this choice, we can find a sequence of integer numbers $\{m_n\}$, $n = 1, 2, \cdots$, $m_n \to \infty$ $(n \to \infty)$ such that

$$\varepsilon_n \max_{a_{m_n} \le x \le a_{m_n-1}} \frac{1}{\rho(x)} \le 1. \quad (16)$$

By Itô formula,

$$u_{m_n}(X_n(t) - X(t)) - u_{m_n}(X_n(0) - X(0))$$
$$= \int_0^t |\sigma_n(s, X_n(s-)) - \sigma(s, X(s-))|^\alpha K_\alpha \varphi_{m_n}(X_n(s) - X(s))ds$$
$$+ M_{m_n}(t),$$

where $M_{m_n}(t)$ is a martingale. By (6)

$$\mathbf{E}[u_{m_n}(X_n(t) - X(t))]$$
$$\le \mathbf{E}[|X_n(0) - X(0)|^{\alpha-1}] + a_{m_n-1}^{\alpha-1}$$
$$+ 2\mathbf{E}\left[\int_0^t |\sigma_n(s, X_n(s-)) - \sigma(s, X_n(s-))|^\alpha K_\alpha \varphi_{m_n}(X_n(s) - X(s))ds\right]$$
$$+ 2\mathbf{E}\left[\int_0^t |\sigma(s, X_n(s-)) - \sigma(s, X(s-))|^\alpha K_\alpha \varphi_{m_n}(X_n(s) - X(s))ds\right]$$
$$= \mathbf{E}[|X_n(0) - X(0)|^{\alpha-1}] + a_{m_n-1}^{\alpha-1} + N_{m_n}^{(1)} + N_{m_n}^{(2)} \quad \text{say.}$$

By the assumption, $\lim_{n\to\infty} \mathbf{E}[|X_n(0) - X(0)|^{\alpha-1}] = 0$ holds. Obviously $\lim_{n\to\infty} a_{m_n-1}^{\alpha-1} = 0$.

For $N_{m_n}^{(1)}$, by (14), (15) and (16) we have

$$N_{m_n}^{(1)} \le 2\mathbf{E}\left[\int_0^t \varepsilon_n K_\alpha \frac{I_{\{a_{m_n} < |X_n(s) - X(s)| < a_{m_n-1}\}}}{m_n \rho(|X_n(s) - X(s)|)} ds\right] \le \frac{2K_\alpha T}{m_n}.$$

From this $\lim_{n\to\infty} N_{m_n}^{(1)} = 0$ follows immediately.

For $N_{m_n}^{(2)}$, by (iii) of (B) and by the definition of φ_m, (14) we have

$$N_{m_n}^{(2)} \leq 2\mathbf{E}\left[\int_0^t \rho(|X_n(s) - X(s)|) K_\alpha \frac{I_{\{a_{m_n} < |X_n(s) - X(s)| < a_{m_n-1}\}}}{m_n \rho(|X_n(s) - X(s)|)} ds\right] \leq \frac{2K_\alpha T}{m_n}.$$

From this $\lim_{n\to\infty} N_{m_n}^{(2)} = 0$ holds.

Note that $\lim_{n\to\infty} u_{m_n}(x) = u(x) = |x|^{\alpha-1}$, then we have $\lim_{n\to\infty} \mathbf{E}[|X_n(t) - X(t)|^{\alpha-1}] = 0$. □

Lemma 5. *Under the same assumption as in Theorem 2,*

[a] $\sup_{0 \leq t \leq T} |X_n(t) - X(t)| \to 0$ *in probability* $(n \to \infty)$,
[b] *the family of random variables*

$$\left\{\sup_{0 \leq t \leq T} |X_n(t) - X(t)|^\beta, \, n = 1, 2, \cdots\right\}$$

is uniformly integrable.

Proof. Let λ be a positive constant. Then, by Giné–Marcus's inequality, there exists a constant $C > 0$ which does not depend on λ such that

$$\mathbf{P}\left(\sup_{0 \leq t \leq T} |X_n(t) - X(t)| > \lambda\right)$$

$$\leq \mathbf{P}\left(|X_n(0) - X(0)| > \frac{\lambda}{2}\right)$$

$$+ \mathbf{P}\left(\sup_{0 \leq t \leq T}\left|\int_0^t \{\sigma_n(s, X_n(s-)) - \sigma(s, X(s-))\} dZ(s)\right| > \frac{\lambda}{2}\right)$$

$$\leq \mathbf{P}\left(|X_n(0) - X(0)| > \frac{\lambda}{2}\right)$$

$$+ \left(\frac{\lambda}{2}\right)^{-\alpha} C \int_0^T \mathbf{E}[|\sigma_n(s, X_n(s-)) - \sigma_n(s, X(s-))|^\alpha] ds$$

$$+ \left(\frac{\lambda}{2}\right)^{-\alpha} C \int_0^T \mathbf{E}[|\sigma_n(s, X(s-)) - \sigma(s, X(s-))|^\alpha] ds$$

$$= J_n^{(1)} + J_n^{(2)} + J_n^{(3)} \quad \text{say}.$$

By the assumption on initial data, $\lim_{n\to\infty} J_n^{(1)} = 0$ is obvious. By (ii) of (B), $\lim_{n\to\infty} J_n^{(3)} = 0$ holds. By (iii) of (B) we have

$$J_n^{(2)} \leq \left(\frac{\lambda}{2}\right)^{-\alpha} C \int_0^T \mathbf{E}[\rho(|X_n(s-) - X(s-)|)] ds.$$

By the result of Lemma 4, for fixed $s \in [0,T]$, $X_n(s-) \to X(s-)$ in probability ($n \to \infty$). Since ρ is bounded continuous and $\rho(0) = 0$, we have $\lim_{n\to\infty} \mathbf{E}[\rho(|X_n(s-) - X(s-)|)] = 0$. Note that $\mathbf{E}[\rho(|X_n(s-) - X(s-)|)]$ is uniformly bounded with respect to $s \in [0,T]$. Then by Lebesgue convergence theorem $\lim_{n\to\infty} J_n^{(2)} = 0$ holds. Thus, we can conclude that [a] follows.

Just the same argument employed in the proof of Lemma 3 implies [b]. □

Proof of Theorem 2. Lemma 5 implies immediately Theorem 2. □

4 Stability of Solutions Under Belfadli–Ouknine Condition

In this section we consider the following sequence of SDEs:

$$X(t) = X(0) + \int_0^t \sigma(X(s-))dZ(s) \tag{17}$$

$$X_n(t) = X_n(0) + \int_0^t \sigma_n(X_n(s-))dZ(s), \quad n = 1, 2, \cdots \tag{18}$$

where $Z(t)$ is a symmetric α stable process ($1 < \alpha < 2$), and coefficient functions σ, σ_n are Borel measurable.

We assume that coefficient functions σ, σ_n, $n = 1, 2, \cdots$ satisfy the following condition.

Condition (C)

(i) there exists two positive constants d, K such that $0 < d < K < \infty$,

$$d \leq \sigma(x) \leq K \quad \forall x \in \mathbb{R}$$
$$d \leq \sigma_n(x) \leq K \quad \forall x \in \mathbb{R}, \, n = 1, 2, \cdots$$

(ii) there exist an increasing function f such that for every real numbers x, y

$$|\sigma(x) - \sigma(y)|^\alpha \leq |f(x) - f(y)|$$
$$|\sigma_n(x) - \sigma_n(y)|^\alpha \leq |f(x) - f(y)|, \quad n = 1, 2, \cdots$$

(iii)

$$\lim_{n\to\infty} \sup_x |\sigma_n(x) - \sigma(x)| = 0.$$

Remark 5. Belfadli and Ouknine [3] show that under the (ii) of (C) the pathwise uniqueness holds for each solutions of (17) and (18).

Let $T > 0$ be fixed.

Theorem 3. *Assume that there exists a positive number $\tilde{M}_0$ such that $|X(0)| \leq \tilde{M}_0$ a.s., $|X_n(0)| \leq \tilde{M}_0$ a.s. $n = 1, 2, \cdots$ and assume also*

$$\lim_{n \to \infty} \mathbf{E}[|X_n(0) - X(0)|^\alpha] = 0.$$

Then, under the condition (C)

$$\lim_{n \to \infty} \mathbf{E}\left[\sup_{0 \leq t \leq T} |X_n(t) - X(t)|^\beta\right] = 0$$

holds for $\beta < \alpha$.

For the proof of Theorem 3, we prepare two lemmas.

Lemma 6. *Under the same assumption as in Theorem 3,*

$$\lim_{n \to \infty} \mathbf{E}[|X_n(t) - X(t)|^{\alpha - 1}] = 0, \quad \text{holds for } t \in [0, T].$$

Proof. Put $u(x) := |x|^{\alpha - 1}$. Choose a sequence $\{a_m\}$ such that $1 = a_0 > a_1 > \cdots$, $\lim_{m \to \infty} a_m = 0$ and $\int_{a_m}^{a_{m-1}} x^{-1} dx = m$. For this choice, we can find a sequence $\{\varphi_m\}$ $m = 1, 2, \cdots$ of sufficiently smooth even functions such that $\int_{-\infty}^{\infty} \varphi_m(x) dx = 1$ and

$$\varphi_m(x) = \begin{cases} 0 & |x| \leq a_m \\ \text{between 0 and } 1/|mx| & a_m < |x| < a_{m-1} \\ 0 & |x| \geq a_{m-1} \end{cases}.$$

Put $u_m := u * \varphi_m$, then $\mathscr{L} u_m = K_\alpha \varphi_m$, where $K_\alpha = -2\pi \alpha^{-1} \cot(\alpha \pi / 2)$. Choose also a sequence $\{\varepsilon_n\}$ of positive numbers such that $\varepsilon_n \downarrow 0$ and

$$\sup_x |\sigma_n(x) - \sigma(x)| \leq \varepsilon_n.$$

For this choice, we can find a sequence $\{m_n\}$ of positive integers such that

$$\varepsilon_n \frac{1}{a_{m_n}} \leq 1.$$

By Itô formula,

$$\mathbf{E}[u_{m_n}(X_n(t) - X(t))]$$
$$= \mathbf{E}[u_{m_n}(X_n(0) - X(0))]$$
$$+ \mathbf{E}\Big[K_\alpha \int_0^t |\sigma_n(X_n(s-)) - \sigma(X(s-))|^\alpha \varphi_{m_n}(X_n(s-) - X(s-))ds\Big]$$
$$\leq \mathbf{E}[|X_n(0) - X(0)|^{\alpha-1}] + a_{m_n-1}^{\alpha-1}$$
$$+ 2K_\alpha \mathbf{E}\Big[\int_0^t |\sigma_n(X_n(s-)) - \sigma_n(X(s-))|^\alpha \varphi_{m_n}(X_n(s-) - X(s-))ds\Big]$$
$$+ 2K_\alpha \mathbf{E}\Big[\int_0^t |\sigma_n(X(s-)) - \sigma(X(s-))|^\alpha \varphi_{m_n}(X_n(s-) - X(s-))ds\Big]$$
$$= \mathbf{E}[|X_n(0) - X(0)|^{\alpha-1}] + a_{m_n-1}^{\alpha-1} + N_n^{(1)} + N_n^{(2)} \text{ say.}$$

Obviously we have $\lim_{n\to\infty} \mathbf{E}[|X_n(0) - X(0)|^{\alpha-1}] = 0$ and $\lim_{n\to\infty} a_{m_n-1}^{\alpha-1} = 0$. For $N_n^{(2)}$, we have

$$N_n^{(2)} \leq 2K_\alpha \mathbf{E}\Big[\int_0^T \varepsilon_n^\alpha \frac{1}{m_n a_{m_n}} ds\Big]$$
$$\leq \frac{2K_\alpha T}{m_n}.$$

So $\lim_{n\to\infty} N_n^{(2)} = 0$.

Finally we will discuss $N_n^{(1)}$. By stopping X and X_n, when one of them first leaves a compact set, we can assume $|X| \vee |X_n| \leq M$ for every $t \geq 0$. Letting $\lambda \downarrow 0$ in the last inequality in the proof of Lemma 2.2 in [3]. We get the following inequality:

$$N_n^{(1)} \leq 2K_\alpha \frac{(M+1)\|f\|_\infty}{d^\alpha m_n} \sup_{|a|\leq M+1} \Big(\int_0^{K_\alpha T} p_s(a)ds\Big)$$

where $\|f\|_\infty = \sup_x |f(x)|$ and $p_s(a) = p_s(a, 0)$ is the transition density function of $Z(s)$. So, we can conclude $\lim_{n\to\infty} N_n^{(1)} = 0$. The proof of Lemma 6 is achieved. □

Lemma 7. *Under the same assumption as in Theorem 3,*

[a] *$\sup_{0\leq t\leq T} |X_n(t) - X(t)| \to 0$ in probability ($n \to \infty$),*
[b] *the family of random variables*

$$\Big\{\sup_{0\leq t\leq T} |X_n(t) - X(t)|^\beta, \ n = 1, 2, \cdots\Big\}$$

is uniformly integrable.

Proof. Let λ be a positive constant. Since Giné–Marcus's inequality, there exists a constant $C > 0$ which does not depend on λ such that

$$\mathbf{P}\left(\sup_{0 \le t \le T} |X_n(t) - X(t)| > \lambda\right)$$

$$\le \mathbf{P}\left(|X_n(0) - X(0)| > \frac{\lambda}{2}\right)$$

$$+ \mathbf{P}\left(\sup_{0 \le t \le T} \left|\int_0^t \{\sigma_n(X_n(s-)) - \sigma(X(s-))\} dZ(s)\right| > \frac{\lambda}{2}\right)$$

$$\le \mathbf{P}\left(|X_n(0) - X(0)| > \frac{\lambda}{2}\right)$$

$$+ \left(\frac{\lambda}{2}\right)^{-\alpha} C \int_0^T \mathbf{E}[|\sigma_n(X_n(s-)) - \sigma_n(X(s-))|^\alpha] ds$$

$$+ \left(\frac{\lambda}{2}\right)^{-\alpha} C \int_0^T \mathbf{E}[|\sigma_n(X(s-)) - \sigma(X(s-))|^\alpha] ds$$

$$= J_n^{(1)} + J_n^{(2)} + J_n^{(3)} \quad \text{say.}$$

By the assumption on initial data, $\lim_{n \to \infty} J_n^{(1)} = 0$ is obvious. By (iii) of (C), $\lim_{n \to \infty} J_n^{(3)} = 0$ holds. By (ii) of (C), we have

$$J_n^{(2)} \le \left(\frac{\lambda}{2}\right)^{-\alpha} C \int_0^T \mathbf{E}[|f(X_n(s-)) - f(X(s-))|] ds.$$

Let D be the countable set of discontinuous points of the function f. The following statement holds (see Lemma 2.3 of [3]):

$$\int_0^T \mathbf{P}\left[\bigcup_{n=1}^\infty \{X_n(s-) \in D\} \cup \{X(s-) \in D\}\right] ds = 0.$$

Therefore, for almost all $s \in [0, T]$ (with respect to Lebesgue measure),

$$\mathbf{E}[|f(X_n(s-)) - f(X(s-))|]$$
$$= \mathbf{E}[|f(X_n(s-)) - f(X(s-))|I_{\{X(s-) \in D\}}]$$
$$+ \mathbf{E}[|f(X_n(s-)) - f(X(s-))|I_{\{X(s-) \notin D\}}]$$
$$= \mathbf{E}[|f(X_n(s-)) - f(X(s-))|I_{\{X(s-) \notin D\}}].$$

Assume that there exists a subsequence $\{X_{n_k}\}$ of $\{X_n\}$ such that

$$\lim_{n\to\infty} \mathbf{E}[|f(X_{n_k}(s-)) - f(X(s-))|I_{\{X(s-)\notin D\}}] = r > 0. \qquad (19)$$

By the result of Lemma 6, we can choose some $\{n'_k\} \subset \{n_k\}$ such that $\lim_{n'_k\to\infty} X_{n'_k}(s-) = X(s-)$ a.s. Note that f is bounded. Then

$$\lim_{n'_k\to\infty} \mathbf{E}[|f(X_{n'_k}(s-)) - f(X(s-))|I_{\{X(s-)\notin D\}}] = 0.$$

It is contradictory to (19). Thus, we obtain the following expression: For almost all $s \in [0, T]$,

$$\lim_{n\to\infty} \mathbf{E}[|f(X_n(s-)) - f(X(s-))|] = 0.$$

By Lebesgue convergence theorem, we have $\lim_{n\to\infty} J_n^{(2)} = 0$. Thus, we can conclude that [a] follows.

Just the same argument employed in the proof of Lemma 3 implies [b]. □

Proof of Theorem 3. This is direct consequence of Lemma 7. □

Acknowledgements This work is motivated by fruitful discussions with Toshio Yamada. The author would like to thank him very much. Professor Shinzo Watanabe reviewed original manuscript and offered some polite suggestions. The author also would like to thank him very much for his valuable comments. The author would like to express his thanks to the anonymous referee for several helpful corrections and suggestions.

References

1. D. Applebaum, *Lévy Processes and Stochastic Calculus*, 1st edn. Cambridge Stud. Adv. Math. 93 (2004)
2. R.F. Bass, Stochastic differential equations driven by symmetric stable processes. *Séminaire de Probabilités, XXXVI*. Lecture Notes in Mathematics, vol. 1801 (Springer, New York, 2004), pp. 302–313
3. R. Belfadli, Y. Ouknine, On the pathwise uniqueness of solutions of Stochastic differential equations driven by symmetric stable Lévy processes. Stochastics **80**, 519–524 (2008)
4. C. Dellacherie, P.A. Meyer, *Probabilités et Potentiel* (Théorie des Martingales, Hermann, 1980)
5. M. Émery, Stabilité des solutions des équations différentielles stochastiques applications aux intégrales multiplicatives stochastiques. Z. Wahr. **41**, 241–262 (1978)
6. E. Giné, M.B. Marcus, The central limit theorem for stochastic integrals with respect to Lévy processes. Ann. Prob. **11**, 58–77 (1983)
7. H. Hashimoto, T. Tsuchiya, T. Yamada, On stochastic differential equations driven by symmetric stable processes of index α. *Stochastic Processes and Applications to Mathematical Finance* (World Scientific Publishing, Singapore, 2006), pp. 183–193
8. A. Janicki, Z. Michna, A. Weron, Approximation of stochastic differential equations driven by α-stable Lévy motion. Appl. Math. **24**, 149–168 (1996)

9. H. Kaneko, S. Nakao, A note on approximation for stochastic differential equations. *Séminaire de Probabilités, XXII*. Lecture Notes in Mathematics, vol. 1321 (Springer, New York, 1998), pp. 155–162
10. Y. Kasahara, K. Yamada, Stability theorem for stochastic differential equations with jumps. Stoch. Process. Appl. **38**, 13–32 (1991)
11. S. Kawabata, T. Yamada, On some limit theorems for solutions of stochastic differential equations. *Séminaire de Probabilités, XVI*. Lecture Notes in Mathematics, vol. 920 (Springer, New York, 1982), pp. 412–441
12. T. Komatsu, On the pathwise uniqueness of solutions of one-dimentional stochastic differential equations of jump type. Proc. Jp. Acad. Ser. A. Math. Sci. **58**, 353–356 (1982)
13. J.F. Le Gall, Applications des temps locaux aux équations différentielles stochastiques unidimensionnelles. *Séminaire de Probabilités, XVII*. Lecture Notes in Mathematics, vol. 986 (Springer, New York, 1983), pp. 15–31
14. S. Nakao, On the pathwise uniqueness of solutions of stochastic differential equations. Osaka J. Math. **9**, 513–518 (1972)
15. P. Protter, *Stochastic Integration and Differential Equations* (Springer, New York, 1992)
16. D. Revuz, M. Yor, *Continuous Martingales and Brownian Motion* (Springer, New York, 1991)
17. T. Tsuchiya, On the pathwise uniqueness of solutions of stochastic differential equations driven by multi-dimensional symmetric α stable class. J. Math. Kyoto Univ. **46**, 107–121 (2006)
18. T. Yamada, Sur une Construction des Solutions d'Équations Différentielles Stochastiques dans le Cas Non-Lipschitzien. *Séminaire de Probabilités*. Lecture Notes in Mathematics, vol. 1771 (Springer, New York, 2004), pp. 536–553
19. P.A. Zanzotto, On solutions of one-dimensional stochastic differential equations driven by stable Lévy motion. Stoch. Process. Appl. **68**, 209–228 (1997)

Path Properties and Regularity of Affine Processes on General State Spaces

Christa Cuchiero and Josef Teichmann

Abstract We provide a new proof for regularity of affine processes on general state spaces by methods from the theory of Markovian semimartingales. On the way to this result we also show that the definition of an affine process, namely as stochastically continuous time-homogeneous Markov process with exponential affine Fourier–Laplace transform, already implies the existence of a càdlàg version. This was one of the last open issues in the fundaments of affine processes.

Keywords Affine processes • Markov semimartingales • Path properties • Regularity

MSC 2000: Primary: 60J25; Secondary: 60G17

1 Introduction

In the last decades affine processes have been of great interest in mathematical finance to model phenomena like stochastic volatility, stochastic interest rates, heavy tails, credit default, etc. Pars pro toto we mention here the one-dimensional short-rate model of Cox, Ingersoll and Ross [8], the stochastic volatility model of Heston [18] and the credit risk model of Lando [24]. In order to accommodate the more and more complex structures in finance, these simple models have

---

C. Cuchiero
Faculty of Mathematics, University of Vienna, Nordbergstrasse 15, 1090 Wien, Austria
e-mail: christa.cuchiero@univie.ac.at

J. Teichmann (✉)
Departement Mathematik, ETH Zürich, Rämistrasse 101, 8092 Zürich, Switzerland
e-mail: josef.teichmann@math.ethz.ch

progressively been extended to higher dimensional affine jump diffusions with values in the so-called canonical state space $\mathbb{R}_+^m \times \mathbb{R}^{n-m}$, or in the cone of positive semidefinite matrices, see, e.g., [11, 13, 14] for affine models on the canonical state space and [4, 10, 17, 26] for multivariate stochastic volatility and interest rate models based on matrix-valued affine processes.

Axiomatically speaking affine processes are stochastically continuous Markov processes on some state space $D \subseteq V$, where V is a finite-dimensional Euclidean vector space with scalar product $\langle \cdot, \cdot \rangle$, such that the Fourier–Laplace transform is of exponential affine form in the initial values. More precisely, this means that there exist functions Φ and ψ such that

$$\mathbb{E}_x \left[e^{\langle u, X_t \rangle} \right] = \Phi(t, u) e^{\langle \psi(t,u), x \rangle},$$

for all $(t, x) \in \mathbb{R}_+ \times D$ and $u \in V + iV$, for which $x \mapsto e^{\langle u, x \rangle}$ is a bounded function on D. From this definition neither the Feller property, nor the existence of a càdlàg version, nor differentiability of the Fourier–Laplace transform with respect to time, a concept called *regularity* (see [12, Definition 2.5]), are immediate. This paper provides a positive answer to the latter two questions, while the Feller property is still an open issue on general state spaces, but can probably be established by building on the results of the present article.

The reasons for the strong interest in affine processes are twofold: first, affine processes are a rich and flexible class of Markov processes containing Lévy processes, Ornstein–Uhlenbeck processes, squared Bessel processes and aggregates of them. Second, affine processes are analytically tractable in the sense that the Fourier–Laplace transform, which is a solution of the backward Kolmogorov equation, a PIDE with affine coefficients, can be calculated by solving a system of ODEs for Φ and ψ, the so-called generalized Riccati equations. Having the Fourier–Laplace transform at hand then means that real-time-calibration is at reach from a numerical point of view. However in order to show that the functions Φ and ψ are solutions of these generalized Riccati differential equations, one first has to prove regularity, in other words the differentiability of Φ and ψ with respect to time.

The theory of affine processes has been developed in several steps: in an article by Kawazu and Watanabe [21] the full classification on the state space $\mathbb{R}_+$ was proved, introducing already the generalized Riccati equations and the related affine technology. A key step in this article is to establish the aforementioned differentiability of the functions Φ and ψ with respect to time. After several seminal papers in finance the classification of affine processes for the so-called canonical state space $D = \mathbb{R}_+^m \times \mathbb{R}^{n-m}$ was done by Duffie, Filipović and Schachermayer [12], although under the standing assumption of regularity. It remained open whether there are affine processes on the canonical state space which are not regular, or if regularity follows in fact from stochastic continuity and the property that the Fourier–Laplace transform is of exponential affine form. Indeed, in [22] it is shown that affine processes on the canonical state space $D = \mathbb{R}_+^m \times \mathbb{R}^{n-m}$ are regular, a reasoning motivated by insights from the solution of Hilbert's fifth problem, see [22] for details. However, this solution depends on the full solution of [12] and thus on the particular polyhedral nature of the canonical state space. It remained open if regularity holds on other "non-polyhedral" state spaces, for instance on sets whose boundary is described by a parabola or on (subsets of) the cone of positive semidefinite $d \times d$ matrices, denoted by S_d^+.

The following example of a possible state space illustrates that affine processes can take values in various types of sets and that particular geometric properties of the state space cannot be taken for granted. Consider the subsets of the cone of positive semidefinite $d \times d$ matrices of the form

$$D_k = \{x \in S_d^+ \mid \text{rank}(x) \leq k\}, \quad k \in \{1, \ldots, d\}.$$

In particular, if $k \in \{1, \ldots, d-1\}$, these sets constitute *non-convex* state spaces of affine processes. The non-convexity of D_k, $k \neq d$, is easily seen by the following argument: If D_k was convex, it would contain all convex combinations of positive semidefinite matrices of rank smaller than or equal to k, thus also matrices of rank strictly greater than k, which contradicts the definition of D_k. Moreover, if $k \in \{1, \ldots, d-2\}$, the sets D_k are maximal state spaces for affine processes in a sense made clear in the sequel. To illustrate this phenomenon by an example, let $\langle x, y \rangle := \text{tr}(xy)$ denote the scalar product on S_d, the vector space of $d \times d$ symmetric matrices, and let $d > 2$ and $k \in \{1, \ldots, d-2\}$. Consider a $k \times d$ matrix of independent Brownian motions $(W_t)_{t \geq 0}$ with initial value $W_0 = y \in \mathbb{R}^{k \times d}$ and define the following process

$$X_t = W_t^\top W_t, \quad X_0 = x := y^\top y. \tag{1}$$

Then the distribution of X_t corresponds to the *non-central Wishart distribution* with shape parameter $\frac{k}{2}$, scale parameter $2tI$ and non-centrality parameter x (see, e.g., [27]). Its Fourier–Laplace transform is given by

$$\mathbb{E}_x\left[e^{\langle u, X_t \rangle}\right] = \det(I - 2tu)^{-\frac{k}{2}} e^{\left\langle \frac{(I-2tu)^{-1}u + u(I-2tu)^{-1}}{2}, x \right\rangle}, \quad u \in -S_d^+ + iS_d, \tag{2}$$

and therefore of exponential affine form in all initial values x with $\text{rank}(x) \leq k$. This implies in particular that (1) is an affine process with state space $D_k = \{x \in S_d^+ \mid \text{rank}(x) \leq k\}$ and functions Φ and ψ given by

$$\Phi(t, u) = \det(I - 2tu)^{-\frac{k}{2}},$$

$$\psi(t, u) = \frac{(I - 2tu)^{-1}u + u(I - 2tu)^{-1}}{2}.$$

Note here that the set $\mathcal{U} := \{u \in S_d + iS_d \mid x \mapsto e^{\langle u, x \rangle} \text{ is bounded on } D_k\}$ corresponds to $-S_d^+ + iS_d$. By differentiating Φ and ψ it is easily seen that these functions are solutions of the following system of Riccati ODEs

$$\partial_t \Phi(t,u) = k\Phi(t,u)\langle I, \psi(t,u)\rangle, \qquad \Phi(0,u) = 1,$$
$$\partial_t \psi(t,u) = 2\psi(t,u)^2, \qquad \psi(0,u) = u.$$

From the characterization of affine processes on S_d^+ via the Riccati equations and the corresponding admissible parameters (see [9, Theorem 2.4 and Condition (2.4)]), we know that (2) is the Fourier–Laplace transform of an affine process with state space S_d^+ (meaning in particular that every starting value in S_d^+ is possible), if and only if $k \geq d - 1$. Hence, for $k \in \{1, \ldots, d-2\}$, the state space D_k cannot be enlarged to its convex hull S_d^+ such that the constructed affine process on D_k can also be extended to an affine process on S_d^+. Further affine processes with state space D_k can be obtained from squares of Ornstein–Uhlenbeck processes (see [3]).

The aim is thus to find a unified treatment which allows to prove regularity for all possible state spaces without relying on particular properties of them. In [23] this general question has been solved: it is shown that affine processes are regular on general state spaces D, however, under the assumption that the affine process admits a càdlàg version. The method of proof is probabilistic in the sense that the "absence of regularity" leads—in a probabilistic way—to a contradiction.

This article now provides a new proof inspired by the theory of Markovian semimartingales as laid down in [5]. In order to apply these reasonings, we first prove one of the last open issues in the basics of affine processes, namely that stochastic continuity and the affine property are already sufficient for the existence of a version with càdlàg trajectories, which can then be defined on the canonical probability space of càdlàg paths with a filtration satisfying the usual conditions for any initial value. Let us remark, that in the existing literature on affine processes, the càdlàg property—if addressed—could directly be deduced from the Feller property, whose proof however strongly depends on the particular choice of the state space. Indeed, the Feller property has been shown—under the regularity condition—by Veerman [31] for state spaces of the form $\mathscr{X} \times \mathbb{R}^{n-m}$, where $\mathscr{X} \subset \mathbb{R}^m$ is a closed convex set such that the boundary of

$$\widetilde{\mathscr{U}} := \{u \in \mathbb{R}^m \mid \sup_{x \in \mathscr{X}} \langle u, x \rangle < \infty\}$$

is described by the zeros of a real-analytic function. In the proof, the regularity assumption is crucial to achieve this result. Otherwise the state spaces considered so far are of type $K \times \mathbb{R}^{n-m}$, where $K \subset \mathbb{R}^m$ denotes some proper convex cone. In these cases, the Feller property and also regularity follow from the fact that the function $\psi(t, \cdot)$ maps $-\overset{\circ}{K}{}^* \times i\mathbb{R}^{n-m}$ to itself[1] for all $t \geq 0$ and that the projection of $\psi(t, \cdot)$ on the components corresponding to the $\mathbb{R}^{n-m}$ part of the state space, denoted by $u \mapsto \Pi_{\mathbb{R}^{n-m}}\psi(t,u)$, is a linear function in u. The first assertion hinges on certain order properties of the function $\operatorname{Re}\psi(t, \cdot)$ on $-K^*$, while the second one builds on

---

[1] Here, K^* denotes the dual cone.

the fact that $\Pi_{\mathbb{R}^{n-m}}\psi(t,\cdot)$ maps $i\mathbb{R}^n$ to $i\mathbb{R}^{n-m}$. Since we lack the mentioned order properties of ψ, a similar result to the first one seems hard to establish on general state spaces. The second one can be extended to a certain degree by considering particular sequences and projections, as done in Lemma 1 below. In this respect the main difficulty arises from the fact that we do not have the specific product structure of the state space at hand. For these reasons, we have to take another route, namely martingale regularization for a lot of "test martingales", to show that affine processes admit a version with càdlàg trajectories.

Having achieved this, we proceed with the proof of regularity and provide a *full* and *complete* class in the sense of [5] by using the process' own harmonic analysis. More precisely, we use the fact that, for all $u \in V + iV$, for which $x \mapsto e^{\langle u,x \rangle}$ is a bounded function on D, the map

$$x \mapsto \int_0^\eta \mathbb{E}_x\left[e^{\langle u,X_s\rangle}\right] ds, \quad \eta > 0$$

always lies in the domain of the extended infinitesimal generator of any time-homogeneous Markov process X. The particular form of $\mathbb{E}_x\left[e^{\langle u,X_s\rangle}\right]$ in the case of affine processes then allows to show that the domain of the extended generator actually contains a full and complete class. This in turn implies on the one hand the semimartingale property (up to the lifetime of the affine process) and on the other hand the absolute continuity of the involved characteristics with respect to the Lebesgue measure. The final proof of regularity then builds to a large extent on these results.

The remainder of the article is organized as follows. In Sect. 2 we define affine processes on general state spaces and derive some fundamental properties of the functions Φ and ψ. Sections 3 and 4 are devoted to show the existence of a càdlàg version and the right-continuity of the appropriately augmented filtration. The results on the semimartingale nature of affine process are established in Sect. 5 and are used in Sect. 6 for the proof of regularity.

2 Affine Processes on General State Spaces

We define affine processes as a particular class of time-homogeneous Markov processes with state space $D \subseteq V$, some closed, non-empty subset of an n-dimensional real vector space V with scalar product $\langle \cdot, \cdot \rangle$. Symmetric matrices and the positive semidefinite matrices on V are denoted by $S(V)$ and $S_+(V)$, respectively. We write $\mathbb{R}_+$ for $[0,\infty)$, $\mathbb{R}_{++}$ for $(0,\infty)$ and $\mathbb{Q}_+$ for nonnegative rational numbers. For the stochastic background and notation we refer to standard text books such as [20] and [29].

To further clarify notation, we find it useful to recall in this section the basic ingredients of the theory of time-homogeneous Markov processes and the particular conventions being made in this article (compare [2, Chap. 1.3], [7, Chap. 1.2],

[15, Chap. 4], [30, Chap. 3, Definition 1.1]). Throughout, D denotes a closed subset of V and $\mathscr{D}$ its Borel σ-algebra. Since we shall not assume the process to be conservative, we adjoin to the state space D a point $\Delta \notin D$, called cemetery state, and set $D_\Delta = D \cup \{\Delta\}$ as well as $\mathscr{D}_\Delta = \sigma(\mathscr{D}, \{\Delta\})$. We make the convention that $\|\Delta\| := \infty$, where $\|\cdot\|$ denotes the norm induced by the scalar product $\langle \cdot, \cdot \rangle$, and we set $f(\Delta) = 0$ for any other function f on D. Moreover, in order to allow for exploding processes we shall also deal with a "point at infinity", denoted by ∞, and $D_\Delta \cup \{\infty\}$ then corresponds to the one-point compactification of D_Δ. If the state space D is compact, we *do not* adjoin $\{\infty\}$, since explosion is anyway not possible.

Consider the following objects on a space Ω:

(i) A stochastic process $X = (X_t)_{t \geq 0}$ taking values in D_Δ such that

$$\text{if } X_s(\omega) = \Delta, \text{ then } X_t(\omega) = \Delta \text{ for all } t \geq s \text{ and all } \omega \in \Omega. \qquad (3)$$

(ii) The filtration generated by X, that is, $\mathscr{F}^0_t = \sigma(X_s, s \leq t)$, where we set $\mathscr{F}^0 = \bigvee_{t \in \mathbb{R}_+} \mathscr{F}^0_t$.

(iii) A family of probability measures $(\mathbb{P}_x)_{x \in D_\Delta}$ on $(\Omega, \mathscr{F}^0)$.

In the course of the article, we shall show that the "point at infinity" ∞ can be identified with Δ, since it will turn out that property (3) also holds true for ∞ in our case.

Definition 1 (Markov process). A *time-homogeneous Markov process*

$$X = \left(\Omega, (\mathscr{F}^0_t)_{t \geq 0}, (X_t)_{t \geq 0}, (p_t)_{t \geq 0}, (\mathbb{P}_x)_{x \in D_\Delta}\right)$$

with state space $(D, \mathscr{D})$ (augmented by Δ) is a D_Δ-valued stochastic process such that, for all $s, t \geq 0$, $x \in D_\Delta$ and all bounded $\mathscr{D}_\Delta$-measurable functions $f : D_\Delta \to \mathbb{R}$,

$$\mathbb{E}_x\left[f(X_{t+s})|\mathscr{F}^0_s\right] = \mathbb{E}_{X_s}[f(X_t)] = \int_D f(\xi) p_t(X_s, d\xi), \quad \mathbb{P}_x\text{-a.s.} \qquad (4)$$

Here, $\mathbb{E}_x$ denotes the expectation with respect to $\mathbb{P}_x$ and $(p_t)_{t \geq 0}$ is a *transition function* on $(D_\Delta, \mathscr{D}_\Delta)$. A transition function is a family of kernels $p_t : D_\Delta \times \mathscr{D}_\Delta \to [0, 1]$ such that

(i) for all $t \geq 0$ and $x \in D_\Delta$, $p_t(x, \cdot)$ is a measure on $\mathscr{D}_\Delta$ with $p_t(x, D) \leq 1$, $p_t(x, \{\Delta\}) = 1 - p_t(x, D)$ and $p_t(\Delta, \{\Delta\}) = 1$;
(ii) for all $x \in D_\Delta$, $p_0(x, \cdot) = \delta_x(\cdot)$, where $\delta_x(\cdot)$ denotes the Dirac measure at x;
(iii) for all $t \geq 0$ and $\Gamma \in \mathscr{D}_\Delta$, $x \mapsto p_t(x, \Gamma)$ is $\mathscr{D}_\Delta$-measurable;
(iv) for all $s, t \geq 0$, $x \in D_\Delta$ and $\Gamma \in \mathscr{D}_\Delta$, the Chapman–Kolmogorov equation holds, that is,

$$p_{t+s}(x, \Gamma) = \int_{D_\Delta} p_s(x, d\xi) p_t(\xi, \Gamma).$$

If $(\mathscr{F}_t)_{t\geq 0}$ is a filtration with $\mathscr{F}_t^0 \subset \mathscr{F}_t$, $t \geq 0$, then X is a *time-homogeneous Markov process relative to* $(\mathscr{F}_t)$ if (4) holds with $\mathscr{F}_s^0$ replaced by $\mathscr{F}_s$.

We can alternatively think of the transition function as inducing a measurable contraction semigroup $(P_t)_{t\geq 0}$ defined by

$$P_t f(x) := \mathbb{E}_x[f(X_t)] = \int_D f(\xi) p_t(x, d\xi), \quad x \in D_\Delta,$$

for all bounded $\mathscr{D}_\Delta$-measurable functions $f : D_\Delta \to \mathbb{R}$.

Remark 1. (i) Note that, in contrast to [12], we do not assume Ω to be the canonical space of all functions $\omega : \mathbb{R}_+ \to D_\Delta$, but work on some general probability space.
(ii) Since we have $p_t(x, \Gamma) = \mathbb{P}_x[X_t \in \Gamma]$ for all $t \geq 0$, $x \in D_\Delta$ and $\Gamma \in \mathscr{D}_\Delta$, property (ii) and (iii) of the transition function, imply $\mathbb{P}_x[X_0 = x] = 1$ for all $x \in D_\Delta$ and measurability of the map $x \mapsto \mathbb{P}_x[X_t \in \Gamma]$ for all $t \geq 0$ and $\Gamma \in \mathscr{D}_\Delta$.

For the following definition of affine processes, let us introduce the set $\mathscr{U}$ defined by

$$\mathscr{U} = \left\{ u \in V + \mathrm{i}\, V \mid \mathrm{e}^{\langle u, x \rangle} \text{ is a bounded function on } D \right\}. \tag{5}$$

Clearly $\mathrm{i}\, V \subseteq \mathscr{U}$. Here, the set $\mathrm{i}\, V$ stands for purely imaginary elements and $\langle \cdot, \cdot \rangle$ is the extension of the real scalar product to $V + \mathrm{i}\, V$, but without complex conjugation. Moreover, we denote by p the dimension of $\mathrm{Re}\,\mathscr{U}$ and write $\langle \mathrm{Re}\,\mathscr{U} \rangle$ for its (real) linear hull and $\langle \mathrm{Re}\,\mathscr{U} \rangle^\perp$ for its orthogonal complement in V. The set $\mathrm{i}\, \langle \mathrm{Re}\,\mathscr{U} \rangle^\perp \subset \mathscr{U}$ corresponds to the purely imaginary directions of $\mathscr{U}$. Finally, for some linear subspace $W \subset V$, $\Pi_W : V \to V$ denotes the orthogonal projection on W, which is extended to $V + \mathrm{i}\, V$ by linearity, i.e., $\Pi_W(v_1 + \mathrm{i}\, v_2) := \Pi_W v_1 + \mathrm{i}\, \Pi_W v_2$.

Furthermore we need the sets

$$\mathscr{U}^m = \left\{ u \in V + \mathrm{i}\, V \mid \sup_{x \in D} \mathrm{e}^{\langle \mathrm{Re}\, u, x \rangle} \leq m \right\}, \quad m \geq 1,$$

and note that $\mathscr{U} = \bigcup_{m \geq 1} \mathscr{U}^m$ and $\mathrm{i}\, V \subseteq \mathscr{U}^m$ for all $m \geq 1$.

Assumption 1. *Recall that* $\dim V = n$. *We suppose that the state space D contains at least $n + 1$ affinely independent elements $x_1, \ldots, x_{n+1}$, that is, the n vectors $(x_1 - x_j, \ldots, x_{j-1} - x_j, x_{j+1} - x_j, \ldots, x_{n+1} - x_j)$ are linearly independent for every $j \in \{1, \ldots, n+1\}$.*

We are now prepared to give our main definition:

Definition 2 (Affine process). A time-homogeneous Markov process X relative to some filtration $(\mathscr{F}_t)$ and with state space $(D, \mathscr{D})$ (augmented by Δ) is called *affine* if

(i) it is stochastically continuous, that is, $\lim_{s \to t} p_s(x, \cdot) = p_t(x, \cdot)$ weakly on D for every $t \geq 0$ and $x \in D$, and

(ii) its Fourier–Laplace transform has exponential affine dependence on the initial state. This means that there exist functions $\Phi : \mathbb{R}_+ \times \mathcal{U} \to \mathbb{C}$ and $\psi : \mathbb{R}_+ \times \mathcal{U} \to V + iV$ such that, for every $x \in D$ and $m \geq 1$, the map $(t, u) \mapsto \langle \psi(t, u), x \rangle$ is locally continuous on the subset of $\mathbb{R}_+ \times \mathcal{U}^m$ where Φ does not vanish, and

$$\mathbb{E}_x \left[e^{\langle u, X_t \rangle} \right] = P_t e^{\langle u, x \rangle} = \int_D e^{\langle u, \xi \rangle} p_t(x, d\xi) = \Phi(t, u) e^{\langle \psi(t, u), x \rangle}, \quad (6)$$

for all $x \in D$ and $(t, u) \in \mathbb{R}_+ \times \mathcal{U}$.

Remark 2. (i) The above definition differs in four crucial details from the definitions given in [12, Definition 2.1, Definition 12.1].[2]

a. First, therein the right hand side of (6) is defined in terms of a function $\phi(t, u)$, namely as $e^{\phi(t,u) + \langle \psi(t,u), x \rangle}$, such that the function $\Phi(t, u)$ in our definition corresponds to $e^{\phi(t,u)}$. Our definition is in line with the one given in [21] and [22, 23] and differs from the one in [12], as we do not require $\Phi(t, u) \neq 0$ a priori. However, since all affine processes on $D = \mathbb{R}_+^m \times \mathbb{R}^{n-m}$ are infinitely divisible (see [12, Theorem 2.15]), it turns out with hindsight that setting $\Phi(t, u) = e^{\phi(t,u)}$ is actually no restriction.

b. Second, we assume that the affine property (6) holds for all $u \in \mathcal{U}$, whereas on the canonical state space $D = \mathbb{R}_+^m \times \mathbb{R}^{n-m}$ it is restricted to $i\mathbb{R}^n$ (see [12]). This however turns out to imply the affine property (6) also on $\mathcal{U}$.

c. Third, in contrast to [12], we take stochastic continuity as part of the definition of an affine process. We remark that there are simple examples of Markov processes which satisfy Definition 2 (ii), but are not stochastically continuous (see [12, Remark 2.11]).

d. Fourth, due to the general structure of the state space D, we decided to assume local continuity of $(t, u) \mapsto \langle \psi(t, u), x \rangle$ on the subset of $\mathbb{R}_+ \times \mathcal{U}^m$ where Φ does not vanish, which we denote by $\mathcal{Q}^m = \{(t, u) \in \mathbb{R}_+ \times \mathcal{U}^m \mid \Phi(s, u) \neq 0, \text{ for all } s \in [0, t]\}$ in the sequel. This condition could be replaced by the following weaker requirement: For every $m \geq 1$ and all $(t_0, u_0, x) \in \mathcal{Q}^m \times D$, there exists some neighborhood U such that for all $(t, u) \in U$

$$|\langle \operatorname{Im} \psi(t, u), x \rangle - \langle \operatorname{Im} \psi(t_0, u_0), x \rangle| < \pi. \quad (7)$$

Indeed, in order to conclude the existence of a unique continuous choice for Φ and ψ on $\mathcal{Q}^m$, this is the only condition needed in the proof of

---

[2]In Definition 2.1 affine processes on the canonical state space $D = \mathbb{R}_+^m \times \mathbb{R}^{n-m}$ are considered, whereas in Definition 12.1 the state space D can be an arbitrary subset of $\mathbb{R}^n$.

Proposition 1 below. Notice that in many cases the mere existence of Φ and ψ satisfying (6) is sufficient for the existence of a continuous selection, e.g., for star shaped spaces D.

(ii) Let us remark that the assumption of a *closed* state space is no restriction. Indeed, if an affine process is defined on some state space D, which is only supposed to be an arbitrary Borel subset of V as done in [23], then the affine property (6) extends automatically to $\overline{D}$: Let $(x_k)_{k\in\mathbb{N}}$ be a sequence in D converging to some $x \in \overline{D}$. Due to the exponential affine form of the characteristic function, we have for all $t \in \mathbb{R}_+$ and $u \in iV$

$$\mathbb{E}_{x_n}\left[e^{\langle u, X_t \rangle}\right] = \Phi(t,u)e^{\langle \psi(t,u), x_n \rangle} \to \Phi(t,u)e^{\langle \psi(t,u), x \rangle}.$$

Since the left hand side is continuous in u, the same holds true for the right hand side. Whence Lévy's continuity theorem implies that the right hand side is a characteristic function of some substochastic measure $p_t(x,\cdot)$ on $\overline{D}$, which is the weak limit of $p_t(x_n,\cdot)$. As stochastic continuity and the Chapman–Kolmogorov equations extend to $\overline{D}$, and since weak convergence implies the convergence of the Fourier–Laplace transforms on $\mathcal{U}$, we thus have constructed an affine process with state space $\overline{D}$.

(iii) Note furthermore that Assumption 1 is no restriction, since we can always pass to a lower dimensional ambient vector space if D does not contain $n+1$ affinely independent elements. Moreover, note also that we do not exclude compact state spaces. For examples of affine processes on compact state spaces we refer to Remark 13.

(iv) We finally remark that in Sect. 3 we consider affine processes on the filtered space $(\Omega, \mathcal{F}^0, \mathcal{F}^0_t)$, where $\mathcal{F}^0_t$ denotes the natural filtration and $\mathcal{F}^0 = \bigvee_{t \in \mathbb{R}_+} \mathcal{F}^0_t$, as introduced above. However, we shall progressively enlarge the filtration by augmenting with the respective null-sets.

Proposition 1. *Let X be an affine process relative to some filtration $(\mathcal{F}_t)$. Then we have the following properties:*

(i) *If we set $\Phi(0, u) = 1$ and $\psi(0, u) = u$ for all $u \in \mathcal{U}$, then there is a unique choice Φ and ψ in (6) such that Φ, ψ are jointly continuous on $\mathcal{Q}^m = \{(t, u) \in \mathbb{R}_+ \times \mathcal{U}^m \mid \Phi(s, u) \neq 0, \text{ for all } s \in [0, t]\}$ for $m \geq 1$.*

(ii) *ψ maps the set $\mathcal{O} = \{(t, u) \in \mathbb{R}_+ \times \mathcal{U} \mid \Phi(t, u) \neq 0\}$ to $\mathcal{U}$.*

(iii) *The functions Φ and ψ satisfy the* semiflow *property: Let $u \in \mathcal{U}$ and $t, s \geq 0$. Suppose that $\Phi(t + s, u) \neq 0$, then also $\Phi(t, u) \neq 0$ and $\Phi(s, \psi(t, u)) \neq 0$ and we have*

$$\begin{aligned}\Phi(t+s, u) &= \Phi(t, u)\Phi(s, \psi(t, u)), \\ \psi(t+s, u) &= \psi(s, \psi(t, u)).\end{aligned} \qquad (8)$$

Proof. Fix $m \geq 1$. It follows e.g. from [1, Lemma 23.7] that stochastic continuity of X implies joint continuity of $(t, u) \mapsto P_t e^{\langle u, x \rangle}$ on $\mathbb{R}_+ \times \mathscr{U}^m$ for all $x \in D$. Hence $(t, u) \mapsto \Phi(t, u) e^{\langle \psi(t,u), x \rangle}$ is jointly continuous on $\mathbb{R}_+ \times \mathscr{U}^m$ for every $x \in D$. By Assumption 1 on the state space D, this in turn yields a unique continuous choice of the functions $(t, u) \mapsto \Phi(t, u)$ and $(t, u) \mapsto \psi(t, u)$ on $\mathscr{Q}^m$. Indeed, by [23, Proposition 2.4], we know that for every $x \in D$ there exists a unique continuous logarithm $g(x; \cdot, \cdot) : \mathscr{Q}^m \to \mathbb{C}$, $(t, u) \mapsto g(x; t, u)$ such that for all $(t, u, x) \in \mathscr{Q}^m \times D$

$$e^{g(x;t,u)} = \Phi(t, u) e^{\langle \psi(t,u), x \rangle},$$

with $g(x; 0, 0) = 0$ holds true. Without loss of generality we suppose $0 \in D$, then it follows that $\Phi(t, u) = e^{g(0;t,u)} =: e^{\phi(t,u)}$, is continuous in (t, u).

Setting $h(x; t, u) := g(x; t, u) - \phi(t, u)$, with $\phi(t, u,) := g(0; t, u)$, we have for all $(t, u, x) \in \mathscr{Q}^m \times D$

$$e^{h(x;t,u)} = e^{\langle \psi(t,u), x \rangle}, \tag{9}$$

whence

$$h(x; t, u) = \langle \psi(t, u), x \rangle + 2\pi i k(t, u, x), \quad k(t, u, x) \in \mathbb{Z}. \tag{10}$$

Moreover, the local continuity assumption on $(t, u) \mapsto \langle \psi(t, u), x \rangle$ implies that for all (t_0, u_0, x) there exists some neighborhood around (t_0, u_0) such that

$$(t, u) \mapsto \langle \psi(t, u), x \rangle$$

is continuous.[3] Since $k(t, u, x) \in \mathbb{Z}$, it follows that $k(t, u, x) = k(x)$ on $\mathscr{Q}^m$. Setting $t = 0$ and $u = 0$ in (10) thus yields for all $x \in D$

$$h(x; 0, 0) - 2\pi i k(x) = -2\pi i k(x) = 0,$$

and in particular a unique continuous specification of $(t, u) \mapsto \psi(t, u)$, since

$$h(x; t, u) = \langle \psi(t, u), x \rangle$$

can be uniquely solved for ψ due to Assumption 1. The choice of Φ and ψ certainly does not depend on m but only on the initial conditions for $t = 0$.

---

[3] Due to relation (9) and the continuity of $(t, u) \mapsto h(x; t, u)$, this is also implied by the weaker condition (7).

Concerning (ii), let $(t, u) \in \mathscr{O} = \{(t, u) \in \mathbb{R}_+ \times \mathscr{U} \mid \Phi(t, u) \neq 0\}$. Since

$$\left| \Phi(t, u) e^{\langle \psi(t,u), x \rangle} \right| = \left| \mathbb{E}_x \left[e^{\langle u, X_t \rangle} \right] \right| \leq \mathbb{E}_x \left[\left| e^{\langle u, X_t \rangle} \right| \right]$$

is bounded on D and as $\Phi(t, u) \neq 0$, we conclude that $\psi(t, u) \in \mathscr{U}$.

Assumption $\Phi(t + s, u) \neq 0$ in (iii) implies

$$\mathbb{E}_x \left[e^{\langle u, X_{t+s} \rangle} \right] = \Phi(t + s, u) e^{\langle \psi(t+s,u), x \rangle} \neq 0. \tag{11}$$

By the law of iterated expectations and the Markov property, we thus have

$$\mathbb{E}_x \left[e^{\langle u, X_{t+s} \rangle} \right] = \mathbb{E}_x \left[\mathbb{E}_x \left[e^{\langle u, X_{t+s} \rangle} \middle| \mathscr{F}_s \right] \right] = \mathbb{E}_x \left[\mathbb{E}_{X_s} \left[e^{\langle u, X_t \rangle} \right] \right]. \tag{12}$$

If $\Phi(t, u) = 0$ or $\Phi(s, \psi(t, u)) = 0$, then the inner or the outer expectation evaluates to 0. This implies that the whole expression is 0, which contradicts (11). Hence $\Phi(t, u) \neq 0$ and $\Phi(s, \psi(t, u)) \neq 0$ and we can write (12) as

$$\mathbb{E}_x \left[e^{\langle u, X_{t+s} \rangle} \right] = \mathbb{E}_x \left[\Phi(t, u) e^{\langle \psi(t,u), X_s \rangle} \right] = \Phi(t, u) \Phi(s, \psi(t, u)) e^{\langle \psi(s, \psi(t,u)), x \rangle}.$$

Comparing with (11) yields the claim by uniqueness of Φ and ψ. □

Remark 3. Henceforth, the symbols Φ and ψ always correspond to the unique choice as established in Proposition 1.

3 Affine Processes Have a Càdlàg Version

The aim of this section is to show that the definition of an affine process already implies the existence of a càdlàg version. This is the core section of this article and of a remarkable subtlety, which is maybe less surprising if one considers the generality of the question. So far we do not know whether general affine processes are Feller processes. If the Feller property held true, this would allow us to conclude the existence of a càdlàg version. Moreover, we also cannot apply the most general standard criteria for the existence of càdlàg versions, as for instance described in [16, Theorem I.6.2].

Our approach to the problem is inspired by martingale regularization for a lot of "test martingales", from which we want to conclude path properties of the original stochastic process. The main difficulty here is that explosions and/or killing might appear.

Indeed, for every fixed $x \in D$, we first establish that for $\mathbb{P}_x$-almost every ω

$$t \mapsto M_t^{T,u}(\omega) := \Phi(T - t, u) e^{\langle \psi(T-t,u), X_t(\omega) \rangle}, \quad t \in [0, T],$$

is the restriction to $\mathbb{Q}_+ \cap [0, T]$ of a càdlàg function for almost all $(T, u) \in (0, \infty) \times \mathscr{U}$, in the sense that $M_t^{T,u} = 0$ if $\Phi(T - t, u) = 0$. This is an application of Doob's regularity theorem for supermartingales, where we can conclude—using Fubini's theorem—that there exists a $\mathbb{P}_x$-null-set outside of which we observe appropriately regular trajectories for almost all (T, u).

Proposition 2. *Let $x \in D$ be fixed and let X be an affine process relative to $(\mathscr{F}_t^0)$. Then*

$$\lim_{\substack{q \in \mathbb{Q}_+ \\ q \downarrow t}} M_q^{T,u} = \lim_{\substack{q \in \mathbb{Q}_+ \\ q \downarrow t}} \Phi(T - q, u) e^{\langle \psi(T-q,u), X_q \rangle}, \quad t \in [0, T],$$

exists $\mathbb{P}_x$-a.s. for almost all $(T, u) \in (0, \infty) \times \mathscr{U}$ and defines a càdlàg function in t.

Proof. In order to prove this result, we adapt parts of the proof of [28, Theorem I.4.30] to our setting. Due to the law of iterated expectations

$$M_t^{T,u} = \Phi(T - t, u) e^{\langle \psi(T-t,u), X_t \rangle} = \mathbb{E}_x \left[e^{\langle u, X_T \rangle} \big| \mathscr{F}_t^0 \right], \quad t \in [0, T],$$

is a (complex-valued) $(\mathscr{F}_t^0, \mathbb{P}_x)$-martingale for every $u \in \mathscr{U}$ and every $T > 0$. From Doob's regularity theorem (see, e.g., [30, Theorem II.65.1]) it then follows that, for any fixed (T, u), the function $t \mapsto M_t^{T,u}(\omega)$, with $t \in \mathbb{Q}_+ \cap [0, T]$, is the restriction to $\mathbb{Q}_+ \cap [0, T]$ of a càdlàg function for $\mathbb{P}_x$-almost every ω. Define now the set

$$\Gamma = \{(\omega, T, u) \in \Omega \times (0, \infty) \times \mathscr{U} \, | \, t \mapsto M_t^{T,u}(\omega), \, t \in \mathbb{Q}_+ \cap [0, T],$$
$$\text{is not the restriction of a càdlàg function}\}. \quad (13)$$

Then Γ is a $\mathscr{F}^0 \otimes \mathscr{B}((0, \infty) \times \mathscr{U})$-measurable set. Due to the above argument concerning regular versions of (super-)martingales, $\int_\Omega 1_\Gamma(\omega, T, u) \mathbb{P}_x(d\omega) = 0$ for any $(T, u) \in (0, \infty) \times \mathscr{U}$. By Fubini's theorem, we therefore have

$$\int_\Omega \int_{(0,\infty) \times \mathscr{U}} 1_\Gamma(\omega, T, u) d\lambda \, \mathbb{P}_x(d\omega) = \int_{(0,\infty) \times \mathscr{U}} \int_\Omega 1_\Gamma(\omega, T, u) \mathbb{P}_x(d\omega) \, d\lambda = 0,$$

where λ denotes the Lebesgue measure. Hence, for $\mathbb{P}_x$-almost every ω, $t \mapsto M_t^{T,u}(\omega)$ with $t \in \mathbb{Q}_+ \cap [0, T]$, is the restriction of a càdlàg function for λ-almost all $(T, u) \in (0, \infty) \times \mathscr{U}$, which proves the result. $\square$

Having established path regularity of the martingales $M^{T,u}$, we want to deduce the same for the affine process X. This is the purpose of the subsequent lemmas and propositions, for which we need to introduce the following sets:

$$\widetilde{\Omega} \text{ is the projection of } \{\Omega \times (0, \infty) \times iV\} \setminus \Gamma \text{ onto } \Omega, \quad (14)$$

$$\mathscr{T} \text{ is the projection of } \{\Omega \times (0, \infty) \times iV\} \setminus \Gamma \text{ onto } (0, \infty), \quad (15)$$

$\mathcal{V}$ is the projection of $\{\Omega \times (0, \infty) \times \mathcal{U}\} \setminus \Gamma$ onto $\mathcal{U}$, (16)

$\mathcal{V}^m$ is the projection of $\{\Omega \times (0, \infty) \times \mathcal{U}^m\} \setminus \Gamma$ onto $\mathcal{U}^m$, (17)

where Γ is given in (13). Denoting by $\mathscr{F}^x$ the completion of $\mathscr{F}^0$ with respect to $\mathbb{P}_x$, let us remark that the measurable projection theorem implies that $\widetilde{\Omega} \in \mathscr{F}^x$ and by the above proposition we have $\mathbb{P}_x[\widetilde{\Omega}] = 1$.

The following lemma is needed to prove Proposition 3 below which is essential for establishing the existence of a càdlàg version of X.

Lemma 1. *Let ψ be given by (6) and assume that there exists some D-valued sequence $(x_k)_{k \in \mathbb{N}}$ such that*

$$\lim_{k \to \infty} \Pi_{\langle \operatorname{Re} \mathcal{U} \rangle} x_k =: \lim_{k \to \infty} y_k \qquad (18)$$

exists finitely valued and

$$\limsup_{k \to \infty} \| \Pi_{\langle \operatorname{Re} \mathcal{U} \rangle^\perp} x_k \| = \infty. \qquad (19)$$

(i) *Then we can choose a subsequence of (x_k) denoted again by (x_k): along this sequence there exist a finite number of mutually orthogonal directions $g_i \in \langle \operatorname{Re} \mathcal{U} \rangle^\perp$ of length 1 such that*

$$x_k - \sum_i \langle x_k, g_i \rangle g_i$$

converges as $k \to \infty$ and $\langle x_k, g_i \rangle$ diverges as $k \to \infty$, where the rates of divergence are non-increasing in i in the sense that

$$\limsup_{k \to \infty} \frac{\langle x_k, g_{i+1} \rangle}{\langle x_k, g_i \rangle} < \infty.$$

(ii) *Moreover, let $T > 0$ be fixed and let $r > 0$ such that $\Phi(t, u) \neq 0$ for all $(t, u) \in [0, T] \times B_r$, where B_r denotes the ball with radius r in iV, i.e.*

$$B_r = \{u \in iV \mid \|u\| < r\}.$$

Then, there exist continuous functions $R: [0, T] \to \mathbb{R}_{++}$ and $\lambda_i: [0, T] \to V$ such that

$$\langle \psi(t, u), g_i \rangle = \langle \lambda_i(t), u \rangle$$

for all $u \in B_{R(t)}$.

Proof. Concerning the first assertion, we define—by choosing appropriate subsequences, still denoted by (x_k)—the directions of divergence in $\langle \mathrm{Re}\,\mathcal{U}\rangle^\perp$ inductively by

$$g_r = \lim_{k\to\infty} \frac{x_k - \sum_{i=1}^{r-1}\langle x_k, g_i\rangle g_i}{\|x_k - \sum_{i=1}^{r-1}\langle x_k, g_i\rangle g_i\|} \tag{20}$$

as long as $\limsup_{k\to\infty} \|x_k - \sum_{i=1}^{r-1}\langle x_k, g_i\rangle g_i\| = \infty$. Notice that we can choose the directions g_i mutually orthogonal and the rates of divergence $\langle g_i, x_k\rangle$ non-increasing in i.

For the second part of the statement, we adapt the proof of [22, Lemma 3.1] to our situation, using in particular the existence of a sequence in D with the properties (18) and (19). As characteristic function, the map $iV \ni u \mapsto \mathbb{E}_x[e^{\langle u, X_t\rangle}]$ is positive definite for any $x \in D$ and $t \geq 0$. Define now for every $u \in B_r$, $x \in D$ and $t \in [0, T]$ the function

$$\Theta(u,t,x) = \frac{\mathbb{E}_x\left[e^{\langle u,X_t\rangle}\right]}{\Phi(t,0)e^{\langle \Pi_{\langle\mathrm{Re}\,\mathcal{U}\rangle}\psi(t,0),\Pi_{\langle\mathrm{Re}\,\mathcal{U}\rangle}x\rangle}} = \frac{\Phi(t,u)e^{\langle \psi(t,u),x\rangle}}{\Phi(t,0)e^{\langle \Pi_{\langle\mathrm{Re}\,\mathcal{U}\rangle}\psi(t,0),\Pi_{\langle\mathrm{Re}\,\mathcal{U}\rangle}x\rangle}}. \tag{21}$$

As $\mathbb{E}_x[e^{\langle 0,X_t\rangle}] = \Phi(t,0)e^{\langle\psi(t,0),x\rangle}$ is real-valued and positive for all $t \in [0,T]$, we conclude—due to Assumption 1 and the continuity of the functions $t \mapsto \Phi(t,0)$ and $t \mapsto \psi(t,0)$—that $\mathrm{Im}\,\Phi(t,0) = 0$ and $\mathrm{Im}\,\psi(t,0) = 0$ for all $t \in [0,T]$. In particular, the denominator in (21) is positive, which implies that $B_r \ni u \mapsto \Theta(u,t,x)$ is a positive definite function for all $t \in [0,T]$ and $x \in D$. Moreover, since $\Pi_{\langle\mathrm{Re}\,\mathcal{U}\rangle^\perp}\psi(t,0)$ is purely imaginary and thus in particular 0 for all $t \in [0,T]$, it follows that

$$\Theta(0,t,x) = \exp\left(\langle \Pi_{\langle\mathrm{Re}\,\mathcal{U}\rangle^\perp}\psi(t,0), \Pi_{\langle\mathrm{Re}\,\mathcal{U}\rangle^\perp}x\rangle\right) = 1$$

for all $t \in [0,T]$ and $x \in D$. This together with the positive definiteness of $u \mapsto \Theta(u,t,x)$ yields

$$|\Theta(u+v,t,x) - \Theta(u,t,x)\Theta(v,t,x)|^2 \leq 1, \quad u,v \in B_{\frac{r}{2}}, t \geq 0, x \in D. \tag{22}$$

Indeed, this inequality is obtained by computing the determinant of the positive semidefinite matrix

$$\begin{pmatrix} \Theta(0,t,x) & \overline{\Theta(u,t,x)} & \overline{\Theta(v,t,x)} \\ \Theta(u,t,x) & \Theta(0,t,x) & \overline{\Theta(u+v,t,x)} \\ \Theta(v,t,x) & \Theta(u+v,t,x) & \Theta(0,t,x) \end{pmatrix}$$

(compare, e.g., [22, Lemma 3.2]). Let us now define $y := \Pi_{\langle \operatorname{Re} \mathscr{U} \rangle} x$ and

$$Z_1(u, v, y, t) = \frac{\Phi(t, u+v) e^{\langle \Pi_{\langle \operatorname{Re} \mathscr{U} \rangle} \psi(t, u+v), y \rangle}}{\Phi(t, 0) e^{\langle \Pi_{\langle \operatorname{Re} \mathscr{U} \rangle} \psi(t, 0), y \rangle}},$$

$$Z_2(u, v, y, t) = \frac{\Phi(t, u) \Phi(t, v) e^{\langle \Pi_{\langle \operatorname{Re} \mathscr{U} \rangle} (\psi(t, u) + \psi(t, v)), y \rangle}}{\Phi(t, 0)^2 e^{2 \langle \Pi_{\langle \operatorname{Re} \mathscr{U} \rangle} \psi(t, 0), y \rangle}},$$

$$\beta_1(u, v, t) = \operatorname{Im}(\Pi_{\langle \operatorname{Re} \mathscr{U} \rangle^\perp} \psi(t, u+v)),$$

$$\beta_2(u, v, t) = \operatorname{Im}(\Pi_{\langle \operatorname{Re} \mathscr{U} \rangle^\perp} \psi(t, u)) + \operatorname{Im}(\Pi_{\langle \operatorname{Re} \mathscr{U} \rangle^\perp} \psi(t, v)),$$

$$r_1(u, v, y, t) = |Z_1| = \left| \frac{\Phi(t, u+v)}{\Phi(t, 0)} \right| e^{\langle \operatorname{Re}(\Pi_{\langle \operatorname{Re} \mathscr{U} \rangle} (\psi(t, u+v) - \psi(t, 0))), y \rangle},$$

$$r_2(u, v, y, t) = |Z_2| = \left| \frac{\Phi(t, u) \Phi(t, v)}{\Phi(t, 0)^2} \right| e^{\langle \operatorname{Re}(\Pi_{\langle \operatorname{Re} \mathscr{U} \rangle} (\psi(t, u) + \psi(t, v) - 2\psi(t, 0))), y \rangle},$$

$$\alpha_1(u, v, y, t) = \arg(Z_1) = \arg\left(\frac{\Phi(t, u+v)}{\Phi(t, 0)} \right)$$
$$\qquad + \langle \operatorname{Im}(\Pi_{\langle \operatorname{Re} \mathscr{U} \rangle} \psi(t, u+v)), y \rangle,$$

$$\alpha_2(u, v, y, t) = \arg(Z_2) = \arg\left(\frac{\Phi(t, u) \Phi(t, v)}{\Phi(t, 0)^2} \right)$$
$$\qquad + \langle \operatorname{Im}(\Pi_{\langle \operatorname{Re} \mathscr{U} \rangle} (\psi(t, u) + \psi(t, v))), y \rangle.$$

Using (22) and the fact that $2 r_1 r_2 \leq r_1^2 + r_2^2$, we then obtain

$$1 \geq \left| r_1 e^{i(\alpha_1 + \langle \beta_1, \Pi_{\langle \operatorname{Re} \mathscr{U} \rangle^\perp} x \rangle)} - r_2 e^{i(\alpha_2 + \langle \beta_2, \Pi_{\langle \operatorname{Re} \mathscr{U} \rangle^\perp} x \rangle)} \right|^2$$
$$= r_1^2 + r_2^2 - 2 r_1 r_2 \cos(\alpha_1 - \alpha_2 + \langle \beta_1 - \beta_2, \Pi_{\langle \operatorname{Re} \mathscr{U} \rangle^\perp} x \rangle)$$
$$\geq 2 r_1 r_2 (1 - \cos(\alpha_1 - \alpha_2 + \langle \beta_1 - \beta_2, \Pi_{\langle \operatorname{Re} \mathscr{U} \rangle^\perp} x \rangle)),$$

whence

$$r_1(u, v, y, t) r_2(u, v, y, t)$$
$$\times (1 - \cos(\alpha_1(u, v, y, t) - \alpha_2(u, v, y, t) + \langle \beta_1(u, v, t) - \beta_2(u, v, t), \Pi_{\langle \operatorname{Re} \mathscr{U} \rangle^\perp} x \rangle)) \leq \frac{1}{2}. \tag{23}$$

Define now

$$R(t, y) = \sup \left\{ \rho \in \left[0, \frac{r}{2}\right] \mid r_1(u, v, y, t) r_2(u, v, y, t) > \frac{3}{4} \text{ for } u, v \in B_{\frac{r}{2}} \right.$$
$$\left. \text{with } \|u\| \leq \rho \text{ and } \|v\| \leq \rho \right\}.$$

Note that $R(t, y) > 0$ for all $(t, y) \in [0, T] \times \Pi_{(\text{Re }\mathcal{U})} D$, which follows from the fact that $r_1(0, 0, y, t) = r_2(0, 0, y, t) = 1$ and the continuity of

$$(u, v) \mapsto r_1(u, v, y, t) r_2(u, v, y, t).$$

Continuity of $(t, y) \mapsto r_1(u, v, y, t) r_2(u, v, y, t)$ also implies that $(t, y) \mapsto R(t, y)$ is continuous. Set now $R(t) := \inf_k R(t, y_k)$ where $y_k = \Pi_{(\text{Re }\mathcal{U})} x_k$. Then (18) implies that $R(t) > 0$ for all $t \in [0, T]$.

Let now t be fixed and g_1 given by (20). Suppose that

$$\langle \beta_1(u^*, v^*, t) - \beta_2(u^*, v^*, t), g_1 \rangle \neq 0$$

for some $u^*, v^* \in B_{R(t)}$. Then due to the continuity of β_1 and β_2, there exists some $\delta > 0$ such that for all u, v in a neighborhood O_δ of (u^*, v^*) defined by

$$O_\delta = \left\{ u, v \in B_{R(t)} \mid \|u - u^*\| < \delta, \|v - v^*\| < \delta \text{ and } \right\},$$

we also have

$$\langle \beta_1(u, v, t) - \beta_2(u, v, t), g_1 \rangle \neq 0. \tag{24}$$

Moreover, there exist some $(u, v) \in O_\delta$ and some $k \in \mathbb{N}$ such that

$$\cos(\alpha_1(u, v, y_k, t) - \alpha_2(u, v, y_k, t) + \langle \beta_1(u, v, t) - \beta_2(u, v, t), \Pi_{(\text{Re }\mathcal{U})^\perp} x_k \rangle)$$

$$= \cos\left(\arg\left(\frac{\Phi(t, u + v)}{\Phi(t, 0)} \right) - \arg\left(\frac{\Phi(t, u) \Phi(t, v)}{\Phi(t, 0)^2} \right) \right.$$

$$+ \langle \text{Im}(\Pi_{(\text{Re }\mathcal{U})}(\psi(t, u + v) - \text{Im } \psi(t, u) - \text{Im } \psi(t, v))), y_k \rangle \tag{25}$$

$$\left. + \langle \beta_1(u, v, t) - \beta_2(u, v, t), \Pi_{(\text{Re }\mathcal{U})^\perp} x_k \rangle \right)$$

$$\leq \frac{1}{3},$$

since y_k stays in a bounded set and $\Pi_{(\text{Re }\mathcal{U})^\perp} x_k$ explodes with highest divergence rate in direction g_1. However, inequality (25) now implies that

$$r_1(u, v, y_k, t) r_2(u, v, y_k, t)$$
$$\times (1 - \cos(\alpha_1(u, v, y_k, t) - \alpha_2(u, v, y_k, t) + \langle \beta_1(u, v, t) - \beta_2(u, v, t), \Pi_{(\text{Re }\mathcal{U})^\perp} x_k \rangle))$$
$$> \frac{1}{2},$$

which contradicts (23). Since g_1 corresponds to the direction of the highest divergence rate, we thus conclude that

$$\langle \beta_1(u,v,t) - \beta_2(u,v,t), g_1 \rangle = \mathrm{Im}(\langle \psi(t,u+v) - \psi(t,u) - \psi(t,v), g_1 \rangle) = 0$$

for all $u, v \in B_{R(t)}$. Continuity of $u \mapsto \psi(t,u)$ therefore implies that $u \mapsto \langle \psi(t,u), g_1 \rangle$ is a linear function. Hence there exists a continuous curve of (real) vectors $\lambda_1(t) \in V$ such that

$$\langle \psi(t,u), g_1 \rangle = \langle \lambda_1(t), u \rangle$$

for all $u \in B_{R(t)}$.

We can now proceed inductively for the remaining directions of divergence g_i. Indeed, assume that $\langle \beta_1(u,v,t) - \beta_2(u,v,t), g_i \rangle = 0$ for all $i \le r-1$ and all $u, v \in B_{R(t)}$. Then repeating the above steps allows us to conclude that $\langle \beta_1(u,v,t) - \beta_2(u,v,t), g_r \rangle = 0$ for all $u, v \in B_{R(t)}$, yielding the assertion. □

As introduced before, we denote by p the dimension of $\mathrm{Re}\,\mathcal{U}$. Let now m^* be fixed such that $\dim(\mathrm{Re}\,\mathcal{U}^{m^*}) = p$. For some $r > 0$, we define K to be the intersection of $\mathcal{V}^{m^*}$ with the closed ball with center 0 and radius r in $\mathcal{U}$, that is,

$$K := \overline{B}(0,r) \cap \mathcal{V}^{m^*} := \{ u \in \mathcal{U} \mid \|\mathrm{Re}\,u\|^2 + \|\mathrm{Im}\,u\|^2 \le r^2 \} \cap \mathcal{V}^{m^*}, \quad (26)$$

where $\mathcal{V}^{m^*}$ is defined in (17). Let now $(u_1, \ldots, u_p)$ be linearly independent vectors in $K \cap \mathrm{Re}\,\mathcal{U}$ and let $(u_{p+1}, \ldots, u_n)$ be linearly independent vectors in $\Pi_{(\mathrm{Re}\,\mathcal{U})^\perp} K$. Then, as a consequence of the fact that $\psi(0,u) = u$ for all $u \in \mathcal{U} \supset K$ and the continuity of $t \mapsto \psi(t,u)$, there exists some $\delta > 0$ such that for every $t \in [0, \delta)$

$$(\psi(t, u_1), \ldots, \psi(t, u_p))$$

and

$$(\Pi_{(\mathrm{Re}\,\mathcal{U})^\perp} \psi(t, u_{p+1}), \ldots, \Pi_{(\mathrm{Re}\,\mathcal{U})^\perp} \psi(t, u_n))$$

are linearly independent.

Moreover, since $(t,u) \mapsto \Phi(t,u) e^{\langle \psi(t,u), x \rangle}$ is jointly continuous on $\mathbb{R}_+ \times \mathcal{U}^{m^*}$, with $\Phi(0,u) = 1$ and $\psi(0,u) = u$ (see Proposition 1), it follows that there exists some $\eta > 0$ such that for all $t \in [0, \eta]$

$$\inf_{u \in K} |\Phi(t,u)| > c \quad \text{and} \quad \sup_{u \in K}(\|\mathrm{Re}\,\psi(t,u)\|^2 + \|\mathrm{Im}\,\psi(t,u)\|^2) < C, \quad (27)$$

with some positive constants c and C. By fixing these constants and some linearly independent vectors in K as described above, we define

$$\varepsilon := \min(\eta, \delta). \quad (28)$$

Furthermore, let $t \geq 0$ be fixed. Then we denote by $I_{t,\varepsilon}^{\mathscr{T}}$ the set

$$I_{t,\varepsilon}^{\mathscr{T}} := (t, t+\varepsilon) \cap \mathscr{T}, \tag{29}$$

where $\mathscr{T}$ is defined in (15).

Proposition 3. *Let K and $I_{t,\varepsilon}^{\mathscr{T}}$ be the sets defined in (26) and (29). Consider the function ψ given in (6) with the properties of Proposition 1. Let $t \geq 0$ be fixed and consider a sequence $(q_k)_{k \in \mathbb{N}}$ with values in $\mathbb{Q}_+$ such that $q_k \uparrow t$. Moreover, let $(x_{q_k})_{k \in \mathbb{N}}$ be a sequence with values in $D_\Delta \cup \{\infty\}$.[4] Then the following assertions hold:*

(i) *If for all $(T, u) \in I_{t,\varepsilon}^{\mathscr{T}} \times K$*

$$\lim_{k \to \infty} N_{q_k}^{T,u} := \lim_{k \to \infty} e^{\langle \psi(T-q_k, u), x_{q_k} \rangle} \tag{30}$$

exists finitely valued and does not vanish, then also $\lim_{k \to \infty} x_{q_k}$ exists finitely valued.

(ii) *If there exists some $(T, u) \in I_{t,\varepsilon}^{\mathscr{T}} \times K$ such that*

$$\lim_{k \to \infty} N_{q_k}^{T,u} := \lim_{k \to \infty} e^{\langle \psi(T-q_k, u), x_{q_k} \rangle} = 0,$$

then we have $\lim_{k \to \infty} \|x_{q_k}\| = \infty$.

Moreover, let $(q_k^T)_{k \in \mathbb{N}, T \in I_{t,\varepsilon}^{\mathscr{T}}}$ be a family of sequences with values in $\mathbb{Q}_+ \cap [t, T]$ such that $q_k^T \downarrow t$ for every $T \in I_{t,\varepsilon}^{\mathscr{T}}$ and the additional property that for every $S, T \in I_{t,\varepsilon}^{\mathscr{T}}$, with $S < T$, there exists some index $N \in \mathbb{N}$ such that, for all $k \geq N$, $q_{k-N}^S = q_k^T$. Then the above assertions hold true for these right limits with q_k replaced by q_k^T.

Remark 4. Concerning assertion (ii) of Proposition 3, note that, e.g. in the case $q_k \uparrow t$, $\lim_{k \to \infty} \|x_{q_k}\| = \infty$ corresponds either to explosion or to the possibility that there exists some index $N \in \mathbb{N}$ such that $x_{q_k} = \Delta$ for all $k \geq N$. In the latter case we also have, due to the convention $\|\Delta\| = \infty$, $\lim_{k \to \infty} \|x_{q_k}\| = \infty$.

Proof. We start by proving the first assertion (i). Let $T \in I_{t,\varepsilon}^{\mathscr{T}}$ be fixed and define for all $u \in K$

$$A(u) := \limsup_{k \to \infty} \langle \operatorname{Re} \psi(T-q_k, u), x_{q_k} \rangle, \quad a(u) := \liminf_{k \to \infty} \langle \operatorname{Re} \psi(T-q_k, u), x_{q_k} \rangle.$$

---

[4] As mentioned at the beginning of Sect. 2, ∞ corresponds to a "point at infinity" and $D_\Delta \cup \{\infty\}$ is the one-point compactification of D_Δ. If the state space D is compact, we *do not* adjoin $\{\infty\}$ and only consider a sequence with values in D_Δ.

Then there exist subsequences $(x_{q_{k_m}})$ and $(x_{q_{k_l}})$ such that[5]

$$A(u) = \lim_{m \to \infty} \langle \operatorname{Re} \psi(T - q_{k_m}, u), x_{q_{k_m}} \rangle,$$

$$a(u) = \lim_{l \to \infty} \langle \operatorname{Re} \psi(T - q_{k_l}, u), x_{q_{k_l}} \rangle.$$

First note that $A(u)$ and $a(u)$ exist finitely valued for all $u \in K$. Indeed, if there is some $u \in K$ such that $A(u) = \pm\infty$ or $a(u) = \pm\infty$, then the limit of $N_{q_k}^{T,u}$ does not exist, or $\lim_{k \to \infty} N_{q_k}^{T,u}$ is either 0 or $+\infty$, which contradicts assumption (30). We now define

$$r_1(u) = \lim_{m \to \infty} \exp\left(\langle \operatorname{Re} \psi(T - q_{k_m}, u), x_{q_{k_m}} \rangle\right),$$

$$r_2(u) = \lim_{l \to \infty} \exp\left(\langle \operatorname{Re} \psi(T - q_{k_l}, u), x_{q_{k_l}} \rangle\right),$$

$$\varphi_m(u) = \langle \operatorname{Im} \psi(T - q_{k_m}, u), x_{q_{k_m}} \rangle,$$

$$\varphi_l(u) = \langle \operatorname{Im} \psi(T - q_{k_l}, u), x_{q_{k_l}} \rangle.$$

Then the limits of $\cos(\varphi_m(u))$, $\cos(\varphi_l(u))$, $\sin(\varphi_m(u))$ and $\sin(\varphi_l(u))$ necessarily exist and

$$r_1(u) \lim_{m \to \infty} \cos(\varphi_m(u)) = r_2(u) \lim_{l \to \infty} \cos(\varphi_l(u)),$$

$$r_1(u) \lim_{m \to \infty} \sin(\varphi_m(u)) = r_2(u) \lim_{l \to \infty} \sin(\varphi_l(u)).$$

This yields $r_1(u) = r_2(u)$ for all $u \in K$, since

$$\lim_{m \to \infty} \left(\cos^2(\varphi_m(u)) + \sin^2(\varphi_m(u))\right) = \lim_{l \to \infty} \left(\cos^2(\varphi_l(u)) + \sin^2(\varphi_l(u))\right) = 1.$$

In particular, we have proved that

$$\lim_{k \to \infty} \langle \operatorname{Re} \psi(T - q_k, u), x_{q_k} \rangle \quad (31)$$

exists finitely valued and does not vanish for all $(T, u) \in I_{t,\varepsilon}^{\mathcal{T}} \times K$. Choosing linear independent vectors $(u_1, \ldots, u_p) \in K \cap \operatorname{Re} \mathcal{U}$ thus implies that

$$\lim_{k \to \infty} \Pi_{\langle \operatorname{Re} \mathcal{U} \rangle} x_{q_k}$$

exists finitely valued.

---

[5]Note that these subsequences depend on u. For notational convenience we however suppress the dependence on u.

Therefore it only remains to focus on $\Pi_{\langle \operatorname{Re} \mathscr{U} \rangle^\perp} x_{q_k}$. From the above, we know in particular that for all $(T, u) \in I_{t,\varepsilon}^{\mathscr{T}} \times K$

$$\lim_{k \to \infty} e^{\left\langle \Pi_{\langle \operatorname{Re} \mathscr{U} \rangle^\perp} \psi(T - q_k, u), \Pi_{\langle \operatorname{Re} \mathscr{U} \rangle^\perp} x_{q_k} \right\rangle} \quad (32)$$

exists finitely valued and does not vanish. This implies that for all $(T, u) \in I_{t,\varepsilon}^{\mathscr{T}} \times K$

$$\operatorname{Im} \left\langle \Pi_{\langle \operatorname{Re} \mathscr{U} \rangle^\perp} \psi(T - q_k, u), \Pi_{\langle \operatorname{Re} \mathscr{U} \rangle^\perp} x_{q_k} \right\rangle = \alpha_k(T, u) + 2\pi z_k(T, u), \quad (33)$$

where $\alpha_k(T, u) \in [-\pi, \pi)$, $\alpha(T, u) := \lim_{k \to \infty} \alpha_k(T, u)$ exists finitely valued and $(z_k(T, u))_{k \in \mathbb{N}}$ is a sequence with values in $\mathbb{Z}$, which a priori does not necessarily have a limit and/or $\lim_{k \to \infty} z_k(T, u) = \pm \infty$.

Let us first assume that

$$\limsup_{k \to \infty} \| \Pi_{\langle \operatorname{Re} \mathscr{U} \rangle^\perp} x_{q_k} \| = \infty. \quad (34)$$

Then we are exactly in the situation of Lemma 1 and the above limit (32) can be written as

$$\lim_{k \to \infty} e^{\left(\sum_i \langle \lambda_i(T - q_k), u \rangle \langle g_i, x_{q_k} \rangle + \langle \Pi_{\langle \operatorname{Re} \mathscr{U} \rangle^\perp} \psi(T - q_k, u), x_{q_k} - \sum_i g_i \langle g_i, x_{q_k} \rangle \rangle \right)}$$

for all $u \in \Pi_{\langle \operatorname{Re} \mathscr{U} \rangle^\perp} K$ with $\| \operatorname{Im} u \| < P(T)$, where $P(T)$ is defined by $P(T) := \inf_k R(T - q_k)$ and R and the directions g_i are given in Lemma 1 after possibly selecting a subsequence such that $x_{q_k} - \sum_i g_i \langle g_i, x_{q_k} \rangle$ converges as $k \to \infty$. Note that due to the strict positivity and continuity of R, $P(T)$ is strictly positive as well. Furthermore, there exists some $T^* \in I_{t,\varepsilon}^{\mathscr{T}}$ and some set $M_{T^*} \subseteq \{u \in \Pi_{\langle \operatorname{Re} \mathscr{U} \rangle^\perp} K \mid \| \operatorname{Im} u \| < P(T^*), \exists i \text{ s.t. } \langle \lambda_i(T^* - t), u \rangle \neq 0\}$ of positive finite measure such that

$$\lim_{k \to \infty} \int_{M_T^*} e^{\left\langle \Pi_{\langle \operatorname{Re} \mathscr{U} \rangle^\perp} \psi(T^* - q_k, u), x_{q_k} - \sum_i g_i \langle g_i, x_{q_k} \rangle \right\rangle} e^{\left(\sum_i \langle \lambda_i(T^* - q_k), u \rangle \langle g_i, x_{q_k} \rangle \right)} du \neq 0. \quad (35)$$

However, it follows from the Riemann–Lebesgue Lemma that the previous limit is zero, whence contradicting (35). We therefore conclude that

$$\limsup_{k \to \infty} \| \Pi_{\langle \operatorname{Re} \mathscr{U} \rangle^\perp} x_{q_k} \| < \infty.$$

This in turn implies that there exists some $(T^*, u^*) \in I_{t,\varepsilon}^{\mathscr{T}} \times K$ and $N \in \mathbb{N}$ such that for all $k \geq N$

$$\operatorname{Im} \left\langle \Pi_{\langle \operatorname{Re} \mathcal{U} \rangle^\perp} \psi(T^* - q_k, u^*), \Pi_{\langle \operatorname{Re} \mathcal{U} \rangle^\perp} x_{q_k} \right\rangle \in (-\pi, \pi).$$

Indeed, this follows from the fact that for every $u \in K$ and $\eta > 0$ there exists some $T^* \in I_{t,\varepsilon}^{\mathcal{T}}$ and $N \in \mathbb{N}$ such that for all $k \geq N$

$$\|\operatorname{Im}(\Pi_{\langle \operatorname{Re} \mathcal{U} \rangle^\perp} \psi(T^* - q_k, u) - \Pi_{\langle \operatorname{Re} \mathcal{U} \rangle^\perp} u)\| \leq \eta. \tag{36}$$

For u^* with $\|\operatorname{Im}(\Pi_{\langle \operatorname{Re} \mathcal{U} \rangle^\perp} u^*)\|$ sufficiently small and k sufficiently large, we thus have

$$\left| \left\langle \Pi_{\langle \operatorname{Re} \mathcal{U} \rangle^\perp} \psi(T^* - q_k, u^*), \Pi_{\langle \operatorname{Re} \mathcal{U} \rangle^\perp} x_{q_k} \right\rangle \right|$$
$$\leq (\|\operatorname{Im}(\Pi_{\langle \operatorname{Re} \mathcal{U} \rangle^\perp} u^*)\| + \|\operatorname{Im}(\Pi_{\langle \operatorname{Re} \mathcal{U} \rangle^\perp} \psi(T^* - q_k, u^*) - \Pi_{\langle \operatorname{Re} \mathcal{U} \rangle^\perp} u^*)\|)$$
$$\times (\limsup_{k \to \infty} \|\Pi_{\langle \operatorname{Re} \mathcal{U} \rangle^\perp} x_{q_k}\| + 1)$$
$$< \pi.$$

Hence,

$$\lim_{k \to \infty} \operatorname{Im} \left\langle \Pi_{\langle \operatorname{Re} \mathcal{U} \rangle^\perp} \psi(T^* - q_k, u^*), \Pi_{\langle \operatorname{Re} \mathcal{U} \rangle^\perp} x_{q_k} \right\rangle = \alpha(T^*, u^*). \tag{37}$$

As we can find $n - p$ linear independent vectors $u_{p+1}, \ldots, u_n$ such that (37) is satisfied, we conclude that

$$\lim_{k \to \infty} \Pi_{\langle \operatorname{Re} \mathcal{U} \rangle^\perp} x_{q_k}$$

exists finitely valued. This proves assertion (i).

Concerning the second statement, observe that we have

$$\lim_{k \to \infty} e^{\langle \psi(T - q_k, u), x_{q_k} \rangle} = 0, \tag{38}$$

if either explosion occurs or if x_{q_N} jumps to Δ for some $N \in \mathbb{N}$ and $x_{q_k} = \Delta$ for all $k \geq N$. (This happens when the corresponding process is killed.) Indeed, since (38) is equivalent to $\lim_{k \to \infty} e^{\langle \operatorname{Re} \psi(T - q_k, u), x_{q_k} \rangle} = 0$ and as $\psi(T - t, u)$ is bounded on K due to the definition of $I_{t,\varepsilon}^{\mathcal{T}}$, we necessarily have

$$\lim_{k \to \infty} \|x_{q_k}\| = \infty.$$

In the case of a jump to Δ, this is implied by the conventions $\|\Delta\| = \infty$ and $f(\Delta) = 0$ for any other function.

Similar arguments yield the assertion concerning right limits. $\square$

Using Propositions 2 and 3 above, we are now prepared to prove Theorem 2 below, which asserts the existence of a càdlàg version of X. Before stating this result, let us recall the notion of the (usual) augmentation of $(\mathscr{F}_t^0)$ with respect to $\mathbb{P}_x$, which guarantees the càdlàg version to be adapted.

Definition 3 (Usual augmentation). We denote by $\mathscr{F}^x$ the *completion* of $\mathscr{F}^0$ with respect to $\mathbb{P}_x$. A sub-σ-algebra $\mathscr{G} \subset \mathscr{F}^x$ is called *augmented* with respect to $\mathbb{P}_x$ if $\mathscr{G}$ contains all $\mathbb{P}_x$-null-sets in $\mathscr{F}^x$. The augmentation of $\mathscr{F}_t^0$ with respect to $\mathbb{P}_x$ is denoted by $\mathscr{F}_t^x$, that is, $\mathscr{F}_t^x = \sigma(\mathscr{F}_t^0, \mathscr{N}(\mathscr{F}^x))$, where $\mathscr{N}(\mathscr{F}^x)$ denotes all $\mathbb{P}_x$-null-sets in $\mathscr{F}^x$.

Theorem 2. *Let X be an affine process relative to $(\mathscr{F}_t^0)$. Then there exists a process $\tilde{X}$ such that, for each $x \in D_\Delta$, $\tilde{X}$ is a $\mathbb{P}_x$-version of X, which is càdlàg in $D_\Delta \cup \{\infty\}$ (in D_Δ respectively if D is compact) and an affine process relative to $(\mathscr{F}_t^x)$. As before, ∞ corresponds to a "point at infinity" and $D_\Delta \cup \{\infty\}$ is the one-point compactification of D_Δ, if D is non-compact.*

Remark 5. We here establish the existence of a càdlàg version $\tilde{X}$ whose sample paths may take ∞ as left limiting value if D is non-compact. A priori, we cannot identify $\tilde{X}_{s-}(\omega)$ with Δ, whenever $\|\tilde{X}_{s-}(\omega)\| = \infty$. Indeed, $\tilde{X}_t(\omega)$ might become finitely valued for some $t \geq s$. This issue is clarified in Theorem 3 below, where we prove that $\mathbb{P}_x$-a.s. $\|\tilde{X}_t\| = \infty$ for all $t \geq s$ and all $s > 0$ if $\|\tilde{X}_{s-}\| = \infty$. In particular, this allows us to identify ∞ with Δ.

In the case $\tilde{X}_s = \Delta$, which happens when the process is killed, Assumption (3) guarantees that $\tilde{X}_t = \Delta$ for all $t \geq s$ and all $s > 0$.

Proof. It follows from Proposition 2 that for every $\omega \in \widetilde{\Omega}$,[6] where $\mathbb{P}_x[\widetilde{\Omega}] = 1$,

$$t \mapsto M_t^{T,u}(\omega) := \Phi(T-t, u)e^{\langle \psi(T-t,u), X_t(\omega) \rangle}, \quad t \in [0, T],$$

is the restriction to $\mathbb{Q}_+ \cap [0, T]$ of a càdlàg function for all $(T, u) \in \mathscr{T} \times \mathscr{V}$. Here, $\widetilde{\Omega}$, $\mathscr{T}$ and $\mathscr{V}$ are defined in (14), (15) and (16). Hence, for every $\omega \in \widetilde{\Omega}$ and all $(T, u) \in \mathscr{T} \times \mathscr{V}$, the limits

$$\lim_{\substack{q \in \mathbb{Q}_+ \\ q \uparrow t}} M_q^{T,u}(\omega), \quad \lim_{\substack{q \in \mathbb{Q}_+ \\ q \downarrow t}} M_q^{T,u}(\omega) \tag{39}$$

exist finitely valued for all $t \in [0, T]$.

Let us now show that the same holds true for X. For notational convenience we first focus on left limits. Consider the sets K and $I_{t,\varepsilon}^{\mathscr{T}}$ defined in (26) and (29) and let $t \geq 0$ be fixed. Take some sequence $(q_k)_{k \in \mathbb{N}}$, as specified in Proposition 3, such that $q_k \uparrow t$. Then there exists some $N \in \mathbb{N}$ such that, for all $k \geq N$ and $(T, u) \in I_{t,\varepsilon}^{\mathscr{T}} \times K$, $\Phi(T - q_k, u) \neq 0$. This is a consequence of the definition of ε (see (28)). Thus we

---

[6]Note that due to the measurable projection theorem, $\widetilde{\Omega} \in \mathscr{F}^x$.

can divide $M_{q_k}^{T,u}(\omega)$ by $\Phi(T - q_k, u)$ for all $k \geq N$ and $(T, u) \in I_{t,\varepsilon}^{\mathcal{T}} \times K$. By the continuity of $t \mapsto \Phi(t, u)$ and (39), it follows that, for every $\omega \in \widetilde{\Omega}$, the limit

$$\lim_{k\to\infty} N_{q_k}^{T,u}(\omega) := \lim_{k\to\infty} e^{\langle \psi(T-q_k, u), X_{q_k}(\omega) \rangle}$$

exists finitely valued for all $(T, u) \in I_{t,\varepsilon}^{\mathcal{T}} \times K$. From Proposition 3 we thus deduce that, for every $\omega \in \widetilde{\Omega}$, the limit

$$\lim_{k\to\infty} X_{q_k}(\omega)$$

exists either finitely valued or $\lim_{k\to\infty} \|X_{q_k}(\omega)\| = \infty$. Using similar arguments yields the same assertion for right limits. Hence we can conclude that $\mathbb{P}_x$-a.s.

$$\tilde{X}_t = \lim_{\substack{q \in \mathbb{Q}_+ \\ q \downarrow t}} X_q \qquad (40)$$

exists for all $t \geq 0$ and defines a càdlàg function in t.

Let now Ω_0 be the set of $\omega \in \Omega$ for which the limit $\tilde{X}_t(\omega)$ exists for every t and defines a càdlàg function in t. Then, as a consequence of [30, Theorem II.62.7, Corollary II.62.12], $\Omega_0 \in \mathscr{F}^0$ and $\mathbb{P}_x[\Omega_0] = 1$ for all $x \in D_\Delta$. For $\omega \in \Omega \setminus \Omega_0$, we set $\tilde{X}_t(\omega) = \Delta$ for all t. Then $\tilde{X}$ is a càdlàg process and $\tilde{X}_t$ is $\mathscr{F}^0$-measurable for every $t \geq 0$. Since X is assumed to be stochastically continuous, we have $X_s \to X_t$ in probability as $s \to t$. Using the fact that convergence in probability implies almost sure convergence along a subsequence, we have

$$\mathbb{P}_x \left[\lim_{\substack{q \in \mathbb{Q}_+ \\ q \downarrow t}} X_q = X_t \right] = 1. \qquad (41)$$

By our definition of $\tilde{X}_t$, the limit in (41) is equal to $\tilde{X}_t$ on Ω_0. Hence, for all $x \in D_\Delta$, we have $\mathbb{P}_x[\tilde{X}_t = X_t]$ for each t, implying that $\tilde{X}$ is a version of X. This then also yields

$$\mathbb{E}_x \left[e^{\langle u, \tilde{X}_t \rangle} \right] = \mathbb{E}_x \left[e^{\langle u, X_t \rangle} \right]$$

and augmentation of $(\mathscr{F}_t^0)$ with respect to $\mathbb{P}_x$ ensures that $\tilde{X}_t \in \mathscr{F}_t^x$ for each t. We therefore conclude that $\tilde{X}$ is an affine process with respect to $(\mathscr{F}_t^x)$. □

If D is non-compact, the càdlàg version (40) on $D_\Delta \cup \{\infty\}$, still denoted by X, can be realized on the space $\Omega' := \mathbb{D}(D_\Delta \cup \{\infty\})$ of càdlàg paths $\omega : \mathbb{R}_+ \to D_\Delta \cup \{\infty\}$ with $\omega(t) = \Delta$ for $t \geq s$, whenever $\omega(s) = \Delta$. However, we still have to prove that we can identify ∞ with Δ, as mentioned in Remark 5. In other words, we have to show that $\|\omega(t)\| = \infty$ for all $t \geq s$ if explosion occurs for some $s > 0$, that is, $\|\omega(s-)\| = \infty$. This is stated in the Theorem 3 below. For its proof let us introduce the following notations:

Due to the convention $\|\Delta\| = \infty$, we define the *explosion time* by (see [6] for a similar definition)

$$T_{\text{expl}} := \begin{cases} T_\Delta, & \text{if } T'_k < T_\Delta \text{ for all } k, \\ \infty, & \text{if } T'_k = T_\Delta \text{ for some } k, \end{cases}$$

where the stopping times T_Δ and T'_k are given by

$$T_\Delta := \inf\{t > 0 \mid \|X_{t-}\| = \infty \text{ or } \|X_t\| = \infty\},$$
$$T'_k := \inf\{t \mid \|X_{t-}\| \geq k \text{ or } \|X_t\| \geq k\}, \quad k \geq 1.$$

Moreover, we denote by relint(C) the *relative interior* of a set C defined by

$$\text{relint}(C) = \{x \in C \mid \overline{B}(x, r) \cap \text{aff}(C) \subseteq C \text{ for some } r > 0\},$$

where aff(C) denotes the affine hull of C.

Lemma 2. *Let X be an affine process with càdlàg paths in $D_\Delta \cup \{\infty\}$ and let $x \in D$ be fixed. If*

$$\mathbb{P}_x[T_{\text{expl}} < \infty] > 0, \tag{42}$$

then relint(Re $\mathcal{U}$) $\neq \emptyset$ *and we have* $\mathbb{P}_x$-*a.s.*

$$\lim_{t \uparrow T_{\text{expl}}} e^{\langle u, X_t \rangle} = 0$$

for all $u \in$ relint(Re $\mathcal{U}$).

Proof. Let us first establish that under Assumption (42), relint Re $\mathcal{U} \neq \emptyset$. To this end, we denote by Ω_{expl} the set

$$\Omega_{\text{expl}} = \{\omega \in \Omega' \mid T_{\text{expl}}(\omega) < \infty\}.$$

Then it follows from Propositions 2 and 3 that, for $\mathbb{P}_x$-almost every $\omega \in \Omega_{\text{expl}}$, there exists some $(T(\omega), v(\omega)) \in (T_{\text{expl}}(\omega), \infty) \times iV$ such that

$$\lim_{t \uparrow T_{\text{expl}}(\omega)} \Phi(T(\omega) - t, v(\omega)) \neq 0$$

and

$$\lim_{t \uparrow T_{\text{expl}}(\omega)} N_t^{T(\omega), v(\omega)}(\omega) = \lim_{t \uparrow T_{\text{expl}}(\omega)} e^{\langle \psi(T(\omega) - t, v(\omega)), X_t(\omega) \rangle} = 0. \tag{43}$$

This implies that

$$\lim_{t \uparrow T_{\text{expl}}(\omega)} \langle \operatorname{Re} \psi(T(\omega) - t, v(\omega)), X_t(\omega) \rangle = -\infty, \tag{44}$$

and in particular that $\mathscr{U} \ni \operatorname{Re} \psi(T(\omega) - T_{\text{expl}}(\omega), v(\omega)) \neq 0$, which proves the claim, since $\operatorname{Re} \mathscr{U} \subseteq \overline{\operatorname{relint}(\operatorname{Re} \mathscr{U})}$.

Furthermore, by (44) we have $\lim_{t \uparrow T_{\text{expl}}(\omega)} \|\Pi_{\langle \operatorname{Re} \mathscr{U} \rangle}(X_t(\omega))\| = \infty$ and an application of Lemma 3 below yields the second assertion. □

Lemma 3. *Assume that* $\operatorname{relint}(\operatorname{Re} \mathscr{U}) \neq \emptyset$ *and that there exists some D-valued sequence $(x_k)_{k \in \mathbb{N}}$ such that*

$$\lim_{k \to \infty} \|\Pi_{\langle \operatorname{Re} \mathscr{U} \rangle} x_k\| = \infty. \tag{45}$$

Then $\lim_{k \to \infty} \langle u, x_k \rangle = -\infty$ *for all* $u \in \operatorname{relint}(\operatorname{Re} \mathscr{U})$.

Proof. Suppose by contradiction that there exists some $u \in \operatorname{relint}(\operatorname{Re} \mathscr{U})$ such that

$$\limsup_{k \to \infty} \langle u, x_k \rangle > -\infty.$$

Then there exists a subsequence, still denoted by (x_k), such that

$$\lim_{k \to \infty} \langle u, x_k \rangle > -\infty. \tag{46}$$

and due to (45) some direction $g \in \langle \operatorname{Re} \mathscr{U} \rangle$ such that

$$\lim_{k \to \infty} \langle g, x_k \rangle = \infty. \tag{47}$$

Moreover, since $u \in \operatorname{relint}(\operatorname{Re} \mathscr{U})$, there exists some $\varepsilon > 0$ such that $u + \varepsilon g \in \operatorname{relint}(\operatorname{Re} \mathscr{U})$. By the definition of $\mathscr{U}$, we have

$$\sup_{x \in D} \langle u + \varepsilon g, x \rangle < \infty.$$

Due to (47), this however implies that

$$\lim_{k \to \infty} \langle u, x_k \rangle = -\infty$$

and contradicts (46). □

Theorem 3. *Let X be an affine process with càdlàg paths in $D_\Delta \cup \{\infty\}$. Then, for every $x \in D$, the following assertion holds $\mathbb{P}_x$-a.s.: If*

$$\|X_{s-}\| = \infty, \tag{48}$$

then $\|X_t\| = \infty$ for all $t \geq s$ and $s \geq 0$. Identifying ∞ with Δ, then yields $X_t = \Delta$ for all $t \geq s$.

Proof. Let $x \in D$ be fixed and let $u \in \operatorname{relint}(\operatorname{Re}\mathcal{U})$. Note that by Lemma 2,

$$\operatorname{relint}(\operatorname{Re}\mathcal{U}) \neq \emptyset$$

and that $\Phi(t, u)$ and $\psi(t, u)$ are real-valued functions with values in $\mathbb{R}_{++}$ and $\operatorname{Re}\mathcal{U}$, respectively. Take now some $T > 0$ and $\delta > 0$ such that

$$\mathbb{P}_x\left[T - \delta < T_{\text{expl}} \leq T\right] > 0,$$

and $\psi(t, u) \in \operatorname{relint}(\operatorname{Re}\mathcal{U})$ for all $t < \delta$. Consider the martingale

$$M_t^{T,u} = \Phi(T - t, u) e^{\langle \psi(T-t,u), X_t \rangle}, \quad t \leq T,$$

which is clearly nonnegative and has càdlàg paths. Moreover, by the choice of δ, it follows from Lemma 2 and the conventions $\|\Delta\| = \infty$ and $f(\Delta) = 0$ for any other function that $\mathbb{P}_x$-a.s.

$$M_{s-}^{T,u} = 0, \quad s \in (T - \delta, T], \tag{49}$$

if and only if $\|X_{s-}\| = \infty$ for $s \in (T - \delta, T]$. We thus conclude using [30, Theorem II.78.1] that $\mathbb{P}_x$-a.s. $M_t^{T,u} = 0$ for all $t \geq s$, which in turn implies that $\|X_t\| = \infty$ for all $t \geq s$. This allows us to identify ∞ with Δ and we obtain $X_t = \Delta$ for all $t \geq s$. Since T was chosen arbitrarily, the assertion follows. □

Combining Theorems 2 and 3 and using Assumption (3), we thus obtain the following statement:

Corollary 1. *Let X be an affine process relative to $(\mathcal{F}_t^0)$. Then there exists a process $\tilde{X}$ such that, for each $x \in D_\Delta$, $\tilde{X}$ is a $\mathbb{P}_x$-version of X, which is an affine process relative to $(\mathcal{F}_t^x)$, whose paths are càdlàg and satisfy $\mathbb{P}_x$-a.s. $\tilde{X}_t = \Delta$ for $t \geq s$, whenever $\|\tilde{X}_{s-}\| = \infty$ or $\|\tilde{X}_s\| = \infty$.*

Remark 6. We will henceforth *always* assume that we are using the càdlàg version of an affine process, given in Corollary 1, which we still denote by X. Under this assumption X can now be realized on the space $\Omega = \mathbb{D}(D_\Delta)$ of càdlàg paths $\omega : \mathbb{R}_+ \to D_\Delta$ with $\omega(t) = \Delta$ for $t \geq s$, whenever $\|\omega(s-)\| = \infty$ or $\|\omega(s)\| = \infty$. The canonical realization of an affine process X is then defined by $X_t(\omega) = \omega(t)$. Moreover, we make the convention that $X_\infty = \Delta$, which allows us to write certain formulas without restriction.

4 Right-Continuity of the Filtration and Strong Markov Property

Using the existence of a right-continuous version of an affine process, we can now show that $(\mathscr{F}_t^x)$, that is, the augmentation of $(\mathscr{F}_t^0)$ with respect to $\mathbb{P}_x$, is right-continuous.

Theorem 4. *Let $x \in D$ be fixed and let X be an affine process relative to $(\mathscr{F}_t^x)$ with càdlàg paths. Then $(\mathscr{F}_t^x)$ is right-continuous.*

Proof. We adapt the proof of [28, Theorem I.4.31] to our setting. We have to show that for every $t \geq 0$, $\mathscr{F}_{t+}^x = \mathscr{F}_t^x$, where $\mathscr{F}_{t+}^x = \bigcap_{s>t} \mathscr{F}_s^x$. Since the filtration is increasing, it suffices to show that $\mathscr{F}_t^x = \bigcap_{n \geq 1} \mathscr{F}_{t+\frac{1}{n}}^x$. In particular, we only need to prove that

$$\mathbb{E}_x \left[e^{\langle u_1, X_{t_1}\rangle + \cdots + \langle u_k, X_{t_k}\rangle} \,\Big|\, \mathscr{F}_t^x \right] = \mathbb{E}_x \left[e^{\langle u_1, X_{t_1}\rangle + \cdots + \langle u_k, X_{t_n}\rangle} \,\Big|\, \mathscr{F}_{t+}^x \right] \qquad (50)$$

for all $(t_1, \ldots, t_k)$ and all $(u_1, \ldots, u_k)$ with $t_i \in \mathbb{R}_+$ and $u_i \in \mathscr{U}$, since this implies $\mathbb{E}_x[Z|\mathscr{F}_t^x] = \mathbb{E}_x[Z|\mathscr{F}_{t+}^x]$ for every bounded $Z \in \mathscr{F}^x$. As both $\mathscr{F}_{t+}^x$ and $\mathscr{F}_t^x$ contain the nullsets $\mathscr{N}(\mathscr{F}^x)$, this then already yields $\mathscr{F}_{t+}^x = \mathscr{F}_t^x$ for all $t \geq 0$.

In order to prove (50), let $t \geq 0$ be fixed and take first $t_1 \leq t_2 \cdots \leq t_k \leq t$. Then we have for all $(u_1, \ldots, u_k)$

$$\mathbb{E}_x \left[e^{\langle u_1, X_{t_1}\rangle + \cdots + \langle u_k, X_{t_k}\rangle} \,\Big|\, \mathscr{F}_t^x \right] = \mathbb{E}_x \left[e^{\langle u_1, X_{t_1}\rangle + \cdots + \langle u_k, X_{t_k}\rangle} \,\Big|\, \mathscr{F}_{t+}^x \right]$$

$$= e^{\langle u_1, X_{t_1}\rangle + \cdots + \langle u_k, X_{t_k}\rangle}.$$

In the case $t_k > t_{k-1} \cdots > t_1 > t$, we give the proof for $k = 2$ for notational convenience. Let $t_2 > t_1 > t$ and fix $u_1, u_2 \in \mathscr{U}$. Then we have by the affine property

$$\mathbb{E}_x \left[e^{\langle u_1, X_{t_1}\rangle + \langle u_2, X_{t_2}\rangle} \,\Big|\, \mathscr{F}_{t+}^x \right] = \lim_{s \downarrow t} \mathbb{E}_x \left[e^{\langle u_1, X_{t_1}\rangle + \langle u_2, X_{t_2}\rangle} \,\Big|\, \mathscr{F}_s^x \right]$$

$$= \lim_{s \downarrow t} \mathbb{E}_x \left[\mathbb{E}_x \left[e^{\langle u_1, X_{t_1}\rangle + \langle u_2, X_{t_2}\rangle} \,\Big|\, \mathscr{F}_{t_1}^x \right] \,\Big|\, \mathscr{F}_s^x \right]$$

$$= \Phi(t_2 - t_1, u_2) \lim_{s \downarrow t} \mathbb{E}_x \left[e^{\langle u_1 + \psi(t_2 - t_1, u_2), X_{t_1}\rangle} \,\Big|\, \mathscr{F}_s^x \right].$$

If $\Phi(t_2 - t_1, u_2) = 0$, it follows by the same step that

$$\mathbb{E}_x \left[e^{\langle u_1, X_{t_1}\rangle + \langle u_2, X_{t_2}\rangle} \,\Big|\, \mathscr{F}_t^x \right] = 0,$$

too. Otherwise, we have by Proposition 1 (ii), $\psi(t_2 - t_1, u_2) \in \mathcal{U}$, and by the definition of $\mathcal{U}$ also $u_1 + \psi(t_2 - t_1, u_2) \in \mathcal{U}$. Hence, again by the affine property and right-continuity of $t \mapsto X_t(\omega)$, the above becomes

$$\mathbb{E}_x \left[e^{\langle u_1, X_{t_1} \rangle + \langle u_2, X_{t_2} \rangle} \mid \mathscr{F}^x_{t+} \right]$$
$$= \Phi(t_2 - t_1, u_2) \lim_{s \downarrow t} \Phi(t_1 - s, u_1 + \psi(t_2 - t_1, u_2)) e^{\langle \psi(t_1 - s, u_1 + \psi(t_2 - t_1, u_2)), X_s \rangle}$$
$$= \Phi(t_2 - t_1, u_2) \Phi(t_1 - t, u_1 + \psi(t_2 - t_1, u_2)) e^{\langle \psi(t_1 - t, u_1 + \psi(t_2 - t_1, u_2)), X_t \rangle}$$
$$= \mathbb{E}_x \left[e^{\langle u_1, X_{t_1} \rangle + \langle u_2, X_{t_2} \rangle} \mid \mathscr{F}^x_t \right].$$

This yields (50) and by the above arguments we conclude that $\mathscr{F}^x_{t+} = \mathscr{F}^x_t$ for all $t \geq 0$. □

Remark 7. A consequence of Theorem 4 is that $(\Omega, \mathscr{F}_t, (\mathscr{F}^x_t), \mathbb{P}_x)$ satisfies the *usual conditions*, since

(i) $\mathscr{F}^x$ is $\mathbb{P}_x$-complete,
(ii) $\mathscr{F}^x_0$ contains all $\mathbb{P}_x$-null-sets in $\mathscr{F}^x$,
(iii) $(\mathscr{F}^x_t)$ is right-continuous.

Let us now set

$$\mathscr{F} := \bigcap_{x \in D_\Delta} \mathscr{F}^x, \quad \mathscr{F}_t := \bigcap_{x \in D_\Delta} \mathscr{F}^x_t. \tag{51}$$

Then $(\Omega, \mathscr{F}, (\mathscr{F}_t), \mathbb{P}_x)$ does not necessarily satisfy the usual conditions, but $\mathscr{F}_t = \mathscr{F}_{t+}$ still holds true. Moreover, it follows e.g. from [29, Proposition III.2.12, III.2.14] that, for each t, X_t is $\mathscr{F}_t$-measurable and a Markov process with respect to $(\mathscr{F}_t)$.

Unless otherwise mentioned, we henceforth *always* consider affine processes on the filtered space $(\Omega, \mathscr{F}, (\mathscr{F}_t))$, where $\Omega = \mathbb{D}(D_\Delta)$, as described in Remark 6, and $\mathscr{F}, \mathscr{F}_t$ are given by (51). Notice that these assumptions on the probability space correspond to the standard setting considered for Feller processes (compare [30, Definition III.7.16, III.9.2]).

Similar as in the case of Feller processes, we can now formulate and prove the strong Markov property for affine processes using the above setting and in particular the right-continuity of the sample paths.

Theorem 5. *Let X be an affine process and let T be a $(\mathscr{F}_t)$-stopping time. Then for each bounded Borel measurable function f and $s \geq 0$*

$$\mathbb{E}_x[f(X_{T+s}) \mid \mathscr{F}_T] = \mathbb{E}_{X_T}[f(X_s)], \quad \mathbb{P}_x\text{-a.s.}$$

Proof. This result can be shown by the same arguments used to prove the strong Markov property of Feller processes (see, e.g., [30, Theorem 8.3, Theorem 9.4]), namely by using a dyadic approximation of the stopping time T and applying the Markov property. Instead of using C_0-functions and the Feller property, we here consider the family of functions $\{x \mapsto e^{\langle u,x \rangle} \mid u \in iV\}$ and the affine property, which asserts in particular that

$$x \mapsto \mathbb{E}_x\left[e^{\langle u,X_t \rangle}\right] = P_t e^{\langle u,x \rangle} = \Phi(t,u) e^{\langle \psi(t,u),x \rangle}$$

is continuous. This together with the right-continuity of paths then implies for every $\Lambda \in \mathscr{F}_T$ and $u \in iV$

$$\mathbb{E}_x\left[e^{\langle u,X_{T+s} \rangle} 1_\Lambda\right] = \mathbb{E}_x\left[P_s e^{\langle u,X_T \rangle} 1_\Lambda\right].$$

The assertion then follows by the same arguments as in [30, Theorem 8.3] or [7, Theorem 2.3.1]. □

5 Semimartingale Property

We shall now relate affine processes to semimartingales, where, for every $x \in D$, semimartingales are understood with respect to the filtered probability space $(\Omega, \mathscr{F}, (\mathscr{F}_t), \mathbb{P}_x)$ defined above. By convention, we call X a semimartingale if $X1_{[0,T_\Delta)}$ is a semimartingale, where—as a consequence of Theorem 3 and Corollary 1—we can now define the *lifetime* T_Δ by

$$T_\Delta(\omega) = \inf\{t > 0 \mid X_t(\omega) = \Delta\}. \tag{52}$$

Let us start with the following definition for general Markov processes (compare [5, Definition 7.1]):

Definition 4 (Extended Generator). An operator $\mathscr{G}$ with domain $\mathscr{D}_\mathscr{G}$ is called *extended generator* for a Markov process X (relative to some filtration $(\mathscr{F}_t)$) if $D_\mathscr{G}$ consists of those Borel measurable functions $f : D \to \mathbb{C}$ for which there exists a function $\mathscr{G}f$ such that the process

$$f(X_t) - f(x) - \int_0^t \mathscr{G}f(X_{s-})ds$$

is a well-defined and $(\mathscr{F}_t, \mathbb{P}_x)$-local martingale for every $x \in D_\Delta$.

In the following lemma we consider a particular class of functions for which it is possible to state the form of the extended generator for a Markov process in terms of its semigroup.

Lemma 4. *Let X be a D_A-valued Markov process relative to some filtration $(\mathscr{F}_t)$. Suppose that $u \in \mathscr{U}$ and $\eta > 0$. Consider the function*

$$g_{u,\eta} : D \to \mathbb{C}, x \mapsto g_{u,\eta}(x) := \frac{1}{\eta} \int_0^\eta P_s e^{\langle u,x \rangle} ds.$$

Then, for every $x \in D$,

$$M_t^u := g_{u,\eta}(X_t) - g_{u,\eta}(X_0) - \int_0^t \frac{1}{\eta} \left(P_\eta e^{\langle u,X_{s-} \rangle} - e^{\langle u,X_{s-} \rangle} \right) ds$$

is a (complex-valued) $(\mathscr{F}_t, \mathbb{P}_x)$-martingale and thus $g_{u,\eta}(X)$ is a (complex-valued) special semimartingale.

Proof. Since $g_{u,\eta}$ and $P_\eta e^{\langle u,\cdot \rangle} - e^{\langle u,\cdot \rangle}$ are bounded, M_t^u is integrable for each t and we have

$$\mathbb{E}_x \left[M_t^u | \mathscr{F}_r \right]$$
$$= M_r^u + \mathbb{E}_x \left[g_{u,\eta}(X_t) - g_{u,\eta}(X_r) - \int_r^t \frac{1}{\eta} \left(P_\eta e^{\langle u,X_{s-} \rangle} - e^{\langle u,X_{s-} \rangle} \right) ds \bigg| \mathscr{F}_r \right]$$
$$= M_r^u + \mathbb{E}_{X_r} \left[g_{u,\eta}(X_{t-r}) - g_{u,\eta}(X_0) - \int_0^{t-r} \frac{1}{\eta} \left(P_\eta e^{\langle u,X_{s-} \rangle} - e^{\langle u,X_{s-} \rangle} \right) ds \right]$$
$$= M_r^u + \frac{1}{\eta} \int_{t-r}^{t-r+\eta} P_s e^{\langle u,X_r \rangle} ds - \frac{1}{\eta} \int_0^\eta P_s e^{\langle u,X_r \rangle} ds$$
$$\quad - \frac{1}{\eta} \int_\eta^{t-r+\eta} P_s e^{\langle u,X_r \rangle} ds + \frac{1}{\eta} \int_0^{t-r} P_s e^{\langle u,X_r \rangle} ds$$
$$= M_r^u.$$

Hence M^u is $(\mathscr{F}_t, \mathbb{P}_x)$-martingale and thus $g_{u,\eta}(X)$ is a special semimartingale, since it is the sum of a martingale and a predictable finite variation process. □

Remark 8. Lemma 4 asserts that the extended generator applied to $g_{u,\eta}$ is given by $\mathscr{G} g_{u,\eta}(x) = \frac{1}{\eta} \left(P_\eta e^{\langle u,x \rangle} - e^{\langle u,x \rangle} \right)$. Note that for general Markov processes and even for affine processes we do not know whether the "pointwise" infinitesimal generator applied to

$$e^{\langle u,x \rangle} = \lim_{\eta \to 0} g_{u,\eta} = \lim_{\eta \to 0} \frac{1}{\eta} \int_0^\eta P_s e^{\langle u,x \rangle} ds,$$

that is,

$$\lim_{\eta \to 0} \frac{1}{\eta} \left(P_\eta e^{\langle u,x \rangle} - e^{\langle u,x \rangle} \right),$$

is well-defined or not.[7] For this reason we consider the family of functions $\{x \mapsto g_{u,\eta}(x) \mid u \in \mathcal{U}, \eta > 0\}$, which exhibits in the case of affine processes similar properties as $\{x \mapsto e^{\langle u,x \rangle} \mid u \in \mathcal{U}\}$ (see Remark 9 (ii) and Lemma 5 below). These properties are introduced in the following definitions (compare [5, Definition 7.7, 7.8]).

Definition 5 (Full Class). A class $\mathscr{C}$ of Borel measurable functions from D to $\mathbb{C}$ is said to be a *full* class if, for all $r \in \mathbb{N}$, there exists a finite family $\{f_1, \ldots, f_N\} \in \mathscr{C}$ and a function $h \in C^2(\mathbb{C}^N, D)$ such that

$$x = h(f_1(x), \ldots, f_N(x)) \tag{53}$$

for all $x \in D$ with $\|x\| \leq r$.

Definition 6 (Complete Class). Let $\beta \in V$, $\gamma \in S_+(V)$, where $S_+(V)$ denotes the positive semidefinite matrices over V, and let F be a nonnegative measure on V, which integrates $(\|\xi\|^2 \wedge 1)$, satisfies $F(\{0\}) = 0$ and $x + \text{supp}(F) \subseteq D_\Delta$ for all $x \in D$. Moreover, let $\chi : V \to V$ denote some truncation function, that is, χ is bounded and satisfies $\chi(\xi) = \xi$ in a neighborhood of 0. A countable subset of functions $\tilde{\mathscr{C}} \subset C_b^2(D)$ is called *complete* if, for any fixed $x \in D$, the countable collection of numbers

$$\kappa(f(x)) = \langle \beta, \nabla f(x) \rangle + \frac{1}{2} \sum_{i,j} \gamma_{ij} D_{ij} f(x)$$
$$+ \int_V (f(x+\xi) - f(x) - \langle \nabla f(x), \chi(\xi) \rangle) F(d\xi), \quad f \in \tilde{\mathscr{C}} \tag{54}$$

completely determines β, γ and F. A class $\mathscr{C}$ of Borel measurable functions from D to $\mathbb{C}$ is said to be *complete class* if it contains such a countable set.

Remark 9. (i) Note that the integral in (54) is well-defined for all $f \in C_b^2(D)$. This is a consequence of the integrability assumption and the fact that $x + \text{supp}(F)$ is supposed to lie in D_Δ for all x.
(ii) The class of functions

$$\mathscr{C}^* := \left\{ D \to \mathbb{C}, x \mapsto e^{\langle u,x \rangle} \mid u \in iV \right\} \tag{55}$$

is a full and complete class. Indeed, for every $x \in D$ with $\|x\| \leq r$, we can find n linearly independent vectors $(u_1, \ldots, u_n)$ such that

$$\text{Im}\langle u_i, x \rangle \in \left[-\frac{\pi}{2}, \frac{\pi}{2}\right].$$

---

[7]In the case of affine processes, this would be implied by the differentiability of Φ and ψ with respect to t, which we only prove in Sect. 6 using the results of this paragraph.

This implies that x is given by

$$x = \left(\arcsin\left(\operatorname{Im} e^{\langle u_1,x\rangle}\right), \ldots, \arcsin\left(\operatorname{Im} e^{\langle u_n,x\rangle}\right)\right) (\operatorname{Im} u_1, \ldots, \operatorname{Im} u_n)^{-1}$$

and proves that $\mathscr{C}^*$ is a full class. Completeness follows by the same arguments as in [20, Lemma II.2.44]. □

Lemma 5. *Let X be an affine process with Φ and ψ given in (6). Consider the class of functions*

$$\mathscr{C} := \left\{ D \to \mathbb{C},\, x \mapsto g_{u,\eta}(x) := \frac{1}{\eta} \int_0^\eta \Phi(s,u) e^{\langle \psi(s,u),x\rangle} ds \,\Big|\, u \in iV,\, \eta > 0 \right\}. \tag{56}$$

Then $\mathscr{C}$ is a full and complete class.

Proof. Let $(u_1, \ldots u_n) \in iV$ be n linearly independent vectors and define a function $f_\eta : D \to \mathbb{C}^n$ by $f_{\eta,i}(x) = g_{u_i,\eta}(x)$. Then the Jacobi matrix $J_{f_\eta}(x)$ is given by

$$\begin{pmatrix} \frac{1}{\eta}\int_0^\eta \Phi(s,u_1)e^{\langle\psi(s,u_1),x\rangle}\psi_1(s,u_1)ds & \cdots & \frac{1}{\eta}\int_0^\eta \Phi(s,u_1)e^{\langle\psi(s,u_1),x\rangle}\psi_n(s,u_1)ds \\ \vdots & \ddots & \vdots \\ \frac{1}{\eta}\int_0^\eta \Phi(s,u_n)e^{\langle\psi(s,u_n),x\rangle}\psi_1(s,u_n)ds & \cdots & \frac{1}{\eta}\int_0^\eta \Phi(s,u_n)e^{\langle\psi(s,u_n),x\rangle}\psi_n(s,u_n)ds \end{pmatrix}.$$

In particular, the imaginary part of each row tends to $(\cos(\operatorname{Im}\langle u_i,x\rangle)\operatorname{Im} u_i)^\top$ for $\eta \to 0$. Hence there exists some $\eta > 0$ such that the rows of $\operatorname{Im} J_{f_\eta}$ are linearly independent. As $\operatorname{Im} f_\eta : D \to \mathbb{R}^n$ is a $C^\infty(D)$-function and as $J_{\operatorname{Im} f_\eta} = \operatorname{Im} J_{f_\eta}$, it follows from the inverse function theorem that, for each $x_0 \in D$, there exists some $r_0 > 0$ such that $\operatorname{Im} f_\eta : B(x_0, r_0) \to W$ has a $C^\infty(W)$ inverse, where $W = \operatorname{Im} f_\eta(B(x_0, r_0))$.

Let now $r \in \mathbb{N}$ and consider $x \in D$ with $\|x\| \leq r$. Assume without loss of generality that $0 \in D$ and let $x_0 = 0$. Since

$$\lim_{\eta \to 0} J_{\operatorname{Im} f_\eta}(x) = (\cos(\operatorname{Im}\langle u_1,x\rangle)\operatorname{Im} u_1, \ldots, \cos(\operatorname{Im}\langle u_n,x\rangle)\operatorname{Im} u_n)^\top,$$

we can assure—by choosing the linearly independent vectors $(u_1, \ldots, u_n)$ such that $|\langle u_i, x\rangle|$ is small enough—that for all $x \in \overline{B}(0,r) \cap D$

$$\|\lim_{\eta\to 0} J_{\operatorname{Im} f_\eta}^{-1}(0) \lim_{\eta\to 0} J_{\operatorname{Im} f_\eta}(x) - I\|$$
$$= \|(\operatorname{Im} u_1, \ldots, \operatorname{Im} u_n)^{-\top}(\cos(\operatorname{Im}\langle u_1,x\rangle)\operatorname{Im} u_1, \ldots, \cos(\operatorname{Im}\langle u_n,x\rangle)\operatorname{Im} u_n)^\top - I\| < 1.$$

By the continuity of the matrix inverse the same holds true for η small enough. The proof of the inverse function theorem (see, e.g., [19, Theorem 4.2] or [25, Lemma XIV.1.3]) then implies that r_0 can be chosen to be r and $\mathscr{C}$ is a full class.

Concerning completeness, note that

$$\kappa(g_{u,\eta}(x)) = \frac{1}{\eta}\int_0^\eta \Phi(s,u)e^{\langle\psi(s,u),x\rangle}\Bigg(\langle\beta,\psi(s,u)\rangle + \frac{1}{2}\langle\psi(s,u)\gamma\psi(s,u)\rangle$$

$$+ \int_V \left(e^{\langle\psi(s,u),\xi\rangle} - 1 - \langle\psi(s,u),\chi(\xi)\rangle\right) F(d\xi)\Bigg)ds. \quad (57)$$

Indeed, by Remark 9 (i), the integral

$$\int_V \int_0^\eta \left|\Phi(s,u)e^{\langle\psi(s,u),x\rangle}\left(e^{\langle\psi(s,u),\xi\rangle} - 1 - \langle\psi(s,u),\chi(\xi)\rangle\right)\right| ds\, F(d\xi)$$

is well-defined, whence by Fubini's theorem we can interchange the integration. From (57) it thus follows that

$$\lim_{\eta\to 0}\kappa(g_{u,\eta}(x)) = \kappa(e^{\langle u,x\rangle})$$

$$= e^{\langle u,x\rangle}\left(\langle\beta,u\rangle + \frac{1}{2}\langle u,\gamma u\rangle + \int_V \left(e^{\langle u,\xi\rangle} - 1 - \langle u,\chi(\xi)\rangle\right) F(d\xi)\right). \quad (58)$$

Moreover, by [20, Lemma II.2.44] or simply as a consequence of the completeness of the class C^*, as defined in (55), the function $u \mapsto \kappa(e^{\langle u,x\rangle})$ admits a unique representation of form (58), that is, if $\kappa(e^{\langle\cdot,x\rangle})$ also satisfies (58) with $(\tilde\beta,\tilde\gamma,\tilde F)$, then $\beta = \tilde\beta$, $\gamma = \tilde\gamma$ and $F = \tilde F$. This property carries over to the class $\mathscr{C}$. Indeed, for every $x \in D$, there exists some $\eta > 0$ such that $\beta = \tilde\beta$, $\gamma = \tilde\gamma$ and $F = \tilde F$ if $u \mapsto \kappa(g_{u,\eta}(x))$ also satisfies (57) with $(\tilde\beta,\tilde\gamma,\tilde F)$. This proves that $\mathscr{C}$ is a complete class. □

In order to establish the semimartingale property of X and to study its characteristics, we need to handle explosions and killing. Similar to [6], we consider again the stopping times T_Δ defined in (52) and T'_k given by

$$T'_k := \inf\{t \mid \|X_{t-}\| \geq k \text{ or } \|X_t\| \geq k\}, \quad k \geq 1.$$

By the convention $\|\Delta\| = \infty$, $T'_k \leq T_\Delta$ for all $k \geq 1$. As a transition to Δ occurs either by a jump or by explosion, we additionally define the stopping times:

$$T_{\text{jump}} = \begin{cases} T_\Delta, & \text{if } T'_k = T_\Delta \text{ for some } k, \\ \infty, & \text{if } T'_k < T_\Delta \text{ for all } k, \end{cases}$$

$$T_{\text{expl}} = \begin{cases} T_\Delta, & \text{if } T'_k < T_\Delta \text{ for all } k, \\ \infty, & \text{if } T'_k = T_\Delta \text{ for some } k, \end{cases} \quad (59)$$

$$T_k = \begin{cases} T'_k, & \text{if } T'_k < T_\Delta, \\ \infty, & \text{if } T'_k = T_\Delta. \end{cases}$$

Note that $\{T_{\text{jump}} < \infty\} \cap \{T_{\text{expl}} < \infty\} = \emptyset$ and $\lim_{k \to \infty} T_k = T_{\text{expl}}$ with $T_k < T_{\text{expl}}$ on $\{T_{\text{expl}} < \infty\}$. Hence T_{expl} is predictable with announcing sequence $T_k \wedge k$. In order to turn X into a semimartingale and to get explicit expressions for the characteristics, we stop X before it explodes, which is possible, since T_{expl} is predictable. Note that we cannot stop X before it is killed, as T_{jump} is totally inaccessible. For this reason we shall concentrate on the process $(X_t^{\tau}) := (X_{t \wedge \tau})$, where τ is a stopping time satisfying $0 < \tau < T_{\text{expl}}$, which exists by the above argument and the càdlàg property of X. Since $X = X^{T_\Delta}$, we have

$$X_t^{\tau} = X_t 1_{\{t < (\tau \wedge T_\Delta)\}} + X_{\tau \wedge T_\Delta} 1_{\{t \geq (\tau \wedge T_\Delta)\}}$$
$$= X_t 1_{\{t < (\tau \wedge T_{\text{jump}})\}} + X_{\tau \wedge T_{\text{jump}}} 1_{\{t \geq (\tau \wedge T_{\text{jump}})\}},$$

which implies that a transition to Δ can only occur through a jump.

Recall that Δ is assumed to be an arbitrary point which does not lie in D. We can thus identify Δ with some point in $V \setminus D$ such that every $C_b^2(D)$-function f can be extended continuously to D_Δ with $f(\Delta) = 0$. Indeed, without loss of generality we may assume that such a point exists, because otherwise we can always embed D_Δ in $V \times \mathbb{R}$.

Theorem 6. *Let X be an affine process and let τ be a stopping time with $\tau < T_{\text{expl}}$, where T_{expl} is defined in (59). Then $X 1_{[0, T_\Delta)}$ and X^{τ} are semimartingales with state space $D \cup \{0\}$ and D_Δ, respectively. Moreover, let (B, C, ν) denote the characteristics of X^{τ} relative to some truncation function χ. Then there exists a version of (B, C, ν), which is of the form*

$$B_{t,i} = \int_0^{t \wedge \tau} b_i(X_{s-}) ds,$$
$$C_{t,ij} = \int_0^{t \wedge \tau} c_{ij}(X_{s-}) ds, \qquad (60)$$
$$\nu(\omega; dt, d\xi) = K(X_t, d\xi) 1_{[0,\tau]} dt,$$

where $b : D \to V$ and $c : D \to S_+(V)$ are measurable functions and $K(x, d\xi)$ is a positive kernel from $(D, \mathscr{D})$ into $(V, \mathscr{B}(V))$, which satisfies $\int_V (\|\xi\|^2 \wedge 1) K(x, d\xi) < \infty$, $K(x, \{0\}) = 0$ and $x + \text{supp}(K(x, \cdot)) \subseteq D_\Delta$ for all $x \in D$.

Proof. We adapt the proof of [5, Theorem 7.9 (ii), (iii)] to our setting. By Lemma 4,

$$g_{u,\eta}(X) = \frac{1}{\eta} \int_0^\eta \Phi(s, u) e^{\langle \psi(s,u), X \rangle} ds$$

is a semimartingale for every $u \in \mathscr{U}$ and $\eta > 0$. Since Lemma 5 asserts that $\mathscr{C}$, as defined in (56), is a full class, an application of Itô's formula to the function h_i appearing in (53) shows that X_i coincides with a semimartingale on each stochastic interval $[0, \tau_r[$, where

$$\tau_r = \inf\{t \geq 0 \mid \|X_t\| \geq r\} \wedge T_\Delta.$$

Since we have $\mathbb{P}_x$-a.s. $\lim_{r \to \infty} \tau_r = T_\Delta$ and since being a semimartingale is a local property (see [20, Proposition I.4.25]), we conclude that $X 1_{[0,T_\Delta)}$ is a semimartingale.

Let now τ denote a stopping time with $\tau < T_{\text{expl}}$. Then X^τ is also a semimartingale with state space D_Δ, since explosion is avoided and the transition to Δ can only occur via killing, that is, a jump to Δ, which is incorporated in the jump characteristic (see [6, Sect. 3]).

By [5, Theorem 6.25], one can find a version of the characteristics (B, C, ν) of X^τ, which is of the form

$$
\begin{aligned}
B_{t,i} &= \int_0^{t \wedge \tau} \tilde{b}_{s-,i} dF_s, \\
C_{t,ij} &= \int_0^{t \wedge \tau} \tilde{c}_{s-,ij} dF_s, \\
\nu(\omega; dt, d\xi) &= 1_{[0,\tau]} dF_t(\omega) \tilde{K}_{\omega,t}(d\xi),
\end{aligned}
\tag{61}
$$

where F is an additive process of finite variation, which is $\mathbb{P}_x$-indistinguishable from an $(\mathscr{F}_t)$-predictable process, $\tilde{b}$ and $\tilde{c}$ are $(\mathscr{F}_t)$-optional processes with values in V and $S_+(V)$, respectively, and $\tilde{K}_{\omega,t}(d\xi)$ is a positive kernel from $(\Omega \times \mathbb{R}_+, \mathcal{O}(\mathscr{F}_t))^8$ into $(V, \mathscr{B}(V))$, which satisfies $\int_V (\|\xi\|^2 \wedge 1) \tilde{K}_{\omega,t}(d\xi) < \infty$, $\tilde{K}_{\omega,t}(\{0\}) = 0$ and $X_t(\omega) + \text{supp}(\tilde{K}_{\omega,t}) \subseteq D_\Delta$ for all $t \in [0, \tau]$ and $\mathbb{P}_x$-almost all ω. Moreover, by [20, Theorem II.2.42], for every $f \in C_b^2(V)$, the process

$$
f(X_t^\tau) - f(x) - \int_0^{t \wedge \tau} \langle \tilde{b}_{s-}, \nabla f(X_{s-}) \rangle dF_s - \frac{1}{2} \int_0^{t \wedge \tau} \sum_{i,j} \tilde{c}_{s-,ij} D_{ij} f(X_{s-}) dF_s
$$
$$
- \int_0^{t \wedge \tau} \int_V (f(X_{s-} + \xi) - f(X_{s-}) - \langle \nabla f(X_{s-}), \chi(\xi) \rangle) \tilde{K}_{\omega,s-}(d\xi) dF_s \tag{62}
$$

is a $(\mathscr{F}_t, \mathbb{P}_x)$-local martingale and the last three terms are of finite variation. Let us denote

$$
\mathscr{L} f(X_{t-}(\omega)) := \langle \tilde{b}_{t-}, \nabla f(X_{t-}(\omega)) \rangle - \frac{1}{2} \sum_{i,j} \tilde{c}_{t-,ij} D_{ij} f(X_{t-}(\omega))
$$
$$
- \int_V (f(X_{t-}(\omega) + \xi) - f(X_{t-}(\omega)) - \langle \nabla f(X_{t-}(\omega)), \chi(\xi) \rangle) \tilde{K}_{\omega,t-}(d\xi).
$$

---

[8] Here, $\mathcal{O}(\mathscr{F}_t)$ denotes the $(\mathscr{F}_t)$-optional σ-algebra.

As proved in Lemma 5, the class of functions $\mathscr{C}$ defined in (56) is complete. Let now $\tilde{\mathscr{C}} \subset \mathscr{C}$ be the countable set satisfying the property stated in Definition 6 and let $g_{\eta,u} \in \tilde{\mathscr{C}}$ for some $u \in iV$ and $\eta > 0$. Then Lemma 4 and Definition 4 imply that

$$g_{\eta,u}(X_t^\tau) - g_{\eta,u}(x) - \int_0^{t\wedge\tau} \mathscr{G} g_{\eta,u}(X_{s-})ds$$

$$= g_{\eta,u}(X_t^\tau) - g_{\eta,u}(x) - \int_0^{t\wedge\tau} \frac{1}{\eta}\left(P_\eta e^{\langle u, X_{s-}\rangle} - e^{\langle u, X_{s-}\rangle}\right)ds \quad (63)$$

is a $(\mathscr{F}_t, \mathbb{P}_x)$-martingale, while $(\int_0^{t\wedge\tau} \mathscr{G} g_{\eta,u}(X_{s-})ds)$ is a predictable finite variation process. Due to (62) and uniqueness of the canonical decomposition of the special semimartingale $g_{\eta,u}(X^\tau)$ (see [20, Definition I.4.22, Corollary I.3.16]), we thus have

$$\int_0^{t\wedge\tau} \tilde{\mathscr{L}} g_{\eta,u}(X_{s-})dF_s = \int_0^{t\wedge\tau} \mathscr{G} g_{\eta,u}(X_{s-})ds \quad \text{up to an evanescent set.} \quad (64)$$

Set now

$$\Lambda = \left\{(\omega, t) : \tilde{\mathscr{L}} g_{\eta,u}(X_{(t\wedge\tau\wedge T_\Delta)-}(\omega)) = 0 \text{ for every } g_{\eta,u} \in \tilde{\mathscr{C}}\right\}.$$

Then the characteristic property (54) of $\tilde{\mathscr{C}}$ implies that Λ is exactly the set where $\tilde{b} = 0$, $\tilde{c} = 0$ and $\tilde{K} = 0$. Hence we may replace F by $1_{\Lambda^c} F$ without altering (61), that is, we can suppose that $1_\Lambda F = 0$. This property together with (64) implies that $dF_t \ll dt \, \mathbb{P}_x$-a.s. Hence we know that there exists a triplet (b', c', K') such that F replaced by t and $(\tilde{b}, \tilde{c}, \tilde{K})$ replaced by (b', c', K') satisfy all the conditions of (61). In particular, we have by [20, Proposition II.2.9 (i)] that X^τ is quasi-left continuous. Due to [5, Theorem 6.27], it thus follows that

$$b'_t = b(X_t) 1_{[0,\tau]},$$
$$c'_t = c(X_t) 1_{[0,\tau]},$$
$$K'_{\omega,t}(d\xi) = K(X_t, d\xi) 1_{[0,\tau]},$$

where the functions b, c and the kernel K have the properties stated in (60). This proves the assertion. □

6 Regularity

By means of the above derived semimartingale property, in particular the fact that the characteristics are absolutely continuous with respect to the Lebesgue measure, we can prove that every affine process is regular in the following sense:

Definition 7 (Regularity). An affine process X is called *regular* if for every $u \in \mathcal{U}$ the derivatives

$$F(u) = \left.\frac{\partial \Phi(t,u)}{\partial t}\right|_{t=0}, \quad R(u) = \left.\frac{\partial \psi(t,u)}{\partial t}\right|_{t=0} \tag{65}$$

exist and are continuous on $\mathcal{U}^m$ for every $m \geq 1$.

Remark 10. In the case of the canonical state space $D = \mathbb{R}_+^m \times \mathbb{R}^{n-m}$, the derivative of $\phi(t,u)$ at $t = 0$ is also denoted by $F(u)$ (see [12, Eq. (3.10)] and Remark 2). Since $\Phi(t,u) = e^{\phi(t,u)}$, we have

$$\partial_t \Phi(t,u)|_{t=0} = e^{\phi(0,u)} \partial_t \phi(t,u)|_{t=0} = \partial_t \phi(t,u)|_{t=0}.$$

Hence our definition of F coincides with the one in [12].

Lemma 6. *Let X be an affine process. Then the functions $t \mapsto \Phi(t,u)$ and $t \mapsto \psi_i(t,u)$, $i \in \{1,\ldots,n\}$, defined in (6) are of finite variation for all $u \in \mathcal{U}$.*

Proof. Due to Assumption 1, there exist $n+1$ vectors such that $(x_1,\ldots,x_n)$ are linearly independent and $x_{n+1} = \sum_{i=1}^n \lambda_i x_i$ for some $\lambda \in V$ with $\sum_{i=1}^n \lambda_i \neq 1$. Let us now take $n+1$ affine processes $X^1, \ldots, X^{n+1}$ such that $\mathbb{P}_{x_i}[X_0^i = x_i] = 1$ for all $i \in \{1,\ldots,n+1\}$. It then follows from Theorem 6 that, for every $i \in \{1,\ldots,n+1\}$, X^i is a semimartingale with respect to the filtered probability space $(\Omega, \mathcal{F}, (\mathcal{F}_t), \mathbb{P}_{x_i})$. We can then construct a filtered probability space $(\Omega', \mathcal{F}', (\mathcal{F}'_t), \mathbb{P}')$, with respect to which $X_1, \ldots, X_{n+1}$ are independent semimartingales such that $\mathbb{P}' \circ (X^i)^{-1} = \mathbb{P}_{x_i}$. One possible construction is the product probability space

$$(\Omega^{n+1}, \otimes_{i=1}^{n+1} \mathcal{F}, (\otimes_{i=1}^{n+1} \mathcal{F}_t), \otimes_{i=1}^{n+1} \mathbb{P}_{x_i}).$$

We write $y_i = (1, x_i)^\top$ and $Y^i = (1, X^i)^\top$ for $i \in \{1,\ldots,n+1\}$. Then the definition of x_i implies that $(y_1,\ldots,y_{n+1})$ are linearly independent. Moreover, as X^i exhibits càdlàg paths for all $i \in \{1,\ldots,n+1\}$, there exists some stopping time $\delta > 0$ such that, for all $\omega \in \Omega'$ and $t \in [0, \delta(\omega))$, the vectors $(Y_t^1(\omega), \ldots, Y_t^{n+1}(\omega))$ are also linearly independent. Let now $T > 0$ and $u \in \mathcal{U}$ be fixed and choose some $0 < \varepsilon(\omega) \leq \delta(\omega)$ such that, for all $t \in [0, \varepsilon(\omega))$, $\Phi(T-t,u) \neq 0$.

Denoting the $(\mathcal{F}'_t, \mathbb{P}')$-martingales $\Phi(T-t,u) e^{\langle \psi(T-t,u), X_t^i \rangle}$ by $M_t^{T,u,i}$ and choosing the right branch of the complex logarithm, we thus have for all $t \in [0, \varepsilon(\omega))$

$$\begin{pmatrix} 1 & X_{t,1}^1(\omega) & \cdots & X_{t,n}^1(\omega) \\ \vdots & \vdots & \ddots & \vdots \\ 1 & X_{t,1}^{n+1}(\omega) & \cdots & X_{t,n}^{n+1}(\omega) \end{pmatrix}^{-1} \begin{pmatrix} \ln M_t^{T,u,1}(\omega) \\ \vdots \\ \ln M_t^{T,u,n+1}(\omega) \end{pmatrix} = \begin{pmatrix} \ln \Phi(T-t,u) \\ \psi_1(T-t,u) \\ \vdots \\ \psi_n(T-t,u) \end{pmatrix}.$$

This implies that $(\Phi(s,u))_s$ and $(\psi(s,u))_s$ coincide on the stochastic interval $(T - \varepsilon(\omega), T]$ with deterministic semimartingales and are thus of finite variation. As this holds true for all $T > 0$, we conclude that $t \mapsto \Phi(t,u)$ and $t \mapsto \psi_i(t,u)$ are of finite variation. □

Using Lemma 6 and Theorem 6, we are now prepared to prove regularity of affine processes. Additionally, our proof reveals that the functions F and R have parameterizations of Lévy–Khintchine type and that the (differential) semimartingale characteristics introduced in (60) depend in an affine way on X.

Theorem 7. *Every affine process is regular. Moreover, the functions F and R, as defined in (65), are of the form*

$$F(u) = \langle u, b \rangle + \frac{1}{2} \langle u, au \rangle - c$$
$$+ \int_V \left(e^{\langle u, \xi \rangle} - 1 - \langle u, \chi(\xi) \rangle \right) m(d\xi), \quad u \in \mathcal{U},$$

$$\langle R(u), x \rangle = \langle u, B(x) \rangle + \frac{1}{2} \langle u, A(x)u \rangle - \langle \gamma, x \rangle$$
$$+ \int_V \left(e^{\langle u, \xi \rangle} - 1 - \langle u, \chi(\xi) \rangle \right) M(x, d\xi), \quad u \in \mathcal{U},$$

where $\chi : V \to V$ denotes some truncation function such that $\chi(\Delta - x) = 0$ for all $x \in D$, $b \in V$, $a \in S(V)$, m is a (signed) measure, $c \in \mathbb{R}$, $\gamma \in V$ and $x \mapsto B(x)$, $x \mapsto A(x)$, $x \mapsto M(x, d\xi)$ are restrictions of linear maps on V such that

$$b(x) = b + B(x),$$
$$c(x) = a + A(x),$$
$$K(x, d\xi) = m(d\xi) + M(x, d\xi) + (c + \langle \gamma, x \rangle)\delta_{(\Delta-x)}(d\xi).$$

Here, the left hand side corresponds to the (differential) semimartingale characteristics introduced in (60).

Furthermore, on the set $\mathcal{Q} = \{(t, u) \in \mathbb{R}_+ \times \mathcal{U} \mid \Phi(s, u) \neq 0, \text{ for all } s \in [0, t]\}$, the functions Φ and ψ satisfy the ordinary differential equations

$$\partial_t \Phi(t, u) = \Phi(t, u) F(\psi(t, u)), \qquad \Phi(0, u) = 1, \qquad (66)$$
$$\partial_t \psi(t, u) = R(\psi(t, u)), \qquad \psi(0, u) = u \in \mathcal{U}. \qquad (67)$$

Remark 11. Recall that without loss of generality we identify Δ with some point in $V \setminus D$ such that every $f \in C_b^2(D)$ can be extended continuously to D_Δ with $f(\Delta) = 0$.

Proof. Let $m \geq 1$ and $u \in \mathcal{U}^m$ be fixed and choose $T_u > 0$ such that $\Phi(T_u-t, u) \neq 0$ for all $t \in [0, T_u]$. As $t \mapsto \Phi(t, u)$ and $t \mapsto \psi(t, u)$ are of finite variation by Lemma 6, their derivatives with respect to t exist almost everywhere and we can write

$$\Phi(T_u - t, u) - \Phi(T_u, u) = \int_0^t -d\Phi(T_u - s, u),$$

$$\psi_i(T_u - t, u) - \psi_i(T_u, u) = \int_0^t -d\psi_i(T_u - s, u),$$

for $i \in \{1, \ldots, n\}$. Moreover, by the semiflow property of Φ and ψ (see Proposition 1 (iii)), differentiability of $\Phi(t, u)$ and $\psi(t, u)$ with respect to t at some $\varepsilon \in (0, T_u]$ implies that the derivatives $\partial_t|_{t=0}\psi(t, \psi(\varepsilon, u))$ and $\partial_t|_{t=0}\Phi(t, \psi(\varepsilon, u))$ exist as well. Let now $(\varepsilon_k)_{k \in \mathbb{N}}$ denote a sequence of points where $\Phi(t, u)$ and $\psi(t, u)$ are differentiable such that $\lim_{k \to \infty} \varepsilon_k = 0$. Then there exists a sequence $(u_k)_{k \in \mathbb{N}}$ given by

$$u_k = \psi(\varepsilon_k, u) \in \mathcal{U} \text{ with } \lim_{k \to \infty} u_k = u \tag{68}$$

such that the derivatives

$$\partial_t|_{t=0}\psi(t, u_k), \quad \partial_t|_{t=0}\Phi(t, u_k) \tag{69}$$

exist for every $k \in \mathbb{N}$. Moreover, since $|\mathbb{E}_x[\exp(\langle u, X_{\varepsilon_k}\rangle)]| < m$, there exists some constant M such that $u_k \in \mathcal{U}^M$ for all $k \in \mathbb{N}$.

Furthermore, due to Theorem 6, the canonical semimartingale representation of X^τ (see [20, Theorem II.2.34]), where τ is a stopping time with $\tau < T_{\text{expl}}$, is given by

$$X_t^\tau = x + \int_0^{t \wedge \tau} b(X_{s-})ds + N_t^\tau + \int_0^{t \wedge \tau} \int_V (\xi - \chi(\xi))\mu^{X^\tau}(\omega; ds, d\xi),$$

where μ^{X^τ} is the random measure associated with the jumps of X^τ and N^τ is a local martingale, namely the sum of the continuous martingale part and the purely discontinuous one, that is,

$$\int_0^{t \wedge \tau} \int_V \chi(\xi)(\mu^{X^\tau}(\omega; ds, d\xi) - K(X_{s-}, d\xi)ds).$$

Let now (u_k) be given by (68). Applying Itô's formula (relative to the measure $\mathbb{P}_x$) to each of the martingales $M_{t \wedge \tau}^{T_{u_k}, u_k} = \Phi(T_{u_k} - (t \wedge \tau), u)e^{\langle \psi(T_{u_k}-(t \wedge \tau), u_k), X_{t \wedge \tau}\rangle}$, $k \in \mathbb{N}$, we obtain

$$M_{t\wedge\tau}^{T_{u_k},u_k}$$
$$= M_0^{T_{u_k},u_k} + \int_0^{t\wedge\tau} M_{s-}^{T_{u_k},u_k} \left(\frac{-d\Phi(T_{u_k}-s,u_k)}{\Phi(T_{u_k}-s,u_k)} + \langle -d\psi(T_{u_k}-s,u_k), X_{s-}\rangle \right)$$
$$+ \int_0^{t\wedge\tau} M_{s-}^{T_{u_k},u_k} \bigg(\langle \psi(T_{u_k}-s,u_k), b(X_{s-})\rangle$$
$$+ \frac{1}{2} \langle \psi(T_{u_k}-s,u_k), c(X_{s-})\psi(T_{u_k}-s,u_k)\rangle$$
$$+ \int_V \left(e^{\langle \psi(T_{u_k}-s,u_k),\xi\rangle} - 1 - \langle \psi(T_{u_k}-s,u_k), \chi(\xi)\rangle \right) K(X_{s-},d\xi) \bigg) ds$$
$$+ \int_0^{t\wedge\tau} M_{s-}^{T_{u_k},u_k} \langle \psi(T_{u_k}-s,u_k), dN_s^\tau\rangle$$
$$+ \int_0^{t\wedge\tau} \int_V M_{s-}^{T_{u_k},u_k} \left(e^{\langle \psi(T_{u_k}-s,u_k),\xi\rangle} - 1 - \langle \psi(T_{u_k}-s,u_k), \chi(\xi)\rangle \right)$$
$$\times \left(\mu^{X^\tau}(\omega;ds,d\xi) - K(X_{s-},d\xi)ds \right).$$

As the last two terms are local martingales and as the rest is of finite variation, we thus have, for almost all $t \in [0, T_{u_k} \wedge \tau]$, $\mathbb{P}_x$-a.s. for every $x \in D$,

$$\frac{d\Phi(T_{u_k}-t,u_k)}{\Phi(T_{u_k}-t,u_k)} + \langle d\psi(T_{u_k}-t,u_k), X_{t-}\rangle$$
$$= \langle \psi(T_{u_k}-t,u_k), b(X_{t-})\rangle dt + \frac{1}{2} \langle \psi(T_{u_k}-t,u_k), c(X_{t-})\psi(T_{u_k}-t,u_k)\rangle dt$$
$$+ \int_V \left(e^{\langle \psi(T_{u_k}-t,u_k),\xi\rangle} - 1 - \langle \psi(T_{u_k}-t,u_k), \chi(\xi)\rangle \right) K(X_{t-},d\xi)dt. \tag{70}$$

By setting $t = T_{u_k}$ on a set of positive measure with $\mathbb{P}_x[\tau \geq T_{u_k}]$ and letting $T_{u_k} \to 0$, we obtain due to (69) for each $k \in \mathbb{N}$ and $x \in D$

$$\partial_t|_{t=0}\Phi(t,u_k) + \langle \partial_t|_{t=0}\psi(t,u_k), x\rangle$$
$$= \langle u_k, b(x)\rangle dt + \frac{1}{2} \langle u_k, c(x)u_k\rangle dt + \int_V \left(e^{\langle u_k,\xi\rangle} - 1 - \langle u_k, \chi(\xi)\rangle \right) K(x,d\xi)dt. \tag{71}$$

Since the right hand side is continuous in u_k, which is a consequence of the support properties of $K(x,\cdot)$ and the fact that $u_k \in \mathcal{U}^M$ for all $k \in \mathbb{N}$, the limit for $u_k \to u$ of the left hand side exists as well. By the affine independence of the $n+1$ elements in D, the coefficients $\partial_t|_{t=0}\Phi(t,u_k)$ and $\partial_t|_{t=0}\psi(t,u_k)$ converge for $u_k \to u$, whence the limit is affine, too. Since $m \geq 1$ and u was arbitrary, it follows that

$$\langle u, b(x)\rangle\, dt + \frac{1}{2}\langle u, c(x)u\rangle\, dt + \int_V \left(e^{\langle u,\xi\rangle} - 1 - \langle u, \chi(\xi)\rangle\right) K(x, d\xi)\, dt$$

is an affine function in x for all $u \in \mathcal{U}$.

By uniqueness of the Lévy–Khintchine representation and the assumption that D contains $n+1$ affinely independent elements, this implies that $x \mapsto b(x)$, $x \mapsto c(x)$ and $x \mapsto K(x, d\xi)$ are affine functions in the following sense:

$$b(x) = b + B(x),$$
$$c(x) = a + A(x),$$
$$K(x, d\xi) = m(d\xi) + M(x, d\xi) + (c + \langle \gamma, x\rangle)\delta_{(\Delta-x)}(d\xi),$$

where $b \in V$, $a \in S(V)$, m a (signed) measure, $c \in \mathbb{R}$, $\gamma \in V$ and $x \mapsto B(x)$, $x \mapsto A(x)$, $x \mapsto M(x, d\xi)$ are restrictions of linear maps on V. Indeed, $c + \langle \gamma, x\rangle$ corresponds to the killing rate of the process, which is incorporated in the jump measure. Here, we explicitly use the convention that $e^{\langle u, \Delta\rangle} = 0$, $b(\Delta) = 0$, $c(\Delta) = 0$, $K(\Delta, d\xi) = 0$ and the fact that $\chi(\Delta - x) = 0$ for all $x \in D$.

Moreover, for t small enough, we have for all $u \in \mathcal{U}$

$$\Phi(t, u) - \Phi(0, u) = \int_0^t \Phi(s, u)\bigg(\langle \psi(s, u), b\rangle + \frac{1}{2}\langle \psi(s, u), a\psi(s, u)\rangle - c$$
$$+ \int_V \left(e^{\langle \psi(s,u),\xi\rangle} - 1 - \langle \psi(s, u), \chi(\xi)\rangle\right) m(d\xi)\bigg) ds,$$

$$\langle \psi(t, u) - \psi(0, u), x\rangle = \int_0^t \bigg(\langle \psi(s, u), B(x)\rangle + \frac{1}{2}\langle \psi(s, u), A(x)\psi(s, u)\rangle - \langle \gamma, x\rangle$$
$$+ \int_V \left(e^{\langle \psi(s,u),\xi\rangle} - 1 - \langle \psi(s, u), \chi(\xi)\rangle\right) M(x, d\xi)\bigg) ds.$$

Note again that the properties of the support of $K(x, \cdot)$ carry over to the measures $M(x, \cdot)$ and $m(\cdot)$ implying that the above integrals are well-defined. Due to the continuity of $t \mapsto \Phi(t, u)$ and $t \mapsto \psi(t, u)$, we can conclude that the derivatives of Φ and ψ exist at 0 and are continuous on $\mathcal{U}^m$ for every $m \geq 1$, since they are given by

$$F(u) = \left.\frac{\partial \Phi(t,u)}{\partial t}\right|_{t=0} = \langle u, b \rangle + \frac{1}{2}\langle u, au \rangle - c$$
$$+ \int_V \left(e^{\langle u,\xi \rangle} - 1 - \langle u, \chi(\xi) \rangle\right) m(d\xi),$$

$$\langle R(u), x \rangle = \left\langle \left.\frac{\partial \psi(t,u)}{\partial t}\right|_{t=0}, x \right\rangle = \langle u, B(x) \rangle + \frac{1}{2}\langle u, A(x)u \rangle - \langle \gamma, x \rangle$$
$$+ \int_V \left(e^{\langle u,\xi \rangle} - 1 - \langle u, \chi(\xi) \rangle\right) M(x, d\xi).$$

This proves the first part of the theorem.

By the regularity of X, we are now allowed to differentiate the semiflow equations (8) on the set $\mathscr{Q} = \{(t,u) \in \mathbb{R}_+ \times \mathscr{U} \mid \Phi(s,u) \neq 0, \text{ for all } s \in [0,t]\}$ with respect to s and evaluate them at $s = 0$. As a consequence, Φ and ψ satisfy (66) and (67). □

Remark 12. The differential equations (66) and (67) are called *generalized Riccati equations*, which is due to the particular form of F and R.

Remark 13. Using the results of Theorem 7, in particular the assertion on the semimartingale characteristics, we aim to construct examples of affine processes on compact state spaces, which justify that we do not restrict ourselves to unbounded sets D. For simplicity, let $n = 1$. Then the pure deterministic drift process with characteristics

$$b(x) = b + Bx, \quad B \leq 0, \; -Br_1 \leq b \leq -Br_2, \quad c(x) = 0, \quad K(x,d\xi) = 0$$

is an affine process on the interval $[r_1, r_2]$. Another example of an affine process on a compact, but discrete, state space of the form

$$\{0, 1, \ldots, k\}$$

can be obtained by a pure jump process X with jump size distribution $\delta_1(d\xi)$ and intensity $k - X$. In terms of the semimartingale characteristics, we thus have $b(x) = 0$, $c(x) = 0$ and $K(x, d\xi) = (k-x)\delta_1(d\xi)$. For such type of jump processes the state space is necessarily discrete and cannot be extended to the whole interval $[0, k]$. In the presence of a diffusion component, the state space is necessarily unbounded, since the stochastic invariance conditions on $c(x) = a + Ax$, which would guarantee that the process remains in some interval $[r_1, r_2]$ imply

$$a + Ar_1 = 0 \quad \text{and} \quad a + Ar_2 = 0,$$

yielding $a = A = 0$.

Acknowledgements We thank Georg Grafendorfer and Enno Veerman for discussions and helpful comments. Both authors gratefully acknowledge the financial support by the ETH Foundation.

References

1. H. Bauer, *Probability Theory*, vol. 23 of *de Gruyter Studies in Mathematics* (Walter de Gruyter, Berlin, 1996). Translated from the fourth (1991) German edition by Robert B. Burckel and revised by the author
2. R.M. Blumenthal, R.K. Getoor, *Markov Processes and Potential Theory*. Pure and Applied Mathematics, vol. 29 (Academic Press, New York, 1968)
3. M.-F. Bru, Wishart processes. J. Theoret. Probab. **4**(4), 725–751 (1991)
4. B. Buraschi, A. Cieslak, F. Trojani, Correlation risk and the term structure of interest rates. Working paper, University St.Gallen, 2007
5. E. Çinlar, J. Jacod, P. Protter, M.J. Sharpe. Semimartingales and Markov processes. Z. Wahrsch. Verw. Gebiete **54**(2), 161–219 (1980)
6. P. Cheridito, D. Filipović, M. Yor, Equivalent and absolutely continuous measure changes for jump-diffusion processes. Ann. Appl. Probab. **15**(3), 1713–1732 (2005)
7. K.L. Chung, J.B. Walsh, *Markov processes, Brownian motion, and time symmetry*, vol. 249 of *Grundlehren der Mathematischen Wissenschaften [Fundamental Principles of Mathematical Sciences]*, 2nd edn. (Springer, New York, 2005)
8. J.C. Cox, J.E.J. Ingersoll, S.A. Ross, A theory of the term structure of interest rates. Econometrica **53**(2), 385–407 (1985)
9. C. Cuchiero, D. Filipović, E. Mayerhofer, J. Teichmann, Affine processes on positive semidefinite matrices. Ann. Appl. Probab. **21**(2), 397–463 (2011)
10. J. Da Fonseca, M. Grasselli, C. Tebaldi, A multifactor volatility Heston model. Quant. Finance **8**(6), 591–604 (2008)
11. Q. Dai, K.J. Singleton, Specification analysis of affine term structure models. J. Finance **55**(5), 1943–1978 (2000)
12. D. Duffie, D. Filipović, W. Schachermayer, Affine processes and applications in finance. Ann. Appl. Probab. **13**(3), 984–1053 (2003)
13. D. Duffie, R. Kan, A yield-factor model of interest rates. Math. Finance **6**(4), 379–406 (1996)
14. D. Duffie, J. Pan, K. Singleton, Transform analysis and asset pricing for affine jump-diffusions. Econometrica **68**(6), 1343–1376 (2000)
15. S.N. Ethier, T.G. Kurtz, *Markov Processes*. Wiley Series in Probability and Mathematical Statistics: Probability and Mathematical Statistics (Wiley, New York, 1986)
16. I.I. Gihman, A.V. Skorohod, *The Theory of Stochastic Processes II*. Grundlehren der Mathematischen Wissenschaften, Bd. 218 (Springer, Berlin, 1983)
17. C. Gourieroux, R. Sufana, Wishart quadratic term structure models. SSRN eLibrary, 2003
18. S.L. Heston, A closed-form solution for options with stochastic volatility with applications to bond and currency options. Rev. Financ. Stud. **6**(2), 327–343 (1993)
19. R. Howard, The inverse function theorem for Lipschitz maps. Lecture Notes, 1997
20. J. Jacod, A.N. Shiryaev, *Limit Theorems for Stochastic Processes*, vol. 288 of *Fundamental Principles of Mathematical Sciences*, 2nd edn. (Springer, Berlin, 2003)
21. K. Kawazu, S. Watanabe, Branching processes with immigration and related limit theorems. Teor. Verojatnost. i Primenen. **16**, 34–51 (1971)
22. M. Keller-Ressel, W. Schachermayer, J. Teichmann, Affine processes are regular. Probab. Theory Relat. Fields **151**(3–4), 591–611 (2011)
23. M. Keller-Ressel, W. Schachermayer, J. Teichmann, Regularity of affine processes on general state spaces. arXiv preprint arXiv:1105.0632 (2011)
24. D. Lando, On Cox processes and credit risky securities. Rev. Derivatives Res. **2**(2–3), 99–120 (1998)

25. S. Lang, *Real and Functional Analysis*, vol. 142 of *Graduate Texts in Mathematics*, 3rd edn. (Springer, New York, 1993)
26. M. Leippold, F. Trojani, Asset pricing with matrix jump diffusions. SSRN eLibrary, 2008
27. G. Letac, H. Massam, The noncentral Wishart as an exponential family, and its moments. J. Multivariate Anal. **99**(7), 1393–1417 (2008)
28. P.E. Protter, *Stochastic Integration and Differential Equations*, vol. 21 of *Stochastic Modelling and Applied Probability*, 2nd edn. Version 2.1, Corrected third printing. (Springer, Berlin, 2005)
29. D. Revuz, M. Yor, *Continuous Martingales and Brownian Motion*, vol. 293 of *Grundlehren der Mathematischen Wissenschaften [Fundamental Principles of Mathematical Sciences]*, 3rd edn. (Springer, Berlin, 1999)
30. L.C.G. Rogers, D. Williams, *Diffusions, Markov Processes, and Martingales. Vol. 1: Foundations*. Wiley Series in Probability and Mathematical Statistics: Probability and Mathematical Statistics, 2nd edn. (Wiley, Chichester, 1994)
31. E. Veerman, Affine Markov processes on a general state space. PhD thesis, University of Amsterdam, 2011

Langevin Process Reflected on a Partially Elastic Boundary II

Emmanuel Jacob

Abstract A particle subject to a white noise external forcing moves according to a Langevin process. Consider now that the particle is reflected at a boundary which restores a portion c of the incoming speed at each bounce. For c strictly smaller than the critical value $c_{crit} = \exp(-\pi/\sqrt{3})$, the bounces of the reflected process accumulate in a finite time, yielding a very different behavior from the most studied cases of perfectly elastic reflection—$c = 1$—and totally inelastic reflection—$c = 0$. We show that nonetheless the particle is not necessarily absorbed after this accumulation of bounces. We define a "resurrected" reflected process as a recurrent extension of the absorbed process, and study some of its properties. We also prove that this resurrected reflected process is the unique solution to the stochastic partial differential equation describing the model, for which well-posedness is nothing obvious. Our approach consists in defining the process conditioned on never being absorbed, via an h-transform, and then giving the Itō excursion measure of the recurrent extension thanks to a formula fairly similar to Imhof's relation.

Keywords Langevin process • Second order reflection • Recurrent extension • Excursion measure • Stochastic differential equation • h-transform

AMS 2010 subject classifications. 60J50, 60H15.

E. Jacob (✉)
Laboratoire de Probabilités et Modèles Aléatoires, Université Pierre et Marie Curie, 4 place Jussieu, 75005 Paris, France
e-mail: emmanuel.jacob@normalesup.org

1 Introduction

A physical particle with a stochastic behavior can often be described by a Brownian motion, for reasonably large time scales. On short time scales, however, a more accurate model is that of a Langevin process, which exhibits C^1 paths, and thus non-exploding velocities. We study in this article reflection problems, which exhibit an infinite number of bounces in a finite time, and hence involve extremely short time scales. For this reason—and others—it seems pertinent to the author to be unsatisfied with the usual models of reflected Brownian motions, and to deepen the investigation of the reflected Langevin processes.

A Langevin equation is simply the usual Newton equation of motion of a physical particle submitted to external forces, when these forces include a random force modeled by a white noise. We restrict ourselves to the case when there is no other external forces, and when the space is one-dimensional. If x is the initial position of the particle and u its initial velocity, its position is then given by

$$X_t = x + ut + \int_0^t B_s \mathrm{d}s,$$

where B is a Brownian motion starting from $B_0 = 0$. We suppose B is standard in the sense that it has variance t at time t, and call X the integrated Brownian motion, or (free) Langevin process. The Langevin process is non-Markov, contrarily to the two-dimensional Kolmogorov process $(X, \dot{X})$, whose first coordinate is a Langevin process, and second its derivative. We refer to Lachal [15] for a detailed account about it.

Further, suppose the particle is constrained to stay in $[0, \infty)$ by a partially elastic barrier at 0, characterized by an elasticity coefficient $c \geq 0$: if the particle hits the barrier with incoming velocity $v < 0$, it will instantly bounce back with velocity $-cv$. Again, write X for the position of the particle and $\dot{X}$ for its (right-continuous) right derivative. The reflection is naturally modeled by second order reflection, expressed by the following second order stochastic differential equation:

(RLP) $\quad \begin{cases} X_t = x + \int_0^t \dot{X}_s \mathrm{d}s \\ \dot{X}_t = u + B_t - (1+c) \sum_{0 < s \leq t} \dot{X}_{s-} \mathbb{1}_{X_s = 0} + N_t, \end{cases}$

where B is the standard Brownian motion driving the motion;

N is a continuous nondecreasing process starting from $N_0 = 0$,
increasing only when the process $(X, \dot{X})$ is at $(0, 0)$,
in the sense $\mathbb{1}_{(X_t, \dot{X}_t) \neq (0,0)} \mathrm{d}N_t \equiv 0$;

(x, u) is the initial or starting condition.

Two particular cases are often considered: a perfectly elastic reflection—$c = 1$—and a totally inelastic reflection—$c = 0$. Whatever the value of c, viewed as a two dimensional first order differential equation, Equations (RLP) enjoy the property of local pathwise existence and uniqueness of solutions, except maybe at $(0, 0)$. This extremal point $(0, 0)$ yields a real obstruction to the well-posedness of the equation. As early as in 1960, Bressan [7] pointed out that multiple solutions may occur, even when the force is $\mathscr{C}^\infty$. Since then, uniqueness results have been shown, but with the requirement the force be analytic (see [1, 20]). But the introduction of a white noise leads to original behaviors of the reflected process and to (weak) uniqueness results, as underlined by some previous works and this one.

In the totally inelastic case $c = 0$, it could be natural to believe that the second order reflection of a Langevin process would mean absorption of the particle at the barrier. In [16], Maury proceeded to numerical simulations, and was the first one to ask whether there could be non-absorbed solutions. The surprising answer that actually the (weak) only solution was non-absorbed is due to Bertoin, see [3, 4]. The solution enjoys the remarkable feature of spending no time at $(0, 0)$ and involving no continuous push of the barrier—it solves Equations (RLP) with $N \equiv 0$. His work is mainly based on a smart construction of a solution, and on Itô excursion theory.

In his thesis, Bect [2] discussed briefly Equations (RLP) for the different values of the elasticity coefficient, observing a phase transition phenomena. When c is greater than the critical value $c_{crit} = \exp(-\pi/\sqrt{3})$, the process $(X, \dot{X})$ starting from $(x, u) \neq (0, 0)$ will never hit $(0, 0)$. This includes the perfectly elastic case $c = 1$, which is particularly simple, as the absolute value of a free Langevin process yields a reflected Langevin process. A perfectly elastic barrier is also considered by Bossy and Jabir [6], who proved the well-posedness of the problem, for more general Langevin type processes. In a former paper [12], we studied the general supercritical regime $c > c_{crit}$, as well as the critical regime $c = c_{crit}$, and showed in particular that the existence of a unique reflected Langevin process stays true when the starting condition is $(0, 0)$, in a weak sense.

Finally, this paper deals with the subcritical regime $0 < c < c_{crit}$, and answers the last questions raised by Bect about the model. In that regime, starting from a nontrivial condition $(x, u) \neq (0, 0)$, the two-dimensional reflected process $(X, \dot{X})$ hits almost surely $(0, 0)$ after an infinite number of bounces. We write ζ_∞ for this hitting time, and $\mathbb{P}^c_{x,u}$ for the law of this reflected process *killed at time* ζ_∞, which is Markov. We prove the existence of a unique excursion law compatible with the semigroup of this Markov process. This defines uniquely a recurrent extension of the Markov process, which spends no time in $(0, 0)$, thanks to Itô's program. Finally we prove that this yields the unique solution, in the weak sense, to Equations (RLP). As in the totally inelastic case, no continuous push of the barrier is involved.

Though the similarity in the results, each regime reveals a very different behavior of the process, and needs a particular study, with dedicated techniques. In this work on the subcritical regime, our guiding line is largely inspired by a paper of Rivero [19], in which he studies the recurrent extensions of a self-similar Markov process with semigroup P_t. First, he recalls that recurrent extensions are equivalent to excursion measures compatible with P_t, thanks to Itô's program. Then a change of probability allows him to define the Markov process conditioned on never hitting 0, where this conditioning is in the sense of Doob, via an h-transform. An inverse h-transform on the Markov process conditioned on never hitting zero *and starting from 0* then gives the construction of the excursion measure.

We will not recall it at each step throughout the paper, but a lot of parallels can be made. However, it is a two-dimensional Markov process that we consider here. Further, its study will rely on an underlying random walk $(S_n)_{n \in \mathbb{N}}$ constructed from the velocities at bouncing times.

In the Preliminaries, we introduce this random walk and use it to estimate the tail of the variable ζ_∞ under $\mathbb{P}^c_{0,1}$. In Sect. 3, we introduce a change of probability, via an h-transform, to define $\tilde{\mathbb{P}}_{x,u}$, law of a process which can be viewed as the reflected Kolmogorov process conditioned on never being killed. We then show in Sect. 3.2 that this law has a weak limit $\tilde{\mathbb{P}}_{0+}$ when (x, u) goes to $(0, 0)$, using the same method that was used in [12] to show that for $c > c_{crit}$, the laws $\mathbb{P}^c_{0,u}$ have the weak limit $\mathbb{P}^c_{0+}$ when u goes to zero. Section 4 contains our main results. The first one is the construction of **n**, unique excursion measure compatible with the semigroup of the killed reflected process. This construction relies on the results of Sect. 3, involving in particular the laws $\mathbb{P}^c_{0+}$. More precisely, the measure **n** is given by a formula similar to Imhof's relation (see [10]) connecting the excursion measure of Brownian motion and the law of a Bessel(3) process. Further, we call *resurrected Kolmogorov process* the Markov process with Itō excursion measure **n** and which spends no time at $(0, 0)$. Then comes the second result: this process, together with $N \equiv 0$, yields the unique solution to (RLP), in the weak sense. The ten last pages contain the proofs.

2 Preliminaries

The preliminaries are partly similar to those of [12], which enter into more details, and to which we refer, if needed.

Context

For the sake of simplicity, we use the same notation (say P) for a probability measure and for the expectation under this measure. We will even authorize ourselves to write $P(f, A)$ for the quantity $P(f \mathbb{1}_A)$, when f is a positive measurable functional and A a measurable event. The set of nonnegative (resp. positive) real number is denoted by $\mathbb{R}_+$ (resp. $\mathbb{R}_+^*$). Introduce $D = (\{0\} \times \mathbb{R}_+^*) \cup (\mathbb{R}_+^* \times \mathbb{R})$ and $D^0 := D \cup \{(0, 0)\}$. Our working space is $\mathscr{C}$, the space of càdlàg trajectories $(x, \dot{x}) : [0, \infty) \to D^0$, which satisfy

$$x(t) = x(0) + \int_0^t \dot{x}(s) ds.$$

That space is endowed with the σ-algebra generated by the coordinate maps and with the topology induced by the following injection:

$$\mathscr{C} \to \mathbb{R}_+ \times \mathbb{D}$$
$$(x, \dot{x}) \mapsto (x(0), \dot{x}),$$

where $\mathbb{D}$ is the space of càdlàg trajectories on $\mathbb{R}_+$, equipped with Skorohod topology.

We denote by $(X, \dot{X})$ the canonical process and by $(\mathfrak{F}_t, t \geq 0)$ its natural filtration, satisfying the usual conditions of right continuity and completeness. For an initial condition $(x, u) \in D$, consider Equations (RLP), which we recall here:

$$\begin{cases} X_t = x + \int_0^t \dot{X}_s ds \\ \dot{X}_t = u + B_t - (1 + c) \sum_{0 < s \leq t} \dot{X}_{s-} \mathbb{1}_{X_s = 0} + N_t, \end{cases}$$

where B is the standard Brownian motion driving the motion;

N is a continuous nondecreasing process starting from $N_0 = 0$,

increasing only when the process $(X, \dot{X})$ is at $(0, 0)$,

in the sense $\mathbb{1}_{(X_t, \dot{X}_t) \neq (0,0)} dN_t \equiv 0$.

In the whole paper, the coefficient c is fixed and satisfies $0 < c < c_{crit} = \exp(-\pi/\sqrt{3})$. Call ζ_1 the first hitting time of zero for the reflected process X, that is $\zeta_1 := \inf\{t > 0, X_t = 0\}$. More generally, the sequence of the successive hitting times of zero $(\zeta_n)_{n \geq 1}$ is defined recursively by $\zeta_{n+1} := \inf\{t > \zeta_n, X_t = 0\}$. Write $(V_n)_{n \geq 1} := (\dot{X}_{\zeta_n})_{n \geq 1}$ for the sequence of the (outgoing) velocities of the process at these hitting times. The limit of the increasing sequence $(\zeta_n)_{n \geq 0}$ coincides almost surely with

$$\zeta_\infty := \inf\{t > 0, X_t = 0, \dot{X}_t = 0\}.$$

Indeed, to ensure this, the only nontrivial thing to prove is that the limit of $(X_t, \dot{X}_t)$ when t increases to $\sup \zeta_n$ is almost surely $(0, 0)$ on the event $\sup \zeta_n < \infty$. This follows, for example, from the almost sure uniform continuity of Brownian motion on compact sets, together with the fact $X_{\zeta_n} = 0$ for all n.

Hence, Equations (RLP) yield a strong unique solution $(X, \dot{X}, N)$ killed at time ζ_∞. It trivially satisfies $N \equiv 0$. We write $\mathbb{P}^c_{x,u}$ for the law of the Markov process $(X, \dot{X})$ killed at time ζ_∞ and call it (killed) reflected Kolmogorov process. We also call its first coordinate—which is no longer Markov—(killed) reflected Langevin process. Finally, when the starting position is $x = 0$ and starting velocity $u > 0$, we simply write $\mathbb{P}^c_u$ for $\mathbb{P}^c_{0,u}$, and we also define $\zeta_0 = 0$ and $V_0 = u$.

The Variable ζ_∞

Now, we will not only prove the almost sure finiteness of ζ_∞—which is the characterization of the subcritical regime—but also estimate its tail. The sequence $\left(\frac{\zeta_{n+1} - \zeta_n}{V_n^2}, \frac{V_{n+1}}{V_n}\right)_{n \geq 0}$ is i.i.d. and of law independent of u, which can be deduced from the law

$$\mathbb{P}_1^c\left((\zeta_1, V_1/c) \in (ds, dv)\right) = ds\, dv \frac{3v}{\pi\sqrt{2s^2}} \exp\left(-2\frac{v^2 - v + 1}{s}\right) \int_0^{4v/s} e^{-\frac{3\theta}{2}} \frac{d\theta}{\sqrt{\pi\theta}}, \quad (1)$$

given by McKean [17]. The second marginal of this law is

$$\mathbb{P}_1^c(V_1/c \in dv) = \frac{3}{2\pi} \frac{v^{\frac{3}{2}}}{1 + v^3} dv. \quad (2)$$

In particular, the sequence $S_n := \ln(V_n)$ is a random walk, with drift

$$\mathbb{P}_1^c(S_1 - S_0) = \ln(c) + \frac{\pi}{\sqrt{3}} = \ln\left(\frac{c}{c_{crit}}\right),$$

which is negative.

Lemma 1. *We have*

$$\mathbb{P}_1^c\left(V_1^x\right) = \frac{c^x}{2\cos\left(\frac{x+1}{3}\pi\right)} \text{ for } x < 1/2. \quad (3)$$

There exists a unique $k = k(c)$ in $(0, 1/4)$ such that $\mathbb{P}_1^c\left(V_1^{2k}\right) = 1$, and

$$\mathbb{P}_1^c(\zeta_\infty > t) \underset{t \to \infty}{\sim} C_1 t^{-k}, \quad (4)$$

where $C_1 = C_1(c) \in (0, \infty)$ is a constant depending only on c, given by

$$C_1 = \frac{\mathbb{P}_1^c\left(\zeta_\infty^k - (\zeta_\infty - \zeta_1)^k\right)}{k\mathbb{P}_1^c(V_1^{2k} \ln(V_1^2))}. \quad (5)$$

In other words, $k(c)$ is given implicitly as the unique solution in $]0, \frac{1}{4}]$ of the equation

$$c = \left[2\cos\left(\frac{2k+1}{3}\pi\right)\right]^{\frac{1}{2k}}. \quad (6)$$

The upper bound $1/4$ stems from the fact that $\mathbb{P}_1^c\left(V_1^{2k}\right)$ becomes infinite for $k = 1/4$. The value of $k(c)$ converges to $1/4$ when c goes to 0, and to 0 when c

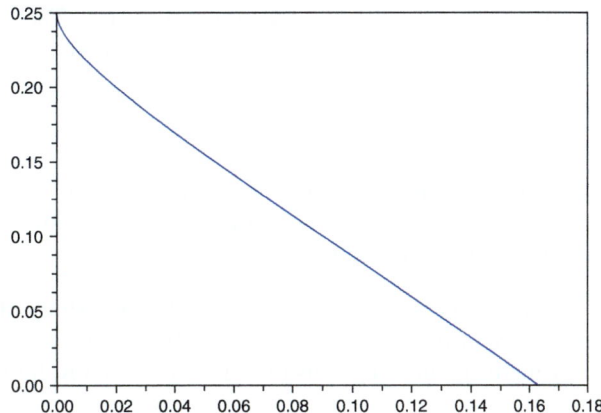

Fig. 1 Graph of the exponent k(c)

goes to c_{crit}, as illustrated by Fig. 1. We may notice that Formula (4) remains true for $c = 0$ and $k = 1/4$.

Proof. Formula (3) is a particular case of formula (12) of [14] and follows easily from Formula (2) and the well-known identity

$$\int_0^\infty \frac{t^{y-1}}{1+t}dt = \frac{\pi}{\sin(\pi y)}.$$

Now, the function $x \mapsto \mathbb{P}_1^c(V_1^x)$ is convex, takes value 1 at $x = 0$ and becomes infinite at $x = 1/2$. Its derivative at 0 is equal to $\mathbb{P}_1^c(\ln(V_1)) = \mathbb{P}_1^c(S_1 - S_0) < 0$. We deduce that there is indeed a unique $k(c)$ in $(0, \frac{1}{4})$ such that $\mathbb{P}_1^c(V_1^{2k}) = 1$.

The variable ζ_∞ can be expressed as the following series:

$$\zeta_\infty = \sum_{n=1}^\infty Q_n \prod_{k=1}^{n-1} M_k,$$

where $Q_n := \frac{\zeta_n - \zeta_{n-1}}{V_{n-1}^2}$ and $M_n := \frac{V_n^2}{V_{n-1}^2}$, so that $\prod_{k=1}^{n-1} M_k = V_{n-1}^2$. The sequence $(Q_n, M_n)_{n \geq 1}$ is i.i.d. with common law that of (ζ_1, V_1^2) under $\mathbb{P}_1^c$. Kesten, in [13], studies such series, using renewal theory. In particular, Theorem 5 of [13] states that $t^k \mathbb{P}_1^c(\zeta_\infty > t)$ converges to a positive and finite constant C_1, provided that the following conditions are satisfied:

$$\mathbb{P}_1^c(\ln(V_1^2)) < 0,$$

$$\mathbb{P}_1^c(V_1^{2k}) = 1,$$

$$\mathbb{P}_1^c(V_1^{2k} \ln(V_1^2) \mathbb{1}_{V_1 \geq 1}) < \infty,$$

$$\mathbb{P}_1^c(\zeta_1^k) < \infty.$$

The first three conditions are straightforward. The last one follows from the inequality $k < 1/4$ and from the following estimate of the queue of the variable ζ_1

$$\mathbb{P}_1^c(\zeta_1 > t) \underset{t\to\infty}{\sim} c' t^{-\frac{1}{4}}, \tag{7}$$

which was already pointed out in Lemma 1 in [12]. Finally, under the same conditions, Goldie, in Theorem (4.1) of [8], provides the explicit formula (5) for the value of the constant C_1.

Next section is devoted to the definition and study of the reflected Kolmogorov process, *conditioned on never hitting* $(0, 0)$. This process will be of great use for studying the recurrent extensions of the reflected Kolmogorov process in Sect. 4.

3 The Reflected Kolmogorov Process Conditioned on Never Hitting $(0, 0)$

3.1 Definition via an h-Transform

Recall that under $\mathbb{P}_1^c$, the sequence $(S_n)_{n\geq 0} = (\ln(V_n))_{n\geq 0}$ is a random walk starting from 0, and write $\mathbf{P}_0$ for its law. The important fact $\mathbb{P}_1^c(V_1^{2k}) = 1$ implies $\mathbb{P}_1^c(V_n^{2k}) = 1$ for any $n > 0$, and can be rewritten $\mathbf{P}_0(\theta^{S_n}) = 1$, with $\theta := \exp(2k)$.

The sequence θ^{S_n} being a martingale, we introduce the change of probability

$$\tilde{\mathbf{P}}_0(S_n \in dt) = \theta^t \mathbf{P}_0(S_n \in dt).$$

Under $\tilde{\mathbf{P}}_0$, $(S_n)_{n\geq 0}$ becomes a random walk drifting to $+\infty$. Informally, it can be viewed as being the random walk of law $\mathbf{P}_0$ and conditioned on hitting arbitrary high levels.

There is a corresponding change of probability for the reflected Kolmogorov process and its law $\mathbb{P}_1^c$. Introduce the law $\tilde{\mathbb{P}}_1$ of a process $(X_t)_{0\leq t < \zeta_\infty}$ determined by

$$\tilde{\mathbb{P}}_1(A) = \mathbb{P}_1^c(\mathbb{1}_A \mathbb{P}_1^c(V_n^{2k}|\mathfrak{F}_T)),$$

for any $n \geq 0$, stopping-time T satisfying $T < \zeta_n$, and $A \in \mathfrak{F}_T$. Note that this definition does not depend on the choices of n and T satisfying $T < \zeta_n$ and $A \in \mathfrak{F}_T$, and that $\tilde{\mathbb{P}}_1$ is a probability measure. Further, write $H(x, u)$ for $\mathbb{P}_{x,u}^c(V_1^{2k})$ and observe that we have $H(0, u) = u^{2k}$ and $\mathbb{P}_1^c(V_n^{2k}|\mathfrak{F}_T) = H(X_T, \dot{X}_T))$, by the strong Markov property. Therefore we have

$$\tilde{\mathbb{P}}_1(A) = \mathbb{P}_1^c(\mathbb{1}_A H(X_T, \dot{X}_T)),$$

which remains true if we just suppose $T < \zeta_\infty$ and $A \in \mathfrak{F}_T$. In other words, the function H is harmonic for the semigroup of the reflected Kolmogorov process, and the law $\tilde{\mathbb{P}}_1$ is the h-transform of $\mathbb{P}_1^c$, in the sense of Doob.

Under $\tilde{\mathbb{P}}_1$, the law of the sequence $(S_n)_{n\geq 0}$ is $\tilde{\mathbf{P}}_0$. This sequence is diverging to $+\infty$, and as a consequence the time ζ_∞ is infinite $\tilde{\mathbb{P}}_1$-almost surely. The law $\tilde{\mathbb{P}}_1$ is that of a process indexed by the whole half-line $[0,\infty)$. We give now a general definition for any starting condition (x,u).

Definition 1. The reflected Kolmogorov process *conditioned on never hitting* $(0,0)$ is the Markov process of law $\tilde{\mathbb{P}}_{x,u}$, for any starting condition $(x,u) \in D$, determined by

$$\tilde{\mathbb{P}}_{x,u}(A, T < \infty) = \frac{1}{H(x,u)} \mathbb{P}^c_{x,u}(\mathbb{1}_A H(X_T, \dot{X}_T), T < \zeta_\infty) \qquad (8)$$

for any stopping-time T and $A \in \mathfrak{F}_T$. We write $\tilde{\mathbf{P}}_t$ for its associated semigroup, and $\tilde{\mathbb{P}}_u$ for $\tilde{\mathbb{P}}_{0,u}$.

This denomination is justified by the following proposition.

Proposition 1. *For any $(x,u) \in D$ and $t > 0$, we have*

$$\tilde{\mathbb{P}}_{x,u}(A) = \lim_{s\to\infty} \mathbb{P}^c_{x,u}(A|\zeta_\infty > s), \qquad (9)$$

for any $A \in \mathfrak{F}_t$.

We stress that in [19], Proposition 2, Rivero defines in a similar way the self-similar Markov process conditioned on never hitting 0. Incidentally, we can find in [11] a thorough study of other h-transforms regarding the Kolmogorov process killed at time ζ_1.

In order to get Formula (9), we first prove the following lemma, which is a slight improvement of (4):

Lemma 2. *For any $(x,u) \in D$,*

$$s^k \mathbb{P}^c_{x,u}(\zeta_\infty > s) \xrightarrow[s\to\infty]{} C_1 H(x,u). \qquad (10)$$

Proof. For $(x,u) = (0,1)$, this is (4). For $x = 0$, the rescaling invariance property immediately yields

$$s^k \mathbb{P}^c_{0,u}(\zeta_\infty > s) = s^k \mathbb{P}^c_{0,1}(\zeta_\infty > su^{-2}) \xrightarrow[s\to\infty]{} C_1 u^{2k} = C_1 H(0,u).$$

For $(x,u) \in D$, the Markov property at time ζ_1 yields

$$s^k \mathbb{P}^c_{x,u}(\zeta_\infty > s) = \mathbb{P}^c_{x,u}(s^k \mathbb{P}^c_{0,V_1}(\zeta_\infty > s - \zeta_1))$$
$$\xrightarrow[s\to\infty]{} C_1 \mathbb{P}^c_{x,u}(H(0,V_1)) = C_1 H(x,u),$$

where the convergence holds by dominated convergence. The lemma is proved.

Formula (9) then results from:

$$\mathbb{P}^c_{x,u}(A|\zeta_\infty > s) = \frac{1}{\mathbb{P}^c_{x,u}(\zeta_\infty > s)} \mathbb{P}^c_{x,u}\left(\mathbb{1}_A \mathbb{P}^c_{X_t,\dot{X}_t}(\zeta_\infty > s - t), \zeta_\infty > t\right)$$

$$\xrightarrow[s\to\infty]{} \frac{1}{H(x,u)} \mathbb{P}^c_{x,u}\left(\mathbb{1}_A H(X_t, \dot{X}_t), \zeta_\infty > t\right)$$

$$= \tilde{\mathbb{P}}_{x,u}(A).$$

3.2 Starting the Conditioned Process from (0, 0)

The study of the reflected Kolmogorov process conditioned on never hitting $(0,0)$ will happen to be very similar to that of the reflected Kolmogorov process in the supercritical case $c > c_{crit}$, done in [12]. Observe the following similarities between the laws $\tilde{\mathbb{P}}_u$, and $\mathbb{P}^c_u$ when $c > c_{crit}$: the sequence $\left(\frac{\zeta_{n+1} - \zeta_n}{V_n^2}, \frac{V_{n+1}}{V_n}\right)_{n \geq 0}$ is i.i.d., we explicitly know its law, and the sequence $S_n = \ln(V_n)$ is a random walk with positive drift. It follows that a major part of [12] can be transcribed *mutatis mutandis*. In particular we will prove in this part a convergence result for the probabilities $\tilde{\mathbb{P}}_u$ when u goes to zero, similar to Theorem 1 of [12].

Under $\tilde{\mathbb{P}}_1$, the sequence $(S_n)_{n \geq 0}$ is a random walk of law $\tilde{\mathbf{P}}_0$. Write μ for its drift, that is the expectation of its jump distribution, which is positive and finite. The associated strictly ascending ladder height process $(H_n)_{n \geq 0}$, defined by $H_k = S_{n_k}$, where $n_0 = 0$ and $n_k = \inf\{n > n_{k-1}, S_n > S_{n_{k-1}}\}$, is a random walk with positive jumps. Its jump distribution also has positive and finite expectation $\mu_H \geq \mu$. The measure

$$m(dy) := \frac{1}{\mu_H} \tilde{\mathbf{P}}_0(H_1 > y)dy \tag{11}$$

is the "stationary law of the overshoot", both for the random walks $(S_n)_{n \geq 0}$ and $(H_n)_{n \geq 0}$. The following proposition holds.

Proposition 2. *The family of probability measures $(\tilde{\mathbb{P}}_{x,u})_{(x,u) \in D}$ on $\mathscr{C}$ has a weak limit when $(x, u) \to (0, 0)$, which we denote by $\tilde{\mathbb{P}}_{0+}$. More precisely, write τ_v for the instant of the first bounce with speed greater than v, that is $\tau_v := \inf\{t > 0, X_t = 0, \dot{X}_t > v\}$. Then the law $\tilde{\mathbb{P}}_{0+}$ satisfies the following properties:*

$$(*) \begin{cases} \lim_{v \to 0^+} \tau_v = 0 \quad \text{almost surely.} \\ \text{For any } u, v > 0, \text{ and conditionally on } \dot{X}_{\tau_v} = u, \text{ the process} \\ (X_{\tau_v + t}, \dot{X}_{\tau_v + t})_{t \geq 0} \text{ is independent of } (X_s, \dot{X}_s)_{s < \tau_v} \text{ and has law } \tilde{\mathbb{P}}_u. \end{cases}$$

$(**)$ *For any $v > 0$, the law of $\ln(\dot{X}_{\tau_v}/v)$ is m.*

In the proof of this proposition we can take $x = 0$ and just prove the convergence result for the laws $\tilde{\mathbb{P}}_u$ when $u \to 0+$. The general result will follow as an application of the Markov property at time ζ_1.

The complete proof follows mainly that of Theorem 1 in [12] and takes many pages. Here, the reader has three choices. Skip this proof and go directly to next section about the resurrected process. Or read the following pages for an overview of the ideas of the proof, with details given only when significantly different from those in [12]. Or, read [12] and the following pages for a complete proof.

Call $T_y(S)$ the hitting time of (y, ∞) for the random walk S starting from $x < y$. Call $\tilde{\mathbf{P}}_\mu$ the law of $(S_n)_{n \geq 0}$ obtained by taking S_0 and $(S_n - S_0)_{n \geq 0}$ independent, with law m and $\tilde{\mathbf{P}}_0$, respectively. That is, we allow the starting position to be nonconstant and distributed according to μ. A result of renewal theory states that the law of the overshoot $(S_{n+T_y} - y)_{n \geq 0}$ under $\tilde{\mathbf{P}}_x$, when x goes to $-\infty$, converges to $\tilde{\mathbf{P}}_m$. Now, for a process indexed by an interval I of $\mathbb{Z}$, we define a spatial translation operator by $\Theta_y^{sp}((S_n)_{n \in I}) = (S_{n+T_y} - y)_{n \in I - T_y}$. We get that under $\tilde{\mathbf{P}}_x$ and when x goes to $-\infty$, the translated process $\Theta_y^{sp}(S)$ converges to a process called the "spatially stationary random walk", a process indexed by $\mathbb{Z}$ which is spatially stationary and whose restriction to $\mathbb{N}$ is $\tilde{\mathbf{P}}_m$ (see [12]). We write $\tilde{\mathbf{P}}$ for the law of this spatially stationary random walk.

There exists a link between the law $\tilde{\mathbf{P}}_x$ and the law $\tilde{\mathbb{P}}_{e^x}$: the first one is the law of the underlying random walk $(S_n)_{n \geq 0} = (\ln V_n)_{n \geq 0}$ for a process $(X, \dot{X})$ following the second one. Now, in a very brief shortcut, we can say that the law $\tilde{\mathbf{P}}$ is linked to a law written $\tilde{\mathbb{P}}_{0+}^*$. And the convergence results of $\tilde{\mathbf{P}}_x \circ \Theta_y^{sp}$ to $\mathbf{P}$ when $x \to -\infty$ provide convergence results of $\tilde{\mathbb{P}}_u$ to $\tilde{\mathbb{P}}_{0+}^*$ when $u \to 0$.

However, this link is different, as the spatially stationary random walk, of law $\tilde{\mathbf{P}}$, is a process indexed by $\mathbb{Z}$. The value S_0 is thus not equal to the logarithm of the velocity of the process at time 0, but at time τ_1 (recall that $\tau_1 = \inf\{t > 0, X_t = 0, \dot{X}_t \geq 1\}$ is the instant of the first bounce with speed not less than one). The sequence $(S_n)_{n \geq 0}$ is then the sequence of the logarithms of the velocities of the process at the bouncing times, starting from that bounce. The sequence $(S_{-n})_{n \geq 0}$ is the sequence of the logarithms of the velocities of the process at the bouncing times happening *before* that bounce.

The law $\tilde{\mathbb{P}}_{0+}^*$ is the law of a process indexed by $\mathbb{R}_+^*$, but we actually construct it "from the random time τ_1". In order for the definition to be clean, we have to prove that the random time τ_1 is finite a.s. In [12], we used the fact that if $(\zeta_{1,k})_{k \geq 0}$ is a sequence of i.i.d random variables, with common law that of ζ_1 under $\mathbb{P}_1^c$, then for any $\varepsilon > 0$ there is almost surely only a finite number of indices k such that $\ln(\zeta_{1,k}) \geq \varepsilon k$. This was based on Formula (7), which, we recall, states

$$\mathbb{P}_1^c(\zeta_1 > t) \underset{t \to \infty}{\sim} c' t^{-\frac{1}{4}},$$

where c' is some positive constant. Here the same results holds with replacing $\mathbb{P}_1^c$ by $\tilde{\mathbb{P}}_1$ and is a consequence from the following lemma.

Lemma 3. We have
$$\tilde{\mathbb{P}}_1(\zeta_1 > t) \underset{t\to\infty}{\sim} c't^{k-\frac{1}{4}}, \qquad (12)$$

where c' is some positive constant.

Proof. From (8) and (1), we get that the density of $(\zeta_1, V_1/c)$ under $\tilde{\mathbb{P}}_1$ is given by

$$f(s,v) := \frac{1}{dsdv}\tilde{\mathbb{P}}_1((\zeta_1, V_1/c) \in dsdv)$$

$$= (cv)^{2k}\frac{3v}{\pi\sqrt{2}s^2}\exp(-2\frac{v^2-v+1}{s})\int_0^{\frac{4v}{s}} e^{-\frac{3\theta}{2}}\frac{d\theta}{\sqrt{\pi\theta}}.$$

Thanks to the inequality

$$4\sqrt{\frac{v}{s\pi}}e^{-\frac{6v}{s}} \leq \int_0^{\frac{4v}{s}} e^{-\frac{3\theta}{2}}\frac{d\theta}{\sqrt{\pi\theta}} \leq 4\sqrt{\frac{v}{s\pi}},$$

we may write

$$f(s,v) = (6\sqrt{2}.\pi^{-\frac{3}{2}}c^{2k})s^{-\frac{5}{2}}v^{\frac{3}{2}+2k}e^{-2\frac{v^2}{s}+\frac{v}{s}K(s,v)},$$

where $(s,v) \mapsto K(s,v)$ is continuous and bounded. The marginal density of ζ_1 is thus given by

$$\frac{1}{ds}\tilde{\mathbb{P}}_1(\zeta_1 \in ds) = \int_{\mathbb{R}_+} f(s,v)dv$$

$$= (3\sqrt{2}.\pi^{-\frac{3}{2}}c^{2k})s^{-\frac{5}{4}+k}\int_{\mathbb{R}_+} w^{\frac{1}{4}+k}e^{-2w+K(s,\sqrt{sw})\sqrt{w/s}}dw$$

$$\underset{s\to\infty}{\sim} (3\sqrt{2}.\pi^{-\frac{3}{2}}c^{2k})s^{-\frac{5}{4}+k}\int_{\mathbb{R}_+} w^{\frac{1}{4}+k}e^{-2w}dw,$$

where we used successively the change of variables $w = v^2/s$ and dominated convergence theorem. Just integrate this equivalence in the neighborhood of $+\infty$ to get

$$\tilde{\mathbb{P}}_1(\zeta_1 > t) \underset{t\to\infty}{\sim} c't^{k-\frac{1}{4}},$$

with the constant

$$c' = \frac{3\sqrt{2}.\pi^{-\frac{3}{2}}c^{2k}}{\frac{1}{4}-k}\int_{\mathbb{R}_+} w^{\frac{1}{4}+k}e^{-2w}dw = \frac{3c^{2k}}{\pi^{\frac{3}{2}}2^{\frac{3}{4}+k}}\cdot\frac{1+4k}{1-4k}\Gamma\left(\frac{1}{4}+k\right).$$

For now, we have introduced $\tilde{\mathbb{P}}_{0+}^*$, law of a process $(X, \dot{X})$ indexed by $\mathbb{R}_+^*$. We keep on following the proof of [12]. First, we get that this law satisfies conditions

(∗) and (∗∗), and that for any $v > 0$, the joint law of τ_v and $(X_{\tau_v+t}, \dot{X}_{\tau_v+t})_{t\geq 0}$ under $\tilde{\mathbb{P}}_u$ converges to that under $\tilde{\mathbb{P}}^*_{0+}$. Then we establish Proposition 2 by controlling the behavior of the process just after time 0, through the two following lemmas:

Lemma 4. *Under $\tilde{\mathbb{P}}^*_{0+}$, we have almost surely $(X_t, \dot{X}_t) \xrightarrow[t \to 0]{} (0,0)$.*

This lemma allows in particular to extend $\tilde{\mathbb{P}}^*_{0+}$ to $\mathbb{R}_+$. We call $\tilde{\mathbb{P}}_{0+}$ this extension. The second lemma is more technical and controls the behavior of the process on $[0, \tau_v[$ under $\tilde{\mathbb{P}}_u$.

Lemma 5. *Write $M_v = \sup\{|\dot{X}_t|, t \in [0, \tau_v[\}$. Then,*

$$\forall \varepsilon > 0, \forall \delta > 0, \exists v_0 > 0, \exists u_0 > 0, \forall 0 < u \leq u_0, \quad \tilde{\mathbb{P}}_u(M_{v_0} \geq \delta) \leq \varepsilon, \qquad (13)$$

In [12], we proved these two results by using the stochastic partial differential equation satisfied by the laws $\mathbb{P}^c$. They are of course not available for the laws $\tilde{\mathbb{P}}$, and we need a new proof. We start by showing a rather simple but really useful inequality:

Lemma 6. *The following inequality holds for any $(x, u) \in D$,*

$$\tilde{\mathbb{P}}_{x,u}\left(V_1/c \geq \frac{|u|}{2}\right) \geq 1 - \frac{\sqrt{3}}{\pi}. \qquad (14)$$

For us, the important fact is that the probability is bounded below by a positive constant, uniformly in x and u. The constant $1 - \sqrt{3}/\pi$ is not intended to be the optimal one. Note that this inequality will also be used again later on in this paper.

Proof (Proof of Lemma 6). For $u = 0$, there is nothing to prove. By a scaling invariance property we may suppose $u \in \{-1, 1\}$, what we do.

The density $f_{x,u}$ of V_1/c under $\mathbb{P}^c_{x,u}$ is given in Gor'kov [9]. If you write $p_t(x, u; y, v)$ for the transition densities of the (free) Kolmogorov process, given by

$$p_t(x, u; y, v) = \frac{\sqrt{3}}{\pi t^2} \exp\left[-\frac{6}{t^3}(y - x - tu)^2 + \frac{6}{t^2}(y - x - tu)(v - u) - \frac{2}{t}(v - u)^2\right],$$

and $\Phi(x, u; y, v)$ for its total occupation time densities, defined by

$$\Phi(x, u; y, v) := \int_0^\infty p_t(x, u; y, v)\,dt,$$

then the density $f_{x,u}$ is given by

$$f_{x,u}(v) = v\left[\Phi(x, u; 0, -v) - \frac{3}{2\pi}\int_0^\infty \frac{\mu^{\frac{3}{2}}}{\mu^3 + 1}\Phi(x, u; 0, \mu v)\,d\mu\right]. \qquad (15)$$

Now, knowing the density of V_1 under $\mathbb{P}^c_{x,u}$, we get that of V_1 under $\tilde{\mathbb{P}}_{x,u}$ by multiplying it by the increasing function $v \mapsto v^{2k}$. This necessarily increases the probability of being greater than $c/2$. Consequently, it is enough to prove

$$\mathbb{P}^c_{x,u}(V_1/c \geq \frac{1}{2}) \geq 1 - \frac{\sqrt{3}}{\pi}$$

as soon as $u \in \{-1, 1\}$. But very rough bounds give

$$f_{x,u}(v) \leq v\Phi(x, u; 0, -v)$$
$$\leq v \int_0^\infty \frac{\sqrt{3}}{\pi t^2} \exp(-\frac{(u+v)^2}{2t}) dt.$$

For $u \in \{-1, 1\}$ and $v \in [0, 1/2]$ we have $|u + v| \geq 1/2$ and thus

$$f_{x,u}(v) \leq \frac{v\sqrt{3}}{\pi} \int_0^\infty \frac{1}{t^2} \exp(-\frac{1}{8t}) dt = \frac{8\sqrt{3}}{\pi} v.$$

Consequently,

$$\mathbb{P}_{x,u}(V_1/c \geq \frac{1}{2}) \geq 1 - \int_0^{1/2} \frac{8\sqrt{3}}{\pi} v \, dv = 1 - \frac{\sqrt{3}}{\pi} > 0.$$

Proof (Proof of Lemma 4). First, observe that conditions (∗) and (∗∗) imply that the variables $\tau_v = \inf\{t > 0, X_t = 0, \dot{X}_t > v\}$ and $\tau_v^- := \sup\{t < \tau_v, X_t = 0\}$ are almost surely strictly positive and go to zero when v goes to zero. Then, observe that it is enough to show the almost sure convergence of $\dot{X}_t$ to 0 when $t \to 0$, and suppose on the contrary that this does not hold.

Then, there would exist a positive x such that $\tilde{\mathbb{P}}^*_{0+}(T_x = 0) > 0$, where we have written $T_x := \inf\{t > 0, |\dot{X}_t| > x\}$. By self-similarity this would be true for any $x > 0$ and in particular we would have

$$K := \tilde{\mathbb{P}}^*_{0+}(T_1 = 0) > 0. \tag{16}$$

Informally, this, together with (14), should induce that $\tau_{c/2}^-$ takes the value zero with probability at least $(1 - \sqrt{3}/\pi)K$, and give the desired contradiction. However it is not straightforward, because we cannot use a Markov property at time T_1, which can take value 0, while the process is still not defined at time 0. Consider the stopping time $T_1^\varepsilon := \inf\{t > \varepsilon, |\dot{X}_t| > x\}$. For any $\eta > 0$, we have

$$\liminf_{\varepsilon \to 0} \tilde{\mathbb{P}}^*_{0+}(T_1^\varepsilon < \eta) \geq \tilde{\mathbb{P}}^*_{0+}(\liminf_{\varepsilon \to 0}\{T_1^\varepsilon < \eta\}) \geq \tilde{\mathbb{P}}^*_{0+}(T_1 < \eta) \geq K,$$

and in particular there is some $\varepsilon_0(\eta)$ such that for any $\varepsilon < \varepsilon_0(\eta)$,

$$\tilde{\mathbb{P}}^*_{0+}(T_1^\varepsilon < \eta) \geq \frac{K}{2}. \tag{17}$$

Now, write θ for the translation operator defined by $\theta_x((X_t)_{t\geq 0}) = (X_{x+t})_{t\geq 0}$, so that $V_1 \circ \theta_{T_1^\varepsilon}$ denotes the velocity of the process at its first bounce after time T_1^ε. From (17) and Lemma 6, a Markov property gives, for $\varepsilon < \varepsilon_0(\eta)$,

$$\tilde{\mathbb{P}}^*_{0+}\left(T_1^\varepsilon < \eta, V_1 \circ \theta_{T_1^\varepsilon} \geq \frac{c}{2}\right) \geq K' := \left(1 - \frac{\sqrt{3}}{\pi}\right)\frac{K}{2}.$$

We have *a fortiori* $\tilde{\mathbb{P}}^*_{0+}(\tau^-_{c/2} \leq \eta) \geq K'$. This result true for any $\eta > 0$ leads to $\tilde{\mathbb{P}}^*_{0+}(\tau^-_{c/2} = 0) \geq K' > 0$, and we get a contradiction. This shows $(X_t, \dot{X}_t) \xrightarrow[t\to 0]{}$ $(0,0)$ under $\tilde{\mathbb{P}}^*_{0+}$, as requested.

Proof (Proof of Lemma 5). We prove (13). Fix $\varepsilon, \delta > 0$. The event $\{M_v \geq \delta\}$ coincides with the event $T_\delta \leq \tau_v$. From a Markov property at time T_δ and (14), we get, for any $v < c\delta/2$, and any u,

$$(1 - \sqrt{3}/\pi)\tilde{\mathbb{P}}_u(M_v \geq \delta) \leq \tilde{\mathbb{P}}_u(\dot{X}_{\tau_v} \geq c\delta/2).$$

Choose v_0 such that $\tilde{\mathbb{P}}_{0+}(\dot{X}_{\tau_{v_0}} \geq c\delta/2) \leq \varepsilon$. Then, from the convergence of the law of $\dot{X}_{\tau_{v_0}}$ under $\tilde{\mathbb{P}}_u$ to that under $\tilde{\mathbb{P}}_{0+}$, we get, for u small enough,

$$\tilde{\mathbb{P}}_u(\dot{X}_{\tau_{v_0}} \geq c\delta/2) \leq 2\varepsilon,$$

and hence

$$\tilde{\mathbb{P}}_v(M_{v_0} \geq \delta) \leq \frac{2}{1 - \sqrt{3}/\pi}\varepsilon.$$

In conclusion, all this suffices to show Proposition 2.

4 The Resurrected Process

4.1 Itô Excursion Measure, Recurrent Extensions, and (RLP) Equations

We finally tackle the problem of interest, that is the recurrent extensions of the reflected Kolmogorov process. A recurrent extension of the latter is a Markov process that behaves like the reflected Kolmogorov process until ζ_∞, the hitting time of $(0,0)$, but that is defined for any positive times and does not stay at $(0,0)$, in the sense that the Lebesgue measure of the set of times when the process is at $(0,0)$ is almost surely 0. More concisely, we will call such a process a resurrected reflected process.

We recall that Itō's program and results of Blumenthal [5] establish an equivalence between the law of recurrent extensions of a Markov process and excursion measures compatible with its semigroup, here P_t^c (where as usually in Itō's excursion theory we identify the measures which are equal up to a multiplicative constant). The *set of excursions* $\mathscr{E}$ is defined by

$$\mathscr{E} := \{(x, \dot{x}) \in \mathscr{C} \mid \zeta_\infty > 0 \text{ and } x_t \mathbb{1}_{t \geq \zeta_\infty} \equiv 0\}.$$

An excursion measure n compatible with the semigroup P_t^c is defined by the three following properties:

1. The measure n is carried by $\mathscr{E}$.
2. For any $\mathfrak{F}_\infty$-measurable function F and any $t > 0$, any $A \in \mathfrak{F}_t$,

$$n(F \circ \theta_t, A \cap \{t < \zeta_\infty\}) = n(\mathbb{P}^c_{X_t, \dot{X}_t}(F), A \cap \{t < \zeta_\infty\}).$$

3. $n(1 - e^{-\zeta_\infty}) < \infty$.

We also say that n is a pseudo-excursion measure compatible with the semigroup P_t^c if only the two first properties are satisfied and not necessarily the third one. We recall that the third property is the necessary condition in Itō's program in order for the lengths of the excursions to be summable, hence in order for Itō's program to succeed. Besides, we are here interested in recurrent extensions which leave $(0, 0)$ continuously. These extensions correspond to excursion measures n which satisfy the additional condition $n((X_0, \dot{X}_0) \neq (0, 0)) = 0$. Our main results are the following:

Theorem 1. *There exists, up to a multiplicative constant, a unique excursion measure $\mathbf{n}$ compatible with the semigroup P_t^c and such that $\mathbf{n}((X_0, \dot{X}_0) \neq (0, 0)) = 0$. We may choose $\mathbf{n}$ such that*

$$\mathbf{n}(\zeta_\infty > s) = C_1 s^{-k}, \qquad (18)$$

where C_1 is the constant defined by (5), and $k = k(c)$ has been introduced in Lemma 1. The measure $\mathbf{n}$ is then characterized by any of the two following formulas:

$$\mathbf{n}(f(X, \dot{X}), \zeta_\infty > T) = \tilde{\mathbb{P}}_{0^+}(f(X, \dot{X}) H(X_T, \dot{X}_T)^{-1}), \qquad (19)$$

for any $\mathfrak{F}_t$-stopping time T and any f positive measurable functional depending only on $(X_t, \dot{X}_t)_{0 \leq t \leq T}$.

$$\mathbf{n}(f(X, \dot{X}), \zeta_\infty > T) = \lim_{(x, u) \to (0, 0)} H(x, u)^{-1} \mathbb{P}^c_{x, u}(f(X, \dot{X}), \zeta_\infty > T), \qquad (20)$$

for any $\mathfrak{F}_t$-stopping time T and any f positive continuous *functional depending only on $(X_t, \dot{X}_t)_{0 \leq t \leq T}$.*

So Itō's program constructs a Markov process with associated Itō excursion measure **n**, which spends no time at $(0, 0)$, that is a recurrent extension, that is a resurrected reflected process. We call its law $\mathbb{P}_0^r$. The second theorem will be the weak existence and uniqueness of a solution to Equations (RLP) with starting condition $(0, 0)$. For any solution $(X, \dot{X}, N, W)$, the law of $(X, \dot{X})$ is $\mathbb{P}_0^r$, and $N \equiv 0$ almost surely. The fact that the continuous push N is degenerate is remarkable and non-obvious. Even if we admit that the process $(X, \dot{X})$ should provide a solution to Equations (RLP), and even knowing that this process spends no time at $(0, 0)$, we could have imagined that the term N be proportional to the local time spent by $(X, \dot{X})$ in $(0, 0)$, for example.

Theorem 2.
- *Consider $(X, \dot{X})$ a process of law $\mathbb{P}_0^r$. Then the jumps of $\dot{X}$ on any finite interval are summable and the process W defined by*

$$W_t = \dot{X}_t + (1+c) \sum_{0 < s \leq t} \dot{X}_{s-} \mathbb{1}_{X_s = 0}$$

is a Brownian motion. As a consequence the quadruplet $(X, \dot{X}, 0, W)$ is a solution to (RLP).
- *For any solution $(X, \dot{X}, N, W)$ to (RLP), the law of $(X, \dot{X})$ is $\mathbb{P}_0^r$ and $N \equiv 0$ almost surely.*

It is implicit in this second theorem and until the end of the paper that the initial condition is $(0, 0)$, though this can be easily generalized to any other initial condition $(x, u) \in D$.

Before tackling the proofs, let us write some comments and consequences of Theorem 1. First, the Itō excursion measure **n** is entirely determined by its entrance law, which is defined by

$$\mathbf{n}_s(dx, du) := n((X_s, \dot{X}_s) \in dx \otimes du, s < \zeta_\infty)$$

for $s > 0$. But Theorem 1 implies that it is characterized by any of the two following formulas:

$$\mathbf{n}_s(f) = \tilde{\mathbb{P}}_{0^+}(f(X_s, \dot{X}_s) H(X_s, \dot{X}_s)^{-1}), \quad s > 0, \tag{21}$$

for $f : D^0 \to \mathbb{R}_+$ measurable.

$$\mathbf{n}_s(f) = \lim_{(x,u) \to (0,0)} H(x, u)^{-1} \mathbb{P}_{x,u}^c(f(X_s, \dot{X}_s), \zeta_\infty > s), \quad s > 0, \tag{22}$$

for $f : D^0 \to \mathbb{R}_+$ continuous.

Formulas similar to these are found in the case of self-similar Markov processes studied by Rivero [19]. This ends the parallel between our works. Rivero underlined that the self-similar Markov process conditioned on never hitting 0 that he introduced plays the same role as the Bessel process for the Brownian motion. In our model, this role is played by the reflected Kolmogorov process conditioned on never hitting $(0,0)$. Here is a short presentation of this parallel. Write P_x for the law of a Brownian motion starting from position x, $\tilde{P}_x$ for the law of the "three-dimensional" Bessel process starting from x. Write n for the Itô excursion measure of the absolute value of the Brownian motion (that is, the Brownian motion reflected at 0), and ζ for the hitting time of 0. Then the inverse function is excessive (i.e. non-negative and superharmonic) for the Bessel process and we have the two well-known formulas

$$\mathbf{n}(f(X), \zeta > T) = \tilde{P}_0(f(X)/X_T)$$

$$\mathbf{n}(f(X), \zeta > T) = \lim_{x \to 0} \frac{1}{x} P_x(f(X), \zeta > T),$$

for any $\mathfrak{F}_t$-stopping time T and any f positive measurable functional (resp. continuous functional for the second formula) depending only on $(X_t)_{0 \leq t \leq T}$.

Now, let us give an application of Formula (18). Write l for the local time spent by X at zero, under $\mathbb{P}_0^c$. Formula (18) implies that the inverse local time l^{-1} is a subordinator with jumping measure Π satisfying $\Pi(\zeta_\infty > s) \propto s^{-k}$. That is, it is a stable subordinator of index k. A well-known result of Taylor and Wendel [21] then gives that the exact Hausdorff function of the closure of its range (the range is the image of $\mathbb{R}_+$ by l^{-1}) is given by $\phi(\varepsilon) = \varepsilon^k (\ln \ln 1/\varepsilon)^{1-k}$ almost surely. The closure of the range of l^{-1} being equal to the zero set $\mathscr{Z} := \{t \geq 0 : X_t = \dot{X}_t = 0\}$, we get the following corollary:

Corollary 1. *The exact Hausdorff function of the set of the passage times to $(0,0)$ of the resurrected reflected Kolmogorov process is $\phi(\varepsilon) = \varepsilon^k (\ln \ln 1/\varepsilon)^{1-k}$ almost surely.*

It is also clear that the set of the bouncing times of the resurrected reflected Langevin process—the moments when the process is at zero with a nonzero speed—is countable. Therefore the zero set of the resurrected reflected Langevin process has the same exact Hausdorff function.

Finally, we should mention that the self-similarity property enjoyed by the Kolmogorov process easily spreads to all the processes we introduced. If a is a positive constant, denote by $(X^a, \dot{X}^a)$ the process $(a^3 X_{a^{-2}t}, a X_{a^{-2}t})_{t \geq 0}$. Then the law of $(X^a, \dot{X}^a)$ under $\mathbb{P}_{x,u}^c$ is simply $\mathbb{P}_{a^3 x, au}^c$. We have $H(a^3 x, au) = a^{2k} H(x, u)$. The law of $(X^a, \dot{X}^a)$ under $\tilde{\mathbb{P}}_{x,u}$, resp. $\tilde{\mathbb{P}}_{0+}$, is simply $\tilde{\mathbb{P}}_{a^3 x, au}$, resp. $\tilde{\mathbb{P}}_{0+}$. Finally, the measure of $(X^a, \dot{X}^a)$ under $\mathbf{n}$ is simply $a^{2k} \mathbf{n}$.

Last two subsections are devoted to the proof of the two theorems.

4.2 The Unique Recurrent Extension Compatible with $\mathbb{P}_t^c$

Construction of the Excursion Measure

The function $1/H$ is excessive for the semigroup $\tilde{P}_t$ and the corresponding h-transform is $\mathbb{P}_t^c$ (see Definition 1). Write $\mathbf{n}$ for the h-transform of $\tilde{\mathbb{P}}_{0+}$ via this excessive function $1/H$. That is, **n** is the unique measure on $\mathscr{C}$ carried by $\{\zeta_\infty > 0\}$ such that under **n** the coordinate process is Markovian with semigroup P_t^c, and for any $\mathfrak{F}_t$-stopping time T and any A_T in $\mathfrak{F}_T$, we have

$$\mathbf{n}(A_T, T < \zeta_\infty) = \tilde{\mathbb{P}}_{0+}(A_T, H(X_T, \dot{X}_T)^{-1}).$$

Then, **n** is a pseudo-excursion measure compatible with semigroup P_t^c, which verifies $\mathbf{n}((X_0, \dot{X}_0) \neq (0,0)) = 0$ and satisfies Formula (19). For f continuous functional depending only on $(X_t, \dot{X}_t)_{t \leq T}$, we have

$$\tilde{\mathbb{P}}_{0+}(f(X_s, \dot{X}_s) H(X_s, \dot{X}_s)^{-1}) = \lim_{(x,u) \to (0,0)} \tilde{\mathbb{P}}_{x,u}(f(X_s, \dot{X}_s) H(X_s, \dot{X}_s)^{-1})$$

$$= \lim_{(x,u) \to (0,0)} \frac{1}{H(x,u)} \mathbb{P}_{x,u}^c(f(X_s, \dot{X}_s), \zeta_\infty > s),$$

so that the pseudo-excursion measure **n** also satisfies Formula (20). In particular, taking $T = s$ and $f = 1$, and considering the limit along the half-line $x = 0$, this gives

$$\mathbf{n}(\zeta_\infty > s) = \lim_{u \to 0} u^{-2k} \mathbb{P}_{0,u}(\zeta_\infty > s).$$

Using Lemma 2 and the scaling invariance property, we get

$$\mathbf{n}(\zeta_\infty > s) = C_1 s^{-k},$$

where C_1 is the constant defined by (5). This is exactly Formula (18). This formula gives, in particular,

$$\mathbf{n}(1 - e^{-\zeta_\infty}) = C_1 \Gamma(1 - k),$$

where Γ denotes the usual Gamma function. Hence, **n** is an excursion measure.

Finally, in order to establish Theorem 1 we just should prove that **n** is the only excursion measure compatible with the semigroup P_t^c such that $\mathbf{n}((X_0, \dot{X}_0) \neq (0,0)) = 0$. That is, we should show the uniqueness of the law of the resurrected reflected process.

Uniqueness of the Excursion Measure

Let **n**′ be such an excursion measure, compatible with the semigroup P_t^c, and satisfying $\mathbf{n}'((X_0, \dot{X}_0) \neq (0,0)) = 0$. We will prove that **n** and **n**′ coincide, up to a multiplicative constant. Recall that ζ_1 is defined as the infimum of $\{t > 0, X_t = 0\}$.

Lemma 7. *The measure* **n**′ *satisfies:*

$$\mathbf{n}'(\zeta_1 \neq 0) = 0$$

Proof. This condition will appear to be necessary to have the third property of excursion measures, that is $\mathbf{n}'(1 - e^{-\zeta_\infty}) < \infty$. Suppose on the contrary that $\mathbf{n}'(\zeta_1 \neq 0) > 0$ and write $\tilde{\mathbf{n}}(\cdot) = \mathbf{n}'(\cdot \mathbb{1}_{\zeta_1 \neq 0})$. The excursion measure $\tilde{\mathbf{n}}$ is compatible with the semigroup $\mathbb{P}_t^c$ and satisfies $\tilde{\mathbf{n}}((X_0, \dot{X}_0) \neq (0,0)) = 0$ and $\tilde{\mathbf{n}}(\zeta_1 = 0) = 0$. Consider $\bar{\mathbf{n}}((X_t, \dot{X}_t)_{t \geq 0}) := \tilde{\mathbf{n}}((X_t \mathbb{1}_{t < \zeta_1}, \dot{X}_t \mathbb{1}_{t < \zeta_1})_{t \geq 0})$ the excursion measure of the process killed at time ζ_1.

The measure $\bar{\mathbf{n}}$ is an excursion measure compatible with the semigroup $\mathbb{P}_t^0$, semigroup of the Kolmogorov process killed at time ζ_1 (the first hitting time of $\{0\} \times \mathbb{R}$). Therefore its first marginal must be the excursion measure of the Langevin process reflected on an inelastic boundary, introduced and studied in [3]. In particular, under $\bar{\mathbf{n}}$, the absolute value of the incoming speed at time ζ_1, or $|\dot{X}_{\zeta_1-}|$, is distributed proportionally to $v^{-\frac{3}{2}} dv$ (see [3], Corollary 2, (ii)). This holds true under $\tilde{\mathbf{n}}$ and implies that $V_1 = c|\dot{X}_{\zeta_1-}|$ is also distributed proportionally to $v^{-\frac{3}{2}} dv$. Now, a Markov property at the stopping time ζ_1 under $\tilde{\mathbf{n}}$ gives

$$\tilde{\mathbf{n}}(\zeta_\infty - \zeta_1 > t | V_1 = v) = \mathbb{P}_v^c(\zeta_\infty > t) = \mathbb{P}_1^c(\zeta_\infty > v^{-2}t) \underset{v^{-2}t \to \infty}{\sim} C v^{2k} t^{-k}$$

As a consequence the function $v \mapsto v^{-\frac{3}{2}} \tilde{\mathbf{n}}(\zeta_\infty - \zeta_1 > t | V_1 = v)$ is not integrable in the neighborhood of 0. That is $\tilde{\mathbf{n}}(\zeta_\infty - \zeta_1 > t) = +\infty$, we get a contradiction.

Recall that we owe to prove that $\mathbf{n}'$ and $\mathbf{n}$ are equal, up to a multiplicative constant. Let us work on the corresponding entrance laws. Take $s > 0$ and f a bounded continuous function. It is sufficient to prove $\mathbf{n}'_s(f) = C\mathbf{n}_s(f)$, where C is a constant independent of s and f.

By reformulating Lemma 7, time ζ_1 is zero $\mathbf{n}'$-almost surely, in the sense that the $\mathbf{n}'$-measure of the complementary event is 0. That is, $\mathbf{n}'$-a.s., the first coordinate of the process comes back to zero just after the initial time, while the second coordinate cannot be zero, for the simple reason that we are working on an excursion outside from $(0, 0)$. This, together with the fact that the velocity starts from $\dot{X}_0 = 0$ and is right-continuous, implies that $\mathbf{n}'$-almost surely, the time τ_v (which, we recall, is the instant of the first bounce with speed greater than v) goes to 0 when v goes to 0.

We deduce, by dominated convergence, from the continuity of f, and, again, from the right-continuity of the paths, that

$$\mathbf{n}'_s(f) = \lim_{v \to 0} \mathbf{n}'(f(X_{s+\tau_v}, \dot{X}_{s+\tau_v}) \mathbb{1}_{\tau_v < \infty, \zeta_\infty > s+\tau_v}). \tag{23}$$

An application of the Markov property gives

$$\mathbf{n}'(f(X_{s+\tau_v}, \dot{X}_{s+\tau_v}) \mathbb{1}_{\tau_v < \infty, \zeta_\infty > s+\tau_v}) = \int_{\mathbb{R}_+} \mathbf{n}'(\dot{X}_{\tau_v} \in du) \mathbb{P}_u^c(f(X_s, \dot{X}_s) \mathbb{1}_{\zeta_\infty > s})$$

$$= \int_{\mathbb{R}_+} \mathbf{n}'(\dot{X}_{\tau_v} \in du) u^{2k} g(u),$$

where $g(u) = u^{-2k}\mathbb{P}^c_u(f(X_s,\dot X_s)\mathbb{1}_{\zeta_\infty>s}) = H(0,u)^{-1}\mathbb{P}^c_u(f(X_s,\dot X_s)\mathbb{1}_{\zeta_\infty>s})$ converges to $\mathbf{n}_s(f)$ when $u \to 0$, by Formula (20). Moreover the function $u^{2k}g(u)$ is bounded by $\|f\|_\infty$, and for any $\varepsilon > 0$ we have $\mathbf{n}'(\dot X_{\tau_v} > \varepsilon) \to 0$ when $v \to 0$. Informally, all this explains that when v is small, all the mass in the integral is concentrated in the neighborhood of 0, where we can replace $g(u)$ by $\mathbf{n}_s(f)$. More precisely, write

$$\int_{\mathbb{R}_+} \mathbf{n}'(\dot X_{\tau_v} \in du) u^{2k} g(u) = I(v) + J(v),$$

where

$$I(v) = \int_0^1 \mathbf{n}'(\dot X_{\tau_v} \in du) u^{2k} \mathbf{n}_s(f),$$

$$J(v) = \int_0^\infty \mathbf{n}'(\dot X_{\tau_v} \in du) u^{2k} (g(u) - \mathbf{n}_s(f)\mathbb{1}_{u\le 1}).$$

For any $\varepsilon \in (0,1)$, let us split the integral defining $J(v)$ on $(0,\varepsilon]$ and on (ε,∞). The absolute value of the sum on (ε,∞) is bounded by

$$\mathbf{n}'(\dot X_{\tau_v} \ge \varepsilon)(\|f\|_\infty + |\mathbf{n}_s(f)|),$$

and therefore goes to 0 when v goes to 0. The absolute value of the sum on $(0,\varepsilon)$ is bounded by

$$\frac{I(v)}{\mathbf{n}_s(f)} \sup_{u\in(0,\varepsilon]} |g(u) - \mathbf{n}_s(f)|.$$

The supremum goes to 0 when ε goes to 0. We deduce that $J(v)$ is negligible compared to $1 \vee I(v)$. Recalling that the sum $I(v) + J(v)$ converges to $\mathbf{n}'_s(f)$ (Formula (23)), we get that $I(v)$ converges to $\mathbf{n}'_s(f)$ when $v \to 0$, while $J(v)$ converges to 0.

We thus have

$$\mathbf{n}'_s(f) = C\mathbf{n}_s(f),$$

where C is independent of s and f and given by

$$C = \lim_{v\to 0} \int_0^1 \mathbf{n}'(\dot X_{\tau_v} \in du) u^{2k}.$$

Uniqueness follows. Theorem 1 is proved.

4.3 The Weak Unique Solution to the (RLP) Equations

We now prove Theorem 2.

Weak Solution

We consider, under $\mathbb{P}_0^r$, the coordinate process $(X, \dot{X})$, and its natural filtration $(\mathfrak{F}_t)_{t \geq 0}$. We first prove that the jumps of $\dot{X}$ are almost surely summable on any finite interval. As there are (a.s.) only finitely many jumps of amplitude greater than a given constant on any finite interval, it is enough to prove that the jumps of amplitude less than a given constant are (a.s.) summable. If a jump occurs at a time s, its amplitude is $\dot{X}_s - \dot{X}_{s-} = -(1+c)\dot{X}_{s-} = (1+1/c)\dot{X}_s$. Besides, write L for a local time of the process $(X, \dot{X})$ in $(0,0)$, L^{-1} for its inverse, and $\mathbf{n}$ for the associated excursion measure. Now, it is sufficient to prove that the expectation of the sum of the jumps of amplitude less than $1 + 1/c$ (i.e. corresponding to an outgoing velocity less than one), and occurring before time $L^{-1}(1)$, is finite. This expectation is equal to

$$(1 + \frac{1}{c})\mathbf{n}\left(\sum_{0 < s < \zeta_\infty} \dot{X}_s \mathbb{1}_{X_s = 0, \dot{X}_s \leq 1} \right)$$

and can be rewritten as

$$(1 + \frac{1}{c})\mathbf{n}\left(\sum_{0 < s < \zeta_\infty} \mathbb{1}_{X_s = 0} \int_0^1 \mathbb{1}_{v \leq \dot{X}_s \leq 1} dv \right) = (1 + \frac{1}{c}) \int_0^1 \mathbf{n}(N_{[v,1]}(X, \dot{X})) dv,$$

where $N_I(X, \dot{X})$ denotes the number of bounces of the process $(X, \dot{X})$ with outgoing speed included in the interval I. For a fixed v, introduce the sequence of stopping times defined by $\tau_0^v = 0$ and $\tau_{n+1}^v = \inf\{t > \tau_n^v, X_t = 0, \dot{X}_t \in [v, 1]\}$ for $n \geq 0$. Then $N_{[v,1]}(X, \dot{X})$ is also equal to $\sup\{n, \tau_n^v < \zeta_\infty\}$. Thanks to formula (19), for any $n > 0$, we have:

$$\mathbf{n}(\zeta_\infty > \tau_n^v) = \tilde{\mathbb{P}}_{0+}(H(X_{\tau_n^v}, \dot{X}_{\tau_n^v})^{-1} \mathbb{1}_{\tau_n^v < \infty})$$
$$= \tilde{\mathbb{P}}_{0+}(\dot{X}_{\tau_n^v}^{-2k} \mathbb{1}_{\tau_n^v < \infty})$$
$$\leq v^{-2k} \tilde{\mathbb{P}}_{0+}(\tau_n^v < \infty).$$

As a consequence, we have

$$\mathbf{n}(N_{[v,1]}(X, \dot{X})) \leq v^{-2k} \tilde{\mathbb{P}}_{0+}(\sup\{n, \tau_n^v < \zeta_\infty\})$$
$$\leq v^{-2k} \tilde{\mathbf{P}}(N_{[\ln v, 0]}^d(S)),$$

where we have written $N^d_{[\ln v,0]}(S)$ for the number of instants $n \in \mathbb{Z}$ such that $S_n \in [\ln v, 0]$. Recall also that $\tilde{\mathbf{P}}$ is the law of the spatially stationary random walk. The stationarity property implies that the expectation $\tilde{\mathbf{P}}(N^d_{[\ln v,0]}(S))$ is proportional to the length of the interval $[\ln(v), 0]$, that is $-\ln v$. Further, it is finite. A simple way to check this it to observe that given $S_M \in [\ln(v), 0]$ for some stopping time M, there is a positive probability, independent of the value of S_M, that the process at this time has a jump of size greater than $-\ln(v)$, and then stay forever above S_{M+1}. Therefore the variable $N^d_{[\ln v,0]}(S)$ is stochastically dominated by a geometric random variable and has finite expectation. As a result,

$$\mathbf{n}(N_{[v,1]}(X, \dot{X})) \underset{v \to 0}{=} O(v^{-2k} \ln(1/v))$$

and (recall $k < 1/4$)

$$\left(1 + \frac{1}{c}\right) \int_0^1 \mathbf{n}(N_{[v,1]}(X, \dot{X}))\,dv < \infty.$$

The jumps are summable.

Now, write

$$W_t = \dot{X}_t + (1+c) \sum_{0 < s \le t} \dot{X}_{s-} \mathbb{1}_{X_s = 0}.$$

We aim to show that the continuous process W is a Brownian motion. The technique we use is the introduction of a sequence of stopping times, which allows us to deal with what happens just after the instants when the process is at $(0, 0)$. This same method will be used several times until the end of the paper. For $\varepsilon > 0$, define the sequence $(T_n^\varepsilon)_{n \ge 0}$ by $T_0^\varepsilon = 0$ and, for $n \ge 0$,

$$\begin{cases} T_{2n+1}^\varepsilon = \inf\{t > T_{2n}^\varepsilon, X_t = 0, \dot{X}_t > \varepsilon\} \\ T_{2n+2}^\varepsilon = \inf\{t > T_{2n+1}^\varepsilon, X_t = \dot{X}_t = 0\} \end{cases}$$

We also introduce $F^\varepsilon = \bigcup_{n \ge 0}[T_{2n}^\varepsilon, T_{2n+1}^\varepsilon]$ and $H_t^\varepsilon = \mathbb{1}_{F^\varepsilon}(t)$. For $0 < \varepsilon' < \varepsilon$, we have $H^{\varepsilon'} \le H^\varepsilon$, or equivalently, $F^{\varepsilon'} \subset F^\varepsilon$. When ε goes to $0+$, F^ε converges to the zero set $\mathscr{Z} = \{t, X_t = \dot{X}_t = 0\}$, and H^ε converges pointwisely to $H^0 = \mathbb{1}_{\mathscr{Z}}$. Note that the processes H^ε and H^0 are $\mathfrak{F}_t$-adapted. Note, also, that Corollary 1 implies in particular that $\mathscr{Z}$ has zero Lebesgue measure. For ease of notations, we will sometimes omit the superscript ε.

Conditionally on $\dot{X}_{T_{2n+1}} = u$, the process $(X_{(T_{2n+1}+t) \wedge T_{2n+2}})_{t \ge 0}$ is independent of $\mathfrak{F}_{T_{2n+1}}$ and has law $\mathbb{P}_u^c$. Hence, the process $(W_{(T_{2n+1}+t) \wedge T_{2n+2}} - W_{T_{2n+1}})_{t \ge 0}$ is a Brownian motion stopped at time $T_{2n+2} - T_{2n+1}$. Write

$$W_t = \int_0^t H_s^\varepsilon dW_s + \int_0^t (1 - H_s^\varepsilon) dW_s.$$

The process $\int_0^t (1 - H_s^\varepsilon) dW_s$ converges almost surely to $\int_0^t (1 - H_s^0) dW_s$. But the process $\int_0^t (1 - H_s^0) dW_s$ is a continuous martingale of quadratic variation $\int_0^t (1 - H_s^0) ds = t$ and thus a Brownian motion. In order to prove that it actually coincides with W, we only need to prove that the term $D_t^\varepsilon := \int_0^t H_s^\varepsilon dW_s$ converges almost surely to 0 when ε goes to 0. Without loss of generality, we only prove it on the event $t \leq L^{-1}(1)$. Rewrite

$$D_t^\varepsilon = \begin{cases} \sum_{k \leq n} (W_{T_{2k+1}} - W_{T_{2k}}) & \text{if } T_{2n+1} \leq t < T_{2n+2}, \\ W_t - W_{T_{2n}} + \sum_{k < n} (W_{T_{2k+1}} - W_{T_{2k}}) & \text{if } T_{2n} \leq t < T_{2n+1}. \end{cases}$$

Now, for any k, we have

$$W_{T_{2k+1}} - W_{T_{2k}} = \dot{X}_{T_{2k+1}} + (1+c) \sum_{T_{2k} < s \leq T_{2k+1}} \dot{X}_{s-} \mathbb{1}_{X_s = 0},$$

and, if $T_{2n} \leq t < T_{2n+1}$,

$$W_t - W_{T_{2n}} = \dot{X}_t + (1+c) \sum_{T_{2n} < s \leq t} \dot{X}_{s-} \mathbb{1}_{X_s = 0}.$$

Hence D_t^ε is a sum of jumps of amplitude less than $(1+c)\varepsilon$, of the fraction $c/(1+c)$ of the jumps occurring at times T_{2k+1}, and of the possible extra term $\dot{X}_t$ if $t \in F^\varepsilon$. Using the summability of the jumps, we just should show that all these terms are small when ε is small. For this, it suffices to show that

$$\sup_{s \leq L^{-1}(1), s \in F^\varepsilon} |\dot{X}_s|$$

converges in probability to 0. Fix $\eta > 0$ and introduce $\tau^\varepsilon = \inf\{s \in F^\varepsilon, |\dot{X}_s| \geq \eta\}$ and $\tau'^\varepsilon = \inf\{s \geq t, X_s = 0\}$. From the Markov property at the stopping time τ^ε and Inequality (14), we get that conditionally on $\tau^\varepsilon < \infty$ and $\mathfrak{F}_{\tau^\varepsilon}$, the probability of the event $\{\dot{X}_{\tau'^\varepsilon} \geq \eta c/2\}$ is bounded below by $1 - \sqrt{3}/\pi$. Hence we just have to prove that the probability of the event

$$\{\tau^\varepsilon < L^{-1}(1), \dot{X}_{\tau'^\varepsilon} \geq \eta c/2\}$$

goes to 0. But this event is equivalent to the existence of an excursion, before time $L^{-1}(1)$, for which there is a bounce with velocity greater than $\eta c/2$ before any bounce with speed greater than ε. The number of such excursions is a Poisson variable of parameter

$$\mathbf{n}(\dot{X}_{T_1^\varepsilon} \geq \eta c/2, \zeta_\infty > T_1^\varepsilon),$$

where T_1^ε is still defined as the time of the first bounce with velocity greater than ε, here for the excursion. We have:

$$\mathbf{n}(\dot{X}_{T_1^\varepsilon} \geq \eta c/2, \zeta_\infty > T_1^\varepsilon) = \tilde{\mathbb{P}}_{0+}(H(0, \dot{X}_{T_1^\varepsilon})^{-1} \mathbb{1}_{\dot{X}_{T_1^\varepsilon} \geq \eta c/2})$$
$$\leq (\eta c/2)^{-2k} \tilde{\mathbb{P}}_{0+}(\dot{X}_{T_1^\varepsilon} \geq \eta c/2)$$
$$\leq (\eta c/2)^{-2k} m(]\ln(\eta c/(2\varepsilon)), \infty[),$$

where we recall that m is the stationary law of the overshoot appearing in Proposition 2. The parameter of the Poisson variable goes to 0 when ε goes to 0, which concludes the proof. The process W is a Brownian motion, and $(X, \dot{X}, 0, W)$ is a solution to Equations (RLP).

Weak Uniqueness

Consider $(X, \dot{X}, N, W)$, with law $\mathbb{P}$, be any solution to (RLP), and its associated filtration $(\mathfrak{F}_t)_{t\geq 0}$. Then we have

$$\dot{X}_t = W_t - (1+c) \sum_{0<s\leq t} \dot{X}_{s-} \mathbb{1}_{X_s=0} + N_t,$$

with W a Brownian motion.

We start with the observation that the process $\dot{X}$ does not explode and that the sum involves only positive terms. Therefore these terms are summable. The process $N_t + \sum_{0<s\leq t} \dot{X}_{s-} \mathbb{1}_{X_s=0}$ is monotone and adapted, hence $\dot{X}$ is a semimartingale. As a consequence, it possesses local times $(L^{(a)})_{a\in\mathbb{R}}$, and we have an occupation formula (see for example [18], Theorem 70 Corollary 1, p. 216):

$$\int_{-\infty}^{+\infty} L_t^{(a)} g(a) da = \int_0^t g(\dot{X}_{s-}) ds,$$

for any g bounded measurable function. Taking $g = \mathbb{1}_{\{0\}}$ shows that $\dot{X}$ spends no time at zero. It follows that the process $(X, \dot{X})$ spends no time at $(0, 0)$.

The next step is to show that the process N has to be constantly equal to 0. We use a stopping time technique. Define $(\hat{T}_n^\varepsilon)_{n\geq 0}$ by $\hat{T}_0^\varepsilon = 0$ and, for $n \geq 0$,

$$\begin{cases} \hat{T}_{2n+1}^\varepsilon = \inf\{t > T_{2n}^\varepsilon, |\dot{X}_t| > \varepsilon\} \\ \hat{T}_{2n+2}^\varepsilon = \inf\{t > T_{2n+1}^\varepsilon, X_t = \dot{X}_t = 0\}, \end{cases}$$

as well as $\hat{F}^\varepsilon = \bigcup_{n\geq 0} [\hat{T}_{2n}^\varepsilon, \hat{T}_{2n+1}^\varepsilon]$ and $\hat{H}_t^\varepsilon = \mathbb{1}_{\hat{F}^\varepsilon}(t)$. When ε goes to $0+$, the closed set $\hat{F}^\varepsilon$ decreases to the zero set $\mathscr{Z} = \{t > 0, X_t = \dot{X}_t = 0\}$. Observe that N

can increase only on $\hat{F}^\varepsilon$, and that for any $s \in \hat{F}^\varepsilon$, we have $|\dot{X}_s| \leq \varepsilon$. Write $n_t^\varepsilon = \sup\{n \geq 0, \hat{T}_{2n}^\varepsilon \leq t\}$. In the case $\hat{T}_{2n_t^\varepsilon} \leq t < \hat{T}_{2n_t^\varepsilon+1}$, we obtain

$$N_t = N_t - N_{\hat{T}_{2n_t^\varepsilon}} + \sum_{n<n_t^\varepsilon}(N_{\hat{T}_{2n+1}} - N_{T_{2n}})$$

$$= -\int_0^t (1-\hat{H}_s^\varepsilon)dW_s + (1+c)\sum_{s\in\hat{F}^\varepsilon, s\leq t}\dot{X}_{s-}\mathbb{1}_{X_s=0} + \dot{X}_t + \sum_{n<n_t^\varepsilon}\dot{X}_{\hat{T}_{2n+1}},$$

by writing the equation on each interval of $\hat{F}^\varepsilon$. In the case $\hat{T}_{2n_t^\varepsilon+1} \leq t < \hat{T}_{2n_t^\varepsilon+2}$, we simply write $N_t = N_{\hat{T}_{2n_t^\varepsilon+1}}$, and we get a similar formula, without the term $\dot{X}_t$. In the last line of the above formula, the first term goes to 0 because $\mathscr{X}$ has zero Lebesgue measure, while the second term goes to 0 because the jumps are summable. The third one is included in $[-\varepsilon, \varepsilon]$ and the last one in $[-\varepsilon n_t^\varepsilon, \varepsilon n_t^\varepsilon]$. In order to deduce $N \equiv 0$ almost surely, it suffices to prove $\varepsilon n_t^\varepsilon \to 0$ in probability. Fix $\eta > 0$. We have

$$\mathbb{P}(\varepsilon n_t^\varepsilon > \eta) = \mathbb{P}(\forall n \leq \eta\varepsilon^{-1}, n+1 \leq n_t^\varepsilon) = \mathbb{P}(\forall n \leq \eta\varepsilon^{-1}, \hat{T}_{2n+2}^\varepsilon \leq t)$$

$$\leq \mathbb{P}(\forall n \leq \eta\varepsilon^{-1}, \hat{T}_{2n+2}^\varepsilon - \hat{T}_{2n+1}^\varepsilon \leq t).$$

But we also have

$$\mathbb{P}(\hat{T}_{2n+2}^\varepsilon - \hat{T}_{2n+1}^\varepsilon > t | \mathfrak{F}_{\hat{T}_{2n+1}^\varepsilon}) = \mathbb{P}^c_{X_{\hat{T}_{2n+1}^\varepsilon}, \dot{X}_{\hat{T}_{2n+1}^\varepsilon}}(\zeta_\infty > t)$$

$$\geq (1 - \frac{\sqrt{3}}{\pi})\mathbb{P}^c_{0,\varepsilon c/2}(\zeta_\infty > t),$$

using Inequality (14), so that

$$\mathbb{P}(\varepsilon n_t^\varepsilon > \eta) \leq \left(1 - (1 - \frac{\sqrt{3}}{\pi})\mathbb{P}^c_{0,\varepsilon c/2}(\zeta_\infty > t)\right)^{\lfloor \eta\varepsilon^{-1}\rfloor}.$$

Finally, by use of (4) and of the scaling invariance property, we have

$$\mathbb{P}^c_{0,\varepsilon c/2}(\zeta_\infty > t) \underset{\varepsilon \to 0}{\sim} C_1 \frac{c^{2k}t^{-k}}{2^{2k}}\varepsilon^{2k}.$$

As $2k < 1$, we conclude

$$\mathbb{P}(\varepsilon n_t^\varepsilon > \eta) \underset{\varepsilon \to 0}{\longrightarrow} 0.$$

Now, exactly as before, introduce, for $\varepsilon > 0$, the sequence of stopping times T_n^ε, defined by $T_0^\varepsilon = 0$ and

$$\begin{cases} T_{2n+1}^\varepsilon = \inf\{t > T_{2n}^\varepsilon, X_t = 0, \dot{X}_t > \varepsilon\} \\ T_{2n+2}^\varepsilon = \inf\{t > T_{2n+1}^\varepsilon, X_t = \dot{X}_t = 0\}, \end{cases}$$

as well as $F^\varepsilon = \bigcup_{n \geq 0}[T_{2n}^\varepsilon, T_{2n+1}^\varepsilon]$ and $H^\varepsilon = \mathbb{1}_{F^\varepsilon}$. Finally, define the closed set $F = \lim_{\varepsilon \to 0} F^\varepsilon$ and the adapted process $H^0 = \mathbb{1}_F$.

Lemma 8. *The set F has almost surely zero Lebesgue measure.*

This result is not immediate. First, observe that the excursions of the process may be of two types. Either an excursion bounces on the boundary just after the initial time, or it doesn't. We call $\mathscr{E}_1$ the set of excursions of the first type, defined by

$$\mathscr{E}_1 := \{(x, \dot{x}) \in \mathscr{E} \mid \zeta_1(x, \dot{x}) := \inf\{t > 0, x_t = 0\} = 0\},$$

and $\mathscr{E}_2 = \mathscr{E} \setminus \mathscr{E}_1$ the set of excursions of the second type. Unlike before, we do not know a priori that all the excursions of the process lie in $\mathscr{E}_1$. If the process starts an excursion at time t, we write e^t for the corresponding excursion. The set F contains not only the zero set $\mathscr{Z}$, but also all the intervals $[t, t + \zeta_1(e^t)]$, where t is the starting time of an excursion $e^t \in \mathscr{E}_2$. Proving Lemma 8 is equivalent to proving that there is actually no excursion in $\mathscr{E}_2$.

Suppose that this fails. Then the process

$$\mathscr{L}(t) = \int_0^t H_s^0 ds$$

is not almost surely constantly equal to zero. We introduce its right-continuous inverse

$$\mathscr{L}^{-1}(t) := \inf\{s > t, \mathscr{L}(s) > t\}.$$

There exists a Brownian motion M such that for $t < \mathscr{L}(\infty)$,

$$M_t = \int_0^{\mathscr{L}^{-1}(t)} H_s^0 dW_s.$$

Introduce the time-changed process stopped at time $\mathscr{L}(\infty)$:

$$(Y_t, \dot{Y}_t) = (X_{\mathscr{L}^{-1}(t)}, \dot{X}_{\mathscr{L}^{-1}(t)}), \qquad 0 \leq t < \mathscr{L}(\infty).$$

This process spends no time at $(0, 0)$ (before being stopped).

Lemma 9. *The quadruplet* $(Y_t, \dot{Y}_t, 0, M_t)_{t < \mathscr{L}(\infty)}$ *under* $\mathbb{P}$ *is a solution of the equations (RLP) with null elasticity coefficient, stopped at time* $\mathscr{L}(\infty)$.

Proof. Write F_d for the set of right extremities of non-trivial intervals of F, that is, the set of those instants $t + \zeta_1(e^t)$, where t is the starting time of an excursion $e^t \in \mathscr{E}_2$. Observe that for $t \in F \setminus F_d$, we have $\dot{X}_{t-} \mathbb{1}_{X_t = 0} = 0$. Fix $t < \mathscr{L}(\infty)$. We have:

$$\dot{Y}_t - M_t = W_{\mathscr{L}^{-1}(t)} - \int_0^{\mathscr{L}^{-1}(t)} H_s^0 dW_s - (1+c) \sum_{0 < s \leq \mathscr{L}^{-1}(t)} \dot{X}_{s-} \mathbb{1}_{X_s = 0}$$

$$= \int_0^{\mathscr{L}^{-1}(t)} \mathbb{1}_{s \notin F} dW_s - (1+c) \sum_{0 < s \leq \mathscr{L}^{-1}(t), s \notin F} \dot{X}_{s-} \mathbb{1}_{X_s = 0}$$

$$-(1+c) \sum_{0 < s \leq \mathscr{L}^{-1}(t), s \in F_d} \dot{X}_{s-} \mathbb{1}_{X_s = 0}.$$

Moreover, for any maximal interval (a, b) of $[0, \mathscr{L}^{-1}(t)) \setminus F$, we have $X_a = X_b = \dot{X}_b = \dot{X}_{b-} = 0$, and $\dot{X}_a = -c\dot{X}_{a-}$. It follows

$$0 = \dot{X}_{b-} - \dot{X}_a - c\dot{X}_{a-}$$
$$= W_b - W_a - (1+c) \sum_{s \in (a,b)} \dot{X}_{s-} \mathbb{1}_{X_s = 0} - c\dot{X}_{a-}.$$

Summing on all the intervals of $[0, \mathscr{L}^{-1}(t)) \setminus F$ yields

$$\int_0^{\mathscr{L}^{-1}(t)} \mathbb{1}_{s \notin F} dW_s - (1+c) \sum_{\substack{s \leq \mathscr{L}^{-1}(t), \\ s \notin F}} \dot{X}_{s-} \mathbb{1}_{X_s = 0} - c \sum_{\substack{s \leq \mathscr{L}^{-1}(t), \\ s \in F_d}} \dot{X}_{s-} \mathbb{1}_{X_s = 0} = 0,$$

and thus

$$\dot{Y}_t - M_t = - \sum_{0 < s \leq \mathscr{L}^{-1}(t), s \in F_d} \dot{X}_{s-} \mathbb{1}_{X_s = 0}. \tag{24}$$

If $s \leq \mathscr{L}^{-1}(t)$ is in F_d, then there exists $r \leq t$ such that $s = \mathscr{L}^{-1}(r-)$ and $\mathscr{L}^{-1}(r-) < \mathscr{L}^{-1}(r)$. Then $\dot{Y}_{r-} = \dot{X}_{\mathscr{L}^{-1}(r-)-} = \dot{X}_{s-}$, and $Y_r = X_{\mathscr{L}^{-1}(r)} = 0 = X_s$. Moreover, for $r \leq t$ such that $\mathscr{L}^{-1}(r-) \notin F_d$, we have $\mathscr{L}^{-1}(r) = \mathscr{L}^{-1}(r-)$ and

$$\dot{Y}_{r-} \mathbb{1}_{Y_r = 0} = \dot{X}_{\mathscr{L}^{-1}(r)-} \mathbb{1}_{X_{\mathscr{L}^{-1}(r)} = 0} = 0.$$

We deduce that (24) can be rewritten as

$$\dot{Y}_t - M_t = -\sum_{0<r\leq t} \dot{Y}_{r-}\mathbb{1}_{Y_r=0}.$$

Besides, it is easy to check $Y_t = \int_0^t \dot{Y}_s ds$. So $(Y, \dot{Y}, 0, M)$ is a solution to (RLP) with null elasticity coefficient, stopped at time $\mathscr{L}(\infty)$.

The article [4], dealing with equations (RLP) with null elasticity coefficient, shows that $(Y, \dot{Y})$ must be a Markov process, with Itō excursion law $\bar{\mathbf{n}}$ (stopped at time $\mathscr{L}(\infty)$). We immediately introduce another change of time, in a very similar way, but without stopping the excursions of $\mathscr{E}_2$ at time ζ_1. Define the random set

$$A := \mathscr{L} \cup \bigcup_{\{t|e^t \in \mathscr{E}_2\}} [t, t+\zeta_\infty(e^t)],$$

and the adapted process $\tilde{H} = \mathbb{1}_A$. Define also

$$\tilde{\mathscr{L}}(t) = \int_0^t \tilde{H}_s ds,$$

and write $\tilde{\mathscr{L}}^{-1}$ for its right-continuous inverse. Then, there exists a Brownian motion $\tilde{M}$ such that

$$\tilde{M}_t = \int_0^{\tilde{\mathscr{L}}^{-1}(t)} \tilde{H}_s dW_s$$

for $t < \tilde{\mathscr{L}}(\infty)$. Finally, the time-changed process

$$(\tilde{Y}_t, \dot{\tilde{Y}}_t) = (X_{\tilde{\mathscr{L}}^{-1}(t)}, \dot{X}_{\tilde{\mathscr{L}}^{-1}(t)}),$$

stopped at time $\tilde{\mathscr{L}}(\infty)$, spends no time at zero and its excursions are the excursions of $(X, \dot{X})$ included in $\mathscr{E}_2$. Remark that we have $\tilde{\mathscr{L}}(\infty) \geq \mathscr{L}(\infty)$ because $A \supset F$. We also get the following lemma, similar to Lemma 9, and whose proof is left to the reader.

Lemma 10. *The quadruplet* $\left(\tilde{Y}_t, \dot{\tilde{Y}}_t, 0, \tilde{M}_t\right)_{t\leq\tilde{\mathscr{L}}(\infty)}$ *under* $\mathbb{P}$ *is a solution of Equations (RLP) (with elasticity coefficient c), stopped at time* $\tilde{\mathscr{L}}(\infty)$.

The process $(\tilde{Y}, \dot{\tilde{Y}})$ spends no time at 0, is a solution to (RLP), and its excursions, stopped at ζ_1, the first return time to $\{0\} \times \mathbb{R}$, are precisely those of $(Y, \dot{Y})$. This induces that $(\tilde{Y}, \dot{\tilde{Y}})$ is a Markov process with Itō excursion measure $\tilde{\mathbf{n}}$ determined by

$$\begin{cases} \tilde{\mathbf{n}}\left((x_{t\wedge\zeta_1})_{t\geq 0} \in \cdot\right) = \bar{\mathbf{n}}(x \in \cdot) \\ \tilde{\mathbf{n}}\left((x_{t+\zeta_1})_{t\geq 0} \in \cdot \,\big|\, \dot{X}_{\zeta_1} = v\right) = \mathbb{P}_v^c(x \in \cdot) \end{cases}$$

Now, the result of uniqueness of the excursion measure implies that $\tilde{\mathbf{n}}$ should be a multiple of $\mathbf{n}$, which is obviously not the case (for example because $\tilde{\mathbf{n}}(\zeta_\infty = 0) = 0$). Therefore $\tilde{\mathscr{L}}(\infty) = 0 = \mathscr{L}(\infty)$ a.s. Lemma 8 is proved.

Finally, we introduce a last time-change, with $\mathscr{L}^\varepsilon(t) := \int_0^t H_s^\varepsilon ds$, and $(\mathscr{L}^\varepsilon)^{-1}(t) := \inf\{s > 0, \mathscr{L}^\varepsilon(s) > t\}$. When ε goes to 0, $(\mathscr{L}^\varepsilon)^{-1}$ goes to $\mathscr{L}^{-1} = Id$. It follows that the process $X^\varepsilon := (X_{(\mathscr{L}^\varepsilon)^{-1}(t)})_{t \geq 0}$ converges uniformly on compacts to X when ε goes to 0, almost surely. In particular the law of X is entirely determined by that of X^ε. The law of X^ε is in turn entirely determined by that of $(\dot{X}_{T_{2n+1}^\varepsilon})_{n \geq 0}$. We will now determine this law, which will prove the uniqueness of the law of X.

In order to avoid complex notations, we only provide the calculation of the law of $\dot{X}_{T_1^1}$, which is not fundamentally different from others. For $\varepsilon > 0$ and $n \geq 0$, a Markov property for the process W applied at time T_{2n+1}^ε shows that conditionally on $\dot{X}_{T_{2n+1}^\varepsilon} = u$, the process $(X_{(T_{2n+1}^\varepsilon + t) \wedge T_{2n+2}^\varepsilon})_{t \geq 0}$ is independent from $\mathfrak{F}_{T_{2n+1}^\varepsilon}$ and has law $\mathbb{P}_u^c$. Write n_1 for the integer satisfying $T_{2n_1+1}^\varepsilon \leq T_1^1 < T_{2n_1+2}^\varepsilon$. Conditionally on $\dot{X}_{T_{2n_1+1}^\varepsilon} = u$, the process $(X_{(T_{2n_1+1}^\varepsilon + t) \wedge T_{2n_1+2}^\varepsilon})_{t \geq 0}$ has the law $\mathbb{P}_u^c$ conditioned on reaching a speed greater than one after a bounce.

In other words, the law of $\dot{X}_{T_1^1}$ under $\mathbb{P}(\cdot | \dot{X}_{T_{2n_1+1}^\varepsilon} = u)$ is equal to that of $\dot{X}_{T_1^1}$ under $\mathbb{P}_u^c(\cdot | T_1^1 < \infty)$. Besides, it should be clear now that $\dot{X}_{T_{2n_1+1}^\varepsilon}$ goes to 0 when ε goes to 0. Recall that ζ_∞, the hitting time of $(0,0)$, is the lifetime of the excursion (under $\mathbb{P}_u^c$ as well as under $\mathbf{n}$). For any positive continuous functional f, we have:

$$\begin{aligned}
\mathbb{P}_u^c(f(\dot{X}_{T_1^1}) | T_1^1 < \zeta_\infty) &= \mathbb{P}_u^c\big(f(\dot{X}_{T_1^1}) \mathbb{1}_{T_1^1 < \zeta_\infty}\big) \big/ \mathbb{P}_u^c(\mathbb{1}_{T_1^1 < \zeta_\infty}) \\
&= \tilde{\mathbb{P}}_u\Big(f(\dot{X}_{T_1^1})(H(0, \dot{X}_{T_1^1}))^{-1}\Big) \big/ \tilde{\mathbb{P}}_u((H(0, \dot{X}_{T_1^1}))^{-1}) \\
&\xrightarrow[u \to 0]{} \tilde{\mathbb{P}}_{0+}\big(f(\dot{X}_{T_1^1})(H(0, \dot{X}_{T_1^1}))^{-1}\big) \big/ \tilde{\mathbb{P}}_{0+}((H(0, \dot{X}_{T_1^1}))^{-1}) \\
&= \mathbf{n}(f(\dot{X}_{T_1^1}) | T_1^1 < \zeta_\infty),
\end{aligned}$$

where we used successively (8), Proposition 2 and (a generalization of) (19). As a consequence, the law of $\dot{X}_{T_1^1}$ under $\mathbb{P}$ is entirely determined, and is equal to that of $\dot{X}_{T_1^1}$ under $\mathbf{n}(\cdot | T_1^1 < \zeta_\infty)$. Uniqueness for the stochastic differential equation follows.

References

1. P. Ballard, The dynamics of discrete mechanical systems with perfect unilateral constraints. Arch. Ration. Mech. Anal. **154**, 199–274 (2000)
2. J. Bect, Processus de Markov diffusifs par morceaux: outils analytiques et numriques. PhD thesis, Supelec, 2007
3. J. Bertoin, Reflecting a Langevin process at an absorbing boundary. Ann. Probab. **35**(6), 2021–2037 (2007)

4. J. Bertoin, A second order SDE for the Langevin process reflected at a completely inelastic boundary. J. Eur. Math. Soc. **10**(3), 625–639 (2008)
5. R.M. Blumenthal, On construction of Markov processes. Z. Wahrsch. Verw. Gebiete **63**(4), 433–444 (1983)
6. M. Bossy, J.-F. Jabir, On confined McKean Langevin processes satisfying the mean no-permeability boundary condition. Stochastic Process. Appl. **121**, 2751–2775 (2011)
7. A. Bressan, Incompatibilit dei teoremi di esistenza e di unicit del moto per un tipo molto comune e regolare di sistemi meccanici. Ann. Scuola Norm. Sup. Pisa Serie III **14**, 333–348(1960)
8. C.M. Goldie, Implicit renewal theory and tails of solutions of random equations. Ann. Appl. Probab. **1**(1), 126–166 (1991)
9. J.P. Gor′kov, A formula for the solution of a certain boundary value problem for the stationary equation of Brownian motion. Dokl. Akad. Nauk SSSR **223**(3), 525–528 (1975)
10. J.-P. Imhof, Density factorizations for Brownian motion, meander and the three-dimensional Bessel process, and applications. J. Appl. Probab. **21**(3), 500–510 (1984)
11. E. Jacob, Excursions of the integral of the Brownian motion. Ann. Inst. H. Poincaré Probab. Stat. **46**(3), 869–887 (2010)
12. E. Jacob, Langevin process reflected on a partially elastic boundary I. Stoch. Process. Appl. **122**(1), 191–216 (2012)
13. H. Kesten, Random difference equations and renewal theory for products of random matrices. Acta Math. **131**, 207–248 (1973)
14. A. Lachal, Les temps de passage successifs de l'intégrale du mouvement brownien. Ann. Inst. H. Poincaré Probab. Stat. **33**(1), 1–36 (1997)
15. A. Lachal, Application de la théorie des excursions à l'intégrale du mouvement brownien. In *Séminaire de Probabilités XXXVII*, vol. 1832 of *Lecture Notes in Math.* (Springer, Berlin, 2003), pp. 109–195
16. B. Maury, Direct simulation of aggregation phenomena. Comm. Math. Sci. **2**(suppl. 1), 1–11 (2004)
17. H.P. McKean, Jr., A winding problem for a resonator driven by a white noise. J. Math. Kyoto Univ. **2**, 227–235 (1963)
18. P.E. Protter, *Stochastic integration and differential equations*, vol. 21 of *Stochastic Modelling and Applied Probability*, 2nd edn, Version 2.1, Corrected third printing (Springer, Berlin, 2005)
19. V. Rivero, Recurrent extensions of self-similar Markov processes and Cramér's condition. Bernoulli **11**(3), 471–509 (2005)
20. M. Schatzman, Uniqueness and continuous dependence on data for one dimensional impact problems. Math. Comput. Model. **28**, 1–18 (1998)
21. S.J. Taylor, J.G. Wendel, The exact Hausdorff measure of the zero set of a stable process. Z. Wahrscheinlichkeitstheorie und Verw. Gebiete **6**, 170–180 (1966)

Windings of Planar Stable Processes

R.A. Doney and S. Vakeroudis

Abstract Using a generalization of the skew-product representation of planar Brownian motion and the analogue of Spitzer's celebrated asymptotic Theorem for stable processes due to Bertoin and Werner, for which we provide a new easy proof, we obtain some limit Theorems for the exit time from a cone of stable processes of index $\alpha \in (0, 2)$. We also study the case $t \to 0$ and we prove some Laws of the Iterated Logarithm (LIL) for the (well-defined) winding process associated to our planar stable process.

AMS 2010 subject classifications. Primary: 60G52, 60G51, 60F05, 60J65; secondary: 60E07, 60B12, 60G18.

Keywords Stable processes • Lévy processes • Brownian motion • Windings • Exit time from a cone • Spitzer's Theorem • Skew-product representation • Lamperti's relation • Law of the Iterated Logarithm (LIL) for small times

---

R.A. Doney
Probability and Statistics Group, School of Mathematics, University of Manchester, Alan Turing Building, Oxford Road, Manchester M13 9PL, UK
e-mail: ron.doney@manchester.ac.uk

S. Vakeroudis (✉)
Probability and Statistics Group, School of Mathematics, University of Manchester, Alan Turing Building, Oxford Road, Manchester M13 9PL, UK

Laboratoire de Probabilités et Modèles Aléatoires (LPMA) CNRS : UMR7599, Université Pierre et Marie Curie - Paris VI, Université Paris-Diderot - Paris VII, 4 Place Jussieu, 75252 Paris Cedex 05, France
e-mail: stavros.vakeroudis@ulb.ac.be

1 Introduction

In this paper, we study the windings of planar isotropic stable processes. More precisely, having as a starting point a work of Bertoin and Werner [7] concerning this subject (following their previous work on windings of planar Brownian motion[1] [5]) and motivated by some works of Shi [37], we attempt to generalize some results obtained recently for the case of planar Brownian motion (see e.g. [40–42] and the references therein). In particular, we are interested in the behaviour of stable processes for small time, an aspect which has already been investigated e.g. by Doney [13] in terms of Spitzer's condition for stable processes (see e.g. [4] and the references therein).

In Sect. 2, we recall some facts about standard isotropic stable processes of index $\alpha \in (0, 2)$ taking values in the complex plane. Then, we follow Bertoin and Werner [7] to define the process of its winding number, we generalize the skew-product representation of planar BM (see e.g. [24, 35, 11]) and we present two Lemmas for the winding process of isotropic stable Lévy processes obtained in [7]. Finally, we mention some properties of the positive and the negative moments of the exit times from a cone of this process.

In Sect. 3, we use some continuity arguments of the composition function due to Whitt [43] and we obtain a new simple proof of the analogue of Spitzer's Theorem for isotropic stable Lévy processes, initially proven by Bertoin and Werner [7]. We reformulate and we extend this result in terms of the exit times from a cone. More precisely, Spitzer's asymptotic Theorem says that, if $(\vartheta_t, t \geq 0)$ denotes the continuous determination of the argument of a planar BM starting away from the origin, then:

$$\frac{2}{\log t} \vartheta_t \xrightarrow[t \to \infty]{(law)} C_1 , \qquad (1)$$

where C_1 is a standard Cauchy variable. For other proofs of (1), see e.g. [44, 17, 31, 33, 5, 46, 41, 42]. Bertoin and Werner state that because an isotropic stable Lévy processes is transient, we expect that it winds more slowly than planar Brownian motion and prove that, with θ now denoting the process of its winding number (appropriately defined, see e.g. Sect. 2), $\theta_t / \sqrt{\log t}$ converges in distribution to some centered Gaussian law as $t \to \infty$ (Theorem 1 in Bertoin and Werner [7], stated here as Theorem 1).

In Sect. 4, and more precisely in Theorems 2 and 4, we study the asymptotics of a symmetric Lévy process and of the winding process of isotropic stable Lévy processes for $t \to 0$, respectively, which are the main results of this article. In particular, we show that $t^{-1/\alpha}\theta_t$ converges in distribution to an α-stable law as $t \to 0$. Using this result, in Proposition 5 we obtain the (weak) limit in distribution

---

[1] When we simply write: Brownian motion, we always mean real-valued Brownian motion, starting from 0. For two-dimensional Brownian motion, we indicate planar or complex BM.

of the process of the exit times from a cone with narrow amplitude and we further obtain several generalizations. We also study the windings of planar stable processes in $(t, 1]$, for $t \to 0$ and we note that, with obvious notation, Spitzer's law is still valid for $\theta_{(t,1]}$ (see Remark 4). Section 5 deals with the Law of the Iterated Logarithm (LIL) for Lévy processes for small times, in the spirit of some well-known LIL for Brownian motion for $t \to \infty$ from Bertoin and Werner [5, 6] and from Shi [36, 37], and for stable subordinators with index $\alpha \in (0, 1)$ for $t \to 0$ from Fristedt [18, 19] and Khintchine [23] (see also [3]). Moreover, we prove a LIL for the winding number process of stable processes, for $t \to 0$.

Finally, in Sect. 6 we discuss the planar Brownian motion case and in Theorem 7 we obtain the asymptotic behaviour of the winding process as $t \to 0$. More precisely, the process $\left(c^{-1/2}\vartheta_{ct}, t \geq 0\right)$ converges in law to a one-dimensional Brownian motion as $c \to 0$.

Notation. In the following text, with the symbol "$\Longrightarrow$" we shall denote the weak convergence in distribution on the appropriate space, endowed with the Skorohod topology.

2 Preliminaries

Following Lamperti [26], a Markov process X with values in $\mathbb{R}^d$, $d \geq 2$ is called isotropic or $O(d)$-invariant ($O(d)$ stands for the group of orthogonal transformations on $\mathbb{R}^d$) if its transition satisfies:

$$P_t(\phi(x), \phi(\mathcal{B})) = P_t(x, \mathcal{B}), \qquad (2)$$

for any $\phi \in O(d)$, $x \in \mathbb{R}^d$ and Borel subset $\mathcal{B} \subset \mathbb{R}^d$.

Moreover, X is said to be α-self-similar if, for $\alpha > 0$,

$$P_{\lambda t}(x, \mathcal{B}) = P_t(\lambda^{-\alpha} x, \lambda^{-\alpha} \mathcal{B}), \qquad (3)$$

for any $\lambda > 0$, $x \in \mathbb{R}^d$ and $\mathcal{B} \subset \mathbb{R}^d$.

We focus now our study on the two-dimensional case ($d = 2$), where (3) holds, and we denote by $(Z_t, t \geq 0)$ a standard isotropic stable process of index $\alpha \in (0, 2)$ taking values in the complex plane and starting from $z_0 + i0, z_0 > 0$. A scaling argument shows that we may assume $z_0 = 1$, without loss of generality, since, with obvious notation:

$$\left(Z_t^{(z_0)}, t \geq 0\right) \stackrel{(law)}{=} \left(z_0 Z_{(t/z_0^\alpha)}^{(1)}, t \geq 0\right). \qquad (4)$$

Thus, from now on, we shall take $z_0 = 1$. More precisely, Z has stationary independent increments, its sample path is right continuous and has left limits (cadlag) and, with $\langle \cdot, \cdot \rangle$ standing for the Euclidean inner product, $E\left[\exp\left(i\langle \lambda, Z_t \rangle\right)\right] =$

$\exp(-t|\lambda|^\alpha)$, for all $t \geq 0$ and $\lambda \in \mathbb{C}$. Z is transient, $\lim_{t \to \infty} |Z_t| = \infty$ a.s. and it a.s. never visits single points. We remark that for $\alpha = 2$, we are in the Brownian motion case.

We are now going to recall some properties of stable processes and Lévy processes (for more details see e.g. [3] or [25]).

To start with, if $\mathscr{Z} = (\mathscr{Z}_t, t \geq 0)$ denotes a planar Brownian motion starting from 1 and $S = (S(t), t \geq 0)$ an independent stable subordinator with index $\alpha/2$ starting from 0, i.e.:

$$E[\exp(-\mu S(t))] = \exp(-t\mu^{\alpha/2}), \qquad (5)$$

for all $t \geq 0$ and $\mu \geq 0$, then the subordinated planar BM $(\mathscr{Z}_{2S(t)}, t \geq 0)$ is a standard isotropic stable process of index α. The Lévy measure of S is:

$$\frac{\alpha}{2\Gamma(1-\alpha/2)} s^{-1-\alpha/2} 1_{\{s>0\}} ds,$$

thus, the Lévy measure ν of Z is:

$$\begin{aligned}
\nu(dx) &= \frac{\alpha}{2\Gamma(1-\alpha/2)} \int_0^\infty s^{-1-\alpha/2} P(\mathscr{Z}_{2s} - 1 \in dx) ds \\
&= \frac{\alpha}{8\pi\Gamma(1-\alpha/2)} \left(\int_0^\infty s^{-2-\alpha/2} \exp(-|x|^2/(4s)) ds \right) dx \\
&= \frac{\alpha \, 2^{-1+\alpha/2} \Gamma(1+\alpha/2)}{\pi \Gamma(1-\alpha/2)} |x|^{-2-\alpha} dx. \qquad (6)
\end{aligned}$$

Contrary to planar Brownian motion, as Z is discontinuous, we cannot define its winding number (recall that, as is well known [21], for planar BM, since it starts away from the origin, it does not visit a.s. the point 0 but keeps winding around it infinitely often. In particular, the winding process is well defined, for further details see also e.g. [33]). However, following [7], we can consider a path on a finite time interval $[0, t]$ and "fill in" the gaps with line segments in order to obtain the curve of a continuous function $f : [0, 1] \to \mathbb{C}$ with $f(0) = 1$. Now, since 0 is polar and Z has no jumps across 0 a.s., we have $f(u) \neq 0$ for every $u \in [0, 1]$. Hence, we can define the process of the winding number of Z around 0, which we denote by $\theta = (\theta_t, t \geq 0)$. It has cadlag paths of absolute length greater than π and, for all $t \geq 0$,

$$\exp(i\theta_t) = \frac{Z_t}{|Z_t|}. \qquad (7)$$

We also introduce the clock:

$$H(t) \equiv \int_0^t \frac{ds}{|Z_s|^\alpha}, \qquad (8)$$

and its inverse:

$$A(u) \equiv \inf\{t \geq 0, H(t) > u\}. \tag{9}$$

Bertoin and Werner following [20] obtained these two Lemmas for $\alpha \in (0, 2)$ (for the proofs see [5]):

Lemma 1. *The time-changed process $(\theta_{A(u)}, u \geq 0)$ is a real-valued symmetric Lévy process. It has no Gaussian component and its Lévy measure has support in $[-\pi, \pi]$.*

We now denote by dz the Lebesgue measure on $\mathbb{C}$. Then, for every complex number $z \neq 0$, $\phi(z)$ denotes the determination of its argument valued in $(-\pi, \pi]$.

Lemma 2. *The Lévy measure of $\theta_{A(\cdot)}$ is the image of the Lévy measure ν of Z by the mapping $z \to \phi(1+z)$. As a consequence, $E[(\theta_{A(u)})^2] = uk(\alpha)$, where*

$$k(\alpha) = \frac{\alpha \, 2^{-1+\alpha/2} \Gamma(1+\alpha/2)}{\pi \Gamma(1-\alpha/2)} \int_{\mathbb{C}} |z|^{-2-\alpha} |\phi(1+z)|^2 dz. \tag{10}$$

Using Lemma 1, we can obtain the analogue of the skew product representation for planar BM which is the Lamperti correspondence for stable processes. Indeed, following [20] and using Lamperti's relation (see e.g. [26, 24, 11, 10] or [35]) and Lemma 1, there exist two real-valued Lévy processes $(\xi_u, u \geq 0)$ and $(\rho_u, u \geq 0)$, the first one non-symmetric whereas the second one symmetric, both starting from 0, such that:

$$\log |Z_t| + i\theta_t = (\xi_u + i\rho_u)\Big|_{u=H_t=\int_0^t \frac{ds}{|Z_s|^\alpha}}. \tag{11}$$

We remark here that $|Z|$ and $Z_{A(\cdot)}/|Z_{A(\cdot)}|$ are NOT independent. Indeed, the processes $|Z_{A(\cdot)}|$ and $Z_{A(\cdot)}/|Z_{A(\cdot)}|$ jump at the same times hence they cannot be independent. Moreover, $A(\cdot)$ depends only upon $|Z|$, hence $|Z|$ and $Z_{A(\cdot)}/|Z_{A(\cdot)}|$ are not independent. For further discussion on the independence, see e.g. [30], where is shown that an isotropic α-self-similar Markov process has a skew-product structure if and only if its radial and its angular part do not jump at the same time.

We also remark that

$$H^{-1}(u) \equiv A(u) \equiv \inf\{t \geq 0 : H(t) > u\} = \int_0^u \exp\{\alpha \xi_s\} \, ds. \tag{12}$$

Hence, (11) may be equivalently written as:

$$\begin{cases} |Z_t| = \exp(\xi(H_t)) \Leftrightarrow |Z_{A(t)}| = \exp(\xi_t), \text{ (extension of Lamperti's identity)} \\ \theta_t = \rho(H_t) \Leftrightarrow \theta(A(t)) = \rho(t). \end{cases}$$
$$\tag{13}$$

We also define the random times $T_c^{|\theta|} \equiv \inf\{t : |\theta_t| \geq c\}$ and $T_c^{|\rho|} \equiv \inf\{t : |\rho_t| \geq c\}$, $(c > 0)$. Using the "generalized" skew-product representation (11) (or (13)), we obtain:

$$T_c^{|\theta|} = H_u^{-1}\Big|_{u=T_c^{|\rho|}} = \int_0^{T_c^{|\gamma|}} ds \exp(\alpha \xi_s) \equiv A_{T_c^{|\rho|}}. \qquad (14)$$

Following [42], for the random times $T_{-d,c}^\theta \equiv \inf\{t : \theta_t \notin (-d,c)\}$, $d,c > 0$, and $T_c^\theta \equiv \inf\{t : \theta_t \geq c\}$, we have:

Remark 1. For $0 < c < d$, the random times $T_{-d,c}^\theta$, $T_c^{|\theta|}$ and T_c^θ satisfy the trivial inequality:

$$T_c^{|\theta|} \leq T_{-d,c}^\theta \leq T_c^\theta. \qquad (15)$$

Hence, with $p > 0$:

$$E\left[\left(T_c^{|\theta|}\right)^p\right] \leq E\left[\left(T_{-d,c}^\theta\right)^p\right] \leq E\left[\left(T_c^\theta\right)^p\right], \qquad (16)$$

and for the negative moments:

$$E\left[\left(T_c^\theta\right)^{-p}\right] \leq E\left[\left(T_{-d,c}^\theta\right)^{-p}\right] \leq E\left[\left(T_c^{|\theta|}\right)^{-p}\right]. \qquad (17)$$

Remark 2. For further details concerning the finiteness of the positive moments of $T_c^{|\theta|}$, see e.g. [12, 2]. Recall also that for the positive moments of the exit time from a cone of planar Brownian motion, Spitzer showed that (with obvious notation) [39, 9]:

$$E\left[\left(T_c^{|\vartheta|}\right)^p\right] < \infty \Leftrightarrow p < \frac{\pi}{4c}, \qquad (18)$$

whereas all the negative moments $E\left[\left(T_c^{|\vartheta|}\right)^{-p}\right]$ are finite [42].

We denote now by $\Psi(u)$ the exponent of the symmetric Lévy process ρ, hence (Lévy–Khintchine formula) $E[e^{iu\rho_t}] = e^{-t\Psi(u)}$, with:

$$\Psi(u) = \int_{(-\infty,\infty)} \left(1 - e^{iux} + iux1_{\{|x|\leq 1\}}\right) \mu(dx), \ u \in \mathbb{R}, \qquad (19)$$

where μ is a Radon measure on $\mathbb{R} \setminus \{0\}$ such that:

$$\int_{(-\infty,\infty)} (x^2 \wedge 1)\mu(dx) < \infty.$$

μ is the Lévy measure of ρ and is symmetric.

3 Large Time Asymptotics

Concerning the clock H, we have the almost sure convergence (see Corollary 1 in Bertoin and Werner [7]):

$$\frac{H(e^u)}{u} \xrightarrow[u \to \infty]{a.s.} 2^{-\alpha} \frac{\Gamma(1-\alpha/2)}{\Gamma(1+\alpha/2)} \equiv K(\alpha) = E\left[|Z_1|^{-\alpha}\right]. \tag{20}$$

Moreover, we have the following:

Proposition 1. *The family of processes*

$$H_x^{(u)} \equiv \left(\frac{H(e^{ux})}{u}, x \geq 0\right)$$

is tight, as $u \to \infty$.

Proof. To prove this, we could repeat some arguments of Pitman and Yor [34] (see the estimates in their proof of Theorem 6.4), however, we give here a straightforward proof, using the definition of tightness:
for every $\varepsilon, \eta > 0$, there exist $\delta > 0$ and $C_\delta > 0$ such that, for every $0 < x < y$:

$$P\left(\sup_{|x-y| \leq \delta} |H(e^{uy}) - H(e^{ux})| \geq u\varepsilon\right) \leq \eta, \quad \text{for } u \geq C_\delta, \tag{21}$$

or equivalently:

$$P\left(\frac{1}{u}|H(e^{u(x+\delta)}) - H(e^{ux})| \geq \varepsilon\right) \leq \eta, \quad \text{for } u \geq C_\delta. \tag{22}$$

First, following Bertoin and Werner [7], we introduce the "Ornstein–Uhlenbeck type" process:

$$\tilde{Z}_u = \exp(-u/\alpha) Z_{\exp(u)}, \quad u \geq 0, \tag{23}$$

which is a stationary Markov process under P_0 (see e.g. [8]). We denote by $p_t(\cdot)$ the semigroup of Z:

$$p_t(\bar{z}) = P_0(Z_t \in d\bar{z})/d\bar{z}, \quad \bar{z} \in \mathbb{C}.$$

We denote by $Z^{(0)}$ another stable process starting at 0. Then, using the scaling property, given that $\tilde{Z}_0 \equiv Z_1 \equiv 1 + Z_1^{(0)} \equiv \bar{x}$, the semigroup $q_u(\cdot)$ of $\tilde{Z}$ is given by:

$$\begin{aligned}
q_u(\bar{x}, \bar{y}) &= p_{\exp(u)-1}\left(e^{u/\alpha}\bar{y} - \bar{x}\right) e^{2u/\alpha} \\
&= (e^u - 1)^{-2/\alpha} e^{2u/\alpha} p_1\left((e^u - 1)^{-1/\alpha}(e^{u/\alpha}\bar{y} - \bar{x})\right) \\
&= (l(u))^2 p_1\left(l(u)(\bar{y} - e^{-u/\alpha}\bar{x})\right), \tag{24}
\end{aligned}$$

where $l(v) = e^{v/\alpha}(e^v - 1)^{-1/\alpha}$. For every $\delta > 0$ and changing variables $s = \exp(v)$, with obvious notation, we have:

$$E\left[|H(e^{u(x+\delta)}) - H(e^{ux})|\right] = \int_{e^{ux}}^{e^{u(x+\delta)}} E\left[|Z_s|^{-\alpha}\right] ds = \int_{ux}^{u(x+\delta)} E_{\tilde{Z}_0}\left[|\tilde{Z}_v|^{-\alpha}\right] dv. \quad (25)$$

We also define $\varepsilon(v) \equiv l(v)e^{-v/\alpha} = (e^v - 1)^{-1/\alpha}$. From (23), using the stability of Z, we have:

$$\tilde{Z}_v = e^{-v/\alpha} Z_{\exp(v)} = e^{-v/\alpha}\left(Z_{\exp(v)-1} + Z_1\right) \stackrel{(law)}{=} e^{-v/\alpha}\left((e^v - 1)^{1/\alpha} Z_1^{(0)} + Z_1\right)$$
$$= (l(v))^{-1}\left(Z_1^{(0)} + \varepsilon(v) Z_1\right). \quad (26)$$

Hence (for simplicity, we use $E \equiv E_0$):

$$E_{\tilde{Z}_0}\left[|\tilde{Z}_v|^{-\alpha}\right] = (l(v))^\alpha E\left[|Z_1^{(0)} + \varepsilon(v)\tilde{Z}_0|^{-\alpha}\right] \equiv (l(v))^\alpha (E_1 + E_2), \quad (27)$$

where, with $\delta' > 0$,

$$E_1 = E\left[|Z_1^{(0)} + \varepsilon(v)\tilde{Z}_0|^{-\alpha} : |Z_1^{(0)} + \varepsilon(v)\tilde{Z}_0| \geq \delta'\right],$$
$$E_2 = E\left[|Z_1^{(0)} + \varepsilon(v)\tilde{Z}_0|^{-\alpha} : |Z_1^{(0)} + \varepsilon(v)\tilde{Z}_0| \leq \delta'\right].$$

We have: $l(v) \stackrel{v\to\infty}{\longrightarrow} 1$ and $\varepsilon(v) \stackrel{v\to\infty}{\longrightarrow} 0$, thus, by Dominated Convergence Theorem:

$$E_1 \stackrel{v\to\infty}{\longrightarrow} E\left[|Z_1^{(0)}|^{-\alpha} : |Z_1^{(0)}| \geq \delta'\right] \stackrel{\delta'\to 0}{\longrightarrow} E\left[|Z_1^{(0)}|^{-\alpha}\right]. \quad (28)$$

Moreover, changing the variables: $\bar{w} = \bar{z} + \varepsilon(v)\bar{x}$, we have:

$$E_2 = \int_{\bar{x},\bar{z}:|\bar{z}+\varepsilon(v)\bar{x}|\leq \delta'} P(\tilde{Z}_0 \in d\bar{x}) \, P(Z_1^{(0)} \in d\bar{z}) \, |\bar{z} + \varepsilon(v)\bar{x}|^{-\alpha}$$
$$= \int_{\bar{x},\bar{w}:|\bar{w}|\leq \delta'} P(\tilde{Z}_0 \in d\bar{x}) \, P(Z_1^{(0)} \in d\bar{w}) \, |\bar{w}|^{-\alpha}.$$

Remarking now that for stable processes: $P(Z_1^{(0)} \in d\bar{y}) \leq C' d\bar{y}$, where C' stands for a positive constant and using $w = (w_1, w_2)$, we have:

$$E_2 \leq C' \int_{\bar{x},\bar{w}:|\bar{w}|\leq \delta'} P(\tilde{Z}_0 \in d\bar{x}) \frac{dw_1 \, dw_2}{|\bar{w}|^\alpha} = C' \int_{\bar{z}:|\bar{w}|\leq \delta'} \frac{dw_1 \, dw_2}{|\bar{w}|^\alpha} \stackrel{\delta'\to 0}{\longrightarrow} 0. \quad (29)$$

Thus, from (27), (28) and (29), invoking again the Dominated Convergence Theorem, we deduce:

$$\lim_{u \to \infty} E_{\tilde{Z}_0}\left[|\tilde{Z}_v|^{-\alpha}\right] = E\left[|Z_1^{(0)}|^{-\alpha}\right], \qquad (30)$$

which is a constant. Hence, for every $\varepsilon, \eta > 0$, there exist $\delta > 0$ and $C_\delta > 0$ such that (22) is satisfied for $u \geq C_\delta$. □

Bertoin and Werner in [7] obtained the analogue of Spitzer's Asymptotic Theorem [39] for isotropic stable Lévy processes of index $\alpha \in (0, 2)$:

Theorem 1. *The family of processes*

$$\Theta_t^{(c)} \equiv \left(c^{-1/2}\theta_{\exp(ct)}, t \geq 0\right)$$

converges in distribution on $D([0, \infty), \mathbb{R})$ endowed with the Skorohod topology, as $c \to \infty$, to $\left(\sqrt{r(\alpha)}B_t, t \geq 0\right)$, where $(B_s, s \geq 0)$ is a real valued Brownian motion and

$$r(\alpha) = \frac{\alpha \, 2^{-1-\alpha/2}}{\pi} \int_{\mathbb{C}} |z|^{-2-\alpha} |\phi(1+z)|^2 dz. \qquad (31)$$

Using some results due to Whitt [43], we can obtain a simple proof of this theorem.

Proof (new proof). Essentially, an argument of continuity of the composition function (Theorem 3.1 in [43]) may replace the martingale argument in the lines of the proof for $t \to \infty$ from Bertoin and Werner. We split the proof in three parts:

(i) Concerning the clock H, we have the almost sure convergence (20):

$$\frac{H(e^u)}{u} \xrightarrow[u \to \infty]{a.s.} K(\alpha) = E\left[|Z_1|^{-\alpha}\right].$$

From this result follows the convergence of the finite dimensional distributions of $v^{-1} H(\exp(vt))$, for $v \to \infty$ and every $t > 0$.

Moreover, from Proposition 1, the family of processes

$$H_x^{(u)} \equiv \left(\frac{H(e^{ux})}{u}, x \geq 0\right)$$

is tight as $u \to \infty$. Hence, from (20) and (21), finally, $H^{(u)}(t) \equiv (u^{-1}H(\exp(ut)), t \geq 0)$ converges weakly to $(tK(\alpha), t \geq 0)$ as $u \to \infty$, i.e.:

$$\left(\frac{H(e^{ut})}{u}, t \geq 0\right) \xRightarrow[u \to \infty]{(d)} (tK(\alpha), t \geq 0), \qquad (32)$$

where the convergence in distribution is viewed on $D([0, \infty), \mathbb{R})$ endowed with the Skorohod topology.

(ii) Using the skew product representation analogue (11) and Lemma 2, we have:

$$\left(\frac{\rho_{tu}}{\sqrt{u}}, t \geq 0\right) \xrightarrow[u \to \infty]{(d)} \left(\sqrt{k(\alpha)}\, B_t, t \geq 0\right), \tag{33}$$

where the convergence in distribution is viewed again on $D([\,0,\infty\,), \mathbb{R})$ endowed with the Skorohod topology and $k(\alpha)$ is given by (10). This follows from the convergence of the finite dimensional distributions:

$$\frac{\rho_u}{\sqrt{u}} = \frac{\theta_{A(u)}}{\sqrt{u}} \xrightarrow[u \to \infty]{(d)} \sqrt{k(\alpha)}\, B_1, \tag{34}$$

a condition which is sufficient for the weak convergence (33), since Lévy processes are semimartingales with stationary independent increments; for further details see e.g. [38] or [22] (Corollary 3.6, Chap. VII, p. 415).

(iii) Theorem 3.1 in [43] states that the composition function on $D([\,0,\infty\,), \mathbb{R}) \times D([\,0,\infty\,), [\,0,\infty\,))$ is continuous at each $(\rho, H) \in (C([\,0,\infty\,), \mathbb{R}) \times D_0([\,0,\infty\,), [\,0,\infty\,)))$, with C denoting the set of continuous functions and D_0 the subset of increasing cadlag functions in D (hence the subset of non-decreasing cadlag functions in D). Hence, from (32) and (33), we have: for every $t > 0$,

$$\frac{\theta_{\exp(ct)}}{\sqrt{c}} = \frac{\rho_{H(\exp(ct))}}{\sqrt{c}} = \frac{\rho_c(H(e^{ct})/c)}{\sqrt{c}}. \tag{35}$$

The result now follows from the continuity of the composition function together with (35) and the weak convergence of $H^{(c)}(\cdot)$ and $c^{-1/2}\rho_c$, as $c \to \infty$.

$\square$

From Theorem 1, we can obtain the asymptotic behaviour of the exit times from a cone for isotropic stable processes which generalizes a recent result in [42]:

Proposition 2. *For $c \to \infty$, for every $x > 0$, we have the weak convergence:*

$$\left(\frac{1}{c}\log\left(T^\theta_{x\sqrt{c}}\right), x \geq 0\right) \xrightarrow[c \to \infty]{(d)} \left(\tau^{(1/2)}_{\sqrt{1/r(\alpha)}}, x \geq 0\right), \tag{36}$$

where for every $y > 0$, $\tau^{(1/2)}_y$ stands for the $\frac{1}{2}$-stable process defined by: $\tau^{(1/2)}_y \equiv \inf\{t : B_t = y\}$.

Proof. We rely now upon Theorem 1, the analogue of Spitzer's Theorem for stable processes by Bertoin and Werner:

$$\Theta^{(c)}_t \equiv \left(c^{-1/2}\theta_{\exp(ct)}, t \geq 0\right) \xrightarrow[c \to \infty]{(d)} \left(B_{r(\alpha)t}, t \geq 0\right). \tag{37}$$

Hence, for every $x > 0$,

$$\frac{1}{c}\log\left(T^\theta_{x\sqrt{c}}\right) = \frac{1}{c}\log\left(\inf\{t : \theta_t > x\sqrt{c}\}\right)$$

$$\stackrel{t=\exp(cs)}{=} \frac{1}{c}\log\left(\inf\left\{e^{cs} : \frac{1}{\sqrt{c}}\theta_{\exp(cs)} > x\right\}\right)$$

$$= \inf\left\{s : \frac{1}{\sqrt{c}}\theta_{\exp(cs)} > x\right\}$$

$$\stackrel{c\to\infty}{\longrightarrow} \inf\{s : B_{r(\alpha)s} > x\}$$

$$= \inf\{s : \sqrt{r(\alpha)}B_s > x\} \equiv \tau^{(1/2)}_{x/\sqrt{r(\alpha)}}. \qquad (38)$$

Moreover, from Theorem 7.1 in [43], we know that the first passage time function mapping is continuous, thus, we deduce (36). □

If we replace c by ac, we can obtain several variants of Proposition 2 for the random times $T^\theta_{-bc,ac}$, $0 < a, b \le \infty$, for $c \to \infty$, and $a, b > 0$ fixed:

Corollary 1. *The following asymptotic results hold:*

$$\frac{1}{c}\log\left(T^\theta_{\sqrt{ac}}\right) \xrightarrow[c\to\infty]{(law)} \tau^{(1/2)}_{\sqrt{a/r(\alpha)}}, \qquad (39)$$

$$\frac{1}{c}\log\left(T^{|\theta|}_{\sqrt{ac}}\right) \xrightarrow[c\to\infty]{(law)} \tau^{|B|}_{\sqrt{a/r(\alpha)}}, \qquad (40)$$

$$\frac{1}{c}\log\left(T^\theta_{-\sqrt{bc},\sqrt{ac}}\right) \xrightarrow[c\to\infty]{(law)} \tau^B_{-\sqrt{b/r(\alpha)},\sqrt{a/r(\alpha)}}, \qquad (41)$$

where for every $x, y > 0$, $\tau^{|B|}_x \equiv \inf\{t : |B_t| = x\}$ and $\tau^B_{-y,x} \equiv \inf\{t : B_t \notin (-y, x)\}$.

Proposition 3. *The following asymptotic result for $\alpha \in (0, 2)$, holds: for every $b > 0$,*

$$P\left(T^\theta_{b\sqrt{\log t}} > t\right) \xrightarrow{t\to\infty} \operatorname{erf}\left(\frac{b}{\sqrt{2r(\alpha)}}\right), \qquad (42)$$

where $\operatorname{erf}(x) \equiv \frac{2}{\sqrt{\pi}}\int_0^x e^{-y^2}\,dy$ *is the error function.*

Proof. Using the notation of Theorem 1, for every $b > 0$, we have:

$$P\left(T^\theta_{b\sqrt{\log t}} > t\right) = P\left(\sup_{u\le t}\theta_u < b\sqrt{\log t}\right) \stackrel{u=t^v}{=} P\left(\sup_{v\le 1}(\log t)^{-1/2}\theta_{t^v} < b\right)$$

$$\stackrel{t=e^c}{=} P\left(\sup_{v\le 1} c^{-1/2}\theta_{\exp(cv)} < b\right)$$

Hence, using Theorem 1 for $t \to \infty$, we deduce:

$$P\left(T^{\theta}_{b\sqrt{\log t}} > t\right) \stackrel{t \to \infty}{\longrightarrow} P\left(\sup_{v \leq 1} \sqrt{r(\alpha)} B_v < b\right) = P\left(|B_1| < \frac{b}{\sqrt{r(\alpha)}}\right)$$

$$= 2 \int_0^{b/\sqrt{r(\alpha)}} \frac{dw}{\sqrt{2\pi}} e^{-w^2/2},$$

and changing the variables $w = y\sqrt{2}$, we obtain (42). □

As mentioned in [7], because an isotropic stable Lévy process Z is transient, the difference between θ and the winding number around an arbitrary fixed $z \neq 1$ is bounded and converges as $t \to \infty$. Hence, with $(\theta_t^i, t > 0)$, $1 \leq i \leq n$ denoting the continuous total angle wound of Z of index $\alpha \in (0, 2)$ around z^i ($z^1, \ldots, z^n$ are n distinct points in the complex plane $\mathbb{C}$) up to time t, we obtain the following concerning the finite dimensional distributions (windings around several points):

Proposition 4. *For isotropic stable Lévy processes of index $\alpha \in (0, 2)$, we have:*

$$\left(\frac{\theta_t^i}{\sqrt{\log t}}, 1 \leq i \leq n\right) \stackrel{(d)}{\underset{t \to \infty}{\Longrightarrow}} \left(\sqrt{r(\alpha)} B_1^i, 1 \leq i \leq n\right), \quad (43)$$

where $(B_s^i, 1 \leq i \leq n, s \geq 0)$ is an n-dimensional Brownian motion and $r(\alpha)$ is given by (31).

4 Small Time Asymptotics

We turn now our study to the behaviour of θ_t for $t \to 0$.

Theorem 2. *For $\alpha \in (0, 2)$, the following convergence in law holds:*

$$\left(t^{-1/\alpha} \rho_{ts}, s \geq 0\right) \stackrel{(d)}{\underset{t \to 0}{\Longrightarrow}} (\zeta_s, s \geq 0), \quad (44)$$

where $(\zeta_s, s \geq 0)$ is a symmetric one-dimensional α-stable process and the convergence in distribution is considered on $D([0, \infty), \mathbb{R})$ endowed with the Skorohod topology.

Proof. From Lemma 2, we use the Lévy measure, say $\tilde{\pi}$, of $\theta_{A(\cdot)}$ (thus the Lévy measure of ρ) and we prove that for $t \to 0$ it converges to the Lévy measure of a one-dimensional α-stable process. Indeed, with

$$L \equiv \frac{\alpha \, 2^{-1+\alpha/2} \Gamma(1 + \alpha/2)}{\pi \Gamma(1 - \alpha/2)},$$

and z denoting a number in $\mathbb{C}$, using polar coordinates, we have:

$$\phi(1+z) = \int_{\mathbb{C}} \frac{dz}{1+z} = 2L \int_0^\pi \int_0^\infty \frac{r\,dr\,d\varphi}{(1+r^2-2r\cos\varphi)^{1+\alpha/2}}.$$

We remark that:

$$1 + r^2 - 2r\cos\varphi = (r - \cos\varphi)^2 + \sin^2\varphi,$$

hence, changing the variables $(r - \cos\varphi)^2 = t^{-1}\sin^2\varphi$ and denoting by:

$$B(y; a, b) = \int_0^y u^{a-1}(1-u)^{b-1}\,du,$$

the incomplete Beta function, for $\varphi > 0$ (we can repeat the same arguments for $\varphi < 0$) we have:

$$\begin{aligned}
\tilde{\pi}(d\varphi) &= d\varphi\, 2L \int_0^\infty \frac{r\,dr}{\left((r-\cos\varphi)^2 + \sin^2\varphi\right)^{1+\alpha/2}} \\
&= d\varphi\, \frac{2L}{2} \left(\frac{2}{\alpha} + \cos\varphi\left(1 - \cos^2\varphi\right)^{-\frac{1}{2}-\frac{\alpha}{2}} \int_0^{1-\frac{1}{\cos^2\varphi}} t^{-\frac{1}{2}+\frac{\alpha}{2}}(1+t)^{-\frac{\alpha}{2}-1}\,dt\right) \\
&\stackrel{u=-t}{=} d\varphi\, L \left(\frac{2}{\alpha} + \cos\varphi\left(-1 + \cos^2\varphi\right)^{-\frac{1}{2}-\frac{\alpha}{2}} \int_0^{1-\frac{1}{\cos^2\varphi}} u^{-\frac{1}{2}+\frac{\alpha}{2}}(1-u)^{-\frac{\alpha}{2}-1}\,du\right) \\
&= d\varphi\, L \left(\frac{2}{\alpha} + \cos\varphi\left(-1 + \cos^2\varphi\right)^{-\frac{1}{2}-\frac{\alpha}{2}} B\left(1 - \frac{1}{\cos^2\varphi}; \frac{1}{2} + \frac{\alpha}{2}, -\frac{\alpha}{2}\right)\right) \\
&\stackrel{\varphi \sim 0}{\sim} \tilde{L}\varphi^{-1-\alpha}\,d\varphi,
\end{aligned} \qquad (45)$$

which is the Lévy measure of an α-stable process. The result now follows by standard arguments. □

Concerning the clock H and its increments, we have:

Theorem 3. *The following a.s. convergence holds:*

$$\left(\frac{H(ux)}{u}, x \geq 0\right) \xrightarrow[u \to 0]{a.s.} (x, x \geq 0). \qquad (46)$$

Proof. From the definition of the clock H we have:

$$\frac{H(ux)}{u} = \frac{1}{u} \int_0^{ux} \frac{ds}{|Z_s|^\alpha}.$$

Hence, for every $x_0 > 0$, we have:

$$\sup_{x \leq x_0} \left| \frac{H(ux) - ux}{u} \right| = \sup_{x \leq x_0} \frac{1}{u} \left| \int_0^{ux} \left(\frac{1}{|Z_s|^\alpha} - 1 \right) ds \right| \leq \frac{1}{u} \int_0^{ux_0} \left| \frac{1}{|Z_s|^\alpha} - 1 \right| ds$$

$$\stackrel{s=uw}{=} \int_0^{x_0} \left| \frac{1}{|Z_{uw}|^\alpha} - 1 \right| dw \xrightarrow[u \to 0]{a.s.} 0. \tag{47}$$

because:

$$|Z_u|^\alpha \xrightarrow[u \to 0]{a.s.} 1. \tag{48}$$

Thus, as (47) is true for every $x_0 > 0$, we obtain (46). □

Remark 3. We remark that this behaviour of the clock is different for the case $t \to \infty$, where (20) can be equivalently stated as:

$$\left(\frac{H(ux)}{\log u}, x \geq 0 \right) \xrightarrow[u \to \infty]{(d)} \left(2^{-\alpha} \frac{\Gamma(1 - \alpha/2)}{\Gamma(1 + \alpha/2)} x, x \geq 0 \right). \tag{49}$$

Using Theorems 2 and 3, we obtain:

Theorem 4. *With $\alpha \in (0, 2)$, the family of processes*

$$\left(c^{-1/\alpha} \theta_{ct}, t \geq 0 \right)$$

converges in distribution on $D([0, \infty), \mathbb{R})$ endowed with the Skorohod topology, as $c \to 0$, to a symmetric one-dimensional α-stable process $(\zeta_t, t \geq 0)$.

Proof. We will use Theorems 2 and 3. More precisely, we shall rely again upon the continuity of the composition function as studied in Theorem 3.1 in [43].

(i) First, concerning the clock H, for every $t > 0$, we have the almost sure convergence (46), which yields the weak convergence (on $D([0, \infty), \mathbb{R})$ endowed with the Skorohod topology) of the family of processes $\tilde{H}^{(u)}(t) \equiv (u^{-1} H(ut), t \geq 0)$ to $(t, t \geq 0)$ as $u \to 0$.

(ii) We use another result of Whitt [43] which states that the composition function on $D([0, \infty), \mathbb{R}) \times D([0, \infty), [0, \infty))$ is continuous at each $(\rho, H) \in (D([0, \infty), \mathbb{R}) \times C_0([0, \infty), [0, \infty)))$, with D denoting the set of cadlag functions and C_0 the subset of strictly-increasing functions in C. Hence, from Theorem 2 and (46), using the weak convergence of $\tilde{H}^{(c)}(\cdot)$ and of $c^{-1/\alpha} \rho_c$, as $c \to 0$, we deduce: for every $t > 0$,

$$\frac{\theta_{ct}}{c^{1/\alpha}} = \frac{\rho_{H(ct)}}{c^{1/\alpha}} = \frac{\rho_c(H(ct)/c)}{c^{1/\alpha}} \xrightarrow[c \to 0]{(d)} \zeta_t, \tag{50}$$

where the convergence in distribution is viewed on $D([0, \infty), \mathbb{R})$ endowed with the Skorohod topology.

□

From the previous results, we deduce the asymptotic behaviour, for $c \to 0$, of the first exit times from a cone for isotropic stable processes of index $\alpha \in (0, 2)$ taking values in the complex plane:

Proposition 5. *For $c \to 0$, we have the weak convergence:*

$$\left(\frac{1}{c} T^\theta_{c^{1/\alpha}x}, \ x \geq 0\right) \xrightarrow[c \to 0]{(d)} \left(T^\zeta_x, \ x \geq 0\right), \tag{51}$$

where for every x, T^ζ_x is the first hitting time defined by: $T^\zeta_x \equiv \inf\{t : \zeta_t = x\}$.

Proof. Using Theorem 4, we have:

$$\frac{1}{c} T^\theta_{c^{1/\alpha}x} = \frac{1}{c} \inf\{t : \theta_t > c^{1/\alpha}x\} \stackrel{t=cs}{=} \frac{1}{c} \inf\{cs : c^{-1/\alpha}\theta_{cs} > x\}$$

$$= \inf\{s : c^{-1/\alpha}\theta_{cs} > x\} \xrightarrow{c \to 0} \inf\{s : \zeta_s > x\} ,$$

which, using again the continuity of the first passage time function mapping (see Theorem 7.1 in [43]), yields (51). □

Finally, we can obtain several variants of Proposition 5 for the random times $T^\theta_{-bc,ac}$, $0 < a, b \leq \infty$ fixed, for $c \to 0$:

Corollary 2. *The following asymptotic results hold:*

$$\frac{1}{c} T^\theta_{ac^{1/\alpha}} \xrightarrow[c \to 0]{(law)} T^\zeta_a , \tag{52}$$

$$\frac{1}{c} T^{|\theta|}_{ac^{1/\alpha}} \xrightarrow[c \to 0]{(law)} T^{|\zeta|}_a , \tag{53}$$

$$\frac{1}{c} T^\theta_{-bc^{1/\alpha}, ac^{1/\alpha}} \xrightarrow[c \to 0]{(law)} T^\zeta_{-b,a} , \tag{54}$$

where, for every $x, y > 0$, $T^{|\zeta|}_x \equiv \inf\{t : |\zeta_t| = x\}$ and $T^\zeta_{-y,x} \equiv \inf\{t : \zeta_t \notin (-y, x)\}$.

Remark 4 (*Windings of planar stable processes in $(t, 1]$ for $t \to 0$*).

(i) We consider now our stable process Z starting from 0 and we want to investigate its windings in $(t, 1]$ for $t \to 0$. We know that it doesn't visit again the origin but it winds a.s. infinitely often around it, hence, its winding process θ in $(t, 1]$ is well-defined. With obvious notation, concerning now the clock $H_{(t,1]} = \int_t^1 du\, |Z_u|^{-\alpha}$, the change of variables $u = tv$ and the stability property, i.e.: $Z_{tv} \stackrel{(law)}{=} t^{1/\alpha} Z_v$, yield:

$$H_{(t,1]} = \int_1^{1/t} \frac{t\, dv}{|Z_{tv}|^\alpha} \stackrel{(law)}{=} \int_1^{1/t} \frac{dv}{|Z_v|^\alpha} = H_{(1,1/t]} .$$

Hence, as before, using Whitt's result [43] on the continuity of the composition function on $D([\,0,\infty\,),\mathbb{R}) \times D([\,0,\infty\,),[\,0,\infty\,))$ at each $(\rho, H) \in (D([\,0,\infty\,),\mathbb{R}) \times C_0([\,0,\infty\,),[\,0,\infty\,)))$, we have (with obvious notation):

$$\theta_{(t,1]} = \rho_{H_{(t,1]}} \overset{(law)}{=} \rho_{H_{(1,1/t]}} = \theta_{(1,1/t]}. \tag{55}$$

The only difference with respect to the "normal" stable case is that the winding process is considered from 1 and not from 0, but this doesn't provoke any problem.

Hence, Bertoin and Werner's Theorem 1 is still valid for Z in $(t, 1], t \to 0$: for $\alpha \in (0, 2)$:

$$\frac{1}{\sqrt{\log(1/t)}} \theta_{(t,1]} \overset{(law)}{=} \frac{1}{\sqrt{\log(1/t)}} \theta_{(1,1/t]} \overset{(d)}{\underset{t \to 0}{\Longrightarrow}} \sqrt{r(\alpha)} N, \tag{56}$$

with $r(\alpha)$ defined in (31) and $N \sim \mathcal{N}(0, 1)$.

(ii) We note that this study is also valid for a planar Brownian motion starting from 0 in $(t, 1]$ for $t \to 0$ and for planar stable processes or planar Brownian motion starting both from a point different from 0 (in order to have an well-defined winding number) in [0, 1]. In particular, for planar Brownian motion $\mathcal{Z}$ with associated winding number ϑ, we obtain that Spitzer's law is still valid for $t \to 0$ (see e.g. [35, 27]):

$$\frac{2}{\log(1/t)} \vartheta_{(t,1]} \overset{(law)}{\underset{t \to 0}{\Longrightarrow}} C_1, \tag{57}$$

where C_1 is a standard Cauchy variable. We also remark that this result could also be obtained from a time inversion argument, that is: with $\mathcal{Z}'$ denoting another planar Brownian motion starting from 0, with winding number ϑ', by time inversion we have: $\mathcal{Z}_u = u\mathcal{Z}'_{1/u}$. Changing now the variables $u = 1/v$, we obtain:

$$\vartheta_{(t,1]} \equiv \operatorname{Im}\left(\int_t^1 \frac{d\mathcal{Z}_u}{\mathcal{Z}_u}\right) = \operatorname{Im}\left(\int_t^1 \frac{d(u\mathcal{Z}'_{1/u})}{u\mathcal{Z}'_{1/u}}\right) = \operatorname{Im}\left(\int_t^1 \frac{d\mathcal{Z}'_{1/u}}{\mathcal{Z}'_{1/u}}\right)$$

$$= \operatorname{Im}\left(\int_1^{1/t} \frac{d\mathcal{Z}'_v}{\mathcal{Z}'_v}\right) \equiv \vartheta'_{(1,1/t]},$$

and we continue as before.

Note that this time inversion argument is NOT valid for planar stable processes.

5 The Law of the Iterated Logarithm (LIL)

In this section, we shall use some notation introduced in [15, 16]. Recall (19); then, for all $x > 0$, because ρ is symmetric, we define:

$$L(x) = 2\hat{\mu}(x) = 2\mu(x, +\infty), \quad U(x) = 2\int_0^x yL(y)dy.$$

We remark that U plays essentially the role of the truncated variance in the random walk case (see e.g. [14]).

Hence, from (44), for $t \to 0$, we have ($\tilde{K}(\alpha)$ is a constant depending on α):

$$U(x) \stackrel{x \sim 0}{\sim} \tilde{K}(\alpha) x^{2-\alpha}. \tag{58}$$

Then, we obtain the following Law of the Iterated Logarithm (LIL) for Lévy processes for small times:

Theorem 5. (LIL for Lévy processes for small times)
For any non-decreasing function $f > 0$,

$$\limsup_{t \to 0} \frac{\rho_t}{f(t)} = \begin{cases} 0; \\ \infty \end{cases} \text{ a.s.} \Leftrightarrow \int_1^\infty (f(t))^{-\alpha} dt \begin{cases} < \infty; \\ = \infty. \end{cases} \tag{59}$$

We can reformulate Theorem 5 by using the skew-product representation (13) stating: $\rho_t = \theta_{A(t)}$, in order to deduce a LIL for the winding process $\theta_{A(\cdot)}$ for small times.

Corollary 3. *For any non-decreasing function $f > 0$,*

$$\limsup_{t \to 0} \frac{\theta_{A(t)}}{f(t)} = \begin{cases} 0; \\ \infty \end{cases} \text{ a.s.} \Leftrightarrow \int_1^\infty (f(t))^{-\alpha} dt \begin{cases} < \infty; \\ = \infty. \end{cases} \tag{60}$$

Proof (Theorem 5).
First, we define:

$$h(y) = y^{-2} U(y), \quad y > 0. \tag{61}$$

Then, we consider $t_n = 2^{-n}$ and we note that (Cauchy's test):

$$I(f) \equiv \int_1^\infty dt\, h(f(t)) < \infty \iff \sum_{n=1}^\infty t_n h(f(t_n)) < \infty.$$

Using now Lemma 2 from Doney and Maller [15], because ρ is symmetric, there exists a positive constant c_1 such that for every $x > 0, t > 0$,

$$P(\rho_t \geq x) \leq P\left(\sup_{0 \leq u \leq t} \rho_u \geq x\right) \leq c_1 t \, h(x) = c_1 t \, \frac{U(x)}{x^2}. \tag{62}$$

Thus:

$$\sum_{n=1}^{\infty} P(\rho_{t_{n-1}} \geq f(t_n)) \leq c_1 \sum_{n=1}^{\infty} t_n \frac{U(f(t_n))}{(f(t_n))^2}.$$

From (58), we have that,

$$\frac{U(f(t_n))}{(f(t_n))^2} \underset{t_n \sim 0}{\sim} (f(t_n))^{-\alpha}.$$

Hence, when $I(f) < \infty$, from Borel–Cantelli Lemma we have that with probability 1, $\rho_{t_{n-1}} \leq f(t_n)$ for all n's, except for a finite number of them. Now, from a monotonicity argument for f, if $t \in [t_n, t_{n-1}]$, we have that: $\rho_{t_{n-1}} \leq f(t_n) \leq f(t)$ for every t sufficiently small. It follows now that $\lim_{t \to 0}(\rho_t/f(t)) \leq 1$ a.s. Finally, we remark that as $I(f) < \infty$, we also have that $I(\varepsilon f) < \infty$, for arbitrarily small $\varepsilon > 0$ and follows that $\rho_t/f(t) \to 0$ a.s.

The proof of the second statement follows from the same kind of arguments. Indeed, using Lemma 2 from Doney and Maller [15] and the fact that ρ is symmetric, there exists a positive constant c_2 such that for every $x > 0, t > 0$,

$$P\left(\sup_{0 \leq u \leq t} \rho_u \leq x\right) \leq \frac{c_2}{t \, h(x)}. \tag{63}$$

Hence:

$$\sum_{n=1}^{\infty} P(\rho_{t_{n-1}} \leq f(t_{n-1})) \leq \sum_{n=1}^{\infty} P\left(\sup_{0 \leq u \leq t_{n-1}} \rho_u \leq f(t_{n-1})\right) \leq \sum_{n=1}^{\infty} \frac{c_2}{t_{n-1} \, h(f(t_{n-1}))}.$$

Thus, for $I(f) = \infty$ (or equivalently $\sum h(f(t_{n-1})) = \infty$), Borel–Cantelli Lemma yields that for every n, a.s. $\rho_{t_{n-1}} > f(t_{n-1})$ infinitely often, which finishes the proof. □

Remark 5. For other kinds of LIL for Lévy processes for small times e.g. of the Chung type, see [1] and the references therein.

Theorem 6 (LIL for the angular part of planar stable processes for small times).
For any non-decreasing function $f > 0$,

$$\limsup_{t\to 0} \frac{\theta_t}{f(t)} = \begin{cases} 0; \\ \infty \end{cases} \text{ a.s.} \Leftrightarrow \int_1^\infty (f(t))^{-\alpha} dt \begin{cases} < \infty; \\ = \infty. \end{cases} \qquad (64)$$

Proof. We use the skew-product representation (13) together with (46), which essentially writes: $t^{-1}H(t) \xrightarrow[t\to 0]{a.s.} 1$. Thus, for every $\varepsilon, \delta > 0$, there exists $t_0 > 0$ such that:

$$P\left(\frac{H(t)}{t} \le 1+\varepsilon\right) \ge 1-\delta, \text{ for } t \le t_0. \qquad (65)$$

We define now the setting:

$$\mathcal{K} \equiv \mathcal{K}(\omega) \equiv \left\{\omega : \frac{H(t)}{t} \le 1+\varepsilon\right\}, \text{ thus : } \overline{\mathcal{K}} \equiv \overline{\mathcal{K}}(\omega) \equiv \left\{\omega : \frac{H(t)}{t} \ge 1+\varepsilon\right\},$$

hence, there exists $t_0 > 0$ such that: for every $t \le t_0$,

$$P(\mathcal{K}) \ge 1-\delta \text{ and } P(\overline{\mathcal{K}}) \le \delta.$$

Hence, choosing $\delta > 0$ small enough, it suffices to restrict our study in the set $\mathcal{K}$ and it follows that:

$$P\left(\sup_{0\le u\le t} \theta_u > x\right) = P\left(\sup_{0\le u\le t} \rho_{H(u)} > x\right) = P\left(\left\{\sup_{0\le u\le t} \rho_{H(u)} > x\right\} \cap \mathcal{K}\right)$$

$$\le P\left(\sup_{0\le u\le t} \rho_{u(1+\varepsilon)} > x\right)$$

Changing now the variables $\tilde{u} = u(1+\varepsilon)$, and invoking (62), there exists another positive constant c_3 such that, for every $x > 0$ and $t > 0$:

$$P\left(\sup_{0\le u\le t} \theta_u > x\right) \le P\left(\sup_{0\le \tilde{u}\le t(1+\varepsilon)} \rho_{\tilde{u}} > x\right) \le c_3 \, t(1+\varepsilon) \frac{U(x)}{x^2}. \qquad (66)$$

Mimicking now the proof of Theorem 5, we obtain the first statement.
For the second statement, we use the settings:

$$\mathcal{K}' \equiv \mathcal{K}'(\omega) \equiv \left\{\omega : \frac{H(t)}{t} \ge 1-\varepsilon\right\},$$

thus:

$$\overline{\mathcal{K}'} \equiv \overline{\mathcal{K}'}(\omega) \equiv \left\{\omega : \frac{H(t)}{t} \le 1-\varepsilon\right\}.$$

Hence, for every $\varepsilon, \delta > 0$, there exists $t_0 > 0$ such that: for every $t \leq t_0$,

$$P(\mathcal{K}') \geq 1 - \delta \text{ and } P(\overline{\mathcal{K}'}) \leq \delta.$$

As before, we choose $\delta > 0$ small enough and we restrict our study in the set $\mathcal{K}'$. The proof finishes by repeating the arguments of the proof of Theorem 5. □

6 The Planar Brownian Motion Case

Before starting, we remark that the notations used in this section are independent from the ones used in the text up to now.

In this section, we state and give a new proof of the analogue of Theorem 4 for the planar Brownian motion case, which is equivalent to a result obtained in [42]. For this purpose, and in order to avoid complexity, we will use the same notation as in the "stable" case. Hence, for a planar BM $\mathcal{Z}$ starting from a point different z_0 from 0 (without loss of generality, let $z_0 = 1$) and with $\vartheta = (\vartheta_t, t \geq 0)$ denoting now the (well defined—see e.g. [21]) continuous winding process, we have the skew product representation (see e.g. [35]):

$$\log|\mathcal{Z}_t| + i\vartheta_t \equiv \int_0^t \frac{d\mathcal{Z}_s}{\mathcal{Z}_s} = (\beta_u + i\gamma_u)\Big|_{u=\mathcal{H}_t=\int_0^t \frac{ds}{|\mathcal{Z}_s|^2}}, \quad (67)$$

where $(\beta_u + i\gamma_u, u \geq 0)$ is another planar Brownian motion starting from $\log 1 + i0 = 0$. The Bessel clock $\mathcal{H}$ plays a key role in many aspects of the study of the winding number process $(\vartheta_t, t \geq 0)$ (see e.g. [45]). We shall also make use of the inverse of $\mathcal{H}$, which is given by:

$$\mathcal{H}_u^{-1} = \inf\{t \geq 0 : \mathcal{H}(t) > u\} = \int_0^u ds \, \exp(2\beta_s) = \mathcal{A}_u. \quad (68)$$

Rewriting (67) as:

$$\log|\mathcal{Z}_t| = \beta_{\mathcal{H}_t}; \quad \vartheta_t = \gamma_{\mathcal{H}_t}, \quad (69)$$

we easily obtain that the two σ-fields $\sigma\{|\mathcal{Z}_t|, t \geq 0\}$ and $\sigma\{\beta_u, u \geq 0\}$ are identical, whereas $(\gamma_u, u \geq 0)$ is independent from $(|\mathcal{Z}_t|, t \geq 0)$, a fact that is in contrast to what happens in the "stable" case.

Theorem 7. *The family of processes*

$$(c^{-1/2}\vartheta_{ct}, t \geq 0)$$

converges in distribution, as $c \to 0$, to a one-dimensional Brownian motion $(\gamma_t, t \geq 0)$.

Proof. We split the proof in two parts:

(i) First, repeating the arguments in the proof of Theorem 3 with $\alpha = 2$, we obtain:

$$\left(\frac{\mathcal{H}(ux)}{u}, x \geq 0\right) \xrightarrow[u \to 0]{a.s.} (x, x \geq 0). \tag{70}$$

which also implies the weak convergence:

$$\left(\frac{\mathcal{H}(ux)}{u}, x \geq 0\right) \xrightarrow[u \to 0]{(d)} (x, x \geq 0). \tag{71}$$

(ii) Using the skew product representation (69) and the scaling property of BM together with (70), we have that for every $s > 0$:

$$t^{-1/2}\vartheta_{st} = t^{-1/2}\gamma_{\mathcal{H}(st)} \stackrel{(law)}{=} \sqrt{\frac{\mathcal{H}(st)}{t}}\gamma_1 \xrightarrow[t \to 0]{a.s.} \sqrt{s}\gamma_1 \stackrel{(law)}{=} \gamma_s, \tag{72}$$

which finishes the proof.

We remark that for part (ii) of the proof, we could also invoke Whitt's Theorem 3.1 concerning the composition function [43], however, the independence in the planar Brownian motion case simplifies the proof. □

From Theorem 7 now, with $T_c^{|\vartheta|} \equiv \inf\{t : |\vartheta_t| = c\}$ and $T_c^{|\gamma|} \equiv \inf\{t : |\gamma_t| = c\}$, $(c > 0)$, we deduce for the exit time from a cone of planar BM (this result has already been obtained in [42], where one can also find several variants):

Corollary 4. *The following convergence in law holds:*

$$\left(\frac{1}{c^2} T_{cx}^{|\vartheta|}, x \geq 0\right) \xrightarrow[c \to 0]{(law)} \left(T_x^{|\gamma|}, x \geq 0\right). \tag{73}$$

Remark 6. We highlight the different behaviour of the clock $\mathcal{H}$ for $t \to 0$ and for $t \to \infty$ (for the second see e.g. [32], followed by [33, 28, 29], a result which is equivalent to Spitzer's Theorem [39] stated in (1)), that is:

$$\frac{\mathcal{H}(t)}{t} \xrightarrow[t \to 0]{a.s.} 1, \tag{74}$$

$$\frac{4\mathcal{H}(t)}{(\log t)^2} \xrightarrow[t \to \infty]{(law)} T_1 \equiv \inf\{t : \beta_t = 1\}, \tag{75}$$

where the latter follows essentially from the classical Laplace argument:

$$\| \cdot \|_p \xrightarrow{p \to \infty} \| \cdot \|_\infty .$$

We also remark that, from Remark 3, the behaviour of the clock for $t \to 0$ is a.s. the same for Brownian motion and for stable processes, whereas it is different for $t \to \infty$. In particular, for $t \to \infty$, compare (49) to (75).

Acknowledgements The author S. Vakeroudis is very grateful to Prof. M. Yor for the financial support during his stay at the University of Manchester as a Post Doc fellow invited by Prof. R.A. Doney.

References

1. F. Aurzada, L. Döring, M. Savov, Small time Chung type LIL for Lévy processes. Bernoulli **19**, 115–136 (2013)
2. R. Bañuelos, K. Bogdan, Symmetric stable processes in cones. Potential Anal. **21**, 263–288 (2004)
3. J. Bertoin, *Lévy Processes* (Cambridge University Press, Cambridge, 1996)
4. J. Bertoin, R.A. Doney, Spitzer's condition for random walks and Lévy processes. Ann. Inst. Henri Poincaré **33**, 167–178 (1997)
5. J. Bertoin, W. Werner, Asymptotic windings of planar Brownian motion revisited via the Ornstein-Uhlenbeck process, in *Séminaire de Probabilités XXVIII*, ed. by J. Azéma, M. Yor, P.A. Meyer. Lecture Notes in Mathematics, vol. 1583 (Springer, Berlin, 1994), pp. 138–152
6. J. Bertoin, W. Werner, Compertement asymptotique du nombre de tours effectués par la trajectoire brownienne plane. *Séminaire de Probabilités XXVIII*, ed. by J. Azéma, M. Yor, P.A. Meyer. Lecture Notes in Mathematics, vol. 1583 (Springer, Berlin, 1994), pp. 164–171
7. J. Bertoin, W. Werner, Stable windings. Ann. Probab. **24**, 1269–1279 (1996)
8. L. Breiman, A delicate law of the iterated logarithm for non-decreasing stable processes. Ann. Math. Stat. **39**, 1818–1824 (1968); correction **41**, 1126
9. D. Burkholder, Exit times of Brownian Motion, harmonic majorization and hardy spaces. Adv. Math. **26**, 182–205 (1977)
10. M.E. Caballero, J.C. Pardo, J.L. Pérez, Explicit identities for Lvy processes associated to symmetric stable processes. Bernoulli **17**(1), 34–59 (2011)
11. O. Chybiryakov, The Lamperti correspondence extended to Lévy processes and semi-stable Markov processes in locally compact groups. Stoch. Process. Appl. **116**, 857–872 (2006)
12. R.D. De Blassie, The first exit time of a two-dimensional symmetric stable process from a wedge. Ann. Probab. **18**, 1034–1070 (1990)
13. R.A. Doney, Small time behaviour of Lévy processes. Electron. J. Probab. **9**, 209–229 (2004)
14. R.A. Doney, R.A. Maller, Random walks crossing curved boundaries: a functional limit theorem, stability and asymptotic distributions for exit positions. Adv. Appl. Probab. **32**, 1117–1149 (2000)
15. R.A. Doney, R.A. Maller, Stability of the overshoot for Lévy processes. Ann. Probab. **30**, 188–212 (2002)
16. R.A. Doney, R.A. Maller, Stability and attraction to normality for Le'vy processes at zero and infinity. J. Theor. Probab. **15**, 751–792 (2002)
17. R. Durrett, A new proof of Spitzer's result on the winding of 2-dimensional Brownian motion. Ann. Probab. **10**, 244–246 (1982)

18. B.E. Fristedt, The behavior of increasing stable processes for both small and large times. J. Math. Mech. **13**, 849–856 (1964)
19. B.E. Fristedt, Sample function behavior of increasing processes with stationary, independent increments. Pacific J. Math. **21**(1), 21–33 (1967)
20. S.E. Graversen, J. Vuolle-Apiala, α-Self-similar Markov processes. Probab. Theor. Relat. Field. **71**, 149–158 (1986)
21. K. Itô, H.P. McKean, *Diffusion Processes and their Sample Paths* (Springer, Berlin, 1965)
22. J. Jacod, A.N. Shiryaev, *Limit Theorems for Stochastic Processes*, 2nd edn. (Springer, Berlin, 2003)
23. A. Khintchine, Sur la croissance locale des processus stochastiques homogènes à accroissements indépendants. (Russian article and French resume) Akad. Nauk. SSSR Izv. Ser. Math. **3**(5–6), 487–508 (1939)
24. S.W. Kiu, Semi-stable Markov processes in $\mathbb{R}^n$. Stoch. Process. Appl. **10**(2), 183–191 (1980)
25. A.E. Kyprianou, *Introductory Lectures on Fluctuations of Lévy Processes with Applications* (Springer, Berlin, 2006)
26. J. Lamperti, Semi-stable Markov processes I. Z. Wahr. Verw. Gebiete, **22**, 205–225 (1972)
27. J.F. Le Gall, Some properties of planar Brownian motion, in *Cours de l'école d'été de St-Flour XX*, ed. by A. Dold, B. Eckmann, E. Takens. Lecture Notes in Mathematics, vol. 1527 (Springer, Berlin, 1992), pp. 111–235
28. J.F. Le Gall, M. Yor, Etude asymptotique de certains mouvements browniens complexes avec drift. Probab. Theor. Relat. Field. **71**(2), 183–229 (1986)
29. J.F. Le Gall, M. Yor, Etude asymptotique des enlacements du mouvement brownien autour des droites de l'espace. Probab. Theor. Relat. Field. **74**(4), 617–635 (1987)
30. M. Liao, L. Wang, Isotropic self-similar Markov processes. Stoch. Process. Appl. **121**(9), 2064–2071 (2011)
31. P. Messulam, M. Yor, On D. Williams' "pinching method" and some applications. J. Lond. Math. Soc. **26**, 348–364 (1982)
32. J.W. Pitman, M. Yor, The asymptotic joint distribution of windings of planar Brownian motion. Bull. Am. Math. Soc. **10**, 109–111 (1984)
33. J.W. Pitman, M. Yor, Asymptotic laws of planar Brownian motion. Ann. Probab. **14**, 733–779 (1986)
34. J.W. Pitman, M. Yor, Further asymptotic laws of planar Brownian motion. Ann. Probab. **17**(3), 965–1011 (1989)
35. D. Revuz, M. Yor, *Continuous Martingales and Brownian Motion*, 3rd edn. (Springer, Berlin, 1999)
36. Z. Shi, Liminf behaviours of the windings and Lévy's stochastic areas of planar Brownian motion. *Séminaire de Probabilités XXVIII*, ed. by J. Azéma, M. Yor, P.A. Meyer. Lecture Notes in Mathematics, vol. 1583 (Springer, Berlin, 1994), pp. 122–137
37. Z. Shi, Windings of Brownian motion and random walks in the plane. Ann. Probab. **26**(1), 112–131 (1998)
38. A.V. Skorohod, *Random Processes with Independent Increments* (Kluwer, Dordrecht, 1991)
39. F. Spitzer, Some theorems concerning two-dimensional Brownian motion. Trans. Am. Math. Soc. **87**, 187–197 (1958)
40. S. Vakeroudis, Nombres de tours de certains processus stochastiques plans et applications à la rotation d'un polymère. (Windings of some planar Stochastic Processes and applications to the rotation of a polymer). Ph.D. Dissertation, Université Pierre et Marie Curie (Paris VI), April 2011.
41. S. Vakeroudis, On hitting times of the winding processes of planar Brownian motion and of Ornstein-Uhlenbeck processes, via Bougerol's identity. SIAM Theor. Probab. Appl. **56**(3), 485–507 (2012) [originally published in Teor. Veroyatnost. i Primenen. **56**(3), 566–591 (2011)]
42. S. Vakeroudis, M. Yor, Integrability properties and limit theorems for the exit time from a cone of planar Brownian motion. arXiv preprint arXiv:1201.2716 (2012), to appear in Bernoulli
43. W. Whitt, Some useful functions for functional limit theorems. Math. Oper. Res. **5**, 67–85 (1980)

44. D. Williams, A simple geometric proof of Spitzer's winding number formula for 2-dimensional Brownian motion. University College, Swansea. Unpublished, 1974
45. M. Yor, Loi de l'indice du lacet Brownien et Distribution de Hartman-Watson. Z. Wahrsch. verw. Gebiete **53**, 71–95 (1980)
46. M. Yor, Generalized meanders as limits of weighted Bessel processes, and an elementary proof of Spitzer's asymptotic result on Brownian windings. Studia Scient. Math. Hung. **33**, 339–343 (1997)

An Elementary Proof that the First Hitting Time of an Open Set by a Jump Process is a Stopping Time

Alexander Sokol

Abstract We give a short and elementary proof that the first hitting time of an open set by the jump process of a càdlàg adapted process is a stopping time.

1 Introduction

For a stochastic process X and a subset B of the real numbers, the random variable $T = \inf\{t \geq 0 | X_t \in B\}$ is called the first hitting time of B by X. A classical result in the general theory of processes is the début theorem, which has as a corollary that under the usual conditions, the first hitting time of a Borel set for a progressively measurable process is a stopping time, see [3], Sect. III.44 for a proof of this theorem, or [1] and [2] for a recent simpler proof. For many purposes, however, the general début theorem is not needed, and weaker results may suffice, where elementary methods may be used to obtain the results. For example, it is elementary to show that the first hitting time of an open set by a càdlàg adapted process is a stopping time, see [4], Theorem I.3. Using somewhat more advanced, yet relatively elementary methods, Lemma II.75.1 of [5] shows that the first hitting time of a compact set by a càdlàg adapted process is a stopping time.

These elementary proofs show stopping time properties for the first hitting times of a càdlàg adapted process X. However, the jump process ΔX in general has paths with neither left limits nor right limits, and so the previous elementary results do not apply. In this note, we give a short and elementary proof that the first hitting time of

A. Sokol (✉)
Institute of Mathematical Sciences, University of Copenhagen, Universitetsparken 5, 2100 Copenhagen, Denmark
e-mail: alexander@math.ku.dk

an open set by ΔX is a stopping time when the filtration is right-continuous and X is càdlàg adapted. This result may be used to give an elementary proof that the jumps of a càdlàg adapted process are covered by the graphs of a countable sequence of stopping times.

2 A Stopping Time Result

Assume given a filtered probability space $(\Omega, \mathcal{F}, (\mathcal{F}_t), P)$ such that the filtration $(\mathcal{F}_t)_{t \geq 0}$ is right-continuous in the sense that $\mathcal{F}_t = \cap_{s > t} \mathcal{F}_s$ for all $t \geq 0$. We use the convention that $X_{0-} = X_0$, so that there is no jump at the timepoint zero.

Theorem 1. *Let X be a càdlàg adapted process, and let U be an open set in $\mathbb{R}$. Define $T = \inf\{t \geq 0 | \Delta X_t \in U\}$. Then T is a stopping time.*

As X has càdlàg, ΔX is zero everywhere except for on a countable set, and so T is identically zero if U contains zero. In this case, T is trivially a stopping time. Thus, it suffices to prove the result in the case where U does not contain zero. Therefore, assume that U is an open set not containing zero. As the filtration is right-continuous, an elementary argument yields that to show the stopping time property of T, it suffices to show $(T < t) \in \mathcal{F}_t$ for $t > 0$, see Theorem I.1 of [4].

To this end, fix $t > 0$ and note that

$$(T < t) = (\exists\, s \in (0, \infty) : s < t \text{ and } X_s - X_{s-} \in U). \tag{1}$$

Let $F_m = \{x \in \mathbb{R} | \forall\, y \in U^c : |x - y| \geq 1/m\}$, F_m is an intersection of closed sets and therefore itself closed. Clearly, $(F_m)_{m \geq 1}$ is increasing, and since U is open, $U = \cup_{m=1}^\infty F_m$. Also, $F_m \subseteq F_{m+1}^\circ$, where F_{m+1}° denotes the interior of F_{m+1}. Let Θ_k be the subset of $\mathbb{Q}^2$ defined by $\Theta_k = \{(p, q) \in \mathbb{Q}^2 | 0 < p < q < t, |p - q| \leq \frac{1}{k}\}$. We will prove the result by showing that

$$\begin{aligned}(\exists\, s \in (0, \infty) &: s < t \text{ and } X_s - X_{s-} \in U) \\ &= \cup_{m=1}^\infty \cup_{n=1}^\infty \cap_{k=n}^\infty \cup_{(p,q) \in \Theta_k} (X_q - X_p \in F_m). \end{aligned} \tag{2}$$

To obtain this, first consider the inclusion towards the right. Assume that there is $0 < s < t$ such that $X_s - X_{s-} \in U$. Take m such that $X_s - X_{s-} \in F_m$. As $F_m \subseteq F_{m+1}^\circ$, we then have $X_s - X_{s-} \in F_{m+1}^\circ$ as well. As F_{m+1}° is open and as X is càdlàg, it holds that there is $\varepsilon > 0$ such that whenever $p, q \geq 0$ with $p \in (s - \varepsilon, s)$ and $q \in (s, s + \varepsilon)$, $X_q - X_p \in F_{m+1}^\circ$. Take $n \in \mathbb{N}$ such that $1/2n < \varepsilon$. We now claim that for $k \geq n$, there is $(p, q) \in \Theta_k$ such that $X_q - X_p \in F_{m+1}$. To prove this, let $k \geq n$ be given. By the density properties of $\mathbb{Q}_+$ in $\mathbb{R}_+$, there are elements

$p, q \in \mathbb{Q}$ with $p, q \in (0, t)$ such that $p \in (s - 1/2k, s)$ and $q \in (s, s + 1/2k)$. In particular, then $0 < p < q < t$ and $|p - q| \le |p - s| + |s - q| \le 1/k$, so $(p, q) \in \Theta_k$. As $1/2k \le 1/2n < \varepsilon$, we have $p \in (s - \varepsilon, s)$ and $q \in (s, s + \varepsilon)$, and so $X_q - X_p \in F_{m+1}^\circ \subseteq F_{m+1}$. This proves the inclusion towards the right.

Now consider the inclusion towards the left. Assume that there is $m \ge 1$ and $n \ge 1$ such that for all $k \ge n$, there exists $(p, q) \in \Theta_k$ with $X_q - X_p \in F_m$. We may use this to obtain sequences $(p_k)_{k \ge n}$ and $(q_k)_{k \ge n}$ with the properties that $p_k, q_k \in \mathbb{Q}, 0 < p_k < q_k < t, |p_k - q_k| \le \frac{1}{k}$ and $X_{q_k} - X_{p_k} \in F_m$. Putting $p_k = p_n$ and $q_k = q_n$ for $k < n$, we then find that the sequences $(p_k)_{k \ge 1}$ and $(q_k)_{k \ge 1}$ satisfy $p_k, q_k \in \mathbb{Q}, 0 < p_k < q_k < t, \lim_k |p_k - q_k| = 0$ and $X_{q_k} - X_{p_k} \in F_m$. As all sequences of real numbers contain a monotone subsequence, we may by taking two consecutive subsequences and renaming our sequences obtain the existence of two monotone sequences (p_k) and (q_k) in $\mathbb{Q}$ with $0 < p_k < q_k < t, \lim_k |p_k - q_k| = 0$ and $X_{q_k} - X_{p_k} \in F_m$. As bounded monotone sequences are convergent, both (p_k) are (q_k) are then convergent, and as $\lim_k |p_k - q_k| = 0$, the limit $s \ge 0$ is the same for both sequences.

We wish to argue that $s > 0$, that $X_{s-} = \lim_k X_{p_k}$ and that $X_s = \lim_k X_{q_k}$. To this end, recall that U does not contain zero, and so as $F_m \subseteq U$, F_m does not contain zero either. Also note that as both (p_k) and (q_k) are monotone, the limits $\lim_k X_{p_k}$ and $\lim_k X_{q_k}$ exist and are either equal to X_s or X_{s-}. As $X_{q_k} - X_{p_k} \in F_m$ and F_m is closed and does not contain zero, $\lim_k X_{q_k} - \lim_k X_{p_k} = \lim_k X_{q_k} - X_{p_k} \ne 0$. From this, we can immediately conclude that $s > 0$, as if $s = 0$, we would obtain that both $\lim_k X_{q_k}$ and $\lim_k X_{p_k}$ were equal to X_s, yielding $\lim_k X_{q_k} - \lim_k X_{p_k} = 0$, a contradiction. Also, we cannot have that both limits are X_s or that both limits are X_{s-}, and so only two cases are possible, namely that $X_s = \lim_k X_{q_k}$ and $X_{s-} = \lim_k X_{p_k}$ or that $X_s = \lim_k X_{p_k}$ and $X_{s-} = \lim_k X_{q_k}$. We wish to argue that the former holds. If $X_s = X_{s-}$, this is trivially the case. Assume that $X_s \ne X_{s-}$ and that $X_s = \lim_k X_{p_k}$ and $X_{s-} = \lim_k X_{q_k}$. If $q_k \ge s$ from a point onwards or $p_k < s$ from a point onwards, we obtain $X_s = X_{s-}$, a contradiction. Therefore, $q_k < s$ infinitely often and $p_k \ge s$ infinitely often. By monotonicity, $q_k < s$ and $p_k \ge s$ from a point onwards, a contradiction with $p_k < q_k$. We conclude $X_s = \lim_k X_{q_k}$ and $X_{s-} = \lim_k X_{p_k}$, as desired.

In particular, $X_s - X_{s-} = \lim_k X_{q_k} - X_{p_k}$. As $X_{q_k} - X_{p_k} \in F_m$ and F_m is closed, we obtain $X_s - X_{s-} \in F_m \subseteq U$. Next, note that if $s = t$, we have $p_k, q_k < s$ for all k, yielding that both sequences must be increasing and $X_s = \lim X_{q_k} = X_{s-}$, a contradiction with the fact that $X_s - X_{s-} \ne 0$ as $X_s - X_{s-} \in U$. Thus, $0 < s < t$. This proves the existence of $s \in (0, \infty)$ with $s < t$ such that $X_s - X_{s-} \in U$, and so proves the inclusion towards the right.

We have now shown (2). Now, as X_s is $\mathscr{F}_t$ measurable for all $0 \le s \le t$, it holds that the set $\bigcup_{m=1}^\infty \bigcup_{n=1}^\infty \bigcap_{k=n}^\infty \bigcup_{(p,q) \in \Theta_k} (X_q - X_p \in F_m)$ is $\mathscr{F}_t$ measurable as well. We conclude that $(T < t) \in \mathscr{F}_t$ and so T is a stopping time.

References

1. R.F. Bass, The measurability of hitting times. Electron. Comm. Probab. **15**, 99–105 (2010)
2. R.F. Bass, Correction to "The measurability of hitting times". Electron. Comm. Probab. **16**, 189–191 (2011)
3. C. Dellacherie, P.-A. Meyer, *Probabilities and Potential*. North-Holland Mathematics Studies, vol. 29 (North-Holland, Amsterdam, 1978)
4. P. Protter, *Stochastic Integration and Differential Equations*, 2nd edn. Version 2.1. Stochastic Modelling and Applied Probability, vol. 21 (Springer, Berlin, 2005)
5. L.C.G. Rogers, D. Williams, *Diffusions, Markov Processes and Martingales*, 2nd edn., vol. 1 (Cambridge University Press, Cambridge, 2000)

Catalytic Branching Processes via Spine Techniques and Renewal Theory

Leif Döring and Matthew I. Roberts

Abstract In this article we contribute to the moment analysis of branching processes in catalytic media. The many-to-few lemma based on the spine technique is used to derive a system of (discrete space) partial differential equations for the number of particles in a variation of constants formulation. The long-time behaviour is then deduced from renewal theorems and induction.

1 Introduction and Results

A classical subject of probability theory is the analysis of branching processes in discrete or continuous time, going back to the study of extinction of family names by Francis Galton. There have been many contributions to the area since, and we present here an application of a recent development in the probabilistic theory. We identify qualitatively different regimes for the longtime behaviour for moments of sizes of populations in a simple model of a branching Markov process in a catalytic environment.

To give some background for the branching mechanism, we recall the discrete-time Galton–Watson process. Given a random variable X with law μ taking values in $\mathbb{N}$, the branching mechanism is modeled as follows: for a deterministic or random initial number $Z_0 \in \mathbb{N}$ of particles, one defines for $n = 1, 2, \ldots$

L. Döring (✉)
Laboratoire de Probabilités et Modèles Aléatoires, Université Paris VI, 4, Place Jussieu, 75005 Paris, France
e-mail: leif.doering@upmc.fr

M.I. Roberts
Weierstrass Institute for Applied Analysis and Stochastics, Mohrenstrasse 39, 10117 Berlin, Germany
e-mail: mattiroberts@gmail.com

$$Z_{n+1} = \sum_{r=0}^{Z_n} X_r(n),$$

where all $X_r(n)$ are independent and distributed according to μ. Each particle in generation n is thought of as giving birth to a random number of particles according to μ, and these particles together form generation $n+1$. For the continuous-time analogue each particle carries an independent exponential clock of rate 1 and performs its breeding event once its clock rings.

It is well-known that a crucial quantity appearing in the analysis is $m = \mathbb{E}[X]$, the expected number of offspring particles. The process has positive chance of long-term survival if and only if $m > 1$. This is known as the supercritical case. The cases $m = 1$ (critical) and $m < 1$ (subcritical) also show qualitatively different behaviour in the rate at which the probability of survival decays. As this paper deals with the moment analysis of a spatial relative to this system, we mention the classical trichotomy for the moment asymptotics of Galton–Watson processes:

$$\lim_{t \to \infty} e^{-k(m-1)t} \mathbb{E}[Z_t^k] \in (0, \infty) \quad \forall k \in \mathbb{N} \quad \text{if } m < 1, \tag{1}$$

$$\lim_{t \to \infty} t^{k-1} \mathbb{E}[Z_t^k] \in (0, \infty) \quad \forall k \in \mathbb{N} \quad \text{if } m = 1, \tag{2}$$

$$\lim_{t \to \infty} e^{-k(m-1)t} \mathbb{E}[Z_t^k] \in (0, \infty) \quad \forall k \in \mathbb{N} \quad \text{if } m > 1, \tag{3}$$

so that all moments increase exponentially to infinity if $m > 1$, increase polynomially if $m = 1$, and decay exponentially fast to zero if $m < 1$.

In the present article we are interested in a simple spatial version of the Galton–Watson process for which a system of branching particles moves in space and particles branch only in the presence of a catalyst. More precisely, we start a particle ξ which moves on some countable set S according to a continuous-time Markov process with Q-matrix $\mathscr{A}$. This particle carries an exponential clock of rate 1 that only ticks if ξ is at the same site as the catalyst, which we assume sits at some fixed site $0 \in S$. If and when the clock rings, then the particle dies and is replaced in its position by a random number of offspring. This number is distributed according to some offspring distribution μ, and all newly born particles behave as independent copies of their parent: they move on S according to $\mathscr{A}$ and branch after an exponential rate 1 amount of time spent at 0.

In recent years several authors have studied such branching systems. Often the first quantities that are analyzed are moments of the form

$$M^k(t, x, y) = \mathbb{E}[N_t(y)^k \mid \xi_0 = x] \quad \text{and} \quad M^k(t, x) = \mathbb{E}[N_t^k \mid \xi_0 = x],$$

where $N_t(y)$ is the number of particles alive at site y at time t, and $N_t = \sum_{y \in S} N_t(y)$ is the total number of particles alive at time t. Under the additional assumption that $\mathscr{A} = \Delta$ is the discrete Laplacian on $\mathbb{Z}^d$, the moment analysis was first carried out in [1–3] via partial differential equations and Tauberian theorems.

More recently, the moment analysis, and moreover the study of conditional limit theorems, was pushed forward to more general spatial movement $\mathscr{A}$ assuming

(**A1**) irreducibility,
(**A2**) spatial homogeneity,
(**A3**) symmetry,
(**A4**) finite variance of jump sizes.

Techniques such as Bellman–Harris branching processes (see [5,6,16,17]), operator theory (see [20]) and renewal theory (see [12]) have been applied successfully. Some of these tools also apply in a non-symmetric framework. We present a purely stochastic approach avoiding the assumptions (**A1**)–(**A4**). In order to avoid many pathological special cases we only assume

(**A**) the motion governed by $\mathscr{A}$ is irreducible.

This assumption is not necessary, and the interested reader may easily reconstruct the additional cases from our proofs.

In order to analyze the moments M^k one can proceed in two steps. First, a set of partial differential equations for M^k is derived. This can be done for instance as in [3] via analytic arguments from partial differential equations for the generating functions $\mathbb{E}_x[e^{-zN_t(y)}]$ and $\mathbb{E}_x[e^{-zN_t}]$ combined with Faà di Bruno's formula of differentiation. The asymptotic properties of solutions to those differential equations are then analyzed in a second step where more information on the transition probabilities corresponding to $\mathscr{A}$ implies more precise results on the asymptotics for M^k. This is where the finite variance assumption is used via the local central limit theorem.

The approach presented in this article is based on the combinatorial spine representation of [11] to derive sets of partial differential equations, in variation of constants form, for the kth moments of $N_t(y)$ and N_t. A set of combinatorial factors can be given a direct probabilistic explanation, whereas the same factors appear otherwise from Faà di Bruno's formula. Those equations are then analyzed via renewal theorems. We have to emphasize that under the assumption (**A**) only, general precise asymptotic results are of course not possible so that we aim at giving a qualitative description. Compared to the fine results in the presence of a local central limit theorem (such as Lemma 3.1 of [12] for finite variance transitions on $\mathbb{Z}^4$) our qualitative description is rather poor. On the other hand, the generality of our results allows for some interesting applications. For example, one can easily deduce asymptotics for moments of the number of particles when the catalyst is not fixed at zero, but rather follows some Markov process of its own, simply by considering the difference walk.

To state our main result, we denote the transition probabilities of $\mathscr{A}$ by $p_t(x, y) = \mathbb{P}_x(\xi_t = y)$ and the Green function by

$$G_\infty(x, y) = \int_0^\infty p_t(x, y)\, dt.$$

Recall that, by irreducibility, the Green function is finite for all $x, y \in S$ if and only if $\mathcal{A}$ is transient. For the statement of the result let us further denote by

$$L_t(y) = \int_0^t \mathbb{1}_{\{\xi_s = y\}} ds$$

the time of ξ spent at site y up to time t.

Theorem 1. *Suppose that μ has finite moments of all orders; then the following regimes occur for all integers $k \geq 1$:*

(i) *If the branching mechanism is **subcritical**, then*

$$\lim_{t \to \infty} M^k(t, x) \in (0, \infty) \quad \text{if } \mathcal{A} \text{ is transient,}$$

$$\lim_{t \to \infty} M^k(t, x) = 0 \quad \text{if } \mathcal{A} \text{ is recurrent,}$$

and

$$\lim_{t \to \infty} M^k(t, x, y) = 0 \quad \text{in all cases.}$$

(ii) *If the branching mechanism is **critical**, then*

$$\lim_{t \to \infty} \frac{M^k(t, x)}{\mathbb{E}_x[L_t(0)^{k-1}]} \in (0, \infty) \quad \text{and} \quad M^1(t, x, y) = p_t(x, y).$$

(iii) *If the branching mechanism is **supercritical**, then there is a critical constant*

$$\beta = \frac{1}{G_\infty(0, 0)} + 1 \geq 1$$

such that

(a) *for $m < \beta$*

$$\lim_{t \to \infty} M^1(t, x) \in (0, \infty) \quad \text{and} \quad \lim_{t \to \infty} M^1(t, x, y) = 0;$$

further, there exist constants c and C such that

$$c \mathbb{E}_x[L_t(0)^{k-1}] \leq M^k(t, x) \leq C t^{k-1}.$$

(b) *for $m = \beta$*

$$\lim_{t \to \infty} M^k(t, x) = \infty,$$

and

$$\lim_{t\to\infty} M^k(t,x,y) = \infty \quad \text{if} \quad \int_0^\infty rp_r(0,0)\,dr = \infty$$

$$\lim_{t\to\infty} \frac{M^k(t,x,y)}{t^{k-1}} \in (0,\infty) \quad \text{if} \quad \int_0^\infty rp_r(0,0)\,dr < \infty.$$

(In both cases the growth is subexponential.)

(c) for $m > \beta$

$$\lim_{t\to\infty} e^{-kr(m)t} M^k(t,x,y) \in (0,\infty) \quad \text{and} \quad \lim_{t\to\infty} e^{-kr(m)t} M^k(t,x) \in (0,\infty)$$

where $r(m)$ equals the unique solution λ to $\int_0^\infty e^{-\lambda t} p_t(0,0)\,dt = \frac{1}{m-1}$.

We did not state all the asymptotics in cases (ii) and (iii)(a). Our methods, see Lemma 3, do allow for investigation of these cases too; in particular they show how $M^k(t,x,y)$ can be expressed recursively by $M^i(t,x,y)$ for $i < k$. However, without further knowledge of the underlying motion, it is not possible to give any useful and general information. If more information on the tail of $p_t(x,y)$ is available then the recursive equations can indeed be analyzed: for instance for kernels on $\mathbb{Z}^d$ with second moments the local central limit theorem can be applied leading to $p_t(x,y) \sim Ct^{-d/2}$, and such cases have already been addressed by other authors.

The formulation of the theorem does not include the limiting constants. Indeed, the proofs give some of those (in an explicit form involving the transition probabilities p_t) in the supercritical regime but they seem to be of little use. The use of spectral theory for symmetric Q-matrices $\mathscr{A}$ allows one to derive the exponential growth rate $r(m)$ as the maximal eigenvalue of a Schrödinger operator with one-point potential and the appearing constants via the eigenfunctions. Our renewal theorem based proof gives the representation of $r(m)$ as the inverse of the Laplace transform of $p_t(0,0)$ at $1/(m-1)$ and the eigenfunction expressed via integrals of $p_t(0,0)$. As $p_t(0,0)$ is rarely known explicitly, the integral form of the constants is not very useful (apart from the trivial case of Example 1 below). Only in case (iii) (b) for $\int_0^\infty rp_r(0,0)\,dr = \infty$ are the proofs unable to give strong asymptotics. This is caused by the use of an infinite-mean renewal theorem which only gives asymptotic bounds up to an unknown factor between 1 and 2. There is basically one example in which $p_t(0,0)$ is trivially known:

Example 1. For the trivial motion $\mathscr{A} = 0$, i.e. branching particles are fixed at the same site as the catalyst, the supercritical cases (iii) (a) and (b) do not occur as $\mathscr{A}$ is trivially recurrent so that $\beta = 1$. Furthermore, in this example $p_t(0,0) = 1$ for all $t \geq 0$ so that $r(m) = m - 1$. In fact by examining the proof of Theorem 1 one recovers (1), (2), (3) with all constants.

Remark 1. We reiterate here that our results can be generalized when the fixed branching source is replaced by a random branching source moving according to a random walk independent of the branching particles. For the proofs the branching particles only have to be replaced by branching particles relative to the branching source.

2 Proofs

The key tool in our proofs will be the many-to-few lemma proved in [11] which relies on modern spine techniques. These emerged from work of Kurtz, Lyons, Pemantle and Peres in the mid-1990s [13–15]. The idea is that to understand certain functionals of branching processes, it is enough to study carefully the behaviour of one special particle, the *spine*. In particular very general *many-to-one* lemmas emerged, allowing one to easily calculate expectations of sums over particles like

$$\mathbb{E}\left[\sum_{v \in N_t} f(v)\right],$$

where $f(v)$ is some well-behaved functional of the behaviour of the particle v up to time t, and N_t here is viewed as the *set* of particles alive at time t, rather than the number. It will always be clear from the context which meaning for N_t is intended.

It is natural to ask whether similar results exist for higher moments of sums over N_t. This is the idea behind [11], wherein it turns out that to understand the kth moment one must consider a system of k particles. The k particles introduce complications compared to the single particle required for first moments, but this is still significantly simpler than controlling the behaviour of the potentially huge random number of particles in N_t.

While we do not need to understand the full spine setup here, we shall require some explanation.

For each $k \geq 0$ let $p_k = \mathbb{P}(X = k)$ and $m_k = \mathbb{E}[X^k]$, the kth moment of the offspring distribution (in particular $m_1 = m$). We define a new measure $\mathbb{Q} = \mathbb{Q}_x^k$, under which there are k distinguished lines of descent known as spines. The construction of $\mathbb{Q}$ relies on a carefully chosen change of measure, but we do not need to understand the full construction and instead refer to [11]. In order to use the technique, we simply have to understand the dynamics of the system under $\mathbb{Q}$. Under $\mathbb{Q}_x^k$ particles behave as follows:

- We begin with one particle at position x which (as well as its position) has a mark k. We think of a particle with mark j as carrying j spines.
- Whenever a particle with mark j, $j \geq 1$, spends an (independent) exponential time with parameter m_j in the same position as the catalyst, it dies and is replaced by a random number of new particles with law A_j.
- The probability of the event $\{A_j = a\}$ is $a^j p_a m_j^{-1}$. (This is the jth size-biased distribution relative to μ.)

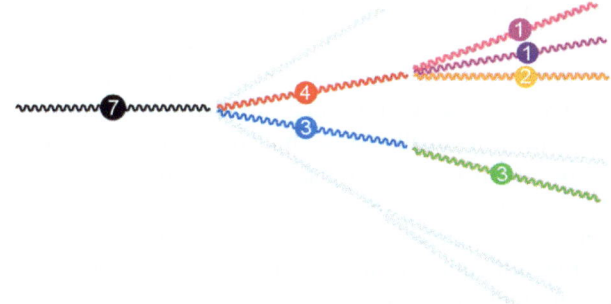

Fig. 1 An impression of the start of the process: each particle in the skeleton is a different color, and particles not in the skeleton are drawn in pale grey. The circles show the number of spines being carried by each particle in the skeleton

- Given that a particles $v_1, \ldots, v_a$ are born, the j spines each choose a particle to follow independently and uniformly at random. Thus particle v_i has mark l with probability $a^{-l}(1-a^{-1})^{j-l}$, $l = 0, \ldots, j$, $i = 1, \ldots, a$. We also note that this means that there are always k spines amongst the particles alive; equivalently the sum of the marks over all particles alive always equals k.
- Particles with mark 0 are no longer of interest (in fact they behave just as under $\mathbb{P}$, branching at rate 1 when in the same position as the catalyst and giving birth to numbers of particles with law μ, but we will not need to use this).

For a particle v, we let $X_v(t)$ be its position at time t and B_v be its mark (the number of spines it is carrying). Let σ_v be the time of its birth and τ_v the time of its death, and define $\sigma_v(t) = \sigma_v \wedge t$ and $\tau_v(t) = \tau_v \wedge t$. Let χ_t^i be the current position of the ith spine. We call the collection of particles that have carried at least one spine up to time t the *skeleton* at time t, and write skel(t). Figure 1 gives an impression of the skeleton at the start of the process.

A much more general form of the following lemma was proved in [11].

Lemma 1 (Many-to-few). *Suppose that $f : \mathbb{R} \to \mathbb{R}$ is measurable. Then, for any $k \geq 1$,*

$$\mathbb{E}\left[\sum_{v_1,\ldots,v_k \in N_t} f(X_{v_1}(t)) \cdots f(X_{v_k}(t)) \right]$$

$$= \mathbb{Q}^k \left[f(\chi_t^1) \cdots f(\chi_t^k) \prod_{v \in \text{skel}(t)} \exp\left((m_{B_v} - 1) \int_{\sigma_v(t)}^{\tau_v(t)} \mathbb{1}_0(X_v(s)) ds \right) \right].$$

Clearly if we take $f \equiv 1$, then the left hand side is simply the kth moment of the number of particles alive at time t. The lemma is useful since the right-hand side depends on at most k particles at a time, rather than the arbitrarily large random number of particles on the left-hand side.

Having introduced the spine technique, we can now proceed with the proof of Theorem 1. We first use Lemma 1 for the case $k = 1$, which is simply the many-to-one lemma, to deduce two convenient representations for the first moments: a Feynman–Kac expression and a variation of constants formula. Indeed, the exponential expression equally works for other random potentials and, hence, is well known for instance in the parabolic Anderson model literature. More interestingly, the variation of constants representation is most useful in the case of a one-point potential: it simplifies to a renewal type equation. Understanding when those are proper renewal equations replaces the spectral theoretic arguments of [1] and explains the different cases appearing in Theorem 1.

Lemma 2. *The first moments can be expressed as*

$$M^1(t, x) = \mathbb{E}_x\left[e^{(m-1)\int_0^t \mathbb{1}_0(\xi_r)\,dr}\right], \tag{4}$$

$$M^1(t, x, y) = \mathbb{E}_x\left[e^{(m-1)\int_0^t \mathbb{1}_0(\xi_r)\,dr}\mathbb{1}_y(\xi_t)\right], \tag{5}$$

where ξ_t is a single particle moving with Q-matrix $\mathscr{A}$. Furthermore, these quantities fulfil

$$M^1(t, x) = 1 + (m - 1)p_t(x, 0) * M^1(t, 0), \tag{6}$$

$$M^1(t, x, y) = p_t(x, y) + (m - 1)p_t(x, 0) * M^1(t, 0, y), \tag{7}$$

where $*$ denotes ordinary convolution in t.

For completeness we include a proof of these well-known relations. First let us briefly mention why the renewal type equations occur naturally. The Feynman–Kac representation can be proved in various ways; we derive it simply from the many-to-few lemma. The Feynman–Kac formula then leads naturally to solutions of discrete-space heat equations with one-point potential:

$$\begin{cases} \frac{\partial}{\partial t}u(t, x) = \mathscr{A}u(t, x) + (m-1)\mathbb{1}_0(x)u(t, x) \\ u(0, x) = \mathbb{1}_y(x) \end{cases}$$

Applying the variation of constants formula for solutions gives

$$u(t, x) = P_t u(0, x) + \int_0^t P_{t-s}(m-1)\mathbb{1}_0(x)u(s, x)\,ds$$

$$= p_t(x, y) + (m-1)\int_0^t p_{t-s}(x, 0)u(s, x)\,ds,$$

where P_t is the semigroup corresponding to $\mathscr{A}$, i.e. $P_t f(x) = \mathbb{E}_x[f(\xi_t)]$.

Proof (of Lemma 2). To prove (4) and (5) we apply the easiest case of Lemma 1: we choose $k = 1$ and $f \equiv 1$ (resp. $f(z) = \mathbb{1}_y(z)$ for (5)). Since there is exactly one spine at all times, the skeleton reduces to a single line of descent. Hence $m_{B_v} - 1 =$

$m-1$ and the integrals in the product combine to become a single integral along the path of the spine up to time t. Thus

$$M^1(t,x) = \mathbb{Q}_x\left[e^{(m-1)\int_0^t \mathbb{1}_0(\xi_r)\,dr}\right] \quad \text{and} \quad M^1(t,x,y) = \mathbb{Q}_x\left[e^{(m-1)\int_0^t \mathbb{1}_0(\xi_r)\,dr}\mathbb{1}_y(\xi_t)\right]$$

which is what we claimed but with expectations taken under $\mathbb{Q}$ rather than the original measure $\mathbb{P}$. However we note that the motion of the single spine is the same (it has Q-matrix $\mathscr{A}$) under both $\mathbb{P}$ and $\mathbb{Q}$, so we may simply replace $\mathbb{Q}$ with $\mathbb{P}$, giving (4) and (5).

The variation of constants formulas can now be derived from the Feynman–Kac formulas. We only prove the second identity, as the first can be proved similarly. We use the exponential series to get

$$\mathbb{E}_x\left[e^{(m-1)\int_0^t \mathbb{1}_0(\xi_r)\,dr}\mathbb{1}_y(\xi_t)\right]$$

$$= \mathbb{E}_x\left[\sum_{n=0}^\infty \frac{(m-1)^n}{n!}\left(\int_0^t \mathbb{1}_0(\xi_r)\,dr\right)^n \mathbb{1}_y(\xi_t)\right]$$

$$= \mathbb{P}_x(\xi_t = y) + \mathbb{E}_x\left[\sum_{n=1}^\infty \frac{(m-1)^n}{n!}\int_0^t\cdots\int_0^t \mathbb{1}_0(\xi_{r_1})\cdots\mathbb{1}_0(\xi_{r_n})\,dr_n\ldots dr_1 \mathbb{1}_y(\xi_t)\right]$$

$$= p_t(x,y)$$

$$+ \mathbb{E}_x\left[\sum_{n=1}^\infty (m-1)^n \int_0^t\int_{r_1}^t\cdots\int_{r_{n-1}}^t \mathbb{1}_0(\xi_{r_1})\cdots\mathbb{1}_0(\xi_{r_n})\,dr_n\ldots dr_2 dr_1 \mathbb{1}_y(\xi_t)\right].$$

The last step is justified by the fact that the function that is integrated is symmetric in all arguments and, thus, it suffices to integrate over a simplex. We can exchange sum and expectation and obtain that the last expression equals

$$p_t(x,y) + (m-1)\int_0^t \sum_{n=1}^\infty$$

$$(m-1)^{n-1}\int_{r_1}^t\cdots\int_{r_{n-1}}^t \mathbb{P}_x[\xi_{r_1}=0,\ldots,\xi_{r_n}=0]\,dr_n\ldots dr_2 dr_1.$$

Due to the Markov property, the last expression equals

$$p_t(x,y) + (m-1)\int_0^t p_{r_1}(x,0)\sum_{n=1}^\infty (m-1)^{n-1}$$

$$\times \int_{r_1}^t\cdots\int_{r_{n-1}}^t \mathbb{P}_0[\xi_{r_2-r_1}=0,\ldots,\xi_{r_n-r_1}=0]\,dr_n\ldots dr_2 dr_1$$

and can be rewritten as

$$p_t(x,y) + (m-1)\int_0^t p_{r_1}(x,0)$$
$$\times \left(\sum_{n=1}^\infty (m-1)^{n-1} \int_0^{t-r_1} \cdots \int_{r_{n-1}}^{t-r_1} \mathbb{P}_0[\xi_{r_2}=0,\ldots,\xi_{r_n}=0]\,dr_n\ldots dr_2\right) dr_1.$$

Using the same line of arguments backwards for the term in parentheses, the assertion follows.

Having derived variation of constants formulas, there are different ways to analyze the asymptotics of the first moments. Assuming more regularity for the transition probabilities, this can be done as sketched in the next remark.

Remark 2. Taking Laplace transforms $\mathscr{L}$ in t, one can transform (6), and similarly (7), into the algebraic equation

$$\mathscr{L} M^1(\lambda, x) = \frac{1}{\lambda} + (m-1)\mathscr{L} M^1(\lambda,0)\mathscr{L} p_\lambda(x,0) \qquad ,\lambda > 0,$$

which can be solved explicitly to obtain

$$\mathscr{L} M^1(\lambda, x) = \frac{1}{\lambda(1-(m-1)\mathscr{L} p_\lambda(x,0))} \qquad ,\lambda > 0. \tag{8}$$

Assuming the asymptotics of $p_t(x,0)$ are known for t tending to infinity (and are sufficiently regular), the asymptotics of $\mathscr{L} p_\lambda(x,0)$ for λ tending to zero can be deduced from Tauberian theorems. Hence, from (8) one can then deduce the asymptotics of $\mathscr{L} M^1(\lambda, x)$ as λ tends to zero. This, using Tauberian theorems in the reverse direction, allows one to deduce the asymptotics of $M^1(t,x)$ for t tending to infinity.

Unfortunately, to make this approach work, ultimate monotonicity and asymptotics of the type $p_t(x,0) \sim Ct^{-\alpha}$ are needed. This motivated the authors of [1] to assume (**A4**) so that by the local central limit theorem

$$p_t(x,0) \sim \left(\frac{d}{2\pi}\right)^{d/2} t^{-d/2}.$$

As we did not assume any regularity for p_t, the aforementioned approach fails in general. We instead use an approach based on renewal theorems recently seen in [7].

Proof (of Theorem 1 for M^1). Taking into account irreducibility and the Markov property of $\mathscr{A}$, we see that the property "$\int_0^\infty \mathbb{1}_0(\xi_r)\,dr = \infty$ almost surely" does not depend on the starting value ξ_0. To prove case (i), we simply apply dominated convergence to (4) and (5). If $\mathscr{A}$ is transient, then $\int_0^\infty \mathbb{1}_0(\xi_r)\,dr < \infty$ almost surely and $M^1(t,x)$ converges to a constant. On the other hand if $\mathscr{A}$ is recurrent, then

$\int_0^\infty 1_0(\xi_r)\, dr = \infty$ almost surely and $M^1(t,x) \to 0$. In both cases $M^1(t,x,y) \to 0$, because if $\mathscr{A}$ is transient then $1_{\{\xi_t = y\}} \to 0$ almost surely, and if $\mathscr{A}$ is recurrent then $M^1(t,x,y) \leq M^1(t,x) \to 0$.

Regime (ii) is trivial as here $M^1(t,x) = 1$ and $M^1(t,x,y) = p_t(x,y)$. Next, for regime (iii) (a) we exploit both the standard and the reverse Hölder inequality for $p > 1$:

$$M^1(t,x,y) \geq \mathbb{E}_x\left[e^{-(1/(p-1))(m-1)\int_0^t 1_0(\xi_r)\, dr}\right]^{-(p-1)} p_t(x,y)^p, \qquad (9)$$

$$M^1(t,x,y) \leq \mathbb{E}_x\left[e^{p(m-1)\int_0^t 1_0(\xi_r)\, dr}\right]^{1/p} p_t(x,y)^{(p-1)/p}. \qquad (10)$$

In the recurrent case $G_\infty(0,0) = \infty$ and thus $\beta = 1$, so this case has already been dealt with in regime (ii). Hence we may assume that $\mathscr{A}$ is transient so that $\int_0^\infty 1_0(\xi_r)\, dr < \infty$ with positive probability. This shows that the expectation in the lower bound (9) converges to a finite constant. By assumption $m - 1 < \beta$ so that there is $p > 1$ satisfying $p(m-1) < \beta$. With this choice of p, part 3) of Theorem 1 of [7] implies that also the expectation in the upper bound (10) converges to a finite constant. In total this shows that

$$C p_t(x,y)^p \leq M^1(t,x,y) \leq C' p_t(x,y)^{(p-1)/p}$$

and the claim for $M^1(t,x,y)$ follows. For $M^1(t,x)$ we can directly refer to Theorem 1 of [7].

For regimes (iii) (b) and (c) we give arguments based on renewal theorems. A closer look at the variation of constants formula (7) shows that only for $x = 0$, $M^1(t,x,y)$ occurs on both sides of the equation. Hence, we start with the case $x = 0$ and afterwards deduce the asymptotics for $x \neq 0$.

Let us begin with the simpler case (iii) (c). As mentioned above, in this case we may assume that $\mathscr{A}$ is transient so that $\int_0^\infty p_r(0,0)\, dr < \infty$. Hence, dominated convergence ensures that the equation $\int_0^\infty e^{-\lambda t} p_t(0,0)\, dt = 1/(m-1)$ has a unique positive root λ, which we call $r(m)$. The definition of $r(m)$ shows that $U(dt) := (m-1)e^{-r(m)t} p_t(0,0)\, dt$ is a probability measure on $[0,\infty)$ and furthermore $e^{-r(m)t} p_t(0,y)$ is directly Riemann integrable. Hence the classical renewal theorem (see page 349 of [9]) can be applied to the (complete) renewal equation

$$f(t) = g(t) + f * U(t),$$

with $f(t) = e^{-r(m)t} M^1(t,0,y)$ and $g(t) = e^{-r(m)t} p_t(0,y)$. The renewal theorem implies that

$$\lim_{t \to \infty} f(t) = \frac{\int_0^\infty g(s)\, ds}{\int_0^\infty U((s,\infty))\, ds} \in (0,\infty) \qquad (11)$$

so that the claim for $M^1(t,0,y)$ follows including the limiting constants.

For (iii) (b), we need to be more careful as the criticality implies that $(m-1)\int_0^\infty p_r(0,0)\, dr = 1$. Hence, the measure U as defined above is already a probability measure so that the variation of constants formula is indeed a proper renewal equation. The renewal measure U only has finite mean if additionally

$$\int_0^\infty r p_r(0,0)\,dr < \infty. \qquad (12)$$

In the case of finite mean the claim follows as above from (11) without the exponential correction (i.e. $r(m) = 0$). Note that $p_t(0, y)$ is directly Riemann integrable as the case $\beta > 0$ implies that $\mathscr{A}$ is transient and $p_t(0, y)$ is decreasing.

If (12) fails, we need a renewal theorem for infinite mean variables. Iterating (7) reveals the representation

$$M^1(t, 0, y) = p_t(0, y) * \sum_{n \geq 0} (m-1)^n p_t(0,0)^{*n}, \qquad (13)$$

where $*n$ denotes n-fold convolution in t and $p_t(0, y) * p_t(0, 0)^{*0} = p_t(0, y)$. Note that convergence of the series is justified by

$$(m-1)^n p_t(0,0)^{*n} \leq \left((m-1)\int_0^t p_r(0,0)\,dr\right)^n$$

and the assumption on m. Lemma 1 of [8] now implies that

$$\sum_{n \geq 0}(m-1)^n p_t(0,0)^{*n} \approx \frac{t}{(m-1)\int_0^t \int_s^\infty p_r(0,0)\,drds} \qquad (14)$$

which tends to infinity as $(m-1)\int_s^\infty p_r(0,0)\,dr \to 0$ for $s \to \infty$ since we assumed that $(m-1)p_r(0,0)$ is a probability density in r. To derive from this observation the result for $M^1(t, 0, y)$, note that the simple bound $p_t(0, y) \leq 1$ gives the upper bound

$$M^1(t,0,y) \leq \int_0^t \sum_{n \geq 0}(m-1)^n p_r(0,0)^{*n}\,dr. \qquad (15)$$

For a lower bound, we use that due to irreducibility and continuity of $p_t(0, y)$ in t, there are $0 < t_0 < t_1$ and $\epsilon > 0$ such that $p_t(0, y) > \epsilon$ for $t_0 \leq t \leq t_1$. This shows that

$$M^1(t,0,y) \geq \epsilon \int_{t-t_0}^{t-t_1} \sum_{n \geq 0}(m-1)^n p_r(0,0)^{*n}\,dr. \qquad (16)$$

Combined with (14) the lower and upper bounds directly prove the claim for $M^1(t, 0, y)$.

It remains to deal with regime (iii) (b) and (c) for $x \neq 0$. The results follow from the asymptotics of the convolutions as those do not vanish at infinity. But this can be deduced from simple upper and lower bounds similar to (15) and (16).

The asymptotic results for the expected total number of particles $M^1(t, x)$ follow from similar ideas: estimating as before

$$1 + \epsilon \int_{t-t_0}^{t-t_1} M^1(r, 0)\, dr \leq M^1(t, x) \leq 1 + \int_0^t M^1(r, 0)\, dr,$$

and applying case (2) of Theorem 1 of [7] to (4) with $x = 0$, the result follows.

We now come to the crucial lemma of our paper. We use the many-to-few lemma to reduce higher moments of N_t and $N_t(y)$ to the first moment. More precisely, a system of equations is derived that can be solved inductively once the first moment is known. This particular useful form is caused by the one-point catalyst. A similar system can be derived in the same manner in the deterministic case if the one-point potential is replaced by a n-point potential. However the case of a random n-point potential is much more delicate as the sources are "attracted" to the particles, destroying any chance of a renewal theory approach.

Lemma 3. *For $k \geq 2$ the kth moments fulfil*

$$M^k(t, x) = M^1(t, x) + M^1(t, x, 0) * g_k\big((M^1(t, 0), \cdots, M^{k-1}(t, 0))\big), \quad (17)$$

$$M^k(t, x, y) = M^1(t, x, y) + M^1(t, x, 0) * g_k\big(M^1(t, 0, y), \cdots, M^{k-1}(t, 0, y)\big), \quad (18)$$

where

$$g_k\big(M^1, \ldots, M^{k-1}\big) = \sum_{j=2}^{k} \mathbb{E}\left[\binom{X}{j}\right] \sum_{\substack{i_1, \ldots, i_j > 0 \\ i_1 + \ldots + i_j = k}} \frac{k!}{i_1! \cdots i_j!} M^{i_1} \cdots M^{i_j}.$$

Proof. We shall only prove (17); the proof of (18) is almost identical. We recall the spine setup and introduce some more notation. To begin with, all k spines are carried by the same particle ξ which branches at rate $m_k = \mathbb{E}[X^k]$ when at 0. Thus the k spines separate into two or more particles at rate $m_k - m$ when at 0 (since it is possible that at a birth event all k spines continue to follow the same particle, which happens at rate m). We consider what happens at this first "separation" time, and call it T.

Let $i_1, \ldots, i_j > 0$, $i_1 + \ldots + i_j = k$, and define $A_k(j; i_1, \ldots, i_j)$ to be the event that at a separation event, i_1 spines follow one particle, i_2 follow another, ..., and i_j follow another. The first particle splits into a new particles with probability $a^k p_a m_k^{-1}$ (see the definition of $\mathbb{Q}^k$). Then given that the first particle splits into a new particles, the probability that i_1 spines follow one particle, i_2 follow another, ..., and i_j follow another is

$$\frac{1}{a^k} \cdot \binom{a}{j} \cdot \frac{k!}{i_1! \cdots i_j!}$$

(the first factor is the probability of each spine making a particular choice from the a available; the second is the number of ways of choosing the j particles to assign the spines to; and the third is the number of ways of rearranging the spines amongst those j particles). Thus the probability of the event $A_k(j; i_1, \ldots, i_j)$ under $\mathbb{Q}^k$ is

$$\frac{1}{m_k} \mathbb{E}\left[\binom{X}{j}\right] \frac{k!}{i_1! \cdots i_j!}.$$

(Note that, as expected, this means that the total rate at which a separation event occurs is

$$m_k \cdot \frac{1}{m_k} \sum_{j=2}^{k} \mathbb{E}\left[\binom{X}{j}\right] \sum_{\substack{i_1,\ldots,i_j > 0 \\ i_1 + \ldots + i_j = k}} \frac{k!}{i_1! \cdots i_j!} = m_k - m$$

since the double sum is just the expected number of ways of assigning k things to X boxes without assigning them all to the same box.)

However, for $j \geq 2$, *given* that we have a separation event, $A_k(j; i_1, \ldots, i_j)$ occurs with probability

$$\frac{1}{m_k} \mathbb{E}\left[\binom{X}{j}\right] \frac{k!}{i_1! \cdots i_j!} \left(\frac{m_k}{m_k - m}\right).$$

Write χ_t for the position of the particle carrying the k spines for $t \in [0, T)$, and define $\mathscr{F}_t$ to be the filtration containing all information (including about the spines) up to time t. Recall that the skeleton $\mathrm{skel}(t)$ is the tree generated by particles containing at least one spine up to time t; let $\mathrm{skel}(s; t)$ similarly be the part of the skeleton falling between times s and t. Using the many-to-few lemma with $f = 1$, the fact that by definition before T all spines sit on the same particle and integrating out T, we obtain

$$\mathbb{E}[N_t^k] = \mathbb{Q}^k \left[\prod_{v \in \mathrm{skel}(t)} e^{(m_{B_v} - 1) \int_{\sigma_v(t)}^{\tau_v(t)} \mathbb{1}_0(X_v(s)) ds} \right]$$

$$= \mathbb{Q}^k \left[e^{(m_k - 1) \int_0^T \mathbb{1}_0(\chi_s) ds} \mathbb{1}_{\{T \leq t\}} \mathbb{Q}^k \left[\prod_{v \in \mathrm{skel}(T; t)} e^{(m_{B_v} - 1) \int_{\sigma_v(t)}^{\tau_v(t)} \mathbb{1}_0(X_v(s)) ds} \bigg| \mathscr{F}_T \right] \right]$$

$$+ \mathbb{Q}^k \left[e^{(m_k - 1) \int_0^t \mathbb{1}_0(\chi_s) ds} \mathbb{1}_{\{T > t\}} \right]$$

$$= \int_0^t \mathbb{Q}^k \left[e^{(m_k-1)\int_0^u \mathbf{1}_0(\chi_s)ds} (m_k - m)\mathbf{1}_0(\chi_u) e^{-(m_k-m)\int_0^u \mathbf{1}_0(\chi_s)ds} \right.$$
$$\cdot \mathbb{Q}^k \left[\prod_{v \in \text{skel}(u;t)} e^{(m_{B_v}-1)\int_{\sigma_v(t)}^{\tau_v(t)} \mathbf{1}_0(X_v(s))ds} \bigg| \mathscr{F}_u; T = u \right] du$$
$$+ \mathbb{Q}^k \left[e^{(m_k-1)\int_0^t \mathbf{1}_0(\chi_s)ds} e^{-(m_k-m)\int_0^t \mathbf{1}_0(\chi_s)ds} \right].$$

To prove (18), the same arguments are used with $f = \mathbf{1}_y$ in place of $f = 1$. Now we split the sample space according to the distribution of the numbers of spines in the skeleton at time T. Since, given their positions and marks at time T, the particles in the skeleton behave independently, we may split the product up into j independent factors. Thus

$$\mathbb{E}[N_t^k] = \int_0^t \sum_{j=2}^k \sum_{\substack{i_1,\ldots,i_j>0 \\ i_1+\ldots+i_j=k}} \mathbb{E}\left[\binom{X}{j}\right] \frac{k!}{i_1!\cdots i_j!} \mathbb{Q}^k \left[e^{(m-1)\int_0^u \mathbf{1}_0(\chi_s)ds} \mathbf{1}_0(\chi_u) \right.$$
$$\cdot \prod_{l=1}^j \mathbb{Q}^{i_l} \left[\prod_{v \in \text{skel}(t-u)} e^{(m_{B_v}-1)\int_{\sigma_v(t-u)}^{\tau_v(t-u)} \mathbf{1}_0(X_v(s))ds} \right] du$$
$$+ \mathbb{Q}^k \left[e^{(m-1)\int_0^t \mathbf{1}_0(\chi_s)ds} \right]$$
$$= \int_0^t \sum_{j=2}^k \sum_{\substack{i_1,\ldots,i_j>0 \\ i_1+\ldots+i_j=k}} \mathbb{E}\left[\binom{X}{j}\right] \frac{k!}{i_1!\cdots i_j!} \mathbb{E}_x[N_u(0)] \cdot \prod_{l=1}^j \mathbb{E}_0\left[N_{t-u}^{i_l}\right] du + \mathbb{E}_x[N_t],$$

where we have used the many-to-few lemma backwards with $f = \mathbf{1}_0$ (first expectation) and $f = 1$ (two last expectations) to obtain the last line. This is exactly the desired equation (17). For (18) we again use $f = \mathbf{1}_y$ in place of $f = 1$ and copy the same lines of arguments.

Remark 3. The factors appearing in g_k are derived combinatorially from splitting the spines. In Lemma 3.1 of [3] they appeared from Faà di Bruno's differentiation formula.

We need the following elementary lemma before we can complete our proof.

Lemma 4. *For any non-negative integer-valued random variable Y, and any integers $a \geq b \geq 1$,*
$$\mathbb{E}[Y^a]\mathbb{E}[Y] \geq \mathbb{E}[Y^b]\mathbb{E}[Y^{a-b+1}].$$

Proof. Assume without loss of generality that $b \geq a/2$. Note that for any two positive integers j and k,

$$j^a k + j k^a - j^b k^{a-b+1} - j^{a-b+1} k^b = jk(j-k)^2 \Big(j^{a-3} + 2 j^{a-4} k + 3 j^{a-5} k^2 + \ldots$$
$$+ (a-b) j^{b-2} k^{a-b-1} + (a-b-1) j^{b-3} k^{a-b}$$
$$+ \cdots + 2 j k^{a-4} + k^{a-3} \Big)$$
$$\geq 0.$$

Thus

$$\mathbb{E}[Y^a]\mathbb{E}[Y] - \mathbb{E}[Y^b]\mathbb{E}[Y^{a-b+1}]$$
$$= \sum_{j \geq 1} j^a \mathbb{P}(Y=j) \sum_{k \geq 1} k \mathbb{P}(Y=k) - \sum_{j \geq 1} j^b \mathbb{P}(Y=j) \sum_{k \geq 1} k^{a-b+1} \mathbb{P}(Y=k)$$
$$= \sum_{j \geq 1} \sum_{k > j} (j^a k + k^a j - j^b k^{a-b+1} - j^{a-b+1} k^b) \mathbb{P}(Y=j) \mathbb{P}(Y=k)$$
$$\geq 0$$

as required.

We can now finish the proof of the main result.

Proof (of Theorem 1 for M^k). Case (i) follows just as for M^1, applying dominated convergence to the $\mathbb{Q}^k$-expectation in Lemma 1. Note that if T is the first split time of the k spines (as in Lemma 3) then $e^{(m_k-1) \int_0^T \mathbf{1}_0(\xi_s) ds}$ is stochastically dominated by $e^{(m_k-1)\tau}$ where τ is an exponential random variable of parameter $m_k - m_1$; this allows us to construct the required dominating random variable.

For case (ii), using Lemmas 2 and 3 we find the lower bound

$$M^k(t,x) \geq 1 + C \int_0^t p_s(x,0) M^{k-1}(t-s,0) ds \geq C \int_0^t p_s(x,0) M^{k-1}(t-s,0) ds. \tag{19}$$

An upper bound can be obtained by additionally using Lemma 4 (to reduce g_k to the leading term $M^1 M^{k-1}$) to obtain

$$M^k(t,x) \leq 1 + C \int_0^t p_s(x,0) M^{k-1}(t-s,0) ds. \tag{20}$$

Using inductively the lower bound (19) and furthermore the iteration

$$\int_0^t \mathbb{P}_x(X_{s_1}=0) \int_0^{t-s_1} \mathbb{P}_0(X_{s_2}=0) \ldots \int_0^{t-s_1-\ldots-s_{k-2}} \mathbb{P}_0(X_{s_{k-1}}=0) ds_{k-1}\ldots ds_2 ds_1$$

$$= \int_0^t \int_{s_1}^t \ldots \int_{s_{k-2}}^t \mathbb{P}_x(X_{s_1}=0, X_{s_2}=0, \ldots, X_{s_{k-1}}=0) ds_{k-1} \ldots ds_2 ds_1$$

$$= \frac{1}{(k-1)!} \int_0^t \int_0^t \ldots \int_0^t \mathbb{P}_x(X_{s_1}=0, X_{s_2}=0, \ldots, X_{s_{k-1}}=0) ds_{k-1} \ldots ds_2 ds_1$$

$$= \frac{1}{(k-1)!} \mathbb{E}_x\left[\left(\int_0^t \mathbb{1}_{\{X_s=0\}} ds\right)^{k-1}\right]$$

$$= \frac{1}{(k-1)!} \mathbb{E}_x\left[L_t(0)^{k-1}\right]$$

(21)

we see that $M^k(t,x)$ goes to infinity if $\mathscr{A}$ is recurrent and to a constant if $\mathscr{A}$ is transient. This implies that the additional summand 1 in (20) can be omitted asymptotically in both cases. The claim follows.

The lower bound of case (iii)(a) follows by the same argument as for case (ii), and the upper bound is a straightforward induction using Lemmas 3 and 4. The cases (iii)(b) and (c) also follow from Lemma 3 and induction based on the asymptotics for M^1.

Acknowledgements Leif Döring would like to thank Martin Kolb for drawing his attention to [1] and Andreas Kyprianou for his invitation to the Bath-Paris workshop on branching processes, where he learnt of the many-to-few lemma from Matthew I. Roberts. The authors would also like to thank Piotr Milos for checking an earlier draft, and a referee for pointing out several relevant articles.

Leif Döring acknowledges the support of the Fondation Sciences Mathématiques de Paris. Matthew I. Roberts thanks ANR MADCOF (grant ANR-08-BLAN-0220-01) and WIAS for their support.

References

1. S. Albeverio, L. Bogachev, E. Yarovaya, Asymptotics of branching symmetric random walk on the lattice with a single source. Compt. Rendus. Acad. Sci. Math. **326**, 975–980 (1998)
2. S. Albeverio, L. Bogachev, E. Yarovaya, Erratum to : Asymptotics of branching symmetric random walk on the lattice with a single source. Compt. Rendus. Acad. Sci. Math. **327**, 585 (1998)
3. S. Albeverio, L. Bogachev, Branching random walk in a catalytic medium. I. Basic equations. Positivity **4**, 41 (2000)
4. K.K. Anderson, K.B. Athreya, A renewal theorem in the infinite mean case. Ann. Probab. **15**, 388–393 (1987)
5. E.V. Bulinskaya, Catalytic branching random walk on three-dimensional lattice. Theory Stoch. Process. **16**(2), 23–32 (2010)

6. E.V. Bulinskaya, Limit distributions arising in branching random walks on integer lattices. Lithuan. Math. J. **51**(3), 310–321 (2011)
7. L. Döring, M. Savov, An application of renewal theorems to exponential moments of local times. Elect. Comm. Probab. **15**, 263–269 (2010)
8. B. Erickson, The strong law of large numbers when the mean is undefined. Trans. Am. Math. Soc. **54**, 371–381 (1973)
9. W. Feller, *An Introduction to Probability Theory and Its Applications*, vol. II (Wiley, New York, 1966)
10. J. Gärtner, M. Heydenreich, Annealed asymptotics for the parabolic Anderson model with a moving catalyst. Stoch. Process. Appl. **116**, 1511–1529 (2006)
11. S. Harris, M. Roberts, The many-to-few lemma and multiple spines. arXiv:1106.4761v1. Preprint (2011)
12. Y. Hu, V.A. Vatutin, V.A. Topchii, Branching random walk in $\mathbb{Z}^4$ with branching at the origin only. Theory Probab. Appl. **56**(2), 224–247 (2011)
13. T. Kurtz, R. Lyons, R. Pemantle, Y. Peres, A conceptual proof of the Kesten-Stigum theorem for multi-type branching processes, in *Classical and Modern Branching Processes (Minneapolis, MN, 1994)*, ed. by K.B. Athreya, P. Jagers. volume 84 of The IMA Volumes in Mathematics and its Applications (Springer, New York, 1997), pp. 181–185
14. R. Lyons, A simple path to Biggins' martingale convergence for branching random walk, in *Classical and Modern Branching Processes (Minneapolis, MN, 1994)*, ed. by K.B. Athreya, P. Jagers. volume 84 of The IMA Volumes in Mathematics and its Applications (Springer, New York, 1997), pp. 217–221
15. R. Lyons, R. Pemantle, Y. Peres, Conceptual proofs of $L \log L$ criteria for mean behavior of branching processes. Ann. Probab. **23**(3), 1125–1138 (1995)
16. V.A. Topchii, V.A. Vatutin, Individuals at the origin in the critical catalytic branching random walk. Discrete Math. Theor. Comput. Sci. **6**, 325–332 (2003)
17. V.A. Vatutin, V.A. Topchii, Limit theorem for critical catalytic branching random walks. Theory Probab. Appl. **49**(3), 498–518 (2005)
18. V.A. Vatutin, V.A. Topchii, E.B. Yarovaya, Catalytic branching random walk and queueing systems with random number of independent servers. Theory Probab. Math. Stat. **69**, 1–15 (2004)
19. E.B. Yarovaya, Use of spectral methods to study branching processes with diffusion in a noncompact phase space. Teor. Mat. Fiz. **88**, 25–30 (1991) (in Russian); English translation: Theor. Math. Phys. **88** (1991)
20. E.B. Yarovaya, The monotonicity of the probability of return into the source in models of branching random walks. Moscow Univ. Math. Bull. **65**(2), 78–80 (2010)

Malliavin Calculus and Self Normalized Sums

Solesne Bourguin and Ciprian A. Tudor

Abstract We study the self-normalized sums of independent random variables from the perspective of the Malliavin calculus. We give the chaotic expansion for them and we prove a Berry–Esséen bound with respect to several distances.

Keywords Chaos expansions • Limit theorems • Malliavin calculus • Multiple stochastic integrals • Self-normalized sums • Stein's method

AMS Subject Classification (2010): 60F05, 60H07, 60H05.

---

*The authors are associate members of the team Samos, Université de Panthéon-Sorbonne Paris 1. They wish to thank Natesh Pillai for interesting discussions and an anonymous referee for useful suggestions. The second author is supported by the CNCS grant PN-II-ID-PCCE-2011-2-0015 (Romania). Support from the ANR grant "Masterie" BLAN 012103 (France) is also acknowledged.

S. Bourguin (✉)
Faculté des Sciences, de la Technologie et de la Communication, Université du Luxembourg, UR en Mathématiques. 6, rue Richard Coudenhove-Kalergi, L-1359 Luxembourg City, Luxembourg
e-mail: solesne.bourguin@gmail.com

C.A. Tudor
Laboratoire Paul Painlevé, Université de Lille 1, F-59655 Villeneuve d'Ascq, Lille, France
and Department of Mathematics, Academy of Economical Studies, Bucharest, Romania
e-mail: tudor@math.univ-lille1.fr

1 Introduction

Let $(\Omega, \mathscr{F}, P)$ be a probability space and $(W_t)_{t \geq 0}$ a one-dimensional Brownian motion on this space. Let F be a random variable defined on Ω which is differentiable in the sense of the Malliavin calculus. Then using the so-called Stein method introduced by Nourdin and Peccati in [13] (see also [14] and [12]), it is possible to measure the distance between the law of F and the standard normal law $\mathscr{N}(0, 1)$. This distance, denoted by d, can be defined in several ways, such as the Kolmogorov distance, the Wasserstein distance, the total variation distance or the Fortet–Mourier distance. More precisely we have, if $\mathscr{L}(F)$ denotes the law of F,

$$d(\mathscr{L}(F), \mathscr{N}(0,1)) \leq c\sqrt{\mathbf{E}\left(1 - \langle DF, D(-L)^{-1}F\rangle_{L^2([0,1])}\right)^2}. \qquad (1)$$

Here D denotes the Malliavin derivative with respect to W, and L is the generator of the Ornstein–Uhlenbeck semigroup. We will explain in the next section how these operators are defined. The constant c is equal to 1 in the case where d is the Kolmogorov distance as well as in the case where d is the Wasserstein distance, $c = 2$ for the case where d is the total variation distance and $c = 4$ in the case where d is the Fortet–Mourier distance.

Our purpose is to apply Stein's method combined with Malliavin calculus and in particular the bound (1) to self-normalized sums. Let us recall some basic facts on this topic. We refer to [7] and the references therein for a more detailed exposition. Let $X_1, X_2, \ldots$ be independent random variables. Set $S_n = \sum_{i=1}^n X_i$ and $V_n^2 = \sum_{i=1}^n X_i^2$. Then $\frac{S_n}{V_n}$ converges in distribution as $n \to \infty$ to the standard normal law $\mathscr{N}(0, 1)$ if and only if $\mathbf{E}(X) = 0$ and X is in the domain of attraction of the standard normal law (see [7], Theorem 4.1). The "if" part of the theorem has been known for a long time (it appears in [11]) while the "only if" part remained open until its proof in [8]. The Berry–Esséen theorem for self-normalized sums has been also widely studied. We refer to [2,9] and [17] (see also [3,4] for the situation where the random variables X_i are non i.i.d.). These results say that the Kolmogorov distance between the law of $\frac{S_n}{V_n}$ and the standard normal law is less than

$$C\left(B_n^{-2}\sum_{i=1}^n \mathbf{E}\left(X_i^2 1_{(|X_i|>B_n)}\right) + B_n^{-3}\sum_{i=1}^n \mathbf{E}\left(X_i^3 1_{(|X_i|\geq B_n)}\right)\right)$$

where $B_n = \sum_{i=1}^n \mathbf{E}(X_i^2)$ and C is an absolute constant. We mention that, as far as we know, these results only exist for the Kolmogorov distance. To use our techniques based on the Malliavin calculus and multiple stochastic integrals, we will put ourselves on a Gaussian space where we will consider the following particular case: the random variables X_i are the increments of the Wiener process $X_i = W_i - W_{i-1}$. The Berry–Esséen bound from above reduces to (see [7], page 53): for $2 < p \leq 3$

$$\sup_{z \in \mathbb{R}} |\mathbf{P}(F_n \leq z) - \Phi(z)| \leq 25 \mathbf{E}(|Z|^p) n^{1-\frac{p}{2}} \qquad (2)$$

where Z is a standard normal random variable and Φ is its distribution function. In particular for $p = 3$ we get

$$\sup_{z \in \mathbb{R}} |\mathbf{P}(F_n \leq z) - \Phi(z)| \leq 25 \mathbf{E}(|Z|^3) n^{-\frac{1}{2}}. \qquad (3)$$

We will compare our result with the above relation (3). The basic idea is as follows: we are able to find the chaos expansion into multiple Wiener–Itô integrals of the random variable $\frac{S_n}{V_n}$ for every $n \geq 2$ and to compute its Malliavin derivative. Note that the random variable $\frac{S_n}{V_n}$ has a decomposition into an infinite sum of multiple integrals in contrast to the examples provided in the papers [5, 13, 14]. Then, we compute the Berry–Esséen bound given by $\sqrt{\mathbf{E}\left(1 - \langle DF, D(-L)^{-1}F\rangle_{L^2([0,1])}\right)^2}$ by using properties of multiple stochastic integrals. Of course, we cannot expect to obtain a rate of convergence better than $c\frac{1}{\sqrt{n}}$, but we have an explicit expression of the constant appearing in this bound and our method is available for several distances between the laws of random variables (not limited to the Kolmogorov distance). This aspect of the problem is new: we provide new bounds with an explicit constant c in the cases of other useful distances. This computation of the Berry–Esséen bound is also interesting in and of itself as it brings to light original relations involving Gaussian measure and Hermite polynomials. It gives an exact expression of the chaos expansion of the self normalized sum and it also shows that the convergence to the normal law of $\frac{S_n}{V_n}$ is uniform with respect to the chaos, in the sense that every chaos of $\frac{S_n}{V_n}$ is weakly convergent to the standard normal law and that the rate is the same for every chaos. Moreover, he hope that this approach which uses Malliavin calculus could lead to another results which may be interesting for a statistician: the bounds in the multidimensional case (see [15]) or exact confidence intervals (see [6]).

We have organized our paper as follows: Sect. 2 contains the elements of the Malliavin calculus needed in the paper and in Sect. 3 we discuss the chaos decomposition of self-normalized sums as well as study the asymptotic behavior of the coefficients appearing in this expansion. Section 4 contains the computation of the Berry–Esséen bound given in terms of the Malliavin calculus. Finally, Sect. 5 is dedicated to the proofs of the results as well as to some technical lemmas needed in those proofs.

2 Preliminaries

We will begin by describing the basic tools of multiple Wiener–Itô integrals and Malliavin calculus that will be needed in our paper. Let $(W_t)_{t \in [0,T]}$ be a classical one-dimensional Wiener process on a standard Wiener space $(\Omega, \mathscr{F}, P)$.

If $f \in L^2([0,T]^n)$ with $n \geq 1$ integer, we introduce the multiple Wiener–Itô integral of f with respect to W. We refer to [16] for a detailed exposition of the construction and the properties of multiple Wiener–Itô integrals.

Let $f \in \mathscr{S}_n$, which means that there exists $n \geq 1$ integers such that

$$f := \sum_{i_1,\ldots,i_n} c_{i_1,\ldots,i_n} 1_{A_{i_1} \times \ldots \times A_{i_n}}$$

where the coefficients satisfy $c_{i_1,\ldots,i_n} = 0$ if two indices i_k and i_ℓ are equal and where the sets $A_i \in \mathscr{B}([0,T])$ are disjoint. For a such step function f we define

$$I_n(f) := \sum_{i_1,\ldots,i_n} c_{i_1,\ldots,i_n} W(A_{i_1}) \ldots W(A_{i_n})$$

where we put $W([a,b]) = W_b - W_a$. It can be seen that the application I_n constructed above from $\mathscr{S}_n$ equipped with the scaled norm $\frac{1}{\sqrt{n!}} \| \cdot \|_{L^2([0,T]^n)}$ to $L^2(\Omega)$ is an isometry on $\mathscr{S}_n$, i.e. for m, n positive integers,

$$\mathbf{E}\left(I_n(f) I_m(g)\right) = n! \langle f, g \rangle_{L^2([0,T]^n)} \quad \text{if } m = n,$$
$$\mathbf{E}\left(I_n(f) I_m(g)\right) = 0 \quad \text{if } m \neq n.$$

It also holds that

$$I_n(f) = I_n(\tilde{f})$$

where $\tilde{f}$ denotes the symmetrization of f defined by

$$\tilde{f}(x_1,\ldots,x_x) = \frac{1}{n!} \sum_{\sigma \in \mathscr{S}_n} f(x_{\sigma(1)},\ldots,x_{\sigma(n)}).$$

Since the set $\mathscr{S}_n$ is dense in $L^2([0,T]^n)$ for every $n \geq 2$, the mapping I_n can be extended to an isometry from $L^2([0,T]^n)$ to $L^2(\Omega)$ and the above properties hold true for this extension. Note also that I_n can be viewed as an iterated stochastic integral (this follows e.g. by Itô's formula)

$$I_n(f) = n! \int_0^1 \int_0^{t_n} \ldots \int_0^{t_2} f(t_1,\ldots,t_n) dW_{t_1} \ldots dW_{t_n}$$

We recall the product for two multiple integrals (see [16]): if $f \in L^2([0,T]^n)$ and $g \in L^2([0,T]^m)$ are symmetric, then it holds that

$$I_n(f) I_m(g) = \sum_{\ell=0}^{m \wedge n} \ell! C_m^\ell C_n^\ell I_{m+n-2\ell}(f \otimes_\ell g) \tag{4}$$

where the contraction $f \otimes_\ell g$ belongs to $L^2([0,T]^{m+n-2\ell})$ for $\ell = 0, 1, \ldots, m \wedge n$ and is given by

$$(f \otimes_\ell g)(s_1, \ldots, s_{n-\ell}, t_1, \ldots, t_{m-\ell})$$
$$= \int_{[0,T]^\ell} f(s_1, \ldots, s_{n-\ell}, u_1, \ldots, u_\ell) g(t_1, \ldots, t_{m-\ell}, u_1, \ldots, u_\ell) du_1 \ldots du_\ell.$$

We recall that any square integrable random variable that is measurable with respect to the σ-algebra generated by W can be expanded into an orthogonal sum of multiple stochastic integrals

$$F = \sum_{n \geq 0} I_n(f_n) \tag{5}$$

where $f_n \in L^2([0,1]^n)$ are (uniquely determined) symmetric functions and $I_0(f_0) = \mathbf{E}(F)$.

Let L be the Ornstein–Uhlenbeck operator

$$LF = -\sum_{n \geq 0} n I_n(f_n) \text{ and } L^{-1} F = -\sum_{n \geq 1} \frac{1}{n} I_n(f_n)$$

if F is given by (5). We denote by D the Malliavin derivative operator that acts on smooth functionals of the form $F = g(W(\varphi_1), \ldots, W(\varphi_n))$ where g is a smooth function with compact support and $\varphi_i \in L^2([0,1])$. For $i = 1, \ldots, n$, the derivative operator is defined by

$$DF = \sum_{i=1}^n \frac{\partial g}{\partial x_i}(W(\varphi_1), \ldots, W(\varphi_n))\varphi_i.$$

The operator D can be extended to the closure $\mathbb{D}^{p,2}$ of smooth functionals with respect to the norm

$$\|F\|_{p,2}^2 = \mathbf{E}(F^2) + \sum_{i=1}^p \mathbf{E}\left(\|D^i F\|_{L^2([0,1]^i)}^2\right)$$

where the ith order Malliavin derivative D^i is defined iteratively.

Let us recall how this derivative acts for random variables in a finite chaos. If $f \in L^2([0,T]^n)$ is a symmetric function, we will use the following rule to differentiate in the Malliavin sense

$$D_t I_n(f) = n I_{n-1}(f(\cdot, t)), \quad t \in \mathbb{R}.$$

Let us also recall how the distances between the laws of random variables are defined. We have

$$d(\mathscr{L}(X), \mathscr{L}(Y)) = \sup_{h \in \mathscr{A}} (|\mathbf{E}(h(X)) - \mathbf{E}(h(Y))|)$$

where $\mathscr{A}$ denotes a set of functions. When $\mathscr{A} = \{h : \|h\|_L \geq 1\}$ (here $\|\cdot\|_L$ is the finite Lipschitz norm) we obtain the Wasserstein distance, when $\mathscr{A} = \{h : \|h\|_{BL} \geq 1\}$ (with $\|\cdot\|_{LB} = \|\cdot\|_L + \|\cdot\|_\infty$) we get the Fortet–Mourier distance, when $\mathscr{A}$ is the set of indicator functions of Borel sets we obtain the total variation distance, and when $\mathscr{A}$ is the set of indicator functions of the form $1_{(-\infty,z)}$ with $z \in \mathbb{R}$, we obtain the Kolmogorov distance that has been presented above.

3 Chaos Decomposition of Self-Normalized Sums

The tools of the Malliavin calculus presented above can be successfully applied in order to study self-normalized sums. Because of the nature of Malliavin calculus, we put ourselves in a Gaussian setting and we consider $X_i = W_i - W_{i-1}$ to be the increments of a classical Wiener process W. We then consider the sums

$$S_n = \sum_{i=1}^n X_i \quad \text{and} \quad V_n^2 = \sum_{i=1}^n X_i^2$$

as well as the *self-normalized sum* F_n defined by

$$F_n = \frac{S_n}{V_n} = \frac{W_n}{\left(\sum_{i=1}^n (W_{i+1} - W_i)^2\right)^{\frac{1}{2}}}. \tag{6}$$

3.1 The Chaos Expansion Theorem

Let us now concentrate our efforts on finding the chaotic decomposition of the random variable F_n. This will be the key to computing Berry–Esséen bounds for the distance between the law of F_n and the standard normal law in the next section.

Proposition 1. *Let F_n be given by (6) and let $f : \mathbb{R}^n \to \mathbb{R}$ be given by*

$$f(x_1, \ldots, x_n) = \frac{x_1 + \ldots + x_n}{(x_1^2 + \ldots + x_n^2)^{\frac{1}{2}}}. \tag{7}$$

Let $\varphi_i = 1_{[i-1,i]}$ for $i = 1, \ldots, n$. Then for every $n \geq 2$, we have

$$F_n = \sum_{k \geq 0} \frac{1}{k!} \sum_{i_1, \ldots, i_k = 1}^n a_{i_1, \ldots, i_k} I_k \left(\varphi_{i_1} \otimes \ldots \otimes \varphi_{i_k}\right)$$

with

$$a_{i_1,\ldots,i_k} \stackrel{def}{=} \mathbf{E}\left(\frac{\partial^k f}{\partial x_{i_1},\ldots,x_{i_k}}(W(\varphi_1),\ldots,W(\varphi_n))\right). \tag{8}$$

Remark 1. The coefficients $a_{i_1,\ldots,i_k}$ also depend on n. We omit n in their notation in order to simplify the presentation.

3.2 Computing the Coefficients in the Chaos Expansion

An important step in the analysis of the decomposition coefficients is to explicitly compute the coefficients $a_{i_1,\ldots,i_k}$ appearing in Proposition 1. Let $\mathbf{H}_n(x)$ denote the nth Hermite polynomial:

$$\mathbf{H}_n(x) = (-1)^n e^{x^2/2} \frac{d^n}{dx^n} e^{-x^2/2}.$$

Define

$$W_n \stackrel{def}{=} W(\varphi_1) + W(\varphi_2) + \cdots + W(\varphi_n)$$

$$V_n \stackrel{def}{=} \left(\sum_{i=1}^n W(\varphi_i)^2\right)^{1/2}.$$

The following proposition is the main result of this subsection and provides an explicit expression for the coefficients of the chaos decomposition proved in Proposition 1.

Proposition 2. *For every $k \geq 0$ and for every $1 \leq i_1, \cdots, i_{2k+1} \leq n$, let $d_r^\star, 1 \leq r \leq n$ be the number of times the integer r appears in the sequence $\{i_1, \cdots, i_{2k+1}\}$. Then,*

$$a_{i_1,\cdots,i_{2k+1}} = \mathbf{E}\left[\frac{1}{V_n} W(\varphi_1) \mathbf{H}_{d_1^\star}(W(\varphi_1)) \mathbf{H}_{d_2^\star}(W(\varphi_2)) \cdots \mathbf{H}_{d_n^\star}(W(\varphi_n))\right] \tag{9}$$

if there is only one odd integer in the sequence $d_r^\star, 1 \leq r \leq n$. If there is more than one odd integer in the sequence $d_r^\star, 1 \leq r \leq n$, we have $a_{i_1,\cdots,i_{2k+1}} = 0$.

Remark 2. Note that in (9), it might be understood that $d_1^\star$ is always the only odd integer in $d_r^\star, 1 \leq r \leq n$. This is obviously not always the case and if $d_1^\star$ is not the odd integer but let's say, $d_i^\star$ with $1 < i \leq n$ is, one can use the equality in law between $W(\varphi_i)$ and $W(\varphi_1)$ to perform an index swap ($i \leftrightarrow 1$) and the equality (9) remains unchanged.

Remark 3. If one is in the case where $a_{i_1,\cdots,i_{2k+1}} \neq 0$, one can rewrite $d_1^\star, d_2^\star, \cdots, d_n^\star$ as $2d_1 + 1, 2d_2, \cdots, 2d_n$ and finally rewrite (9) as

$$a_{i_1,\cdots,i_{2k+1}} = \mathbf{E}\left[\frac{1}{V_n}W(\varphi_1)\mathbf{H}_{2d_1+1}\left(W(\varphi_1)\right)\mathbf{H}_{2d_2}\left(W(\varphi_2)\right)\cdots\mathbf{H}_{2d_n}\left(W(\varphi_n)\right)\right]. \tag{10}$$

3.3 Asymptotic Behavior of the Coefficients

Having an explicit expression of the coefficients in the chaos decomposition will allow us to finally determine their asymptotic behavior as $n \to \infty$. We established the following result.

Proposition 3. *For every* $1 \leq i_1, \cdots, i_{2k+1} \leq n$, *let* $a_{i_1,\cdots,i_{2k+1}}$ *be as defined in (8). As in (10), let* $2d_1 + 1, 2d_2, \cdots, 2d_r, \cdots, 2d_n$ *denote the number of times the integers* $1, 2, \ldots, n$ *appears in the sequence* $\{i_1, i_2, \cdots, i_{2k+1}\}$ *with* $\sum_{r=1}^n d_r = k$ *(see Remark 3). Then when* $n \to \infty$, *(the symbol $\sim$ means that the ratio of the two sides converges to 1 as* $n \to \infty$)

$$a_{i_1,\cdots,i_{2k+1}} \sim \frac{1}{k!}(2k-1)!! \frac{(2d_1+1)!(2d_2)!\cdots(2d_n)!}{(d_1!d_2!\cdots d_n!)^2}$$

$$\times 2^{-2k}(-1)^k \left(\prod_{j=0}^n \sum_{l_j=0}^{d_j} (-1)^{l_j} C_{d_j}^{l_j} l_j^{d_j}\right) \frac{1}{n^{\frac{1}{2}+|A|}} \tag{11}$$

where

$$A := \{2d_1 + 1, 2d_2, \cdots, 2d_n\} \setminus \{0, 1\}$$

and $|A|$ *is the cardinal of A.*

4 Computation of the Berry–Esséen Bound

Let us first recall the following result (see [7], page 53): for $2 < p \leq 3$,

$$\sup_{z \in \mathbb{R}} |P(F_n \leq z) - \Phi(z)| \leq 25\mathbf{E}\left(|Z|^p\right) n^{1-\frac{p}{2}} \tag{12}$$

where Z is a standard normal random variable and Φ is its distribution function. In particular for $p = 3$ we get

$$\sup_{z \in \mathbb{R}} |P(F_n \leq z) - \Phi(z)| \leq 25\mathbf{E}\left(|Z|^3\right) n^{-\frac{1}{2}}.$$

We now compute the Berry–Esséen bound obtained via Malliavin calculus in order to compare it with (12). This is the objet of the main result of this section.

Theorem 1. *For any integer $n \geq 2$,*

$$\mathbf{E}\left(\left(\langle DF_n, D(-L)^{-1} F_n\rangle - 1\right)^2\right) \leq \frac{c_0}{n}$$

with

$$c_0 = \sum_{m \geq 1} (2m)! \left(\sum_{k=0}^{2m} \frac{1}{2k!} \frac{1}{(2m-2k)!} \sum_{r \geq 0} \frac{1}{(2r)!} \frac{1}{2m-2k+2r+1} c(k,r,m)\right)^2 \quad (13)$$

$$+ \left(\sum_{k=0}^{2m} \frac{1}{(2k+1)!} \frac{1}{(2m-2k-1)!} \sum_{r \geq 0} \frac{1}{(2r-1)!} \frac{1}{2m-2k+2r+1} c(k,r,m)\right)^2$$

and where $c(k,r,m)$ is given by (22).

An immediate consequence of this theorem is the fact that the self-normalized sequence F_n converges uniformly to its limit in the sense that each element of the chaos decomposition converges weakly to a standard normal random variable.

Corollary 1. *Let $J_m(F_n)$ denotes the projection on the mth Wiener chaos of the random variable F_n. Then for every $m \geq 1$ the sequence $J_m(F_n)$ converges as $n \to \infty$ to a standard normal random variable.*

5 Proofs and Technical Lemmas

5.1 Proofs of Propositions 1, 2 and 3

Proof (Proof of Proposition 1). First note that F_n can be written as

$$F_n = f(W(\varphi_1), \cdots, W(\varphi_n)).$$

We can also write

$$f(x_1, \cdots, x_n) = \sum_{i=1}^{n} f_i(x_1, \cdots, x_n),$$

where $f_i(x_1, \cdots, x_n)$ is defined by

$$f_i(x_1, \cdots, x_n) = \frac{x_i}{(x_1^2 + \cdots + x_n^2)^{\frac{1}{2}}}.$$

The chaotic decomposition of $f_i(W(\varphi_1), \cdots, W(\varphi_n))$ was obtained (in a slightly different setting) by Hu and Nualart in the proof of Proposition 10 in [10]. They proved that

$$f_i(W(\varphi_1), \cdots, W(\varphi_n)) = \sum_{k=0}^{\infty} \sum_{j_1,\cdots,j_k=1}^{n} b_{i,j_1,\cdots,j_k} I_k\left(\varphi_{j_1} \otimes \cdots \otimes \varphi_{j_k}\right),$$

where

$$b_{i,j_1,\cdots,j_k} = \frac{(-1)^k}{(2\pi)^{n/2}} \int_{\mathbb{R}^d} \left[\frac{\partial^k}{\partial x_{j_1} \cdots \partial x_{j_k}} e^{-\frac{(x_1+\cdots+x_n)^2}{2}}\right] f_i(x_1,\cdots,x_n) dx_1 \cdots dx_n.$$

Define $b_{j_1,\cdots,j_k} = \sum_{i=1}^{n} b_{i,j_1,\cdots,j_k}$. By the above result, we have

$$b_{j_1,\cdots,j_k} = \frac{(-1)^k}{(2\pi)^{n/2}} \int_{\mathbb{R}^d} \left[\frac{\partial^k}{\partial x_{j_1} \cdots \partial x_{j_k}} e^{-\frac{(x_1+\cdots+x_n)^2}{2}}\right] f(x_1,\cdots,x_n) dx_1 \cdots dx_n.$$

Thus,

$$F_n = \sum_{k \geq 0} \sum_{i_1,\cdots,i_k=1}^{n} b_{i_1,\cdots,i_k} I_k\left(\varphi_{i_1} \otimes \cdots \otimes \varphi_{i_k}\right).$$

Finally, using the Gaussian integration by part formula yields $b_{i_1,\cdots,i_k} = a_{i_1,\cdots,i_k}$, where $a_{i_1,\cdots,i_k}$ is defined by (8), which concludes the proof.

Proof (Proof of Proposition 2). Since $\sum_{r=1}^{n} d_r^{\star} = 2k+1$, there is an odd number of odd integers in the sequence $d_r^{\star}$, $1 \leq r \leq n$. Recall that by Lemma 1 (see the technical lemmas subsection), we have

$$a_{i_1,\cdots,i_{2k+1}} = \sum_{u=1}^{n} \mathbf{E}\left(\frac{W(\varphi_u)}{V_n} \prod_{r=1}^{n} \mathbf{H}_{d_r^{\star}}(W(\varphi_r))\right)$$

$$= \mathbf{E}\left[\frac{1}{V_n} W(\varphi_1) \mathbf{H}_{d_1^{\star}}(W(\varphi_1)) \mathbf{H}_{d_2^{\star}}(W(\varphi_2)) \ldots \mathbf{H}_{d_n^{\star}}(W(\varphi_n))\right]$$

$$+ \mathbf{E}\left[\frac{1}{V_n} W(\varphi_2) \mathbf{H}_{d_1^{\star}}(W(\varphi_1)) \mathbf{H}_{d_2^{\star}}(W(\varphi_2)) \cdots \mathbf{H}_{d_n^{\star}}(W(\varphi_n))\right]$$

$$\vdots$$

$$+ \mathbf{E}\left[\frac{1}{V_n} W(\varphi_n) \mathbf{H}_{d_1^{\star}}(W(\varphi_1)) \mathbf{H}_{d_2^{\star}}(W(\varphi_2)) \cdots \mathbf{H}_{d_n^{\star}}(W(\varphi_n))\right]. \quad (14)$$

Because of Lemma 3 (see the technical lemmas section), for each i, the term

$$\mathbf{E}\left[\frac{1}{V_n} W(\varphi_i) \mathbf{H}_{d_1^{\star}}(W(\varphi_1)) \mathbf{H}_{d_2^{\star}}(W(\varphi_2)) \cdots \mathbf{H}_{d_n^{\star}}(W(\varphi_n))\right]$$

is non null if d_i^* is the only odd integer in $\{d_r^*\}$, $1 \leq r \leq n$. Thus, $a_{i_1,\cdots,i_{2k+1}} \neq 0$ if there is only one odd integer in $\{d_r^*\}$, $1 \leq r \leq n$. Let d_i^* with $1 \leq i \leq n$ be this only odd integer. Then, if $j \neq i$, by Lemma 3,

$$\mathbf{E}\left[\frac{1}{V_n} W(\varphi_j) \mathbf{H}_{d_1^*}(W(\varphi_1)) \mathbf{H}_{d_2^*}(W(\varphi_2)) \cdots \mathbf{H}_{d_n^*}(W(\varphi_n))\right] = 0.$$

Thus, using (14) yields

$$a_{i_1,\cdots,i_{2k+1}} = \mathbf{E}\left[\frac{1}{V_n} W(\varphi_i) \mathbf{H}_{d_1^*}(W(\varphi_1)) \mathbf{H}_{d_2^*}(W(\varphi_2)) \cdots \mathbf{H}_{d_n^*}(W(\varphi_n))\right]$$

if there is only one odd integer in the sequence $\{d_r^*\}$, $1 \leq r \leq n$ and $a_{i_1,\cdots,i_{2k+1}} = 0$ if there is more than one odd integer in the sequence $\{d_r^*\}$, $1 \leq r \leq n$. Using the equality in law between $W(\varphi_i)$ and $W(\varphi_1)$, one can perform an index swap ($i \leftrightarrow 1$) to finally obtain the desired result.

Proof (Proof of Proposition 3). We recall the following explicit formula for the Hermite polynomials (see for example [1], p. 775):

$$\mathbf{H}_d(x) = d! \sum_{l=0}^{[\frac{d}{2}]} \frac{(-1)^l}{2^l l! (d-2l)!} x^{d-2l}. \tag{15}$$

Using (15) and (10) we can write

$$a_{i_1,\cdots,i_{2k+1}} = \mathbf{E}\left[\frac{1}{V_n} W(\varphi_1) \mathbf{H}_{2d_1+1}(W(\varphi_1)) \mathbf{H}_{2d_2}(W(\varphi_2)) \cdots \mathbf{H}_{2d_n}(W(\varphi_n))\right]$$

$$= (2d_1+1)!(2d_2)!\cdots(2d_n)! \sum_{l_1=0}^{d_1} \sum_{l_2=0}^{d_2} \cdots \sum_{l_n=0}^{d_n} \frac{(-1)^{l_1+l_2+\cdots+l_n}}{2^{l_1+l_2+\cdots+l_n} l_1! \cdots l_n!}$$

$$\times \frac{\mathbf{E}\left[\frac{1}{V_n} W(\varphi_1)^{2d_1+2-2l_1} W(\varphi_2)^{2d_2-2l_2} \cdots W(\varphi_n)^{2d_n-2l_n}\right]}{(2d_1+1-2l_2)!(2d_2-2l_2)!\cdots(2d_n-2l_n)!}.$$

At this point, we use Lemma 5 (see the technical lemmas section) to rewrite the expectation in the last equation:

$$\mathbf{E}\left[\frac{1}{V_n} W(\varphi_1) \mathbf{H}_{2d_1+1}(W(\varphi_1)) \mathbf{H}_{2d_2}(W(\varphi_2)) \cdots \mathbf{H}_{2d_n}(W(\varphi_n))\right]$$

$$= (2d_1+1)!(2d_2)!\cdots(2d_n)! \sum_{l_1=0}^{d_1} \sum_{l_2=0}^{d_2} \cdots \sum_{l_n=0}^{d_n} \frac{(-1)^{l_1+l_2+\cdots+l_n}}{2^{l_1+l_2+\cdots+l_n} l_1! \cdots l_n!}$$

$$\times \frac{2^{d_1+1+d_2+\cdots+d_n-(l_1+l_2+\cdots+l_n)+\frac{n-1}{2}}}{(2\pi)^{\frac{n}{2}}(2d_1+1-2l_2)!(2d_2-2l_2)!\cdots(2d_n-2l_n)!}$$

$$\times \frac{\Gamma\left(d_1+1+d_2+\cdots+d_n-(l_1+l_2+\cdots+l_n)+\frac{n-1}{2}\right)}{\Gamma\left(d_1+1+d_2+\cdots+d_n-(l_1+l_2+\cdots+l_n)+\frac{n}{2}\right)}$$

$$\times \Gamma\left(d_1+1-l_1+\frac{1}{2}\right)\Gamma\left(d_2-l_2+\frac{1}{2}\right)\cdots\Gamma\left(d_n-l_n+\frac{1}{2}\right)$$

$$= (2d_1+1)!(2d_2)!\cdots(2d_n)! \sum_{l_1=0}^{d_1}\sum_{l_2=0}^{d_2}\cdots\sum_{l_n=0}^{d_n} \frac{(-1)^{l_1+l_2+\cdots+l_n}}{2^{2(l_1+l_2+\cdots+l_n)}l_1!\cdots l_n!}$$

$$\times \frac{2^{d_1+1+d_2+\cdots+d_n-\frac{1}{2}}}{\pi^{\frac{n}{2}}(2d_1+1-2l_2)!(2d_2-2l_2)!\cdots(2d_n-2l_n)!}$$

$$\times \frac{\Gamma\left(d_1+1+d_2+\cdots+d_n-(l_1+l_2+\cdots+l_n)+\frac{n-1}{2}\right)}{\Gamma\left(d_1+1+d_2+\cdots+d_n-(l_1+l_2+\cdots+l_n)+\frac{n}{2}\right)}$$

$$\times \Gamma\left(d_1+1-l_1+\frac{1}{2}\right)\Gamma\left(d_2-l_2+\frac{1}{2}\right)\cdots\Gamma\left(d_n-l_n+\frac{1}{2}\right).$$

We claim that for any integers $d \geq l$,

$$\frac{(-1)^l}{2^{-2l}l!(2d-2l)!}\Gamma\left(d-l+\frac{1}{2}\right) = \sqrt{\pi}\frac{2^{-2d}(-1)^l}{d!}C_d^l. \tag{16}$$

Recall the relation satisfied by the Gamma function: for every $z > 0$,

$$\Gamma(z+1) = z\Gamma(z) \text{ and } \Gamma(z)\Gamma\left(z+\frac{1}{2}\right) = \sqrt{\pi}2^{1-2z}\Gamma(2z). \tag{17}$$

Then

$$\frac{(-1)^l}{2^{-2l}l!(2d-2l)!}\Gamma\left(d-l+\frac{1}{2}\right) = \frac{(-1)^l}{2^{-2l}l!(2d-2l)!}\frac{\Gamma\left(d-l+1+\frac{1}{2}\right)}{d-l-\frac{1}{2}}$$

$$= \frac{(-1)^l}{2^{-2l}l!(2d-2l)!}\frac{\Gamma(2d-2l+2)}{\Gamma(d-l+1)}\sqrt{\pi}2^{1-2(d-l+1)}$$

$$= \sqrt{\pi}2^{-2d}\frac{(-1)^l}{l!(2d-2l)!}\frac{(2d-2l+1)!}{(d-l)!(2d-2l+1)}$$

$$= \sqrt{\pi}\frac{2^{-2d}(-1)^l}{d!}C_d^l$$

and (16) is proved. In the same way, using only the second relation in (17), we obtain

$$\frac{(-1)^{l_1}}{2^{-2l_1}l_1!(2d_1+1-2l_1)!}\Gamma\left(d_1+1-l_1+\frac{1}{2}\right) = \sqrt{\pi}\frac{2^{-1-2d_1}(-1)^{l_1}}{d_1!}C_{d_1}^{l_1}. \quad (18)$$

Putting together (16) and (18) we find

$$\mathbf{E}\left[\frac{1}{V_n}W(\varphi_1)\mathbf{H}_{2d_1+1}\left(W(\varphi_1)\right)\mathbf{H}_{2d_2}\left(W(\varphi_2)\right)\cdots\mathbf{H}_{2d_n}\left(W(\varphi_n)\right)\right]$$

$$= \frac{(2d_1+1)!(2d_2)!\cdots(2d_n)!}{d_1!d_2!\cdots d_n!}2^{-(d_1+\cdots+d_n)-\frac{1}{2}}$$

$$\times \sum_{l_1=0}^{d_1}\sum_{l_2=0}^{d_2}\cdots\sum_{l_n=0}^{d_n}(-1)^{l_1+l_2+\cdots+l_n}C_{d_1}^{l_1}\cdots C_{d_n}^{l_n}$$

$$\times \frac{\Gamma\left(d_1+1+d_2+\cdots+d_n-(l_1+l_2+\cdots+l_n)+\frac{n-1}{2}\right)}{\Gamma\left(d_1+1+d_2+\cdots+d_n-(l_1+l_2+\cdots+l_n)+\frac{n}{2}\right)}.$$

By Stirling's formula, when n goes to infinity, we have

$$\frac{\Gamma\left(d_1+1+d_2+\cdots+d_n-(l_1+l_2+\cdots+l_n)+\frac{n-1}{2}\right)}{\Gamma\left(d_1+1+d_2+\cdots+d_n-(l_1+l_2+\cdots+l_n)+\frac{n}{2}\right)}$$

$$\sim \frac{1}{\sqrt{k+1-(l_1+\cdots+l_n)+\frac{n}{2}}}.$$

Therefore we need to study the behavior of the sequence

$$t_n := \sum_{l_1=0}^{d_1}\sum_{l_2=0}^{d_2}\cdots\sum_{l_n=0}^{d_n}(-1)^{l_1+l_2+\cdots+l_n}C_{d_1}^{l_1}\cdots C_{d_n}^{l_n}\frac{1}{\sqrt{k+1-(l_1+\cdots+l_n)+\frac{n}{2}}}$$

as $n \to \infty$. We can write

$$t_n = \frac{1}{\sqrt{n}}\sqrt{2}g\left(\frac{1}{n}\right)$$

where

$$g(x) = \sum_{l_1=0}^{d_1}\sum_{l_2=0}^{d_2}\cdots\sum_{l_n=0}^{d_n}(-1)^{l_1+l_2+\cdots+l_n}C_{d_1}^{l_1}\cdots C_{d_n}^{l_n}\frac{1}{\sqrt{2k+2-(l_1+\cdots+l_n)x+1}}.$$

Since for every $d \geq 1$
$$\sum_{l=0}^{d}(-1)^{l}C_{d}^{l}=0,$$
we clearly have $g(0) = 0$. The qth derivative of g at zero is
$$g^{(q)}(0) = (-1)^{q}\frac{(2q-1)!!}{2^{q}}\,[2k+2-(l_{1}+\cdots+l_{n})]^{q}.$$
Repeatedly using the relation $C_{n}^{k} = \frac{n}{k}C_{n-1}^{k-1}$ we can prove that
$$\sum_{l=0}^{d}(-1)^{l}C_{d}^{l}l^{q} = 0$$
for every $q = 0, 1, \cdots, d - 1$. Therefore the first non-zero term in the Taylor decomposition of the function g around zero is
$$\sum_{l_{1}=0}^{d_{1}}\sum_{l_{2}=0}^{d_{2}}\cdots\sum_{l_{n}=0}^{d_{n}}(-1)^{l_{1}+l_{2}+\cdots+l_{n}}C_{d_{1}}^{l_{1}}\cdots C_{d_{n}}^{l_{n}}l_{1}^{d_{1}}\cdots l_{n}^{d_{n}}$$
which appears when we take the derivative of order $d_{1} + d_{2} + \cdots + d_{n}$. We obtain that, for x close to zero,
$$g(x) \sim (-1)^{d_{1}+\cdots+d_{n}}\frac{(2(d_{1}+\cdots+d_{n})-1)!!}{2^{d_{1}+\cdots+d_{n}}}\prod_{j=0}^{n}\sum_{l_{j}=0}^{d_{j}}(-1)^{l_{j}}C_{d_{j}}^{l_{j}}l_{j}^{d_{j}} \times H(d_{1},\cdots,d_{n})x^{|A|}$$
where
$$A = \{d_{1},\cdots,d_{n}\} \setminus \{0\} = \{2d_{1}+1, 2d_{2},\cdots, 2d_{n}\} \setminus \{0, 1\}$$
and $H(d_{1}, \cdots, d_{n})$ is the coefficient of $l_{1}^{d_{1}}\cdots l_{n}^{d_{n}}$ in the expansion of $(l_{1} + \cdots + l_{n})^{d_{1}+\cdots+d_{n}}$. That is
$$H(d_{1},\cdots,d_{n}) = C_{d_{1}+\cdots+d_{n}}^{d_{1}}C_{d_{2}+\cdots+d_{n}}^{d_{2}}\cdots C_{d_{n-1}+d_{n}}^{d_{n-1}} = \frac{(d_{1}+\cdots+d_{n})!}{d_{1}!\cdots d_{n}!}.$$
We finally have

$$a_{i_1,\cdots,i_{2k+1}}$$
$$= \frac{(2d_1+1)!(2d_2)!\cdots(2d_n)!}{(d_1!d_2!\cdots d_n!)^2} 2^{-(d_1+\cdots+d_n)}(-1)^{d_1+\cdots+d_n}\frac{(2(d_1+\cdots+d_n)-1)!!}{2^{d_1+\cdots+d_n}}$$
$$\times \left(\prod_{j=0}^{n}\sum_{l_j=0}^{d_j}(-1)^{l_j}C_{d_j}^{l_j}l_j^{d_j}\right)\frac{(d_1+\cdots+d_n)!}{d_1!\cdots d_n!}\frac{1}{n^{\frac{1}{2}+|A|}}$$
$$= k!(2k-1)!!\frac{(2d_1+1)!(2d_2)!\cdots(2d_n)!}{(d_1!d_2!\cdots d_n!)^2}2^{-2k}(-1)^k$$
$$\times \left(\prod_{j=0}^{n}\sum_{l_j=0}^{d_j}(-1)^{l_j}C_{d_j}^{l_j}l_j^{d_j}\right)\frac{1}{n^{\frac{1}{2}+|A|}}$$
$$= k!(2k-1)!!\frac{(2d_1+1)!(2d_2)!\cdots(2d_n)!}{(d_1!d_2!\cdots d_n!)^2}2^{-2k}(-1)^k\left(\prod_{j=0}^{n}t(d_j)\right)\frac{1}{n^{\frac{1}{2}+|A|}}$$

with for $i=1,\cdots,n$

$$t(d_j):=\sum_{l_j=0}^{d_j}(-1)^{l_j}C_{d_j}^{l_j}l_j^{d_j}. \tag{19}$$

5.2 Proof of Theorem 1

Before proving our main result, let us discuss a particular case as an example in order to better understand the general phenomenon. This is both useful and important in order to have a good overview of how a simple case works. Assume that $k=0$ and $l=1$. The corresponding summand in (27) reduces to

$$\frac{1}{3!}\sum_{u=1}^{n}a_u\sum_{j_1,j_2=1}^{n}a_{u,j_1,j_2}I_2\left(\varphi_{j_1}\otimes\varphi_{j_2}\right).$$

Its L^2-norm is

$$\frac{1}{3}\sum_{j_1,j_2=1}^{n}\left(\sum_{u=1}^{n}a_u a_{u,j_1,j_2}\right)^2 = \frac{1}{3}\sum_{j_1=1}^{n}\left(\sum_{u=1}^{n}a_u a_{u,j_1,j_1}\right)^2$$

because $a_{u,j_1,j_2}=0$ if $j_1\neq j_2$. Using (11), it reduces to a quantity equivalent to

$$\frac{1}{3}(na_1^2 a_{1,1,1}^2 + n((n-1)a_1 a_{1,1,2})^2)$$

which, using (11) again, is of order

$$n\left(\frac{1}{\sqrt{n}}\right)^2\left(\frac{1}{n^{\frac{3}{2}}}\right)^2 + n\left((n-1)\frac{1}{\sqrt{n}}\frac{1}{n^{\frac{3}{2}}}\right)^2 \sim n^{-1}.$$

Proof (Proof of Theorem 1). Observe that the integers $r+1+k$ and $r+1+2m-k$ both have to be odd numbers (otherwise the coefficients $a_{u_1,u_2,\cdots,u_{r+1},i_1,\cdots,i_k}$ and $a_{u_1,u_2,\cdots,u_{r+1},i_{k+1},\cdots,i_{2m}}$ vanish). This implies two cases: either r is even and k is even or r is odd and k is odd. Thus, we can write

$$\mathbf{E}\left(\left(\langle DF_n, D(-L)^{-1}F_n\rangle - 1\right)^2\right)$$

$$= \sum_{m\geq 1}(2m)! \sum_{i_1,\cdots,i_{2m}=1}^{n}\left(\sum_{k=0}^{2m}\frac{1}{2k!}\frac{1}{(2m-2k)!}\sum_{r\geq 0}\frac{1}{(2r)!}\frac{1}{2m-2k+2r+1}\right.$$

$$\left.\times \sum_{u_1,\cdots,u_{2r+1}=1}^{n} a_{u_1,u_2,\cdots,u_{2r+1},i_1,\cdots,i_{2k}} a_{u_1,u_2,\cdots,u_{2r+1},i_{2k+1},\cdots,i_{2m}}\right)^2$$

$$+ \sum_{m\geq 1}(2m)! \sum_{i_1,\cdots,i_{2m}=1}^{n}\left(\sum_{k=0}^{2m}\frac{1}{(2k+1)!}\frac{1}{(2m-2k-1)!}\right.$$

$$\times \sum_{r\geq 0}\frac{1}{(2r-1)!}\frac{1}{2m-2k+2r+1}$$

$$\left.\times \sum_{u_1,\cdots,u_{2r}=1}^{n} a_{u_1,u_2,\cdots,u_{2r},i_1,\cdots,i_{2k+1}} a_{u_1,u_2,\cdots,u_{2r},i_{2k+2},\cdots,i_{2m}}\right)^2. \quad (20)$$

Let us treat the first part of the sum (20). Assume that the number of common numbers occurring in the sets $\{u_1,\cdots,u_{2r+1}\}$ and $\{i_1,\cdots,i_{2k}\}$ is x and the number of common numbers occurring in the sets $\{u_1,\cdots,u_{2r+1}\}$ and $\{i_{2k+1},\cdots,i_{2m-2k}\}$ is y. This can be formally written as

$$|\{u_1,\cdots,u_{2r+1}\}\cap\{\{i_1,\cdots,i_{2k}\}| = x$$

and

$$|\{u_1,\cdots,u_{2r+1}\}\cap\{i_{2k+1},\cdots,i_{2m-2k}\}| = y.$$

It is clear that

$$x \leq (2r+1) \wedge 2k \text{ and } y \leq (2r+1) \wedge 2m-2k.$$

This also implies $x + y \leq 2m$. According to the definitions of x and y, it can be observed that x and y must be even. We will denote them by $2x$ and $2y$ from now on.

The next step in the proof is to determine how many distinct sequences of numbers can occur in the set

$$\{u_1, \cdots, u_{2r+1}, i_1, \cdots, i_{2k}\}.$$

We can have sequences of lengths (all of the lengths that we consider from now on are greater or equal to one) $2c_1, 2c_2, \cdots, 2c_{l_1}$ with $2(c_1 + \cdots + c_{l_1}) = 2x$ in the set $\{u_1, \cdots, u_{2r+1}\} \cap \{i_1, \cdots, i_{2k}\}$ but also sequences of lengths $2d_1, 2d_2, \cdots, 2d_{l_2}$ with $2(d_1 + \cdots + d_{l_2}) = 2k - 2x$ in the set $\{i_1, \cdots, i_{2k}\} \setminus \{u_1, \cdots, u_{2r+1}\}$ as well as sequences of lengths $2e_1 + 1, 2e_2, \cdots, 2e_{l_3}$ with $1 + 2(e_1 + \cdots + e_{l_3}) = 2r + 1 - 2x$ in the set $\{u_1, \cdots, u_{2r+1}\} \setminus \{i_1, \cdots, i_{2k}\}$. In this last sequence we have one (and only one) length equal to 1 (because we are allowed to choose only one odd number in the set $\{u_1, \cdots, u_{2r+1}\} \setminus \{i_1, \cdots, i_{2k}\}$). We will have, if we have a configuration as above,

$$a_{u_1, u_2, \cdots, u_{2r+1}, i_1, \cdots, i_{2k}} \leq c(r, c, e) n^{-\frac{1}{2} - l_1 - l_2 - l_3}$$

where

$$c(r, c, e) = r!(2r-1)!! \frac{(2c_1)! \cdots (2c_{l_1})!(2e_1 + 1)!(2e_2)! \cdots (2e_{l_3})!}{(c_1! \cdots c_{l_1}! e_1! \cdots e_{l_3}!)^2}$$
$$\times t(c_1) \cdots t(c_{l_1}) t(e_1) \cdots t(e_{l_3}) \qquad (21)$$

and the constants $t(\cdot)$ are given by (19).

In the same way, assuming that we have sequences of lengths $2f_1, 2f_2, \cdots, 2f_{l_4}$ with $2(f_1 + \cdots + f_{l_4}) = 2m - 2k - 2y$ in the set $\{i_{2k+1}, \cdots, i_{2m}\} \setminus \{u_1, \cdots, u_{2r+1}\}$ and sequences of lengths $2g_1 + 1, 2g_2, \cdots, 2g_{l_5}$ with $1 + 2(g_1 + \cdots + g_5) = 2r + 1 - 2y$ in the set $\{u_1, \cdots, u_{2r+1}\} \setminus \{i_{2k+1}, \cdots, i_{2m}\}$. We will obtain

$$a_{u_1, u_2, \cdots, u_{2r+1}, i_{2k+1}, \cdots, i_{2n}} \leq c(k, c, d) n^{-\frac{1}{2} - l_1 - l_4 - l_5 + 1}$$

with $c(k, c, d)$ defined as in (21). The sum over $u_1, \cdots, u_{r+1}$ from 1 to n reduces to a sum of $l_1 + l_3 + l_5 - 1$ distinct indices from 1 to n. Therefore we get

$$\sum_{u_1, \cdots, u_{2r+1} = 1}^{n} a_{u_1, u_2, \cdots, u_{2r+1}, i_1, \cdots, i_{2k}} a_{u_1, u_2, \cdots, u_{2r+1}, i_{2k+1}, \cdots, i_{2n}}$$
$$\leq c(k, r, m) n^{-l_1 - l_2 - l_4}$$

with

$$c(k, r, m) = \sum_{x+y=2m} \sum_{c_1 + \cdots + c_{l_1} = x} \sum_{d_1 + \cdots + d_{l_2} = y} \sum_{e_1 + \cdots + e_{l_3} = r - x} c(r, c, e) c(k, c, d).$$
$$(22)$$

We need to consider the sum $i_1, \cdots, i_{2m}$ from 1 to n. It reduces to a sum over $l_2 + l_4$ distinct indices. Thus

$$\sum_{i_1,\cdots,i_{2m}=1}^{n} \left(\sum_{k=0}^{2m} \frac{1}{2k!} \frac{1}{(2m-2k)!} \sum_{r\geq 0} \frac{1}{(2r)!} \frac{1}{2m-2k+2r+1} \sum_{u_1,\cdots,u_{2r+1}=1}^{n} \sum_{u_1,\cdots,u_{2r+1}=1}^{n} \right.$$

$$a_{u_1,u_2,\cdots,u_{2r+1},i_1,\cdots,i_{2k}} a_{u_1,u_2,\cdots,u_{2r+1},i_{2k+1},\cdots,i_{2m}} \Big)^2$$

$$\leq n^{l_2+l_4} \left(\frac{1}{n^{2l_1+l_2+l_4}} \right)^2 \sum_{k=0}^{2m} \frac{1}{2k!} \frac{1}{(2m-2k)!} \sum_{r\geq 0} \frac{1}{(2r)!} \frac{1}{2m-2k+2r+1} c(k,r,m)$$

$$= \frac{1}{n^{2l_1+l_2+l_4}} \left(\sum_{k=0}^{2m} \frac{1}{2k!} \frac{1}{(2m-2k)!} \sum_{r\geq 0} \frac{1}{(2r)!} \frac{1}{2m-2k+2r+1} c(k,r,m) \right)^2.$$

Note that either $l_1 + l_2 \geq 1$ or $l_1 + l_4 \geq 1$ (this is true because $m \geq 1$). Then this term is at most of order of n^{-1}.

Let us now look at the second part of the sum in (20). Suppose that in the sets $\{u_1, \cdots, u_{2r}\} \cap \{i_1, \cdots, i_{2k+1}\}$, $\{i_1, \cdots, i_{2k+1}\} \setminus \{u_1, \cdots, u_{2r}\}$, $\{u_1, \cdots, u_{2r}\} \setminus \{i_1, \cdots, i_{2k+1}\}$, $\{i_{2k+2}, \cdots, i_{2m-2k}\} \setminus \{u_1, \cdots, u_{2r}\}$, $\{u_1, \cdots, u_{2r}\} \setminus \{i_{2k+2}, \cdots, i_{2m-2k}\}$ we have sequences with lengths

$$p_1, \ p_2, \ p_3, \ p_4, \ p_5$$

respectively (the analogous of $l_1, \cdots, l_5$ above). Then the behavior with respect to n of

$$\sum_{u_1,\cdots,u_{2r}=1}^{n} a_{u_1,u_2,\cdots,u_{2r},i_1,\cdots,i_{2k+1}} a_{u_1,u_2,\cdots,u_{2r},i_{2k+2},\cdots,i_{2m}}$$

is of order of $n^{p_1+p_3} \frac{1}{n^{2p_1+p_3+p_4}}$. Therefore the behavior with respect to n of the second sum in (20) is of order

$$n^{p_2+1+p_4+1} \left(\frac{1}{n^{1+2p_1+p_2+p_4}} \right)^2 = \frac{1}{n^{2p_1+p_2+p_4}}.$$

Again, since either $p_1 + p_2 \geq 1$ or $p_1 + p_4 \geq 1$, the behavior of the term is at most of order n^{-1}. Therefore

$$\mathbf{E}\left((\langle DF_n, D(-L)^{-1} F_n \rangle - 1)^2 \right) \leq \frac{c_0}{n}$$

where the constant c_0 is given by (13). The fact that the sum over m is finite is a consequence of the following argument: $\langle DF_n, D(-L)^{-1} F_n \rangle$ belongs to $\mathbb{D}^{\infty,2}(\Omega)$

(which is true based on the derivation rule—Exercise 1.2.13 in [16]—and since F_n belongs to $\mathbb{D}^{\infty,2}$ as a consequence of Proposition 1.2.3 in [16]), this implies that $\sum_m m! m^k \|h_m^{(n)}\|_2 < \infty$ for every k where $h_m^{(n)}$ is given by (28). Therefore, the constant $c(m, k, r)$ defined in (22) behaves at most as a power function with respect to m.

5.3 Some Technical Lemmas

Let us first give the following lemma that can be proved using integration by part.

Lemma 1. *For every* $1 \leq i_1, ..., i_k \leq n$, *let* $a_{i_1,\cdots i_k}$ *be as defined in* (8). *Let* $d_r, 1 \leq r \leq n$ *denote the number of times the integer r appears in the sequence* $\{i_1, i_2, \cdots, i_k\}$ *with* $\sum_{r=1}^n d_r = k$. *Then we have*

$$a_{i_1,\cdots i_k} = \mathbf{E}\left(\frac{W_n}{V_n} \prod_{r=1}^n \mathbf{H}_{d_r}(W(\varphi_r))\right).$$

Proof. If $X \sim \mathcal{N}(0, 1)$, then for any $g \in C^{(n)}(\mathbb{R})$ with g and its derivatives having polynomial growth at infinity, we have the Gaussian integration by part formula

$$\mathbf{E}(g^{(n)}(X)) = \mathbf{E}(g(X)\mathbf{H}_n(X))$$

where $g^{(n)}(x) \stackrel{\text{def}}{=} \frac{d^n}{dx^n} g(x)$.

Notice that the function f defined in (7) satisfies $|f(x)| \leq C|x|, \forall x \in \mathbb{R}^n$ for a constant C, and thus applying the above integration by part formula recursively yields

$$\begin{aligned}
a_{i_1,\cdots i_k} &= \frac{1}{(\sqrt{2\pi})^n} \int_{\mathbb{R}^n} \left(\frac{\partial^k f}{\partial x_1^{d_1}, \cdots, x_n^{d_n}}\right)(x_1, \cdots, x_n) e^{-\frac{x_1^2}{2}} \cdots e^{-\frac{x_n^2}{2}} dx_1 \cdots dx_n \\
&= \frac{1}{(\sqrt{2\pi})^n} \int_{\mathbb{R}^n} \left(\frac{\partial^k f}{\partial x_1^{d_1}, \cdots, x_{n-1}^{d_{n-1}}}\right)(x_1, \cdots, x_n) \mathbf{H}_{d_n}(x_n) e^{-\frac{x_1^2}{2}} \cdots e^{-\frac{x_n^2}{2}} dx_1 \cdots dx_n \\
&= \frac{1}{(\sqrt{2\pi})^n} \int_{\mathbb{R}^n} f(x_1, \cdots, x_n) \prod_{r=1}^n \mathbf{H}_{d_r}(x_r) e^{-\frac{x_1^2}{2}} \cdots e^{-\frac{x_n^2}{2}} dx_1 \cdots dx_n \\
&= \mathbf{E}\left(\frac{W_n}{V_n} \prod_{r=1}^n \mathbf{H}_{d_r}(W(\varphi_r))\right).
\end{aligned}$$

This concludes the proof.

The next lemma ensures that $a_{i_1,\cdots i_k} = 0$ when k is even.

Lemma 2. *If k is even, then*
$$a_{i_1,\cdots i_k} = 0.$$

Proof. Let k be an even number and $d_1, d_2, \cdots, d_n$ be as defined in Lemma 1. By Lemma 1, we have

$$a_{i_1,\cdots i_k} = \sum_{u=1}^n \mathbf{E}\left(\frac{W(\varphi_u)}{V_n} \prod_{r=1}^n \mathbf{H}_{d_r}\left(W(\varphi_r)\right)\right). \tag{23}$$

Note that the product $\prod_{r=1}^n \mathbf{H}_{d_r}(W(\varphi_r))$ is an even function of $(W(\varphi_1), W(\varphi_2), \cdots, W(\varphi_n))$. Indeed, since k is even and $\sum_{r=1}^n d_r = k$, either all of the integers $d_r, r \leq n$ are even or there is an even number of odd integers in $d_r, r \leq n$. In either case the product $\prod_{r=1}^n \mathbf{H}_{d_r}(W(\varphi_r))$ is an even function of $(W(\varphi_1), W(\varphi_2), \cdots, W(\varphi_n))$, since $\mathbf{H}_m(x) = \mathbf{H}_m(-x)$ for all even $m \in \mathbb{N}$ and $\mathbf{H}_m(x) = -\mathbf{H}_m(-x)$ for all odd $m \in \mathbb{N}$.

Thus for each $u \leq n$, the expression $\frac{W(\varphi_u)}{V_n} \prod_{r=1}^n \mathbf{H}_{d_r}(W(\varphi_r))$ is an odd function of $W(\varphi_u)$ and thus has expectation zero since $W(\phi_u)$ is a standard Gaussian random variable. The fact that (23) is a sum of such expectations concludes the proof.

Remark 4. As a consequence of Lemma 2, we have

$$F_n = \sum_{k \geq 0} \frac{1}{(2k+1)!} \sum_{i_1,\cdots,i_{2k+1}=1}^n a_{i_1,\cdots,i_{2k+1}} I_{2k+1}\left(\varphi_{i_1} \otimes \cdots \otimes \varphi_{i_{2k+1}}\right). \tag{24}$$

This implies that in order to compute the coefficients $a_{i_1,\cdots i_k}$, it suffices to focus on the case where k is odd.

Based on that last remark, we state the following lemma.

Lemma 3. *Let $k \geq 0$ be a positive integer and let $d_r, 1 \leq r \leq n$ denote the number of times the integer r appears in the sequence $\{i_1, i_2, \cdots, i_{2k+1}\}$ with $\sum_{r=1}^n d_r = 2k+1$. Then, if there is more than one odd integer in the sequence $d_r, 1 \leq r \leq n$, for each $1 \leq i \leq n$,*

$$\mathbf{E}\left[\frac{1}{V_n} W(\varphi_i) \mathbf{H}_{d_1}(W(\varphi_1)) \mathbf{H}_{d_2}(W(\varphi_2)) \cdots \mathbf{H}_{d_n}(W(\varphi_n))\right] = 0.$$

Proof. Note that the equality $\sum_{r=1}^n d_r = 2k+1$ implies that there can only be an odd number of odd integers in the sequence d_r, otherwise the sum $\sum_{r=1}^n d_r$ could not be odd. Therefore, more than one odd integer in the sequence d_r means that there are at least three of them. We will prove the lemma for this particular case of three odd integers in the sequence d_r for the sake of readability of the proof, as the other cases follow with the exact same arguments. Hence, assume that there are three odd integers d_j, d_k and d_l in the sequence $d_r, 1 \leq r \leq n$. We will first consider the case where i is different than j, k, l. Then,

$$\mathbf{E}\left[\frac{1}{V_n}W(\varphi_i)\mathbf{H}_{d_1}(W(\varphi_1))\mathbf{H}_{d_2}(W(\varphi_2))\cdots\mathbf{H}_{d_n}(W(\varphi_n))\right]$$

$$=\frac{1}{(2n)^{\frac{n}{2}}}\int_{\mathbb{R}^n}\frac{x_i\mathbf{H}_{d_1}(x_1)\cdots\mathbf{H}_{d_n}(x_n)}{\sqrt{x_1^2+\cdots+x_n^2}}e^{-\frac{1}{2}(x_1^2+\cdots+x_n^2)}dx_1\cdots dx_n$$

$$=\frac{1}{(2n)^{\frac{n}{2}}}\int_{\mathbb{R}^{n-1}}x_i\mathbf{H}_{d_1}(x_1)\cdots\mathbf{H}_{d_{j-1}}(x_{j-1})\mathbf{H}_{d_{j+1}}(x_{j+1})\cdots\mathbf{H}_{d_n}(x_n)$$

$$\times\left(\int_{\mathbb{R}}\frac{\mathbf{H}_{d_j}(x_j)}{\sqrt{x_1^2+\cdots+x_n^2}}e^{-\frac{x_j^2}{2}}dx_j\right)\exp\left[-\frac{1}{2}\sum_{\substack{p=1\\p\neq j}}^n x_p^2\right]dx_1\cdots dx_{j-1}dx_{j+1}\cdots dx_n,$$

d_j being odd, $\mathbf{H}_{d_j}$ is an odd function of x_j and $x_j\mapsto\frac{\mathbf{H}_{d_j}(x_j)}{\sqrt{x_1^2+\cdots+x_n^2}}e^{-\frac{x_j^2}{2}}$ is also an odd function of x_j. Thus, $\int_{\mathbb{R}}\frac{\mathbf{H}_{d_j}(x_j)}{\sqrt{x_1^2+\cdots+x_n^2}}e^{-\frac{x_j^2}{2}}dx_j=0$ and finally

$$\mathbf{E}\left[\frac{1}{V_n}W(\varphi_i)\mathbf{H}_{d_1}(W(\varphi_1))\mathbf{H}_{d_2}(W(\varphi_2))\cdots\mathbf{H}_{d_n}(W(\varphi_n))\right]=0.$$

The other cases one could encounter is when $i=j$ or $i=k$ or $i=l$ and the proof follows based on the exact same argument.

In the following lemma, we compute the L^2 norm of F_n. This technical result is needed in the computation of the Berry–Esséen bound.

Lemma 4. *Let $a_{i_1,\cdots,i_{2k+1}}$ be as given in (24). Then, for every $n\in\mathbb{N}$, we have*

$$\|F_n\|_{L^2(\Omega)}^2=\sum_{k\geq 0}\frac{1}{(2k+1)!}\sum_{i_1,\cdots,i_{2k+1}=1}^n a_{i_1,\cdots,i_{2k+1}}^2=1.$$

Proof. Firstly, using the isometry of multiple stochastic integrals and the orthogonality of the kernels φ_i, one can write

$$\mathbf{E}(F_n^2)=\sum_{k\geq 0}\left(\frac{1}{(2k+1)!}\right)^2(2k+1)!\sum_{\substack{i_1,\cdots,i_{2k+1}=1\\j_1,\cdots,j_{2k+1}=1}}^n a_{i_1,\cdots,i_{2k+1}}a_{j_1,\cdots,j_{2k+1}}$$

$$\times\langle\varphi_{i_1}\otimes\cdots\otimes\varphi_{i_{2k+1}},\varphi_{j_1}\otimes\cdots\otimes\varphi_{j_{2k+1}}\rangle_{L^2([0,1]^{2k})}$$

$$=\sum_{k\geq 0}\frac{1}{(2k+1)!}\sum_{i_1,\cdots,i_{2k+1}=1}^n a_{i_1,\cdots,i_{2k+1}}^2.$$

Secondly, using the fact that $F_n^2 = \frac{W_n^2}{V_n^2}$, we have

$$\mathbf{E}\left(F_n^2\right) = \frac{1}{(2\pi)^{\frac{n}{2}}} \int_{\mathbb{R}^n} \frac{(x_1 + \cdots + x_n)^2}{x_1^2 + \cdots + x_n^2} e^{-\frac{1}{2}(x_1^2 + \cdots + x_n^2)} dx_1 \cdots dx_n$$

$$= \frac{1}{(2\pi)^{\frac{n}{2}}} \int_{\mathbb{R}^n} \frac{x_1^2 + \cdots + x_n^2}{x_1^2 + \cdots + x_n^2} e^{-\frac{1}{2}(x_1^2 + \cdots + x_n^2)} dx_1 \cdots dx_n = 1$$

because the mixed terms vanish as in the proof of Lemma 2.

The following lemma is the second key result for the computation of the coefficients.

Lemma 5. *Let $\{a_1, a_2, \cdots a_n\}$ be non-negative numbers. Then it holds that*

$$\mathbf{E}\left(\frac{W(\varphi_1)^{2a_1} W(\varphi_2)^{2a_2} \cdots W(\varphi_n)^{2a_n}}{V_n}\right)$$

$$= \frac{1}{(2\pi)^{\frac{n}{2}}} 2^{a_1 + \cdots + a_n + \frac{n-1}{2}} \frac{\Gamma(a_1 + \cdots + a_n + \frac{n-1}{2})}{\Gamma(a_1 + \cdots + a_n + \frac{n}{2})} \Gamma\left(a_1 + \tfrac{1}{2}\right) \cdots \Gamma\left(a_n + \tfrac{1}{2}\right).$$

Proof. Recall that if X is a Chi-squared random variable with n degrees of freedom (denoted by χ_n^2) then for any $m \geq 0$,

$$\mathbf{E}(X^m) = 2^m \frac{\Gamma(m + \frac{n}{2})}{\Gamma(\frac{n}{2})}.$$

where $\Gamma(\cdot)$ denotes the standard Gamma function.

When $k = 0$, the coefficients $a_{i_1, \ldots, i_{2k+1}}$ can be easily computed. Indeed, noticing that V_n^2 has a χ_n^2 distribution, we obtain

$$\sum_{i=1}^n a_i = \mathbf{E}\left(\sum_{i=1}^n \frac{1}{V_n} W(\varphi_i)^2\right) = \mathbf{E}\left((V_n^2)^{\frac{1}{2}}\right) = 2^{\frac{1}{2}} \frac{\Gamma(\frac{1}{2} + \frac{n}{2})}{\Gamma(\frac{n}{2})}.$$

Since $a_1 = a_2 = \cdots = a_n$ we obtain that for every $i = 1, \ldots, n$

$$a_i = \frac{2^{\frac{1}{2}}}{n} \frac{\Gamma(\frac{1}{2} + \frac{n}{2})}{\Gamma(\frac{n}{2})}.$$

By definition, we have

$$\mathbf{E}\left(\frac{W(\varphi_1)^{2a_1} W(\varphi_2)^{2a_2} \cdots W(\varphi_n)^{2a_n}}{V_n}\right)$$

$$= \frac{1}{(2\pi)^{\frac{n}{2}}} \int_{\mathbb{R}^n} \frac{x_1^{2a_1} x_2^{2a_2} \cdots x_n^{2a_n}}{\sqrt{x_1^2 + x_2^2 + \cdots + x_n^2}} e^{-\frac{1}{2}(x_1^2 + x_2^2 + \cdots + x_n^2)} dx_1 dx_2 \cdots dx_n$$

$$= \frac{1}{(2\pi)^{\frac{n}{2}}} I.$$

To compute the above integral I, we introduce n-dimensional polar coordinates. Set

$$x_1 = r \cos \theta_1$$

$$x_j = r \cos \theta_j \prod_{i=1}^{j-1} \sin \theta_i, \quad j = 2, \cdots, n-2$$

$$x_{n-1} = r \sin \psi \prod_{i=1}^{n-2} \sin \theta_i, \quad x_n = r \cos \psi \prod_{i=1}^{n-2} \sin \theta_i$$

with $0 \leq r < \infty$, $0 \leq \theta_i \leq \pi$ and $0 \leq \psi \leq 2\pi$. It can be easily verified that $x_1^2 + x_2^2 + \cdots + x_n^2 = r^2$. The Jacobian of the above transformation is given by

$$J = r^{n-1} \prod_{k=1}^{n-2} \sin^k \theta_{n-1-k}.$$

Therefore our integral denoted by I becomes

$$\int_0^\infty r^{2(a_1 + \cdots + a_n) + n - 2} e^{-\frac{r^2}{2}} dr \int_0^{2\pi} (\sin \psi)^{2a_{n-1} + 2a_n} (\cos \psi)^{2a_n} d\psi$$

$$\prod_{k=2}^{n-1} \int_0^\pi (\sin \theta_{n-k})^{2a_n + 2a_{n-1} + \cdots + 2a_{n-k+1} + k - 1} (\cos \theta_{n-k})^{2a_{n-k}} d\theta_{n-k}.$$

Let us compute the first integral with respect to dr. Using the change of variables $\frac{r^2}{2} = y$, we get

$$\int_0^\infty r^{2(a_1 + \cdots + a_n) + n - 2} e^{-\frac{r^2}{2}} dr = 2^{a_1 + \cdots + a_n + \frac{n-1}{2} - 1} \int_0^\infty dy\, y^{a_1 + \cdots + a_n + \frac{n-1}{2} - 1} e^{-y}$$

$$= 2^{a_1 + \cdots + a_n + \frac{n-1}{2} - 1} \Gamma\left(a_1 + \cdots + a_n + \frac{n-1}{2}\right).$$

Let us now compute the integral with respect to $d\psi$. We use the following formula: for every $a, b \in \mathbb{Z}$, it holds that

$$\int_0^{2\pi} (\sin\theta)^a (\cos\theta)^b \, d\theta = 2\beta\left(\tfrac{a+1}{2}, \tfrac{b+1}{2}\right) \text{ if } a \text{ and } b \text{ are even}$$
$$= 0, \quad \text{if } a \text{ or } b \text{ are odd}.$$

This implies that

$$\int_0^{2\pi} (\sin\psi)^{2a_n-1+2a_n} (\cos\psi)^{2a_n} \, d\psi = 2\beta\left(a_n + \tfrac{1}{2}, a_{n-1} + \tfrac{1}{2}\right).$$

Finally, we deal with the integral with respect to $d\theta_i$ for $i = 1$ to $n-2$. Using the fact that, for $a, b > -1$, it holds that

$$\int_0^{\frac{\pi}{2}} (\sin\theta)^a (\cos\theta)^b \, d\theta = \tfrac{1}{2}\beta\left(\tfrac{a+1}{2}, \tfrac{b+1}{2}\right)$$

yields

$$\int_0^{\pi} (\sin\theta_{n-k})^{2a_n+2a_{n-1}+\cdots+2a_{n-k+1}+k-1} (\cos\theta_{n-k})^{2a_{n-k}} \, d\theta_{n-k}$$
$$= \int_0^{\frac{\pi}{2}} (\sin\theta_{n-k})^{2a_n+2a_{n-1}+\cdots+2a_{n-k+1}+k-1} (\cos\theta_{n-k})^{2a_{n-k}} \, d\theta_{n-k}$$
$$+ \int_{\frac{\pi}{2}}^{\pi} (\sin\theta_{n-k})^{2a_n+2a_{n-1}+\cdots+2a_{n-k+1}+k-1} (\cos\theta_{n-k})^{2a_{n-k}} \, d\theta_{n-k}$$
$$= \tfrac{1}{2}\beta\left(a_n + \cdots + a_{n-k+1} + \tfrac{k}{2}, a_{n-k} + \tfrac{1}{2}\right)$$
$$+ \int_0^{\frac{\pi}{2}} (\sin(\theta_{n-k} + \tfrac{\pi}{2}))^{2a_n+2a_{n-1}+\cdots+2a_{n-k+1}+k-1} (\cos(\theta_{n-k} + \tfrac{\pi}{2}))^{2a_{n-k}} \, d\theta_{n-k}$$
$$= \beta\left(a_n + \cdots + a_{n-k+1} + \tfrac{k}{2}, a_{n-k} + \tfrac{1}{2}\right)$$

because $\sin(\theta + \tfrac{\pi}{2}) = \cos\theta$ and $\cos(\theta + \tfrac{\pi}{2}) = -\sin(\theta)$. By gathering the above calculations, the integral I becomes

$$I = 2^{a_1+\cdots+a_n+\tfrac{n-1}{2}} \Gamma\left(a_1 + \cdots + a_n + \tfrac{n-1}{2}\right) \beta\left(a_n + \tfrac{1}{2}, a_{n-1} + \tfrac{1}{2}\right)$$
$$\times \prod_{k=2}^{n-1} \beta\left(a_n + \cdots + a_{n-k+1} + \tfrac{k}{2}, a_{n-k} + \tfrac{1}{2}\right)$$
$$= 2^{a_1+\cdots+a_n+\tfrac{n-1}{2}} \Gamma\left(a_1 + \cdots + a_n + \tfrac{n-1}{2}\right) \frac{\Gamma\left(a_n + \tfrac{1}{2}\right) \Gamma\left(a_{n-1} + \tfrac{1}{2}\right)}{\Gamma(a_n + a_{n-1} + 1)}$$

$$\times \prod_{k=2}^{n-1} \frac{\Gamma\left(a_n + \cdots + a_{n-k+1} + \frac{k}{2}\right) \Gamma\left(a_{n-k} + \frac{1}{2}\right)}{\Gamma\left(a_n + a_{n-1} + \cdots + a_{n-k} + \frac{k+1}{2}\right)}$$

$$= 2^{a_1 + \cdots + a_n + \frac{n-1}{2}} \frac{\Gamma\left(a_1 + \cdots + a_n + \frac{n-1}{2}\right)}{\Gamma\left(a_1 + \cdots + a_n + \frac{n}{2}\right)} \Gamma\left(a_1 + \frac{1}{2}\right) \cdots \Gamma\left(a_n + \frac{1}{2}\right).$$

This concludes the proof.

The following lemma gives an expression of the bound appearing in the right hand side of (1) in terms of the coefficients of the chaos decomposition of the self-normalized sequence.

Lemma 6. *For every $n \geq 2$,*

$$\mathbf{E}\left(1 - \langle DF_n, D(-L)^{-1} F_n \rangle\right)^2$$

$$= \sum_{m \geq 1} (2m)! \sum_{i_1, \cdots, i_{2m}=1}^{n} \left(\sum_{k=0}^{2m} \frac{1}{k!} \frac{1}{(2m-k)!} \sum_{r \geq 0} \frac{1}{r!} \frac{1}{2m - k + r + 1} \right.$$

$$\left. \sum_{u_1, \cdots, u_{r+1}=1}^{n} a_{u_1, u_2, \cdots, u_{r+1}, i_1, \cdots, i_k} a_{u_1, u_2, \cdots, u_{r+1}, i_{k+1}, \cdots, i_{2m}} \right)^2.$$

Proof. Formula (24) yields

$$D_\alpha F_n = \sum_{k \geq 0} \frac{2k+1}{(2k+1)!} \sum_{i_1, \cdots, i_{2k+1}=1}^{n} a_{i_1, \cdots, i_{2k+1}} I_{2k}\left((\varphi_{i_1} \otimes \cdots \otimes \varphi_{i_{2k-1}})^\sim\right)(\cdot, \alpha) \tag{25}$$

(here $(\varphi_{i_1} \otimes \cdots \otimes \varphi_{i_{2k-1}})^\sim$ denotes the symmetrization of the function $\varphi_{i_1} \otimes \cdots \otimes \varphi_{i_k}$ with respect to its k variables) and

$$D_\alpha (-L)^{-1} F_n = \sum_{k} \frac{1}{(2k+1)!} \sum_{i_1, \cdots, i_{2k+1}=1}^{n} a_{i_1, \cdots, i_{2k+1}} I_{2k}\left((\varphi_{i_1} \otimes \cdots \otimes \varphi_{i_{2k+1}})^\sim\right)(\cdot, \alpha). \tag{26}$$

Using (25) and (26), we can calculate the following quantity.

$$\langle DF_n, D(-L)^{-1} F_n \rangle$$

$$= \sum_{k,l \geq 0} \frac{1}{(2k)!} \frac{1}{(2l+1)!} \sum_{i_1, \cdots, i_{2k+1}=1}^{n} a_{i_1, \cdots, i_{2k+1}} \sum_{j_1, \cdots, j_{2l+1}=1}^{n} a_{j_1, \cdots, j_{2l+1}}$$

$$\times \int_0^\infty d\alpha \, I_{2k}\left((\varphi_{i_1} \otimes \cdots \otimes \varphi_{i_{2k+1}})^\sim\right)(\cdot, \alpha) I_{2l}\left((\varphi_{j_1} \otimes \cdots \otimes \varphi_{j_{2l+1}})^\sim\right)(\cdot, \alpha)$$

$$= \sum_{k,l \geq 0} \frac{1}{(2k)!} \frac{1}{(2l+1)!} \sum_{u=1}^{n} \sum_{i_1,\cdots,i_{2k}=1}^{n} a_{u,i_1,\cdots,i_{2k}} \sum_{j_1,\cdots,j_{2l}=1}^{n} a_{u,j_1,\cdots,j_{2l}}$$
$$\times I_{2k}\left((\varphi_{i_1} \otimes \cdots \otimes \varphi_{i_{2k}})\right) I_{2l}\left((\varphi_{j_1} \otimes \cdots \otimes \varphi_{j_{2l}})\right).$$

The product formula (4) applied to the last equality yields

$$\sum_{i_1,\cdots,i_{2k}=1}^{n} a_{u,i_1,\cdots,i_{2k}} \sum_{j_1,\cdots,j_{2l}=1}^{n} a_{u,j_1,\cdots,j_{2l}} I_{2k}\left((\varphi_{i_1} \otimes \cdots \otimes \varphi_{i_{2k}})\right) I_{2l}\left((\varphi_{j_1} \otimes \cdots \otimes \varphi_{j_{2l}})\right)$$
$$= \sum_{r=0}^{(2k)\wedge(2l)} r! C_{2k}^{r} C_{2l}^{r}$$
$$\times \sum_{u_1,\cdots,u_r=1}^{n} \sum_{i_1,\cdots,i_{2k-r}=1}^{n} \sum_{j_1,\cdots,j_{2l-r}=1}^{n} a_{u,u_1,\cdots,u_r,i_1,\cdots,i_{2k-r}} a_{u,u_1,\cdots,u_r,j_1,\cdots,j_{2l-r}}$$
$$\times I_{2k+2l-2r}\left(\varphi_{i_1} \otimes \cdots \otimes \varphi_{i_{2k-r}} \otimes \varphi_{j_1} \otimes \cdots \otimes \varphi_{j_{2l-r}}\right)$$

and therefore we obtain

$$\langle DF_n, D(-L)^{-1} F_n \rangle$$
$$= \sum_{k,l \geq 0} \frac{1}{(2k)!} \frac{1}{(2l+1)!} \sum_{r=0}^{(2k)\wedge(2l)} r! C_{2k}^{r} C_{2l}^{r}$$
$$\times \sum_{u_1,\cdots,u_{r+1}=1}^{n} \sum_{i_1,\cdots,i_{2k-r}=1}^{n} \sum_{j_1,\cdots,j_{2l-r}=1}^{n} a_{u_1,u_2,\cdots,u_{r+1},i_1,\cdots,i_{2k-r}} a_{u_1,u_2,\cdots,u_{r+1},j_1,\cdots,j_{2l-r}}$$
$$\times I_{2k+2l-2r}\left(\varphi_{i_1} \otimes \cdots \otimes \varphi_{i_{2k-r}} \otimes \varphi_{j_1} \otimes \cdots \otimes \varphi_{j_{2l-r}}\right). \tag{27}$$

Remark 5. The chaos of order zero in the above expression is obtained for $k = l$ and $r = 2k$. It is therefore equal to

$$\sum_{k \geq 0} \frac{1}{(2k)!} \frac{1}{(2k+1)!} (2k)! \sum_{i_1,\cdots,i_{2k+1}=1}^{n} a_{i_1,\cdots,i_{2k+1}}^{2}$$

which is also equal to 1 as follows from Lemma 4. Therefore it will vanish when we consider the difference $1 - \langle DF_n, D(-L)^{-1} F_n \rangle$. This difference will have only chaoses of even orders.

By changing the order of summation and by using the changes of indices $2k - r = k'$ and $2l - r = l'$, we can write

$$\langle DF_n, D(-L)^{-1} F_n \rangle$$

$$= \sum_{r \geq 0} r! \sum_{2k \geq r} \sum_{2l \geq r} \frac{1}{(2k)!} \frac{1}{(2l+1)!} C_{2k}^r C_{2l}^r$$

$$\times \sum_{u_1, \cdots, u_{r+1}=1}^{n} \sum_{i_1, \cdots, i_{2k-r}=1}^{n} \sum_{j_1, \cdots, j_{2l-r}=1}^{n} a_{u_1, u_2, \cdots, u_{r+1}, i_1, \cdots, i_{2k-r}} a_{u_1, u_2, \cdots, u_{r+1}, j_1, \cdots, j_{2l-r}}$$

$$\times I_{2k+2l-2r} \left(\varphi_{i_1} \otimes \cdots \otimes \varphi_{i_{2k-r}} \otimes \varphi_{j_1} \otimes \cdots \otimes \varphi_{j_{2l-r}} \right)$$

$$= \sum_{r \geq 0} \sum_{k,l \geq 0} \frac{1}{(k+r)!} \frac{1}{(l+r+1)!} C_{k+r}^r C_{l+r}^r$$

$$\times \sum_{u_1, \cdots, u_{r+1}=1}^{n} \sum_{i_1, \cdots, i_k=1}^{n} \sum_{j_1, \cdots, j_l=1}^{n} a_{u_1, u_2, \cdots, u_{r+1}, i_1, \cdots, i_k} a_{u_1, u_2, \cdots, u_{r+1}, j_1, \cdots, j_l}$$

$$\times I_{2k+2l-2r} \left(\varphi_{i_1} \otimes \cdots \otimes \varphi_{i_k} \otimes \varphi_{j_1} \otimes \cdots \otimes \varphi_{j_l} \right)$$

$$= \sum_{k,l \geq 0} \sum_{r \geq 0} r! \frac{1}{(k+r)!} \frac{1}{(l+r+1)!} C_{k+r}^r C_{l+r}^r$$

$$\times \sum_{u_1, \cdots, u_{r+1}=1}^{n} \sum_{i_1, \cdots, i_k=1}^{n} \sum_{j_1, \cdots, j_l=1}^{n} a_{u_1, u_2, \cdots, u_{r+1}, i_1, \cdots, i_k} a_{u_1, u_2, \cdots, u_{r+1}, j_1, \cdots, j_l}$$

$$\times I_{k+l} \left(\varphi_{i_1} \otimes \cdots \otimes \varphi_{i_k} \otimes \varphi_{j_1} \otimes \cdots \otimes \varphi_{j_l} \right).$$

Once again using a change of indices ($k + l = m$), we obtain

$$\langle DF_n, D(-L)^{-1} F_n \rangle$$

$$= \sum_{m \geq 0} \sum_{k=0}^{m} \sum_{r \geq 0} r! \frac{1}{(k+r)!} \frac{1}{(m-k+r+1)!} C_{k+r}^r C_{m-k+r}^r$$

$$\times \sum_{u_1, \cdots, u_{r+1}=1}^{n} \sum_{i_1, \cdots, i_k=1}^{n} \sum_{j_1, \cdots, j_{m-k}=1}^{n} a_{u_1, u_2, \cdots, u_{r+1}, i_1, \cdots, i_k} a_{u_1, u_2, \cdots, u_{r+1}, j_1, \cdots, j_{m-k}}$$

$$\times I_m \left(\varphi_{i_1} \otimes \cdots \otimes \varphi_{i_k} \otimes \varphi_{j_1} \otimes \cdots \otimes \varphi_{j_{m-k}} \right)$$

$$= \sum_{m \geq 0} \sum_{k=0}^{m} \frac{1}{k!} \frac{1}{(m-k)!} \sum_{r \geq 0} \frac{1}{r!} \frac{1}{m-k+r+1} \sum_{u_1, \cdots, u_{r+1}=1}^{n} \sum_{i_1, \cdots, i_m=1}^{n}$$

$$\times a_{u_1, u_2, \cdots, u_{r+1}, i_1, \cdots, i_k} a_{u_1, u_2, \cdots, u_{r+1}, i_{k+1}, \cdots, i_m} I_m \left(\varphi_{i_1} \otimes \cdots \otimes \varphi_{i_k} \otimes \varphi_{i_{k+1}} \otimes \cdots \otimes \varphi_{i_m} \right)$$

where at the end we renamed the indices $j_1, \cdots, j_{m-k}$ as $i_{k+1}, \cdots, i_m$. We obtain

$$\langle DF_n, D(-L)^{-1} F_n \rangle = \sum_{m \geq 0} I_m(h_m^{(n)})$$

where

$$h_m^{(n)} = \sum_{k=0}^{m} \frac{1}{k!} \frac{1}{(m-k)!} \sum_{r \geq 0} \frac{1}{r!} \frac{1}{m-k+r+1} \sum_{u_1, \cdots, u_{r+1}=1}^{n} \sum_{i_1, \cdots, i_m=1}^{n}$$
$$a_{u_1, u_2, \cdots, u_{r+1}, i_1, \cdots, i_k} a_{u_1, u_2, \cdots, u_{r+1}, i_{k+1}, \cdots, i_m} \varphi_{i_1} \otimes \cdots \otimes \varphi_{i_k} \otimes \varphi_{i_{k+1}} \otimes \cdots \otimes \varphi_{i_m}$$
(28)

Let us make some comments about this result before going any further. These remarks will simplify the expression that we have just obtained. As follows from Lemma 1, the coefficients $a_{i_1, \cdots, i_k}$ are zero if k is even. Therefore, the numbers $r+1+k$ and $r+1+m-k$ must be odd. This implies that m must be even and this is coherent with our previous observation (see Remark 5) that the chaos expansion of $\langle DF_n, D(-L)^{-1} F_n \rangle$ only contains chaoses of even orders. The second comment concerns the chaos of order zero. If $m = 0$ then $k = 0$ and we obtain

$$h_0^{(n)} = \sum_{r \geq 0} \sum_{u_1, \cdots, u_{r+1}=1}^{n} \frac{1}{r!} \frac{1}{r+1} a_{u_1, \cdots, u_{r+1}}^2 = \sum_{r \geq 1} \frac{1}{r!} \sum_{u_1, \cdots, u_r=1}^{n} a_{u_1, \cdots, u_r}^2.$$

Thus, because the summand $\sum_{r \geq 1} \frac{1}{r!} \sum_{u_1, \cdots, u_r=1}^{n} a_{u_1, \cdots, u_r}^2 - 1$ is zero by using Lemma 4,

$$\langle DF_n, D(-L)^{-1} F_n \rangle - 1 = \left(\sum_{r \geq 1} \frac{1}{r!} \sum_{u_1, \cdots, u_r=1}^{n} a_{u_1, \cdots, u_r}^2 - 1 \right) + \sum_{m \geq 1} I_{2m}(h_{2m}^{(n)})$$
$$= \sum_{m \geq 1} I_{2m}(h_{2m}^{(n)})$$

with $h_{2m}^{(n)}$ given by (28).

Using the isometry formula of multiple integrals in order to compute the L^2 norm of the above expression and noticing that the function $h_{2m}^{(n)}$ is symmetric, we find that

$$\mathbf{E}\left(\left(\langle DF_n, D(-L)^{-1} F_n \rangle - 1 \right)^2 \right) = \sum_{m \geq 1} (2m)! \langle h_{2m}^{(n)}, h_{2m}^{(n)} \rangle_{L^2([0,1]^{2m})}$$

$$= \sum_{m \geq 1} (2m)! \sum_{k,l=0}^{2m} \frac{1}{k!} \frac{1}{l!} \frac{1}{(2m-k)!} \frac{1}{(2m-l)!} \sum_{r,q \geq 0} \frac{1}{r!} \frac{1}{q!} \frac{1}{2m-k+r+1} \frac{1}{2m-l+q+1}$$

$$\times \sum_{u_1,\cdots,u_{r+1}=1}^{n} \sum_{v_1,\cdots,v_{q+1}=1}^{n} \sum_{i_1,\cdots,i_{2m}=1}^{n}$$

$$\times a_{u_1,u_2,\cdots,u_{r+1},i_1,\cdots,i_k} a_{u_1,u_2,\cdots,u_{r+1},i_{k+1},\cdots,i_{2m}} a_{v_1,v_2,\cdots,v_{q+1},i_1,\cdots,i_k} a_{v_1,v_2,\cdots,v_{q+1},i_{k+1},\cdots,i_{2m}}$$

$$= \sum_{m\geq 1}(2m)! \sum_{i_1,\cdots,i_{2m}=1}^{n} \left(\sum_{k=0}^{2m} \frac{1}{k!}\frac{1}{(2m-k)!} \sum_{r\geq 0}\frac{1}{r!}\frac{1}{2m-k+r+1} \sum_{u_1,\cdots,u_{r+1}=1}^{n} \right.$$

$$\left. a_{u_1,u_2,\cdots,u_{r+1},i_1,\cdots,i_k} a_{u_1,u_2,\cdots,u_{r+1},i_{k+1},\cdots,i_{2m}} \right)^2,$$

which is the desired result.

References

1. M. Abramowitz, I.A. Stegun (eds). *Handbook of Mathematical Functions with Formulas, Graphs, and Mathematical Tables*. Reprint of the 1972 edition (Dover, New York, 1992)
2. V. Bentkus, F. Götze, The Berry-Essén bound for Student's statistics. Ann. Probab. **24**, 491–503 (1996)
3. V. Bentkus, F. Götze, A. Tikhomirov, Berry-Esseen bounds for statistics of weakly dependent samples. Bernoulli **3**, 329–349 (1997)
4. V. Bentkus, M. Bloznelis, F. Götze, A Berry-Esséen bound for Student's statistics in the non i.i.d. case. J. Theor. Probab. **9**, 765–796 (1996)
5. S. Bourguin, C.A. Tudor, Berry-Esséen bounds for long memory moving averages via Stein's method and Malliavin calculus. Stoch. Anal. Appl. **29**, 881–905 (2011)
6. J.-C. Breton, I. Nourdin, G. Peccati, Exact confidence intervals for the Hurst parameter of a fractional Brownian motion. Electron. J. Stat. **3**, 415–425 (2009)
7. V.H. de la Pena, T.L. Lai, Q.M.Shao, *Self Normalized Processes* (Springer, New York, 2009)
8. E. Giné, E. Götze, D. Mason, When is the t-Student statistics asymptotically standard normal? Ann. Probab. 25, 1514–1531 (1997)
9. P. Hall, Q. Wang, Exact convergence rate and leading term in central limit theorem for Student's t statistic. Ann. Probab. **32**(2), 1419–1437 (2004)
10. Y. Hu, D. Nualart, Some processes associated with fractional Bessel processes. J. Theor. Probab. **18**(2), 377–397 (2005)
11. R. Maller, On the law of the iterated logarithm in the infinite variance cas. J. Aust. Math. Soc. **30**, 5–14 (1981)
12. I. Nourdin, G. Peccati, Stein's method meets Malliavin calculus: a short survey with new estimates. To appear in *Recent Advances in Stochastic Dynamics and Stochastic Analysis* (World Scientific, Singapore, 2010), pp. 207–236
13. I. Nourdin, G. Peccati, Stein's method on Wiener chaos. Probab. Theor. Relat. Field. **145**(1–2), 75–118 (2009)
14. I. Nourdin, G. Peccati, Stein's method and exact Berry-Esséen asymptotics for functionals of Gaussian fields. Ann. Probab. **37**(6), 2200–2230 (2009)
15. I. Nourdin, G. Peccati, A. Réveillac, Multivariate normal approximation using Steins method and Malliavin calculus. Ann. Inst. H. P. Probab. Stat. **46**(1), 45–58 (2010)
16. D. Nualart, *Malliavin Calculus and Related Topics*, 2nd edn (Springer, New York, 2006)
17. Q.M. Shao, An explicit Berry-Esseen bound for Student's t-statistic via Stein's method. Stein's method and applications, pp. 143–155. Lect. Notes Ser. Inst. Math. Sci. Natl. Univ. Singap., 5 (Singapore University Press, Singapore, 2005)

A Note on Stochastic Calculus in Vector Bundles*

Pedro J. Catuogno, Diego S. Ledesma, and Paulo R. Ruffino

Abstract The aim of these notes is to relate covariant stochastic integration in a vector bundle E [as in Norris (*Séminaire de Probabilités, XXVI*, vol. 1526, Springer, Berlin, 1992, pp. 189–209)] with the usual Stratonovich calculus via the connector $\mathscr{K}_\nabla : TE \to E$ [cf. e.g. Paterson (Canad. J. Math. 27(4):766–791, 1975) or Poor (*Differential Geometric Structures*, McGraw-Hill, New York, 1981)] which carries the connection dependence.

Keywords Global analysis • Stochastic calculus • Vector bundles

AMS Subject Classification (2010): 58J65 (60J60, 60H05).

1 Introduction

Stochastic calculus on vector bundles has been studied by several authors, among others, Arnaudon and Thalmaier [1], Norris [6], Driver and Thalmaier [3]. In these articles, the stochastic integral of a semimartingale v_t in a vector bundle $\pi : E \to M$ is defined by decomposing v_t into horizontal and vertical (covariant) components

---

*P. J. Catuogno was partially supported by CNPq, grant no. 302704/2008-6, 480271/2009-7 and FAPESP, grant no. 07/06896-5. D. S. Ledesma was supported by FAPESP, grant no. 10/20347-7. P. R. Ruffino was partially supported by CNPq, grant no. 306264/2009-9, 480271/2009-7 and FAPESP, grant no. 7/06896-5.

P.J. Catuogno · D.S. Ledesma · P.R. Ruffino (✉)
Departamento de Matemática, Universidade Estadual de Campinas,
13083-859-Campinas, SP, Brazil
e-mail: pedrojc@ime.unicamp.br; dledesma@ime.unicamp.br; ruffino@ime.unicamp.br

according to a given connection in E. The aim of these notes is to relate covariant stochastic integral in vector bundles (Norris [6]) with the usual Stratonovich calculus using an appropriate operator, the connector $\mathcal{K}_\nabla$ (cf. e.g. Paterson [7] and Poor [8]), from the tangent space TE to E which carries the connection dependence.

We denote by M a smooth differentiable manifold. Let E be an n-dimensional vector bundle over M endowed with a connection ∇. This connection induces a natural projection $\mathcal{K}_\nabla : TE \to E$ called the associated connector (cf. Paterson [7] and Poor [8]) which projects into the vertical subspace of TE identified with E. More precisely: Given a differentiable curve $v_t \in E$, decompose $v_t = u_t f_t$, where u_t is the unique horizontal lift of $\pi(v_t)$ in the principal bundle $Gl(E)$ of frames in E starting at a certain u_0 with $\pi(u_0) = \pi(v_0)$ and $f_t \in \mathbf{R}^n$. Then

$$\mathcal{K}_\nabla(v_0') := u_0 f_0'.$$

Norris [6] defines the covariant Stratonovich integration of a section θ in the dual vector bundle E^* along a process $v_t \in E$ by:

$$\int \theta D v_t := \int \theta u_t \circ df_t,$$

where $v_t = u_t f_t$; and the corresponding covariant Itô version:

$$\int \theta D^I v_t := \int \theta u_t \, df_t.$$

2 Main Results

Initially, observe that using the connector $\mathcal{K}_\nabla$, the covariant integral above reduces to a classical Stratonovich integral of 1-forms:

Proposition 1. *Let v_t be a semimartingale in E and $\theta \in \Gamma(E^*)$. Then*

$$\int \theta \, Dv_t = \int \theta \, \mathcal{K}_\nabla \circ dv_t.$$

Proof. Let $\phi : Gl(E) \times \mathbf{R}^n \to E$ be the action map $\phi(u, f) = uf$. The right hand side in the equation above is

$$\int \theta \, \mathcal{K}_\nabla \circ d\phi(u_t, f_t) = \int \phi_{u_t}^* \theta \mathcal{K}_\nabla \circ df_t + \int \phi_{f_t}^* \theta \, \mathcal{K}_\nabla \circ du_t. \tag{1}$$

The second term on the right hand side vanishes since $\phi_{f_t}^* \theta \, \mathcal{K}_\nabla = 0$. The formula holds because $\phi_{u_t}^* \mathcal{K}_\nabla(z) = u_t z$ for all $z \in \mathbf{R}^n$. □

Remark 1. In the special case of $E = TM$, one can compare the classical integration of 1-forms with the covariant integration: Let Y_t be a semimartingale in M and v_t be a semimartingale in E. If $v_t = u_t f_t$ such that u_t is a horizontal lift of Y_t and f_t is the antidevelopment of Y_t, then for any 1-form θ, the classical integration in M and the covariant integration in E coincide:

$$\int \theta \circ dY_t = \int \theta \mathcal{K}_\nabla \circ dv_t.$$

Local Coordinates

Let $\{\delta_1, \ldots \delta_n\}$ be local sections in E which is a basis in a coordinate neighbourhood $(U, \varphi = (x^1, \ldots, x^d))$, where d is the dimension of M. For $1 \leq \alpha, \beta \leq n$ and $1 \leq i \leq d$, we write

$$\nabla_{\frac{\partial}{\partial x^i}} \delta_\alpha = \Gamma^\beta_{i\alpha} \delta_\beta,$$

then

$$\mathcal{K}_\nabla \left(\frac{\partial \delta_\alpha}{\partial x^i} \right) = \Gamma^\beta_{i\alpha} \delta_\beta.$$

Let γ_t be a differentiable curve in M and u_t be a horizontal lift of γ_t in $Gl(E)$, we write $u_t^\beta = u_t(e_\beta) = u_t^{\beta\alpha} \delta_\alpha(\gamma_t)$. Naturally

$$\nabla_{\gamma_t'} u_t^\beta = 0,$$

and the parallel transport equation is given by

$$\frac{du_t^{\alpha\beta}}{dt} + \frac{d\gamma^j}{dt} u_t^{\alpha\gamma} \Gamma^\beta_{j\gamma}(\gamma_t) = 0.$$

For $\theta \in \Gamma(E^*)$, write $\theta = \theta^\alpha \delta_\alpha^*$, where $\theta^\alpha = \theta(\delta_\alpha)$. We have, for each $1 \leq \alpha \leq n$

$$\left(\nabla_{\frac{\partial}{\partial x^j}} \theta \right) \delta^\alpha = \frac{\partial \theta^\alpha}{\partial x^j} - \theta(\nabla_{\frac{\partial}{\partial x^j}} \delta^\alpha) = \frac{\partial \theta^\alpha}{\partial x^j} - \Gamma^\beta_{j\alpha} \theta^\beta.$$

That is,

$$\nabla \theta = \left(\frac{\partial \theta^\alpha}{\partial x^j} - \Gamma^\beta_{j\alpha} \theta^\beta \right) dx^j \otimes \delta_\alpha^*.$$

Cross Quadratic Variation in Sections of $TM^* \otimes E^*$

In order to find a covariant conversion formula for Itô–Stratonovich integrals we introduce stochastic integration formulae for sections of $TM^* \otimes E^*$, which is the space where the covariant derivative $\nabla \theta$ lives. Let v_t be a semimartingale in E. Denoting $x_t = \pi(v_t)$, we have the following identities:

1. For $\alpha \in \Gamma(TM^*)$ and $\theta \in \Gamma(E^*)$,

$$\int \alpha \otimes \theta \, (dx_t, Dv_t) = \left\langle \int \alpha \circ d\pi(v_t), \int \theta Dv_t \right\rangle.$$

2. For $b \in \Gamma(TM^* \otimes E^*)$ and $f \in C^\infty(M)$,

$$\int fb \, (dx_t, Dv_t) = \int f(\pi(v_t)) \circ d \int b \, (dx_t, Dv_t).$$

This is well defined (similarly to Emery [5, p. 23]). In particular, for $b = \nabla \theta$, in local coordinates:

$$\int \nabla \theta \, (dx_t, Dv_t) = \int (\frac{\partial \theta^\alpha}{\partial x^j} - \Gamma^\beta_{j\alpha} \theta^\beta) \circ d \int dx^j \otimes \delta^*_\alpha \, (dx_t, Dv_t)$$

$$= \int (\frac{\partial \theta^\alpha}{\partial x^j} - \Gamma^\beta_{j\alpha} \theta^\beta)(x_t) \circ d < x_t^j, \int u^{\gamma\alpha} df^\gamma >_t$$

$$= \int (\frac{\partial \theta^\alpha}{\partial x^j} - \Gamma^\beta_{j\alpha} \theta^\beta)(x_t) u_t^{\gamma\alpha} \circ d < x^j, f^\gamma >_t.$$

We write in the language of stochastic integration on manifolds the Itô–Stratonovich covariant conversion formula (26) of J. R. Norris [6].

Proposition 2. *Let v_t be a semimartingale in E and $\theta \in \Gamma(E^*)$. Then*

$$\int \theta \, Dv_t = \int \theta D^I v_t + \frac{1}{2} \int \nabla \theta \, (dx_t, Dv_t). \tag{2}$$

Proof. In local coordinates we have that

$$\int \theta \, Dv_t = \int \theta_{x_t}(u_t e_\alpha) \circ df_t^\alpha = \int \theta_{x_t}(u_t e_\alpha) \, df_t^\alpha + \frac{1}{2} < \theta(ue_\alpha), f^\alpha >.$$

We have to show that

$$\int \nabla \theta \, (dx_t, Dv_t) = < \theta(ue_\alpha), f^\alpha >.$$

But

$$\begin{aligned}
<\theta(ue_\alpha), f^\alpha> &= <\theta_x^\beta \delta_\beta^*(ue_\alpha), f^\alpha> \\
&= <\theta_x^\beta(ue_\alpha)^\beta, f^\alpha> \\
&= \int (ue_\alpha)^\beta \, d<\theta_x^\beta, f^\alpha> + \int \theta_x^\beta \, d<(ue_\alpha)^\beta, f^\alpha> \\
&= \int u^{\alpha\beta} \, d<\theta_x^\beta, f^\alpha> + \int \theta_x^\beta \, d<u^{\alpha\beta}, f^\alpha> \\
&= \int u^{\alpha\beta} \frac{\partial \theta^\beta}{\partial x^j} d<x^j, f^\alpha> + \int \theta_x^\beta \, d<u^{\alpha\beta}, f^\alpha> \\
&= \int u^{\alpha\beta} \frac{\partial \theta^\beta}{\partial x^j} d<x^j, f^\alpha> - \int \theta_x^\beta u^{\alpha\gamma} \Gamma_{j\gamma}^\beta(x) <x^j, f^\alpha> \\
&= \int \left(u^{\alpha\beta} \frac{\partial \theta^\beta}{\partial x^j} - \theta_x^\beta u^{\alpha\gamma} \Gamma_{j\gamma}^\beta(x) \right) <x^j, f^\alpha> \\
&= \int \left(u^{\alpha\gamma} \frac{\partial \theta^\gamma}{\partial x^j} - \theta_x^\beta u^{\alpha\gamma} \Gamma_{j\gamma}^\beta(x) \right) <x^j, f^\alpha> \\
&= \int \left(\frac{\partial \theta^\gamma}{\partial x^j} - \theta_x^\beta \Gamma_{j\gamma}^\beta(x) \right) u^{\alpha\gamma} <x^j, f^\alpha> \\
&= \int \nabla \theta \, (dx_t, Dv_t).
\end{aligned}$$

□

Itô Representation

The vertical lift of an element $w \in E$ to the tangent space $T_e E$, with e and w in the same fiber is given by

$$w^v = \frac{d}{dt}[e + tw]_{t=0} \in T_e E. \tag{3}$$

Let r, s be sections of E and X, Y be vector fields of M. We shall consider a connection ∇^h in E, a prolongation of ∇, which satisfies the following:

$$\nabla_{r^v}^h s^v = 0, \quad \nabla_{X^h}^h s^v = (\nabla_X s)^v,$$

$$\nabla_{r^v}^h Y^h = 0, \quad \nabla_{X^h} Y^h \text{ is horizontal.}$$

Remark 2. An example of this connections is the horizontal connection defined by Arnaudon and Thalmaier [1], where, considering a connection $\tilde{\nabla}$ in M, the extra condition $\nabla^h_{X^h} Y^h = (\tilde{\nabla}_X Y)^h$ characterizes this connection.

Next proposition shows a geometrical characterization of the covariant Itô integral.

Proposition 3. *Let v_t be a semimartingale in E and $\theta \in \Gamma(E^*)$. Then*

$$\int \theta \, D^I v_t = \int \theta \, \mathcal{K}_\nabla \, d^{\nabla^h} v_t,$$

where d^{∇^h} is the Itô differential with respect to ∇^h.

Proof. We have to calculate each component of $\nabla^h \theta \, \mathcal{K}_\nabla$. Using that for A, B vector fields in E we have that

$$\nabla^h_A \theta \, \mathcal{K}_\nabla(B) = A(\theta \, \mathcal{K}_\nabla(B)) - \theta \, \mathcal{K}_\nabla(\nabla^h_A B),$$

we obtain the components

$$\nabla^h_{r^v} \theta \, \mathcal{K}_\nabla(s^v) = 0, \qquad \nabla^h_{r^v} \theta \, \mathcal{K}_\nabla(Y^h) = 0,$$

$$\nabla^h_{X^h} \theta \, \mathcal{K}_\nabla(s^v) = \nabla_X \theta(s) \circ \pi, \qquad \nabla^h_{X^h} \theta \, \mathcal{K}_\nabla(Y^h) = 0.$$

Hence, using Itô–Stratonovich conversion formula for classical 1-form integration, see e.g. Catuogno and Stelmastchuk [2]:

$$\int \theta \, Dv_t = \int \theta \, \mathcal{K}_\nabla \circ dv_t$$
$$= \int \theta \, \mathcal{K}_\nabla \, d^{\nabla^h} v_t + \frac{1}{2} \int \nabla^h \theta \, \mathcal{K}_\nabla(dv_t, dv_t).$$

For the correction term, we have that:

$$\nabla^h \theta \, \mathcal{K}_\nabla = \nabla \theta \, (\pi_* \times \mathcal{K}_\nabla),$$

in the sense that $\nabla^h_{\pi_* A} \theta \mathcal{K}_\nabla(B) = \nabla \theta (\pi_* \times \mathcal{K}_\nabla)(A, B)$. But

$$\int \nabla^h \theta \, \mathcal{K}_\nabla(dv_t, dv_t) = \int \nabla \theta(dx_t, Dv_t).$$

Combining with Eq. (2), we have that

$$\int \theta \, D^I v_t = \int \theta \, \mathcal{K}_\nabla \, d^{\nabla^h} v_t. \qquad \square$$

Vector Bundle Mappings

Consider two vector bundles $\pi : E \to M$, $\pi' : E' \to M'$ and a differentiable fibre preserving mapping $F : E \to E'$ over a differentiable map $\tilde{F} : M \to M'$, i.e. $\pi' \circ F = \tilde{F} \circ \pi$.

Let $\mathcal{K}_\nabla$ and $\mathcal{K}'_\nabla$ be connectors in E and in E' respectively. We define the vertical derivative (or derivative in the fibre) of F in the direction of w by:

$$D^v F(e)(w) = \mathcal{K}'_\nabla F_*(w^v),$$

where the vertical component w^v is given by Eq. (3). For $Z \in T_{\pi(e)}M$, the horizontal (or parallel) derivative is:

$$D^h F(e)(Z) = \mathcal{K}'_\nabla F_*(Z^h).$$

For a vector field X in E, we have that

$$X = (\pi_* X)^h + \mathcal{K}_\nabla(X),$$

hence

$$\mathcal{K}'_\nabla F_*(X) = D^v F(\mathcal{K}_\nabla(X)) + D^h F(\pi_*(X)). \tag{4}$$

The Itô formula for the Stratonovich covariant integration includes an usual 1-form integration, compare with Norris [6, Eq. (20)]:

Proposition 4. *Given a fibre preserving map F as above,*

$$\int \theta DF(v_t) = \int (D^v F)^* \theta Dv_t + \int (D^h F)^* \theta \circ d(\pi v_t). \tag{5}$$

Proof. We just have to use the decomposition of Eq. (4).

$$\int \theta DF(v_t) = \int \theta \mathcal{K}'_\nabla F_* \circ dv_t$$
$$= \int (\theta D^v F \mathcal{K}_\nabla + \theta D^h F \pi_*) \circ dv_t$$
$$= \int (D^v F)^* \theta Dv_t + \int (D^h F)^* \theta \circ d(\pi v_t). \quad \square$$

Proposition 5. *For a section b' in $(TM')^* \otimes (E')^*$ and a fibre preserving map $F : E \to E'$ over $\tilde{F} : M \to M'$ we have that*

$$\int b'(d\pi' F(v_t), DF(v_t)) = \int (\tilde{F}_* \otimes D^v F)^* b'(d\pi v_t, Dv_t) + \int (\tilde{F}_* \otimes D^h F)^* b'(d\pi v_t, d\pi v_t).$$

Proof. We have

$$\int b'(d\pi' F(v_t), DF(v_t)) = \int b'(\pi'_* \otimes \mathcal{K}'_\nabla)(dF(v_t), dF(v_t))$$
$$= \int b'(\pi'_* \otimes \mathcal{K}'_\nabla)(F_* \otimes F_*)(dv_t, dv_t).$$

Using that

$$(\pi'_* \otimes \mathcal{K}'_\nabla)(F_* \otimes F_*) = \tilde{F}_* \pi_* \otimes (D^v F \mathcal{K}_\nabla + D^h F \pi_*)$$

yields

$$\int b'(d\pi' F(v_t), DF(v_t)) = \int (\tilde{F}_* \otimes D^h F)^* b'(d\pi v_t, Dv_t) + \int (\tilde{F}_* \otimes D^h F)^* b'(d\pi v_t, d\pi v_t).$$

$\square$

Itô version of Formula (5) is given by:

Proposition 6. *Given a fibre preserving map F as above,*

$$\int \theta D^I F(v_t) = \int (D^v F)^* \theta D^I v_t + \int (D^h F)^* \theta \circ d\pi v_t +$$

$$\frac{1}{2} \int \left(\nabla (D^v F^* \theta) - (\tilde{F}_* \otimes D^V F)^* \nabla' \theta \right) (d\pi v_t, Dv_t) +$$

$$\frac{1}{2} \int \left(\tilde{F}_* \otimes D^h F \right)^* \nabla' \theta (d\pi v_t, d\pi v_t).$$

Proof. By Proposition 2 we have that

$$\int \theta D^I F(v_t) = \int \theta DF(v_t) - \frac{1}{2} \int \nabla' \theta (d\pi' F(v_t), DF(v_t))$$

and

$$\int (D^v F)^* \theta Dv_t = \int (D^v F)^* \theta D^I v_t + \frac{1}{2} \nabla (D^v F)^* (d\pi v_t, Dv_t).$$

But, Proposition 4 says that:

$$\int \theta DF(v_t) = \int (D^v F)^* \theta Dv_t + \int (D^h F)^* \theta \circ d(\pi v_t).$$

Finally, by Proposition 5, we have that

$$\int \nabla'\theta(d\pi' F(v_t), DF(v_t)) = \int (\tilde{F}_* \otimes D^v F)^* \nabla'\theta(d\pi v_t, Dv_t) +$$

$$\int (\tilde{F}_* \otimes D^h F)^* \nabla'\theta(d\pi v_t, d\pi v_t),$$

which implies the formula. □

3 Applications

Commutation Formulae

Given a differentiable map $(a, b) \in \mathbf{R}^2 \mapsto E$, let $s_E : TTE \to TTE$ be the symmetry map given by $s_E(\partial_a \partial_b s(a, b)) = \partial_b \partial_a s(a, b)$. Let $C = \mathcal{K}\mathcal{K}_* - \mathcal{K}\mathcal{K}_* s_E : TTE \to E$ be the curvature of $\mathcal{K}$. If $u, v \in TM$ and $s \in \Gamma(E)$ then the relation between the curvature of $\mathcal{K}$ with the curvature of the connection ∇ is given by $R^E(u, v)s = C(uvs)$, see Paterson [7].

Let $I \subset \mathbf{R}$ be an open interval and consider $a \in I \mapsto J(a)$ a differentiable 1-parameter family of semimartingales in E. Then

$$\int \theta \, D\nabla_a J = \int \theta \mathcal{K}_\nabla \circ d(\nabla_a J)$$

$$= \int \theta \mathcal{K}_\nabla \circ d\mathcal{K}_\nabla \partial_a J$$

$$= \int \theta \mathcal{K}_\nabla \mathcal{K}_{\nabla*} \circ d\partial_a J$$

$$= \int \theta \mathcal{K}_\nabla \mathcal{K}_{\nabla*} \circ d\partial_a J - \int \theta \mathcal{K}_\nabla \mathcal{K}_{\nabla*} s_E \circ d\partial_a J$$

$$+ \int \theta \mathcal{K}_\nabla \mathcal{K}_{\nabla*} s_E \circ d\partial_a J$$

$$= \int C \circ d\partial_a J + \int \theta \nabla_a DJ.$$

Where we have used that

$$\int \theta \, \mathcal{K}_\nabla \, \mathcal{K}_{\nabla *} \, s_E \circ d\partial_a J = \int \theta \, \mathcal{K}_\nabla \, \mathcal{K}_{\nabla *} \, \partial_a \circ dJ$$
$$= \int \theta \, \mathcal{K}_\nabla \, \partial_a DJ$$
$$= \int \theta \, \nabla_a DJ.$$

Compare with Arnaudon and Thalmaier [1, Eq.(4.13)]. An Itô version, as in [1] can be obtained by conversion formulae.

Harmonic Sections

Let M be a Riemannian manifold and $\pi : V \to M$ be a Riemannian vector bundle with a connection ∇ which is compatible with its metric. We denote by E^p the vector bundle $\bigwedge^p T^*M \otimes V$ over M. In this context, we shall consider three differential geometric operators. The exterior differential operator $d : \Gamma(E^p) \to \Gamma(E^{p+1})$ is defined by

$$d\sigma(X_1, \ldots, X_{p+1}) := (-1)^k (\nabla_{X_k} \sigma)(X_0, \ldots, \hat{X}_k, \ldots, X_p).$$

The co-differential operator $\delta : \Gamma(E^p) \to \Gamma(E^{p-1})$ is defined by

$$\delta\sigma(X_1, \ldots, X_{p-1}) := -(\nabla_{e_k} \sigma)(e_k, X_1, \ldots, X_{p-1}),$$

where $\{e_i\}$ is a local orthonormal frame field. And the Hodge–Laplace operator $\Delta : \Gamma(E^p) \to \Gamma(E^p)$ is given by

$$\Delta = (d\delta + \delta d).$$

One of the cornerstones of modern geometric analysis is the Weitzenböck formula which states that

$$\Delta \sigma = -\nabla^2 \sigma + \Phi(\sigma),$$

for a $\Phi \in \text{End}(E^p)$, see e.g. Eells and Lemaire [4, p.11] or Xin [9, p.21].

Let B_t be a Brownian motion in M and $e_t \in \text{End}(E^p)$ be the solution of

$$D^I e_t = e_t \circ \Phi(B_t) \, dt.$$

Theorem 1. A section $\sigma \in \Gamma(E^p)$ is harmonic (i.e. $\Delta\sigma = 0$) if and only if for any $\theta \in \Gamma(E^{p*})$

$$\int \theta \, D^I \sigma_t$$

is a local martingale, where $\sigma_t = e_t \sigma(B_t)$.

The result now is consequence of Weitzenböck formula and the following

Lemma 1. Consider $\sigma \in \Gamma(E^p)$, $\theta \in \Gamma(E^{p*})$ and a semimartingale x_t in M. Given $V \in \text{End}(E^p)$, let $e_t \in \text{End}(E^p)$ be the solution of

$$D^I e_t = e_t \circ V(x_t) \, g(dx_t, dx_t).$$

Write $\sigma_t = e_t \sigma(x_t)$. Then

$$\int \theta D^I \sigma_t = \int (\theta \circ \nabla \sigma) \, d^{\nabla^M} x_t + \int (\theta \circ e_t) \left(\frac{1}{2}\nabla^2 + V(\sigma(x_t))g\right) \sigma \, (dx_t, dx_t).$$

Proof. By covariant Itô–Stratonovich conversion formula, Eq. (2), we have that

$$\int \theta D^I \sigma_t = \int \theta \, D^I e_t(\sigma(x_t)) + \int (\theta \circ e_t) \, D^I(\sigma(x_t))$$

$$= \int (\theta \circ e_t) \, V(\sigma(x_t)) \, g(dx_t, dx_t) + \int (\theta \circ e_t) \, D^S(\sigma(x_t))$$

$$+ \frac{1}{2} \int \nabla(\theta \circ e_t) \, (dx_t, D\sigma(x_t)). \tag{6}$$

Now, by usual Itô–Stratonovich conversion formula:

$$\int (\theta \circ e_t) \, D^S(\sigma(x_t)) = \int (\theta \circ e_t) \mathcal{H}_\nabla \sigma_* \, dx_t$$

$$= \int (\theta \circ e_t) \nabla \sigma \, d^{\nabla^M} x_t$$

$$- \frac{1}{2} \int \nabla^M (\theta \circ e_t \circ \nabla \sigma)(dx_t, dx_t). \tag{7}$$

We have that

$$\int \nabla(\theta \circ e_t) \, (dx_t, D\sigma(x_t)) = \int \nabla(\theta \circ e_t) \circ (I \otimes \nabla \sigma) \, (dx_t, dx_t) \tag{8}$$

substituing (7) and (8) in (6) one finds:

$$\int \theta D^I \sigma_t = \int (\theta \circ e_t)\, V(\sigma(x_t))\, g(dx_t, dx_t) + \int (\theta \circ e_t) \nabla \sigma\, d^{\nabla^M} x_t$$
$$- \frac{1}{2} \int \nabla^M (\theta \circ e_t \circ \nabla \sigma)(dx_t, dx_t)$$
$$+ \frac{1}{2} \int \nabla(\theta \circ e_t) \circ (I \otimes \nabla \sigma)\, (dx_t, dx_t).$$

The result follows using that for all $\theta \in \Gamma(E^*)$,

$$\nabla \theta \circ (I \otimes \nabla \sigma) - \nabla^M(\theta \circ \nabla \sigma) = \theta(\nabla^2 \sigma). \qquad \square$$

References

1. M. Arnaudon, A. Thalmaier, Horizontal martingales in vector bundles. *Séminaire de Probabilités, XXXVI*. Lecture Notes in Math., vol. 1801 (Springer, Berlin, 2003), pp. 419–456
2. P.J. Catuogno, S. Stelmastchuk, Martingales on frame bundles. Potential Anal. **28**(1), 61–69 (2008)
3. B.K. Driver, A. Thalmaier, Heat equation derivative formulas for vector bundles. J. Funct. Anal. **183**(1), 42–108 (2001)
4. J. Eells, L. Lemaire, Selected topics in harmonic maps. *CBMS Regional Conference Series in Mathematics*, vol. 50 (American Mathematical Society, Providence, RI, 1983)
5. M. Emery, *Stochastic Calculus in Manifolds*. With an appendix by P. A. Meyer. Universitext (Springer, Berlin, 1989)
6. J.R. Norris, A complete differential formalism for stochastic calculus in manifolds. *Séminaire de Probabilités, XXVI*. Lecture Notes in Math., vol. 1526 (Springer, Berlin, 1992), pp. 189–209
7. L.N. Patterson, Connexions and prolongations. Canad. J. Math. **27**(4), 766–791 (1975)
8. W.A. Poor, *Differential Geometric Structures* (McGraw-Hill, New York, 1981)
9. Y. Xin, Geometry of harmonic maps. *Progress in Nonlinear Differential Equations and their Applications*, vol. 23 (Birkhäuser Boston, Boston, MA, 1996)

Functional Co-monotony of Processes with Applications to Peacocks and Barrier Options

Gilles Pagès

Abstract We show that several general classes of stochastic processes satisfy a functional co-monotony principle, including processes with independent increments, Brownian bridge, Brownian diffusions, Liouville processes, fractional Brownian motion. As a first application, we recover and extend some recent results about peacock processes obtained by Hirsch et al. in (*Peacocks and Associated Martingales, with Explicit Constructions*, Bocconi & Springer, 2011, 430p) [see also (Peacocks sous l'hypothèse de monotonie conditionnelle et caractérisation des 2-martingales en termes de peacoks, thèse de l'Université de Lorraine, 2012, 169p)] which were themselves motivated by a former work of Carr et al. in (Finance Res. Lett. 5:162–171, 2008) about the sensitivities of Asian options with respect to their volatility and residual maturity (seniority). We also derive semi-universal bounds for various barrier options.

Keywords Antithetic simulation method • Asian options • Barrier options • Co-monotony • Fractional Brownian motion • Liouville processes • Processes with independent increments • Sensitivity

1 Introduction

The aim of this paper is to show that the classical co-monotony principle for real-valued random variables also holds for large classes of stochastic processes like Brownian diffusion processes, processes with independent increments, Liouville processes, fractional Brownian motion(s), etc, if one considers the natural partial order on the space of real-valued functions defined on an interval. We also provide

---

G. Pagès (✉)
Laboratoire de Probabilités et Modèles aléatoires, UMR 7599, UPMC, case 188, 4, pl. Jussieu, F-75252 Paris Cedex 5, France
e-mail: gilles.pages@upmc.fr

several examples of application, with a special emphasis on peacocks (English quasi-acronym for "processus croissants pour l'ordre convexe") inspired by recent works by Hirsch et al. in [9], which find themselves their original motivation in [4] by Carr et al. about the sensitivities of Asian options in a Black–Scholes model. We also derive (semi-)universal upper or lower bounds for various barrier options when the dynamics of the underlying asset price satisfies an appropriate functional co-monotony principle.

The starting point of what can be called *co-monotony principle* finds its origin in the following classical proposition dealing with one-dimensional real-valued random variables.

Proposition 1 (One dimensional co-monotony principle). *Let $X : (\Omega, \mathscr{A}, \mathbb{P}) \to \mathbb{R}$ be a random variable and let $f, g : \mathbb{R} \to \mathbb{R}$ be two monotone functions sharing the same monotony property.*

(a) *If $f(X), g(X), f(X)g(X) \in L^1(\mathbb{P})$, then $\mathrm{Cov}(f(X), g(X)) \geq 0$ i.e.*

$$\mathbb{E}\, f(X)g(X) \geq \mathbb{E}\, f(X) \mathbb{E}\, g(X).$$

Furthermore, the inequality holds as an equality if and only if $f(X)$ or $g(X)$ is $\mathbb{P}$-a.s. constant.
If f and g are monotone with opposite monotony then the reverse inequality holds.

(b) *If f and g have the same constant sign, then integrability is no longer requested. As a consequence, if f and g have opposite monotony, then*

$$\|f(X)g(X)\|_1 = \mathbb{E} f(X)g(X) \leq \mathbb{E} f(X)\mathbb{E} g(X) = \|f(X)\|_1 \|g(X)\|_1.$$

These inequalities are straightforward consequences of Fubini's Theorem applied on $(\mathbb{R} \times \mathbb{R}, \mathscr{B}or(\mathbb{R})^{\otimes 2}, \mathbb{P}_X^{\otimes 2})$ to the function $(x, x') \mapsto (f(x) - f(x'))(g(x) - g(x'))$ where $\mathbb{P}_X$ denotes the distribution of X.

Typical applications of this scalar co-monotony principle are, among others, the antithetic simulation method for variance reduction and more recently a priori sign results for the sensitivity of derivatives in Finance.

▷ *Antithetic simulation.* Let $X : (\Omega, \mathscr{A}, \mathbb{P}) \to \mathbb{R}$ be a random variable and let $\varphi : \mathbb{R} \to \mathbb{R}$ be a non-increasing function such that $\varphi(X) \stackrel{\mathscr{L}}{\sim} X$. Then, for every monotone function $f : \mathbb{R} \to \mathbb{R}$ such that $f(X) \in L^2(\mathbb{P})$ and $\mathbb{P}(f(X) \neq \mathbb{E} f(X)) > 0$, we have

$$\mathrm{Var}\left(\frac{f(X) + f \circ \varphi(X)}{2}\right) = \frac{2(\mathrm{Var}(f(X)) + \mathrm{Cov}(f(X), f \circ \varphi(X)))}{4}$$

$$< \frac{\mathrm{Var}(f(X))}{2}$$

since $\mathrm{Cov}(f(X), f \circ \varphi(X)) < 0$.

The variance is reduced by more than a 2-factor whereas the complexity of the simulation of $\frac{f(X)+f\circ\varphi(X)}{2}$ is only twice higher than that of $f(X)$ (if one neglects the additional cost of the computation of $\varphi(x)$ compared to that of x).

▷ *Sensitivity (vega of an option).* Let $\varphi : (0, \infty) \to \mathbb{R}$ be a convex function with (at most) polynomial growth at 0 and $+\infty$ in the sense that there exists a real constant $C > 0$ such that

$$\forall x \in (0, +\infty), \quad |\varphi(x)| \leq C(x^r + x^{-r})$$

and let $Z : (\Omega, \mathscr{A}, \mathbb{P}) \to \mathbb{R}$ be an $\mathscr{N}(0; 1)$-distributed random variable. Set

$$f(\sigma) = \mathbb{E}\,\varphi\!\left(e^{\sigma Z - \frac{\sigma^2}{2}}\right), \quad \sigma > 0.$$

Although it does not appear as a straightforward consequence of its definition, one easily derives from the above proposition that f is a non-decreasing function of σ on $(0, \infty)$. In fact, φ is differentiable outside an at most countable subset of $(0, +\infty)$ (where its right and left derivatives differ) and its derivative φ' is non-decreasing, with polynomial growth as well since

$$|f'(x)| \leq \max\left(|f(x+1) - f(x)|, 2x^{-1}|f(x) - f(x/2)|\right), \quad x \in (0, +\infty).$$

Since Z has no atom, one easily checks that one can interchange derivative and expectation to establish that f is differentiable with derivative

$$f'(\sigma) = \mathbb{E}\!\left(\varphi'\!\left(e^{\sigma Z - \frac{\sigma^2}{2}}\right) e^{\sigma Z - \frac{\sigma^2}{2}} (Z - \sigma)\right), \quad \sigma > 0.$$

A Cameron–Martin change of variable then yields

$$f'(\sigma) = \mathbb{E}\!\left(\varphi'\!\left(e^{\sigma Z + \frac{\sigma^2}{2}}\right) Z\right)$$

so that, applying the co-monotony principle to the two non-decreasing (square integrable) functions $z \mapsto \varphi'\!\left(e^{\sigma z - \frac{\sigma^2}{2}}\right)$ and $z \mapsto z$, implies

$$f'(\sigma) \geq \mathbb{E}\!\left(\varphi'\!\left(e^{\sigma Z + \frac{\sigma^2}{2}}\right)\right) \mathbb{E}(Z) = \mathbb{E}\!\left(\varphi'\!\left(e^{\sigma Z + \frac{\sigma^2}{2}}\right)\right) \times 0 = 0.$$

(However, note that a shortest proof is of course to apply Jensen's inequality to $e^{W_{\sigma^2} - \frac{\sigma^2}{2}} \stackrel{\mathscr{L}}{\sim} e^{\sigma Z - \frac{\sigma^2}{2}}$, where W is a standard Brownian motion).

Extensions of the above co-monotony principle to functions on $\mathbb{R}^d$, $d \geq 2$, are almost as classical as the one dimensional case. They can be established by induction when both functions $\Phi(x_1, \ldots, x_d)$ and $\Psi(x_1, \ldots, x_d)$ defined on $\mathbb{R}^d$ are co-monotone in each variable x_i (*i.e.* having the same monotony property) and when the $\mathbb{R}^d$-valued random vector X has independent marginals (see Sect. 2).

Our aim in this paper is to show that this co-monotony principle can be again extended into a *functional co-monotony principle* satisfied by various classes of stochastic processes $X = (X_t)_{t \in [0,T]}$ whose paths lie in a sub-space E of the vector space $\mathscr{F}([0, T], \mathbb{R})$ of real-valued functions defined on the interval $[0, T]$, $T > 0$, equipped with the pointwise (*partial*) order on functions, defined by

$$\forall \alpha, \beta \in \mathscr{F}([0, T], \mathbb{R}), \quad \alpha \le \beta \text{ if } \forall t \in [0, T], \ \alpha(t) \le \beta(t).$$

Then a functional $F : E \to \mathbb{R}$ is said to be non-decreasing if

$$\forall \alpha, \beta \in E, \quad \alpha \le \beta \implies F(\alpha) \le F(\beta).$$

The choice of E will be motivated by the pathwise regularity of the process X. The space E will also be endowed with a metric topology (and its Borel σ-field) so that X can be seen as an E-valued random vector. The functionals F and G involved in the co-monotony principle will be assumed to be continuous on E (at least $\mathbb{P}$-a.s.). Typical choices for E will be $E = \mathscr{C}([0, T], \mathbb{R})$, $\mathscr{C}([0, T], \mathbb{R}^d)$, $\mathbb{D}([0, T], \mathbb{R})$ or $\mathbb{D}([0, T], \mathbb{R}^d)$ and occasionally $L^p_{\mathbb{R}^d}([0, T], dt)$ (in this case we will switch to the dt-a.e. pointwise order instead of the pointwise order). Then by co-monotony principle, we mean that for every non-decreasing functionals F and G defined on E, $\mathbb{P}_X$-a.s. continuous,

$$\mathbb{E} F(X)G(X) \ge \mathbb{E} F(X)\mathbb{E} G(X).$$

(The case of non-increasing functionals follows by considering the opposite functionals and the opposite monotony case by considering the opposite of only one of the functionals). Among the (classes of) processes of interest, we will consider continuous Gaussian processes with nonnegative covariance function (like the standard and the fractional Brownian motion, "nonnegative" Liouville processes), the Markov processes with monotony preserving transitions (which includes of course Brownian diffusions), processes with independent increments, etc.

The main problem comes from the fact that the naive pointwise order on functional spaces is not total so that the formal one-dimensional proof based on Fubini's theorem no longer applies.

As applications of such functional results, we will be able to extend the above sign property for the *vega* of a "vanilla" option (whose payoff function is a function of the risky asset S_T at the maturity T) to "exotic" options. By "exotic", we classically mean that their payoff is typically a path-dependent functional $F\big((S_t)_{t \in [0,T]}\big)$ of the risky asset $(S_t)_{t \in [0,T]}$. The dynamics of this risky asset is still a Black–Scholes model where $S_t^\sigma = s_0 e^{\sigma W_t + (r - \frac{\sigma^2}{2})t}$, $s_0, \sigma > 0$. Doing so we will retrieve Carr et al. results about the sensitivity of Asian type options in a Black–Scholes model with respect to the volatility (see [4]). We will also emphasize the close connection between co-monotony and the theory of *peacocks*[1] characterized by Kellerer in [13] and recently put back into light in the book [9] (see also the references therein). Let us briefly recall that an integrable process $(X_\lambda)_{\lambda \geq 0}$ is a peacock if for every convex function $\varphi : \mathbb{R} \to \mathbb{R}$ such that $\mathbb{E}|\varphi(X_\lambda)| < +\infty, \lambda \geq 0$, the function $\lambda \mapsto \mathbb{E}\varphi(X_\lambda)$ is non-decreasing. Kellerer's characterization theorem says that a process is a peacock if and only if there exists a martingale $(M_\lambda)_{\lambda \geq 0}$ such that $X_\lambda \stackrel{\mathscr{L}}{\sim} M_\lambda, \lambda \in \mathbb{R}_+$. Moreover, the process $(M_\lambda)_{\lambda \geq 0}$ can be chosen to be Markovian. This proof being non-constructive, it does not help at all establishing whether or not a process is a peacock. See also a new proof of Kellerer's Theorem due to Hirsch and Roynette in [8]. By contrast, one can find in [9] a huge number of peacocks with an explicit marginal martingale representation characterized through various tools from the theory of stochastic processes.

More generally, when applied in its "opposite" version, the co-monotony principle between nonnegative function simply provides a significant improvement of the Hölder inequality since it makes the L^1-norm sub-multiplicative. It can be used to produce less conservative bounds in various fields of applied probability, like recently in [14, 20] where to provide bounds depending on functionals of a Brownian diffusion process, in the spirit of the inequalities proposed in Sect. 6.4 for barrier options.

The paper is organized as follows: Sect. 2 is devoted to the finite-dimensional co-monotony principle, Sect. 3 to the definition of functional co-monotony principle and some first general results. Section 4 deals with continuous processes, Sect. 5 with càdlàg processes like Lévy processes. Section 6 deals with examples of applications, to peacocks and to exotic options for which we establish universal bounds (among price dynamics sharing the functional co-monotony principle).
NOTATIONS:

- $x_{0:n} = (x_0, \ldots, x_n) \in \mathbb{R}^{n+1}$, $x_{1:n} = (x_1, \ldots, x_n) \in \mathbb{R}^n$, etc. $(x|y) = \sum_{0 \leq k \leq n} x_k y_k$ denotes the canonical inner product on $\mathbb{R}^{n+1}$.
- We denote by $\leq$ the componentwise order on $\mathbb{R}^{n+1}$ defined by $x_{0:n} \leq x'_{0:n}$ if $x_i \leq x'_i, i = 0, \ldots, n$.
- $\mathscr{M}(d, r)$ denotes the vector space of matrices with d rows and r columns. M^* denotes the transpose of matrix M.
- $\perp\!\!\!\perp$ will emphasize in formulas the (mutual) independence between processes.
- $\|\alpha\|_{\sup} = \sup_{t \in [0,T]} |\alpha(t)|$ for any function $\alpha : [0, T] \to \mathbb{R}$.
- u_+ denotes the positive part of the real number u. λ_d denotes the Lebesgue measure on $(\mathbb{R}^d, \mathscr{B}or(\mathbb{R}^d))$ where $\mathscr{B}or(\mathbb{R}^d)$ denotes the Borel σ-field on $\mathbb{R}^d$ (we will often denote λ instead of λ_1).
- $X \stackrel{\mathscr{L}}{\sim} \mu$ means that the random vector X has distribution μ.

---

[1] Stands for the French acronym PCOC (Processus Croissant pour l'Ordre Convexe).

2 Finite-Dimensional Co-monotony Principle

2.1 Definitions and Main Results

Let $(P(x, dy))_{x \in \mathbb{R}}$ be a probability transition, *i.e.* a family of probability measures such that for every $x \in \mathbb{R}$, $P(x, dy)$ is a probability measure on $(\mathbb{R}, \mathscr{B}or(\mathbb{R}))$ and for every $B \in \mathscr{B}or(\mathbb{R})$, $x \mapsto P(x, B)$ is a Borel function.

Definition 1. (*a*) The transition $(P(x, dy))_{x \in \mathbb{R}}$ is monotony preserving if, for every bounded or nonnegative monotone function $f : \mathbb{R} \to \mathbb{R}$, the function Pf defined for every real number $x \in \mathbb{R}$ by $Pf(x) = \int f(y) P(x, dy)$ is monotone with the same monotony.
(*b*) Two Borel functions $\Phi, \Psi : \mathbb{R}^d \to \mathbb{R}$ are componentwise co-monotone if, for every $i \in \{1, \ldots, d\}$ and $(x_1, \ldots, x_{i-1}, x_{i+1}, \ldots, x_d) \in \mathbb{R}^{d-1}$, both section functions $x_i \mapsto \Phi(x_1, \ldots, x_i, \ldots, x_d)$ and $x_i \mapsto \Psi(x_1, \ldots, x_i, \ldots, x_d)$ have the same monotony, not depending on the $(d-1)$-tuple $(x_1, \ldots, x_{i-1}, x_{i+1}, \ldots, x_d)$ nor on i.
(*c*) If Φ and $-\Psi$ are co-monotone, Φ and Ψ are said to be anti-monotone.

Remark 1. If P is monotony preserving and $f : \mathbb{R} \to \mathbb{R}$ is a monotone function such that $f \in \cap_{x \in \mathbb{R}} L^1(P(x, dy))$, then Pf has the same monotony as f. This is an easy consequence of the Lebesgue dominated convergence theorem and the approximation of f by the "truncated" bounded functions $f_N = (-N) \vee (f \wedge N)$ which have the same monotony as f.

Definition 2 (Componentwise co-monotony principle). An $\mathbb{R}^d$-valued random vector X satisfies a componentwise co-monotony principle if, for every pair of Borel componentwise co-monotone functions $\Phi, \Psi : \mathbb{R}^d \to \mathbb{R}$ such that $\Phi(X), \Psi(X), \Phi(X)\Psi(X) \in L^1(\mathbb{P})$,

$$\mathbb{E}\,\Phi(X)\Psi(X) \geq \mathbb{E}\,\Phi(X)\mathbb{E}\,\Psi(X). \tag{1}$$

Remark 2. • This property is also known as the *positive association* of the components $X_1, \ldots, X_d$ of X (see *e.g.* [17]).
- If X satisfies a componentwise co-monotony principle, then, for every pair of componentwise anti-monotone functions $\Phi, \Psi : \mathbb{R}^{n+1} \to \mathbb{R}$, the reverse inequality holds. In both cases, if Φ and Ψ both take values in $\mathbb{R}_+$ or $\mathbb{R}_-$ then the inequalities remain true (in $\overline{\mathbb{R}}$) without integrability assumption.
- Owing to elementary approximation arguments, it is clear that it suffices to check (1) for bounded or nonnegative componentwise co-monotone functions.

As a straightforward consequence of the fact that the functions $x_{0:n} \mapsto x_k$ and $x_{0:n} \mapsto x_\ell$ are co-monotone, we derive the following necessary condition for the componentwise co-monotony property.

Proposition 2. Let $X = (X_k)_{1 \leq k \leq d}$ be a random vector. If $X_k \in L^2$, $k = 1, \ldots, d$, then,

$$\forall k, \ell \in \{1, \ldots, d\}, \quad \mathrm{Cov}(X_k, X_\ell) = \mathbb{E} X_k X_\ell - \mathbb{E} X_k \mathbb{E} X_\ell \geq 0$$

i.e. the covariance matrix of X has nonnegative entries.

In finite dimension, the main result on componentwise co-monotony is the following.

Proposition 3. *(a) Let $X = (X_k)_{0 \leq k \leq n}$ be an $\mathbb{R}$-valued Markov chain defined on a probability space $(\Omega, \mathscr{A}, \mathbb{P})$ having a (regular) version of its transitions*

$$P_{k-1,k}(x, dy) = \mathbb{P}(X_k \in dy \,|\, X_{k-1} = x), \; k = 1, \ldots, n$$

which are monotony preserving in the above sense. Then X satisfies a componentwise co-monotony principle.
(b) If the random variables $X_0, \ldots, X_n$ are independent, the conclusion remains true under the following weak co-monotony assumption: there exists a permutation τ of the index set $\{0, \ldots, n\}$ such that for every $i \in \{0, \ldots, n\}$ and every $(x_0, \ldots, x_{i-1}, x_{i+1}, \ldots, x_n) \in \mathbb{R}^n$, $x_i \mapsto \Phi(x_0, \ldots, x_i, \ldots, x_n)$ and $x_i \mapsto \Psi(x_0, \ldots, x_i, \ldots, x_n)$ have the same monotony, possibly depending on $(x_{\tau(0)}, \ldots, x_{\tau(i-1)})$. Then the same conclusion as in (a) holds true.

Proof (Proof of (a)). One proceeds by induction on $n \in \mathbb{N}$. If $n = 0$, the result follows from the scalar co-monotony principle applied to X_0 (with distribution μ_0). $(n) \Longrightarrow (n+1)$: We may assume that Φ and Ψ are bounded and, by changing if necessary the functionals into their opposite, that they are both componentwise nondecreasing. Put $\mathscr{F}_k^X = \sigma(X_0, \ldots, X_k)$, $k = 0, \ldots, n$. It follows from the Markov property that

$$\mathbb{E}\big(\Phi(X_{0:n+1}) \,|\, \mathscr{F}_n^X\big) = \Phi^{(n)}(X_{0:n})$$

where

$$\Phi^{(n)}(x_{0:n}) = P_{n,n+1}\big(\Phi(x_{0:n}, .)\big)(x_n).$$

In particular, we have $\mathbb{E}\big(\Phi(X_{0:n+1})\big) = \mathbb{E}\big(\Phi^{(n)}(X_{0:n})\big)$. Let $x_{0:n} \in \mathbb{R}^{n+1}$. Applying the one dimensional co-monotony principle with the probability distribution $P_{n,n+1}(x_n, dy)$ to $\Phi(x_{0:n}, .)$ and $\Psi(x_{0:n}, .)$ we get

$$\begin{aligned}
(\Phi\,\Psi)^{(n)}(x_{0:n}) &= P_{n,n+1}\big(\Phi\,\Psi(x_{0:n}, .)\big)(x_n) \\
&\geq P_{n,n+1}\big(\Phi(x_{0:n}, .)\big)(x_n) P_{n,n+1}\big(\Psi(x_{0:n}, .)\big)(x_n) \\
&= \Phi^{(n)}(x_{0:n}) \Psi^{(n)}(x_{0:n}) \quad (2)
\end{aligned}$$

so that, considering $X_{0:n+1}$ and taking expectation, we get

$$\mathbb{E}\,(\Phi\,\Psi)(X_{0:n+1}) = \mathbb{E}\big((\Phi\,\Psi)^{(n)}(X_{0,n})\big)$$
$$\geq \mathbb{E}\big(\Phi^{(n)}(X_{0:n})\Psi^{(n)}(X_{0:n})\big).$$

It is clear that for every $i \in \{0,\ldots,n-1\}$, $x_i \mapsto \Phi^{(n)}(x_0,\ldots,x_n)$ is non-decreasing since the transition $P_{n,n+1}$ is a nonnegative operator. Now let $x_n, x'_n \in \mathbb{R}$, $x_n \leq x'_n$. Then

$$P_{n,n+1}\big(\Phi(x_0,\ldots,x_n,.)\big)(x_n) \leq P_{n,n+1}\big(\Phi(x_0,\ldots,x'_n,.)\big)(x_n)$$
$$\leq P_{n,n+1}\big(\Phi(x_0,\ldots,x'_n,.)\big)(x'_n)$$

where the first inequality follows from the non-negativity of the operator $P_{n,n+1}$ and the second follows from its monotony preserving property since $x_{n+1} \mapsto \Phi(x_{0:n+1})$ is non-decreasing. The function $\Psi^{(n)}$, defined likewise, shares the same properties.

An induction assumption applied to the Markov chain $(X_k)_{0 \leq k \leq n}$ completes the proof since

$$\mathbb{E}\Big(\Phi^{(n)}(X_{0:n})\Psi^{(n)}(X_{0:n})\Big) \geq \mathbb{E}\,\Phi^{(n)}(X_{0:n})\mathbb{E}\,\Psi^{(n)}(X_{0:n})$$
$$= \mathbb{E}\,\Phi(X_{0:n+1})\mathbb{E}\,\Psi(X_{0:n+1}).$$

Proof (Proof of (b)). By renumbering the $(n+1)$-tuple $(X_0,\ldots,X_n)$ we may assume $\tau = id$. Then the transition $P_{k-1,k}(x_{k-1},dy) = \mathbb{P}X_k(dy)$ does not depend upon x_{k-1} so that $P_{k-1,k}f$ is a constant function. Then (2) holds as an equality and the monotony of $\Phi^{(n)}$ in each of its variable $x_0,\ldots,x_n$ is that of Φ for the same variables. A careful inspection of the proof of claim (a) then shows that the weak co-monotony is enough to conclude. $\square$

Example 1. Let $A, B \in \mathscr{B}or(\mathbb{R}^{n+1})$ be two Borel sets such that, for every $x = x_{0:n} \in A$, $x + te_i \in A$ for every $t \in \mathbb{R}_+$ and every $i \in \{0,\ldots,n\}$ (where e_i denotes the ith vector of the canonical basis of $\mathbb{R}^{n+1}$), *idem* for B. Then for any $\mathbb{R}^d$-Markov chain $X = (X_k)_{0 \leq k \leq n}$, having monotony preserving transitions (in the sense of Proposition 3(a)), we have

$$\mathbb{P}\big((X_0,\ldots,X_n) \in A \cap B\big) \geq \mathbb{P}\big((X_0,\ldots,X_n) \in A\big)\mathbb{P}\big((X_0,\ldots,X_n) \in B\big).$$

The monotony preserving property of the transitions $P_{k-1,k}$ cannot be relaxed as emphasized by the following easy counter-example.

Counter-example 1. Let $X = (X_0, X_1)$ be a Gaussian bi-variate random vector with distribution $\mathscr{N}\Big(0; \begin{bmatrix} 1 & \rho \\ \rho & 1 \end{bmatrix}\Big)$ where the correlation $\rho \in (-1,0)$). One checks that the transition $P_{0,1}(x_0,dx_1)$ reads on bounded or nonnegative Borel functions

$$P_{0,1}(f)(x_0) := \mathbb{E}(f(X_1) \mid X_0 = x_0) = \mathbb{E}f\left(\rho x_0 + \sqrt{1-\rho^2}\, Z\right), \quad Z \overset{\mathscr{L}}{\sim} \mathscr{N}(0;1).$$

This shows that $P_{0,1}$ is monotony... *inverting*. In particular we have $\mathbb{E}\, X_0 X_1 = \rho < 0 = \mathbb{E}\, X_0 \mathbb{E}\, X_1$. In fact it is clear that (X_0, X_1) satisfies the co-monotony principle if and only if $\rho \geq 0$. In the next section we extend this result to higher dimensional Gaussian vectors.

2.2 More on the Gaussian Case

Let $X = (X_1, \ldots, X_d)$ be a Gaussian vector with covariance matrix $\Sigma = [\sigma_{ij}]_{1 \leq i,j \leq d}$ (its mean obviously plays no role here and can be assumed to be zero). This covariance matrix characterizes the distribution of X so it characterizes as well whether or not X shares a co-monotony property in the sense of (1). But can we *read easily* this property on Σ?

As mentioned above, a necessary condition for co-monotony is obviously that

$$\forall\, i,j \in \{1,\ldots,d\}, \quad \sigma_{ij} = \mathrm{Cov}(X_i, X_j) \geq 0.$$

In fact this simple condition does characterize co-monotony: this result, due to L. Pitt, is established in [17].

Theorem 1 (Pitt (1982)). *A Gaussian random vector X with covariance matrix $\Sigma = [\sigma_{ij}]_{1 \leq i,j \leq d}$ satisfies a componentwise co-monotony principle if and only if*

$$\forall\, i,j \in \{1,\ldots,d\}, \quad \sigma_{ij} \geq 0.$$

Remark 3. • Extensions have been proved in [11]. Typically, if $Z \sim \mathscr{N}(0; I_d)$, under appropriate regularity and integrability assumptions on a function $h : \mathbb{R}^d \to \mathbb{R}$, one has

$$\left(\forall\, x \in \mathbb{R}^d,\ \frac{\partial^2 h}{\partial x_i \partial x_j}(x) \geq 0\right) \implies \left(\sigma_{ij} \mapsto \mathbb{E}\bigl(h(\sqrt{\Sigma}\, Z)\bigr) \text{ is non-decreasing}\right).$$

- Another natural criterion for co-monotony—theoretically straightforward although not easy to "read" in practice on the covariance matrix itself—is to make the assumption that there exists a matrix $A = [a_{ij}]_{1 \leq i \leq d, 1 \leq j \leq r}$, $r \in \mathbb{N}^*$, with *nonnegative entries* $a_{ij} \geq 0$ such that $\Sigma = AA^*$. Then $X \overset{\mathscr{L}}{\sim} AZ$, $Z \overset{\mathscr{L}}{\sim} \mathscr{N}(0; I_r)$. So every component is a linear combination with nonnegative coefficients of the components of Z and Proposition 3(b) straightforwardly implies that X shares the co-monotony property (1).

However, surprisingly, this criterion is not a characterization in general: if $d \leq 4$, symmetric matrices Σ with non-negative entries can always be decomposed as $\Sigma = AA^*$ where A has non negative entries. But if $d \geq 5$, this is no longer true. The negative answer is inspired by a former counter-example—originally due to Horn—when $d = r \geq 5$, reported and justified in [7] (see Eqs. (15.39) and (15.53) and the lines that follow, see also [5] for an equivalent formulation). To be precise, the nonnegative symmetric 5×5 matrix Σ (with rank 4) defined by

$$\Sigma = \begin{bmatrix} 1 & 0 & 0 & 1/2 & 1/2 \\ 0 & 1 & 3/4 & 0 & 1/2 \\ 0 & 3/4 & 1 & 1/2 & 0 \\ 1/2 & 0 & 1/2 & 1 & 0 \\ 1/2 & 1/2 & 0 & 0 & 1 \end{bmatrix}$$

has nonnegative entries but cannot be written AA^* where A has nonnegative entries. Another reference of interest about this question is [1], especially concerning more geometrical aspects connected with this problem.

2.3 Application to the Euler Scheme

The Euler scheme of a diffusion is an important example of Markov chain to which one may wish to apply the co-monotony principle. Let $X = (X_t)_{t \in [0,T]}$ be a Brownian diffusion assumed to be solution to the stochastic differential equation

$$SDE \equiv dX_t^x = b(t, X_t^x)dt + \sigma(t, X_t^x)dW_t, \quad t \in [0, T], \quad X_0 = x.$$

Its Euler scheme with step $h = T/n$ and Brownian increments is entirely characterized by its transitions

$$P_{k,k+1}(f)(x) = \mathbb{E} f\left(x + hb(t_k^n, x) + \sigma(t_k^n, x)\sqrt{h} Z\right), \quad Z \overset{\mathcal{L}}{\sim} \mathcal{N}(0; 1),$$

$$k = 0, \ldots, n-1,$$

where $t_k^n = kh = \frac{k}{n}T$, $k = 0, \ldots, n$. One easily checks that if the function b is Lipschitz continuous in x uniformly in $t \in [0, T]$ and if $\sigma(t, x) = \sigma(t)$ is deterministic and lies in $L^2([0, T], dt)$, then, for large enough n, the Euler transitions $P_{k,k+1}$ are monotony preserving.

This follows from the fact that $x \mapsto x + hb(t, x)$ is non-decreasing provided $h \in (0, \frac{1}{[b]_{\text{Lip}}})$, where $[b]_{\text{Lip}}$ is the uniform Lipschitz coefficient of b.

3 Functional Co-monotony Principle

The aim of this section is to extend the above co-monotony principle to continuous time processes relying on the above multi-dimensional co-monotony result. To do so, we will view processes as random variables taking values in a path vector subspace $E \subset \mathscr{F}([0, T], \mathbb{R})$ endowed with the (trace of the) Borel σ-field of pointwise convergence topology on $\mathscr{F}([0, T], \mathbb{R})$, namely $\sigma(\pi_t, t \in [0, T])$ where $\pi_t(\alpha) = \alpha(t)$, $\alpha \in E$. Consequently, a process X having E-valued paths can be seen as an E-valued random variable if and only if for every $t \in [0, T]$, X_t is an $\mathbb{R}$-valued random variable (which is in some sense a tautology since it is the lightest definition of a stochastic process).

We consider on E the (partial) order induced by the natural partial "pointwise order" on $\mathscr{F}([0, T], \mathbb{R})$ defined by

$$\forall \alpha, \beta \in \mathscr{F}([0, T], \mathbb{R}), \quad \alpha \leq \beta \text{ if } \forall t \in [0, T], \ \alpha(t) \leq \beta(t).$$

Definition 3. (*a*) A measurable functional $F : E \to \mathbb{R}$ is *monotone* if it is either non-decreasing or non-increasing with respect to the order on E.

(*b*) A *pair* of measurable functionals are co-monotone if they are both monotone, with the same monotony.

Then, in order to establish a functional co-monotony principle (see the definition below) our approach will be transfer a finite dimensional co-monotony principle satisfied by appropriate converging (time) discretizations of the process X of interest. Doing so we will need to equip E with a topology ensuring the above convergence for the widest class of ($\mathbb{P}_X$-*a.s.* continuous) functionals. That is why we will consider as often as we can the sup-norm topology, not only on $\mathscr{C}([0, T], \mathbb{R})$, but also on the Skorokhod space $\mathbb{D}([0, T], \mathbb{R})$ of càdlàg (French acronym for right continuous left limited) functions defined on $[0, T]$ since there are more continuous functionals for this topology than for the classical J_1-Skorokhod topology (having in mind that, furthermore, $\mathbb{D}([0, T], \mathbb{R})$ is not a topological space for the latter). We recall that $\mathscr{D}_T := \sigma(\pi_t, t \in [0, T])$ is the Borel σ-field related to both the $\|.\|_{\sup}$-norm and the J_1-topologies on the Skorokhod space.

We will also consider (see Sect. 5.2) the space $L^p([0, T], \mu)$, $0 < p < +\infty$, equipped with its usual $L^p(\mu)$-norm where μ is a finite measure on $[0, T]$. In the latter case (which is not—strictly speaking—a set of functions), we will consider the "μ-*a.e.*" (partial) order

$$\alpha \leq_\mu \beta \text{ if } \alpha(t) \leq \beta(t) \ \mu(dt)\text{-}a.e.$$

A functional F which is monotone for the order $\leq_\mu$ is called μ-monotone. The definition of μ-co-monotony follows likewise.

A formal definition of the co-monotony property on a partially ordered normed vector space is the following.

Definition 4. A random variable X whose paths take values in a partially ordered normed vector space $(E, \|.\|_E, \leq_E)^2$ satisfies a co-monotony principle on E if, for every bounded, co-monotone, $\mathbb{P}_X$-a.s. continuous, measurable functionals F, $G : E \to \mathbb{R}$,

$$\mathbb{E}\, F(X)G(X) \geq \mathbb{E}\, F(X)\mathbb{E}\, G(X).$$

When X is a stochastic process and E is its natural path space, we will often use the term *functional co-monotony principle*.

Extensions 1. • The extension to square integrable or nonnegative $\mathbb{P}_X$-a.s. continuous functionals is canonical by a standard truncation procedure: replace F by $F_N := (-N) \vee (F \wedge N)$, $N > 0$, and let N go to infinity.
• More generally, the inequality also holds for pairs of co-monotone functionals F, G whose truncations F_K and G_K are limits in $L^2(\mathbb{P}_X)$ of $\mathbb{P}_X$-a.s. continuous co-monotone functionals.

3.1 A Stability Result for Series of Independent Random Vectors

We will rely several times on the following proposition which shows that series of independent E-valued random vectors satisfying the co-monotony principle also share this property.

Proposition 4. (a) *Let $(X_n)_{n \geq 1}$ be a sequence of independent E-valued random vectors defined on $(\Omega, \mathscr{A}, \mathbb{P})$ where $(E, \|.\|_E, \leq)$ is a partially ordered normed vector space. Assume that, for every $n \geq 1$, X_n satisfies a co-monotony principle on E. Let $(a_n)_{n \geq 1}$ be a sequence of real numbers such that the series $X = \sum_{n \geq 1} a_n X_n$ converges a.s. for the norm $\|.\|_E$. Then X satisfies a co-monotony principle on E.*
(b) *Assume furthermore that $(E, \|.\|_E)$ is a Banach space with an unconditional norm, that the X_n are nonnegative random vectors for the order on E and that $\sum_{n \geq 1} X_n$ converges in $L^1_E(\mathbb{P})$. Then, for every sequence of independent random variables $(A_n)_{n \geq 1}$ taking values in a fixed compact interval of $\mathbb{R}$ and independent of $(X_n)_{n \geq 1}$, the series $X = \sum_{n \geq 1} A_n X_n$ satisfies a co-monotony principle on E.*

Remark 4. When $E = \mathscr{C}([0, T], \mathbb{R})$, Lévy–Itô–Nisio's Theorem (see *e.g.* [15], Theorem 6.1, p. 151) shows to some extent the equivalence between functional convergence in distribution and *a.s.* convergence for series of independent processes as above.

---

[2]For every $\alpha, \beta, \gamma \in E$ and every $\lambda \in \mathbb{R}_+, \alpha \leq \beta \Rightarrow \alpha + \gamma \leq \beta + \gamma$ and $\lambda \geq 0 \Rightarrow \lambda\alpha \leq \lambda\beta$.

Proof. (a) We may assume without loss of generality that the two functionals F and G are non-decreasing. We first show the result for the sum of two independent processes, *i.e.* we assume $a_k = 0$, $k \geq 3$ (and $a_1 a_2 \neq 0$). By Fubini's Theorem

$$\mathbb{E}\Big(F(a_1 X_1 + a_2 X_2) G(a_1 X_1 + a_2 X_2)\Big)$$
$$= \mathbb{E}\Big(\big[\mathbb{E}(F(a_1 X_1 + a_2 \alpha) G(a_1 X_1 + a_2 \alpha))\big]_{|\alpha = X_2}\Big).$$

Let $\mathrm{Cont}(F)$ denote the set of elements of E at which F is continuous. It follows, still from Fubini's Theorem, that

$$1 = \mathbb{P}(a_1 X_1 + a_2 X_2 \in \mathrm{Cont}(F))) = \int \mathbb{P} X_2(d\alpha_2) \mathbb{P} X_1(F(a_1 \cdot + a_2 \alpha_2))$$

so that $\mathbb{P} X_2(d\alpha_2)$-a.s. $\alpha_1 \mapsto F(a_1 \alpha_1 + a_2 \alpha_2)$ and $\alpha_1 \mapsto G(a_1 \alpha_1 + a_2 \alpha_2)$ are $\mathbb{P} X_1$-a.s. continuous. Noting that these functionals are co-monotone (non-decreasing if $a_1 \geq 0$, non-increasing if $a_1 \leq 0$), this implies

$$\mathbb{E}\big(F(a_1 X_1 + a_2 \alpha_2) G(a_1 X_1 + a_2 \alpha_2)\big) \geq \mathbb{E}\big(F(a_1 X_1 + a_2 \alpha_2)\big) \mathbb{E}\big(G(a_1 X_1 + a_2 \alpha_2)\big).$$

Now, both

$$\alpha_2 \mapsto \mathbb{E} F(a_1 X_1 + a_2 \alpha_2) = \int \mathbb{P} X_1(d\alpha_1) F(a_1 \alpha_1 + a_2 \alpha_2)$$

and

$$\alpha_2 \mapsto \mathbb{E} G(a_1 X_1 + a_2 \alpha_2) = \int \mathbb{P} X_1(d\alpha_1) G(a_1 \alpha_1 + a_2 \alpha_2)$$

are co-monotone (non-decreasing if $a_2 \geq 0$, non-increasing if $a_2 \leq 0$) and one checks that both are $\mathbb{P} X_2(d\alpha_2)$-a.s. continuous which implies in turn that

$$\mathbb{E}\Big(\big[\mathbb{E} F(a_1 X_1 + a_2 \alpha)\big]_{|\alpha=X_2} \big[\mathbb{E} G(a_1 X_1 + a_2 \alpha)\big]_{|\alpha=X_2}\Big)$$
$$\geq \mathbb{E}\Big(\big[\mathbb{E} F(a_1 X_1 + a_2 \alpha)\big]_{|\alpha=X_2}\Big)$$
$$\times \mathbb{E}\Big(\big[\mathbb{E} G(a_1 X_1 + a_2 \alpha)\big]_{|\alpha=X_2}\Big)$$
$$= \mathbb{E}\big(F(a_1 X_1 + a_2 X_2)\big) \mathbb{E}\big(G(a_1 X_1 + a_2 X_2)\big)$$

where we used again Fubini's Theorem in the second line.
One extends this result by induction to the case where $X = X_1 + \cdots + X_n$.

To make n go to infinity, we proceed as follows: let $\mathscr{G}_n = \sigma(X_k, k \geq n+1)$. By the reverse martingale convergence theorem, we know that for any bounded measurable functional $\Phi : E \to \mathbb{R}$,

$$\mathbb{E}\left(\Phi(X) \,|\, \mathscr{G}_n\right) = [\mathbb{E}\,\Phi(X_1 + \cdots + X_n + \tilde{\alpha}_n)]_{\tilde{\alpha}_n = \tilde{X}_n}$$

where $\tilde{X}_n = X - (X_1 + \cdots + X_n)$. We know from the above case $n = 2$ that, for $\Phi = F, G$, one has $\mathbb{P}\tilde{X}_n(d\tilde{\alpha}_n)$-a.s., $\alpha \mapsto \Phi(\alpha + \tilde{\alpha}_n)$ is $\mathbb{P}X_1 + \cdots + X_n(d\alpha)$-continuous so that

$$\mathbb{E}\,FG(X_1 + \cdots + X_n + \tilde{\alpha}_n) \geq \mathbb{E}\,F(X_1 + \cdots + X_n + \tilde{\alpha}_n)\mathbb{E}\,G(X_1 + \cdots + X_n + \tilde{\alpha}_n).$$

This equality also reads

$$\mathbb{E}\,(FG(X) \,|\, \mathscr{G}_n) \geq \mathbb{E}\,(F(X) \,|\, \mathscr{G}_n)\mathbb{E}\,(G(X) \,|\, \mathscr{G}_n)$$

which in turn implies by letting $n \to \infty$

$$\mathbb{E}\,F(X)G(X) \geq \mathbb{E}\,F(X)\mathbb{E}\,G(X)$$

owing to the reverse martingale convergence theorem.

(b) For every bounded sequence $(a_n)_{n \geq 1}$, it follows from the unconditionality of the norm $\|\cdot\|_E$ that $\sum_{n \geq 1} a_n X_n$ a.s. converges in $L^1_E(\mathbb{P})$. Then it follows from (a) that, for every $n \geq 1$,

$$\mathbb{E}\left(F\left(\sum_{k=1}^n a_k X_k\right) G\left(\sum_{k=1}^n a_k X_k\right)\right) \geq \mathbb{E}\,F\left(\sum_{k=1}^n a_k X_k\right) \mathbb{E}\,G\left(\sum_{k=1}^n a_k X_k\right).$$

Now for every $k \in \{1, \ldots, n\}$, the function defined on the real line by $a_k \mapsto \mathbb{E}\,F\left(\sum_{i=1}^n a_i X_i\right)$ has the same monotony as F since $X_k \geq 0$ and is bounded. Consequently for any pair F, G of bounded co-monotone Borel functionals,

$$\mathbb{E}\Big([\mathbb{E}\,F(\sum_{k=1}^n a_k X_k)]_{|a_{1:n} = A_{1:n}}[\mathbb{E}\,G(\sum_{k=1}^n a_k X_k)]_{|a_{1:n} = A_{1:n}}\Big)$$

$$\geq \mathbb{E}\Big([\mathbb{E}\,F(\sum_{k=1}^n a_k X_k)]_{|a_{1:n} = A_{1:n}}\Big)$$

$$\times \mathbb{E}\Big([\mathbb{E}\,G(\sum_{k=1}^n a_k X_k)]_{|a_{1:n} = A_{1:n}}\Big).$$

The conclusion follows for a fixed $n \geq 1$ by preconditioning. One concludes by letting n go to infinity since F and G are continuous. $\square$

A First Application to Gaussian Processes. Let $X = (X_t)_{t \in [0,T]}$ be a continuous centered Gaussian process with a covariance operator C_X defined on the Hilbert space $L_T^2 := L^2([0, T], dt)$ into itself by

$$C_X(f) = \mathbb{E}(\langle f, X \rangle_{L_T^2} X) = \int_0^T \mathbb{E}(X_s X_.) f(s) ds \in L_T^2.$$

The process X can be seen as a random vector taking values in the separable Banach space $\mathscr{C}([0, T], \mathbb{R})$ (equipped with the sup-norm). Assume that C_X admits a decomposition as follows

$$C_X = AA^*, \quad A : (K, |.|_K) \longrightarrow \mathscr{C}([0, T], \mathbb{R}), \ A \text{ continuous linear mapping,}$$

where $(K, |.|_K)$ is a separable Hilbert space.

Then, we know from Proposition 1 (and Theorem 1) in [16], that for any orthonormal basis (or even any *Parseval frame*, see [16]) $(e_n)_{n \geq 1}$ of K that the sequence $(A(e_n))_{n \geq 1}$ is *admissible* for the process X in the following sense: for any i.i.d. sequence $(\xi_n)_{n \geq 1}$ of normally distributed random variables defined on a probability space $(\Omega, \mathscr{A}, \mathbb{P})$

$$\begin{cases} (i) \ \sum_{n \geq 1} \xi_n A(e_n) \text{ a.s. converges in } (\mathscr{C}([0, T], \mathbb{R}), \|.\|_{\sup}) \\ (ii) \ \sum_{n \geq 1} \xi_n A(e_n) \stackrel{\mathscr{L}}{\sim} X. \end{cases}$$

Assume furthermore that *all the continuous functions $A(e_n)$ are nonnegative*. Then, for every $n \geq 1$, the continuous stochastic process $X_n = \xi_n A(e_n)$ satisfies a co-monotony principle (for the natural pointwise partial order on $\mathscr{C}([0, T], \mathbb{R})$). This makes up a sequence of independent random elements of $\mathscr{C}([0, T], \mathbb{R})$. It follows from Proposition 4(a) that the process X satisfies a co-monotony principle.

Example 2. Let us consider the standard Brownian motion W with covariance function $\mathbb{E} W_t W_s = s \wedge t$. One checks that $C_W = AA^*$ where $A : L_T^2 \to \mathscr{C}([0, T], \mathbb{R})$ is defined by

$$Af \equiv \left(t \mapsto \int_0^t f(s) ds \right) \in \mathscr{C}([0, T], \mathbb{R}).$$

Applied to the orthonormal basis $e_n(t) = \sqrt{\frac{2}{T}} \sin\left(\pi n \frac{t}{T}\right)$, $n \geq 1$, we get

$$A(e_n)(t) = \sqrt{2T} \frac{1 - \cos\left(\frac{\pi n t}{T}\right)}{\pi n} \geq 0, \ t \in [0, T], \ n \geq 1,$$

so that

$$\tilde{W} = \sqrt{2T} \sum_{n \geq 1} \frac{\xi_n}{\pi n}\left(1 - \cos\left(\pi n \frac{\cdot}{T}\right)\right), \quad (\xi_n)_{n \geq 1} \text{ i.i.d.}, \quad \xi_1 \stackrel{\mathcal{L}}{\sim} \mathcal{N}(0;1),$$

is an *a.s.* converging series for the sup-norm which defines a standard Brownian motion. As a consequence, the standard Brownian motion satisfies a co-monotony principle (in the sense of Definition 4).[3]

As we will see further on in Sect. 4.1, the above approach is clearly neither the most elementary way nor the most straightforward to establish the co-monotony principle for the Wiener process. Furthermore, the above criterion is not an equivalence as emphasized in a finite dimensional setting: a continuous Gaussian process X may satisfy a functional co-monotony principle albeit its covariance operator C_X admits representation of the form $C_X = AA^*$ for which there exists an orthonormal basis (or even Parseval frame, see [16]) whose image by A is made of nonnegative functions. Thus, no such decomposition is known to us for the fractional Brownian motion (with Hurst constant $H \neq \frac{1}{2}$) although it satisfies a co-monotony principle (see Sect. 4.1.4 further on).

3.2 From $[0, T]$ to $\mathbb{R}_+$

We state our results for processes defined on a finite interval $[0, T]$. However they can be extended canonically on $\mathbb{R}_+$, provided that there exists a sequence of positive real constants $T_N \uparrow +\infty$ such that $\mathbb{P}(d\alpha)$-*a.s.* on $E \subset \mathscr{F}(\mathbb{R}_+, \mathbb{R})$, $\alpha^{T_n} \equiv (t \mapsto \alpha(t \wedge T_n))$ converges in E toward α for the topology on E. Such a sequence does exist the topology of convergence on compact sets but also for the Skorokhod topology on the positive real line. Then the transfer of co- and anti-monotony property (if any) from the stopped process X^{T_n} to X is straightforward for bounded functionals. The extension to square integrable or nonnegative functionals follows by the usual truncation arguments.

4 Application to Pathwise Continuous Processes

In this section functional co-monotony principle is always considered *on the normed vector space* $(\mathscr{C}([0, T], \mathbb{R}), \|.\|_{\sup})$.

We will use implicitly that its Borel σ-field of is $\sigma(\pi_t, t \in [0, T])$, where $\pi_t(\alpha) = \alpha(t)$ for every $\alpha \in \mathscr{C}([0, T], \mathbb{R})$ and every $t \in [0, T]$ (see [2], Chap. 2).

---

[3]The fact that $A(L_T^2)$ is the Cameron–Martin space *i.e.* the reproducing space of the covariance operator, which is obvious here, is a general fact for any such decomposition (see [16]).

Proposition 5. *Let $X = (X_t)_{t \in [0,T]}$ be a pathwise continuous process defined on a probability space $(\Omega, \mathscr{A}, \mathbb{P})$ sharing the following finite dimensional co-monotony property: for every integer $n \geq 1$ and every subdivision $t_1, \ldots, t_n \in [0, T]$, $0 \leq t_1 < t_2 < \cdots < t_n \leq T$, the random vector $(X_{t_k})_{1 \leq k \leq n}$ satisfies a componentwise co-monotony principle. Then X satisfies a functional co-monotony principle on its path space $\mathscr{C}([0, T], \mathbb{R})$.*

Proof. We may assume that F and G are both non-decreasing for the natural order on $\mathscr{C}([0, T], \mathbb{R})$. Let $n \in \mathbb{N}$, $n \geq 1$. We introduce the uniform mesh $t_k^n = \frac{kT}{n}$, $k = 0, \ldots, n$ and, for every function $\alpha \in \mathscr{C}([0, T], \mathbb{R})$, the canonical linear interpolation approximation

$$\alpha^{(n)}(t) = \frac{t_{k+1}^n - t}{t_{k+1}^n - t_k^n} \alpha(t_k^n) + \frac{t - t_k^n}{t_{k+1}^n - t_k^n} \alpha(t_{k+1}^n), \ t \in [t_k^n, t_{k+1}^n], \ k = 0, \ldots, n-1.$$

One checks that $\|\alpha - \alpha^{(n)}\|_{\sup} \leq w(\alpha, T/n)$ goes to 0 as $n \to \infty$ where $w(\alpha, .)$ denotes the uniform continuity modulus of α. As a consequence, X having a.s. continuous paths by assumption, the sequence $(X^{(n)})_{\geq 1}$ of interpolations of X a.s. uniformly converges toward X.

Then set for every $n \geq 1$ and every $x = x_{0:n} \in \mathbb{R}^{n+1}$,

$$\chi^n(x, t) = \frac{t_{k+1}^n - t}{t_{k+1}^n - t_k^n} x_k + \frac{t - t_k^n}{t_{k+1}^n - t_k^n} x_{k+1}, \ t \in [t_k^n, t_{k+1}^n], \ k = 0, \ldots, n-1$$

and

$$F_n(x) = F(\chi^n(x, .)).$$

It is clear that if $x \leq x'$ in $\mathbb{R}^{n+1}$ (in the componentwise sense) then $F_n(x) \leq F_n(x')$ since $\chi^n(x, .) \leq \chi^n(x', .)$ as functions. This is equivalent to the fact that F_n is non-decreasing in each of its variables.

On the other hand $X^{(n)} = \chi((X_{t_k^n})_{0 \leq k \leq n}, .)$ so that $F_n\big((X_{t_k^n})_{0 \leq k \leq n}\big) = F(X^{(n)})$. The sequence $(X_{t_k^n})_{0 \leq k \leq n}$ satisfies a componentwise co-monotony principle. As a consequence, it follows from Proposition 3 that if F and G are bounded, for every $n \in \mathbb{N}$,

$$\mathbb{E}\, F(X^{(n)}) G(X^{(n)}) \geq \mathbb{E}\, F(X^{(n)}) \mathbb{E}\, G(X^{(n)}).$$

One derives the expected inequality by letting n go to infinity since F and G are continuous with respect to the sup-norm. The extension to unbounded functionals F or G follows by the usual truncation arguments. □

4.1 Continuous Gaussian Processes

Let $X = (X_t)_{t \in [0,T]}$ be a continuous centered Gaussian process. Its (continuous) covariance function C_X is defined on $[0, T]^2$ as follows

$$\forall\, s, t \in [0, T], \qquad C_X(s,t) = \mathbb{E}\, X_s X_t.$$

We establish below the functional counterpart of Pitt's Theorem.

Theorem 2 (Functional Pitt's Theorem). *A continuous Gaussian process $X = (X_t)_{t \in [0,T]}$ with covariance operator C_X satisfies a functional co-monotony principle on $\mathscr{C}([0, T], \mathbb{R})$ if and only if, for every $s, t \in [0, T]$, $C_X(s,t) \geq 0$.*

Proof. For every $n \geq 1$, $(X^n_{t^n_k})_{0 \leq k \leq n}$ is a Gaussian random vector satisfying a componentwise co-monotony principle since its covariance matrix $\Sigma^n = \left[C_X(t^n_k, t^n_\ell)\right]_{0 \leq k, \ell \leq n}$ has nonnegative entries. One concludes by the above Proposition 5. □

We briefly inspect below several classical classes of Gaussian processes.

4.1.1 Brownian Motion, Brownian Bridge, Wiener Integrals

The covariance of the standard Brownian motion W is given by $C_W(s,t) = s \wedge t \geq 0$, $s, t \in [0, T]$ and that of the Brownian bridge over $[0, T]$, defined by $X_t = W_t - \frac{t}{T}W_T$, $t \in [0, T]$, is given by for every $s, t \in [0, T]$ by

$$C_X(s,t) = s \wedge t - \frac{st}{T} \geq 0.$$

As for Wiener integrals, let $X_t = \int_0^t f(s)dW_s$, $t \in [0, T]$, where $f \in L^2([0, T], dt)$. The process X admits a continuous modification and its covariance function is given by $C_X(s,t) = \int_{s \wedge t}^{s \vee t} f^2(u)du \geq 0$.

They all satisfy a satisfies a co-monotony principle on $\mathscr{C}([0, T], \mathbb{R})$.

4.1.2 Liouville Processes

Definition 5. Let $f : [0, T] \to \mathbb{R}$ be a locally ρ-Hölder function, $\rho \in (0, 1]$, in the following sense: there exists $\varphi \in \mathscr{L}^2([0, T], dt)$, $\rho \in (0, 1]$, $a \in (0, +\infty)$ such that

$$(L_{\rho,a}) \equiv \begin{cases} (i) & \forall\, t, t' \in [0, T], \quad |f(t) - f(t')| \leq [f]_{\rho,\varphi} |t - t'|^\rho \varphi(t \wedge t') \\ (ii) & \int_0^t f^2(s)\,ds = O(t^a). \end{cases}$$

(3)

Then the Gaussian process defined for every $t \in [0, T]$ by

$$X_t = \int_0^t f(t - s)\,dW_s$$

admits a continuous modification called *Liouville process* (related to f) with covariance function

$$C_X(s, t) = \int_0^{s \wedge t} f(t - u) f(s - u)\,du.$$

JUSTIFICATION: First note that $f \in \mathcal{L}^2(dt)$ since $f(t)| \leq |f(0)| + t^\rho |\varphi(0)|$. Then, for every $t, t' \in [0, T]$, $t \leq t'$,

$$X_{t'} - X_t = \int_t^{t'} f(t' - s)\,dW_s + \int_0^t (f(t' - s) - f(t - s))\,dW_s$$

so that

$$\begin{aligned} \mathbb{E}|X_{t'} - X_t|^2 &= \int_t^{t'} f^2(t' - s)\,ds + \int_0^t (f(t' - s) - f(t - s))^2\,ds \\ &\leq \int_0^{t'-t} f^2(s)\,ds + [f]_{\rho,\varphi}^2 |t' - t|^{2\rho} \int_0^T \varphi^2(s)\,ds \\ &\leq C_f(|t' - t|^a + |t' - t|^{2\rho}) \\ &\leq C_f |t' - t|^{(2\rho) \wedge a} \end{aligned}$$

so that, using the Gaussian feature of the process X

$$\mathbb{E}|X_{t'} - X_t|^p \leq C_{f,p} |t' - t|^{p(\rho \wedge a/2)}$$

for every $p \geq 2$ which in turn implies, owing to Kolmogorov's continuity criterion, that $(X_t)_{t \in [0,T]}$ admits a version $(\rho \wedge \frac{a}{2}) - \eta)$-Hölder continuous for any small enough $\eta > 0$.

Proposition 6. *Let* $X_t = \int_0^t f(t-s)dW_s$, $t \in [0,T]$, *be a continuous Liouville process where W is a standard B.M. defined on a probability space $(\Omega, \mathscr{A}, \mathbb{P})$ and f satisfies $(L_{\rho,a})$ for a couple $(\rho, a) \in (0,1] \times (0, +\infty)$. If furthermore f is λ-a.e. nonnegative, then X satisfies a co-monotony principle on $\mathscr{C}([0,T], \mathbb{R})$.*

The proof of the proposition is straightforward since $f \geq 0$ λ-a.e. implies that the covariance function C_X is nonnegative.

Example 3. If $f(u) = u^{H-1/2}$ with $H \in (0,1]$, then f satisfies $(L_{a,\rho})$ with $a = 2H$ and $\rho \in (0, \frac{1}{2} - H)$ if $H < \frac{1}{2}$, $\rho = H - \frac{1}{2}$ if $H > \frac{1}{2}$ (and $\rho = 1$ if $H = \frac{1}{2}$). This corresponds to the pseudo-fractional Brownian motion with Hurst constant H.

4.1.3 Wiener Integrals Depending on a Parameter

Now we consider a class of processes which is wider than Liouville's class and for which we provide a slightly less refined criterion of existence (as a pathwise continuous process).

$$X_t = \int_0^\infty f(t,s)dW_s, \quad t \in [0,T],$$

where $f : [0,T] \times \mathbb{R}_+ \to \mathbb{R}$ satisfies a dominated ρ-Hölder assumption reading as follows: there exists $\varphi \in \mathscr{L}^2(\mathbb{R}, dt)$, non-increasing, and $\rho \in (0,1]$ such that

$$(L'_\rho) \equiv \forall t, t' \in [0,T], \forall s \in \mathbb{R}_+, \quad |f(t',s) - f(t,s)| \leq [f]_\rho |t'-t|^\rho \varphi(s).$$

Such a process has a continuous modification since $t \mapsto X_t$ is ρ-Hölder from $[0,T]$ to $L^2(\mathbb{P})$ and Gaussian (still owing to Kolmogorov's continuity criterion).

As for Liouville processes, if furthermore, for every $t \in [0,T]$, $f(t,.)$ is λ-a.e. nonnegative, then the process X satisfies a co-monotony principle.

Example 4. Let $f_H(t,s) = (t+s)^{H-\frac{1}{2}} - s^{H-\frac{1}{2}}$, $H \in (0,1]$.

- If $H \geq 1/2$, f_H satisfies $(L'_{H-\frac{1}{2}})$ with $\varphi(s) = 1$.
- If $H \in (0, \frac{1}{2}]$, f_H satisfies $(L'_{\frac{1}{2}-H})$ with $\varphi(s) = s^{2H-1}$.

4.1.4 Fractional Brownian Motion with Hurst Constant $H \in (0,1]$

The fractional Brownian motion is a continuous Gaussian process characterized by its covariance function defined by

$$\forall s, t \in [0,T], \quad C^H(s,t) = \frac{1}{2}\left(t^{2H} + s^{2H} - |t-s|^{2H}\right).$$

Since $u \mapsto u^H$ is H-Hölder and $|t-s|^2 \le t^2 + s^2$, it is clear that $C(s,t) \ge 0$. Consequently, *the fractional Brownian motion satisfies a co-monotony principle on* $\mathscr{C}([0,T], \mathbb{R})$.

Remark 5. An alternative approach could be to rely on the celebrated Mandelbrot–Van Ness representation of the fractional Brownian motion with Hurst constant $H \in (0,1]$, given by

$$B_t^H = \underbrace{\int_0^{+\infty} \left((t+s)^{H-\frac{1}{2}} - s^{H-\frac{1}{2}}\right) dW_s^1}_{=:B_t^{H,1}} + \underbrace{\int_0^t |t-s|^{H-\frac{1}{2}} dW_s^2}_{=:B_t^{H,2}}$$

where W^1 and W^2 are independent standard Brownian motions. These two Wiener integrals define pathwise continuous independent processes, both satisfying the co-monotony principle for $\mathbb{P}B^{H,i}$-a.s. $\|.\|_{\sup}$-continuous functionals, consequently their sum satisfies a co-monotony principle for $\mathbb{P}B^H$-a.s. $\|.\|_{\sup}$-continuous functionals owing to Lemma 4(*b*).

4.2 Continuous Markov Processes, Brownian Diffusion Processes

Proposition 7. *Let* $X = (X_t)_{t \in [0,T]}$ *be a pathwise continuous Markov process defined on a probability space* $(\Omega, \mathscr{A}, \mathbb{P})$ *with transition operators* $(P_{s,t})_{0 \le s \le t \le T}$ *satisfying the monotony preserving property. Then X satisfies a functional co-monotony principle on its path space* $\mathscr{C}([0,T], \mathbb{R})$.

Proof. The sequence $(X_{t_k^n})_{0 \le k \le n}$ is a discrete time Markov chain whose transition operators $P_{t_k^n, t_{k+1}^n}$, $k=0,\ldots,n-1$ satisfy the monotony preserving property. □

The main application of this result is the *co-monotony principle for Brownian diffusions* (*i.e.* solutions of stochastic differential equations driven by a standard Brownian motion). We consider the Brownian diffusion

$$(SDE) \equiv dX_t^x = b(t, X_t^x)dt + \sigma(t, X_t^x)dW_t, \quad X_0^x = x$$

where $b: [0,T] \times \mathbb{R} \to \mathbb{R}$ is continuous, Lipschitz continuous in x, uniformly in $t \in [0,T]$ and $\sigma : [0,T] \times \mathbb{R} \to \mathbb{R}$ is continuous with linear growth in x, uniformly in $t \in [0,T]$ and satisfies

$$\forall x, y \in \mathbb{R}, \ \forall t \in [0,T], \quad |\sigma(t,x) - \sigma(t,y)| \le \rho(|x-y|),$$

where

$$\rho : \mathbb{R} \to \mathbb{R} \text{ is increasing, } \rho(0) = 0 \text{ and } \int_{0+} \frac{du}{\rho^2(u)} = +\infty.$$

Then the equation *(SDE)* satisfies a weak existence property since b and σ are continuous with linear growth: its Euler scheme weakly functionally converges to a weak solution of *(SDE)* as its step T/n goes to 0 (see Theorem 5.3 in [10]). It also satisfies a strong uniqueness property (see Proposition 2.13 in [12]) hence a strong existence-uniqueness property. This implies the existence of (Feller) Markov transitions $(P_{s,t}(x, dy))_{t \geq s \geq 0}$ such that, *a.s.* for every $x \in \mathbb{R}$, $P_{s,t}(f)(x) = \mathbb{E}(f(X_t) \mid X_s = x)$ (see *e.g.* Theorem 1.9 in [18]). Furthermore the flow $(X_t^x)_{x \in \mathbb{R}, t \in [0,T]}$ satisfies a comparison principle (Yamada–Watanabe's Theorem) *i.e.* for every $x, x' \in \mathbb{R}$, $x \leq x'$, $\mathbb{P}$-*a.s.*, for every $t \in [0, T]$, $X_t^x \leq X_t^{x'}$. The functional co-monotony principle follows immediately since it implies that the Markov transitions $P_{s,t}$ are monotony preserving.

Remark 6.
- It is to be noticed that the Euler scheme does not share the componentwise co-monotony principle as a Markov chain in full generality, in particular, when σ does depend on x. So, the above result for diffusions is not a simple transfer from the discrete time case. However when $\sigma(x, t) = \sigma(t)$ with $\sigma : [0, T] \to \mathbb{R}$ continuous, the result can be transferred from the Euler scheme (see Sect. 2) since this scheme functionally weakly converges toward X^x as its step T/n goes to 0.
- This result is strongly related to strong uniqueness of solutions of (SDE). The above conditions are not minimal, see *e.g.* [6] for more insights on these aspects.

5 Functional Co-monotony Principle for Càdlàg Processes

The most natural idea is to mimick Proposition 5 by simply replacing $(\mathscr{C}([0, T], \mathbb{R}), \|\cdot\|_{\sup})$ by the space $\mathbb{D}([0, T], \mathbb{R})$ of càdlàg functions endowed with the J_1-Skorokhod topology (see [2], Chap. 3) whose Borel σ-field $\mathscr{D}_T$ is still $\sigma(\pi_t, t \in [0, T])$. Although this is not a topological vector space (which make some results fail like Proposition 5), this approach yields non-trivial results. To be precise, let us consider, instead of the interpolation operator of a continuous function on the uniform subdivision $(t_k^n)_{0 \leq k \leq n}$, the *stepwise constant approximation* operator defined on every function $\alpha \in \mathbb{D}([0, T], \mathbb{R})$ by

$$\tilde{\alpha}^{(n)} = \sum_{k=1}^{n} \alpha(t_{k-1}^n) \mathbf{1}_{[t_{k-1}^n, t_k^n)} + \alpha(T) \mathbf{1}_{\{T\}}, \quad t_k^n = \frac{kT}{n}, \; k = 0, \ldots, n. \quad (4)$$

It follows from Proposition 6.37 in [10], Chap. VI (second edition), that $\tilde{\alpha}^{(n)} \to \alpha$ for the Skorokhod topology as $n \to +\infty$. Then by simply mimicking the proof of Proposition 5, we get the following result.

Proposition 8. *Let $X = (X_t)_{t \in [0,T]}$ be a pathwise càdlàg process defined on a probability space $(\Omega, \mathscr{A}, \mathbb{P})$ sharing the finite dimensional co-monotony property. Then X satisfies a functional co-monotony principle on its path space $\mathbb{D}([0,T], \mathbb{R})$ for the J_1-Skorokhod topology.*

Thus if one considers now a general càdlàg process with independent increments (PII) $(X_t)_{t \geq 0}$ defined on a probability space $(\Omega, \mathscr{A}, \mathbb{P})$, it is clear, *e.g.* from Proposition 3(a), that X shares the finite dimensional co-monotony property since $(X_{t_k^n})_{0 \leq k \leq n}$ is a Markov chain whose transitions $P_{t_{k-1}^n, t_k^n}(x_{k-1}, dy) = \mathscr{L}(x_{k-1} + X_{t_k^n} - X_{t_{k-1}^n})$, $k = 1, \ldots, n$, are clearly monotony preserving.

Consequently *any càdlàg process with independent increments (PII) $(X_t)_{t \geq 0}$ satisfies a functional co-monotony principle on its path space $\mathbb{D}([0,T], \mathbb{R})$ for the J_1-Skorokhod topology.*

Note that a continuous process which satisfies the functional co-monotony property on its path space for the sup-norm will always satisfy the functional co-monotony principle on $\mathbb{D}([0,T], \mathbb{R})$ for the J_1-topology since, when α is continuous, convergence of a sequence (α_n) to α for the sup-norm and the Skorokhod topology coincide.

However this result is not fully satisfactory since there are not so many functionals which are continuous or even $\mathbb{P}_X$-*a.s.* continuous with respect to the Skorokhod topology. Thus the partial maxima functional $\beta \mapsto \sup_{s \in [0,t]} |\beta(s)|$ is not Skorokhod continuous at α if α is not continuous at t (except when $t = T$) and this functional co-monotony principle will fail, *e.g.* for any process X having a fixed discontinuity at t. This is the reason why we establish in the next subsection a functional co-monotony principle for general *PII* on $\mathbb{D}([0,T], \mathbb{R})$ *endowed with the sup-norm topology.*

5.1 Sup-norm Co-monotony for Processes with Independent Increments

We consider a general càdlàg process with independent increments (*PII*) $(X_t)_{t \geq 0}$ defined on a probability space $(\Omega, \mathscr{A}, \mathbb{P})$. We rely on its Lévy–Khintchine decomposition as exposed in [10], Chap. II, Sect. 3. First one can decompose X as the sum

$$X = X^{(1)} \stackrel{\perp\!\!\!\perp}{+} X^{(2)}$$

where $X^{(1)}$ and $X^{(2)}$ are two independent PII: $X^{(1)}$ is a *PII without fixed discontinuities* and $X^{(2)}$ is a pure jump PII, possibly jumping only at a deterministic sequence of times, namely

$$X_t^{(2)} = \sum_{n\geq 1} U_n \mathbf{1}_{\{t_n \leq t\}}, \ t \in [0, T],$$

where $(t_n)_{n\geq 1}$ is a sequence of $[0, T]$-valued real numbers and $(U_n)_{n\geq}$ is a sequence of independent random variables satisfying the usual assumption of the three series theorem

$$\sum_n \mathbb{P}(|U_n| \geq 1) < +\infty,$$

$$\sum_n \mathbb{E} U_n \mathbf{1}_{\{|U_n| \leq 1\}} < +\infty,$$

$$\sum_n \mathbb{E}\big(U_n^2 \mathbf{1}_{\{|U_n|\leq 1\}} - (\mathbb{E} U_n \mathbf{1}_{\{|U_n|\leq 1\}})^2\big) < +\infty.$$

Proposition 9. *A càdlàg PII satisfies a co-monotony principle on* $(\mathbb{D}([0, T], \mathbb{R}), \|\cdot\|_{\sup})$.

Proof. Owing to Lemma 4(a), we will inspect successively the cases of *PII* without fixed discontinuities and of pure jumps.

STEP 1. *X is a PII without fixed discontinuities*: This means that $X^{(2)} \equiv 0$. The classical (pathwise) Lévy–Khintchine formula for *PII* without fixed discontinuities says that, a truncation level $\varepsilon > 0$ being fixed, X reads as the sum of three mutually independent processes as follows

$$\forall\, t \in [0, T], \qquad X_t = b^\varepsilon(t) + W_{c(t)} \overset{\perp\!\!\!\perp}{+} \sum_{s\leq t} \Delta X_s \mathbf{1}_{\{\Delta X_s| > \varepsilon\}} \overset{\perp\!\!\!\perp}{+} M_t^\varepsilon$$

where b^ε is a continuous function on $[0, T]$, c is a nonnegative non-decreasing continuous function on $[0, T]$ with $c(0) = 0$ and M_t^ε is a pure jump martingale satisfying

$$\mathbb{E}\left(\sup_{s\in[0,t]} |M_s^\varepsilon|^2\right) \leq 4 \int_{\mathbb{R}\setminus\{0\}} x^2 \mathbf{1}_{\{|x|\leq\varepsilon\}} \nu^X([0, t] \times dx)$$

where the measure ν^X is the Lévy measure of X, *i.e.* the dual predictable projection of the jump measure $\mu^X(ds, dx) = \sum_{s\in[0,T]} \mathbf{1}_{\{\Delta X_s \neq 0\}} \Delta X_s$. The Lévy measure is characterized by the fact that, for every bounded Borel function $g : \mathbb{R} \to \mathbb{R}$ null in the neighbourhood of 0,

$$\int_{[0,t]} \int_{\mathbb{R}\setminus\{0\}} g(x)(\mu^X(ds, dx) - \nu^X(ds, dx)), \ t \geq 0, \quad \text{is a local martingale.}$$

In particular for any such function we get the compensation formula

$$\mathbb{E}\Big(\sum_{t\le T}g(\Delta X_t)\Big) = \int g(x)v^X([0,T]\times dx)$$

which extends to any nonnegative function g or satisfying $\int_{\mathbb{R}\setminus\{0\}}|g(x)|v^X([0,T]\times dx]) < +\infty$. The Lévy measure v^X satisfies

$$v^X(\{0\}\times\mathbb{R}) = v^X(\{t\}\times dx) = 0, \quad \int_{\mathbb{R}}(x^2\wedge 1)v^X([0,t]\times dx) < +\infty,\ t\in\mathbb{R}_+.$$

In what follows, we make the convention that $\Delta\alpha_\infty = 0$ for any càdlàg function α defined on $\mathbb{R}_+$.

First, owing to Lemma 4(a) and the result about the standard Brownian in Sect. 3, we can assume that $c \equiv 0$ in what follows, i.e. that there is no Brownian component. Then we define two independent marked Poisson processes with positive jumps as follows

$$\tilde{X}_t^{\varepsilon,\pm} = \sum_{s\le t}(\Delta X_s)_\pm \mathbf{1}_{\{(\Delta X_s)_\pm > \varepsilon\}}$$

and $\tilde{X}^\varepsilon = \tilde{X}^{\varepsilon,+} - \tilde{X}^{\varepsilon,-}$. For each process, we define their inter jump sequence $(\tilde{\Theta}_n^{\varepsilon,\pm})_{n\ge 0}$ i.e., with the convention $\tilde{\Theta}_0^{\varepsilon,\pm} = 0$,

$$\tilde{\Theta}_{n+1}^{\varepsilon,\pm} = \min\{s > \tilde{S}_n^\pm \mid (\Delta X_{\tilde{S}_n^\pm + s})_\pm > \varepsilon\} \in (0,+\infty],\quad n\ge 0,$$

where $\tilde{S}_n^{\varepsilon,\pm} = \tilde{\Theta}_1^{\varepsilon,\pm} + \cdots + \tilde{\Theta}_n^{\varepsilon,\pm}$.

Both processes $\tilde{X}^{\varepsilon,+}$ and $\tilde{X}^{\varepsilon,-}$ are independent since they have no common jumps. Furthermore the four sequences $(\tilde{\Theta}_n^{\varepsilon,\pm})_{n\ge 1}$ and $(\Delta\tilde{X}_{\tilde{S}_n^\pm}^{\varepsilon,\pm})_{n\ge 1}$ are mutually independent and made of mutually independent terms.

Let F be a bounded measurable non-decreasing defined on functional on $\mathbb{D}([0,T],\mathbb{R})$. Now, for every $n\ge 1$, we define on $\mathbb{R}^{2n}$ the function F_n by

$$F_n(\xi_1,\theta_1,\ldots,\xi_n,\theta_n) = F\Big(\Big(\sum_{k=1}^n(\xi_k)_+\mathbf{1}_{\{\theta_1+\cdots+\theta_n\le t\}}\Big)_{t\in[0,T]}\Big),$$

$$\xi_1,\ldots,\xi_n\in\mathbb{R},\ \theta_1,\ldots,\theta_n\in\overline{\mathbb{R}}.$$

It is straightforward that the functions F_n are non-decreasing in each variable ξ_i and non-increasing in each variable $\theta_i\in\overline{\mathbb{R}}_+$.

For every $n\ge 1$, set $\tilde{X}_t^{\varepsilon,n,\pm} = \sum_{k=1}^n(\Delta X_{\tilde{S}_k^{\varepsilon,\pm}})_\pm \mathbf{1}_{\{\tilde{S}_k^{\varepsilon,\pm}\le t\}}$ so that

$$F(\tilde{X}^{\varepsilon,n,\pm}) = F_n\big(((\Delta X_{\tilde{S}_k^{\varepsilon,\pm}})_\pm,\tilde{\Theta}_k^{\varepsilon,\pm})_{k=1,\ldots,n}\big).$$

Consequently, if F and G are co-monotone (measurable) functionals on $\mathbb{D}([0,T],\mathbb{R})$, it follows from Proposition 3(b) (co-monotony principle for mutually independent random variables) that

$$\mathbb{E}\, F(\tilde{X}^{\varepsilon,n,\pm})G(\tilde{X}^{\varepsilon,n,\pm}) \geq \mathbb{E}\, F(\tilde{X}^{\varepsilon,n,\pm})\mathbb{E}\, G(\tilde{X}^{\varepsilon,n,\pm}).$$

Now

$$\sup_{t\in[0,T]} |\tilde{X}_t^{\varepsilon,\pm} - \tilde{X}_t^{\varepsilon,n,\pm}| \leq \sum_{k\geq n+1} (\Delta X_{\tilde{S}_k^{\varepsilon,\pm}})_\pm \mathbf{1}_{\{\tilde{S}_k^{\varepsilon,\pm}\leq T\}}$$

so that

$$\mathbb{P}\Big(\sup_{t\in[0,T]} |\tilde{X}_t^{\varepsilon,\pm} - \tilde{X}_t^{\varepsilon,n,\pm}| > 0\Big) \leq \mathbb{P}(\tilde{S}_{n+1}^\pm \leq T) \to 0 \quad \text{as} \quad n\to\infty$$

since the process X has finitely many jumps of size greater than ε on any bounded time interval. The continuity of F and G transfers the co-monotony inequality to $\tilde{X}^{\varepsilon,\pm}$. In turn, the independence of these two processes, combined with Lemma 4(a), propagates co-monotony to the global Poisson process $\tilde{X}^\varepsilon$.

Noting that F and $F(b^\varepsilon + .)$ have the same monotony (if any), one derives that $X - M^\varepsilon$ satisfies the co-monotony principle for every $\varepsilon > 0$. One concludes by noting that $\|M^\varepsilon\|_{\sup} \to 0$ as $\varepsilon \to 0$ in L^2. □

5.2 Càdlàg Markov Processes and $\|\cdot\|_{L_T^p(\mu)}$-Continuous Functionals

It is often convenient to consider some path spaces of the form $L^p([0,T],\mu)$ where μ is a σ-finite measure and $p \in [1,+\infty)$, especially because of the properties of differentiation on these spaces which allow the natural introduction of gradient fields. Of course, less functionals are continuous for such a topology than with the $\|\cdot\|_{\sup}$-norm topology when the process X has continuous (or even càdlàg) paths.

Then, following the lines of the proof of Proposition 7 but with a new canonical approximation procedure of a function α, this time by a stepwise constant function, one shows the following property.

Proposition 10. *Let $(X_t)_{t\in[0,T]}$ be a càdlàg Markov process defined on a probability space $(\Omega,\mathscr{A},\mathbb{P})$ with transitions operators $(P_{s,t})_{t\geq s\geq 0}$ satisfying the monotony property. Let μ be a finite measure on $([0,T],\mathscr{B}or([0,T]))$ and let $p\in[1,+\infty)$. Let $F,G : \mathbb{D}([0,T],\mathbb{R}) \to \mathbb{R}$ be two μ-co-monotone functionals, $\mathbb{P}_X$-a.s. continuous with respect to the $L_T^p(\mu)$-norm on $\mathbb{D}([0,T],\mathbb{R})$. If $F(X)$, $G(X)$ and $F(X)G(X)$ are integrable or have $\mathbb{P}_X$-a.s. a common constant sign, then*

$$\mathbb{E}\, F(X)G(X) \geq \mathbb{E}\, F(X)\mathbb{E}\, G(X).$$

Proof. For every $\alpha \in \mathbb{D}([0,T],\mathbb{R})$ and very integer $n \geq 1$ we define the stepwise constant approximation operator $\tilde{\alpha}^{(n)}$ defined by (4). It is clear that $\alpha^{(n)}(t) \to \alpha(t)$ at every $t \in [0,T]$ and that the sequence $(\alpha^{(n)})_{n \geq 1}$ is bounded by $\|\alpha\|_{\sup}$. Hence $\alpha^{(n)}$ converges to α in every $L^p_T(\mu)$, $1 \leq p < +\infty$. The rest of the proof is similar to that of Proposition 7. □

6 Examples of Applications

6.1 Functional Antithetic Simulation Method

Of course, the first natural application is a functional version of the antithetic simulation method which was briefly described in the introduction. In fact, if a process X taking values in a vector subspace $E \subset \mathscr{F}([0,T],\mathbb{R})$ (partially ordered by the pointwise order) satisfies a functional co-monotony principle in the sense of Definition 4 and is invariant in distribution under a continuous non-increasing mapping $T : E \to E$ (by T non-increasing we mean that $\alpha \leq \beta \Rightarrow T(\alpha) \geq T(\beta)$, $\alpha, \beta \in E$) that for any sup-norm continuous monotone functional $F : E \to \mathbb{R}$ (square integrable or with constant sign)

$$\mathrm{Cov}(F(X), F \circ T(X)) = \mathbb{E}\, F(X) F \circ T(X) - (\mathbb{E}\, F(X))^2 \leq 0.$$

As a consequence, in order to compute $\mathbb{E}\, F(X)$ by a Monte Carlo simulation, it follows that the computation of (independent copies of) $\dfrac{F(X) + F \circ T(X)}{2}$ will induce, for a prescribed simulation budget, a lower variance than a simulation only computing (independent copies of) $F(X)$ like in the scalar framework. In practice such simulations rely on discretization schemes of X for which the co-monotony principle is only true asymptotically (when the discretization step will go to zero). So is the case for the Euler scheme of a Brownian diffusion with non deterministic diffusion coefficient.

It remains that this strongly suggests, in order to compute $\mathbb{E}\, F(X)$ where $X = (X_t)_{t \in [0,T]}$ is a Brownian diffusion and F is a monotone $\mathbb{P}_X$-a.s. sup-norm continuous functional, to simulate systematically two coupled paths of an Euler scheme (with a small enough step): one with a sequence of Brownian increments $(W_{t^n_{k+1}} - W_{t^n_k})_{k \geq 0}$ and one with its opposite $-(W_{t^n_{k+1}} - W_{t^n_k})_{k \geq 0}$. In fact we know, e.g. from [18] (Chap. IX, p. 341) that $X = \mathcal{E}(W)$. Although we do not know whether $F \circ \mathcal{E}$ is monotone (and $\mathbb{P}_W$-a.s. sup-norm continuous) the sign of the covariance can be roughly tested on a small simulated sample.

6.2 A First Application to Peacocks

The aim of this section is to prove that the (centered) antiderivative of an integrable process satisfying a co-monotony principle is a peacock in the sense of the definition given in the introduction.

Proposition 11. *Let $X = (X_t)_{t \geq 0}$ be an integrable càdlàg process satisfying a co-monotony principle for the sup norm on every interval $[0, T]$, $T > 0$, and let μ be a Borel measure on $(\mathbb{R}_+, \mathscr{B}or(\mathbb{R}_+))$. Assume that $\sup_{[0,t]} \mathbb{E}|X_s| < +\infty$ for every $t > 0$ and that $t \mapsto \mathbb{E} X_t$ is càdlàg. Then the process*

$$\left(\int_{[0,t]} (X_s - \mathbb{E} X_s) \mu(ds) \right)_{t \geq 0} \text{ is a peacock.}$$

Remark 7. If $\sup_{t \in [0,T]} \mathbb{E}|X_t|^{1+\varepsilon} < +\infty$ for an $\varepsilon > 0$, then $t \mapsto \mathbb{E} X_t$ is càdlàg by a uniform integrability argument.

Proof. First we may assume without loss of generality that the process X is centered since $(X_t - \mathbb{E} X_t)_{t \in [0,T]}$ clearly satisfies a co-monotony principles on $\mathbb{D}([0, T], \mathbb{R})$ for every $T > 0$ if X does (since $t \mapsto \mathbb{E} X_t$ is càdlaàg). Set for convenience $Y_t = \int_0^t X_s \mu(ds)$. It is clear from the assumption and Fubini's Theorem that $Y_t \in L^1(\mathbb{P})$ and $\mathbb{E} Y_t = 0$.

▷ STEP 1 : Let $\varphi : \mathbb{R} \to \mathbb{R}$ be a convex function with linear growth (so that $\varphi(Y_t) \in L^1(\mathbb{P})$ for every $t \geq 0$). Its right derivative φ'_r is a non-decreasing bounded function. The convexity of the function φ implies, for every $x, y \in \mathbb{R}$,

$$\varphi(y) - \varphi(x) \geq \varphi'_r(x)(y - x)$$

so that, if $t_1 < t_2$

$$\varphi(Y_{t_2}) - \varphi(Y_{t_1}) \geq \Phi_{t_1}(X) \int_{(t_1, t_2]} X_s \mu(ds)$$

where $\Phi_{t_1}(\alpha) = \varphi'_r \left(\int_{[0,t_1]} \alpha(s) \mu(ds) \right)$ is bounded, continuous for the sup norm topology on $\mathbb{D}([0, t_2], \mathbb{R})$ and non-decreasing for the pointwise order. The functional $\alpha \mapsto \int_{(t_1, t_2]} \alpha(s) \mu(ds)$ is also continuous for the sup norm topology, pointwise non-decreasing and

$$\left| \mathbb{E} \int_{(t_1, t_2]} X_s \mu(ds) \right| \leq \mathbb{E} \int_{(t_1, t_2]} |X_s| \mu(ds) \leq \mu((t_1, t_2]) \sup_{s \in [t_1, t_2]} \mathbb{E}|X_s| < +\infty.$$

Consequently, owing to the co-monotony principle, we get

$$\mathbb{E}\Big(\Phi_{t_1}(X)\int_{(t_1,t_2]} X_s\mu(ds)\Big) \geq \mathbb{E}\,\Phi_{t_1}(X)\mathbb{E}\Big(\int_{(t_1,t_2]} X_s\mu(ds)\Big)$$
$$= \mathbb{E}\,\Phi_{t_1}(X) \times \int_{(t_1,t_2]} \mathbb{E}\,X_s\mu(ds) = 0$$

so that $\mathbb{E}\,\varphi(Y_{t_2}) \geq \mathbb{E}\,\varphi(Y_{t_1}) \in L^1(\mathbb{P})$.

STEP 2: Assume now that φ is simply convex. For every $A > 0$, we define the following convex function φ_A with linear growth:

$$\varphi_A(y) = \begin{cases} \varphi(y) & \text{if } |y| \leq A \\ \varphi(A) + \varphi'_r(A)(y-A) & \text{if } y \geq A \\ \varphi(-A) + \varphi'_r(-A)(y+A) & \text{if } y \geq A. \end{cases}$$

It is clear that $\varphi_A \uparrow \varphi$ as $A \uparrow +\infty$ and that $\mathbb{E}\,\varphi_A(Y_{t_2}) \geq \mathbb{E}\,\varphi_A(Y_{t_1})$ by Step 1.

Now φ_A has linear growth so that $\varphi_A(Y_t) \in L^1(\mathbb{P})$. Consequently it follows from the monotone convergence theorem that $\mathbb{E}\,\varphi_A(Y_t) \uparrow \mathbb{E}\,\varphi(Y_t) \in (-\infty, +\infty]$ as $A \uparrow +\infty$. This completes the proof. □

6.3 From the Sensitivity of Asian Path-Dependent Options to Peacocks

Let μ be a finite measure on $([0,T], \mathscr{B}or([0,T]))$ and, for every $p \in [1,+\infty)$, let q denote its Hölder conjugate. Note that, of course, $\mathbb{D}([0,T], \mathbb{R}) \subset L^\infty_T(\mu) \subset \cap_{p \geq 1} L^p_T(\mu)$.

Definition 6. (a) Let $p \in [1,+\infty)$. A measurable functional $F : L^p_T(\mu) \to \mathbb{R}$ is *regularly differentiable* on $\mathbb{D}([0,T], \mathbb{R})$ if, for every $\alpha \in \mathbb{D}([0,T], \mathbb{R})$, there exists a measurable "gradient" functional $\nabla F : ([0,T] \times \mathbb{D}([0,T], \mathbb{R}),$ $\mathscr{B}or([0,T]) \otimes \mathscr{D}_T) \to \mathbb{R}$ such that

$$\begin{cases} (i) \ \nabla F(.,\alpha) \in L^q_T(\mu) \\ (ii) \ \lim_{\|h\|_{L^p_T(\mu)} \to 0, h \in L^p_T(\mu)} \dfrac{\left|F(\alpha+h) - F(\alpha) - \int_0^T \nabla F(s,\alpha) h(s)\mu(ds)\right|}{\|h\|_{L^p_T(\mu)}} = 0. \end{cases} \quad (5)$$

(b) Furthermore, a gradient functional ∇F is *monotone* if, for every $t \in [0,T]$, $\nabla F(t,.)$ is monotone on $\mathbb{D}([0,T], \mathbb{R})$ and if this monotony does not depend on $t \in [0,T]$.

Proposition 12. *Let $X = (X_t)_{t \in [0,T]}$ be a (càdlàg) PII such that, for every $u \in \mathbb{R}$, $L(u,t) = \mathbb{E}\,e^{uX_t}$ is bounded and bounded away from 0 over $[0,T]$ so that, in particular, the function $\Psi(u,t) = \log \mathbb{E}\,e^{uX_t}$ can be defined as a real valued function. Let $F : L^p_T(\mu) \to \mathbb{R}$ be a measurable functional, regularly differentiable* with a monotone gradient ∇F on $[0, T] \times \mathbb{D}([0, T], \mathbb{R})$ satisfying the following Lipschitz continuity assumption

$$\forall \alpha, \beta \in \mathbb{D}([0,T], \mathbb{R}), \quad |F(\alpha) - F(\beta)| \leq [F]_{\text{Lip}} \|\alpha - \beta\|_{L^p_T(\mu)}.$$

Set, for every $\sigma > 0$,

$$f(\sigma) = \mathbb{E}\left(F\left(e^{\sigma X_\cdot - \Psi(\sigma, \cdot)}\right)\right). \tag{6}$$

Then, under the above assumptions, the function f is (differentiable and) non-decreasing.

Remark 8. At least for Lévy processes, the assumption $\sup_{t \in [0,T]} \mathbb{E} e^{u X_t} < +\infty$, $u \in \mathbb{R}$, is satisfied as soon as $\mathbb{E} e^{u X_t} < +\infty$ for every $u \in \mathbb{R}$ (see [19], Theorem 25.18, p. 166).

Before proving the proposition, we need the following technical lemma about the regularity of function L whose details of proof are left to the reader.

Lemma 1. *Under the assumption made on the function L in Proposition 12, the function Λ defined on $\mathbb{R}^2_+$ by $\Lambda(a, t) = \mathbb{E} e^{a|X_t|}$ is finite. Then for every $a \in (0, +\infty)$, L is Lipschitz continuous in u on $[-a, a]$, uniformly in $t \in [0, T]$, with Lipschitz coefficient (upper-bounded by) $\Lambda(a, T)$. Furthermore, for every $u \in \mathbb{R}$, there exists $\kappa_{u,T} > 0$ and $\varepsilon = \varepsilon(u, T) > 0$ such that*

$$\forall t \in [0, T], \ \forall u' \in [u - \varepsilon, u + \varepsilon], \ L(u, t) \geq \kappa_{u,T}.$$

Proof (Proof of Proposition 12). Formally, the derivative of f reads

$$f'(\sigma) = \mathbb{E}\left(\int_0^T \nabla F\left(e^{\sigma X_\cdot - \Psi(\sigma, \cdot)}, t\right) e^{\sigma X_t - \Psi(\sigma, t)} (X_t - \Psi'_\sigma(\sigma, t)) \mu(dt)\right)$$

$$= \int_0^T \mathbb{E}\left(\nabla F\left(e^{\sigma X_\cdot - \Psi(\sigma, \cdot)}, t\right) e^{\sigma X_t - \Psi(\sigma, t)} (X_t - \Psi'_\sigma(\sigma, t))\right) \mu(dt).$$

To justify that this interchange of differentiation and expectation in the first line is valid we need to prove that the ratio

$$\frac{F\left(e^{\sigma' X_\cdot - \Psi(\sigma', \cdot)}\right) - F\left(e^{\sigma X_\cdot - \Psi(\sigma, \cdot)}\right)}{\sigma' - \sigma}, \quad \sigma' \neq \sigma, \ \sigma, \sigma' \in [\epsilon_0, 1/\epsilon_0], \ \epsilon_0 > 0,$$

is $L^{1+\eta}$-bounded for an $\eta > 0$. Without loss of generality, we may assume that $p = 1 + \eta > 1$ since $\|\cdot\|_{L^p_T} \leq \mu([0,T])^{\frac{1}{p} - \frac{1}{p'}} \|\cdot\|_{L^{p'}_T}$ if $1 \leq p \leq p'$. This follows from the Lipschitz continuity of F and from the properties of the Laplace transform L established in Lemma 1.

Let $\mathcal{G}_t := \sigma(X_s - X_t, \ s \in [t, T])$. This σ-field is independent of $\mathcal{F}_t^X$. Elementary computations show that, for every $t \in [0, T]$,

$$\mathbb{E}\left(\nabla F\left(e^{\sigma X_\cdot - \Psi(\sigma,\cdot)}, t\right) \mid \mathcal{G}_t\right) = \Phi\left(X_t - \frac{\Psi(\sigma,\cdot)}{\sigma}, \left(X_s - \frac{\Psi(\sigma,\cdot)}{\sigma}\right)_{s \in [0,t]}, t\right)$$

where, for every $\beta \in \mathbb{D}([0,t], \mathbb{R})$,

$$\Phi(\xi, \beta, t) = \mathbb{E}\left(\nabla F\left(e^{\sigma(X_\cdot - X_t) - (\Psi(\sigma,\cdot) - \Psi(\sigma,t)) + \sigma\beta(t)}\mathbf{1}_{(t,T]} + e^{\sigma\beta}\mathbf{1}_{[0,t]}, t\right)\right).$$

Note that, for every $t \in [0, T]$, the function $\Phi(\cdot, \cdot, t)$ is non-decreasing in both remaining arguments. Now

$$f'(\sigma) = \int_0^T \mathbb{E}\left(\Phi\left(\left(X_s - \frac{\Psi(\sigma,\cdot)}{\sigma}\right)_{s \in [0,t]}, t\right) e^{\sigma\left(X_t - \frac{\Psi(\sigma,t)}{\sigma}\right)} (X_t - \Psi'_\sigma(\sigma, t))\right)\mu(dt). \quad (7)$$

Set $\mathbb{Q}_{(t)} = e^{\sigma X_t - \Psi(\sigma,t)} \cdot \mathbb{P}$. It is classical that $(X_s)_{s \in [0,t]}$ is still a PII under $\mathbb{Q}_{(t)}$ with exponential moment at any order and a log-Laplace transform $\Psi_{(t)}$ given by

$$\Psi_{(t)}(u, s) = \Psi(\sigma + u, s) - \Psi(\sigma, s).$$

Note that $\Psi_{(t)}$ does not depend on t but on σ. Consequently, for every $s \in [0, t]$,

$$\mathbb{E}_{\mathbb{Q}_{(t)}}(X_s) = \frac{\partial \Psi_{(t)}}{\partial u}(0, s) = \Psi'_\sigma(\sigma, s)$$

where $\Psi'_\sigma(\sigma, s)$ denotes the partial derivative of Ψ with respect to σ. Putting $\tilde{X}_s = X_s - \Psi'_\sigma(\sigma, s)$, we get

$$f'(\sigma) = \int_0^T \mathbb{E}_{\mathbb{Q}_{(t)}}\left(\Phi\left(\left(\tilde{X}_s + \Psi'_\sigma(\sigma, s) - \frac{\Psi(\sigma, s)}{\sigma}\right)_{s \in [0,t]}, t\right) \tilde{X}_t\right) \mu(dt). \quad (8)$$

Applying the co-monotony principle to the PII $\tilde{X}$ and to the two non-decreasing $L_T^p(\mu)$-continuous functionals $F(\alpha) = \Phi\left((\alpha(s) + \Psi'_\sigma(\sigma, s) - \frac{\Psi(\sigma,s)}{\sigma})_{s \in [0,t]}\right)$ and $G(\alpha) = \alpha(t)$ yields that, for every $t \in [0, T]$,

$$\mathbb{E}_{\mathbb{Q}_{(t)}}\left(\Phi\left(\left(\tilde{X}_s + \Psi'_\sigma(\sigma, s) - \frac{\Psi(\sigma, s)}{\sigma}\right)_{s \in [0,t]}, t\right) \tilde{X}_t\right) \geq 0$$

since $\mathbb{E}_{\mathbb{Q}_{(t)}} \tilde{X}_t = 0$. As a consequence, f is a non-decreasing function. $\square$

Corollary 1 (see [9]). *Under the assumptions of Proposition 12 on the càdlàg PII X, the process $\sigma \mapsto \int_0^T e^{\sigma X_t - \Psi(\sigma, t)} \mu(dt)$, $\sigma \in \mathbb{R}_+$, is a peacock (with the definition recalled in the introduction).*

Proof. Let $\varphi : \mathbb{R} \to \mathbb{R}$ be a convex function and, for every $A > 0$, let φ_A be defined by $\varphi'_A(x) = \varphi(x)$ if $x \in [-A, A]$ and φ_A affine and differentiable on $(-\infty, -A] \cup [A, +\infty)$. It is clear that $\varphi_A \uparrow \varphi$ since φ takes values in $(-\infty, +\infty]$. Then set $\varphi_{A,\varepsilon}(x) = \mathbb{E}\,\varphi_A(x + \varepsilon Z)$ where $Z \overset{\mathscr{L}}{\sim} \mathcal{N}(0; 1)$. The function $\varphi_{A,\varepsilon}$ is (finite) convex, infinitely differentiable, Lipschitz continuous and converges uniformly to φ_A when $\varepsilon \to 0$. The functional $F_{A,\varepsilon}(\alpha) = \varphi_{A,\varepsilon}\big(\int_0^T \alpha(t)\mu(dt)\big)$ satisfies the assumptions of the above Proposition 12 so that the function $f_{A,\varepsilon}$ defined by (6) is non-decreasing. Letting $\varepsilon \to 0$ and $A \to +\infty$ successively implies that the function f related (still through (6)) to the original functional $F(\alpha) = \varphi\big(\int_0^T \alpha(t)\mu(dt)\big)$ is non-decreasing which completes the proof (for $A \uparrow +\infty$ the arguments are those of the proof of Proposition 11. □

Remark 9. • In fact this proof remains close in spirit to that proposed in [9]. Roughly speaking we replace the notion of *conditional monotony* used in [9] by a functional co-monotony argument (which also spares a time discretization phase). The notion of conditional monotony and its applications have been extensively investigated in the recent PhD thesis of A. Bogso (see [3]). Conditional monotony has been developed on the basis of finite dimensional distributions of a process but it is clear that a functional version can be derived for (continuous) functionals. Then, when the parameter of interest is time, the connection with functional co-monotony looks clear since it corresponds to a *weak form* of the functional co-monotony principle restricted to couples of functionals of the form $F(\alpha^t)$ and $G(\alpha) = g(\alpha(t))$ (α^t denotes the stopped function α at t).

• As already noticed in [9], specifying μ into δ_T or $\frac{1}{T}\lambda_{|[0,T]}$ in the corollary provides the two main results for peacocks devised from $e^{\sigma X_t - \Psi(\sigma,t)}$. When $\mu = \frac{1}{T}\lambda_{|[0,T]}$ one can combine the above results with some self-similarity property of the *PII* process $(X_t)_{t \in [0,T]}$ (if any) to produce other peacocks. So is the case with the seminal example investigated in [4] where the original aim was, for financial purposes, to prove that

$$\left(\frac{1}{t}\int_0^t e^{B_s - \frac{s}{2}}ds\right)_{t \in (0,T]} \text{ is a peacock.}$$

Many other examples of this type are detailed in [3, 9].

APPLICATION TO A CLASS OF ASIAN OPTIONS. As concerns the sensitivity of exotic derivatives, one can derive or retrieve classical results in a Black–Scholes model for the class of Asian options with convex payoff. To be precise, we consider payoff functionals of the form $\Phi_T = \varphi\big(\frac{1}{T}\int_0^T S_s ds\big)$ where φ is a nonnegative convex function (with linear growth) and $S_t = s_0 e^{(r - \frac{\sigma^2}{2})t + \sigma W_t}$, $t \in [0, T]$, where $s_0 > 0$, $\sigma > 0$ and W is a standard Brownian motion (r is a possibly negative interest rate). The holder of an option contract "written" on this payoff receives in cash at the *maturity* $T > 0$ the value of the payoff Φ_T. Classical arbitrage arguments yield that the premium or price at time 0 of such an option is given by

$$\text{Premium}_0(s_0, \sigma, r, T) = e^{-rT} \mathbb{E}\,\varphi\left(\frac{1}{T}\int_0^T S_s ds\right).$$

By considering the measure $\mu(dt) = e^{rt}\frac{1}{T}\lambda_{|[0,T]}(dt)$, one derives from Corollary 1 that $\sigma \mapsto \text{Premium}_0(s_0, \sigma, r, T)$ is non-decreasing. When $r = 0$ a change of variable shows that the premium is also non-decreasing as a function of the maturity T.

6.4 Examples of Bounds for Barrier Options

Let $S = (S_t)_{t \in [0,T]}$ be a càdlàg nonnegative stochastic process defined on a probability space $(\Omega, \mathscr{A}, \mathbb{P})$, modeling the price dynamics of a risky asset. We will assume that $\mathbb{P}$ is a pricing measure in the sense that derivatives products "written" on the asset S are priced under $\mathbb{P}$. In particular we do not ask $\mathbb{P}$ to be risk-neutral. We assume for convenience that the zero-coupon bond (also known as the riskless asset) is constant equal to 1 (or equivalently that all interest rates are constant equal to 0) but what follows remains true if the price dynamics of this bond is deterministic.

For notational convenience, for a càdlàg function $\alpha : [0, T] \to \mathbb{R}$, we will denote by $_*\alpha_t := \inf_{s \leq t} \alpha_s$, $(t \in [0, T])$, the running minimum of the function α and by $\alpha_t^* := \sup_{s \leq t} \alpha_s$ its running maximum process. In what follows we will extensively use the following classical facts: $\alpha \mapsto {_*\alpha_T} T$ and $\alpha \mapsto \alpha_T^* T$ are Skorokhod continuous on $\mathbb{D}([0, T], \mathbb{R})$ and, for every $t \in [0, T)$, $\alpha \mapsto {_*\alpha_t}$ and $\alpha \mapsto \alpha_t^*$ are sup-norm continuous on $\mathbb{D}([0, T], \mathbb{R})$ (and Skorokhod continuous at every α continuous at t).

We assume throughout this section that *the asset price dynamics* $(S_t)_{t \in [0,T]}$ *satisfies a functional co-monotony principle*. This seems is a quite natural and general assumption given the various classes of examples detailed above.

We will focus on *Down-and-In Call* and *Down-and-Out Call* with maturity T. The payoff functional of a *Down-and-In Call* with maturity T is defined for every strike price $K > 0$ and every barrier $L \in (0, S_0)$ by

$$F_{D\&I}(\alpha) = (\alpha(T) - K)_+ \mathbf{1}_{\{_*\alpha_T \leq L\}}.$$

This means that the holder of the *Down-and-In Call* written on the risky asset S contract receives $F_{D\&I}(S)$ at the maturity T, namely $S_T - K$ at the maturity $T > 0$ provided this flow is positive and the asset attained at least once the (low) barrier L between 0 and $T > 0$.

The premium of such a contract at time 0 is defined by

$$\text{Call}_{D\&In}(K, L, T) = \mathbb{E}\big(F_{D\&I}(S)\big).$$

We will denote by $\text{Call}(K, T) = \mathbb{E}(S_T - K)_+$ the premium of the regular (or vanilla) *Call* option with strike K (and maturity T).

Proposition 13. *If the nonnegative càdlàg process* $(S_t)_{t\in[0,T]}$ *satisfies a finite dimensional co-monotony principle (hence functional co-monotony principle on* $\mathbb{D}([0, T], \mathbb{R})$ *for the Skorokhod topology), then the following semi-universal bound holds:*

$$\mathrm{Call}_{D\&In}(K, L, T) \leq \mathrm{Call}(K, T)\mathbb{P}(_*S_T \leq L).$$

Proof. For every $\alpha \in \mathbb{D}([0, T], \mathbb{R})$ and every $\varepsilon > 0$, we have

$$F_{D\&I}(\alpha) \leq \bigl(\alpha(T) - K\bigr)_+ \left(\left(1 - \frac{{}_*\alpha_T - L}{\varepsilon}\right)_+ \wedge 1\right).$$

The two functionals involved in the product of the right hand side of the above equation are clearly anti-monotone, nonnegative and continuous with respect to the Skorokhod topology, consequently

$$\mathrm{Call}_{D\&In}(K, L, T) \leq \mathrm{Call}(K, T)\mathbb{E}\left(\left(1 - \frac{{}_*S_T - L}{\varepsilon}\right)_+ \wedge 1\right).$$

The result follows by letting $\varepsilon \to 0$ owing to Fatou's Lemma. $\square$

As concerns the *Down-and-Out Call* with payoff

$$F_{D\&O}(\alpha) = \bigl(\alpha(T) - K\bigr)_+ \mathbf{1}_{\{_*\alpha_T > L\}}$$

for which the holder of the option receives $S_T - K$ at the maturity $T > 0$ if this flow is positive and if the asset did not attain the (low) level L between 0 and $T > 0$, one gets, either by a direct approach or by using the obvious parity equation $\mathrm{Call}_{D\&In}(K, L, T) + \mathrm{Call}_{D\&Out}(K, L, T) = \mathrm{Call}(K, T)$,

$$\mathrm{Call}_{D\&Out}(K, H, T) \geq \mathrm{Call}(K, T)\mathbb{P}(_*S_T > L).$$

Similar bounds can be derived for *Up-and-In* and *Up-and-Out Calls* with barrier $L > S_0$ (and strike K), namely

$$\mathrm{Call}_{U\&In}(K, L, T) \geq \mathrm{Call}(K, T)\mathbb{P}(S_T^* > L)$$

and

$$\mathrm{Call}_{U\&Out}(K, L, T) \leq \mathrm{Call}(K, T)\mathbb{P}(S_T^* \leq L).$$

If one considers extensions of the above payoffs in which the barrier needs to be (un-)knocked strictly prior to T, at a time $T' < T$, similar semi-universal bounds can be obtained provided one of the following assumption is true $\mathbb{P}(S_{T'} = S_{T'-}) = 1$ or $(S_t)_{t\in[0,T]}$ satisfies a functional co-monotony principle with respect to the sup-norm on $\mathbb{D}([0, T], \mathbb{R})$.

6.5 A Remark on Running Extrema

If a càdlàg process $X = (X_t)_{t \in [0,T]}$ satisfies a co-monotony principle and X_T and $\sup_{t \in [0,T]} X_t$ have no atom so that, for every $x, y \in \mathbb{R}$, $x \leq y$, the functional $\alpha \mapsto \left(1_{\{\alpha(T) \geq x\}}, 1_{\{\sup_{t \in [0,T]} \alpha(t) \geq y\}}\right)$ is $\mathbb{P}_X$-a.s. $\|.\|_{\sup}$-continuous, then

$$\forall\, y \in \mathbb{R}, \quad \mathbb{P}\Big(\sup_{t \in [0,T]} X_t \geq y\Big) = \inf_{x \leq y} \mathbb{P}\Big(\sup_{t \in [0,T]} X_t \geq y \mid X_T \geq x\Big).$$

Of course the list of possible applications is not exhaustive. In more specific problems, one can take advantage of the functional co-monotony principle to establish less conservative inequalities and bounds on parameters of a problem. A typical example is provided by [14] devoted to optimal order execution on a financial market, cited here since it was partially at the origin of the present work.

Acknowledgements The author thanks the anonymous referee for constructive suggestions and F. Panloup for his help.

References

1. P. Bauman, M. Émery, Peut-on "voir" dans l'espace à n dimensions, L'ouvert **116**, 1–8 (2008)
2. P. Billingsley, *Convergence of Probability Measures*, 2nd edn. Wiley Series in Probability and Statistics: Probability and Statistics. A Wiley-Interscience Publication (Wiley, New York, 1999), 277p
3. A.M. Bogso, Peacocks sous l'hypothèse de monotonie conditionnelle et caractérisation des 2-martingales en termes de peacocks, thèse de l'Université de Lorraine, 2012, 169p
4. P. Carr, C.-O. Ewald, Y. Xiao, On the qualitative effect of volatility and duration on prices of Asian options. Finance Res. Lett. **5**, 162–171 (2008)
5. P.H. Diananda, On nonnegative forms in real variables some or all of which are nonnegative. Proc. Cambridge Philos. Soc. **58**, 17–25 (1962)
6. H.J. Engelbert, On the theorem of T. Yamada and S. Watanabe. Stochast. Stochast. Rep. **36**(3–4), 205–216 (1991)
7. M. Hall, Discrete problems, in *A Survey of Numerical Analysis*, ed. by John Todd (McGraw-Hill, New York, 1962)
8. F. Hirsch, B. Roynette, A new proof of Kellerer's Theorem. ESAIM: P&S **16**, 48–60 (2012)
9. F. Hirsch, C. Profeta, B. Roynette, M. Yor, *Peacocks and Associated Martingales, with Explicit Constructions* Bocconi and Springer Series, 3 (Springer, Milan; Bocconi University Press, Milan, 2011), 430p
10. J. Jacod, A.N. Shiryaev, *Limit Theorems for Stochastic Processes*, 2nd edn. Grundlehren der Mathematischen Wissenschaften [Fundamental Principles of Mathematical Sciences], vol. 288 (Springer, Berlin, 2003), 661p
11. K. Joag-Dev, M.D. Perlman, L.D. Pitt, Association of normal random variables and Slepian's inequality. Ann. Probab. **11**(2), 451–455 (1983)
12. I. Karatzas, S.E. Shreve, *Brownian Motion and Stochastic Calculus*, 2nd edn. Graduate Texts in Mathematics, vol. 113 (Springer, New York, 1991), 470p
13. H.G. Kellerer, Markov-Komposition und eine Anwendung auf Martingale. (German) Math. Ann. **198**, 99–122 (1972)

14. S. Laruelle, C.A. Lehalle, Optimal posting price of limit orders: learning by trading. arXiv preprint arXiv:1112.2397 (2011)
15. M. Ledoux, M. Talagrand, *Probability in Banach Spaces. Isoperimetry and Processes*, Ergebnisse der Mathematik und ihrer Grenzgebiete (3) [Results in Mathematics and Related Areas (3)], vol. 23 (Springer, Berlin, 1991), 480p
16. H. Luschgy, G. Pagès, Expansions for Gaussian processes and Parseval frames. Electron. J. Probab. **14**(42), 1198–1221 (2009)
17. L.D. Pitt, Positively correlated normal variables are associated. Ann. Probab. **10**(2), 496–499 (1982)
18. D. Revuz, M. Yor, *Continuous Martingales and Brownian Motion*, 3rd edn. Grundlehren der Mathematischen Wissenschaften [Fundamental Principles of Mathematical Sciences], vol. 293 (Springer, Berlin, 1999), 560p
19. K.I. Sato, *Lévy Distributions and Infinitely Divisible Distributions*. Cambridge Studies in Advanced Mathematics (Cambridge University Press, London, 1999), 486p (first japanese edition in 1990)
20. G. Zbaganu, M. Radulescu, Trading prices when the initial wealth is random. Proc. Rom. Acad. Ser. A Math. Phys. Tech. Sci. Inf. Sci. **10**(1), 1–8 (2009)

Fluctuations of the Traces of Complex-Valued Random Matrices

Salim Noreddine

Abstract The aim of this paper is to provide a central limit theorem for complex random matrices $(X_{i,j})_{i,j \geq 1}$ with i.i.d. entries having moments of any order. Tao and Vu (Ann. Probab. 38(5):2023–2065, 2010) showed that for large renormalized random matrices, the spectral measure converges to a circular law. Rider and Silverstein (Ann. Probab. 34(6):2118–2143, 2006) studied the fluctuations around this circular law in the case where the imaginary part and the real part of the random variable $X_{i,j}$ have densities with respect to Lebesgue measure which have an upper bound, and their moments of order k do not grow faster than $k^{\alpha k}$, with $\alpha > 0$. Their result does not cover the case of real random matrices. Nourdin and Peccati (ALEA 7:341–375, 2008) established a central limit theorem for real random matrices using a probabilistic approach. The main contribution of this paper is to use the same probabilistic approach to generalize the central limit theorem to complex random matrices.

Keywords Central limit theorems • Invariance principles • Normal approximation • Nualart–Peccati criterion of asymptotic normality • Random matrices

1 Introduction

This paper provides a central limit theorem for complex random matrices with i.i.d entries having moments of any order. It generalizes the results of Rider and Silverstein [11] which do not include the case of discrete random variables, and random variables with moments of order k that grow faster than $k^{\alpha k}$, with $\alpha > 0$. We use the probabilistic approach of Nourdin and Peccati [7] based on the principle of

---

S. Noreddine (✉)
Laboratoire de Probabilités et Modèles Aléatoires, Université Paris 6, Boîte courrier 188, 4 Place Jussieu, 75252 Paris Cedex 5, France
e-mail: salim.noreddine@polytechnique.org

universality for a homogeneous sum of i.i.d random variables, which is an instance of the Linderberg principle, see Rotar [12] and Mossel et al. [6]. This universality principle states that, under some conditions, the convergence of a homogeneous sum of a family of i.i.d. random variables is equivalent to the convergence of this homogeneous sum subject to the substitution of the family of random variables by a Gaussian i.i.d. family. Furthermore, the homogeneous sum of a Gaussian family belongs to the Wigner chaos. Thus, we can use the fourth moment theorem of Nualart and Peccati [9] and Peccati Tudor [10] to establish our main result.

In the particular case of symmetric band matrices, Anderson and Zeitouni [2] have shown a central limit theorem using a version of the classical method of moments based on graph enumerations. These techniques require the estimation of all joint moments of traces whereas our approach merely requires the computation of variances and fourth moments. For the case of symmetric matrices, we refer to Sinai and Soshnikov [13] or Guionnet [5] or Anderson et al.[1].

The general statement proved by Chatterjee [3] (Theorem 3.1) concerns the normal approximation of linear statistics of possibly non-Hermitian random matrices. However, the techniques used by the author require that the entries can be re-written as smooth transformations of Gaussian random variables. In particular, the results of Chatterjee [3] cannot be used for discrete distributions.

Let X be a centered random variable with unit variance, taking its values in $\mathbb{C}$ and admitting moments of all orders. Let $\{X_{i,j}\}_{i,j\geq 1}$ be a family of independent and identically distributed copies of X. We denote by X_n the random matrix defined as

$$X_n = \left\{\frac{X_{i,j}}{\sqrt{n}}\right\}_{1\leq i,j\leq n}.$$

In this paper, we aim to find the limit (in law) of

$$\text{trace}(X_n^d) - E[\text{trace}(X_n^d)], \tag{1}$$

where X_n^d denotes the d-th power of X_n. To achieve this goal, we will make use of the following identity:

$$\text{trace}(X_n^d) = n^{-\frac{d}{2}} \sum_{i_1,\ldots,i_d=1}^{n} X_{i_1,i_2} X_{i_2,i_3} \ldots X_{i_d,i_1}.$$

In the case where X is real-valued, the problem was solved by Nourdin and Peccati [7]. The present study extends [7] to the more general case of a complex-valued random variable X.

When $d = 1$, the expression (1) is very simple; we have indeed

$$\text{trace}(X_n) - E[\text{trace}(X_n)] = \frac{1}{\sqrt{n}} \sum_{i=1}^{n} X_{i,i}.$$

As a consequence, if X is real-valued then a straightforward application of the standard cental limit theorem (CLT) yields the convergence in law to $N(0, 1)$. The case where X is complex-valued is not much more difficult, as one only needs to use the bidimensional CLT to get the convergence in law to $Z = Z_1 + iZ_2$, where (Z_1, Z_2) is a Gaussian vector with the same covariance matrix as that of $(\text{Re}(X), \text{Im}(X))$.

When $d \geq 2$, we have:

$$\text{trace}(X_n^d) - E[\text{trace}(X_n^d)] = n^{-\frac{d}{2}} \sum_{i_1,\ldots,i_d=1}^{n} \left(X_{i_1,i_2}\ldots X_{i_d,i_1} - E[X_{i_1,i_2}\ldots X_{i_d,i_1}]\right). \tag{2}$$

If X is real-valued, it is shown in [7] that there is convergence in law of (2) to the centered normal law with variance d. The idea behind the proof is to separate the sum in the right-hand side of (2) into two parts: a first part consisting of the sum over the diagonal terms, i.e. the terms with indices $i_1, \ldots, i_d$ such that there is at least two distinct integers p and q satisfying $(i_p, i_{p+1}) = (i_q, i_{q+1})$; and a second part consisting of the sum over non-diagonal terms, i.e. the sum over the remaining indices. Using combinatorial arguments, it is possible to show that the sum over diagonal terms converges to 0 in L^2. Thus, the contribution to the limit comes from the non-diagonal terms only. In order to tackle the corresponding sum, the idea [7] is to focus first on the particular case where the entries $X_{i,j}$ are Gaussian. Indeed, in this context calculations are much simpler because we then deal with a quantity belonging to the d-th Wiener chaos, so that the Nualart–Peccati [9] criterion of asymptotic normality may be applied. Then, we conclude in the general case (that is, when the entries are no longer supposed to be Gaussian) by extending the invariance principle of Nourdin, Peccati and Reinert [8], so to deduce that it was actually not a loss of generality to have assumed that the entries were Gaussian.

In this paper we study the more general case of complex-valued entries. As we will see, the obtained limit law is now that of a random variable $Z = Z_1 + iZ_2$, where (Z_1, Z_2) is a Gaussian vector whose covariance matrix is expressed by means of the limits of the expectations of the square of (1), as well as the modulus of the square of (1). To show our result, our strategy consists to adapt, to the complex case, the same method used in the real case. Specifically, we show the following theorem.

Theorem 1. *Let $\{X_{ij}\}_{i,j\geq 1}$ be a family of centered, complex-valued, independent and identically distributed random variables, with unit variance and admitting moments of all orders. Set*

$$X_n = \left\{\frac{X_{i,j}}{\sqrt{n}}\right\}_{1\leq i,j\leq n}.$$

Then, for any integer $k \geq 1$,

$$\{\text{trace}(X_n^d) - E[\text{trace}(X_n^d)]\}_{1\leq d\leq k} \xrightarrow{\text{law}} \{Z_d\}_{1\leq d\leq k}.$$

The limit vector $\{Z_d\}_{1 \le d \le k}$ takes its values in $\mathbb{C}^k$, and is characterized as follows: the random variables $Z_1, \ldots, Z_k$ are independent and, for any $1 \le d \le k$, we have $Z_d = Z_d^1 + i Z_d^2$, where (Z_d^1, Z_d^2) denotes a Gaussian vector with covariance matrix equal to

$$\sqrt{d} \begin{pmatrix} a & c \\ c & b \end{pmatrix},$$

with $a + b = 1$ and $a - b + i2c = E(X_{1,1}^2)^d$.

The closest result to ours in the existing literature, other than the previously quoted reference by Nourdin and Peccati [7], is due to Rider and Silverstein [11]. At this stage of the exposition, we would like to stress that Theorem 1 already appears in the paper [11], but under the following additional assumption on the law of $X_{1,1}$: $\mathrm{Re}(X_{1,1})$ and $\mathrm{Im}(X_{1,1})$ must have a joint density with respect to Lebesgue measure, this density must be bounded, and there exists a positive α such that $E((X_{1,1})^k) \le k^{\alpha k}$ for every $k > 2$. These assumptions can sometimes be too restrictive, typically when one wants to deal with discrete laws. Nevertheless, it is fair to mention that Rider and Silverstein focus more generally on Gaussian fluctuations of $\mathrm{trace}(f(X_n)) - E[\mathrm{trace}(f(X_n))]$, when $f : \mathbb{C} \to \mathbb{C}$ is holomorphic and satisfies some additional technical assumptions (whereas, in our paper, we "only" discuss the polynomial case $f \in \mathbb{C}[X]$).

The rest of the paper is devoted to the proof of Theorem 1. To be in position to do so in Sect. 4, we need to establish some preliminary combinatorial results in Sect. 2, as well as some results related to the Gaussian approximation in Sect. 3.

2 Some Preliminary Combinatorial Results

As we said, before giving the proof of Theorem 1 we need to present some combinatorial results. In what follows, we assume that $d \ge 2$ is fixed. Note that the pairs $(i_1, i_2), \ldots, (i_d, i_1)$ appearing in formula (2) are completely determined by the d-tuple $(i_1, \ldots, i_d)$. Indeed, it is straightforward that the set C_n of elements $((i_1, i_2), \ldots, (i_d, i_1))$ in $([1,n]^2)^d$ is in bijection with $[1,n]^d$ via the application $((i_1, i_2), \ldots, (i_d, i_1)) \mapsto (i_1, \ldots, i_d)$. The cardinality of C_n is therefore equal to n^d. We denote by D_n the set of *diagonal* terms of C_n, i.e. the set of elements of C_n such that there exist (at least) two distinct integers j and k such that $(i_j, i_{j+1}) = (i_k, i_{k+1})$, with the convention that $i_{d+1} = i_1$. We denote by $ND_n = C_n \setminus D_n$ the set of *non-diagonal* terms. If $i_1, \ldots, i_d$ are pairwise distinct, then $((i_1, i_2), \ldots, (i_d, i_1))$ belongs to ND_n. Thus, the cardinality of ND_n is greater or equal to $n(n-1) \ldots (n-d+1)$, or, equivalently, the cardinality of D_n is less or equal to $n^d - n(n-1) \ldots (n-d+1)$. In particular, the cardinality of D_n is $O(n^{d-1})$.

For any integer $p \ge 1$, any integers $\alpha, \beta \in [1, n]$ and any element I_p having the following form

$$I_p = ((x_1, y_1), \ldots, (x_p, y_p)) \in ([1,n]^2)^p, \qquad (3)$$

we denote by $mc_{I_p}(\alpha, \beta)$ the number of times that the pair (α, β) appears in (3). Furthermore, we denote by $mp_{I_p}(\alpha)$ the number of occurrences of α in $\{x_i, y_j\}_{1 \le i,j \le p}$. For example, if $I_4 = ((1, 3), (3, 4), (1, 3), (5, 7))$ then $mp_{I_4}(3) = 3$, $mp_{I_4}(1) = 2$, $mc_{I_4}(1, 3) = 2$ and $mc_{I_4}(3, 4) = 1$. For r elements

$$J_k = ((i_1^{(k)}, i_2^{(k)}), \ldots, (i_d^{(k)}, i_1^{(k)})) \in C_n, \quad k = 1, \ldots, r,$$

we define the concatenation $J_1 \sqcup \ldots \sqcup J_r$ as being

$$\left((i_1^{(1)}, i_2^{(1)}), \ldots, (i_d^{(1)}, i_1^{(1)}), (i_1^{(2)}, i_2^{(2)}), \ldots, (i_d^{(2)}, i_1^{(2)}), \ldots, (i_1^{(r)}, i_2^{(r)}), \ldots, (i_d^{(r)}, i_1^{(r)})\right).$$

As such, $J_1 \sqcup \ldots \sqcup J_r$ is an element of $(C_n)^r$.

From now on, we denote by $\#A$ the cardinality of a finite set A. The following technical lemma will allow us to estimate the moments of (2). More precisely, $(i), (ii), (iii), (iv)$ will imply that the variance of the sum of the diagonal terms converges in L^2 to 0, (v) and (vi) will allow us to show that the variance of the sum of the non-diagonal terms converges to d, and (vii) and $(viii)$ will be used in the computation of the fourth moment of that sum.

Lemma 1. *Let the notations previously introduced prevail, and consider the following sets:*

$$A_n = \{I_{2d} = J_1 \sqcup J_2 \in (D_n)^2 : mc_{I_{2d}}(\alpha, \beta) \ne 1 \text{ for every } \alpha, \beta \in [1, n]\}$$

$$B_n = \{I_{2d} \in A_n : mp_{I_{2d}}(\alpha) \in \{0, 4\} \text{ for every } \alpha \in [1, n]\}$$

$$= \{I_{2d} \in A_n : mc_{I_{2d}}(\alpha, \beta) \in \{0, 2\} \text{ for every } \alpha, \beta \in [1, n]\}$$

$$E_n = \{I_d \in D_n : mc_{I_d}(\alpha, \beta) \ne 1 \text{ for every } \alpha, \beta \in [1, n]\}$$

$$F_n = \{I_d \in E_n : mp_{I_d}(\alpha) \in \{0, 4\} \text{ for every } \alpha \in [1, n]\}$$

$$= \{I_d \in E_n : mc_{I_d}(\alpha, \beta) \in \{0, 2\} \text{ for every } \alpha, \beta \in [1, n]\}$$

$$G_n = \{I_{2d} = J_1 \sqcup J_2 \in (ND_n)^2 : mc_{I_{2d}}(\alpha, \beta) \in \{0, 2\} \text{ for every } \alpha, \beta \in [1, n]\}$$

$$H_n = \{I_{2d} \in G_n : mp_{I_{2d}}(\alpha) \in \{0, 4\} \text{ for every } \alpha \in [1, n]\}$$

$$= \{I_{2d} \in G_n : mc_{I_{2d}}(\alpha, \beta) \in \{0, 2\} \text{ for every } \alpha, \beta \in [1, n]\}$$

$$K_n = \{I_{4d} = J_1 \sqcup J_2 \sqcup J_3 \sqcup J_4 \in (ND_n)^4 : mc_{I_{4d}}(\alpha, \beta) \in \{0, 2, 4\}$$
$$\text{for every } \alpha, \beta \in [1, n]\}$$

$$L_n = \{I_{4d} \in H_n : mp_{I_{4d}}(\alpha) \in \{0, 4\} \text{ for every } \alpha \in [1, n]\}$$

$$= \{I_{4d} \in H_n : mc_{I_{4d}}(\alpha, \beta) \in \{0, 2\} \text{ for every } \alpha, \beta \in [1, n]\}.$$

As $n \to \infty$, we have:

(i) $\#(A_n \setminus B_n) = O(n^{d-1})$.
(ii) If d is even, $\#B_n = n \ldots (n - d + 1)$; if d is odd, $\#B_n = 0$.
(iii) $\#(E_n \setminus F_n) = O(n^{\frac{d-1}{2}})$.
(iv) If d is even, $\#F_n = n \ldots (n - \frac{d}{2} + 1)$; if d is odd, $\#F_n = 0$.
(v) $\#G_n \setminus H_n = O(n^{d-1})$.
(vi) $\#H_n = d \times n \ldots (n - d + 1)$.
(vii) $\#(K_n \setminus L_n) = O(n^{2d-1})$.
(viii) $\#L_n = 3d^2 \times n \ldots (n - 2d + 1)$.

Proof. Before giving the proof of the lemma we will present some examples of element belonging to A_4, B_4 in order to understand the construction of these sets. Then $J_1 = ((1, 2), (2, 1), (1, 2), (2, 1))) \in D_4$ because $(1, 2)$ appear 2 times in J_1. We have also $J_2 = ((5, 6), (6, 5), (5, 6), (6, 5)) \in D_4$ then the concatenation of J_1 and J_2 belong to A_4

$$J_1 \sqcup J_2 = ((1, 2), (2, 1), (1, 2), (2, 1), (5, 6), (6, 5), (5, 6), (6, 5)) \in A_4$$

and every element $(1, 2), (2, 1), (5, 6)$ and $(6, 5)$ appear exactly 2 times, this is why it is also an element of B_4.

Also $((1, 1), (1, 1), (1, 1), (1, 1), (5, 6), (6, 5), (5, 6), (6, 5))$ is an element of A_4, but does not belong to B_4 because $(1, 1)$ appear 4 times.

(i) Let $I_{2d} = ((i_1, i_2), \ldots, (i_d, i_1), (i_{d+1}, i_{d+2}), \ldots, (i_{2d}, i_{d+1})) \in A_n \setminus B_n$. By definition of A_n, we have $mp_{I_{2d}}(i_j) \geq 4$ for any $j = 1, \ldots, 2d$. Furthermore, the fact that $I_{2d} \notin B_n$ ensures the existence of at least one integer j_0 between 1 and $2d$ such that $mp_{I_{2d}}(i_{j_0}) > 4$. Let $\sigma : [1, 2d] \to [1, 2d]$ be defined by $j \mapsto \sigma(j) = \min\{k : i_k = i_j\}$. It is readily checked that

$$4d = \sum_{\alpha \in \mathrm{Im}(\sigma)} mp_{I_{2d}}(i_\alpha).$$

We conclude that $\#\mathrm{Im}(\sigma) < d$. Therefore, $\#(A_n \setminus B_n) = O(n^{d-1})$.

(ii) Assume that B_n is non-empty. Let

$$I_{2d} = ((i_1, i_2), \ldots, (i_d, i_1), (i_{d+1}, i_{d+2}), \ldots, (i_{2d}, i_{d+1})) \in B_n.$$

For every integer $j \in [1, 2d]$, we have $mp_{I_{2d}}(i_j) = 4$. Defining σ and proceeding as in point (i) above, we obtain that $\#\mathrm{Im}(\sigma) = d$. We set $m = \min\{l \in \mathrm{Im}(\sigma) | l + 1 \notin \mathrm{Im}(\sigma)\}$. Since $\#\mathrm{Im}(\sigma) = d$, it follows that $m \leq d$. In fact $m \leq d - 1$, otherwise the elements of the d-tuple $(i_1, \ldots, i_d)$ would be all distinct, and $((i_1, i_2), \ldots, (i_d, i_1))$ could not be in D_n, which would yield a contradiction. In the case $d = 2$, $I_{2d} = ((i_1, i_2), (i_2, i_1), (i_3, i_4), (i_4, i_3)) \in B_n$ if and only if $i_1 = i_2$, $i_3 = i_4$ and $i_1 \neq i_3$. Thus, the cardinality of B_n is equal to $n(n-1)$.

In what follows we suppose that $d \geq 3$.

Let us show that d is even (which will prove that B_n is empty if d is odd) and that I_{2d} can be written as

$$\Big((l_1, l_2), \ldots, (l_{\frac{d}{2}-1}, l_{\frac{d}{2}}), (l_{\frac{d}{2}}, l_1), (l_1, l_2), \ldots, (l_{\frac{d}{2}-1}, l_{\frac{d}{2}}), (l_{\frac{d}{2}}, l_1),$$
$$(j_1, j_2) \ldots, (j_{\frac{d}{2}-1}, j_{\frac{d}{2}}), (j_{\frac{d}{2}}, j_1), (j_1, j_2) \ldots, (j_{\frac{d}{2}-1}, j_{\frac{d}{2}}), (j_{\frac{d}{2}}, j_1)\Big),$$

where $l_1, \ldots, l_{\frac{d}{2}}, j_1, \ldots, j_{\frac{d}{2}}$ are pairwise distinct integers in $[1, n]$, which will prove that the formula for $\#B_n$ given in (ii) holds true. The proof is divided in several parts.

(a) Using a proof by contradiction, let us assume that there exists an integer q in $[m+1, d]$ such that i_q does not belong to $\{i_1, \ldots, i_m\}$. We denote by γ the smallest element verifying this. Note that $\gamma \geq m + 2$ necessarily, and that there exists an integer $p \leq m$ such that $i_{\gamma-1} = i_p$. Therefore, $i_{\gamma-1}$ appears in the four pairs

$$(i_{p-1}, i_p), (i_p, i_{p+1}), (i_{\gamma-2}, i_{\gamma-1}), (i_{\gamma-1}, i_\gamma).$$

Note that for the four pairs above, it is possible that the two pairs in the middle are the same. By definition of B_n, we have $mp_{I_{2d}}(i_{\gamma-1}) = 4$ so these pairs are the only pairs of I_{2d} containing the integer $i_{\gamma-1}$. Moreover, by definition of A_n, we have $mc_{I_{2d}}(i_{\gamma-1}, i_\gamma) \geq 2$. Thus, we necessarily have either $(i_{\gamma-1}, i_\gamma) = (i_p, i_{p+1})$; or $(i_{\gamma-1}, i_\gamma) = (i_{\gamma-2}, i_{\gamma-1})$; or $(i_{\gamma-1}, i_\gamma) = (i_{p-1}, i_p)$. If we had $(i_{\gamma-1}, i_\gamma) = (i_{\gamma-2}, i_{\gamma-1})$, then we would have $i_{\gamma-2} = i_{\gamma-1} = i_\gamma$ and i_γ would appear at least six times in the writing of I_{2d}, which is not possible. Similarly, $(i_{\gamma-1}, i_\gamma) = (i_{p-1}, i_p)$ is impossible. Thus, it must hold that $(i_{\gamma-1}, i_\gamma) = (i_p, i_{p+1})$. We can therefore state that $i_\gamma = i_{p+1}$. Since we also have that $p + 1 \leq m + 1$ and $i_{m+1} \in \{i_1, \ldots, i_m\}$, we can conclude that $i_\gamma \in \{i_1, \ldots, i_m\}$, which yields the desired contradiction. Hence,

$$\{i_{m+1}, \ldots, i_d\} \subset \{i_1, \ldots, i_m\}. \tag{4}$$

(b) Let us show that if $l, k \leq d - 1$ are two distinct integers satisfying $i_k = i_l$, then $(i_k, i_{k+1}) = (i_l, i_{l+1})$. Let $l, k \leq d - 1$ be two integers such that $l \neq k$ et $i_k = i_l$. The integer i_l appears in the four pairs $\{(i_{l-1}, i_l), (i_l, i_{l+1}), (i_{k-1}, i_k), (i_k, i_{k+1})\}$ (or only in three pairs, if both pairs in the middle are the same, which happens whether $l = k - 1$). As $mp_{I_{2d}}(i_k) = 4$, these pairs are the only pairs of I_{2d} containing the integer i_k. By definition of A_n, all pairs of I_{2d} must have at least two occurrences in I_{2d}. If we have $(i_k, i_{k+1}) = (i_{k-1}, i_k)$ then we have $i_k = i_{k+1} = i_{k-1}$ and i_k appears at least six times in I_{2d}, which cannot be true. Similarly, $(i_k, i_{k+1}) = (i_{l-1}, i_l)$ is impossible. Therefore, it must hold that $(i_k, i_{k+1}) = (i_l, i_{l+1})$.

(c) It follows from the definition of m that there exists an integer $r \in [1, m]$ satisfying $i_{m+1} = i_r$. Let us show that

$$(i_1, \ldots, i_d) = (i_1, \ldots, i_m, i_r, \ldots, i_{r+d-m-1}). \tag{5}$$

If $m = d - 1$, then $(i_1, \ldots, i_d) = (i_1, \ldots, i_m, i_r)$ and (5) is verified. If $m \leq d - 2$ then, being given that $i_{m+1} = i_r$ and that we already showed in (b) that if $l, k \leq d - 1$ are two distinct integers satisfying $i_k = i_l$ then $i_{k+1} = i_{l+1}$, we can state that $i_{m+2} = i_{r+1}$. Thus, if $m = d - 2$ then $(i_1, \ldots, i_d) = (i_1, \ldots, i_m, i_r, i_{r+1})$, and (5) is once again verified. Finally, if $m \leq d - 3$, we iterate this process as many times as necessary until we get (5).

(d) Let us now prove that the elements of $(i_r, \ldots, i_{r+d-m-1})$ are all distinct. Once again, we use a proof by contradiction. Thus, let us assume that there exists an integer p in $[1, n]$ which appears at least twice in the uplet $(i_r, \ldots, i_{r+d-m-1})$. We then have $\{i_r, \ldots, i_{r+d-m-1}\} = \{i_{m+1}, \ldots, i_d\} \subset \{i_1, \ldots, i_m\}$, see (4) and (5). Thus, p appears at least three times overall in the uplet $(i_1, \ldots, i_d)$. This latter fact implies $mp_{I_{2d}}(p) \geq 6$, which contradicts the assumption $mp_{I_{2d}}(p) = 4$.

(e) Finally, let us establish that $2m = d$ and $r = 1$. The elements of $(i_r, \ldots, i_{r+d-m-1})$ being all distinct, the couple $(i_{r+d-m-1}, i_1) = (i_d, i_1)$ cannot belong to the set of pairs

$$\{(i_r, i_{r+1}), \ldots, (i_{r+d-m-2}, i_{r+d-m-1})\} = \{(i_{m+1}, i_{m+2}), \ldots, (i_{d-1}, i_d)\}.$$

(because, by (d), no pair of this set can have $i_{r+d-m-1}$ as a first coordinate.) Moreover, since i_1 does not belong to $\{i_2, \ldots, i_m\}$ then the pair $(i_{r+d-m-1}, i_1)$ cannot belong to the set of pairs $\{(i_1, i_2), \ldots, (i_{m-1}, i_m)\}$ (because no pair of this set can have i_1 as a second coordinate). Also, the integer i_d appearing at least twice in the uplet $(i_1, \ldots, i_d)$, it cannot belong to the uplet $(i_{d+1}, \ldots, i_{2d})$ (otherwise, i_d would appear at least six times in the pairs of I_{2d}). Thus, the only way for the occurrence of the pair (i_d, i_1) in I_{2d} to be greater or equal than 2 is that $(i_{r+d-m-1}, i_1) = (i_m, i_{m+1})$. Therefore, $i_1 = i_{m+1} = i_r$. As $i_1, \ldots, i_m$ are all distinct and $r \leq m$, it must hold that $r = 1$. Hence, $r + d - m - 1 = d - m$ and $i_{d-m} = i_m$. Since $i_1, \ldots, i_{d-m}$ are all distinct, see indeed (d), it must be true that $m \geq d - m$. Since $i_{d-m} = i_m$, we conclude that $d - m = m$, that is, $d = 2m$. As such, we establish that $(i_{d+1}, \ldots, i_{2d}) = (i_{d+1}, \ldots, i_{\frac{3d}{2}}, i_{d+1}, \ldots, i_{\frac{3d}{2}})$. Let us finally note that $i_1, \ldots, i_{\frac{d}{2}}, i_{d+1}, \ldots, i_{\frac{3d}{2}}$ are necessarily distinct because $mp_{I_{2d}}(i_j) = 4$. This completes the proof of part (ii).

(iii) Consider

$$I = ((i_1, i_2), \ldots, (i_d, i_1)) \in E_n \setminus F_n.$$

Let $\xi : [1,d] \to [1,d]$ be defined by $j \mapsto \xi(j) = min\{k \mid i_k = i_j\}$. From the equation $2d = \sum_{\alpha \in Im(\xi)} mp_I(i_\alpha)$, and using the fact that all $mp_I(i_\alpha)$ are greater or equal than 4, as well as there exists an α in $Im(\xi)$ satisfying $mp_I(i_\alpha) > 4$, we conclude that $\#Im(\xi) < \frac{d}{2}$. Therefore, the cardinality of $E_n \setminus F_n$ is equal to $O(n^{\frac{d-1}{2}})$.

(iv) Let us assume that F_n is not empty. Consider
$$I_d = \big((i_1, i_2), \ldots, (i_d, i_1)\big) \in F_n.$$

Proceeding as in point (ii) above, we conclude that F_n is empty in the case where d is odd, and that the elements of F_n have the following form when d is even:
$$\Big((l_1, l_2), \ldots, (l_{\frac{d}{2}-1}, l_{\frac{d}{2}})(l_{\frac{d}{2}}, l_1), (l_1, l_2), \ldots, (l_{\frac{d}{2}-1}, l_{\frac{d}{2}})(l_{\frac{d}{2}}, l_1)\Big).$$

Here, $l_1, \ldots, l_{\frac{d}{2}}$ are pairwise distinct integers in $[1, n]$. The formula of $\#F_n$ given in (iv) follows directly from that.

(v) Consider
$$I = \big((i_1, i_2), \ldots, (i_d, i_1), (i_{d+1}, i_{d+2}), \ldots, (i_{2d}, i_{d+1})\big) \in G_n \setminus H_n.$$

Let $\zeta : [1, 2d] \to [1, 2d]$ be defined by $j \mapsto \zeta(j) = min\{k \mid i_k = i_j\}$. From the identity $4d = \sum_{\alpha \in Im(\zeta)} mp_I(i_\alpha)$, and using the fact that $mp_I(i_\alpha)$ are greater or equal than 4, as well as that there exists an α in $Im(\zeta)$ satisfying $mp_I(i_\alpha) > 4$, we conclude as in (i) that $\#Im(\zeta) < d$. Therefore, the cardinality of $G_n \setminus H_n$ is $O(n^{d-1})$.

(vi) Consider
$$I = \big((i_1, i_2), \ldots, (i_d, i_1), (i_{d+1}, i_{d+2}), \ldots, (i_{2d}, i_{d+1})\big) \in H_n.$$

(a) By definition of ND_n, there is no redundancy neither among the pairs $(i_1, i_2), \ldots, (i_d, i_1)$ nor among the pairs $(i_{d+1}, i_{d+2}), \ldots, (i_{2d}, i_{d+1})$. Therefore, to satisfy the constraint defining H_n, it is necessary and sufficient that each couple of $(i_1, i_2), \ldots, (i_d, i_1)$ matches one and only one couple among $(i_{d+1}, i_{d+2}), (i_{d+2}, i_{d+3}), \ldots, (i_{2d}, i_{d+1})$.

(b) Using a proof by contradiction, let us show that the elements of $\{i_1, \ldots, i_d\}$ are pairwise distinct. If p and q were two distinct integers in $[1, d]$ such that $i_p = i_q$ then, according to (a), there would exist $k \in [d+1, 2d]$ satisfying $(i_p, i_{p+1}) = (i_k, i_{k+1})$, which would yield $i_p = i_q = i_k$ and, consequently, $mp_I(i_p) \geq 6$. This would contradict the fact that $mp_I(i_p) = 4$.

(c) Let us establish that, for every $p \in [1, d]$ and $q \in [d+1, 2d]$ such that $i_p = i_q$, we have $i_{p+1} = i_{q+1}$. Using a proof by contradiction, let us assume that there exists an integer $q' \in [d+1, 2d]$ different from q such that $(i_p, i_{p+1}) = (i_{q'}, i_{q'+1})$. Then it must hold that $i_p = i_q = i_{q'}$ and $mp_I(i_p) \geq 6$, which contradicts the fact that $mp_I(i_p) = 4$.

The results (a), (b) et (c) allow us to conclude that there exists an integer $k \in [1, d]$ satisfying $(i_1, \ldots, i_d) = (i_{d+k}, i_{d+k+1}, \ldots, i_{2d}, i_{d+1}, \ldots, i_{d+k-1})$. Thus, the elements of G_n are completely characterized by a given integer $k \in [1, d]$ and a given set $\{i_1, \ldots, i_d\}$ where i_j are pairwise distinct integers in $[1, n]$. We can therefore conclude that $\#H_n = d \times n \ldots (n - d + 1)$.

(vii) Consider

$$I = \big((i_1, i_2), \ldots, (i_d, i_1), \ldots, (i_{3d+1}, i_{3d+2}), \ldots, (i_{4d}, i_{3d+1})\big) \in K_n \setminus L_n.$$

Let $\eta : [1, 4d] \longrightarrow [1, 4d]$ be the application defined by $\eta(j) = min\{k | i_k = i_j\}$. From the identity $8d = \sum_{\alpha \in Im(\eta)} mp_I(i_\alpha)$, and using the fact that $mp_I(i_\alpha)$ are all greater or equal than 4, as well as for at least one $\alpha \in Im(\eta)$ it must hold that $mp_I(i_\alpha) > 4$, we conclude that $\#Im(\eta) < 2d$. Therefore, the cardinality of $K_n \setminus L_n$ is $O(n^{2d-1})$.

(viii) Consider

$$I = \big((i_1, i_2), \ldots, (i_d, i_1), \ldots, (i_{3d+1}, i_{3d+2}), \ldots, (i_{4d}, i_{3d+1})\big) \in L_n.$$

For every $j \leq 4d$, we have $mp_I(i_j) = 4$. Then $2d = \#Im(\eta)$, with η as in point (vii).

(a) Using a proof by contradiction, let us show that, for every $k \in [0, 3]$, the integers $i_{kd+1}, \ldots, i_{(k+1)d}$ are all distinct. Assume that there exist two distinct integers l and h in $[1, d]$, as well as an integer k in $[0, 3]$, satisfying $i_{kd+l} = i_{kd+h}$. By definition of the set L_n, we have $\big((i_{kd+1}, i_{kd+2}), \ldots, (i_{(k+1)d}, i_{kd+1})\big) \in ND_n$. Then, the pairs

$$\{(i_{kd+1}, i_{kd+2}), \ldots, (i_{(k+1)d}, i_{kd+1})\}$$

are all distinct, and we have $mc_I(i_{kd+h}, i_{kd+h+1}) = 2$, which implies that there exists $k' \in [0, 3]$, different from k, and $h' \in [1, d]$ satisfying $i_{kd+h} = i_{k'd+h'}$. It follows that i_{kd+h} appears at least six times in I, which contradicts the fact that $mp_I(i_{kd+h}) = 4$.

(b) For any $p = 0, \ldots, 3$, let us introduce $M_p = \{i_{pd+1}, \ldots, i_{(p+1)d}\}$. For any integers p, q in $[0, 3]$, we have either $M_p \cap M_q = \emptyset$ or $M_p = M_q$. Otherwise there would exist an integer j such that $i_{qd+j} \in M_p$ and $i_{qd+j+1} \notin M_p$ and, since $mc_I(i_{qd+j}, i_{qd+j+1}) = 2$, there would exist $q' \in [0, 3]$, different from p and q, and $j' \in [1, d]$ such that $(i_{qd+j}, i_{qd+j+1}) = (i_{q'd+j'}, i_{q'd+j'+1})$; therefore i_{qd+j} would appear at least six times in I, which would yield a contradiction.

(c) If $M_p = M_q$, then proceeding as in point (vi), we show that there exists $j \in [1, d]$ such that

$$(i_{pd+1}, \ldots, i_{(p+1)d}) = (i_{qd+j}, \ldots, i_{(q+1)d}, i_{qd+1}, \ldots, i_{dq+j-1}).$$

The results (a), (b) et (c) allow us to conclude that a generic element of L_n is characterized by:

- The choice of one case among the following three cases: either $M_0 = M_1$ and $M_2 = M_3$; or $M_0 = M_2$ and $M_1 = M_3$; or $M_0 = M_3$ and $M_1 = M_2$. In what follows, we consider the case $M_0 = M_1$ and $M_2 = M_3$ (we can proceed similarly in the other two cases).
- The choice of $2d$ integers $i_1, \ldots, i_d, i_{2d+1}, \ldots, i_{3d}$ that are pairwise distinct in $[1, n]$.
- The choice of an integer $k \in [1, d]$ such that:

$$(i_{d+1}, \ldots, i_{2d}) = (i_k, \ldots, i_d, i_1, \ldots, i_{k-1}).$$

- The choice of an integer $k' \in [1, d]$ such that:

$$(i_{3d+1}, \ldots, i_{4d}) = (i_{2d+k'}, \ldots, i_{3d}, i_{2d+1}, \ldots, i_{2d+k'-1}).$$

It is now easy to deduce that $\#L_n = 3d^2 n \ldots (n - 2d + 1)$.

3 Gaussian Approximations

Let $X = \{X^i\}_{i \geq 1}$ be a family of centered independent random variables taking values in $\mathbb{R}^r$ and having pairwise uncorrelated components with unit variance. Let $G = \{G^i\}_{i \geq 1}$ be a family of independent standard Gaussian random variables taking values in $\mathbb{R}^r$ and having independent components. Suppose also that X and G are independent, and set

$$X = (X_1^1, \ldots, X_r^1, X_1^2, \ldots, X_r^2, \ldots) = (X_1, \ldots, X_r, X_{r+1}, \ldots, X_{2r}, \ldots).$$

i.e., $X_{j+(i-1)r} = X_j^i$.

Consider integers $m \geq 1$, $d_m \geq \ldots \geq d_1 \geq 2$, $N_1, \ldots, N_m$, as well as real symmetric functions $f_1, \ldots, f_m$ such that each function f_i is defined on $[1, rN_i]^{d_i}$ and vanishes at the points $(i_1, \ldots, i_{d_i})$ such that $\exists j \neq k$ for which $\lceil i_j/r \rceil = \lceil i_k/r \rceil$ (we remind that $\lceil x \rceil$ means the unique integer k such that $k < x \leq k + 1$). Let us define

$$Q^i(X) = Q_{d_i}(f_i, X) = \sum_{i_1, \ldots, i_d = 1}^{rN_i} f_i(i_1, \ldots, i_{d_i}) X_{i_1} \ldots X_{i_{d_i}}.$$

In the case of complex-valued matrices, the real and imaginary parts of the entries $X_{i,j}$ are not necessarily independent. Therefore, we will need to modify the results used by Nourdin and Peccati in the paper [9]. The following lemma is a variant, weaker in terms of assumptions, of the hypercontractivity property.

Lemma 2. *Let the notations previously introduced prevail. Assume that $\alpha = \sup_i E(|X_i|^4) < \infty$ and set $K = 36 \times 25^r \times (1 + 2\alpha^{\frac{3}{4}})^2$. Then*

$$E(Q_d(X)^4) \leq K^d E(Q_d(X)^2)^2. \tag{6}$$

Proof. Set

$$\begin{cases} U = \sum_{\forall k: i_k \notin \{(N-1)r+1,\ldots,Nr\}} f(i_1,\ldots,i_d) X_{i_1} \ldots X_{i_d} \\ V_j = \sum_{\exists! k: i_k = (N-1)r+j} f(i_1,\ldots,i_d) X_{i_1} \ldots \overline{X_{(N-1)r+j}} \ldots X_{i_d} \end{cases}$$

The notation $\overline{X_{(N-1)r+j}}$ means that this term is removed from the product. Observe that $X_{(N-1)r+j} = X_j^N$ according to the notation that we adopted previously, and that the quantity $Q_d(X)$ is given by:

$$Q_d(X) = U + \sum_{j=1}^{r} X_j^N V_j$$

(as f vanishes at the points $(i_1,\ldots,i_{d_i})$ such that there exist $j \neq k$ for which $\lceil i_j/r \rceil = \lceil i_k/r \rceil$). Note that, for every $p \leq N$ and every $i,j \in [1,r]$, X_j^p is independent from U and V_i. Thus, by choosing $p = N$, we have

$$E(Q_d(X)^4) = \sum_{s_0+\ldots+s_r=4} \frac{24}{s_0!\ldots s_r!} E(U^{s_0} \prod_{j=1}^{r} (V_j X_j^N)^{s_j})$$

$$= E(U^4) + \sum_{s_1+\ldots+s_r=2} \frac{12}{s_1!\ldots s_r!} E(U^2 \prod_{j=1}^{r} V_j^{s_j}) E(\prod_{j=1}^{r} (X_j^N)^{s_j})$$

$$+ \sum_{s_1+\ldots+s_r=3} \frac{24}{s_1!\ldots s_r!} E(U \prod_{j=1}^{r} V_j^{s_j}) E(\prod_{j=1}^{r} (X_j^N)^{s_j})$$

$$+ \sum_{s_1+\ldots+s_r=4} \frac{24}{s_1!\ldots s_r!} E(\prod_{j=1}^{r} V_j^{s_j}) E(\prod_{j=1}^{r} (X_j^N)^{s_j}).$$

In the equation above, we used that

$$\sum_{s_1+\ldots+s_r=1} \frac{4}{s_1!\ldots s_r!} E(U^3 \prod_{j=1}^{r} (V_j X_j^N)^{s_j}) = 0$$

since X_j^N are centered. By using the generalized Hlder inequality, we obtain:

$$E(U^{s_0}\prod_{j=1}^r V_j^{s_j}) \le E(U^4)^{\frac{s_0}{4}} \prod_{j=1}^r E(V_j^4)^{\frac{s_j}{4}}.$$

Since the terms $E(V_j^4)^{\frac{s_j}{4}}$ are upper bounded by $\left(\sum_{j=1}^r E(V_j^4)^{\frac{1}{2}}\right)^{\frac{s_j}{2}}$, we obtain:

$$\sum_{s_1+\ldots+s_r=4-s_0} E(U^{s_0}\prod_{j=1}^r V_j^{s_j}) \le 5^r E(U^4)^{\frac{s_0}{4}} \left(\sum_{j=1}^r E(V_j^4)^{\frac{1}{2}}\right)^{\frac{4-s_0}{2}}.$$

Using the generalized Hlder inequality again, we have

$$E(\prod_{j=1}^r (X_j^N)^{s_j}) \le \prod_{j=1}^r E\left((X_j^N)^4\right)^{\frac{s_j}{4}} \le \alpha^{\frac{\sum s_j}{4}}.$$

Therefore:

$$E(Q_d(X)^4) \le E(U^4) + 12 \times 5^r E(U^4)^{\frac{1}{2}} \sum_{j=1}^r E(V_j^4)^{\frac{1}{2}} \qquad (7)$$

$$+24 \times 5^r \alpha^{\frac{3}{4}} E(U^4)^{\frac{1}{4}} \left(\sum_{j=1}^r E(V_j^4)^{\frac{1}{2}}\right)^{\frac{3}{2}}$$

$$+24 \times 5^r \alpha \left(\sum_{j=1}^r E(V_j^4)^{\frac{1}{2}}\right)^2.$$

Note that α does not appear in the second term of the right-hand side of the inequality above because X_j^N are random variables with unit variance and zero covariance. Using the inequality $x^{\frac{1}{4}}y^{\frac{3}{2}} \le x^{\frac{1}{2}}y + y^2$, obtained by separating the cases $x \le y^2$ and $x \ge y^2$, we get:

$$E(U^4)^{\frac{1}{4}}\left(\sum_{j=1}^r E(V_j^4)^{\frac{1}{2}}\right)^{\frac{3}{2}} \le E(U^4)^{\frac{1}{2}}\sum_{j=1}^r E(V_j^4)^{\frac{1}{2}} + \left(\sum_{j=1}^r E(V_j^4)^{\frac{1}{2}}\right)^2.$$

Then

$$E(Q_d(X)^4) \le E(U^4) + 12 \times 5^r(1+2\alpha^{\frac{3}{4}})E(U^4)^{\frac{1}{2}}\sum_{j=1}^r E(V_j^4)^{\frac{1}{2}} \qquad (8)$$

$$+24 \times 5^r(\alpha^{\frac{3}{4}}+\alpha)\left(\sum_{j=1}^r E(V_j^4)^{\frac{1}{2}}\right)^2.$$

To prove the hypercontractivity property (6), we will use an induction on N. When $N = 1$, because f vanishes at the points $(i_1, \ldots, i_d)$ such that $\exists j \neq k$ for which $\lceil i_j/r \rceil = \lceil i_k/r \rceil$, then the only case where the value taken by $Q_d(X)$ is not zero is when $d = 1$, that is, when $Q_d(X)$ has the form $\sum_{j=1}^{r} a_j X_j^1$. In this case, $U = 0$ and $V_j = a_j$. Thus, by (7), we have $E(Q_d(X)^4) \leq 24 \times 5^r \alpha \left(\sum_{j=1}^{r} a_j^2 \right)^2$. It follows that $E(Q_d(X)^4) \leq K E(Q_d(X)^2)^2$. Let us now assume that the result holds for $N - 1$. Then, because U and V_j are functions of $X^1, \ldots, X^{N-1}$, we can apply the recursive hypothesis to $E(U^4)$ and $E(V_j^4)$, and obtain that:

$$E(Q_d(X)^4) \leq K^d \left[E(U^2)^2 + \frac{12 \times 5^r (1 + 2\alpha^{\frac{3}{4}})}{K^{\frac{1}{2}}} E(U^2) \sum_{j=1}^{r} E(V_j^2) \right]$$

$$+ K^d \frac{24 \times 5^r (\alpha + \alpha^{\frac{3}{4}})}{K} \left(\sum_{j=1}^{r} E(V_j^2) \right)^2$$

$$\leq K^d \left[E(U^2)^2 + 2 E(U^2) \sum_{j=1}^{r} E(V_j^2) + \left(\sum_{j=1}^{r} E(V_j^2) \right)^2 \right]$$

$$= K^d \left[E(U^2) + \sum_{j=1}^{r} E(V_j^2) \right]^2.$$

Furthermore, since the X_j^N are centered, unit-variance and independent of U and of V_j, we have

$$E(Q_d(X)^2) = E((U + \sum_{j=1}^{r} X_j^N V_j)^2)$$

$$= E(U^2) + 2 \sum_{j=1}^{r} E(UV_j) E(X_j^N) + \sum_{i,j=1,\ldots,r} E(V_i V_j) E(X_i^N X_j^N)$$

$$= E(U^2) + \sum_{j=1}^{r} E(V_j^2),$$

which completes the proof.

The following two lemmas will be used to prove the convergence in law of the sum of the non-diagonal terms in (2), and to show that the limit does not depend on the common law of $X_{i,j}$.

Lemma 3. *Let $\{X^i\}_{i \geq 1}$, $\{G^i\}_{i \geq 1}$ and $Q^i(X)$ be as in the beginning of Sect. 3. Let us assume that $\beta = \sup_i E(|X_i|^3) < \infty$, $E(Q^i(X)^2) = 1$, and that V is* the symmetric matrix defined as $V(i,j) = E(Q^i(X)Q^j(X))$. Consider $Z_V = (Z_V^1, \ldots, Z_V^m) \sim N_m(0, V)$ (i.e. Z_V is a Gaussian vector with a covariance matrix equal to V).

1. If $\varphi : \mathbb{R}^m \to \mathbb{R}$ is a function of class C^3 such that $\|\varphi'''\|_\infty < \infty$ then

$$|E(\varphi(Q^1(X), \ldots, Q^m(X))) - E(\varphi(Q^1(G), \ldots, Q^m(G)))|$$
$$\leq \|\varphi'''\|_\infty \left(\beta + \sqrt{\frac{8}{\pi}}\right) K^{\frac{3}{4}(d_m-1)} r^3 m^4 \frac{d_m!^3}{d_1!(d_1-1)!} \sqrt{\max_{1 \leq k \leq m} \max_{1 \leq j \leq N_k} \inf_j f_k},$$

where

$$\inf_j f_k = \sum_{i_1, \ldots, i_{d_k-1}=1}^{rN_k} f_k(j, i_1, \ldots, i_{d_k-1})^2.$$

2. If $\varphi : \mathbb{R}^m \to \mathbb{R}$ is a function of class C^3 such that $\|\varphi'''\|_\infty < \infty$ then

$$|E(\varphi(Q^1(X), \ldots, Q^m(X))) - E(\varphi(Z_V))| \leq \|\varphi''\|_\infty \left(\sum_{i=1}^m \Delta_{i,i} + 2 \sum_{1 \leq i < j \leq m} \Delta_{i,j}\right)$$

$$+ \frac{\|\varphi'''\|_\infty m^4 d_m!^3}{d_1!(d_1-1)!} \left(\left(\beta + \sqrt{\frac{8}{\pi}}\right) K^{\frac{3}{4}(d_m-1)} r^3 + \sqrt{\frac{32}{\pi}} \left(\frac{64}{\pi}\right)^{d_m-1}\right)$$

$$\times \sqrt{\max_{1 \leq k \leq m} \max_{1 \leq j \leq N_k} \inf_j f_k}$$

where $\inf_j f_k$ as above and $\Delta_{i,j}$ given by

$$d_j \sum_{s=1}^{d_i-1} (s-1)! \binom{d_i-1}{s-1} \binom{d_j-1}{s-1} \sqrt{(d_i+d_j-2s)!} \left(\|f_i \star_{d_i-s} f_i\|_2 + \right.$$

$$\|f_j \star_{d_j-s} f_j\|_2) + 1_{d_i < d_j} \sqrt{d_j! \binom{d_j}{d_i}} \|f_j \star_{d_j-d_i} f_j\|_2,$$

with

$$f_j \star_r f_j(i_1, \ldots, i_{2d_j-2r}) = \sum_{k_1, \ldots, k_r=1}^{rN_j} f_j(k_1, \ldots, k_r, i_1, \ldots, i_{d_j-r})$$
$$\times f_j(k_1, \ldots, k_r, i_{d_j-r+1}, \ldots, i_{2d_j-2r}).$$

Proof. Set $Q(X) = (Q^1(X), \ldots, Q^m(X))$ and, for any $1 \leq p \leq N + 1$, consider

$$\begin{cases} Z^{(p)} = (G_1, \ldots, G_{(p-1)r}, X_{(p-1)r+1}, \ldots, X_{rN}) \\ U_p^{(i)} = \sum_{\forall k: i_k \notin \{(p-1)r+1, \ldots, pr\}} f_i(i_1, \ldots, i_d) Z_{i_1}^{(p)} \ldots Z_{i_d}^{(p)} \\ V_{p,j}^{(i)} = \sum_{\exists! k: i_k = (p-1)r+j} f_i(i_1, \ldots, i_d) Z_{i_1}^{(p)} \ldots \widetilde{Z_{(p-1)r+j}^{(p)}} \ldots Z_{i_d}^{(p)} \end{cases}$$

The notation $\widetilde{Z_{(p-1)r+j}^{(p)}}$ means that this term is removed from the product. Let us set $U_p = (U_p^{(1)}, \ldots, U_p^{(m)})$ and $V_{p,j} = (V_{p,j}^{(1)}, \ldots, V_{p,j}^{(m)})$. Note that $Q(Z^{(p)})$ can be written as

$$Q(Z^{(p)}) = U_p + \sum_{j=1}^{r} X_j^p V_{p,j}.$$

Similarly, we have: $Q(Z^{(p+1)}) = U_p + \sum_{j=1}^{r} G_j^p V_{p,j}$.

For a vector $Y = (Y_1, \ldots, Y_m)$ in $\mathbb{R}^m$ and a vector $s = (s_1, \ldots, s_m)$ in $\mathbb{N}^m$, we set $Y^s = \prod_{i=1}^{m} Y_i^{s_i}$.

1. Let φ be a function of class C^3. The Taylor formula gives:

$$\left| E\left(\varphi(Q(Z^{(p)}))\right) - E\left(\sum_{|s| \leq 2} \frac{1}{s!} \partial^s \varphi(U_p) \left(\sum_{j=1}^{r} X_j^p V_{p,j}\right)^s\right) \right|$$

$$\leq \|\varphi'''\|_\infty \left| E\left(\sum_{|s|=3} \left(\sum_{j=1}^{r} X_j^p V_{p,j}\right)^s\right) \right|.$$

Note that, for every p, X_j^p is independent from U_p and from $V_{p,i}$. Thus, we have:

$$\left| E\left(\sum_{|s|=3} \left(\sum_{j=1}^{r} X_j^p V_{p,j}\right)^s\right) \right|$$

$$= \left| E\left(\sum_{k,l,q=1}^{m} \sum_{j_1=1}^{r} X_{j_1}^p V_{p,j_1}^{(k)} \sum_{j_2=1}^{r} X_{j_2}^p V_{p,j_2}^{(l)} \sum_{j_3=1}^{r} X_{j_3}^p V_{p,j_3}^{(q)}\right) \right|$$

$$= \left| \sum_{k,l,q=1}^{m} \sum_{j_1=1}^{r} \sum_{j_2=1}^{r} \sum_{j_3=1}^{r} E\left(X_{j_1}^p X_{j_2}^p X_{j_3}^p\right) E\left(V_{p,j_1}^{(k)} V_{p,j_2}^{(l)} V_{p,j_3}^{(q)}\right) \right|.$$

The Hlder inequality ensures that:

$$\left|E\left(X_{j_1}^p X_{j_2}^p X_{j_3}^p\right)\right| \leq E\left(\left|X_{j_1}^p\right|^3\right)^{\frac{1}{3}} E\left(\left|(X_{j_2}^p)\right|^3\right)^{\frac{1}{3}} E\left(\left|(X_{j_3}^p)\right|^3\right)^{\frac{1}{3}} \leq \beta.$$

Using the Hlder inequality, as well as the hypercontractivity property stated in Lemma 2 and the relation $E\left((V_{p,n}^{(k)})^2\right) = d_k! ^2 \inf_{pr+n} f_k$, we obtain

$$\left|E\left(V_{p,j_1}^{(k)} V_{p,j_2}^{(l)} V_{p,j_3}^{(q)}\right)\right| \leq E\left(\left|V_{p,j_1}^{(k)}\right|^4\right)^{\frac{1}{4}} E\left(\left|V_{p,j_2}^{(l)}\right|^4\right)^{\frac{1}{4}} E\left(\left|V_{p,j_3}^{(q)}\right|^4\right)^{\frac{1}{4}}$$

$$\leq K^{\frac{3}{4}(d_m-1)} E\left(\left|V_{p,j_1}^{(k)}\right|^2\right)^{\frac{1}{2}} E\left(\left|V_{p,j_2}^{(l)}\right|^2\right)^{\frac{1}{2}} E\left(\left|V_{p,j_3}^{(q)}\right|^2\right)^{\frac{1}{2}}$$

$$\leq K^{\frac{3}{4}(d_m-1)} \left(d_m!^2 \max_{1\leq j\leq r}\max_{1\leq k\leq m} \inf_{pr+j} f_k\right)^{\frac{3}{2}}.$$

Then,

$$\left|E\left(\varphi(Q(Z^{(p)}))\right) - E\left(\sum_{|s|\leq 2}\frac{1}{s!}\partial^s\varphi(U_p)\left(\sum_{j=1}^r X_j^p V_{p,j}\right)^s\right)\right|$$

$$\leq \|\varphi'''\|_\infty \beta K^{\frac{3}{4}(d_m-1)}(r\,m)^3 \left(d_m!^2 \max_{1\leq j\leq r}\max_{1\leq k\leq m} \inf_{pr+j} f_k\right)^{\frac{3}{2}}.$$

By writing the same formula for $Q(Z^{(p+1)})$ we obtain this time

$$\left|E\left(\varphi(Q(Z^{(p+1)}))\right) - E\left(\sum_{|s|\leq 2}\frac{1}{s!}\partial^s\varphi(U_p)\left(\sum_{j=1}^r G_j^p V_{p,j}\right)^s\right)\right|$$

$$\leq \|\varphi'''\|_\infty \sqrt{\frac{8}{\pi}} K^{\frac{3}{4}(d_m-1)}(r\,m)^3 \left(d_m!^2 \max_{1\leq j\leq r}\max_{1\leq k\leq m} \inf_{pr+j} f_k\right)^{\frac{3}{2}}.$$

In the last inequality, the term $\sqrt{\frac{8}{\pi}}$ comes from the fact that G_j^p are standard Gaussian which implies that $E\left(\left|G_j^p\right|^3\right) = \sqrt{\frac{8}{\pi}}$. Since the vectors X^p and G^p are centered, have the same covariance matrix and are independent from U_p and from V_j^p, then by putting the two inequalities together, we obtain:

$$\left|E\left(\varphi\left(Q(Z^{(p+1)})\right)\right) - E\left(\varphi\left(Q(Z^{(p)})\right)\right)\right|$$

$$\leq \|\varphi'''\|_\infty \left(\beta + \sqrt{\frac{8}{\pi}}\right) K^{\frac{3}{4}(d_m-1)}(r\,m)^3 \left(d_m!^2 \max_{1\leq j\leq r}\max_{1\leq k\leq m} \inf_{pr+j} f_k\right)^{\frac{3}{2}}. \quad (9)$$

Since $\sum_{j=1}^{r \max_i N_i} \inf_j f_k = \frac{E\left((Q^k(X))^2\right)}{d_k!(d_k-1)!}$ then

$$\sum_{p=1}^{\max_i N_i} \max_{1 \leq j \leq r} \max_{1 \leq k \leq m} \inf_{pr+j} f_k \leq \sum_{p=1}^{\max_i N_i} \sum_{j=1}^{r} \sum_{k=1}^{m} \inf_{pr+j} f_k$$

$$\leq \sum_{k=1}^{m} \sum_{j=1}^{r \max_i N_i} \inf_j f_k$$

$$\leq \sum_{k=1}^{m} \frac{E\left((Q^k(X))^2\right)}{d_k!(d_k-1)!} \leq \frac{m}{d_1!(d_1-1)!}.$$

By summing over p in (9), we finally obtain that:

$$|E(\varphi(Q(X))) - E(\varphi(Q(G)))|$$
$$\leq \|\varphi'''\|_\infty \left(\beta + \sqrt{\frac{8}{\pi}}\right) K^{\frac{3}{4}(d_m-1)} r^3 m^4 \frac{d_m!^3}{d_1!(d_1-1)!} \sqrt{\max_{1 \leq k \leq m} \max_{1 \leq j \leq N_k} \inf_j f_k}.$$

2. Let φ be a function of class C^3. We have

$$|E(\varphi(Q(X))) - E(\varphi(Z_V))| \leq |E(\varphi(Q(X))) - E(\varphi(Q(G)))|$$
$$+ |E(\varphi(Q(G))) - E(\varphi(Z_V))|.$$

For the first term we use the point 1 of Lemma 3 to find an upper bound. For the second term we observe that the vector G have independent components, which allows us to use Theorem 7.2 in [8] to get the following inequality:

$$|E(\varphi(Q^1(X), \ldots, Q^m(X))) - E(\varphi(Z_V))| \leq \|\varphi''\|_\infty \left(\sum_{i=1}^{m} \Delta_{i,i} + 2 \sum_{1 \leq i < j \leq m} \Delta_{i,j}\right)$$
$$+ C \|\varphi'''\|_\infty \sqrt{\frac{32}{\pi}} \left[\sum_{j=1}^{m} (\frac{64}{\pi})^{\frac{d_j-1}{3}} d_j!\right]^3 \sqrt{\max_{1 \leq k \leq m} \max_{1 \leq j \leq N_k} \inf_j f_k}.$$

The constant C is such that $\sum_{i=1}^{\max_k N_k} \max_{1 \leq j \leq m} \inf_i f_j \leq C$ and since

$$\sum_{i=1}^{\max N_k} \max_{1\leq j\leq m} \inf_i f_j \leq \sum_{j=1}^{m} \sum_{i=1}^{\max N_k} \inf_i f_j \leq \sum_{j=1}^{m} \frac{E\left((Q^j(X))^2\right)}{d_j!(d_j-1)!} \leq \frac{m}{d_1!(d_1-1)!}$$

then we can choose the constant C equal to $\frac{m}{d_1!(d_1-1)!}$. Thus, we obtain

$$\left|E(\varphi(Q^1(X),\ldots,Q^m(X))) - E(\varphi(Z_V))\right| \leq \|\varphi''\|_\infty \left(\sum_{i=1}^{m} \Delta_{i,i} + 2\sum_{1\leq i<j\leq m} \Delta_{i,j}\right)$$

$$+ \|\varphi'''\|_\infty \sqrt{\frac{32}{\pi}} \left(\frac{64}{\pi}\right)^{d_m-1} m^4 \frac{(d_m!)^3}{d_1!(d_1-1)!} \sqrt{\max_{1\leq k\leq m} \max_{1\leq j\leq N_k} \inf_j f_k}.$$

Lemma 4. *Let the notations used in Lemma 3 prevail. Consider the class $\mathcal{H}$ of indicator functions on measurable convex sets in $\mathbb{R}^m$. Let us define*

$$B_1 = \left(\sum_{i=1}^{m} \Delta_{i,i} + 2 \sum_{1\leq i<j\leq m} \Delta_{i,j}\right)$$

$$B_2 = \frac{m^4 d_m!^3}{d_1!(d_1-1)!} \left(\left(\beta + \sqrt{\frac{8}{\pi}}\right) K^{\frac{3}{4}(d_m-1)} r^3 + \sqrt{\frac{32}{\pi}} \left(\frac{64}{\pi}\right)^{d_m-1}\right)$$

$$\times \sqrt{\max_{1\leq k\leq m} \max_{1\leq j\leq N_k} \inf_j f_k}$$

1. *Let us assume that the covariance matrix V is the m-dimensional identity matrix. Then*

$$\sup_{h\in\mathcal{H}(\mathbb{R}^m)} \left|E\left[h(Q^1(X),\ldots,Q^m(X))\right] - E\left[h(Z_V)\right]\right|$$

$$\leq \left(\frac{8}{3^{\frac{6}{7}}} + \frac{4}{3^{\frac{13}{7}}}\right) (5B_1 + 5B_2)^{\frac{1}{7}} m^{\frac{3}{7}}.$$

2. *Let us assume that the covariance matrix V is invertible and let $\Lambda = \mathrm{diag}(\lambda_1,\ldots,\lambda_k)$ be the diagonal matrix of the eigenvalues of V. Let B be an orthogonal matrix (i.e. $B^T B = I_m$ and $BB^T = I_m$) such that $V = B\Lambda B^T$, and let $b = \max_{i,j}(\Lambda^{-\frac{1}{2}} B^T)$. Then*

$$\sup_{h\in\mathcal{H}(\mathbb{R}^m)} \left|E\left[h(Q^1(X),\ldots,Q^m(X))\right] - E\left[h(Z_V)\right]\right|$$

$$\leq \left(\frac{8}{3^{\frac{6}{7}}} + \frac{4}{3^{\frac{13}{7}}}\right) (5b^2 B_1 + 5b^3 B_2)^{\frac{1}{7}} m^{\frac{3}{7}}.$$

Proof. 1. Let us assume that the covariance matrix V is the m-dimensional identity matrix. Denote by Φ the standard normal distribution in $\mathbb{R}^m$, and by ϕ the corresponding density function. Consider $h \in \mathcal{H}(\mathbb{R}^m)$ and define the following function: $h_t(x) = \int_{\mathbb{R}^m} h(\sqrt{t}y + \sqrt{1-t}x)\Phi(dy)$, $0 < t < 1$. The key result is Lemma 2.11 in [4] which states that, for every probability measure Q on $\mathbb{R}^m$, every random variables $W \sim Q$ and $Z \sim \Phi$, and any $0 < t < 1$, we have

$$\sup_{h \in \mathcal{H}(\mathbb{R}^m)} |E[h(W)] - E[h(Z_V)]|$$
$$\leq \frac{4}{3}\left[\sup_{h \in \mathcal{H}(\mathbb{R}^m)} |E[h_t(W)] - E[h_t(Z_V)]| + 2\sqrt{m}\sqrt{t}\right]. \quad (10)$$

Let us define $u(x,t,z) = (2\pi t)^{-\frac{m}{2}} \exp\left(-\sum_{i=1}^{m} \frac{(z_i - \sqrt{1-t}x_i)^2}{2t}\right)$. Using the change of variable $z = \sqrt{t}y + \sqrt{1-t}x$ in $h_t(x)$ leads to

$$h_t(x) = \int_{\mathbb{R}^m} h(z)u(x,t,z)dz.$$

By the dominated convergence theorem, we may differentiate under the integral sign and obtain

$$\frac{\partial^2 h_t}{\partial x_i^2}(x) = -\frac{1-t}{t}\int_{\mathbb{R}^m} h(z)u(x,t,z)dz + \frac{1-t}{t^2}\int_{\mathbb{R}^m} h(z)(z_i - \sqrt{1-t}x_i)^2 u(x,t,z)dz.$$

Since $\|h\|_\infty \leq 1$ then we have

$$\left|\frac{\partial^2 h_t}{\partial x_i^2}(x)\right| \leq \frac{1-t}{t} + \frac{1-t}{t^2}\int_{\mathbb{R}^m} (z_i - \sqrt{1-t}x_i)^2 u(x,t,z)dz.$$

If $(Y_1, \ldots, Y_m)$ is a Gaussian vector with covariance matrix tI_m then $\int_{\mathbb{R}^m}(z_i - \sqrt{1-t}x_i)u(x,t,z)dz = E(Y_i^2) = t$. Therefore, we have

$$\left|\frac{\partial^2 h_t}{\partial x_i^2}(x)\right| \leq 2\frac{1-t}{t}.$$

Furthermore, for $i \neq j$ we have

$$\frac{\partial^2 h_t}{\partial x_i \partial x_j}(x) = \frac{1-t}{t^2}\int_{\mathbb{R}^m} h(z)(z_i - \sqrt{1-t}x_i)(z_j - \sqrt{1-t}x_j)u(x,t,z)dz,$$

So that $\left|\frac{\partial^2 h_t}{\partial x_i \partial x_j}(x)\right| \leq \frac{1-t}{t^2} E(|Y_i|) E(|Y_i|) = \frac{2(1-t)}{\pi t}$. We conclude that $\|h''\|_\infty \leq \frac{2}{t} \leq \frac{5}{t^3}$. Similarly, for i, j, k in $[1, m]$ it holds that:

$$\left|\frac{\partial^3 h_t}{\partial x_i \partial x_j \partial x_k}(x)\right| \leq \frac{(1-t)^{\frac{3}{2}}}{t^3} \max\left(3E(|Y_i|)t + E(|Y_i|^3); E(|Y_j|)t + E(|Y_i|^2)E(|Y_j|); \right.$$

$$\left. ; E(|Y_i|)E(|Y_j|)E(|Y_k|) \right).$$

Therefore $\|h'''\|_\infty \leq \frac{5}{t^3}$. Combining the latter inequality with the result (10) and point 2 of Lemma 3, we obtain

$$\sup_{h \in \mathcal{H}(\mathbb{R}^m)} |E[h(Q(X))] - E[h(Z_V)]|$$

$$\leq \frac{4}{3}\left[\sup_{h \in \mathcal{H}(\mathbb{R}^m)} |E[h_t(Q(X))] - E[h_t(Z_V)]| + 2\sqrt{m}\sqrt{t} \right]$$

$$\leq \frac{8}{3}\sqrt{m}\sqrt{t} + \frac{4}{3}(5B_1 + 5B_2)t^{-3}.$$

The function in the right-hand side of the inequality reaches its minimum at $t = \left(\frac{15(B_1 + B_2)}{\sqrt{m}}\right)^{\frac{2}{7}}$, hence

$$\sup_{h \in \mathcal{H}(\mathbb{R}^m)} |E[h(Q(X))] - E[h(Z_V)]| \leq \left(\frac{8}{3^{\frac{6}{7}}} + \frac{4}{3^{\frac{13}{7}}}\right)(5B_1 + 5B_2)^{\frac{1}{7}} m^{\frac{3}{7}}.$$

2. Set $Q(X) = (Q^1(X), \ldots, Q^m(X))$. For any $h \in \mathcal{H}(\mathbb{R}^m)$, we have

$$E(h(Q(X))) - E(h(Z_v)) = E(h(B\Lambda^{\frac{1}{2}} \Lambda^{-\frac{1}{2}} B^T Q(X))) - E(h(B\Lambda^{\frac{1}{2}} \Lambda^{-\frac{1}{2}} B^T Z_v)).$$

Define $g(x) = h(B\Lambda^{\frac{1}{2}} x)$, $x \in \mathbb{R}^m$. Since $g \in \mathcal{H}(\mathbb{R}^m)$ then, using inequality (10), we get

$$\sup_{h \in \mathcal{H}(\mathbb{R}^m)} |E[h(Q(X))] - E[h(Z_V)]|$$

$$\leq \sup_{g \in \mathcal{H}(\mathbb{R}^m)} \left| E\left[g(\Lambda^{-\frac{1}{2}} B^T Q(X))\right] - E\left[g(\Lambda^{-\frac{1}{2}} B^T Z_V)\right] \right|$$

$$\leq \frac{4}{3}\left[\sup_{g \in \mathcal{H}(\mathbb{R}^m)} \left| E\left[g_t(\Lambda^{-\frac{1}{2}} B^T Q(X))\right] - E\left[g_t(\Lambda^{-\frac{1}{2}} B^T Z_V)\right] \right| + 2\sqrt{m}\sqrt{t} \right].$$

We can find an upper bound for the second and third derivatives of $f_t(x) = g_t(\Lambda^{-\frac{1}{2}} B^T x)$. Indeed, $\|f_t''\|_\infty \leq 5b^2 t^{-3}$ and $\|f_t'''\|_\infty \leq 5b^3 t^{-3}$. By using the same reasoning as in point 1 and replacing B_1 by $b^2 B_1$ and B_2 by $b^3 B_2$ in (10), we obtain the result.

The main difference between the proof of Nourdin Peccati and our proof is that we consider a family of variables X_i not necessary independent. We need this relaxation of the hypothesis because the real part and the imaginary part of random variables $X_{i,j}$ are not independent, so we modified the Lemma 3 by including a dependance in certain block of random variables. In the sketch proof of this lemma we use the hypercontractity property for homogeneous sum of a family of variables not independent, which lead us to relax the independent hypothesis in the Lemma 2.

4 Proof of Theorem 1

We use hereafter the notation adopted in the beginning of Sect. 2. If we separate the diagonal terms from the non-diagonal terms in (2), we obtain

$$\text{trace}(X_n^d) - E(\text{trace}(X_n^d))$$
$$= \frac{1}{n^{\frac{d}{2}}} \sum_{((i_1,i_2),\ldots,(i_d,i_1)) \in D_n} (X_{i_1,i_2} \ldots X_{i_d,i_1} - E(X_{i_1,i_2} \ldots X_{i_d,i_1}))$$
$$+ \frac{1}{n^{\frac{d}{2}}} \sum_{((i_1,i_2),\ldots,(i_d,i_1)) \in ND_n} X_{i_1,i_2} \ldots X_{i_d,i_1}.$$

The expectation in the second sum is equal to zero because the $X_{i,j}$ are independent and centered. The variance of the term containing the diagonal terms is upper bounded by $O\left(\frac{1}{\sqrt{n}}\right)$ and, therefore, goes to 0 as n goes to infinity. Indeed, if we set $M = \sup_{i,j} E(|X_{i,j}|^{2d})$, then

$$\text{Var}\left(\frac{1}{n^{\frac{d}{2}}} \sum_{((i_1,i_2),\ldots,(i_d,i_1)) \in D_n} X_{i_1,i_2} \ldots X_{i_d,i_1}\right)$$
$$= \frac{1}{n^d}\left[E\left(\left(\sum_{((i_1,i_2),\ldots,(i_d,i_1)) \in D_n} X_{i_1,i_2} \ldots X_{i_d,i_1}\right)^2\right)\right.$$
$$\left. - \left(E\left(\sum_{(i_1,i_2),\ldots,(i_d,i_1)) \in D_n} X_{i_1,i_2} \ldots X_{i_d,i_1}\right)\right)^2\right].$$

Keeping the notation introduced in Lemma 1, we have:

$$E\left(\left(\sum_{((i_1,i_2),\ldots,(i_d,i_1))\in D_n} X_{i_1,i_2}\ldots X_{i_d,i_1}\right)^2\right)$$

$$= \sum_{((i_1,i_2),\ldots,(i_d,i_1),(i_{d+1},i_{d+2}),\ldots,(i_{2d},i_{d+1}))\in A_n} E\left(X_{i_1,i_2}\ldots X_{i_d,i_1} X_{i_{d+1},i_{d+2}}\ldots X_{i_{2d},i_{d+1}}\right).$$

Since $E\left(X_{i_1,i_2}\ldots X_{i_d,i_1} X_{i_{d+1},i_{d+2}}\ldots X_{i_{2d},i_{d+1}}\right)$ is equal to 1 over the subset B_n of A_n, and is upper bounded by M over the subset $A_n \setminus B_n$, then we can state that:

$$\left| E\left(\left(\sum_{((i_1,i_2),\ldots,(i_d,i_1))\in D_n} X_{i_1,i_2}\ldots X_{i_d,i_1}\right)^2\right) - \#B_n \right| \leq M\#(A_n \setminus B_n). \quad (11)$$

Furthermore, since the $X_{i,j}$ are centered and independent, then $E(X_{i_1,i_2}\ldots X_{i_d,i_1}) = 0$ if $((i_1,i_2),\ldots,(i_d,i_1)) \in D_n \setminus E_n$. Thus,

$$E\left(\sum_{((i_1,i_2),\ldots,(i_d,i_1))\in D_n} X_{i_1,i_2}\ldots X_{i_d,i_1}\right) = \sum_{((i_1,i_2),\ldots,(i_d,i_1))\in E_n} E(X_{i_1,i_2}\ldots X_{i_d,i_1}).$$

On the other hand, $E(X_{i_1,i_2}\ldots X_{i_d,i_1})$ is equal to 1 over the subset F_n of E_n, and bounded by $\sqrt{M}$ over $E_n \setminus F_n$. Then,

$$\left| E\left(\sum_{((i_1,i_2),\ldots,(i_d,i_1))\in D_n} X_{i_1,i_2}\ldots X_{i_d,i_1}\right) - \#F_n \right| \leq \sqrt{M}\#(E_n \setminus F_n). \quad (12)$$

Finally, by combining the estimations (11) and (12), and using points (i) to (iv) of Lemma 1 and the fact that $\#D_n = O(n^{d-1})$, we get the following result, with Z_n defined by $Z_n = \sum_{((i_1,i_2),\ldots,(i_d,i_1))\in D_n} X_{i_1,i_2}\ldots X_{i_d,i_1}$:

$$\text{Var}(Z_n) = E(Z_n^2) - E(Z_n)^2$$
$$= \#B_n + \left(E(Z_n^2) - \#B_n\right) - (\#F_n)^2 + \left((\#F_n)^2 - E(Z_n)^2\right)$$
$$= \#B_n - (\#F_n)^2 + \left(E(Z_n^2) - \#B_n\right)$$
$$+ \left(\#F_n - E(Z_n)\right)\left(\#F_n + E(Z_n)\right).$$

From points (*ii*) and (*iv*) of Lemma 1, it follows that $\#B_n - (\#F_n)^2 = O(n^{d-1})$. Using point (*i*) and the relation (11), we obtain the estimation $E(Z_n^2) - \#B_n = O(n^{d-1})$. Finally, using points (*iii*)–(*iv*) and the relation (12), we get the following estimations: $E(Z_n) - \#F_n = O(n^{\frac{d-1}{2}})$ and $E(Z_n) + \#F_n = O(n^{\frac{d}{2}})$. From these estimations, we conclude that:

$$\mathrm{Var}\left(\frac{1}{n^{\frac{d}{2}}} \sum_{((i_1,i_2),\ldots,(i_d,i_1))\in D_n} X_{i_1,i_2} \ldots X_{i_d,i_1}\right) = O\left(\frac{1}{\sqrt{n}}\right).$$

Consider now a bijection $\sigma : [1, n^2] \to [1, n] \times [1, n]$. Let us define $X_i = X_{\sigma(i)}$ and $R = E(Re(X_i)^2)$. When $R = 1$, the X_i are real-valued, which corresponds exactly to the result of Nourdin et Peccati [7] (there is then nothing more to prove). By contrast, when $R = 0$, the X_i are purely imaginary-valued; factoring out by i^d in the trace formula shows that the result in this case can be derived from the case $R = 1$. In what follows, we can then freely assume that $R \in (0, 1)$. Set $\rho = \frac{E(Re(X_i)Im(X_i))}{R\sqrt{1-R}}$, and define:

$$\begin{cases} X_i^0 = Re(X_i) - \rho\sqrt{\frac{R}{1-R}}\, Im(X_i) \\ X_i^1 = Im(X_i) \\ X_{i,j}^0 = Re(X_{i,j}) - \rho\sqrt{\frac{R}{1-R}}\, Im(X_{i,j}) \\ X_{i,j}^1 = Im(X_{i,j}) \end{cases},$$

$$f_n(i_1, \ldots, i_d) = \frac{1}{n^{\frac{d}{2}}} \mathbf{1}_{\{(\sigma(i_1),\ldots,\sigma(i_d))\in ND_n\}},$$

and

$$Q_d(f_n, X) = \sum_{i_1,\ldots,i_d=1}^{n^2} f_n(i_1, \ldots, i_d)\, X_{i_1} \ldots X_{i_d}.$$

We have:

$$X_{i_1} \ldots X_{i_d} = \prod_{k=1}^{d}\left(X_{i_k}^0 + \rho\sqrt{\frac{R}{1-R}} X_{i_k}^1 + i X_{i_k}^1\right) = \prod_{k=1}^{d}\left(X_{i_k}^0 + \left(\rho\sqrt{\frac{R}{1-R}} + i\right) X_{i_k}^1\right).$$

Hence

$$X_{i_1} \ldots X_{i_d} = \sum_{j_1,\ldots,j_d \in \{0,1\}^d} \left(i + \rho\sqrt{\frac{R}{1-R}}\right)^{\sum j_k} X_{i_1}^{j_1} \ldots X_{i_d}^{j_d},$$

which yields

$$Q_d(f_n, X)$$
$$= \sum_{k=0}^{d} \left(i + \rho\sqrt{\frac{R}{1-R}}\right)^k \sum_{\substack{(j_1,\ldots,j_d)\in\{0,1\}^d \\ j_1+\ldots+j_d=k}} \sum_{(i_1,\ldots,i_d)\in[1,n^2]^d} f_n(i_1,\ldots,i_d) X_{i_1}^{j_1} \ldots X_{i_d}^{j_d}$$

$$= \sum_{k=0}^{d} \left(i + \rho\sqrt{\frac{R}{1-R}}\right)^k \sum_{\substack{(j_1,\ldots,j_d)\in\{0,1\}^d \\ j_1+\ldots+j_d=k}} \sum_{((i_1,i_2),\ldots,(i_d,i_1))\in ND_n} \frac{1}{n^{\frac{d}{2}}} X_{i_1,i_2}^{j_1} \ldots X_{i_d,i_1}^{j_d}.$$

We define for any two elements $(i_1,\ldots,i_d)$, $(j_1,\ldots,j_d)$ of $[1,n]^d$, and $(p_1,\ldots,p_d) \in \{0,1\}^d$ the quantity $g_n^k\left[((i_1,j_1),p_1),\ldots((i_d,j_d),p_d)\right]$ as follows: $g_n^k\left[((i_1,j_1),p_1),\ldots((i_d,j_d),p_d)\right] = \frac{1}{n^{\frac{d}{2}}}$ if $((i_1,j_1),\ldots,(i_d,j_d)) \in ND_n$ and $\sum_{l=1}^{d} p_l = k$, and $g_n^k\left[((i_1,j_1),p_1),\ldots((i_d,j_d),p_d)\right] = 0$ otherwise. Set $R_0 = \sqrt{\mathrm{Var}(X_i^0)}$, $R_1 = \sqrt{\mathrm{Var}(X_i^1)}$, and $Y = (Y_{i,j}^k)_{\substack{(i,j)\in[1,n]^2 \\ k\in\{0,1\}}}$ a family of random variables defined by $Y_{i,j}^k = \frac{X_{i,j}^k}{R_k}$. Then

$$Q_d(g_n^k, Y)$$
$$= \sum_{\substack{(x_1,\ldots,x_d)\in[1,n]^d \\ (y_1,\ldots,y_d)\in[1,n]^d \\ (p_1,\ldots,p_d)\in\{0,1\}^d}} g_n^k\left[((x_1,y_1),p_1),\ldots,((x_d,y_d),p_d)\right] Y_{x_1,y_1}^{p_1} \ldots Y_{x_d,y_d}^{p_d}$$

$$= \frac{1}{(R_0)^{d-k}(R_1)^k} \sum_{\substack{(j_1,\ldots,j_d)\in\{0,1\}^d \\ \sum j_p = k}} \sum_{((i_1,i_2),\ldots,(i_d,i_1))\in ND_n} \frac{1}{n^{\frac{d}{2}}} X_{i_1,i_2}^{j_1} \ldots X_{i_d,i_1}^{j_d}.$$

We can then conclude that

$$Q_d(f_n, X) = \sum_{k=0}^{d} \left(i + \rho\sqrt{\frac{R}{1-R}}\right)^k (R_0)^{d-k}(R_1)^k Q_d(g_n^k, Y).$$

If $\widetilde{g_n^k}$ stands for the symmetrization of g_n^k then $Q_d(\widetilde{g_n^k}, Y) = Q_d(g_n^k, Y)$, where $\widetilde{g_n^k} = \sum_{\sigma\in S_d} g_n^{k,\sigma}$ and

$$g_n^{k,\sigma}\left[((x_1,y_1),p_1),\ldots,((x_d,y_d),p_d)\right]$$
$$= g_n^k\left[((x_{\sigma(1)},y_{\sigma(1)}),p_{\sigma(1)}),\ldots,((x_{\sigma(d)},y_{\sigma(d)}),p_{\sigma(d)})\right].$$

To establish that $Q_d(f_n, X)$ converges in law to the variable $Z_d^1 + i Z_d^2$ where $Z_d = (Z_d^1, Z_d^2)$ is a Gaussian vector, it is sufficient to show that the $Q_d(g_n^k, Y), k = 0, \ldots, d$, converge in law to a Gaussian vector having independent components. Using part 2 of Lemma 4 (in the particular case $r = 2$), we show that $Q_d(g_n^k, Y)$ converges in law to a Gaussian vector whose covariance matrix V is given by $V(k, k') = \lim_{\infty} E(Q_d(g_n^k, Y) Q_d(g_n^{k'}, Y))$. To do so, it is sufficient to check the assumptions of Lemma 4, that is:

(i) $\max_{i=1,\ldots,2N} \inf_{(a,b),p} \widetilde{g_N^k} \to 0$.

(ii) for every $1 \leq s \leq d-1$, $\left\| \widetilde{g_N^k} \star_s \widetilde{g_N^k} \right\|_2 \to 0$.

(iii) $E(Q_d(g_n^k, Y) Q_d(g_n^{k'}, Y)) \to \delta_{i,j}$ (with $\delta_{i,j}$ the Kronecker symbol).

(iv) $E(Q_d(g_n^k, Y)^2) \to \sigma^2$.

We can rewrite $Q_d(g_n^k, Y)$ as

$$Q_d(g_n^k, Y) = \frac{1}{n^{\frac{d}{2}}} \sum_{\substack{(j_1,\ldots,j_d) \in \{0,1\}^n \\ j_1+\ldots+j_d=k}} \sum_{(i_1,\ldots,i_d) \in ND_n} Y_{i_1,i_2}^{j_1} \ldots Y_{i_d,i_1}^{j_d}.$$

The second-order moment of $Q_d(g_n^k, Y)$ is equal to

$$\frac{1}{n^d} \sum_{\substack{(j_1,\ldots,j_{2d}) \in \{0,1\}^n \\ j_1+\ldots+j_d=j_{d+1}+\ldots+j_{2d}=k}} \sum_{\substack{((i_1,i_2),\ldots,(i_d,i_1)) \in ND_n \\ ((i_{d+1},i_{d+2}),\ldots,(i_{2d},i_{d+1})) \in ND_n}} E(Y_{i_1,i_2}^{j_1} \ldots Y_{i_d,i_1}^{j_d} Y_{i_{d+1},i_{d+2}}^{j_{d+1}} \ldots Y_{i_{2d},i_{d+1}}^{j_{2d}}). \quad (13)$$

For the expectation corresponding to the indices $i_1, \ldots, i_{2d}, j_1, \ldots, j_{2d}$ in (13) to be different from zero, it must hold that $(i_1, \ldots, i_{2d})$ belongs to G_n, where G_n has been defined in Lemma 1. Furthermore, since the subset $G_n \setminus H_n$ is of cardinality $O(n^{d-1})$, its contribution to the moment of order 2 of $Q_d(g_n^k, Y)$ is $O(\frac{1}{n})$. It remains then to see what happens when $(i_1, \ldots, i_{2d})$ belongs to H_n. In this case, let us recall from the proof of point (vi) of Lemma 1 that the elements of the set H_n are completely characterized by d given pairwise distinct integers $i_1, \ldots, i_d \in [1, n]$ and a given integer $k \in [1, d]$ such that $(i_{d+1}, \ldots, i_{2d}) = (i_k, \ldots, i_d, i_1, \ldots, i_{k-1})$. Moreover, if the expectation corresponding to the indices $i_1, \ldots, i_{2d}, j_1, \ldots, j_{2d}$ in (13) is different from zero, then it must hold that $(j_{d+1}, \ldots, j_{2d}) = (j_k, \ldots, j_d, j_1, \ldots, j_{k-1})$ and this expectation is equal to 1. Thus,

$$E\left(Q_d(g_n^k, Y)^2\right) = \frac{1}{n^d} \sum_{\substack{(j_1,\ldots,j_d) \in \{0,1\}^n \\ j_1+\ldots+j_d=k}} d \times n \ldots \times (n-d+1) + O\left(\frac{1}{n}\right)$$

$$= \frac{d C_d^k \times n \ldots \times (n-d+1)}{n^d} + O(\frac{1}{n}),$$

which yields $E\left(Q_d(g_n^k, Y)^2\right) \xrightarrow[n\to\infty]{} dC_d^k$. Moreover, $E(Q_d(g_N^k, Y)Q_d(g_N^j, Y))$ is equal to

$$\frac{1}{n^d} \sum_{\substack{(j_1,\ldots,j_{2d})\in\{0,1\}^n \\ j_1+\cdots+j_d=k, j_{d+1}+\cdots+j_{2d}=j}} \sum_{\substack{((i_1,i_2),\ldots,(i_d,i_1))\in ND_n \\ ((i_{d+1},i_{d+2}),\ldots,(i_{2d},i_{d+1}))\in ND_n}} E(Y_{i_1,i_2}^{j_1}\cdots Y_{i_d,i_1}^{j_d} Y_{i_{d+1},i_{d+2}}^{j_{d+1}} \cdots Y_{i_{2d},i_{d+1}}^{j_{2d}}).$$
(14)

Similarly to the computation of the second-order moment of $Q_d(g_n^k, Y)$, the set of elements for which the expectation in (14) is different from zero is the set G_n of Lemma 1. The subset $G_n \setminus H_n$ is of cardinality $O(n^{d-1})$, which implies that its contribution to $E(Q_d(g_N^k, Y)Q_d(g_N^j, Y))$ is $O(\frac{1}{n})$. Furthermore, the elements of the set H_n are characterized by d given pairwise distinct integers $i_1,\ldots,i_d \in [1,n]$ and a given integer $k \in [1,d]$ such that $(i_{d+1},\ldots,i_{2d}) = (i_k,\ldots,i_d,i_1,\ldots,i_{k-1})$. Moreover, for $E(Y_{i_1,i_2}^{j_1}\cdots Y_{i_d,i_1}^{j_d} Y_{i_{d+1},i_{d+2}}^{j_{d+1}} \cdots Y_{i_{2d},i_{d+1}}^{j_{2d}})$ to be different from zero, it must hold that $(j_{d+1},\ldots,j_{2d}) = (j_k,\ldots,j_d,j_1,\ldots,j_{k-1})$, which is impossible in the case $j \neq k$. We conclude that $E(Q_d(g_N^k, Y)Q_d(g_N^j, Y)) \to 0$ for every $j \neq k$.

From the definition of $\widetilde{g_n^k}$, it is clear that $\left\|\widetilde{g_n^k}\right\|_\infty \leq \left\|g_n^k\right\|_\infty \leq \frac{1}{n^{\frac{d}{2}}}$. Then,

$$\inf_{(a,b),p} \widetilde{g_N^k}$$

$$= \sum_{\substack{(x_1,\ldots,x_{d-1})\in[1,n]^{d-1} \\ (y_1,\ldots,y_{d-1})\in[1,n]^{d-1} \\ (p_1,\ldots,p_{d-1})\in\{0,1\}^{d-1}}} \widetilde{g_n^k}\left(((a,b),p),((x_1,y_1),p_1),\ldots,((x_{d-1},y_{d-1}),p_{d-1})\right)^2$$

$$\leq \sum_{\substack{(i_1,\ldots,i_d)\in[1,n]^d \\ (p_1,\ldots,p_d)\in\{0,1\}^d \\ \sigma\in\mathfrak{S}_d}} g_n^k\left(((i_{\sigma(1)},i_{\sigma(1)+1}),p_{\sigma(1)}),\ldots,((i_{\sigma(d)},i_{\sigma(d)+1}),p_{\sigma(d)})\right)^2$$

$$\times 1_{\{a=i_{\sigma(1)}\}} \times 1_{\{b=i_{\sigma(1)+1}\}} \times 1_{\{p=p_{\sigma(1)}\}}$$

$$\leq 2^{d-1}(d)! n^{d-2} \left\|\widetilde{g_n^k}\right\|_\infty^2 \leq \frac{2^{d-1}(d)!}{n^2}.$$

Therefore $\max_{i=1,\ldots,2N} \inf_{(a,b),p} \widetilde{g_N^k} \leq \frac{2^{d-1}(d)!}{n^2} \to 0.$

Now, let $1 \leq s \leq d-1$ and $\sigma_1, \sigma_2 \in \mathfrak{S}_d$. Then

$$g_n^{k,\sigma_1} \star_s g_n^{k,\sigma_2} \left[((x_1, y_1), p_1), \ldots, ((x_{d-s}, y_{d-s}), p_{d-s}), ((x'_1, y'_1), p'_1), \ldots, \right.$$
$$\left. ((x'_{d-s}, y'_{d-s}), p'_{d-s}) \right]$$

$$= \sum_{\substack{(x_{d-s+1},\ldots,x_d)\in[1,n]^s \\ (y_{d-s+1},\ldots,y_d)\in[1,n]^s \\ (p_{d-s+1},\ldots,p_d)\in\{0,1\}^s}} g_n^{k,\sigma_1} \left[((x_1, y_1), p_1), \ldots, ((x_{d-s}, y_{d-s}), p_{d-s}), \right.$$
$$\left. ((x_{d-s+1}, y_{d-s+1}), p_{d-s+1}), \ldots, ((x_d, y_d), p_d) \right]$$

$$\times g_n^{k,\sigma_2} \left[((x'_1, y'_1), p'_1), \ldots, ((x'_{d-s}, y'_{d-s}), p'_{d-s}), \right.$$
$$\left. ((x_{d-s+1}, y_{d-s+1}), p_{d-s+1}), \ldots, ((x_d, y_d), p_d) \right]$$

so that

$$\left\| \widetilde{g_n^k} \star_s \widetilde{g_n^k} \right\|_2^2$$

$$= \frac{1}{(d!)^4} \sum_{\sigma_1,\sigma_2,\sigma_3,\sigma_4 \in \mathfrak{S}_d} \sum_{\substack{(x_1,\ldots,x_{d-s})\in[1,n]^{d-s} \\ (y_1,\ldots,y_{d-s})\in[1,n]^{d-s} \\ (p_1,\ldots,p_{d-s})\in\{0,1\}^{d-s}}} \sum_{\substack{(x'_1,\ldots,x'_{d-s})\in[1,n]^{d-s} \\ (y'_1,\ldots,y'_{d-s})\in[1,n]^{d-s} \\ (p'_1,\ldots,p'_{d-s})\in\{0,1\}^{d-s}}}$$

$$\sum_{\substack{(x_{d-s+1},\ldots,x_d)\in[1,n]^s \\ (y_{d-s+1},\ldots,y_d)\in[1,n]^s \\ (p_{d-s+1},\ldots,p_d)\in\{0,1\}^s}} \sum_{\substack{(x'_{d-s+1},\ldots,x'_d)\in[1,n]^s \\ (y'_{d-s+1},\ldots,y'_d)\in[1,n]^s \\ (p'_{d-s+1},\ldots,p'_d)\in\{0,1\}^s}}$$

$$\times g_n^{k,\sigma_1} \left[((x_1, y_1), p_1), \ldots, ((x_{d-s}, y_{d-s}), p_{d-s}), ((x_{d-s+1}, y_{d-s+1}), p_{d-s+1}), \ldots, \right.$$
$$\left. ((x_d, y_d), p_d) \right]$$
$$\times g_n^{k,\sigma_2} \left[((x'_1, y'_1), p'_1), \ldots, ((x'_{d-s}, y'_{d-s}), p'_{d-s}), ((x_{d-s+1}, y_{d-s+1}), p_{d-s+1}), \ldots, \right.$$
$$\left. ((x_d, y_d), p_d) \right]$$
$$\times g_n^{k,\sigma_3} \left[((x_1, y_1), p_1), \ldots, ((x_{d-s}, y_{d-s}), p_{d-s}), ((x'_{d-s+1}, y'_{d-s+1}), p'_{d-s+1}), \ldots, \right.$$
$$\left. ((x'_d, y'_d), p'_d) \right]$$
$$\times g_n^{k,\sigma_4} \left[((x'_1, y'_1), p'_1), \ldots, ((x'_{d-s}, y'_{d-s}), p'_{d-s}), ((x'_{d-s+1}, y'_{d-s+1}), p'_{d-s+1}), \ldots, \right.$$
$$\left. ((x'_d, y'_d), p'_d) \right].$$

For the sake of notational simplicity and because this case is representative of the difficulty, in the rest of the proof we assume that $\sigma_1 = \sigma_2 = \sigma_3 = \sigma_4 = I_d$, where

I_d stands for the identity permutation over $[1, d]$. Since g_n^k is equal to zero at point $[((x_1, y_1), p_1), \ldots, ((x_d, y_d), p_d)]$ if $y_i \neq x_{i+1}$ or $y_d \neq x_1$, then

$$\sum_{\substack{(x_1,\ldots,x_{d-s})\in[1,n]^{d-s} \\ (y_1,\ldots,y_{d-s})\in[1,n]^{d-s} \\ (p_1,\ldots,p_{d-s})\in\{0,1\}^{d-s}}} \sum_{\substack{(x'_1,\ldots,x'_{d-s})\in[1,n]^{d-s} \\ (y'_1,\ldots,y'_{d-s})\in[1,n]^{d-s} \\ (p'_1,\ldots,p'_{d-s})\in\{0,1\}^{d-s}}} \sum_{\substack{(x_{d-s+1},\ldots,x_d)\in[1,n]^s \\ (y_{d-s+1},\ldots,y_d)\in[1,n]^s \\ (p_{d-s+1},\ldots,p_d)\in\{0,1\}^s}} \sum_{\substack{(x'_{d-s+1},\ldots,x'_d)\in[1,n]^s \\ (y'_{d-s+1},\ldots,y'_d)\in[1,n]^s \\ (p'_{d-s+1},\ldots,p'_d)\in\{0,1\}^s}}$$

$\times g_n^{k,I_d}\,[\,((x_1,y_1),p_1),\ldots,((x_{d-s},y_{d-s}),p_{d-s}),((x_{d-s+1},y_{d-s+1}),p_{d-s+1}),\ldots,$
$\quad ((x_d,y_d),p_d)\,]$

$\times g_n^{k,I_d}\,[\,((x'_1,y'_1),p'_1),\ldots,((x'_{d-s},y'_{d-s}),p'_{d-s}),((x_{d-s+1},y_{d-s+1}),p_{d-s+1}),\ldots,$
$\quad ((x_d,y_d),p_d)\,]$

$\times g_n^{k,I_d}\,[\,((x_1,y_1),p_1),\ldots,((x_{d-s},y_{d-s}),p_{d-s}),((x'_{d-s+1},y'_{d-s+1}),p'_{d-s+1}),\ldots,$
$\quad ((x'_d,y'_d),p'_d)\,]$

$\times g_n^{k,I_d}\,[\,((x'_1,y'_1),p'_1),\ldots,((x'_{d-s},y'_{d-s}),p'_{d-s}),((x'_{d-s+1},y'_{d-s+1}),p'_{d-s+1}),\ldots,$
$\quad ((x'_d,y'_d),p'_d)\,]$

$=$

$$\sum_{\substack{(\alpha_1,\ldots,\alpha_{d-s+1})\in[1,n]^{d-s+1} \\ (p_1,\ldots,p_{d-s})\in\{0,1\}^{d-s}}} \sum_{\substack{(\alpha'_1,\ldots,\alpha'_{d-s+1})\in[1,n]^{d-s+1} \\ (p'_1,\ldots,p'_{d-s})\in\{0,1\}^{d-s}}} \sum_{\substack{(i_1,\ldots,i_{s-1})\in[1,n]^{s-1} \\ (p_{d-s+1},\ldots,p_d)\in\{0,1\}^s}} \sum_{\substack{(i'_1,\ldots,i'_{s-1})\in[1,n]^{s-1} \\ (p'_{d-s+1},\ldots,p'_d)\in\{0,1\}^s}}$$

$\times 1_{\{\alpha_1=\alpha'_1\}} \times 1_{\{\alpha_{d-s+1}=\alpha'_{d-s+1}\}}$

$\times g_n^k\,[\,((\alpha_1,\alpha_2),p_1),\ldots,((\alpha_{d-s},\alpha_{d-s+1}),p_{d-s}),((\alpha_{d-s+1},i_1),p_{d-s+1}),\ldots,$
$\quad ((i_{s-1},x_1),p_d)\,]$

$\times g_n^k\,[\,((\alpha'_1,\alpha'_2),p'_1),\ldots,((\alpha'_{d-s},\alpha'_{d-s+1}),p'_{d-s}),((\alpha'_{d-s+1},i_1),p_{d-s+1}),\ldots,$
$\quad ((i_{s-1},x'_1),p_d)\,]$

$\times g_n^k\,[\,((\alpha_1,\alpha_2),p_1),\ldots,((\alpha_{d-s},\alpha_{d-s+1}),p_{d-s}),((\alpha_{d-s+1},i'_1),p'_{d-s+1}),\ldots,$
$\quad ((i'_{s-1},x_1),p'_d)\,]$

$\times g_n^k\,[\,((\alpha'_1,\alpha'_2),p'_1),\ldots,((\alpha'_{d-s},\alpha'_{d-s+1}),p'_{d-s}),((\alpha'_{d-s+1},i'_1),p'_{d-s+1}),\ldots,$
$\quad ((i'_{s-1},x'_1),p'_d)\,]$

$\leq 2^{2d} n^{2d-2}\,\|g_n^k\|_\infty^4 \leq 2^{2d} n^{-2}.$

We conclude that $\left\|\widetilde{g_n^k} \star_s \widetilde{g_n^k}\right\|_2^2 \to 0$.

Then, all the assumptions of Lemma 4 are fulfilled by $Q_d(g_N^k, Y)$. Therefore,

$$Q_d(f_N, Y) \xrightarrow{law} \sum_{k=1}^{d} \left(i + \rho\sqrt{\frac{R}{1-R}}\right)^k R_0^k R_1^{d-k} \sqrt{dC_d^k} G_k,$$

where the G_k's are independent standard Gaussian random variables. We can rewrite this result as:

$$Q_d(f_N, Y) \xrightarrow{law} Z_d^1 + iZ_d^2,$$

where $Z_d = (Z_d^1, Z_d^2)$ is a Gaussian vector; its covariance matrix is

$$\begin{pmatrix} \sigma_1^2 & \sigma_{1,2} \\ \sigma_{1,2} & \sigma_2^2 \end{pmatrix},$$

with $d = \sigma_1^2 + \sigma_2^2$ and $dE(X_1^2)^d = \sigma_1^2 - \sigma_2^2 + i\, 2\sigma_{1,2}$.
This completes the proof of Theorem 1. □

Acknowledgements This work is part of my forthcoming PhD dissertation. I am extremely grateful to my advisor Ivan Nourdin for suggesting this topic, as well as for many advices and encouragements.

References

1. G. Anderson, A. Guionnet, O. Zeitouni, *An Introduction to Random Matrices*, vol. 118 of Cambridge Studies in Advanced Mathematics (Cambridge University Press, Cambridge, 2009)
2. G. Anderson, O. Zeitouni, A CLT for a band matrix model. Probab. Theory Relat. Field **134**(2), 283–338 (2006)
3. S. Chatterjee, Fluctuations of eigenvalues and second order Poincar inequalities. Probab. Theory Relat. Field **143**(1–2), 1–40 (2009)
4. F. Goetze, On the rate of convergence in the multivariate CLT. Ann. Probab. **19**, 724–739 (1991)
5. A. Guionnet, *Large Random Matrices: Lectures on Macroscopic Asymptotics*, vol. 1957 of Lecture Notes in Mathematics (Springer, Berlin, 2009)
6. E. Mossel, R. O'Donnell, K. Oleszkiewicz, Noise stability of functions with low inluences: invariance and optimality. Ann. Math. (2) **171**(1), 295–341 (2010)
7. I. Nourdin, G. Peccati, Universal Gaussian fluctuations of non-Hermitian matrix ensembles. ALEA **7**, 341–375 (2008)
8. I. Nourdin, G. Peccati, G. Reinert, Invariance principles for homogeneous sums: universality of Gaussian Wiener chaos. Ann. Probab. **38**(5), 1947–1985 (2010)
9. D. Nualart, G. Peccati, Central limit theorems for sequences of multiple stochastic integrals. Ann. Probab. **33**(1), 177–193 (2005)
10. G. Peccati, C.A. Tudor, Gaussian limits for vector-valued multiple stochastic integrals. *Séminaire de Probabilités XXXVIII*, Lecture Notes in Math. (Springer, Berlin, 2004), pp. 247–262
11. B. Rider, J.W. Silverstein, Gaussian fluctuations for non-hemitian random matrix ensembles. Ann. Probab. **34**(6), 2118–2143 (2006)
12. I. Rotar, Limit theorems for polylinear forms. J. Multivariate Anal. **9**(4), 511–530 (1979)

13. Ya. Sinai, A. Soshnikov, Central limit theorem for traces of large random symmetric matrices with independent matrix elements. Bull. Braz. Math. Soc. (Bol. Soc. Bras. Mat., Nova Sr.) 29(1), 1–24 (1998)
14. T. Tao, V. Vu, M. Krishnapur, Random matrices: Universality of ESDs and the circular law. Ann. Probab. **38**(5), 2023–2065 (2010)

Functionals of the Brownian Bridge

Janosch Ortmann

Abstract We discuss the distributions of three functionals of the free Brownian bridge: its L^2-norm, the second component of its signature and its Lévy area. All of these are freely infinitely divisible. Two representations of the free Brownian bridge as series of free semicircular random variables are introduced and used. These are analogous to the Fourier representations of the classical Brownian bridge due to Lévy and Kac and the latter extends to all semicircular processes.

1 Introduction

In this note we discuss the distributions of three non-commutative random variables defined in terms of a free Brownian bridge.

In his paper [11], Lévy introduces the following representation of the Brownian bridge. Let ξ_n, η_n be independent standard Gaussian random variables then the process defined by

$$\beta_{2\pi}(t) = \sum_{n=1}^{\infty} \frac{\cos(nt) - 1}{n\sqrt{\pi}} \xi_n + \sum_{n=1}^{\infty} \frac{\sin(nt)}{n\sqrt{\pi}} \eta_n \qquad (1)$$

defines a Brownian bridge on $[0, 2\pi]$. Another representation is given by Kac [9]. Retaining the notation for the η_n it is a consequence of Mercer's theorem that the Gaussian process defined by

$$\beta_1(t) = \sum_{n=1}^{\infty} \frac{\sqrt{2} \sin(n\pi t)}{n\pi} \eta_n \qquad (2)$$

J. Ortmann (✉)
Warwick Mathematics Institute, University of Warwick, CV4 7AL, UK
e-mail: j.ortmann@warwick.ac.uk

has the covariance kernel of a Brownian bridge. The analogue of the Gaussian distribution and processes in non-commutative probability theory are the semicircle law and semicircular processes. It turns out that the crucial properties of the Gaussian distribution needed for the observations above are shared by the semicircular law. Therefore if we replace ξ_n, η_n by free standard semicirculars then (1) and (2) define free Brownian bridges on $[0, 2\pi]$ and $[0, 1]$ respectively. We will use this fact to prove various properties of the square norm, the second component of the signature and the Lévy area of the free Brownian bridge.

The L^2-norm of the classical Brownian bridge was first considered by Kac who used his representation (2) to compute its Fourier transform. Further calculations were performed using Kac's work, see Tolmatz [18] and the references therein. We will compute the R-transform of the free analogue of this object and use the fact that its law is freely infinitely divisible to prove that it has a smooth density for which we give an implicit equation.

In [6] Capitaine and Donati-Martin construct the second component Z of the signature of the free Brownian motion. This process plays a role in the theory of rough paths, see [6], Lyons [12] and Victoir [19] for details. The second component of the signature is a process taking values in the tensor product of the underlying non-commutative probability space with itself. Equipped with the product expectation this is a probability space in its own right and we compute the R-transform of Z. A connection between the cumulants of Z and the number of *2-irreducible meanders*, a combinatorial object introduced by Lando–Zvonkin [10] and further analysed by Di Francesco–Golinelli–Guitter [7] is pointed out.

Finally we apply the Lévy-type representation to compute the R-transform of the Lévy area corresponding to the free Brownian bridge. This random variable is also freely infinitely divisible. Once again this allows us to deduce that the law in question has a smooth density. Again we obtain an implicit equation.

From the considerations involving free infinite divisibility it also follows that the support of the law of both Lévy area and square norm is a single interval, in the former case symmetric about the origin, in the latter strictly contained in the positive half-line. In [15] a large deviations principle is established for the blocks of a uniformly random non-crossing partition. This result allows us to determine the maximum of the support from the free cumulants. We obtain implicit equations that determine the essential suprema of Lévy area and square norm. In particular we show that the support of the law of the square norm is contained in $\left[0, \frac{1}{2}\right]$.

2 Free Probability Theory

We recall here some definitions and properties from free probability theory. For an introduction to the subject see for example [8, 20, 21].

2.1 Freeness, Distributions, and Transforms

Throughout let $(\mathscr{A}, \phi)$ be a *non-commutative probability space*, i.e. a unital *-algebra equipped with a tracial state ϕ on $\mathscr{A}$. All non-commutative probability spaces considered in this paper are assumed to have additional topological structure. Namely we assume that $\mathscr{A}$ is a C*-algebra. That is, there exists a Hilbert space $\mathscr{H}$ such that $\mathscr{A}$ is a closed subalgebra of $B(\mathscr{H})$ and $\phi(a) = \langle \Omega, a(\Omega) \rangle_{\mathscr{H}}$ for some fixed unit vector $\Omega \in \mathscr{H}$. See also Sect. 2.3.

We think of elements $a \in \mathscr{A}$ as non-commutative random variables and consider $\phi(a)$ to be the expectation of $a \in \mathscr{A}$. We will only consider self-adjoint $a \in \mathscr{A}$. Then there exists a compactly supported measure μ_a on $\mathbb{R}$, called the *distribution* of a, such that

$$\phi(a^n) = \int t^n \mu_a(\mathrm{d}t) \qquad \forall\, n \in \mathbb{N}.$$

Recall that the *Cauchy transform* of μ_a is defined to be

$$G_{\mu_a}(z) = \int_{\mathbb{R}} \frac{\mu_a(\mathrm{d}t)}{z-t} = \sum_{n=0}^{\infty} \phi(a^n) z^{-n-1}.$$

Since μ_a is compactly supported the first equality defines an analytic map G_{μ_a} from $\mathbb{C}^+$ to $\mathbb{C}^-$, and the power series expansion is valid on a neighbourhood U_a of infinity. We will also write G_a for G_{μ_a}.

Definition 1. Subalgebras $\mathscr{B}_1, \ldots, \mathscr{B}_N$ of $\mathscr{A}$ are said to be *free* if for every set of indices $\{r_j\}_{j=1}^m \subseteq \{1, \ldots, N\}$ and collection $\{a_j \in \mathscr{B}_{r_j} : 1 \leq j \leq m\}$ such that $r_j \neq r_{j+1}$ and $\phi(a_j) = 0\ \forall\, j$ we already have

$$\phi(a_1, \ldots, a_m) = 0.$$

Random variables $a_1, \ldots, a_N$ are said to be free if the unital algebras generated by the a_j are free.

If a and b are free then the distribution of $a + b$ is uniquely determined by those of a and b (see Remark 1(2) below). Denote the laws of a, b by μ_1, μ_2 respectively. Then the *free convolution* of μ_1 and μ_2 is defined to be the distribution of $a + b$. Because self-adjoint elements of $\mathscr{A}$ are determined by their distribution this induces a binary operation on the space of compactly supported probability measures, denoted $\boxplus$.

A partition π of the set $\underline{n} = \{1, \ldots, n\}$ is said to be *crossing* if there exist distinct blocks V_1, V_2 of π and $x_j, y_j \in V_j$ such that $x_1 < x_2 < y_1 < y_2$. Otherwise π is said to be *non-crossing*. Equivalently, arrange the numbers $1, \ldots, n$ clockwise on a circle and connect any two elements of the same block of π by a straight line. Then π is non-crossing if and only if the lines drawn are pairwise disjoint. Let $\mathrm{NC}(n)$ denote the set of non-crossing partitions on $\underline{n}$ (Fig. 1).

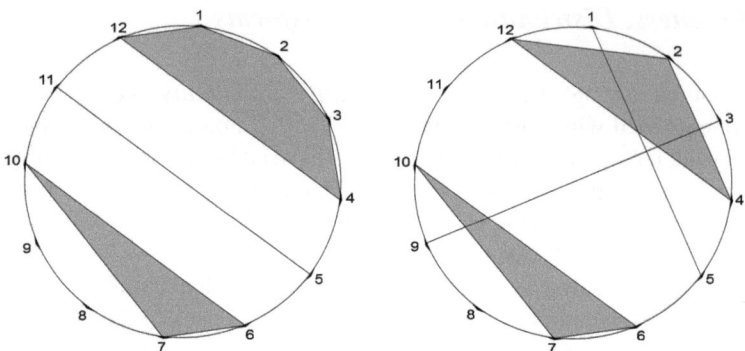

Fig. 1 The partition $\{\{8\}, \{9\}, \{10, 7, 6\}, \{11, 5\}, \{12, 4, 3, 2, 1\}\}$ is non-crossing, $\{\{5, 1\}, \{8\}, \{9, 3\}, \{10, 7, 6\}, \{12, 4, 2\}\}$ is crossing

Definition 2. The *free cumulants* of $\mathscr{A}$ are the maps $k_n \colon \mathscr{A}^n \longrightarrow \mathbb{C}$ (where $n \in \mathbb{N}$), defined indirectly by the following system of equations:

$$\phi(a_1, \ldots, a_n) = \sum_{\pi \in \mathrm{NC}(n)} k_\pi [a_1, \ldots, a_n] \qquad (3)$$

where k_π denotes the product of cumulants according to the block structure of π. That is, if $V_1, \ldots, V_r$ are the components of $\pi \in \mathrm{NC}(n)$ then

$$k_\pi [a_1, \ldots, a_n] = k_{V_1}[a_1, \ldots, a_n] \ldots k_{V_n}[a_1, \ldots, a_n]$$

where, for $V = (v_1, \ldots, v_r)$ we just have $k_V[a_1, \ldots, a_n] = k_{|V|}(a_{v_1}, \ldots, a_{v_r})$.

Note that (3) has the form $\phi(a_1, \ldots, a_n) = k_n[a_1, \ldots, a_n] +$ lower order terms, so that we can find the k_n inductively. Alternatively, (3) defines the k_n by Möbius inversion. See [14] for details.

We will write $k_n(a)$ for $k_n[a, \ldots, a]$. The *R-transform* of a random variable $a \in \mathscr{A}$ is defined to be the formal power series

$$R_a(z) = \sum_{n=0}^{\infty} k_{n+1}(a) z^n. \qquad (4)$$

If the law of a has compact support then Eq. (4) defines an analytic function on a neighbourhood of zero [8, Theorem 3.2.1]. Moreover the Cauchy transform G_a of a is locally invertible on a neighbourhood of infinity and the inverse K_a satisfies

$$K_a(z) = R_a(z) + \frac{1}{z}.$$

Remark 1. The following three properties of the R-transform are easy to check using the continuity of ϕ and multilinearity of the cumulants.

1. If a_n converges to a in the operator topology of $\mathscr{A}$ then there exists a neighbourhood U of zero where R_n, R are defined for all $n \in \mathbb{N}$ and $R_{a_n}(z) \longrightarrow R_a(z)$ as $n \to \infty$ for every $z \in U$.
2. If $a, b \in \mathscr{A}$ are free then $R_{a+b}(z) = R_a(z) + R_b(z)$
3. For $\lambda \in \mathbb{C}$ we have $R_{\lambda a}(z) = \lambda R_a(\lambda z)$.

2.2 Semicircular Processes

Definition 3. A collection $\mathscr{S} = (s_j)_{j \in I}$ of non-commutative variables on $\mathscr{A}$ is said to be a *semicular family* with *covariance* $(c(i,j))_{i,j \in I}$ if the cumulants are given by

$$k_\pi[s_{j_1}, \ldots s_{j_n}] = \begin{cases} \prod_{p \sim_\pi q} c(j_p, j_q) & \text{if } \pi \text{ is a pair partition} \\ 0 & \text{otherwise.} \end{cases}$$

If $\mathscr{S}$ consists of a singleton s_1 and $r = 2\sqrt{c(1,1)}$ then the distribution of s_1 is the *centred semicircle law of radius* r, that is the measure σ_r on $\mathbb{R}$ given by

$$\sigma_r(\mathrm{d}t) = \frac{2}{\pi r^2}\sqrt{r^2 - t^2}\, \mathbf{1}_{[-r,r]}(t)\, \mathrm{d}t.$$

In particular σ_2 is also called the *standard semicircle law* and non-commutative random variables with law σ_r (σ_2) are referred to as *(standard) semicirculars*.

The semicircle law plays a similar role to the Gaussian distribution on classical probability theory. In particular there exists a central limit theorem [21, Theorem 3.5.1], and a collection of random variables with a joint semicircular law is determined by its covariance. To be more precise we recall the following result, which is stated as Proposition 8.19 in Nica–Speicher [14].

Proposition 1. *Let $(s_i)_{i \in I}$ be a semicircular family of covariance $(c(i,j))_{i,j \in I}$ and suppose I is partitioned by $I_1, \ldots, I_d$. Then the following are equivalent:*

1. *The collections $\{s_j : j \in I_1\}, \ldots, \{s_j : j \in I_d\}$ are free*
2. *We have $c(r, j) = 0$ whenever $r \in I_p$ and $j \in I_q$ with $p \neq q$.*

In particular $\{s_j : j \in I\}$ is a free family if and only if $C = (c(r, j))_{r,j \in I}$ is diagonal.

Definition 4. A process $(X(t))_{t \geq 0}$ on $\mathscr{A}$ is said to be a *semicircular process* if for every $t_1, \ldots, t_n \in [0, \infty)$, the set $(X(t_1), \ldots, X(t_n))$ is a semicircular family.

By the considerations above the finite-dimensional distributions of a semicircular process are determined by the *covariance structure* of the process, i.e. by the function $C: [0,\infty)^2 \longrightarrow \mathbb{C}$ defined by

$$C(s,t) = \phi(X(s)X(t)).$$

2.3 The Full Fock Space, Creation and Annihilation

In order to deal with convergence issues it will be useful to choose a specific non-commutative probability space. Let $\mathcal{H}_0$ be an infinite-dimensional separable complex Hilbert space and define the *full Fock space* to be

$$\mathcal{H} = \bigoplus_{n=0}^{\infty} \mathcal{H}_0^{\otimes n}. \tag{5}$$

where by convention $\mathcal{H}_0^{\otimes 0} = \mathbb{C}\Omega$ for a distinguished unit vector Ω. Equip the C*-algebra $B(\mathcal{H})$ of continuous linear functionals on $\mathcal{H}$ with the tracial state ϕ given by

$$\phi(a) = \langle a(\Omega), \Omega \rangle. \tag{6}$$

Definition 5. For $h \in \mathcal{H}_0$ define the *creation* and *annihilation operators* to be $l(h)$ and $l^*(h)$ respectively where

$$l(h)(h_1 \otimes \ldots \otimes h_n) = h \otimes h_1 \otimes \ldots \otimes h_n \tag{7}$$

$$l^*(h)(h_1 \otimes \ldots \otimes h_n) = \langle h, h_1 \rangle h_2 \otimes \ldots \otimes h_n. \tag{8}$$

Let $s(h)$ be the self-adjoint element of $B(\mathcal{H})$ defined by $s(h) = l(h) + l^*(h)$. The following result is Theorem 2.6.2 in [21].

Lemma 1. *Let $(e_n)_{n \in \mathbb{N}}$ be an orthonormal sequence in $\mathcal{H}_0$ and put $\xi_n = s(e_n)$.*

(i) *If $\mathscr{A}$ denotes the sub-von Neumann algebra of $B(\mathcal{H})$ generated by $(\xi_n)_{n \in \mathbb{N}}$ then ϕ is a faithful normal trace on $\mathscr{A}$.*
(ii) *The set $\{s(e_n): n \in \mathbb{N}\}$ forms a semicircular family in $\mathscr{A}$ with covariance kernel $C(m,n) = \delta_{mn}$.*

Since all of the results in this paper are only concerned with the distributions of non-commutative probability spaces we can, and will, assume throughout that $\mathscr{A}$ is a C*-subalgebra of $B(\mathcal{H})$, the space of bounded linear operators on the full Fock space, and that ϕ is as in (6). In particular all semicircular random variables that appear will be defined in terms of the creation and annihilation operators.

2.4 The Lévy Representation of the Free Brownian Bridge

Definition 6. A centred semicircular process $(\beta_T(t))_{t\in[0,T]}$ on $\mathscr{A}$ is said to be a *free Brownian bridge* on $[0, T]$ if its covariance structure is given by

$$\phi(\beta_T(s)\beta_T(t)) = s \wedge t - \frac{st}{T}.$$

Remark 2. In analogy with classical probability it can be easily checked that if β is a free Brownian bridge on $[0, 1]$ and ξ_0 is a free standard semicircular free from $\{\beta(t): t \in [0, 1]\}$, then $X(t) = \xi_0 t + \beta(t)$ defines a *free Brownian motion*, that is

(i) the distribution of $X(t)$ is a centred semicircular law with radius t;
(ii) $X(t) - X(s)$ is free from $\{X(r): r \leq s\}$
(iii) $X(t) - X(s)$ has the same distribution as $X(t - s)$.

The following proposition is the analogue of Lévy's representation of the classical Brownian bridge [11]. Let $(e_n, f_n : n \in \mathbb{N})$ be an orthonormal sequence in the full Fock space $\mathscr{H}$ and define $\xi_n = s(e_n)$ and $\eta_n = s(f_n)$, so that $\{\xi_n, \eta_m : (n, m) \in \mathbb{N}^2\}$ is a set of free standard semicircular variables in $\mathscr{A}$.

Proposition 2. *The process $\beta_{2\pi}$ defined by*

$$\beta_{2\pi}(t) = \sum_{n=1}^{\infty} \frac{\cos(nt) - 1}{n\sqrt{\pi}} \xi_n + \sum_{n=1}^{\infty} \frac{\sin(nt)}{n\sqrt{\pi}} \eta_n \qquad (9)$$

is a free Brownian bridge on $[0, 2\pi]$.

Proof. By continuity and linearity of the operator $s(\cdot)$ it follows that the right-hand side of (9) converges in $\mathscr{A}$ and that $\beta_{2\pi}(t)$ is a centred semicircular variable. A direct computation verifies that $\beta_{2\pi}$ has the right covariance kernel. □

2.5 A Representation for Centred Semicircular Processes

In this section we show how Kac's representation [9] for the classical Brownian bridge on the unit interval can be translated into the setting of free probability. His method extends to all centred semicircular (or indeed Gaussian) processes, as follows. Everything relies on the following classical result from functional analysis, see Bollobas [5].

Theorem 1 (Mercer's theorem). *Let $K : [0, 1] \times [a, b] \longrightarrow \mathbb{R}$ be a non-negative definite symmetric kernel. Let T_K be the operator on $\mathscr{H}$ associated to K, that is,*

$$T_K(f)(s) = \int_0^1 K(s, t) f(t) \, dt. \qquad (10)$$

Then there exists an orthonormal basis $(f_n)_{n\in\mathbb{N}}$ of $L^2[0,1]$ consisting of eigenfunctions of T_K such that the corresponding eigenvalues λ_n are non-negative, $f_n \in \mathscr{C}[0,1]$ whenever $\lambda_n \neq 0$ and

$$K(s,t) = \sum_{n=1}^{\infty} \lambda_n f_n(s) f_n(t) \tag{11}$$

where the convergence is absolute and uniform, and hence also in $L^2[0,1]$.

We can use Mercer's theorem to represent any centred semicircular process as a series of free standard semicircular random variables, noting that if Y is a centred semicircular process indexed by $[0,1]$ then its covariance function K defined by $K(s,t) = \phi(Y(s)Y(t))$ is a non-negative kernel on $[0,1]$ which is also symmetric, by traciality of ϕ.

Corollary 1. *Let $K, \mathscr{H}, (\lambda_n, f_n)_{n\in\mathbb{N}}$ be as in Mercer's theorem and let $(\eta_n)_{n\in\mathbb{N}}$ be a sequence of free standard semicirculars, defined in terms of creation and annihilation operators as in Sect. 2.4. Then the process Y defined by*

$$Y(t) = \sum_{n=1}^{\infty} \sqrt{\lambda_n} f_n(t) \eta_n \tag{12}$$

is a centred semicircular process of covariance K.

Proof. As before convergence in the operator topology of $\mathscr{A}$ follows from linearity and continuity of the operator $s(\cdot)$. Further it is once more immediate that Y is a centred semicircular process. Its covariance kernel is given by

$$\phi(Y(s)Y(t)) = \sum_{m,n=1}^{\infty} \sqrt{\lambda_m \lambda_n} f_m(s) f_n(t) \phi(\eta_m \eta_n)$$

$$= \sum_{n=1}^{\infty} \lambda_n f_n(s) f_n(t) = K(s,t)$$

by Mercer's theorem. □

For the free Brownian bridge on $[0,1]$ we have $K(s,t) = s \wedge t - st$. Solving the corresponding eigenvalue-eigenvector equation we obtain Kac's representation in the free setting.

$$\beta_1(t) = \sum_{n=1}^{\infty} \frac{\sqrt{2}\sin(n\pi t)}{n\pi} \eta_n. \tag{13}$$

3 Square Norm of the Free Brownian Bridge

In this section we consider the square-norm of a free Brownian bridge β on interval. Recall that $\mathscr{A}$ is a C*-algebra so that we can consider β as a map from $[0, 1]$ into a Banach space which is easily seen to be continuous. We can therefore use Riemann integration to define

$$\Gamma = \int_0^1 \beta(t)^2 \, dt$$

where β is a free Brownian bridge on $[0, 1]$. In this section we discuss the distribution of the non-commutative random variable Γ, using the representation (13). Kac[9] showed that the Laplace transform of the commutative analogue of Γ is given by

$$\hat{f}(p) = \left(\frac{\sqrt{2p}}{\sinh \sqrt{2p}} \right)^{(1/2)}.$$

Other properties, in particular the density function f, were computed, most recently by Tolmatz[18].

We give here the R-transform of Γ and an expression for its moments involving a sum over non-crossing partitions. Further below we show that the distribution μ_Γ of Γ is freely infinitely divisible. This gives us some analytic tools to show that there exist $a, b \in [0, \infty)$ with $a < b$ such that the support of μ_Γ is $[a, b]$ and that μ_Γ has a smooth positive density on $[a, b]$. We give an implicit equation and a sketch for the density.

Finally we use a result from [15] to characterise the maximum b of the support of μ_Γ. In particular we show that $b < \frac{1}{2}$.

3.1 The R-Transform

The Kac representation of semicircular process is well suited for computing quadratic functionals. Let Y be a semicircular process with covariance kernel K and series representation as in Corollary 1. By orthonormality of the eigenfunctions,

$$\int_0^1 Y^2(s) \, ds = \sum_{n=1}^\infty \lambda_n \eta_n^2.$$

Now the distribution of η_n^2 is well-known: the square of a standard semicircular random variable is a free Poisson element of unit rate and jump size (Nica–Speicher [14], Proposition 12.13). So the free cumulants of η_n^2 are all equal to 1 and hence its R-transform is given by $R_n(z) = \frac{1}{1-z}$, see [14, p. 205].

Using the properties of the R-transform mentioned in Remark 1 we can now compute the R-transform of $\int_0^1 Y^2(s)\,ds$. In the case where Y is a free Brownian bridge we obtain the following

Proposition 3. *The R-transform of the square norm Γ of the free Brownian bridge is given by*

$$R_\Gamma(z) = \frac{1 - \sqrt{z}\cot(\sqrt{z})}{2z}. \tag{14}$$

Proof. The eigenvalues of K are given by $\lambda_n = \frac{1}{n\pi}$. So for $|z| < \pi^2$ we have

$$R_\Gamma(z) = \sum_{n=1}^\infty \frac{1}{\pi^2 n^2} R_n\left(\frac{z}{\pi^2 n^2}\right) = \sum_{n=1}^\infty \frac{1}{n^2\pi^2 - z} = \frac{1 - \sqrt{z}\cot(\sqrt{z})}{2z}$$

as claimed. □

The free cumulants of Γ are therefore given by

$$k_m = \frac{\zeta(2m)}{\pi^{2m}} = (-4)^{m+1}\frac{B_{2m}}{2(2m)!}$$

where B_n is the nth Bernoulli number and ζ the Riemann zeta function. With (3) we obtain a formula for the moments involving a sum over non-crossing partitions:

$$\phi(\Gamma^n) = \frac{1}{\pi^{2n}} \sum_{\sigma \in NC(n)} \prod_{r=1}^{m_\sigma} \zeta(2l_r^\sigma) = (-4)^{n+1} \sum_{\sigma \in NC(n)} \prod_{r=1}^{m_\sigma} \frac{B_{2l_r^\sigma}}{2(2l_r^\sigma)!}$$

where m_σ denotes the number of equivalence classes of a non-crossing partition σ and l_r^σ is the size of the rth equivalence class of σ.

While there does not seem to exist a closed-form expression for the inverse of $K_\Gamma(z) = R_\Gamma(z) - \frac{1}{z}$ (and hence, by the Stieltjes inversion formula, for the density) we will describe some properties of the law μ_Γ of Γ. We will prove that μ_Γ is freely infinitely divisible, has a positive analytic density on a single interval and give an equation for the right end point of that interval.

3.2 Free Infinite Divisibility

The concept of infinite divisibility has a natural analogue in free probability theory. Noting that the square norm of the free Brownian bridge is freely infinitely divisible we will use the approach of P. Biane in his appendix to the paper [2] to prove that the law of Γ has a smooth density on its support and give an implicit formula for that density.

Definition 7. A compactly supported probability measure μ is said to be *freely infinitely divisible* (or $\boxplus$-*infinitely divisible*) if for every $n \in \mathbb{N}$ there exists a compactly probability measure μ_n such that

$$\mu = \mu_n^{\boxplus n} = \underbrace{\mu_n \boxplus \ldots \boxplus \mu_n}_{n \text{ times}}$$

where $\boxplus$ denotes free convolution (Sect. 2).

Since for each n the free random variable η_n^2 has a free Poisson distribution and is therefore freely infinitely divisible it follows that Γ is also $\boxplus$-infinitely divisible.

Recall that the Cauchy transform G_Γ of Γ is an analytic map from the upper half plane $\mathbb{C}^+$ into the lower half plane $\mathbb{C}^-$, which is locally invertible on a neighbourhood of infinity, and that its local inverse is given by the K-transform K_Γ where

$$K_\Gamma(z) = R_\Gamma(z) + \frac{1}{z} = \frac{3 - \sqrt{z}\cot(\sqrt{z})}{2z}.$$

From Proposition 5.12 in Bercovici–Voiculescu [3] and the infinite divisibility of Γ it is straightforward to deduce the following result.

Lemma 2. *The law μ_Γ of the square norm of the free Brownian bridge can have at most one atom. Moreover its Cauchy transform G_Γ is an analytic injection from $\mathbb{C}^+$ whose image is the connected component Ω in $\mathbb{C}^-$ of*

$$\hat{\Omega} = \{z \in \mathbb{C}^- : \Im(K_\Gamma(z)) > 0\}$$

that contains iy for small values of y.

It will be useful to characterise the boundary $\partial\Omega$.

Lemma 3. *For every $t \in (\pi, 2\pi)$ there exists unique $r(t) > 0$ such that*

$$\Im\left[(K_\Gamma(r(t)e^{it}))\right] = 0.$$

Moreover we have

$$\left.\frac{\partial}{\partial z}\Im K_\Gamma(z)\right|_{z=r(t)e^{it}} \neq 0 \qquad \forall\, t \in (\pi, 2\pi). \tag{15}$$

Proof. Fix $t \in (\pi, 2\pi)$. The imaginary part of K_Γ can be written in polar co-ordinates by

$$h_t(r) := \Im K_\Gamma\left(r e^{it}\right) = -\frac{3\sin(t)}{2r} + \frac{\gamma \sinh(\sigma\sqrt{r})\cosh(\sigma\sqrt{r}) + \sigma \sin(\gamma\sqrt{r})\cos(\gamma\sqrt{r})}{2\sqrt{r}\left(\sin^2(\gamma\sqrt{r}) + \sinh^2(\sigma\sqrt{r})\right)}$$

where $\sigma = \sin(t/2)$ and $\gamma = \cos(t/2)$. Define $g_t(r) = 2r\, h_t(r^2)$. Then

$$g_t(r) = -\frac{6\sigma\gamma}{r} + \frac{\sigma \sin(2\gamma r) + \gamma \sinh(2\sigma r)}{2\left[\sin^2(\gamma\sqrt{r}) + \sinh^2(\sigma\sqrt{r})\right]}.$$

The function g_t blows up to $+\infty$ as $r \downarrow 0$. In particular g_t is strictly positive on $(0, R_2(t))$ for some $R_2(t) > 0$. Further there must be $R_1(t) > 0$ with $g'_t(r)$ negative on $(0, R_1(t))$. Splitting into the three cases whether $t \in \left(\pi, \frac{3\pi}{2}\right)$ or $t \in \left[\frac{3\pi}{2}, \frac{5\pi}{3}\right]$ or $t \in \left(\frac{5\pi}{3}, 2\pi\right)$ we can check directly that there exists $R_3(t) \in (0, R_1(t))$ such that $g_t(r) < 0$ for all $r > R_3(t)$. Further details on this lengthy but elementary computation can be found in [16].

Hence g_t has a unique zero ρ_t, which must lie in $(R_2(t), R_3(t)) \subset (0, R_1(t))$. Hence $g'_t(\rho(t)) < 0$ and the result follows. □

Therefore $\hat{\Omega}$ is actually simply connected: it is given by the area enclosed by the real axis and the curve $\lambda = \{r_t e^{it} : t \in (\pi, 2\pi)\}$. In particular $\Omega = \hat{\Omega}$ and $\partial\Omega$ is a continuous simple curve. So Carathéodory's theorem applies, wherefore the analytic bijection $G_\Gamma : \mathbb{C}^+ \longrightarrow \Omega$ extends to a homeomorphism (denoted $\hat{G}_\Gamma$) from $\mathbb{C}^+ \cup \mathbb{R} \cup \{\infty\}$ to the closure $\overline{\Omega}$ of Ω in $\mathbb{C} \cup \{\infty\}$.

Since Ω is bounded, so is its closure, whence $\hat{G}_\Gamma$ is finite on $\mathbb{C}^+ \cup \mathbb{R} \cup \{\infty\}$. The set of isolated points of the support of μ_Γ is exactly the set of $t \in \mathbb{R}$ such that $\hat{G}_\Gamma(t) = \infty$ so $\mathrm{supp}(\mu_\Gamma)$ must be an interval $[a, b]$. From the Stieltjes inversion formula (see for example [8], p.93) it now follows that if we put for $x \in [a, b]$

$$\Phi(x) = -\frac{1}{\pi} \lim_{y \to 0} \mathfrak{Im}\left(G_\Gamma(x + iy)\right) = -\frac{1}{\pi} \mathfrak{Im}\left(\hat{G}_\Gamma(x)\right) \tag{16}$$

then μ_Γ has density Φ with respect to Lebesgue measure on $[a, b]$. Since K_Γ is the inverse of G_Γ and because of (15) the implicit function theorem applies and hence Φ is smooth on $[a, b]$. Moreover it follows that

$$\mathrm{supp}\,\mu_\Gamma = K_\Gamma\left(\partial\Omega \cap \mathbb{C}^-\right) = [K_\Gamma(r_{\pi+}) \wedge K_\Gamma(r_{2\pi-}), K_\Gamma(r_{\pi+}) \vee K_\Gamma(r_{2\pi-})].$$

where $r_{\pi+} = \lim_{s \downarrow 0} r_{\pi+s}$ and $r_{2\pi-} = \lim_{s \downarrow 0} r_{2\pi-s}$.

The operator Γ is positive so the support of μ_Γ must be contained in $[0, \infty)$. (We will show below that in fact the support is contained in $[0, 1/2]$.) Let us summarise the results of this section.

Proposition 4. *There exist $b > a \geq 0$ and a positive smooth function $\Phi : [a, b] \longrightarrow \mathbb{R}$ such that*

$$\mu_\Gamma(dt) = \Phi(t)\mathbf{1}_{[a,b]}(t). \tag{17}$$

The function Φ is given by $\Phi(x) = -\frac{1}{\pi} r(\tau_x) \sin(\tau_x)$ where $\tau_x \in (\pi, 2\pi)$ is the unique solution to $K_\Gamma\left(r(\tau_x)e^{i\tau_x}\right) = x$ (Fig. 2).

Fig. 2 Sketch of the density of the L^2-norm of the free Brownian bridge, based on numerical computations

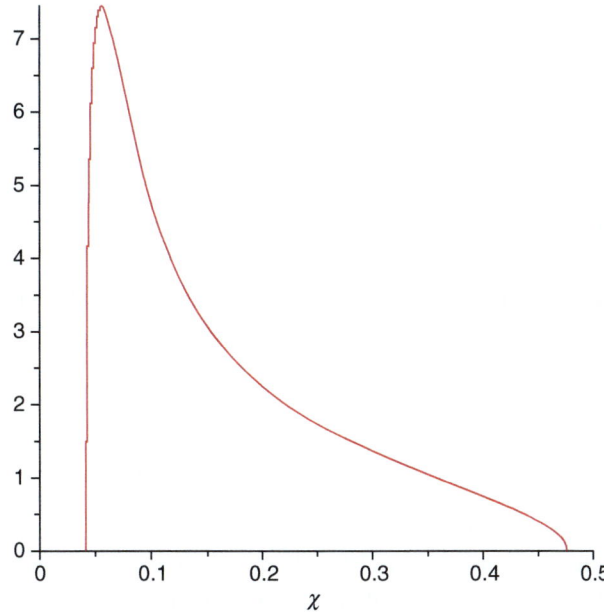

3.3 The Maximum of the Support

We now study the maximum of the support of μ_Γ. We will need Theorem 5.4 from [15]:

Theorem 2. *Let μ be a compactly supported probability measure on $[0, \infty)$ such that its free cumulants $(k_j)_{j \in \mathbb{N}}$ are all positive. Then the right edge ρ_μ of the support of μ is given by*

$$\log \rho_\mu = \sup \left\{ \frac{1}{m_1(p)} \sum_{m=1}^{\infty} p_m \log\left(\frac{k_m}{p_m}\right) + \frac{\Theta(m_1(p))}{m_1(p)} : p \in \mathfrak{M}_1^1(\mathbb{N}) \right\} \quad (18)$$

where $\mathfrak{M}_1^1(\mathbb{N}) = \{p \in \mathfrak{M}_1(\mathbb{N}) : m_1(p) < \infty\}$ is the set of probability measures on $\mathbb{N}$ with finite mean and $\Theta(m) = \log(m-1) - m \log\left(1 - \frac{1}{m}\right)$.

It turns out that this variational problem can be solved using the method of Lagrange multipliers. There exists a unique maximiser p^* for the supremum on the right-hand side of (18). Using the series expansion of $\zeta(2n)$ and interchanging summation we obtain

$$p_n^* = \frac{1}{m^* - 1} \zeta(2n) \left(\frac{\gamma}{\pi}\right)^{2n}$$

where γ is a rational function of m^* and m^* is the unique solution on $\left(\frac{3}{2},\infty\right)$ of the equation

$$m - 3 = \sqrt{4m^2 - 2m - 6} \cot\left(\frac{\sqrt{4m^2 - 2m - 6}}{m - 1}\right) \tag{19}$$

Details of the computations can be found in [16]. In the end we obtain an implicit equation for the right edge of the support of μ_Γ:

Proposition 5. *The number b from Proposition 4 is given by*

$$b = \frac{(m^*)^2 - m^*}{4(m^*)^2 - 2m^* - 6}$$

where m^* is the unique solution of (19) on $\left(\frac{3}{2},\infty\right)$.

Remark 3. The function $B: m \longmapsto \frac{m^2-m}{4m^2-2m-6}$ is strictly decreasing on $\left(\frac{3}{2},2\right)$. Since the left-hand side of (19) is bigger than the right-hand side for $m = \frac{8}{5}$ but smaller for $m = 2$ it follows that $m^* \in \left(\frac{8}{5},2\right)$ and hence $b \leq B\left(\frac{8}{5}\right) < \frac{1}{2}$. It follows that the support of μ_Γ is contained in $\left[0,\frac{1}{2}\right]$.

4 The Signature of the Free Brownian Bridge

4.1 Signature and Rough Paths

In T. Lyons's paper [12] a new approach to differential equations driven by rough paths is proposed. For a general Banach-valued path $p: \mathbb{R}_+ \longrightarrow E$ we define, when this makes sense, the *signature* of p to be the process $S(p)$ taking values in the tensor algebra $T((E)) = \bigoplus_{n=0}^\infty E^{\otimes n}$ whose nth component is given by the n-times iterated integral against p:

$$S(p)_n(t) = \int_{0 < t_1 < \ldots < t_n < t} dp(t_1) \otimes \ldots \otimes dp(t_n).$$

The signature is then used to solve general differential equations of the form

$$dS(t) = S(t) \otimes dp(t).$$

In order to show that this works if the path in question is a free Brownian motion X, Capitaine-Donati-Martin [6] define an integral of a class of suitable processes $\mathfrak{P}$ against X that yields a process taking values in the tensor product $\mathscr{A} \otimes \mathscr{A}$ and prove that X itself is contained in $\mathfrak{P}$. The integral is defined taking Riemann-type approximations, so it is straightforward to extend it to processes with finite variation. Using Remark 2 we can therefore define the *second component of the signature* of a free Brownian bridge β on $[0, 2\pi]$ by

$$Z(t) = \int_0^t \beta \otimes d\beta \qquad t \in [0, 2\pi]$$

where the integral is in the sense of [6], see also Victoir [19].

If $\mathscr{A}$ is a von Neumann algebra and ϕ a faithful tracial state on $\mathscr{A}$ then its tensor product $\phi \otimes \phi$ is a faithful tracial state on the von Neumann tensor product $\mathscr{A} \otimes \mathscr{A}$ of $\mathscr{A}$ with itself, see for example [19], p. 109. So we can consider $(\mathscr{A} \otimes \mathscr{A}, \phi \otimes \phi)$ as a non-commutative probability space in its own right. We will discuss here the law of $Z(2\pi)$ with respect to this space.

We will also use the notation $\hat{\mathscr{A}}$, $\hat{\phi}$ for $\mathscr{A} \otimes \mathscr{A}$, $\phi \otimes \phi$ respectively.

4.2 Using the Lévy Representation

The representation (9) and a straightforward calculation using orthogonality of the trigonometric functions yield

Proposition 6. *The Lévy area of the free Brownian bridge at time 2π has the same law as the random variable*

$$Z(2\pi) = \sum_{n=1}^{\infty} \frac{1}{n} (\xi_n \otimes \eta_n - \eta_n \otimes \xi_n). \tag{20}$$

Lemma 4. *Let ξ, η be two freely independent standard semicircular random variables. Then $\{\xi \otimes \eta, \eta \otimes \xi\}$ is a free set.*

Proof. Denote by $\mathscr{A}_1, \mathscr{A}_2$ respectively the algebras generated by $\xi \otimes \eta$ and $\eta \otimes \xi$. By Theorem 11.20 in [14] it is enough to show that the free cumulant $k_n(a_1, \ldots, a_n) = 0$ whenever $n \geq 2$ and $a_j = \alpha_j \otimes \beta_j$ where $\{\alpha_j, \beta_j\} = \{\xi, \eta\}$ for each j, $a_1 = \xi \otimes \eta$ and for some $j > 1$ we have $a_j = \eta \otimes \xi$. We will establish this by induction, in analogy to the proof of Theorem 11.15 in [14]. If $n = 2$ we must have $a_2 = \eta \otimes \xi$. Because ξ, η are free and centred we have

$$k_2(a_1, a_2) = \hat{\phi}(a_1 a_2) = \hat{\phi}(\xi\eta \otimes \eta\xi) = \phi(\xi\eta)\phi(\eta\xi) = 0.$$

Suppose now the claim holds for all $l < n$. Recall [14, (11.5)] that

$$k_n(a_1, \ldots, a_n) = \sum_{\sigma \in NC(n)} \hat{\phi}_\sigma[a_1, \ldots, a_n] \mu(\sigma, 1_n)$$

$$= \sum_{\sigma \in NC(n)} \phi_\sigma[\beta_1, \ldots, \beta_n]^2 \mu(\sigma, 1_n)$$

where 1_n denotes the partition of $\underline{n}$ consisting only of singletons, μ is the Möbius function on NC(n) and $\phi_\sigma[a_1,\ldots,a_n] = \prod_{V\in\pi}\phi(a_{j_1}\ldots a_{j_{s_V}})$, using the notation $\{j_1,\ldots,j_{s_V}\}$ for V. For a definition of the Möbius function see Chap. 10 of [14], but we will not need it here since we will show that each $\phi_\sigma[a_1,\ldots,a_n] = 0$. Suppose first that the a_j are alternating (i.e. $a_j = \xi \otimes \eta$ if and only if j is odd). Since each $\sigma \in $ NC(n) contains an interval, i.e. a block of the form $\{j,\ldots,j+p\}$ (allowing for the possibility $p = 0$), each $\phi_\sigma[\beta_1,\ldots,\beta_n]^2$ must vanish because of freeness of ξ and η.

Turning to the general case we write $a_1\ldots a_n = A_1\ldots A_N$ so that each $A_j \in \mathscr{A}_1$ if j is odd and $A_j \in \mathscr{A}_2$ if j is even. That is, we group the a_j into products, alternating between powers of $\xi \otimes \eta$ and $\eta \otimes \xi$. Note that we must have $N > 1$ and that $k_N(A_1,\ldots,A_N) = 0$ because of the above. Denote by σ the non-crossing partition corresponding to this grouping, i.e. two elements i, j of $\underline{n}$ are in the same block of σ if and only if a_i and a_j form part of the same A_r. By Theorem 11.12 in [14],

$$0 = k_N(A_1,\ldots,A_N) = \sum_{\substack{\pi\in\text{NC}(n)\\ \pi\vee\sigma=1_n}} k_\pi[a_1,\ldots a_n]$$

$$= k_n(a_1,\ldots,a_n) + \sum_{\substack{\pi\in\text{NC}(n)\setminus\{1_n\}\\ \pi\vee\sigma=1_n}} k_\pi[a_1,\ldots a_n]$$

By the inductive hypothesis, the only non-zero terms in the sum on the right-hand side are those where each block of π contains only elements from either $\mathscr{A}_1$ or $\mathscr{A}_2$, but not both. By the way we have defined σ, the same is true for σ. By Remark 11.14 in [14] it now follows that a non-zero term can only appear if *all* appearing elements are from the same $\mathscr{A}_j$. But we started out by assuming that both $\xi \otimes \eta$ and $\eta \otimes \xi$ appears at least once. Hence the second summand on the right-hand side vanishes and we have $k_n(a_1,\ldots,a_n) = 0$. This completes the proof. □

Since ξ_n, η_n have symmetric distributions, so do $\xi_n \otimes \eta_n$ and $\eta_n \otimes \xi_n$. Hence the R-transform of $Z(2\pi)$ is given by

$$R_{Z(2\pi)}(z) = 2\sum_{n=1}^\infty \frac{1}{n} R_{\xi\otimes\eta}\left(\frac{z}{n}\right). \qquad (21)$$

Remark 4. By the definition of $\hat\phi$ we have $\hat\phi\left((\xi\otimes\eta)^k\right) = \phi(\xi^k)^2$ for $k \in \mathbb{N}$. Recall that $R_a(z) = \sum_{m=0}^\infty k_{m+1}(a)z^m$ where $k_m(a)$ denotes the mth cumulant of a. In particular $k_1(\xi\otimes\eta) = \phi(\xi)^2 = 0$ so that (on a neighbourhood of zero) $R_{\xi\otimes\eta}(z) = zP(z)$ for some analytic function P. Rewriting (21) yields

$$R_{Z(2\pi)}(z) = 2z\sum_{n=1}^\infty \frac{1}{n^2} P\left(\frac{z}{n}\right), \qquad (22)$$

in particular the right hand side of (21) converges in a neighbourhood of zero.

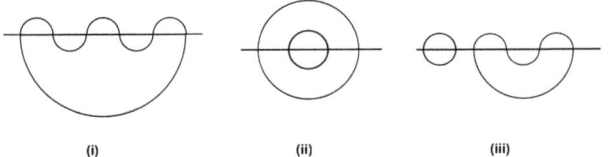

Fig. 3 (a) 1-component meander of order 3; (b) order 2, 2 components; (c) order 3, 2 components

4.3 The Distribution of $\xi \otimes \eta$ and Meanders

We proceed to compute the R-transform of $\zeta := \xi \otimes \eta$ with ξ, η free standard semicirculars. Recall that the odd moments of ξ vanish and that $\phi(\xi^{2n})$ is given by the nth *Catalan number*

$$\phi(\xi^{2n}) = C_n := \frac{1}{2n+1}\binom{2n}{n}. \qquad (23)$$

Since ξ, η are self-adjoint, so is ζ. Hence its law is a probability measure ν with compact support in $\mathbb{R}$. In particular ν is determined by its moments which are given by

$$\int t^m \nu(\mathrm{d}t) = \phi((\xi \otimes \eta)^m) = \phi(\xi^m)\phi(\eta^m) = \begin{cases} (C_k)^2 & \text{if } m = 2k \\ 0 & \text{if } m \text{ is odd} \end{cases} \qquad (24)$$

i.e. ν is the law of $\zeta_1 \zeta_2$ where the ζ_i are independent commutative random variables with standard semicircular distribution. Therefore ν is absolutely continuous with respect to Lebesgue measure with density ϕ given by

$$\phi(u) = \frac{1}{4\pi^2} \int_{-2}^{2} \sqrt{4-s^2}\sqrt{4-\left(\frac{u}{s}\right)^2}\, \mathbf{1}_{[-2,2]}\left(\frac{u}{s}\right) \frac{\mathrm{d}s}{s}. \qquad (25)$$

The Catalan numbers C_n are well-known in combinatorics. They give, for example, the number of Dyck paths of length $2n$. Similarly there is a combinatorial interpretation of the squares of the Catalan numbers, as detailed in Lando–Zvonkin [10] and Di Francesco–Golinelli–Guitter [7]: consider an infinite line in the plane and call it the *river*. A *meander* of order n is a closed self-avoiding connected loop intersecting the line through $2n$ points (the *bridges*). Two meanders are said to be *equivalent* if they can be deformed into each other by a smooth transformation without changing the order of the bridges. If a meander of order n consists of k closed connected non-intersecting (but possibly interlocking) loops it is said to have k *components* (Fig. 3).

A multi-component meander is said to be *k-reducible* if a proper non-trivial collection of its connected components can be detached from the meander by

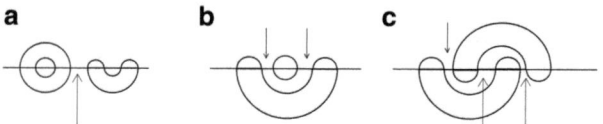

Fig. 4 Meanders that are (**a**) 1-reducible but 2-irreducible; (**b**) 1- and 2-reducible but 3-irreducible (**c**) 3-reducible

cutting the river k times between the bridges. Otherwise the meander is said to be *k-irreducible* (Fig. 4).

The 2-irreducible meanders have been studied extensively in [10] (where they are called irreducible meanders). Our connection to these objects is the following

Proposition 7. *Let q_n denote the number of 2-irreducible meanders of order $2n$ and $k_n = k_n(\xi \otimes \eta)$ the nth cumulant of $\xi \otimes \eta$. Then*

$$k_n(\xi \otimes \eta) = \begin{cases} q_m & \text{if } n = 2m \\ 0 & \text{if } n \text{ is odd} \end{cases} \qquad (26)$$

Proof. We first prove by induction that $k_n = 0$ if n is odd, which will follow from the fact that $\hat{\phi}((\xi \otimes \eta)^n) = 0$ for n odd. Assume that $k_m = 0$ whenever $m < n$ is odd. From (3) it follows that

$$k_n = - \sum_{\substack{\pi \in NC(n) \\ \pi \neq \mathbf{1}}} k_\pi$$

where $k_\pi = k_{V_1} \ldots k_{V_r}$ if $V_1, \ldots, V_r$ are the equivalence classes of π and $\mathbf{1}$ denotes the identity partition, i.e. $[k]_\mathbf{1} = \underline{n}$. Every $\pi \in NC(n) \setminus \{\mathbf{1}\}$ must contain at least one equivalence class of size m for some odd integer $m < n$. Since k_m is a factor of k_π and $k_m = 0$, the inductive hypothesis implies $k_n = 0$ as required. Hence

$$R_{\xi \otimes \eta}(z) = \sum_{n=1}^{\infty} k_{2n} z^{2n-1}.$$

Define the *moment series* of $\xi \otimes \eta$ by

$$M(z) = \frac{1}{z} G\left(\frac{1}{z}\right) = 1 + \sum_{n=1}^{\infty} \hat{\phi}\left((\xi \otimes \eta)^n\right) z^n.$$

It is a consequence of the relationship between Cauchy and R-transform that

$$M(z) = 1 + zM(z) R(zM(z)). \qquad (27)$$

Functionals of the Brownian Bridge

We will introduce one more generating series. Put

$$\rho(z) = \sum_{n=1}^{\infty} q_n z^{2n-1}.$$

From (7.10) in [7] we have

$$M(z) = 1 + zM(z)\,\rho(zM(z)). \tag{28}$$

Combining (27) and (28) yields $\rho = R$ as power series. That $k_{2n} = q_n$ now follows from comparing coefficients. □

4.4 The Distribution of $Z(2\pi)$

So we have an explicit expression for the R-transform of $\xi \otimes \eta$. We will use this to obtain the R-transform of $Z(2\pi)$.

Recall that all odd cumulants of $\xi_n \otimes \eta_n$ and $\eta_n \otimes \xi_n$ vanish, hence the same is true of $Z(2\pi)$.

Proposition 8. *The $2n$th cumulant of $Z(2\pi)$ is $2\zeta(2n)q_n$ where ζ is the Riemann zeta function.*

Proof. Recall that $\zeta(m) = \sum_{n=1}^{\infty} n^{-m}$. So

$$R_{Z(2\pi)}(z) = 2\sum_{n=1}^{\infty} \frac{1}{n} R_{\xi \otimes \eta}\left(\frac{z}{n}\right) = 2\sum_{n=1}^{\infty} \frac{1}{n} \sum_{m=1}^{\infty} k_m \left(\frac{z}{n}\right)^{m-1}$$

$$= 2\sum_{n=1}^{\infty} \sum_{m=1}^{\infty} n^{-2m} q_m z^{2m-1}$$

$$= \sum_{m=1}^{\infty} 2\zeta(2m) q_m z^{2m-1}$$

where interchanging the sums over m and n is justified by absolute convergence.

Definition 8. Let $(a_n)_{n\in\mathbb{N}}$, $(b_n)_{n\in\mathbb{N}}$ be two sequences with generating functions f, g respectively. The *Hadamard product* of f, g is defined to be the generating function of $(a_n b_n)$, denoted $f \boxasterisk g$. That is

$$f \boxasterisk g(z) = \sum_{n=1}^{\infty} a_n b_n z^n.$$

See [17]. So $R_{Z(2\pi)}$ is twice the Hadamard product of the generating functions of the 2-irreducible meanders and that of the sequence $\{\zeta(2m): m \in \mathbb{N}\}$. From (6.3.14) in Abramowitz–Stegun [1] we have for $|z| < 1$,

$$\sum_{n=2}^{\infty} \zeta(n+1)z^n = -\gamma - \Psi(1-z)$$

where γ is the Euler constant and Ψ is the *Digamma function* defined by

$$\Psi(z) = \frac{d}{dz} \log \Gamma(z) = \frac{\Gamma'(z)}{\Gamma(z)}.$$

Since the generating series can be considered as functions inside their radius of convergence, we can use complex analysis to compute their Hadamard product. Namely

Lemma 5. *Let f, g be generating functions of $(a_n)_{n\in\mathbb{N}}$, $(b_n)_{n\in\mathbb{N}}$ and suppose that they are analytic on a neighbourhood of 0. Then*

$$(f \boxasterisk g)(z^2) = \frac{1}{2\pi i} \int_\Gamma f(zw) \, g\left(\frac{z}{w}\right) \frac{dw}{w} \tag{29}$$

on a neighbourhood U of 0, where γ is a smooth closed curve around 0 and contained in U.

Proof. Let U_1, U_2 be neighbourhoods of 0 on which f and g respectively are analytic. Then for $z \in U = U_1 \cap U_2$,

$$\frac{1}{2\pi i} \int_\Gamma f(zw) \, g\left(\frac{z}{w}\right) \frac{dw}{w} = \left[f(z\eta) \, g\left(\frac{z}{\eta}\right) \right]_{\eta^0}$$

$$= \left[\sum_{n=0}^{\infty} a_n (z\eta)^n \sum_{m=0}^{\infty} b_m \left(\frac{z}{w}\right)^m \right]_{\eta^0}$$

$$= \left[\sum_{m,n} a_n b_m z^{n+m} \eta^{n-m} \right]_{\eta^0}$$

$$= \sum_{n=0}^{\infty} a_n b_n z^{2n} = f \boxasterisk g(z^2)$$

where $[\cdot]_{\eta^0}$ denotes the constant term in a Laurent series in η.

Corollary 2. *Let $\epsilon \in (0, \rho)$ where ρ is the radius of convergence of $R_{Z(2\pi)}$ and choose the canonical branch of the square root on $B(0, \epsilon)$. Then for $z \in B(0, \epsilon)$*

$$R_{Z(2\pi)}(z) = -\frac{z^{1/2}}{\pi i} \int_\Gamma \Psi(1 - z^{1/2}w) \mathscr{Q}\left(\frac{z}{w}\right) dw \qquad (30)$$

where $\Gamma = \partial B(0, \epsilon)$ and $\mathscr{Q}$ is the generating series of the q_m (recall that q_m denotes the number of 2-irreducible meanders of order $2n$).

Proof. By Proposition 8 we have $R_{Z(2\pi)} = 2 \mathscr{Q} \boxplus \Lambda$ where, using (29)

$$\Lambda(z) = \sum_{n=1}^{\infty} \zeta(m) z^m = -z\Psi(1 - z) - \gamma z.$$

Lemma 5 now yields

$$(\mathscr{Q} \boxplus \Lambda)(z^2) = \frac{1}{2\pi i} \int_\Gamma \Lambda(zw) \mathscr{Q}\left(\frac{z}{w}\right) \frac{dw}{w}$$

$$= -\frac{1}{2\pi i} \int_\Gamma zw \left(\Psi(1 - zw) + \gamma\right) \mathscr{Q}\left(\frac{z}{w}\right) \frac{dw}{w}$$

$$= -\frac{1}{2\pi i} \int_\Gamma z\Psi(1 - zw) \mathscr{Q}\left(\frac{z}{w}\right) dw$$

$$- \frac{\gamma z}{2\pi i} \int_\Gamma \mathscr{Q}\left(\frac{z}{w}\right) dw.$$

The argument of the integral in the second summand has a power series with only even powers of w so the integral itself must vanish. We therefore have

$$(q \boxplus \Phi)(z^2) = \frac{z}{2\pi i} \int_\Gamma \Psi(1 - zw) q\left(\frac{z}{w}\right) dw$$

Remark 5. In [7] it has been shown that the radius of convergence of $\mathscr{Q}$ is $\frac{4}{\pi} - 1$. Since $\zeta(m) \to 1$ as $m \to \infty$, it follows that the radius of convergence of $R_{Z(2\pi)}$ is also $\frac{4}{\pi} - 1$. It also follows that the R-transform of each $\xi_n \otimes \eta_n$ extends to a Pick function on $(1 - \frac{4}{\pi}, \frac{4}{\pi}, 1)$, see Sect. 5 below. Hence by Theorem 3 the law of $\xi_n \otimes \eta_n$ is $\boxplus$-infinitely divisible. Since free infinite divisibility is preserved by free linear combinations and weak limits, it follows that $Z(2\pi)$ is also $\boxplus$-infinitely divisible.

Unfortunately it seems that there is no explicit formula for $\mathscr{Q}$. It is therefore not apparent how a similar analysis to that for the square norm could be applied in order to obtain further details about the distribution of $Z(2\pi)$.

5 Lévy Area of the Free Brownian Bridge

In this section we use the Lévy representation

$$\beta(t) = \sum_{n=1}^{\infty} \frac{\cos(nt) - 1}{n\sqrt{\pi}} \xi_n + \sum_{n=1}^{\infty} \frac{\sin(nt)}{n\sqrt{\pi}} \eta_n \quad (31)$$

of the free Brownian bridge to compute the distribution of the free analogue of the classical Lévy area process defined by

$$\mathscr{L}(t) = \frac{i}{2} \int_0^t [\beta(s), d\beta(s)] = \frac{i}{2} \int_0^t (\beta(s) d\beta(s) - d\beta(s) \beta(s)). \quad (32)$$

When β is a two-dimensional commutative Brownian motion this is very similar to the object studied by Lévy [11]. By standard properties of the non-commutative integral [4] and self-adjointness of β we have

$$\int_0^t \beta(s) d\beta(s) = \left(\int_0^t d\beta(s) \beta(s) \right)^*.$$

A straightforward calculation yields that the left hand side equals, for $t = 2\pi$,

$$\int_0^{2\pi} \beta(s) d\beta(s) = \sum_{n=1}^{\infty} \frac{1}{n} (\xi_n \eta_n - \eta_n \xi_n) \quad (33)$$

which is easily seen to be anti-self-adjoint. This is the reason for the factor of i in (32): multiplying an anti-self-adjoint operator by i yields a self-adjoint random variable whose distribution is therefore supported in $\mathbb{R}$. Thus $\mathscr{L} := \mathscr{L}(2\pi)$ is equal to either side of (33) multiplied by i.

The summands are *commutators* of free semicircular random variables. Commutators have been studied by Nica–Speicher [13], where the semicircle distribution is discussed in Example 1.5(2). If $c_n = i(\xi_n \eta_n - \eta_n \xi_n)$, then the support of μ_{c_n} is $[-r, r]$ where $r = \sqrt{\frac{11+5\sqrt{5}}{2}}$ and

$$R_{c_n}(z) = \frac{2z}{1 - z^2} = 2 \sum_{m=1}^{\infty} z^{2m-1}. \quad (34)$$

From this we can now compute the R-transform of the classical Lévy area. Let that function be denoted $R_{\mathscr{L}}$ then

$$R_{\mathscr{L}} = \sum_{n=1}^{\infty} \frac{1}{n} R_{c_n}\left(\frac{z}{n}\right) = \sum_{n=1}^{\infty} \frac{2z}{n^2 - z^2}$$

$$= \frac{1}{z} - \pi \cot(\pi z). \quad (35)$$

We can deduce the free cumulants of $\mathscr{L}$, either from the Taylor series of (35) or by calculating

$$R_{\mathscr{L}} = \sum_{n=1}^{\infty} \frac{2}{n} \sum_{m=1}^{\infty} \left(\frac{z}{n}\right)^{2m-1} = \sum_{m=1}^{\infty} 2 \left(\sum_{n=1}^{\infty} n^{-2m}\right) z^{2m-1}$$

$$= \sum_{m=1}^{\infty} 2\zeta(2m) z^{2m-1}$$

where the interchanging of the infinite sums is justified by absolute convergence. The free cumulants of $\mathscr{L}$ are therefore given by

$$k_m(\mathscr{L}) = \begin{cases} 2\zeta(m) & \text{if } m \text{ is even} \\ 0 & \text{otherwise.} \end{cases} \tag{36}$$

Free infinite divisibility is characterised by an analytic property of the R-transform. An analytic function $f: \mathbb{C}^+ \longrightarrow \mathbb{C}^+$ is called a *Pick function*. For $a, b \in \mathbb{R}$ with $a < b$ we denote by $\mathscr{P}(a, b)$ the set of Pick functions f which have an analytic continuation $g: \mathbb{C} \setminus \mathbb{R} \cup (a, b) \longrightarrow \mathbb{C}$ such that $g(\bar{z}) = \overline{g(z)}$. The following result is Theorem 3.3.6 of Hiai-Petz [8]:

Theorem 3. *A compactly supported probability measure μ is $\boxplus$-infinitely divisible if and only if its R-transform extends to a Pick function in $\mathscr{P}(-\epsilon, \epsilon)$ for some $\epsilon > 0$.*

It is easy to see that the common R-transform of the c_n extends to a Pick function in $\mathscr{P}(-1, 1)$. Therefore each c_n is $\boxplus$-infinitely divisible.

Corollary 3. *The distribution of $\mathscr{L}$ is $\boxplus$-infinitely divisible.*

As in Sect. 3 we can use free infinite divisibility together with the analytic properties of the R-transform and the formula for the maximum of the support from [15] to describe further the distribution in question.

The variational formula of Sect. 3 (Theorem 2) assumed that all free cumulants are positive, which is not the case for $\mathscr{L}$ (which is symmetric and therefore has vanishing odd free cumulants). However non-negativity of all free cumulants is actually enough [15, Theorem 5.9]:

Theorem 4. *Let $a \in \mathscr{A}$ be a self-adjoint non-commutative random variable with distribution μ and free cumulants $k_m \geq 0$ for all m. Denote by L the set of $m \in \mathbb{N}$ such that $k_m > 0$. Then the right edge ρ_μ of the support of μ is given by*

$$\log(\rho_\mu) = \sup \left\{ \frac{1}{m_1(p)} \sum_{n \in L} p_n \log\left(\frac{k_n}{p_n}\right) - \frac{\Theta(m_1(p))}{m_1(p)} : p \in \mathfrak{M}_1^1(L) \right\} \tag{37}$$

where $\mathfrak{M}_1^1(L)$ denotes the set of $p \in \mathfrak{M}_1^1(\mathbb{N})$ such that $p(L^c) = 0$ and Θ was defined in Theorem 2.

The inverse of the Cauchy transform of $\mathscr{L}$ is given by

$$K_{\mathscr{L}} = R_{\mathscr{L}} + \frac{1}{z} = \frac{2}{z} - \pi \cot(\pi z).$$

Similarly to the situation in Sect. 3.2 there exists, for every $t \in (\pi, 2\pi)$, unique $r(t) > 0$ such that $\Im \left[K_{\mathscr{L}} \left(r(t) e^{it} \right) \right] = 0$ and

$$\frac{\partial}{\partial z} \Im [K_{\mathscr{L}}(z)] \bigg|_{z=r(t)e^{it}} \neq 0 \qquad \forall t \in (\pi, 2\pi). \tag{38}$$

Summarising, we obtain the following characterisation of the distribution of $\mathscr{L}$:

Proposition 9. *The non-commutative random variable $\mathscr{L}$ is distributed according to $\mu_{\mathscr{L}}(dt) = \Phi_{\mathscr{L}}(t) 1_{[-\rho_{\mathscr{L}}, \rho_{\mathscr{L}}]} dt$ where $\Phi_{\mathscr{L}}(x) = -\frac{1}{\pi} r(\tau_x) \sin(\tau_x)$ and τ_x is the unique solution on $(\pi, 2\pi)$ to*

$$\frac{2}{r(\tau_x) e^{i\tau_x}} - \pi \cot \left(\pi r(\tau_x) e^{i\tau_x} \right) = x. \tag{39}$$

for every $x \in (-\rho_{\mathscr{L}}, \rho_{\mathscr{L}})$. The number $\rho_{\mathscr{L}}$ is given by

$$\rho_{\mathscr{L}} = \frac{m_* \pi}{\sqrt{m_*^2 - 2}} \tag{40}$$

where m_* is the unique solution on $(\sqrt{2}, \infty)$ of

$$m - 2 = \sqrt{m^2 - 2} \cot \left(\frac{\sqrt{m^2 - 2}}{m - 1} \right) \quad \text{(See Fig. 5)}. \tag{41}$$

Proof of Proposition 9. The law $\mu_{\mathscr{L}}$ of $\mathscr{L}$ is symmetric about 0. Together with the analytic arguments of Sect. 3.2, suitably modified, this implies the existence of $\rho_{\mathscr{L}} > 0$ such that the density $\Phi_{\mathscr{L}}$ of $\mu_{\mathscr{L}}$ is smooth, positive on $(-\rho_{\mathscr{L}}, \rho_{\mathscr{L}})$ and zero everywhere else. The function $\Phi_{\mathscr{L}}$ is given by $\Phi_{\mathscr{L}}(x) = -\frac{1}{\pi} r(\tau_x) \sin(\tau_x)$ where τ_x is characterised by (39).

For the remainder of the statement we apply Theorem 4. Only the free cumulants of even order are nonzero, so that the set L from Theorem 4 is given by $\{2n : n \in \mathbb{N}\}$. Otherwise the calculations are very similar to those in the proof of Proposition 5: we apply the methods of Lagrange multipliers and deduce that the supremum on the right-hand side of (37) is attained by a unique maximiser which is characterised by Eq. (41). The argument of the supremum evaluated at this maximiser yields the right edge of the support, and is given by (40). This completes the proof of the proposition. $\square$

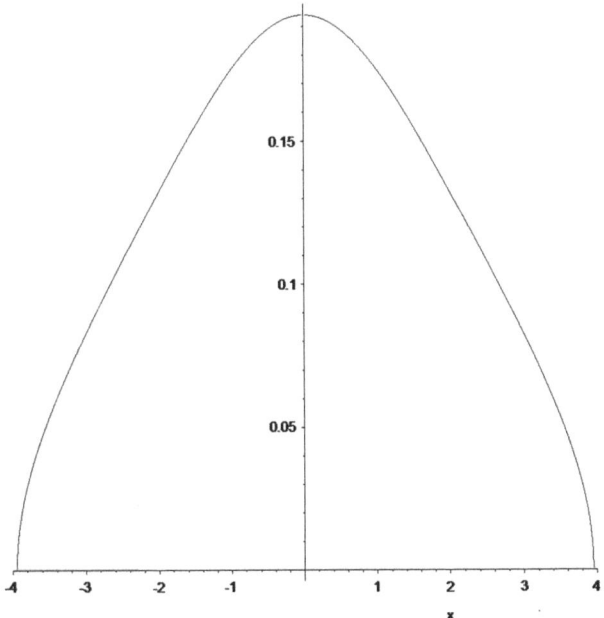

Fig. 5 Density of the free Lévy area

Acknowledgements The author would like to thank his PhD advisor, Neil O'Connell for his advice and support in the preparation of this paper. We also thank Philippe Biane for helpful discussions and suggestions and the anonymous referee whose detailed comments have lead to a much improved version of the paper.

References

1. M. Abramowitz, I.A. Stegun (eds.) *Handbook of Mathematical Functions with Formulas, Graphs, and Mathematical Tables* Reprint of the 1972 edition (Dover, New York, 1992)
2. H. Bercovici, V. Pata, Stable Laws and domains of attraction in free probability theory. With an appendix by Philippe Biane. Ann. Math. **149**, 1023–1060 (1999)
3. H. Bercovici, D. Voiculescu, Free convolution of measures with unbounded support. Indiana Univ. Math. J. **42**, 733–773 (1993)
4. P. Biane, R. Speicher, Stochastic calculus with respect to free Brownian motion and analysis on Wigner space. Prob. Theory Relat. Field **112**, 373–409 (1998)
5. B. Bollobás, *Linear Analysis*, 2nd edn. (Cambridge University Press, Cambridge, 1999)
6. M. Capitaine, C. Donati-Martin, The Lévy area process for the free Brownian motion. J. Funct. Anal. **179**, 153–169 (2001)
7. P. Di Francesco, O. Golinelli, E. Guitter, Meander, folding and arch statistics. Math. Comput. Modelling **26**(8), 97–147 (1997)
8. F. Hiai, D. Petz, *The Semicircle Law, Free Random Variables and Entropy. Mathematical Surveys and Monographs*, Vol. 77 (American Mathematical Society, Providence, RI, 2000)

9. M. Kac, On some connections between probability theory and differential and integral equations. *Proceedings of the Second Berkeley Symposium on Mathematical Statistics and Probability*, pp. 189–215 (1950)
10. S.K. Lando, A. Zvonkin, Plane and projective meanders. Theor. Comput. Sci. **117**, 227–241 (1993)
11. P. Lévy, Wiener's random function, and other laplacian random functions. *Proceedings of the Second Berkeley Symposium on Mathematical Statistics and Probability*, pp. 171–187 (1950)
12. T. Lyons, Differential equations driven by rough signals. Revista Matemática Iberoamericana **14**, 215–310 (1998)
13. A. Nica, R. Speicher, Commutators of free random variables. Duke Math. J., **92**(3), 553–592 (1998)
14. A. Nica, R. Speicher, *Lectures on the Combinatorics of Free Probability* (Cambridge University Press, Cambridge, 2006)
15. J. Ortmann, Large deviations for non-crossing partitions. arXiv preprint arXiv:1107.0208 (2011)
16. J. Ortmann, Random matrices, large deviations and reflected Brownian motion. PhD thesis, Warwick Mathematics Institute, 2011
17. R.P. Stanley, *Enumerative Combinatorics, vol. 1*, vol. 49 of *Cambridge Studies in Advanced Mathematics* (Cambridge University Press, Cambridge, 1997)
18. L. Tolmatz, On the distribution of the square integral of the Brownian bridge. Ann. Probab. **30**(1), 253–269 (2002)
19. N. Victoir, Lévy area for the free Brownian motion: existence and non-existence. J. Funct. Anal. **208**, 107–121 (2004)
20. D.V. Voiculescu, *Lectures on Free Probability Theory*. No. 1738 in Lecture Notes in Mathematics (Lectures on Probability and Theory and Statistics) (Springer, 2000), pp. 283–349
21. D.V. Voiculescu, K.J. Dykema, A. Nica, *Free Random Variables*. CRM Monograph Series, vol. 1 (American Mathematical Society, Providence, RI, 1992)

Étude spectrale minutieuse de processus moins indécis que les autres

Laurent Miclo et Pierre Monmarché

Résumé. On cherche ici à quantifier la convergence à l'équilibre de processus de Markov non réversibles, en particulier en temps court. La simplicité des modèles considérés nous permet de donner une expression assez explicite de l'évolution temporelle de l'erreur L^2 en norme opérateur et de la comparer avec celle des cas réversibles correspondants.

1 Introduction : un processus de volte-face

Le recours à la réversibilité peut parfois limiter les performances des algorithmes stochastiques (voir par exemple [3,4,8]), ce qui nous motive à mieux comprendre la convergence vers l'équilibre des processus non-réversibles. Dans ce papier nous étudierons en détail un modèle, pour lequel on verra comment se quantifie le fait que les processus non-réversibles ont d'abord tendance à aller moins vite à l'équilibre que leur équivalent réversibles, avant d'atteindre des taux asymptotiques de convergence bien meilleurs. On retrouvera notamment pour une chaîne de Markov en temps discret et à espace d'état fini (étudiée dans [4] d'un point de vue asymptotique) les phénomènes d'amorce lente de convergence mis en évidence dans [7], dans un contexte continu d'équations d'évolutions cinétiques simples.

Plus précisément, soit $(P_t)_{t\geq 0}$ un semi-groupe markovien admettant une probabilité invariante μ. Sous des conditions d'ergodicité, P_t converge, en divers sens, vers μ pour de grands temps $t \geq 0$. Considérons la convergence forte dans $L^2(\mu)$: en interprétant μ comme l'opérateur $f \mapsto (\int f \, d\mu)\mathbb{1}$, on s'intéresse à la norme opérateur $\|P_t - \mu\|$ dans $L^2(\mu)$.

L. Miclo (✉) · P. Monmarché
Institut Mathématiques de Toulouse, 118 route de Narbonne F-31062 Toulouse Cedex 9, France
e-mail: laurent.miclo@math.univ-toulouse.fr; pierre.monmarche@ens-cachan.org

Sous hypothèse de réversibilité, le générateur $\mathscr{L}$ du semi-groupe se diagonalise dans une base orthonormée (ou plus généralement, relativement à une résolution de l'identité formée d'une famille monotone de projections), ce qui permet de voir que

$$\forall\, t \geq 0, \qquad \|P_t - \mu\| = \exp(-\lambda t)\,,$$

où $-\lambda \leq 0$ est la borne supérieure du spectre de $\mathscr{L}_{|\mathbb{1}^\perp}$, la restriction de $\mathscr{L}$ à l'espace orthogonal aux fonctions constantes dans $L^2(\mu)$ (s'il est non nul, λ est appelé le trou spectral de $\mathscr{L}$).

Dans les cas non-réversibles, il peut en être autrement, même si la fonction $\mathbb{R}_+ \ni t \mapsto \|P_t - \mu\|$ est toujours décroissante (il s'agit d'une conséquence de l'inégalité de Jensen). Ainsi dans [7], pour la diffusion constituée du couple d'un processus d'Ornstein-Uhlenbeck linéaire et de son intégrale sur le cercle, la décroissance de $\ln(\|P_t - \mu\|)$ pour $t \geq 0$ petit commence par être d'ordre t^3.

Pour mieux appréhender ce phénomène, on va s'intéresser ici à un modèle très simple, analogue en temps continu de la marche persistante d'ordre 2 de [4] : une particule se déplaçant à vitesse constante sur un cercle et faisant brusquement volte-face à taux constant. Autrement dit, on considère $(Y_t)_{t \geq 0}$ un processus sur $\{-1, 1\}$ qui change de signe avec un taux exponentiel $a > 0$, et on pose pour tout $t \geq 0$, $X_t := \int_0^t Y(s)ds$ sur $\mathbb{T} = \mathbb{R}/2\pi\mathbb{Z}$, de sorte que (X_t, Y_t) représente le couple position-vitesse de la particule au temps $t \geq 0$. Ce modèle est cité comme exemple simple d'hypocoercivité dans [5]. Le processus $(X_t, Y_t)_{t \geq 0}$ est caractérisé par son générateur infinitésimal, qui agit sur des fonctions tests convenables f par

$$\forall\, (x, y) \in \mathbb{T} \times \{-1, 1\}, \qquad \mathscr{L}_a f(x, y) := y \partial_x f(x, y) + a\left(f(x, -y) - f(x, y)\right)$$

ou par le semi-groupe $(P_t^a)_{t \geq 0}$ qu'il engendre sur $L^2(\mu)$: pour tout $f \in L^2(\mu)$,

$$\forall\, t \geq 0,\, \forall\, (x, y) \in \mathbb{T} \times \{-1, 1\}, \quad P_t^a f(x, y) := \mathbb{E}\left(f(X_t, Y_t) | X_0 = x, Y_0 = y\right)\,.$$

La mesure invariante μ correspondante est la loi uniforme sur $\mathbb{T} \times \{-1, 1\}$. Il est connu que P_t (pour alléger les notations, le paramètre $a > 0$ sera souvent sous-entendu) converge fortement dans $L^2(\mu)$ vers μ et que la vitesse finit par être exponentielle (voir la section 1.4 de [5], bien que le taux optimal n'y soit pas obtenu). Comme ce serait le cas pour des opérateurs de dimension finie, on suspecte que

$$\lim_{t \to +\infty} \frac{1}{t} \log \|P_t - \mu\| = -\lambda \qquad (1)$$

avec

$$\lambda := \inf\{-\mathfrak{R}(\theta),\ \theta \text{ valeur propre de } \mathscr{L}_{|\mathbb{1}^\perp}\} \qquad (2)$$

Étude spectrale minutieuse de processus moins indécis que les autres

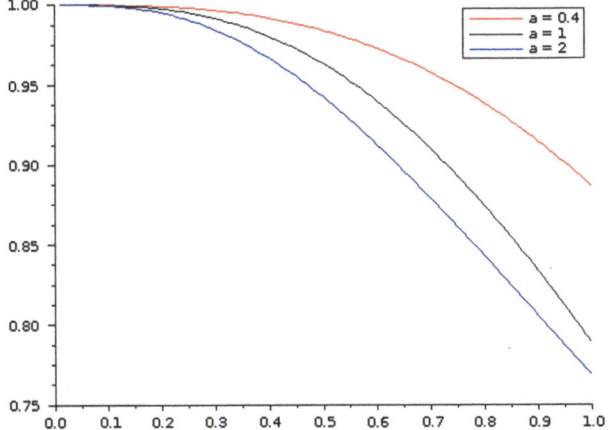

Fig. 1 Norme du semi-groupe pour différentes valeurs de a au cours du temps (ici $t \in [0, 1]$). Au début la décroissance est d'autant plus rapide que a est grand

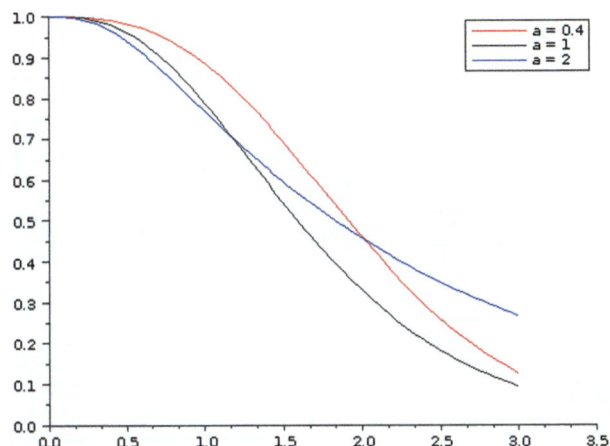

Fig. 2 Cependant la tendance finit par s'inverser (ici $t \in [0, 3]$)

On va vérifier que ceci est juste, mais on cherche surtout des résultats plus quantitatifs, en estimant précisément la norme $\|P_t - \mu\|$ en tout temps $t \geq 0$, car en pratique des renseignements asymptotiques tels que (1) ne sont pas très exploitables. Voilà l'essentiel des résultats obtenus (illustrés par les figures 1, 2 et 3) sur ce modèle :

Théorème 1. *Pour $a \geq 1$, on a $\lambda = a - \sqrt{a^2 - 1}$ et pour $a \leq 1$, $\lambda = a$. Plus précisément, pour tout $t > 0$,*

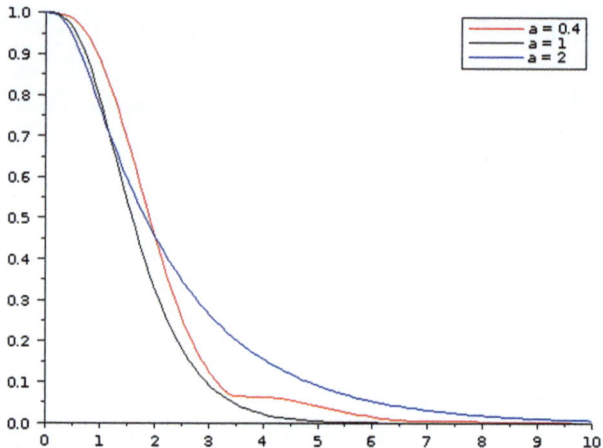

Fig. 3 La meilleure vitesse asymptotique est obtenue pour $a = 1$. Pour $a < 1$ il arrive que la dérivée de la norme s'annule presque (ici $t \in [0, 5]$)

- Si $a > 1$ alors, en notant $\omega = \sqrt{a^2 - 1}$ et $\gamma = e^{-2\omega t}$,

$$\|P_t - \mu\| = e^{(-a+\sqrt{a^2-1})t}\sqrt{1 + \frac{2}{\omega^2\left(\frac{1+\gamma}{1-\gamma}\right) + a\sqrt{1 + \omega^2\left(\frac{1+\gamma}{1-\gamma}\right)^2} - 1}}$$

$$= 1 - \frac{t^3}{3} + \underset{t \to 0}{o}(t^3)$$

$$\underset{t \to +\infty}{\sim} \frac{a^2}{a^2 - 1} e^{\lambda t}.$$

- Si $a = 1$ alors

$$\|P_t - \mu\| = e^{-t}\sqrt{1 + \frac{2}{\sqrt{1 + \frac{1}{t^2}} - 1}}$$

$$= 1 - \frac{t^3}{3} + \underset{t \to 0}{o}(t^3)$$

$$\underset{t \to +\infty}{\sim} 2te^{-t}.$$

- Si $a < 1$ alors

$$\|P_t - \mu\| = e^{-at}\sqrt{g(t)}$$

$$= 1 - \frac{at^3}{3} + \underset{t \to 0}{o}(t^3),$$

avec g telle que

$$\limsup_{t\to+\infty} g(t) = \frac{1+a}{1-a}$$

$$\liminf_{t\to+\infty} g(t) = 1$$

et, en notant $v = 2\sqrt{1-a^2}$, *si* $t \in \left[0, \frac{\pi}{v}\right]$ *alors*

$$g(t) = \left(1 + \frac{2}{\sqrt{\frac{v^2}{a^2}\frac{1}{2(1-\cos(vt))}+1}-1}\right).$$

Comme dans [7], on observe une décroissance initiale en t^3. Dans ce contexte non-réversible, la norme opérateur $\|P_t - \mu\|$ se comporte donc différemment du rayon spectral de $P_t - \mu$, qui n'est autre que $\exp(-\lambda t)$, avec λ défini en (2). Comme nous l'a fait remarquer le referee, ceci traduit aussi l'aspect anormal des opérateurs P_t, pour $t > 0$. Par ailleurs, le choix optimal de a (au sens du meilleur taux asymptotique de convergence exponentielle) correspond à $a = 1$ et voit le facteur pré-exponentiel exploser linéairement en temps grand.

Le processus $(X_t, Y_t)_{t\geq 0}$ précédent est un exemple de processus de Markov déterministe par morceaux, famille de plus en plus étudiée dans la littérature, notamment pour ce qui concerne les processus de type TCP (voir par exemple les articles [2, 1] et les références qu'ils contiennent). Actuellement les méthodes de couplage semblent les plus efficaces pour étudier leur convergence, au sens de la distance de Wasserstein ou de la variation totale. Pourtant nous nous demandons si l'un au moins de ces processus, la version du TCP à taux de saut constant, ne pourrait pas être étudié par le biais d'une variante de l'approche spectrale que nous allons suivre dans ce papier. En effet, il s'agit du processus sur $\mathbb{R}_+$ dont le générateur $\mathscr{L}$ agit sur des fonctions tests f par

$$\forall\, x \in \mathbb{R}_+, \qquad \mathscr{L}f(x) := f'(x) + l(f(rx) - f(x)),$$

où $l > 0$ et $r \in (0, 1)$ sont des constantes. Même si la probabilité invariante associée μ est difficile à décrire explicitement, ses moments se calculent immédiatement (en faisant agir $\mathscr{L}$ sur les monômes). La diagonalisation de $\mathscr{L}$ est facile à obtenir, car les vecteurs propres sont des polynômes. On en déduit également une formule pour leurs produits scalaires. On dispose donc de toute l'information spectrale nécessaire théoriquement pour calculer les normes opérateurs. Malheureusement nous n'avons toujours pas réussi à mener à bien les calculs. Une autre caractéristique spectrale curieuse de $\mathscr{L}$ est que bien que son spectre soit formé de valeurs propres de multiplicité 1 et bornées par l, $\mathscr{L}$ n'est pas borné en tant qu'opérateur dans $L^2(\mu)$, du fait de sa composante différentielle.

Le théorème 1 sera démontré au cours de la partie 2. La partie 3 s'attache au lien entre le modèle discret de la marche persistante et son analogue continu du volte-face. Lorsque la fréquence de changement de vitesse devient grande ce processus continu tend vers le mouvement brownien, ce qui est étudié en partie 4. La partie 5 quant à elle discute des généralisations de ces premiers résultats à des potentiels quelconques et à la dimension supérieure. Enfin, l'appendice regroupe quelques lemmes techniques utilisés dans le reste du texte.

2 Calcul exact de la norme

Remarquons une fois encore que si le processus était réversible, le travail serait simple puisque $\mathscr{L}_a$ serait diagonalisable en base orthonormée (dans $L^2(\mu)$). Ce n'est pas le cas ici mais on va tout de même pouvoir décomposer l'espace en plans stables orthogonaux ce qui nous ramènera à calculer des normes d'opérateurs en dimension 2, qu'il faudra ensuite comparer entre elles.

Lemme 1. *Les plans $V_n = \{f : (x, y) \mapsto e^{inx}g(y), g \in \mathbb{C}^{\{-1,1\}}\}$, pour $n \in \mathbb{Z}$, sont invariants par $\mathscr{L}_a$, orthogonaux et totaux dans $L^2(\mu)$. L'action de P_t^a sur V_n est donnée par $e^{tK_n^{(a)}}$, où pour toute fonction test g,*

$$\forall\, y \in \{\pm 1\}, \qquad K_n^{(a)}g(y) := inyg(y) + a(g(-y) - g(y))$$

(à l'instar du générateur et du semi-groupe, le paramètre a sera généralement omis par la suite).

Preuve. L'orthogonalité et le caractère total découlent directement de ceux de $(x \mapsto e^{inx})_{n \in \mathbb{N}}$ dans $L^2(\mathbb{T})$. On s'assure ensuite directement que pour $f(x, y) = e^{inx}g(y)$ on a bel et bien $\mathscr{L}f(x,y) = e^{inx}K_n g(y)$.

On est donc ramené à calculer la norme d'une matrice 2×2. Notons

$$R(t, a, n) \stackrel{def}{=} \|P_t^a - \mu\|_{V_n}^2.$$

Notons que pour tout $n \neq 0$ on a $V_n \subset Ker(\mu)$. Le cas $n = 0$ est un peu à part et facile à régler : K_0 est diagonalisable avec deux valeurs propres, 0 (associées aux constantes, que l'on retranche ici) et $-2a$. Ainsi

$$R(t, a, 0) = e^{-4at}.$$

Cette restriction ne réalisera en fait jamais la norme globale (sauf $t = 0$ bien sûr) : en effet on va voir que, quelque soit a, $\mathscr{L}$ possède des valeurs propres de parties réelles $-a$; ainsi sur une droite propre pour une telle valeur propre $\|P_t\| = e^{-at} > e^{-2at}$. D'autre part $K_n = \bar{K}_{-n}$ et on se restreindra donc dans la suite à $n > 0$. Finalement,

$$\|P_t - \mu\| = \sup_{n \geq 1}(\|P_t\|_{V_n}) = \sup_{n \geq 1}\left(\sqrt{R(t,a,n)}\right).$$

Étude spectrale minutieuse de processus moins indécis que les autres

Calcul des normes des restrictions

Lemme 2. *Si $a > n$ alors pour tout $t > 0$*

$$R(t,a,n) = e^{-2(a-\sqrt{a^2-n^2})t} \times \left(1 + \frac{2}{\omega^2\left(\frac{1+\gamma}{1-\gamma}\right) + \frac{a}{n}\sqrt{1+\omega^2\left(\frac{1+\gamma}{1-\gamma}\right)^2} - 1}\right),$$

avec $\omega = \sqrt{\left(\frac{a}{n}\right)^2 - 1}$ *et* $\gamma = e^{-2\sqrt{a^2-n^2}t}$.

Preuve. Les deux valeurs propres de K_n, réelles, sont $\lambda_1 = -a + n\omega > \lambda_2 = -a - n\omega$. On calcule que (e_1, e_2) sont des vecteurs propres correspondants unitaires ils vérifient $|<e_1,e_2>| = \frac{n}{a}$ (les vecteurs propres sont « d'autant plus orthogonaux » que a est loin de n), on peut donc choisir (e_1, e_2) unitaires tels que $<e_1,e_2> = \frac{n}{a}$. En posant $u = re^{i\theta}e_1 + e_2$ on a ainsi

$$e^{tK_n}u = re^{i\theta}e^{\lambda_1 t}e_1 + e^{\lambda_2 t}e_2$$

$$\|u\|^2 = r^2 + 1 + 2r\frac{n}{a}\cos(\theta)$$

$$\|e^{tK_n - \lambda_1 t}u\|^2 = r^2 + \gamma^2 + 2r\gamma\frac{n}{a}\cos(\theta)$$

$$= \|u\|^2 + (\gamma - 1) \times \left[\gamma + 1 + 2r\frac{n}{a}\cos(\theta)\right].$$

En conséquence

$$\frac{\|e^{tK_n - \lambda_1 t}u\|^2}{\|u\|^2} = \frac{r^2 + \gamma^2 + 2r\gamma\frac{n}{a}\cos(\theta)}{r^2 + 1 + 2r\frac{n}{a}\cos(\theta)}$$

$$= \gamma + \frac{r^2 + \gamma^2 - \gamma r^2 - \gamma}{r^2 + 1 + 2r\frac{n}{a}\cos(\theta)},$$

quantité qui, à r fixé, est monotone en $\cos(\theta)$. Les valeurs extrémales sont donc obtenues avec $\cos(\theta) = 1$ (quitte à prendre $r < 0$). On a alors

$$\frac{\|e^{tK_n - \lambda_1 t}u\|^2}{\|u\|^2} = 1 + (\gamma - 1) \times \frac{\gamma + 1 + 2r\frac{n}{a}}{r^2 + 1 + 2r\frac{n}{a}}$$

$$= 1 - 2\frac{n}{a}(1 - \gamma) \times \frac{(r + \frac{n}{a}) - \frac{n}{a} + \frac{a}{2n}(1+\gamma)}{(r + \frac{n}{a})^2 + 1 - \left(\frac{n}{a}\right)^2}.$$

D'après le lemme 11, les valeurs extrêmales sont

$$\frac{\|e^{tK_n - \lambda_1 t} u\|^2}{\|u\|^2} = 1 - \frac{\left(\frac{n}{a}\right)^2 (1 - \gamma)}{\left(\frac{n}{a}\right)^2 - \left(\frac{1+\gamma}{2}\right) \pm \sqrt{\left(\frac{1+\gamma}{2}\right)^2 - \gamma \left(\frac{n}{a}\right)^2}}.$$

Le maximum est obtenu pour $\pm = -$, et l'on obtient

$$\|e^{tK_n - \lambda_1 t}\|^2 = 1 + \frac{\left(\frac{n}{a}\right)^2 (1 - \gamma)}{\left(\frac{1+\gamma}{2}\right) - \left(\frac{n}{a}\right)^2 + \sqrt{\left(\frac{1+\gamma}{2}\right)^2 - \gamma \left(\frac{n}{a}\right)^2}}$$

$$= 1 + \frac{2}{\omega^2 \left(\frac{1+\gamma}{1-\gamma}\right) + \frac{a}{n} \sqrt{1 + \omega^2 \left(\frac{1+\gamma}{1-\gamma}\right)^2} - 1}.$$

Lemme 3. *Si $a < n$ alors pour tout $t > 0$*

$$R(t, a, n) = e^{-2at} \times \left(1 + \frac{2}{\sqrt{\frac{v_n^2}{a^2} \frac{1}{2(1 - \cos(v_n t))} + 1} - 1}\right),$$

avec $v_n = 2\sqrt{n^2 - a^2}$.

Preuve. Dans ce cas les valeurs propres de K_n sont complexes conjuguées, $\lambda_1 = \bar{\lambda}_2 = \lambda = -a + i\sqrt{n^2 - a^2}$, de partie réelle a. On trouve des vecteurs propres normés associés e_1 et e_2 vérifiant $<e_1, e_2> = \frac{a}{n}$ (là encore le produit scalaire des vecteurs propres tend vers 0 à mesure que a et n s'éloignent).

Posons $u = e_1 + re^{i\theta} e_2$ avec $r \in \mathbb{R}$ et $\theta \in]-\pi, \pi]$. On a alors $e^{tK_n} u = e^{\lambda t} \left(e_1 + re^{i\theta} e^{-2i\sqrt{n^2 - a^2}} e_2\right)$, et ainsi

$$\|u\|^2 = r^2 + 1 + 2r \frac{a}{n} \cos(\theta)$$

$$\|e^{tK_n - t\lambda} u\|^2 = r^2 + 1 + 2r \frac{a}{n} \cos(\theta - 2t\sqrt{n^2 - a^2}).$$

Par le lemme 11 on obtient que le rapport entre les deux est extrémal pour $r = \pm 1$, on est donc ramené à

$$\|e^{tK_n - t\lambda}\|^2 = \sup_{\theta \in \mathbb{T}} \frac{\alpha_n + \cos(\theta - v_n t)}{\alpha_n + \cos(\theta)},$$

avec $\alpha_n = \frac{n}{a} > 1$. Le lemme 12 de l'appendice conclut.

Lemme 4. *Si $a = n$ alors pour tout $t > 0$*

$$R(t,a,n) = e^{-2at} \times \left(1 + \frac{2}{\sqrt{1 + \frac{1}{n^2 t^2}} - 1}\right).$$

Preuve. Dans ce cas $-n$ est valeur propre double de K_n. Considérons la base $g_1(y) = 1 + iy$ et $g_2(y) = \frac{1}{n}$ de $\mathbb{C}^{\{-1,1\}}$. La matrice de K_n dans cette base est alors un bloc de Jordan, d'exponentielle $e^{-nt} \begin{pmatrix} 1 & t \\ 0 & 1 \end{pmatrix}$. En renormalisant g_1 et g_2, on obtient des vecteurs de base unitaires e_1 et e_2 avec $<e_1, e_2> = \frac{1}{\sqrt{2}}$, $e^{tK_n} e_1 = e^{-nt} e_1$ et $e^{tK_n} e_2 = e^{-nt}(e_2 + \sqrt{2}nt e_1)$. En posant $u = (x + iy)e_1 + e_2$, on a ainsi

$$e^{tK_n} u = e^{-nt}(u + \sqrt{2}nt e_1)$$
$$\|u\|^2 = x^2 + y^2 + 1 + \sqrt{2}x$$
$$\|e^{tK_n + nt} u\|^2 = \|u\|^2 + 2n^2 t^2 + 2\sqrt{2}nt\left(x + \frac{1}{\sqrt{2}}\right).$$

Le rapport $\frac{\|e^{tK_n + nt} u\|^2}{\|u\|^2}$ est donc optimal pour $y = 0$. Reste à choisir x.

$$\frac{\|e^{tK_n + nt} u\|^2}{\|u\|^2} = 1 + 2\sqrt{2}nt \times \frac{x + \frac{1}{\sqrt{2}} + \frac{nt}{\sqrt{2}}}{(x + \frac{1}{\sqrt{2}})^2 + \frac{1}{2}}.$$

D'après le lemme 11, les valeurs extrêmales sont

$$\frac{\|e^{tK_n + nt} u\|^2}{\|u\|^2} = 1 + \sqrt{2}nt \times \frac{1}{-\frac{nt}{\sqrt{2}} \pm \sqrt{\frac{n^2 t^2}{2} + \frac{1}{2}}}$$

et le maximum est obtenu pour $\pm = +$, ce qui donne le résultat escompté.

Remarquons qu'on aurait pu obtenir ce résultat par continuité à partir des cas $a \lessgtr n$.

Comparaison des $R(t, a, n)$

Il s'agit maintenant de comparer les normes de ces restrictions entre elles. Un développement limité en $t = 0$ montre que $R(t, a, n) = 1 - \frac{n^3}{3} t^3 + o(t^3)$ pour $a \geq n$ et $R(t, a, n) = 1 - \frac{an^2}{3} t^3 + o(t^3)$ pour $a \leq n$, ce qui laisse penser qu'au moins au début $R(t, a, 1)$ prévaut (autrement dit que l'erreur décroît lentement sur V_1 les fonctions de grande longueur d'onde en x). D'autre part, si $a > 1$, c'est aussi sur V_1

que se trouve la droite propre associée à la valeur propre de $\mathscr{L}$ de plus grande partie réelle, c'est donc également $R(t, a, 1)$ qui devrait prévaloir asymptotiquement. En fait nous allons voir que, pour l'essentiel, seule compte cette norme sur V_1. Notons que les expressions calculés pour $R(t, a, n)$ permettent d'étendre leur définition à n non entier et qu'alors $n \in]0, +\infty[\mapsto R(t, a, n)$ est continue.

Dans un premier temps, on peut dériver $R(t, a, n)$ pour $n \in]0, a[$. Le lemme 14 de l'annexe montre que cette dérivée est négative et ainsi $\max_{1 \leq n < a} R(t, a, n) = R(t, a, 1)$ pour tout $t > 0$. Par continuité on a même $\max_{1 \leq n \leq a} R(t, a, n) = R(t, a, 1)$. Ainsi a-t-on réglé les cas $a \geq 1$ du théorème 1, puisqu'alors $\|P_t - \mu\| = \max_{n \in \mathbb{Z}^*} R(t, a, n) = R(t, a, 1)$.

Le cas des $n > a$ est un peu plus délicat, pour qui

$$R(t, a, n) = e^{-ta} \sqrt{g_n(t)}$$

avec, si $v_n = 2\sqrt{n^2 - a^2}$,

$$g_n(t) = 1 + \frac{2}{\sqrt{\frac{v_n^2}{a^2} \frac{1}{2(1-\cos(v_n t))} + 1} - 1},$$

qui est $2\pi/v_n$ périodique. Calculer le supremum des g_n pour tout t est à peu près impossible du fait des périodes incommensurables (cf. figure 4). Cependant on peut penser (d'après le développement limité en 0) qu'en temps petit la norme prépondérante correspond à n minimal et qu'elle le reste jusqu'à ce que g_n atteigne son maximum. C'est effectivement le cas, comme on va le montrer dans un instant. Ensuite le suprémum des g_k oscillera entre ce maximum et 1.

Lemme 5. *Si $k < n$ alors pour tout $t \in \left[0, \frac{\pi}{v_k}\right]$ on a $g_k(t) \geq g_n(t)$.*

Preuve.

$$g_n(t) \leq g_k(t) \Leftrightarrow 1 + \frac{2}{\sqrt{\frac{v_n^2}{a^2} \frac{1}{2(1-\cos(v_n t))} + 1} - 1} \leq 1 + \frac{2}{\sqrt{\frac{v_k^2}{a^2} \frac{1}{2(1-\cos(v_k t))} + 1} - 1}$$

$$\Leftrightarrow \frac{1 - \cos(v_n t)}{v_n^2} \leq \frac{1 - \cos(v_k t)}{v_k^2}.$$

Ces deux termes sont égaux et de dérivées égales en $t = 0$, pour les comparer il suffit donc de comparer leurs dérivées secondes. Or, si $v_n \geq v_k$ alors $\cos(v_n t) \leq \cos(v_k t)$ pour $t \in \left[0, \frac{\pi}{v_n}\right]$, et donc $g_n(t) \leq g_k(t)$ pour ces t. Puisque g_k est croissante sur $\left[0, \frac{\pi}{v_k}\right]$ on a pour $t \in \left[\frac{\pi}{v_n}, \frac{\pi}{v_k}\right]$

Fig. 4 Plus l est grand plus l'amplitude et la longueur d'onde de g_l sont faibles

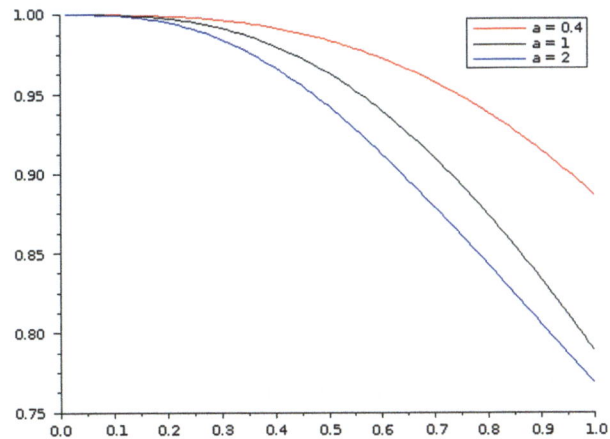

$$g_k(t) \geq g_k(\frac{\pi}{v_n}) \geq g_n(\frac{\pi}{v_n}) \geq g_n(t) \ .$$

On achève en constatant que v_n est croissante en n.

Lemme 6. *Si $n > a$ alors pour tout $t > 0$ on a $R(t, a, n) \leq R(t, a, a)$.*

Preuve. D'après le lemme précédent, pour tout $\varepsilon > 0$ on a $R(t, a, n) \leq R(t, a, a + \varepsilon)$ pour $t \leq \frac{\pi}{v_{a+\varepsilon}}$; or $v_{a+\varepsilon} \underset{\varepsilon \to 0}{\longrightarrow} 0$ et la continuité de R conclut.

En particulier si $a \geq 1$ pour tout t on aura $||P_t - \mu|| = R(t, a, 1)$, ce qui démontre les deux tiers du théorème 1. Pour $a < 1$ on peut comparer plus finement les g_n :

Lemme 7. *Soit $g(t) = \sup_{n \in \mathbb{N}} g_n(t)$. Si $t \leq \frac{\pi}{v_1}$ alors $g(t) = g_1(t)$, et d'autre part*

$$\limsup_{t \to +\infty} g(t) = \frac{1 + a}{1 - a} \quad (= \sup g)$$

$$\liminf_{t \to +\infty} g(t) = 1 \quad (= \inf g) \ .$$

Preuve. La première assertion a déjà été démontrée, et le résultat pour la limite supérieure découle directement de la périodicité de g_1. Pour la limite inf, considérons $\varepsilon > 0$, et soit $N \in \mathbb{N}$ tel que $\frac{1 + \frac{a}{N}}{1 - \frac{a}{N}} \leq 1 + \varepsilon$. On a ainsi, pour tout $k \geq N$ et pour tout $t > 0$, $g_k(t) \leq 1 + \varepsilon$. On cherche ensuite un temps où les fonctions restantes (en nombre fini) sont simultanément proches de leur minimum. Fixons $\delta > 0$ tel que pour tout $n < N$ et tout $k \in \mathbb{Z}$, on ait

$$|t - \frac{2k\pi}{v_n}| \leq \delta \Rightarrow g_n(t) \leq 1 + \varepsilon.$$

Le lemme 13 de l'appendice nous fournit $t \geq 1$ et des entiers $k_1, \ldots, k_{N-1} \in \mathbb{N}$ tels que $|\frac{2\pi}{v_n} k_n - t| < \delta$ pour tout $n < N$; on obtient que $g_n(t) \leq 1 + \varepsilon$ pour tout $n < N$, et donc pour tout $n \in \mathbb{N}$. Soit ε_0 le minimum sur $[1/2, t+1]$ de $g - 1$ (fonction continue). Si $\varepsilon_0 = 0$ alors g est périodique et son minimum est sa limite inférieure. Sinon on peut recommencer l'argument ci-dessus pour obtenir un temps $t_2 \geq 1$ tel que pour tout $n \in \mathbb{N}$ on ait $g_n(t_2) \leq 1 + \varepsilon_0/2$, donc nécessairement $t_2 > t + 1$; finalement en itérant le procédé on peut trouver des temps arbitrairement grand où g est arbitrairement proche de 1, ce qui conclut.

Ce lemme finit de démontrer le théorème 1.

3 Du discret au continu

L'étude du volte-face a initialement été motivée par celle de la marche considérée dans [4] : Y_n est une chaîne de Markov sur $\{-1, +1\}$ qui change de signe avec probabilité $(1-\alpha)/2$, et $X_{n+1}^N = X_n^N + Y_n$ dans $\mathbb{Z}/N\mathbb{Z} =: \mathbb{Z}_N$, avec $N \in \mathbb{N} \setminus \{0, 1\}$. Ainsi pour son $n^{\text{ième}}$ saut la particule (dont la position est X_n^N) persiste dans le même sens qu'au coup précédent avec une probabilité supérieure à $1/2$, c'est bien l'analogue discret du processus continu des sections précédentes. Notons que la chaîne $(X_n^N)_{n \in \mathbb{N}}$ est markovienne d'ordre 2.

Pour peu que N soit impair la chaîne est irréductible apériodique et converge donc en loi vers son unique probabilité invariante μ_N, qui est la mesure uniforme sur $\mathbb{Z}_N \times \{\pm 1\}$. L'opérateur $M_\alpha f(x, y) = \mathbb{E}(f(X_1, Y_1) | X_0 = x, Y_0 = y)$ associé agit sur les fonctions de $L^2(\mu_N)$ et la norme d'opérateur $\|M_\alpha^n - \mu_N\|_{L^2(\mu_N)} \xrightarrow[n \to +\infty]{} 0$ (en voyant à nouveau μ_N comme l'opérateur $f \mapsto (\int f \, d\mu_N)\mathbb{1}$). On a même

$$\lim_{n \to +\infty} \frac{1}{n} \log\left(\|M_\alpha^n - \mu_N\|\right) = \log(\lambda_\alpha)$$

où, en notant $\sigma(M_\alpha)$ le spectre de M_α, $\lambda_\alpha = \sup(|\sigma(M_\alpha) \setminus \{1\}|)$. Ce taux exponentiel de convergence $\log(\lambda_\alpha)$ est de valeur absolue maximale (et donc de vitesse asymptotique la meilleure) pour $\alpha_{opt} = \frac{1 - \sin(\pi/N)}{1 + \sin(\pi/N)}$, pour lequel $\lambda_{opt} = \sqrt{\alpha_{opt}}$ (cf. [4]). En comparaison, pour la marche isotrope ($\alpha = 0$), on a $\lambda_0 = \cos(\pi/N)$. On a donc amélioré la convergence en temps long car

$$\cos(\pi/N) = \sqrt{(1 - \sin(\pi/N))(1 + \sin(\pi/N))} \geq \sqrt{\frac{1 - \sin(\pi/N)}{1 + \sin(\pi/N)}}.$$

L'étude du volte-face a permis de mieux comprendre l'amorce de convergence en temps petit, et nous pouvons maintenant faire le lien avec la marche discrète. D'abord constatons que des calculs identiques aux précédents nous permettent de calculer la norme de M. Pour $k \in [\![1, N]\!]$ on notera $e^{2ik\pi/N} = C_k + iS_k$, $\alpha_l = \frac{1 - |S_l|}{1 + |S_l|}$, $C_0^2 = \frac{4\alpha}{(1+\alpha)^2}$ et $S_0^2 = \left(\frac{1-\alpha}{1+\alpha}\right)^2$.

Lemme 8. *Les plans $W_k = \{(x,y) \mapsto e^{2ik\pi x/N} g(y), g \in \mathbb{C}^{\pm 1}\}$ sont stables par M. Notons $R_N(n,\alpha,k) \stackrel{\text{def}}{=} \|M_\alpha^n - \mu\|_{W_k}^2$.*

– *si $\alpha < \alpha_k$ alors*

$$R_N(n,\alpha,k) = \lambda_+^{2n} \times \left(1 + \frac{2}{\omega^2 \left(\frac{1+\gamma}{1-\gamma}\right) + \frac{S_0}{S_k}\sqrt{1 + \omega^2 \left(\frac{1+\gamma}{1-\gamma}\right)^2 - 1}}\right),$$

avec $\lambda_\pm = \sqrt{\alpha}\left(\frac{|C_k|}{C_0} \pm \sqrt{\left(\frac{C_k}{C_0}\right)^2 - 1}\right)$, $\gamma = \left(\frac{\lambda_-}{\lambda_+}\right)^n$ *et* $\omega^2 = \left(\frac{S_0}{S_k}\right)^2 - 1$.

– *si $\alpha > \alpha_k$ alors*

$$R_N(n,\alpha,k) = \alpha^n \times \left(1 + \frac{2}{\sqrt{2\frac{\left(\frac{S_k}{S_0}\right)^2 - 1}{1-\cos(2n\psi)} + 1} - 1}\right),$$

où $\tan \psi = \sqrt{\left(\frac{C_0}{C_k}\right)^2 - 1}$.

– *si enfin $\alpha = \alpha_k$ alors*

$$R_N(n,\alpha,k) = \alpha^n \times \left(1 + \frac{2}{\sqrt{1 + \frac{C_0^2}{S_0^2 n^2}} - 1}\right).$$

Preuve. La démarche et les calculs sont quasiment les mêmes que dans le cas continu et n'amènent aucune difficulté nouvelle.

Lorsqu'on veut passer du modèle discret au continu, plutôt que $X_n^N \in \mathbb{Z}_N$ il vaut mieux regarder $U_t^N = \frac{2\pi}{N} X_n^N \in \mathbb{T}$ si $t = n\frac{2\pi}{N}$ que l'on prolonge de façon affine à $t \geq 0$ et $V_t^N = Y_n$ si $t \in \frac{2\pi}{N}[n, n+1[$. Si la probabilité de changer de sens $\frac{1-\alpha_N}{2}$ est de l'ordre de $\frac{1}{N}$, la convergence des temps entre deux changements vers une loi exponentielle donne la convergence en loi de (U^N, V^N) vers le processus continu. Remarquons que pour $u = \frac{2\pi}{N} x$ on peut réécrire $e^{i\frac{2k\pi}{N} x} = e^{iku}$, l'espace V_k correspond donc à W_k :

Lemme 9. *Pour tout $t > 0$ et $k \in \mathbb{Z}$, si $\alpha^{(N)} \in [0,1]$ est tel que $\frac{N}{2\pi} \times \frac{1-\alpha^{(N)}}{2} \xrightarrow[N \to +\infty]{} a$ alors*

$$R_N\left(\left\lfloor \frac{Nt}{2\pi} \right\rfloor, \alpha^{(N)}, k\right) \xrightarrow[N \to +\infty]{} R(t,a,k).$$

Preuve. On le vérifie sans difficulté particulière sur les expressions analytique données dans le lemme 8 et la partie 2.

Cependant, contrairement au cas continu, dans la marche discrète la plus grande valeur propre (associée au $|\cos(\frac{2k\pi}{N})|$ maximal) ne correspond pas à $k = 1$ mais à $k = \pm \lfloor \frac{N}{2} \rfloor$. Pour avoir la convergence des normes globales d'opérateurs il faut ignorer les deux plans $W_{\pm \lfloor \frac{N}{2} \rfloor}$. En un sens le caractère fini des positions prises par la particule entraîne l'existence d'observables qui convergent mal, ce qui disparaît à la limite des processus, mais pas dans le passage à la limite des normes.

Lemme 10. *Pour tout $t > 0$ et $k \in \mathbb{Z}$, si $\alpha^{(N)} \in [0,1]$ est tel que $\frac{N}{2\pi} \times \frac{1-\alpha^{(N)}}{2} \xrightarrow[N \to +\infty]{} a$ alors*

$$R_N\left(\left\lfloor \frac{Nt}{2\pi} \right\rfloor, \alpha^{(N)}, \left\lfloor \frac{N}{2} \right\rfloor - k\right) \xrightarrow[N \to +\infty]{} R\left(t, a, k + \frac{1}{2}\right).$$

Preuve. Les calculs sont les mêmes que précédemment ; le $1/2$ apparaît avec

$$\sin\left(\frac{2\pi}{N}\left(\left\lfloor \frac{N}{2} \right\rfloor - k\right)\right) = \sin\left(\pi - \frac{2\pi}{N}\left(\left\lfloor \frac{N}{2} \right\rfloor - k\right)\right) = \sin\left(\frac{2\pi}{N}\left(k + \frac{1}{2}\right)\right)$$

Le travail de comparaison des $R(t, a, n)$ englobait déjà les n non-entiers, et en notant pour tout $t \geq 0$,

$$\forall u \in \mathbb{Z}_N, \ \forall v \in \{\pm 1\}, \qquad P_t^N f(u, v) := \mathbb{E}(f(X_n^N, Y_n^N)|X_0^N = u, Y_0^N = v)$$

avec $n = \lfloor Nt/(2\pi) \rfloor$, on obtient *in fine*

Théorème 2. *Si $\frac{N}{2\pi} \times \frac{1-\alpha^{(N)}}{2} \xrightarrow[N \to +\infty]{} a \geq \frac{1}{2}$ alors*

$$\|P_t^N - \mu_N\| \xrightarrow[N \to +\infty]{} R\left(t, a, \frac{1}{2}\right).$$

D'autre part si l'on note $\mathscr{V}_N := \text{Vect}(W_{\lfloor N/2 \rfloor}, W_{-\lfloor N/2 \rfloor})^{\perp}$ et si $a \geq 1$ alors

$$\|P_t^N - \mu_N\|_{\mathscr{V}_N} \xrightarrow[N \to +\infty]{} \|P_t^a - \mu\|.$$

Les convergences sont uniformes en t.

Preuve. Tout est déjà démontré sauf le caractère uniforme en t ; les fonctions en présence étant toutes décroissantes et les limites continues, il découle du théorème de Dini.

Remarque: Notons que, grosso modo, les choses se passent bien également pour $a < \frac{1}{2}$ dans le premier cas et pour $a < 1$ dans le second mais avec de très légères subtilités : par exemple, dans le deuxième cas et pour reprendre les notations de la partie 2, la fonction $g(t)$ limite n'est pas le supremum des $g_n(t)$ pour n entier mais pour n entier ou demi-entier, ce qui peut éventuellement légèrement changer la valeur exacte de la norme lors d'un « creux » de $R(t, a, 1)$.

Un constat particulier sur ce défaut de convergence du discret vers le continu est que si l'on prend pour tout N la probabilité optimale (au sens du trou spectral maximal) de changer de sens dans la marche persistante, alors on converge vers un taux $1/2$ de saut pour Y_t, qui n'est pas optimal pour le processus continu, et qui donne le même taux exponentiel $1/2$ de convergence que le mouvement brownien sur le tore.

Cependant le phénomène de décroissance initiale en t^3, lui, n'est pas affecté par cette subtilité ; c'est normal car son origine n'est pas dans la prise du supremum des normes des restrictions mais, déjà localement, sur chacun des plans W_k. Une interprétation possible est que prendre, au lieu d'un processus réversible, l'intégrale d'un processus réversible retarde initialement l'effet de mélange du hasard ; ou bien que la particule commence par se déplacer de façon déterministe et brouille donc moins bien les pistes qu'une diffusion au moins initialement.

Si pour N grand, on compare (en oubliant le défaut de convergence et les fonctions de $\text{Vect}(W_{\lfloor N/2 \rfloor}, W_{-\lfloor N/2 \rfloor})$) la marche simple et la marche persistante pour $a = 1$ à la limite, pour un nombre n d'itérations fixé, l'écart L^2 à l'équilibre de la marche réversible est environ $1 - \frac{t}{2}$ avec $t = n\left(\frac{2\pi}{N}\right)^2$ (si cette quantité est petite) et celle de la marche persistante est $1 - \frac{t^3}{3}$ avec $t = n\frac{2\pi}{N}$ (si $n \ll N$), qui devient meilleure que la précédente pour $n \approx \sqrt{\frac{3}{4\pi}N}$ (qui assure aussi la validité des asymptotiques précédentes): c'est le nombre d'itérations à partir duquel la marche d'ordre 2 est plus proche de la mesure uniforme que la réversible.

4 Du continu au mouvement brownien

Lorsque $a \to +\infty$, la vitesse du processus continu saute de plus en plus vite de -1 en 1 ; à la limite, les vitesses en deux temps distincts devraient donc être décorrélées. Le processus devrait en conséquence être l'intégrale d'un bruit blanc, autrement dit un mouvement brownien. Avec la bonne renormalisation, c'est effectivement le cas :

Théorème 3. $X^a = (X_{ta})_{t>0}$ *converge en loi vers un mouvement brownien standard sur* $\mathbb{T}$ *quand* $a \to +\infty$.

Preuve. Notons $\tilde{Y}_t = (-1)^{N_t}$ où N_t est un processus de de Poisson de paramètre 1. Ainsi X suit la même loi que $\int_0^{\cdot} \tilde{Y}_{as} ds$

$$X^a(t) \stackrel{\mathscr{L}}{=} \int_0^{ta} \tilde{Y}_{as} ds$$

$$= \frac{1}{a} \int_0^{ta^2} \tilde{Y}_u du,$$

ce qui nous ramène à l'exemple 3 p. 360 de [6] où l'on nous indique la marche à suivre.

Détaillons : on montre d'abord que $M(t) = \tilde{Y}_t + 2\int_0^t \tilde{Y}_u du$ est une martingale. En effet le nombre de changement de signes de $\tilde{Y}_t$ dans une période $t-s$ suit une loi de Poisson de paramètre $t-s$, et ainsi

$$\mathbb{P}(\tilde{Y}_t = \tilde{Y}_s) = \sum_{k \text{ pair}} \frac{(t-s)^k}{k!} e^{-(t-s)} = \cosh(t-s) e^{-(t-s)}$$

$$\mathbb{P}(\tilde{Y}_t = -\tilde{Y}_s) = \sum_{k \text{ impair}} \frac{(t-s)^k}{k!} e^{-(t-s)} = \sinh(t-s) e^{-(t-s)}.$$

Ainsi $\mathbb{E}(\tilde{Y}_t | \mathscr{F}_s) = \tilde{Y}_s e^{-2(t-s)}$ et

$$\mathbb{E}(M(t)|\mathscr{F}_s) = \tilde{Y}_s e^{-2(t-s)} + 2\int_0^s \tilde{Y}_u du + 2\int_s^t \tilde{Y}_s e^{-2(u-s)} du$$

$$= \tilde{Y}_s + 2\int_0^s \tilde{Y}_u du$$

$$= M(s).$$

Si l'on montre la convergence de la martingale $\frac{1}{n} M(n^2 t) = 2 X_t^n + \frac{1}{n} \tilde{Y}_t$ vers le brownien, on aura celle de X^n ; or la première s'obtient de la convergence des crochets. La variation quadratique de $\int_0^s \tilde{Y}_u du$, processus 1-lipschitzien, est nulle, donc

$$<M>_t = \lim_{\delta \to 0}^{\mathbb{P}} \sum_{t_i \in \pi} (\tilde{Y}_{t_{i+1}} - \tilde{Y}_{t_i})^2$$

où la limite en proba a lieu lorsque le pas δ de la partition π de $[0,t]$ tend vers 0. Notons Z_t le nombre de saut de $\tilde{Y}$ sur cet intervalle.

$$\mathbb{P}(\sum_{t_i \in \pi} (\tilde{Y}_{t_{i+1}} - \tilde{Y}_{t_i})^2 \neq 4 Z_t) \leq \mathbb{P}(\text{deux sauts sont distants de moins de } \delta)$$

$$\underset{\delta \to 0}{\to} 0.$$

Ainsi $<M>_t = 4Z_t$, et $<\frac{1}{n}M(n^2\cdot)>_t = \frac{4}{n^2}Z_{n^2t} \xrightarrow[n\to+\infty]{} 4t$ (par la loi des grands nombres), ce qui donne la convergence de $\frac{1}{2}\frac{1}{n}M(n^2t)$ (et donc de X_t^n) vers le mouvement brownien standard (cf [6]).

Qu'en est-il de la norme ? Celle du modèle irréversible converge-t-elle vers celle du brownien ? C'est effectivement le cas. Le générateur du mouvement brownien est $\frac{1}{2}\partial_x^2$, diagonalisable dans la base orthonormée des $x \mapsto e^{inx}$ pour les valeurs propres $-\frac{n^2}{2}$. Rappelons la norme du semi-groupe associé à (X_t, Y_t) sur le plan V_n, quand $a > n$:

$$\|P_t\|_{V_n}^2 = e^{2\lambda_1 t} \left(1 + \frac{2}{\frac{\omega^2}{n^2}\left(\frac{1+\gamma}{1-\gamma}\right) + \frac{a}{n}\sqrt{1 + \frac{\omega^2}{n^2}\left(\frac{1+\gamma}{1-\gamma}\right)^2} - 1}\right),$$

avec $\lambda_1 = -a + \sqrt{a^2 - n^2}$, $\omega = \sqrt{a^2 - n^2}$ et $\gamma = e^{-2t\sqrt{a^2-n^2}}$. On observe que $a\lambda_1 \to -\frac{1}{2}n^2$, $\omega \to +\infty$ et que $\gamma^a \to 0$ quand $a \to +\infty$; Au final, en notant P_t^a le semi-groupe associé à (X_{at}, Y_{at}), on récupère

$$\|P_t^a\|_{V_n} \xrightarrow[a\to+\infty]{} e^{-\frac{1}{2}n^2 t},$$

ce qui est la norme du semi-groupe Q_t associé au mouvement brownien sur la droite $\text{Vect}\{x \mapsto e^{inx}\}$. En particulier la convergence pour $n = 1$ donne la convergence de la norme globale $\|P_t^a - \mu\| \longrightarrow \|Q_t - \lambda\|$.

5 Généralisations

5.1 Avec un potentiel général

En fait le cas précédent, où la mesure invariante pour X_t est la loi uniforme sur le cercle, est immédiatement généralisable à des processus admettant pour loi limite n'importe quelle mesure de la forme $\nu = e^{-V(x)}dx/(2\pi)$, où le potentiel V est supposé normalisé de sorte que $\nu(\mathbb{T}) = 1$. En effet, considérons comme précédemment $Y_t \in \{-1, 1\}$ qui, avec taux a, change de signe. Soit $X_t \in \mathbb{T}$ la solution de

$$dX_t = Y_t e^{V(X_t)} dt . \tag{3}$$

Autrement dit X_t représente la position d'une particule se déplaçant à vitesse (déterministe) inversement proportionnelle à la densité $e^{-V(x)}$ (les zones « peu intéressantes » sont parcourues plus vite) et changeant de sens de parcours selon des temps exponentiels. Montrons qu'alors la mesure invariante pour (X_t, Y_t) est $\mu = \nu \otimes \mathcal{U}_{\{-1,1\}}$, et que la norme 2 du semi-groupe associé se calcule exactement comme précédemment. Le générateur markovien associé au processus est

$$\mathscr{L} f(x,y) = e^{V(x)} y \partial_x f(x,y) + a \left(f(x,-y) - f(x,y) \right),$$

et l'on vérifie

$$\nu \otimes \mathcal{U}_{\{-1,1\}} [\mathscr{L} f(x,y)]$$
$$= \int_{x \in \mathbb{T}} \int_{y=\pm} \left[e^{V(x)} y \partial_x f(x,y) + a (f(x,-y) - f(x,y)) \right] e^{-V(x)} dx dy$$
$$= \int_{y=\pm} y \left(\int_{x \in \mathbb{T}} \partial_x f(x,y) dx \right) dy$$
$$= 0.$$

Considérons pour $n \in \mathbb{N}$, $g_n(x) = \exp\left(in \int_0^x e^{-V(u)} du \right)$ (on a bien $g_n(0) = g_n(2\pi)$ de par la normalisation de V) et des fonctions de la forme $f(x,y) = g_n(x) h(y)$. On a alors

$$\mathscr{L} f(x,y) = g_n(x) [inyh(y) + a(h(-y) - h(y))]$$
$$= g_n(x) K_n^{(a)} h(y), \qquad (4)$$

où $K_n^{(a)}$ a été défini dans le lemme 1 pour le cas uniforme. On parvient donc là encore à décomposer l'espace en plans stables V_n, et ces plans sont à nouveau orthogonaux entre eux dans $L^2(\mu)$:

$$< g_n, g_k >_{L^2(\mu)} = \int_0^{2\pi} \exp\left(i(n-k) \int_0^x e^{-V(u)} du \right) e^{-V(x)} dx$$
$$= \int_0^{2\pi} e^{i(n-k)u} du$$
$$= 2\pi \delta_{nk}.$$

Finalement, si P_t^V est le semi-groupe associé au processus (et P_t est toujours celui associé au potentiel nul), on a exactement

$$\|P_t^V - \mu\|_{L^2(\mu)} = \|P_t - \lambda \otimes \mathcal{U}_{\{-1,1\}}\|_{L^2(\lambda \otimes \mathcal{U}_{\{-1,1\}})}.$$

D'après la section 1, le meilleur taux de convergence asymptotique est donc obtenu en choisissant $a = 1$. Remarquons que lorsque V n'est connu qu'à une constante additive près et que l'on veut garder le bénéfice de l'écriture (3), il faut modifier en conséquence la définition de ν et des g_n, pour $n \in \mathbb{N}$, et on doit remplacer $K_n^{(a)}$ par $Z^{-1}K_n^{(aZ)}$, avec $Z := \int_0^{2\pi} e^{-V(x)}\,dx/(2\pi)$ dans (4). Le choix optimal de a est alors Z^{-1}, qui malheureusement n'est pas connu en pratique.

5.2 Remarque sur les dimensions supérieures

Remarquons que, dans l'optique d'un algorithme de Monte-Carlo non réversible, les résultats s'adaptent à la dimension supérieure. Ainsi en définissant $Y_t^1, \ldots, Y_t^d$ et $X_t^1, \ldots, X_t^d$ comme précédemment, dans le cas où $V(x) = \sum V_i(x_i)$, on construit un semi-groupe P_t^V sur $\mathbb{T}^d$ de mesure invariante μ proportionnelle à $e^{-V(x)}dx \otimes \mathcal{U}_{\{-1,1\}}^{\otimes d}$ et de norme

$$\|P_t^V\|_{L^2(\nu)} = \prod_{i=1}^d \|P_t^{V_i}\|_{L^2(Z_i e^{-V_i(x_i)}dx_i \otimes U_{\{-1,1\}})}$$

où les Z_i, $i \in [\![1, d]\!]$, sont les constantes de normalisation. On aurait pu imaginer un autre processus, construit en gardant l'idée d'une particule dont la vitesse scalaire dépendrait de façon déterministe de la position mais dont la direction changerait aléatoirement à taux constant. Cela donnerait un générateur du type :

$$\mathscr{L}f(x,y) = e^{V(x)}\nabla_x f(x,y).y + a\int_{\mathbb{S}^d} (f(x,z) - f(x,y))\,dz$$

pour des fonctions tests f régulières. Ci-dessus les vitesses sont prises uniformément sur la sphère mais on aurait pu les choisir différemment sans que les remarques à suivre ne s'en trouvent modifiées. La mesure invariante est alors $Ze^{-V(x)}dx \otimes \mathcal{U}_{\mathbb{S}^d}$, avec $Z = Z_1 \cdots Z_d$, ce qui semble bien parti. Néanmoins, à part pour un potentiel nul, on ne va pas pouvoir se ramener à l'étude d'un opérateur sur les vitesses par la même méthode qu'avant, c'est-à-dire en trouvant des fonctions propres de la famille d'opérateurs $K_y : f(x) \mapsto e^{V(x)}\nabla f(x).y$ sous la forme $f(x) = e^{u(x)}$, qui permettaient jusqu'ici de se ramener à des opérateurs n'agissant que sur les vitesses. En effet on a alors

$$K_y f(x) = e^{V(x)} f(x) \nabla u(x).y\ .$$

Il s'agirait donc de trouver une fonction $u : \mathbb{R}^d \to \mathbb{R}$ de différentielle $x \mapsto e^{-V(x)}(c_1 dx_1 + c_2 dx_2 + \cdots + c_d dx_d)$, avec $c_1, \ldots, c_d$ des constantes. Or si $d > 1$, cette 1-forme linéaire n'est pas exacte (n'étant pas fermée), et un tel u ne saurait exister. En comparaison, pour le processus avec des coordonnées indépendantes du début de ce paragraphe, la 1-forme linéaire qui apparaît est $x \mapsto \sum e^{-V_i(x_i)}dx_i$, qui est bel et bien exacte.

6 Appendice

Lemme 11. *Si* $f(R) = \frac{R-a}{R^2+b}$ *avec* $b > 0$, *alors* f *admet ses valeurs extrémales en* $R_\pm = a \pm \sqrt{a^2+b}$, *et ces valeurs sont* $f(R_\pm) = \frac{1}{2R_\pm}$.

Lemme 12. *Si* $g(\theta) = \frac{\alpha + \cos(\theta - s)}{\alpha + \cos(\theta)}$ *avec* $\alpha > 1$, *alors*

$$\max_{\theta \in \mathbb{T}} g(\theta) = 1 + \frac{2}{\sqrt{\frac{2(\alpha^2-1)}{1-\cos(s)} + 1} - 1}.$$

De plus ce maximum est majoré par $\frac{\alpha+1}{\alpha-1}$, *borne atteinte uniquement pour* $s = \pi \, [2\pi]$.

Preuve. Le premier lemme ne présente aucune difficulté. Pour le second, remarquons tout d'abord pour $s = 0 \, [2\pi]$ que g est alors constante égale à 1 et son max l'est également, le lemme est donc vrai dans ce cas. Supposons dans la suite que $1 - \cos(s) \neq 0$. Réécrivons maintenant

$$g(\theta) = \frac{\alpha + \cos(\theta + s)}{\alpha + \cos(\theta)} = \cos(s) + \frac{\alpha(1 - \cos(s)) - \sin(\theta)\sin(s)}{\alpha + \cos(\theta)},$$

$g(\theta)$ étant continue périodique il suffit de déterminer ses points critiques. Or $g'(\theta) = 0$ équivaut à

$$0 = -\cos(\theta)\sin(s)(\alpha + \cos(\theta)) + \sin(\theta)(\alpha(1 - \cos(s)) - \sin(\theta)\sin(s))$$
$$= -\sin(s) + \sin(\theta)(\alpha(1 - \cos(s)) - \alpha\cos(\theta)\sin(s),$$

équation affine dont les solutions sont

$$\begin{pmatrix} \cos(\theta) \\ \sin(\theta) \end{pmatrix} = \begin{pmatrix} -\beta \\ 0 \end{pmatrix} + t \begin{pmatrix} 1 - \cos(s) \\ \sin(s) \end{pmatrix},$$

pour $t \in \mathbb{R}$ et où l'on note $\beta = \frac{1}{\alpha}$. La condition $\cos^2 + \sin^2 = 1$ équivaut à

$$t^2 - \beta t + \frac{\beta^2 - 1}{1 - \cos(s)},$$

qui admet nécessairement deux solutions réelles puisque g est périodique non constante donc possède au moins deux points critiques. Ces solutions sont données par

$$t(1 - \cos(s)) = \frac{1}{2}\beta(1 - \cos(s)) + \frac{1}{2}\varepsilon\sqrt{2(1 - \cos(s)) - \beta^2\sin^2(s)},$$

où $\varepsilon = \pm 1$. On obtient ainsi les valeurs extrêmales de g :

$$g(\theta_\varepsilon) = \cos(s) + \frac{\alpha(1-\cos(s)) - t\sin^2(s)}{\alpha - \beta + t(1-\cos(s))}$$

$$= \frac{\alpha - t\sin^2(s) - \beta\cos(s) + t\cos(s) - t\cos^2(s)}{\alpha - \beta + t(1-\cos(s))}$$

$$= \frac{\alpha - \beta\cos(s) - t(1-\cos(s))}{\alpha - \beta + t(1-\cos(s))}$$

$$= \frac{\alpha - \frac{1}{2}\beta(1+\cos(s)) - \frac{1}{2}\varepsilon\sqrt{2(1-\cos(s)) - \beta^2\sin^2(s)}}{\alpha - \frac{1}{2}\beta(1+\cos(s)) + \frac{1}{2}\varepsilon\sqrt{2(1-\cos(s)) - \beta^2\sin^2(s)}}.$$

Puisque $\alpha > 1 > \beta$, on a $\alpha - \frac{1}{2}\beta(1+\cos(s)) > 0$ et la valeur ci-dessus est maximale pour $\varepsilon = -1$, et ainsi

$$\max_{\theta \in \mathbb{T}} g(\theta) = \frac{\alpha - \beta\left(\frac{1+\cos(s)}{2}\right) + \sqrt{\left(\frac{1-\cos(s)}{2}\right)\left(1 - \beta^2\frac{1+\cos(s)}{2}\right)}}{\alpha - \beta\left(\frac{1+\cos(s)}{2}\right) - \sqrt{\left(\frac{1-\cos(s)}{2}\right)\left(1 - \beta^2\frac{1+\cos(s)}{2}\right)}} \times \frac{\alpha}{\alpha}$$

$$= \frac{\alpha^2 - \left(\frac{1+\cos(s)}{2}\right) + \sqrt{\left(\frac{1-\cos(s)}{2}\right)\left(\alpha^2 - \frac{1+\cos(s)}{2}\right)}}{\alpha^2 - \left(\frac{1+\cos(s)}{2}\right) - \sqrt{\left(\frac{1-\cos(s)}{2}\right)\left(\alpha^2 - \frac{1+\cos(s)}{2}\right)}} \times \frac{\sqrt{\alpha^2 - \frac{1+\cos(s)}{2}}}{\sqrt{\alpha^2 - \frac{1+\cos(s)}{2}}}$$

$$= \frac{\sqrt{\alpha^2 - \frac{1+\cos(s)}{2}} + \sqrt{\frac{1-\cos(s)}{2}}}{\sqrt{\alpha^2 - \frac{1+\cos(s)}{2}} - \sqrt{\frac{1-\cos(s)}{2}}} \times \frac{\sqrt{\frac{1-\cos(s)}{2}}}{\sqrt{\frac{1-\cos(s)}{2}}}$$

$$= \frac{\sqrt{\frac{2(\alpha^2-1)}{1-\cos(s)} + 1} + 1}{\sqrt{\frac{2(\alpha^2-1)}{1-\cos(s)} + 1} - 1}$$

$$= 1 + \frac{2}{\sqrt{\frac{2(\alpha^2-1)}{1-\cos(s)} + 1} - 1}.$$

Lemme 13. *Considérons $M \in \mathbb{N}^*$, $T_i > 0$, pour $1 \leq i \leq M$, et $\delta > 0$ donnés. Il existe $t \geq 1$ et des entiers $k_1, \ldots, k_M$ tels que pour tout $1 \leq i \leq M$,*

$$|k_i T_i - t| < \delta.$$

Preuve. Considérons le réseau de $\mathbb{R}^{M+1}$ engendré par les $(0,\ldots,0,T_n, 0,\ldots,0)$ (avec T_n en $n^{\text{ième}}$ position) pour $n \leq M$ et par $(1,1,\ldots,1)$, de volume fondamental V le produit des T_n. Ainsi, en considérant le pavé $[-\delta,\delta]\times\cdots\times[-\delta,\delta]\times[-V\delta^{-M},V\delta^{-M}]$, de volume $V2^d$, on sait par le théorème de Minkowski qu'il contient au moins un point du réseau autre que l'origine. Les M premières coordonnées de ce point sont de la forme $h_n T_n + h_{M+1}$ avec $h_j \in \mathbb{Z}$ pour $j \in [\![1, M+1]\!]$. Si $\delta < \min(T_1,\ldots,T_M,1)$ (et quitte à réduire δ, nous supposons ceci satisfait), aucun de ces coefficients h_j ne peut être nul, et nécessairement h_{M+1} est de signe opposé aux autres h_j. Il suffit donc de prendre $k_n = |h_n|$ et $t = |h_{M+1}|$.

Lemme 14. *Notons, pour $s > 0$ et $p \in]0,1[$,*

$$h(p) = \frac{p}{1-p^2}\frac{1+e^{-ps}}{1-e^{-ps}}$$

$$\phi(p) = e^{ps}\left(1 + \frac{2}{ph(p) + \sqrt{h(p)^2 + \frac{1}{1-p^2}} - 1}\right).$$

Alors pour tout s, $p \mapsto \phi(p)$ est croissante.

En prenant $p = \sqrt{1-\left(\frac{n}{a}\right)^2}$ et $s = 2at$ on obtient en particulier que pour $t > 0$ et $a > 0$, $n \in]0, a[\mapsto R(t, a, n)$ est décroissante.

Preuve. Le calcul de la dérivée est effectué *via* Maple :

```
h:=p->p/(1-p^2)*(1+exp(-s*p))/(1-exp(-s*p)):
phi:=p->exp(p*s)*(1+2/(p*h(p)+sqrt(h(p)^2+1/(1-p^2))-
1)):
resultat := simplify(exp(-p*s)*diff(phi(p),p)):
```

Le résultat est de la forme $\frac{\text{numérateur}(p)}{(\text{un terme})^2(p^2-1)(e^{-ps}-1)}$; il s'agit donc de vérifier que le numérateur est positif. À l'instruction

```
solve(numerateur(p)=0,p);
```

la réponse est

```
-RootOf(_Z exp(_Z) + _Z + 2 - 2 exp(_Z))/s
```

Autrement dit le numérateur s'annule en p si $e^p = \frac{2+p}{2-p}$, équation dont la seule solution est $p = 0$: en effet, s'il y avait une autre solution p^*, la dérivée de $e^p\frac{2-p}{2+p}$ s'annulerait entre 0 et p^*, or celle-ci est $\frac{-z^2 e^z}{(2+z)^2}$. Ainsi le numérateur est de signe constant pour $p \in]0,1[$ ϕ est monotone. Les limites de ϕ en 0 et 1 sont respectivement $1 + \frac{2}{\sqrt{4s^{-2}+1}-1}$ et e^s, dont l'égalité est équivalente à $2e^s - 2 - s^2 = 0$, d'unique solution $s = 0$; vu leurs équivalents pour $s \to +\infty$ on a donc $\phi(1) > \phi(0)$ pour $s > 0$, donc ϕ est croissante.

Remerciements. Nous sommes reconnaissant à Jérémy Leborgne pour l'élégant argument du lemme 13.

References

1. J.-B. Bardet, A. Christen, A. Guillin, F. Malrieu, P.-A. Zitt, Total variation estimates for the TCP process. Consultable sur http://hal.archives-ouvertes.fr/hal-00655462, December 2011
2. D. Chafaï, F. Malrieu, K. Paroux, On the long time behavior of the TCP window size process. Stoch. Proc. Appl. **120**(8), 1518–1534 (2010)
3. P. Diaconis, S. Holmes, R.M. Neal, Analysis of a nonreversible Markov chain sampler. Ann. Appl. Probab. **10**(3), 726–752 (2000)
4. P. Diaconis, L. Miclo, On the spectral analysis of second-order Markov chains. Consultable sur http://hal.archives-ouvertes.fr/hal-00719047, 2009
5. J. Dolbeault, C. Mouhot, C. Schmeiser, Hypocoercivity for linear kinetic equations conserving mass. Consultable sur http://hal.archives-ouvertes.fr/ccsd-00482286, 2010
6. S.N. Ethier, T.G. Kurtz, *Markov Processes: Characterization and Convegence*. Wiley Series in Probability and Mathematical Statistics: Probability and Mathematical Statistics (Wiley, New York, 1986)
7. S. Gadat, L. Miclo, Spectral decompositions and $\mathbb{L}^2$-operator norms of toy hypocoercive semigroups. Consultable sur http://hal.archives-ouvertes.fr/hal-00717653, 2011
8. R.M. Neal, Improving asymptotic variance of MCMC estimators: non-reversible chains are better. Technical Report No. 0406, Department of Statistics, University of Toronto. Consultable sur arXiv:math/0407281, 2004

Combinatorial Optimization Over Two Random Point Sets

Franck Barthe and Charles Bordenave

Abstract Let $(\mathscr{X}, \mathscr{Y})$ be a pair of random point sets in $\mathbb{R}^d$ of equal cardinal obtained by sampling independently $2n$ points from a common probability distribution μ. In this paper, we are interested by functions L of $(\mathscr{X}, \mathscr{Y})$ which appear in combinatorial optimization. Typical examples include the minimal length of a matching of $\mathscr{X}$ and $\mathscr{Y}$, the length of a traveling salesperson tour constrained to alternate between points of each set, or the minimal length of a connected bipartite r-regular graph with vertex set $(\mathscr{X}, \mathscr{Y})$. As the size n of the point sets goes to infinity, we give sufficient conditions on the function L and the probability measure μ which guarantee the convergence of $L(\mathscr{X}, \mathscr{Y})$ under a suitable scaling. In the case of the minimal length matching, we extend results of Dobrić and Yukich, and Boutet de Monvel and Martin.

Keywords Combinatorial optimization • Geometric probability • Minimal matching

1 Introduction

This work pertains to the probabilistic study of Euclidean combinatorial optimization problems. The starting point in this field is the celebrated theorem of Beardwood, Halton, and Hammersley [2] about the traveling salesperson problem. It ensures that given a sequence $(X_i)_{i\geq 1}$ of independent random variables on $\mathbb{R}^d$, $d \geq 2$ with common law μ of bounded support, then almost surely

F. Barthe · C. Bordenave (✉)
Institut de Mathématiques (CNRS UMR 5219). Université Paul Sabatier. 31062 Toulouse cedex 09, France
e-mail: barthe@math.univ-toulouse.fr; bordenave@math.univ-toulouse.fr

$$\lim_{n\to\infty} n^{\frac{1}{d}-1} T(X_1,\ldots,X_n) = \beta(d) \int f^{1-\frac{1}{d}}. \tag{1}$$

Here $\beta(d)$ is a constant depending only on the dimension, f is the density of the absolutely continuous part of μ and

$$T(X_1,\ldots,X_n) = \inf_{\sigma \in \mathscr{S}_n} \sum_{i=1}^{n-1} |X_{\sigma(i+1)} - X_{\sigma(i)}| + |X_{\sigma(1)} - X_{\sigma(n)}|$$

is the length (for the canonical Euclidean distance) of the shortest tour through the points $X_1,\ldots,X_n$. In the above formula $\mathscr{S}_n$ stands for the set of permutations of $\{1,2,\ldots,n\}$. Very informally, this result supports the following interpretation: when the number of points n is large, for μ almost every x, if the salesperson is at $X_i = x$ then the distance to the next point in the optimal tour is comparable to $\beta(d)(nf(x))^{-1/d}$ if $f(x) > 0$ and of lower order otherwise. This should be compared to the fact that the distance from $X_i = x$ to $\{X_j, j \le n \text{ and } j \ne i\}$ also stabilizes at the same rate.

Later, Papadimitriou [9] and Steele [14] have initiated a general theory of Euclidean functionals $F(\{X_1,\ldots,X_n\})$ that satisfy almost sure limits of this type. We refer the reader to the monographs of Steele [15] and Yukich [19] for a full treatment of this now mature theory, and present a short outline. It is convenient to consider multisets rather than sets, so throughout the paper $\{x_1,\ldots,x_n\}$ will stand for a multiset (the elements are unordered but may be repeated). The umbrella theorem in [19] puts forward the following three features of a functional F on finite multisets of $\mathbb{R}^d$:

- F is 1-homogeneous if it is translation invariant and dilation covariant:

$$F(a + \lambda \mathscr{X}) = \lambda F(\mathscr{X})$$

for all finite multisets $\mathscr{X}$, all $a \in \mathbb{R}^d$ and $\lambda \in \mathbb{R}^+$.
- The key assumption is subadditivity: F is subadditive if there exists a constant $C > 0$ such that for all multisets $\mathscr{X}, \mathscr{Y}$ in the unit cube $[0,1]^d$,

$$F(\mathscr{X} \cup \mathscr{Y}) \le F(\mathscr{X}) + F(\mathscr{Y}) + C.$$

By an inductive argument, Rhee in [12] has noticed that these two assumptions imply that there is another constant C' such that for all multiset in $[0,1]^d$,

$$|F(\mathscr{X})| \le C' (\text{card}(\mathscr{X}))^{1-\frac{1}{d}}. \tag{2}$$

Hence the worst case for n points is at most in $n^{1-\frac{1}{d}}$ and the above mentioned theorems show that the average case is of the same order.

- The third important property is smoothness (or regularity). A functional F on finite multisets $\mathbb{R}^d$ is smooth if there is a constant C'' such that for all multisets $\mathscr{X}, \mathscr{Y}, \mathscr{Z}$ in $[0,1]^d$, it holds

$$|F(\mathscr{X} \cup \mathscr{Y}) - F(\mathscr{X} \cup \mathscr{Z})| \leq C'' \left(\operatorname{card}(\mathscr{Y})^{1-\frac{1}{d}} + \operatorname{card}(\mathscr{Z})^{1-\frac{1}{d}} \right).$$

As in the model of the Beardwood, Halton, Hammersley theorem, these three properties are enough to prove that almost surely,

$$\limsup n^{\frac{1}{d}-1} F(X_1, \ldots, X_n) \leq \beta(d) \int f^{1-\frac{1}{d}},$$

where $\beta(d)$ is constant. To have the full limits, the umbrella theorem of [19] also requires to check a few more properties of a so-called boundary functional associated with F. They are more complicated to state in a general framework.

Next, let us present a classical optimization problem which does not enter the above picture. Given two multi-subsets of $\mathbb{R}^d$ with the same cardinality, $\mathscr{X} = \{X_1, \ldots, X_n\}$ and $\mathscr{Y} = \{Y_1, \ldots, Y_n\}$, the cost of the minimal bipartite matching of $\mathscr{X}$ and $\mathscr{Y}$ is defined as

$$M_1(\mathscr{X}, \mathscr{Y}) = \min_{\sigma \in \mathscr{S}_n} \sum_{i=1}^{n} |X_i - Y_{\sigma(i)}|,$$

where the minimum runs over all permutations of $\{1, \ldots, n\}$. It is well-known that $n^{-1} M_1(\{X_i\}_{i=1}^n, \{Y_i\}_{i=1}^n)$ coincides with the power of the L_1-Wasserstein distance between the empirical distributions

$$W_1\left(\frac{1}{n} \sum_i \delta_{X_i}, \frac{1}{n} \sum_i \delta_{Y_i}\right),$$

(see e.g. [10, Theorem 13.3]). Hence it is easily seen to tend to 0, for example, when μ has bounded support. Recall that given two finite measures μ_1, μ_2 on $\mathbb{R}^d$ with the same total mass,

$$W_1(\mu_1, \mu_2) = \inf_{\pi \in \Pi(\mu_1, \mu_2)} \int_{\mathbb{R}^d \times \mathbb{R}^d} |x - y| \, d\pi(x, y),$$

where $\Pi(\mu_1, \mu_2)$ is the set of measures on $(\mathbb{R}^d)^2$ having μ_1 as first marginal and μ_2 as second marginal (see e.g. [11, 18] for more background). Note that for all finite multisets $\mathscr{X}, \mathscr{Y}$ in $[0,1]^d$ with $\operatorname{card}(\mathscr{X}) = \operatorname{card}(\mathscr{Y})$,

$$M_1(\mathscr{X}, \mathscr{Y}) \leq \sqrt{d} \, \operatorname{card}(\mathscr{X}),$$

and equality holds for some well-chosen configurations of any cardinal (all elements in $\mathscr{X}$ at $(0,\cdots,0)$ and all elements in $\mathscr{Y}$ at $(1,\cdots,1)$). Hence, an interesting feature of M_1 (as well as others bipartite Euclidean optimization functionals) is that the growth bound assumption (2) fails, hence it is not subadditive in the above sense. However Dobrić and Yukich have stated the following theorem:

Theorem 1 ([4]). *Let $d \geq 3$ be an integer. Assume that μ is a probability measure on $\mathbb{R}^d$ having a bounded support. Consider mutually independent random variables $(X_i)_{i\geq 1}$ and $(Y_j)_{j\geq 1}$ having distribution μ. Then, almost surely,*

$$\lim_n n^{\frac{1}{d}-1} M_1(\{X_1,\ldots,X_n\},\{Y_1,\ldots,Y_n\}) = \beta_1(d) \int_{\mathbb{R}^d} f^{1-\frac{1}{d}},$$

where $f(x)\,dx$ is the absolutely continuous part of μ and $\beta_1(d)$ is a constant depending only on the dimension d.

When f is not the uniform measure on the unit cube, there is an issue in the proof of [4] that apparently cannot be easily fixed (the problem lies in their Lemma 4.2 which is used for proving that the lim inf is at least $\beta_1(d)\int_{\mathbb{R}^d} f^{1-\frac{1}{d}}$). In any case, the proof of Dobrić and Yukich is very specific to the bipartite matching as it uses from the start the Kantorovich–Rubinstein dual representation of the optimal transportation cost. It is not adapted to a general treatment of bipartite functionals. The starting point of our work was a recent paper of Boutet de Monvel and Martin [3] which (independently of [4]) establishes the convergence of the bipartite matching for uniform variables on the unit cube, without using the dual formulation of the transportation cost. Building on their approach we are able to propose a soft approach of bipartite functionals, based on appropriate notions of subadditivity and regularity. These properties allow to establish upper estimates on upper limits. In order to deal with lower limits we adapt to the bipartite setting the ideas of boundary functionals exposed in [19]. We are able to explicitly construct such functionals for a class of optimization problems involving families of graphs with good properties, and to establish full convergence for absolutely continuous laws. Finally we introduce a new notion of inverse subadditivity which allows to deal with singular parts.

This viewpoint sheds a new light on the result of Dobrić and Yukich, that we extend in other respects, by considering power distance costs, and unbounded random variables satisfying certain tail assumptions. Note that in the classical theory of Euclidean functionals, the analogous question for unbounded random variables was answered in Rhee [13] and generalized in [19].

Let us illustrate our results in the case of the bipartite matching with power distance cost: given $p > 0$ and two multi-subsets of $\mathbb{R}^d$, $\mathscr{X} = \{X_1,\ldots,X_n\}$ and $\mathscr{Y} = \{Y_1,\ldots,Y_n\}$, define

$$M_p(\mathscr{X},\mathscr{Y}) = \min_{\sigma \in \mathscr{S}_n} \sum_{i=1}^n |X_i - Y_{\sigma(i)}|^p,$$

where the minimum runs over all permutations of $\{1,\ldots,n\}$. Note that we have the same result for the bipartite traveling salesperson problem, and that our generic approach puts forward key properties that allow to establish similar facts for other functionals. As mentioned in the title, our results apply to relatively high dimension. More precisely, if the length of edges are counted to a power p, our study applies to dimensions $d > 2p$ only.

Theorem 2. *Let $0 < 2p < d$. Let μ be a probability measure on $\mathbb{R}^d$ with absolutely continuous part $f(x)\,dx$. We assume that for some $\alpha > \frac{4dp}{d-2p}$,*

$$\int |x|^\alpha d\mu(x) < +\infty.$$

Consider mutually independent random variables $(X_i)_{i\geq 1}$ and $(Y_j)_{j\geq 1}$ having distribution μ. Then there are positive constants $\beta_p(d), \beta'_p(d)$ depending only on (p,d) such that the following convergence holds almost surely

$$\limsup_n n^{\frac{p}{d}-1} M_p(\{X_1,\ldots,X_n\},\{Y_1,\ldots,Y_n\}) \leq \beta_p(d) \int_{\mathbb{R}^d} f^{1-\frac{p}{d}},$$

$$\liminf_n n^{\frac{p}{d}-1} M_p(\{X_1,\ldots,X_n\},\{Y_1,\ldots,Y_n\}) \geq \beta'_p(d) \int_{\mathbb{R}^d} f^{1-\frac{p}{d}}.$$

Moreover, almost surely,

$$\lim_n n^{\frac{p}{d}-1} M_p(\{X_1,\ldots,X_n\},\{Y_1,\ldots,Y_n\}) = \beta_p(d) \int_{\mathbb{R}^d} f^{1-\frac{p}{d}}$$

provided one of the following hypothesis is verified:

- μ *is the uniform distribution over a bounded set $\Omega \subset \mathbb{R}^d$ with positive Lebesgue measure.*
- $d \in \{1,2\}$, $p \in (0,d/2)$ *or* $d \geq 3$, $p \in (0,1]$, *and f is up to a multiplicative constant the indicator function over a bounded set $\Omega \subset \mathbb{R}^d$ with positive Lebesgue measure.*

Our constant $\beta'(d)$ has an explicit expression in terms of the cost of an optimal boundary matching for the uniform measure on $[0,1]^d$ (see Lemma 11). We strongly suspect that $\beta_p(d) = \beta'_p(d)$ but we have not been able to solve this important issue. Also, assuming only $\alpha > \frac{2dp}{d-2p}$, we can establish convergence in probability. As we shall check, the bounded differences inequality will imply that if μ has bounded support the convergence holds also in L^q for all $q \geq 1$.

Note that this result again supports the following heuristic interpretation: when the number of points n is large, for μ almost every x, given that $X_i = x$, the i-th point is matched to a point $Y_{\sigma(i)}$ at distance of order $\beta_p(d)^{1/p}(nf(x))^{-1/d}$ if $f(x) > 0$ and of lower order otherwise. This can be compared to the fact that the distance from $X_i = x$ to $\{Y_j, 1 \leq j \leq n\}$ also stabilizes at the same rate. This holds as long as $0 < 2p < d$ (see Section 7 for a more detailed discussion).

The paper is organized as follows: Section 2 presents the key properties for bipartite functionals (homogeneity, subadditivity and regularity) and gathers useful preliminary statements. Section 3 establishes the convergence for uniform samples on the cube. Section 4 proves upper bounds on the upper limits. These two sections essentially rely on classical subadditive methods, nevertheless a careful analysis is needed to control the differences of cardinalities of the two samples in small domains. In Section 5, we introduce some examples of bipartite functionals. The lower limits are harder to prove and require a new notion of penalized boundary functionals. It is however difficult to build an abstract theory there, so in Section 6, we will first present the proof for bipartite matchings with power distance cost, and put forward a few lemmas which will be useful for other functionals. We then check that for a natural family of Euclidean combinatorial optimization functionals defined in Section 5.3, the lower limit also holds. This family includes the bipartite traveling salesman tour. Finally, Section 7 mentions possible variants and extensions.

2 A General Setting

Let $\mathcal{M}_d$ be the set of all finite multisets contained in $\mathbb{R}^d$. We consider a bipartite functional:

$$L : \mathcal{M}_d \times \mathcal{M}_d \to \mathbb{R}^+.$$

Let $p > 0$. We shall say that L is p-homogeneous if for all multisets $\mathcal{X}, \mathcal{Y}$, all $a \in \mathbb{R}^d$ and all $\lambda \geq 0$,

$$L(a + \lambda \mathcal{X}, a + \lambda \mathcal{Y}) = \lambda^p L(\mathcal{X}, \mathcal{Y}). \qquad (\mathcal{H}_p)$$

Here $a + \lambda \{x_1, \ldots, x_k\}$ is by definition $\{a + \lambda x_1, \ldots, a + \lambda x_k\}$. For the sake of brevity, we call the above property $(\mathcal{H}_p)$. Note that a direct consequence is that $L(\emptyset, \emptyset) = 0$.

The functional L satisfies the regularity property $(\mathcal{R}_p)$ if there exists a number C such that for all multisets $\mathcal{X}, \mathcal{Y}, \mathcal{X}_1, \mathcal{Y}_1, \mathcal{X}_2, \mathcal{Y}_2$, denoting by Δ the diameter of their union, the following inequality holds

$$L(\mathcal{X} \cup \mathcal{X}_1, \mathcal{Y} \cup \mathcal{Y}_1) \qquad (\mathcal{R}_p)$$
$$\leq L(\mathcal{X} \cup \mathcal{X}_2, \mathcal{Y} \cup \mathcal{Y}_2) + C \Delta^p \big(\text{card}(\mathcal{X}_1) + \text{card}(\mathcal{X}_2) + \text{card}(\mathcal{Y}_1) + \text{card}(\mathcal{Y}_2) \big).$$

The above inequality implies in particular an easy size bound: $L(\mathcal{X}, \mathcal{Y}) \leq C \Delta^p (\text{card}(\mathcal{X}) + \text{card}(\mathcal{Y}))$ when $L(\emptyset, \emptyset) = 0$.

Eventually, L verifies the subadditivity property $(\mathcal{S}_p)$ if there exists a number C such that for every $k \geq 2$ and all multisets $(\mathcal{X}_i, \mathcal{Y}_i)_{i=1}^k$, denoting by Δ the diameter of their union, the following inequality holds

$$L\left(\bigcup_{i=1}^{k} \mathscr{X}_i, \bigcup_{i=1}^{k} \mathscr{Y}_i\right) \leq \sum_{i=1}^{k} L(\mathscr{X}_i, \mathscr{Y}_i) + C\Delta^p \sum_{i=1}^{k} \Big(1 + |\mathrm{card}(\mathscr{X}_i) - \mathrm{card}(\mathscr{Y}_i)|\Big). \quad (\mathscr{S}_p)$$

Remark 1. A less demanding notion of "geometric subadditivity" could be introduced by requiring the above inequality only when the multisets $\mathscr{X}_i \cup \mathscr{Y}_i$ lie in disjoint parallelepipeds (see [19] where such a notion is used in order to encompass more complicated single sample functionals). It is clear from the proofs that some of our results hold assuming only geometric subadditivity (upper limit for bounded absolutely continuous laws for example). We will not push this idea further in this paper.

We will see in Sect. 5 that suitable extensions of the bipartite matching, of the bipartite traveling salesperson problem, and of the minimal bipartite spanning tree with bounded maximal degree satisfy all these properties. Our main generic result on bipartite functionals is the following.

Theorem 3. *Let $d > 2p > 0$ and let L be a bipartite functional on $\mathbb{R}^d$ with the properties $(\mathscr{H}_p)$, $(\mathscr{R}_p)$ and $(\mathscr{S}_p)$. Consider a probability measure μ on $\mathbb{R}^d$ such that there exists $\alpha > \frac{4dp}{d-2p}$ with*

$$\int |x|^\alpha d\mu(x) < +\infty.$$

Consider mutually independent random variables $(X_i)_{i\geq 1}$ and $(Y_j)_{j\geq 1}$ having distribution μ. Let f be a density function for the absolutely continuous part of μ, then, almost surely,

$$\limsup_{n\to\infty} \frac{L(\{X_1, \cdots, X_n\}, \{Y_1, \cdots, Y_n\})}{n^{1-\frac{p}{d}}} \leq \beta_L \int f^{1-\frac{p}{d}},$$

for some constant β_L depending only on L. Moreover, if μ is the uniform distribution over a bounded set Ω with positive Lebesgue measure, then there is equality: almost surely,

$$\lim_{n\to\infty} \frac{L(\{X_1, \cdots, X_n\}, \{Y_1, \cdots, Y_n\})}{n^{1-\frac{p}{d}}} = \beta_L \mathrm{Vol}(\Omega)^{\frac{p}{d}}.$$

Beyond uniform distributions, lower limits are harder to obtain. In Sect. 6, we will state a lower bound for a subclass of bipartite functionals which satisfy the properties $(\mathscr{H}_p)$, $(\mathscr{R}_p)$ and $(\mathscr{S}_p)$ (see the forthcoming Theorem 10 and, for the bipartite traveling salesperson tour, Theorem 11).

Remark 2. Let $B(1/2) = \{x \in \mathbb{R}^d : |x| \leq 1/2\}$ be the Euclidean ball of radius $1/2$ centered at the origin. It is immediate that the functional L satisfies the regularity property $(\mathscr{R}_p)$ if it satisfies property $(\mathscr{H}_p)$ and if for all multisets $\mathscr{X}, \mathscr{Y}, \mathscr{X}_1, \mathscr{Y}_1, \mathscr{X}_2, \mathscr{Y}_2$ in $B(1/2)$,

$$L(\mathcal{X} \cup \mathcal{X}_1, \mathcal{Y} \cup \mathcal{Y}_1) \qquad (\mathcal{R})$$
$$\leq L(\mathcal{X} \cup \mathcal{X}_2, \mathcal{Y} \cup \mathcal{Y}_2) + C\bigl(\operatorname{card}(\mathcal{X}_1) + \operatorname{card}(\mathcal{X}_2) + \operatorname{card}(\mathcal{Y}_1) + \operatorname{card}(\mathcal{Y}_2)\bigr).$$

Similarly, L will enjoy the subadditivity property $(\mathcal{S}_p)$ if it satisfies property $(\mathcal{H}_p)$ and if for every $k \geq 2$ and all multisets $(\mathcal{X}_i, \mathcal{Y}_i)_{i=1}^k$ in $B(1/2)$,

$$L\left(\bigcup_{i=1}^k \mathcal{X}_i, \bigcup_{i=1}^k \mathcal{Y}_i\right) \leq \sum_{i=1}^k L(\mathcal{X}_i, \mathcal{Y}_i) + C \sum_{i=1}^k \bigl(1 + |\operatorname{card}(\mathcal{X}_i) - \operatorname{card}(\mathcal{Y}_i)|\bigr).$$
$$(\mathcal{S})$$

The set of assumptions $(\mathcal{H}_p)$, $(\mathcal{R}_p)$, $(\mathcal{S}_p)$ is thus equivalent to the set of assumptions $(\mathcal{H}_p)$, $(\mathcal{R})$, $(\mathcal{S})$.

2.1 Consequences of Regularity

2.1.1 Poissonization

The proof of Theorem 3 will use partitions of $[0,1]^d$ into subcubes. In order to obtain independence and scaling properties of the point sets in each partition, it is much more convenient to consider the Poissonized version of the above problem. Let $(X_i)_{i \geq 1}$, $(Y_i)_{i \geq 1}$ be mutually independent variables with distribution μ. Considering independent variables N_1, N_2 with Poisson distribution $\mathcal{P}(n)$, the random sets $\{X_1, \ldots, X_{N_1}\}$ and $\{Y_1, \ldots, Y_{N_2}\}$ are independent Poisson point processes with intensity measures $n\mu$. For the sake of brevity, we set

$$L(n\mu) := L\bigl(\{X_1, \ldots, X_{N_1}\}, \{Y_1, \ldots, Y_{N_2}\}\bigr).$$

When $d\mu(x) = f(x)\,dx$ we write $L(nf)$ instead of $L(n\mu)$. Note that whenever we are dealing with Poisson processes, $n \in (0, +\infty)$ is not necessarily an integer. More generally $L(\nu)$ may be defined for any finite measure, as the value of the functional L for two independent Poisson point processes with intensity ν.

Assume for a moment that the measure μ has a bounded support, of diameter Δ. The regularity property ensures that

$$|L(\{X_1, \ldots, X_n\}, \{Y_1, \ldots, Y_n\}) - L(\{X_1, \ldots, X_{N_1}\}, \{Y_1, \ldots, Y_{N_2}\})|$$
$$\leq C\Delta^p \bigl(|N_1 - n| + |N_2 - n|\bigr).$$

Note that $\mathbb{E}|N_i - n| \leq \bigl(\mathbb{E}(N_i - n)^2\bigr)^{1/2} = \operatorname{Var}(N_i) = \sqrt{n}$. Hence the difference between $\mathbb{E}L(\{X_i\}_{i=1}^n, \{Y_i\}_{i=1}^n)$ and $\mathbb{E}L(n\mu)$ is at most a constant times $\sqrt{n} = o(n^{1-p/d})$ when $d > 2p$. Hence in this case, the original quantity and the Poissonized version are the same in average at the relevant scale $n^{1-p/d}$. The boundedness assumption can actually be relaxed. To show this, we need a lemma.

Lemma 1. *Let $\alpha > 0$, $n > 0$ and let μ be a probability measure on $\mathbb{R}^d$ such that for all $t > 0$, $\mu(\{x; |x| \geq t\}) \leq ct^{-\alpha}$. Let $\mathscr{X}$, $\mathscr{Y}$ be two independent Poisson point processes of intensity $n\mu$ and $T_n = \max\{|Z| : Z \in \mathscr{X} \cup \mathscr{Y}\}$. Then, for all $0 < \gamma < \alpha$ there exists a constant $K = K(c, \alpha, \gamma)$ such that for all $n \geq 1$,*

$$\mathbb{E}[T_n^\gamma]^{\frac{1}{\gamma}} \leq K n^{\frac{1}{\alpha}}.$$

Moreover the same conclusion holds if $\mathscr{X} = \{X_1, \ldots, X_n\}$, $\mathscr{Y} = \{Y_1, \ldots, Y_n\}$ are two mutually independent sequences of n variables with distribution μ.

Proof. For $t \geq 0$, let $A_t = \{x \in \mathbb{R}^d : |x| \geq t\}$ and $g(t) = \int_{A_t} d\mu$. By assumption, $\mu(A_t) \leq ct^{-\alpha}$. We start with the Poisson case. Since $\mathscr{X}$, $\mathscr{Y}$ are independent, we have $\mathbb{P}(T_n < t) = \mathbb{P}(\mathscr{X} \cap A_t = \emptyset)^2 = e^{-2n\mu(A_t)}$. Therefore, using $1 - e^{-u} \leq \min(1, u)$,

$$\mathbb{E}[T_n^\gamma] = \gamma \int_0^\infty t^{\gamma-1} \mathbb{P}(T_n \geq t) dt$$

$$= \gamma \int_0^\infty t^{\gamma-1}(1 - e^{-2n\mu(A_t)}) dt$$

$$\leq \gamma \int_0^{n^{1/\alpha}} t^{\gamma-1} dt + \int_{n^{1/\alpha}}^\infty 2nct^{\gamma-\alpha-1} dt$$

$$= n^{\gamma/\alpha} + \frac{2c}{\alpha - \gamma} n^{\gamma/\alpha},$$

For the second case, since $\mathbb{P}(T_n \geq t) = 1 - (1 - \mu(A_t))^{2n} \leq \min(1, 2n\mu(A_t))$ the same conclusion holds. □

The next proposition implies that our original problem is well approximated by its Poissonized version.

Proposition 1. *Let $d > 2p > 0$. Let μ be a probability measure on $\mathbb{R}^d$ such that $\int |x|^\alpha d\mu(x) < +\infty$ for some $\alpha > \frac{2dp}{d-2p}$. Let $(X_i)_{i \geq 1}$, $(Y_i)_{i \geq 1}$ be mutually independent variables with distribution μ. If L satisfies the regularity property $(\mathscr{R}_p)$ then*

$$\lim_{n \to \infty} \frac{\mathbb{E}\left|L(\{X_i\}_{i=1}^n, \{Y_i\}_{i=1}^n) - L(n\mu)\right|}{n^{1-\frac{p}{d}}} = 0.$$

Remark 3. We have not proved the finiteness of $\mathbb{E}L(\{X_i\}_{i=1}^n, \{Y_i\}_{i=1}^n$ and $\mathbb{E}L(n\mu)$ yet. This will be done later. With the convention that $\infty - \infty = 0$, the above statement establishes nevertheless that $n^{\frac{p}{d}-1}(\mathbb{E}L(\{X_i\}_{i=1}^n, \{Y_i\}_{i=1}^n) - \mathbb{E}L(n\mu))$ converges to 0.

Proof. Let N_1 and N_2 be Poisson random variables with mean value n. Let $T = \max\{|Z| : Z \in \{X_1, \cdots, X_{N_1}\} \cup \{Y_1, \cdots, Y_{N_2}\}\}$ and $S = \max\{|Z| : Z \in \{X_1, \cdots, X_n\} \cup \{Y_1, \cdots, Y_n\}\}$, with the convention that the maximum over an empty set is 0. The regularity property ensures that

$$\left| L(\{X_1, \ldots, X_n\}, \{Y_1, \ldots, Y_n\}) - L(\{X_1, \ldots, X_{N_1}\}, \{Y_1, \ldots, Y_{N_2}\}) \right|$$
$$\leq C(T+S)^p \left(|N_1 - n| + |N_2 - n|\right).$$

Taking expectation gives, using Cauchy–Schwarz inequality and the bound $(a+b)^q \leq \max(1, 2^{q-1})(a^q + b^q)$ valid for $a, b, q > 0$

$$\mathbb{E}\left| L(\{X_1, \ldots, X_n\}, \{Y_1, \ldots, Y_n\}) - L(\{X_1, \ldots, X_{N_1}\}, \{Y_1, \ldots, Y_{N_2}\}) \right|$$
$$\leq c_p \left(\mathbb{E}[T^{2p}] + \mathbb{E}[S^{2p}]\right)^{\frac{1}{2}} \left(\mathbb{E}[|N_1 - n|^2] + \mathbb{E}[|N_2 - n|^2]\right)^{\frac{1}{2}}$$
$$= c_p \sqrt{2n} \left(\mathbb{E}[T^{2p}] + \mathbb{E}[S^{2p}]\right)^{\frac{1}{2}}$$

Since $\alpha > 2p$, by Lemma 1, for some $c > 0$ and all $n \geq 1$, $\mathbb{E}[T^{2p}] \leq cn^{2p/\alpha}$ and $\mathbb{E}[S^{2p}] \leq cn^{2p/\alpha}$. Hence the above difference of expectations is at most a constant times $n^{\frac{p}{\alpha}+\frac{1}{2}}$, which is negligible with respect to $n^{1-\frac{p}{d}}$ since α is assumed to be large enough. □

2.1.2 Approximations

We now study the continuity of $\mathbb{E}L(\mu)$ as a function of the finite measure μ. We first look at the regularity of $\mathbb{E}L(\mu)$ under scaling.

Proposition 2. *Assume that a bipartite functional L satisfies the regularity property $(\mathscr{R}_p)$. Let $m, n > 0$ and μ be a probability measure with support included in a set Q. Then*

$$\mathbb{E}L(n\mu) \leq \mathbb{E}L(m\mu) + C \operatorname{diam}(Q)^p |m - n|.$$

Proof. Assume $n < m$ (the other case is treated in the same way). Let $(X_i)_{i \geq 1}$, $(Y_i)_{i \geq 1}$, N_1, N_2, K_1, K_2 be mutually independent random variables, such that for all $i \geq 1$, X_i and Y_i have law μ, and for $j \in \{1, 2\}$, the law of N_j is $\mathscr{P}(n)$ and the law of K_j is $\mathscr{P}(m - n)$. Then $M_i = N_i + K_i$ is $\mathscr{P}(m)$-distributed. Then $\{X_1, \ldots, X_{N_1}\}$ and $\{Y_1, \ldots, Y_{N_2}\}$ are independent Poisson point processes of intensity $n\mu$, while $\{X_1, \ldots, X_{M_1}\}$ and $\{Y_1, \ldots, Y_{M_2}\}$ are independent Poisson point processes of intensity $m\mu$. By the regularity property,

$$L(\{X_1, \ldots, X_{N_1}\}, \{Y_1, \ldots, Y_{N_2}\})$$
$$\leq L(\{X_1, \ldots, X_{N_1+K_1}\}, \{Y_1, \ldots, Y_{N_2+K_2}\}) + C \operatorname{diam}(Q)^p (K_1 + K_2).$$

Taking expectations gives the claim. □

Applying the above inequality for $m = 0$ gives a weak size bound on $\mathbb{E}L(\nu)$.

Corollary 1. *Assume that L satisfies $(\mathcal{R}_p)$ and $L(\emptyset, \emptyset) = 0$ (a consequence of e.g. $(\mathcal{H}_p)$), then if ν is a finite measure with support included in a set Q,*

$$\mathbb{E}L(\nu) \leq C \operatorname{diam}(Q)^p \nu(Q).$$

We now look at the regularity of $\mathbb{E}L(\mu)$ under small perturbations of μ. Recall the total variation distance between two probability measures on $\mathbb{R}^d$ is defined as

$$d_{\mathrm{TV}}(\mu, \mu') = \sup\{|\mu(A) - \mu'(A)| : A \text{ Borel set of } \mathbb{R}^d\}.$$

Proposition 3. *Assume that L satisfies $(\mathcal{R}_p)$. Let μ, μ' be two probability measures on $\mathbb{R}^d$ with bounded supports. Set Δ the diameter of the union of their supports. Then*

$$\mathbb{E}L(n\mu) \leq \mathbb{E}L(n\mu') + 4C\Delta^p n \, d_{\mathrm{TV}}(\mu, \mu').$$

Proof. The difference of expectations is estimated thanks to a proper coupling argument. Let π be a probability measure on $\mathbb{R}^d \times \mathbb{R}^d$ having μ as its first marginal and μ' as its second marginal. We consider mutually independent random variables $N_1, N_2, (X_i, X'_i)_{i\geq 1}, (Y_i, Y'_i)_{i\geq 1}$ such that N_1, N_2 are $\mathcal{P}(n)$ distributed and for all $i \geq 1$, (X_i, X'_i) and (Y_i, Y'_i) are distributed according to π. Then the random multisets

$$\mathcal{X} = \{X_1, \ldots, X_{N_1}\} \quad \text{and} \quad \mathcal{Y} = \{Y_1, \ldots, Y_{N_2}\}$$

are independent Poisson point processes with intensity measure $n\mu$. Similarly $\mathcal{X}' = \{X'_1, \ldots, X'_{N_1}\}$ and $\mathcal{Y}' = \{Y'_1, \ldots, Y'_{N_2}\}$ are independent Poisson point processes with intensity measure $n\mu'$.

The regularity property ensures that

$$L(\{X_1, \ldots, X_{N_1}\}, \{Y_1, \ldots, Y_{N_2}\})$$

$$\leq L(\{X'_1, \ldots, X'_{N_1}\}, \{Y'_1, \ldots, Y'_{N_2}\}) + 2C\Delta^p \left(\sum_{i=1}^{N_1} \mathbf{1}_{X_i \neq X'_i} + \sum_{j=1}^{N_2} \mathbf{1}_{Y_j \neq Y'_j}\right).$$

Taking expectations yields

$$\mathbb{E}L(n\mu) \leq \mathbb{E}L(n\mu') + 2C\Delta^p \mathbb{E}\left(\sum_{i=1}^{N_1} \mathbb{P}(X_i \neq X'_i) + \sum_{j=1}^{N_2} \mathbb{P}(Y_j \neq Y'_j)\right)$$

$$= \mathbb{E}L(n\mu') + 4C\Delta^p n \, \pi(\{(x, y) \in (\mathbb{R}^d)^2; \, x \neq y\}).$$

Optimizing the later term on the coupling π yields the claimed inequality involving the total variation distance. □

Corollary 2. *Assume that the functional L satisfies the regularity property $(\mathscr{R}_p)$. Let $m > 0$, $Q \subset \mathbb{R}^d$ be measurable with positive Lebesgue measure and let f be a nonnegative locally integrable function on $\mathbb{R}^d$. Let $\alpha = \int_Q f / \mathrm{vol}(Q)$ be the average value of f on Q. It holds*

$$\mathbb{E}L(m\, f\mathbf{1}_Q) \leq \mathbb{E}L(m\alpha\mathbf{1}_Q) + 2Cm \operatorname{diam}(Q)^p \int_Q |f(x) - \alpha|\, dx.$$

Proof. We simply apply the total variation bound of the previous lemma with $n = m\int_Q f = m\alpha \mathrm{vol}(Q)$, $d\mu(x) = f(x)\mathbf{1}_Q(x)dx / \int_Q f$ and $d\mu'(x) = \mathbf{1}_Q(x)dx/\mathrm{vol}(Q)$. Note that

$$2d_{TV}(\mu, \mu') = \int \left| \frac{f(x)\mathbf{1}_Q(x)}{\int_Q f} - \frac{\mathbf{1}_Q(x)}{\mathrm{vol}(Q)} \right| dx = \frac{\int_Q |f(x) - \alpha|\, dx}{\int_Q f}.$$

□

2.1.3 Average is Enough

It is known since the works of Rhee and Talagrand that concentration inequalities often allow to deduce almost sure convergence from convergence in average. Without much surprise, this is also the case in our general setting.

Proposition 4. *Let L be a bipartite functional on multisets of $\mathbb{R}^d$, satisfying the regularity property $(\mathscr{R}_p)$. Assume $d > 2p > 0$. Let μ be a probability measure μ on $\mathbb{R}^d$ with $\int |x|^\alpha d\mu(x) < +\infty$. Consider independent variables $(X_i)_{i \geq 1}$ and $(Y_i)_{i \geq 1}$ with distribution μ.*

If $\alpha > 2dp/(d-2p)$ then the following convergence holds in probability:

$$\lim_{n \to \infty} \frac{L(\{X_i\}_{i=1}^n, \{Y_i\}_{i=1}^n) - \mathbb{E}L(\{X_i\}_{i=1}^n, \{Y_i\}_{i=1}^n)}{n^{1-\frac{p}{d}}} = 0.$$

Moreover if $\alpha > 4dp/(d-2p)$, the convergence happens almost surely, and if μ has bounded support, then it also holds in L^q for any $q \geq 1$.

Proof. This is a simple consequence of Azuma's concentration inequality. It is convenient to define $Z(n) = (X_1, \ldots, X_n, Y_1, \ldots, Y_n)$, $Z(n)$ is a vector of dimension $2n$ and its i-th coordinate is denoted by Z_i. Assume first that the support of μ is bounded and let Δ denote its diameter. By the regularity property, modifying one point changes the value of the functional by at most a constant:

$$|L(Z_1, \ldots, Z_{2n}) - L(Z_1, \ldots, Z_{i-1}, Z_i', Z_{i+1}, \ldots, Z_{2n})| \leq 2C\Delta^p.$$

By conditional integration, we deduce that the following martingale difference:

$$d_i := \mathbb{E}\big(L(Z(n)) \mid Z_1, \ldots, Z_i\big) - \mathbb{E}\big(L(Z(n)) \mid Z_1, \ldots, Z_{i-1}\big)$$

is also bounded $|d_i| \leq 2C\Delta^p$ almost surely. Recall that Azuma's inequality states that

$$\mathbb{P}\left(\left|\sum_{i=1}^k d_i\right| > t\right) \leq 2e^{-\frac{t^2}{2\sum_i \|d_i\|_\infty^2}}.$$

Therefore, we obtain that

$$\mathbb{P}\big(|L(\{X_i\}_{i=1}^n, \{Y_i\}_{i=1}^n) - \mathbb{E}L(\{X_i\}_{i=1}^n, \{Y_i\}_{i=1}^n)| > t\big) \leq 2e^{-\frac{t^2}{16nC^2\Delta^{2p}}}, \quad (3)$$

and there is a number C' (depending on Δ only) such that

$$\mathbb{P}\left(\frac{|L(\{X_i\}_{i=1}^n, \{Y_i\}_{i=1}^n) - \mathbb{E}L(\{X_i\}_{i=1}^n, \{Y_i\}_{i=1}^n)|}{n^{1-\frac{p}{d}}} > t\right) \leq 2e^{-C't^2 n^{1-\frac{2p}{d}}}.$$

When $d > 2p$, we may conclude by the Borel–Cantelli lemma.

If μ is not assumed to be of bounded support, a conditioning argument allows to use the above method. Let $S := \max\{|Z_i|;\ i \leq 2n\}$, $s > 0$ and $B(s) = \{x;\ |x| \leq s\}$. Given $\{S \leq s\}$, the variables $\{X_1, \cdots, X_n\}$ and $\{Y_1, \cdots, Y_n\}$ are mutually independent sequences with distribution $\mu_{|B(s)}/\mu(B(s))$. Hence, applying (3) for $\mu_{|B(s)}/\mu(B(s))$ instead of μ and $2s$ instead of Δ, for any $t > 0$,

$$\mathbb{P}\left(\left|\frac{L(\{X_i\}_{i=1}^n, \{Y_i\}_{i=1}^n)}{n^{1-\frac{p}{d}}} - \frac{\mathbb{E}L(\{X_i\}_{i=1}^n, \{Y_i\}_{i=1}^n)}{n^{1-\frac{p}{d}}}\right| > t \mid S \leq s\right)$$

$$\leq 2\exp\left(-\frac{n^{1-\frac{2p}{d}} t^2}{c_p s^{2p}}\right).$$

Hence for $\delta > 0$ to be chosen later,

$$u_n := \mathbb{P}\left(\left|\frac{L(\{X_i\}_{i=1}^n, \{Y_i\}_{i=1}^n)}{n^{1-\frac{p}{d}}} - \frac{\mathbb{E}L(\{X_i\}_{i=1}^n, \{Y_i\}_{i=1}^n)}{n^{1-\frac{p}{d}}}\right| > t\right)$$

$$\leq \mathbb{P}(S > n^{\frac{1}{\delta}}) + 2\exp\left(-\frac{n^{1-\frac{2p}{d}-\frac{2p}{\delta}} t^2}{c_p}\right).$$

Since $\mathbb{P}(S > s) = 1 - (1 - \mu(B(s)))^{2n} \leq 2n\mu(B(s)) \leq 2n(\int |x|^\alpha d\mu(x))/s^\alpha$, we get that for some constant c and any $\delta > 0$,

$$u_n \leq cn^{1-\frac{\alpha}{\delta}} + 2\exp\left(-\frac{n^{1-\frac{2p}{d}-\frac{2p}{\delta}}t^2}{c_p}\right).$$

Since $\alpha > 2dp/(d-2p)$ we may choose $\delta \in [2dp/(d-2p), \alpha]$, which ensures that the latter quantities tend to zero as n increases. This shows the convergence in probability to 0 of

$$\frac{L(\{X_i\}_{i=1}^n, \{Y_i\}_{i=1}^n)}{n^{1-\frac{p}{d}}} - \frac{\mathbb{E}L(\{X_i\}_{i=1}^n, \{Y_i\}_{i=1}^n)}{n^{1-\frac{p}{d}}}.$$

If $\alpha > 4dp/(d-2p)$ we may choose $\delta \in [2dp/(d-2p), \alpha/2]$, which ensures that $\sum_n u_n < +\infty$. The Borel–Cantelli lemma yields the almost sure convergence to 0. □

2.2 Consequences of Subadditivity

We start with a very general statement, which is however not very precise when the measures do not have disjoint supports.

Proposition 5. *Let L satisfy $(\mathscr{S}_p)$. Let μ_1, μ_2 be finite measures on $\mathbb{R}^d$ with supports included in a set Q. Then*

$$\mathbb{E}L(\mu_1+\mu_2) \leq \mathbb{E}L(\mu_1) + \mathbb{E}L(\mu_2) + 2C\operatorname{diam}(Q)^p\left(1+\sqrt{\mu_1(Q)}+\sqrt{\mu_2(Q)}\right).$$

Proof. Consider four independent Poisson point processes $\mathscr{X}_1, \mathscr{Y}_1, \mathscr{X}_2, \mathscr{Y}_2$ such that for $i \in \{1,2\}$, the intensity of $\mathscr{X}_i$ and of $\mathscr{Y}_i$ is μ_i. It is classical [8] that the random multiset $\mathscr{X}_1 \cup \mathscr{X}_2$ is a Poisson point process with intensity $\mu_1 + \mu_2$. Also, $\mathscr{Y}_1 \cup \mathscr{Y}_2$ is an independent copy of the latter process. Applying the subadditivity property,

$$L(\mathscr{X}_1 \cup \mathscr{X}_2, \mathscr{Y}_1 \cup \mathscr{Y}_2) \leq L(\mathscr{X}_1, \mathscr{Y}_1) + L(\mathscr{X}_2, \mathscr{Y}_2)$$
$$+ C\operatorname{diam}(Q)^p\left(1+|\operatorname{card}(\mathscr{X}_1)-\operatorname{card}(\mathscr{Y}_1)|+1+|\operatorname{card}(\mathscr{X}_2)-\operatorname{card}(\mathscr{Y}_2)|\right).$$

Since $\operatorname{card}(\mathscr{X}_i)$ and $\operatorname{card}(\mathscr{Y}_i)$ are independent with Poisson law of parameter $\mu_i(Q)$ (the total mass of μ_i),

$$\mathbb{E}|\operatorname{card}(\mathscr{X}_i)-\operatorname{card}(\mathscr{Y}_i)|$$
$$\leq \left(\mathbb{E}(\operatorname{card}(\mathscr{X}_i)-\operatorname{card}(\mathscr{Y}_i))^2\right)^{\frac{1}{2}} = \sqrt{2\operatorname{var}(\operatorname{card}(\mathscr{X}_i))} = \sqrt{2\mu_i(Q)}.$$

Hence, taking expectations in the former estimate leads to the claimed inequality. □

Partition techniques are essential in the probabilistic theory of Euclidean functionals. The next statement allows to apply them to bipartite functionals. In what follows, given a multiset $\mathscr{X}$ and a set P, we set $\mathscr{X}(P) := \operatorname{card}(\mathscr{X} \cap P)$. If μ is a measure and f a nonnegative function, we write $f \cdot \mu$ for the measure having density f with respect to μ.

Proposition 6. *Assume that the functional L satisfies $(\mathscr{S}_p)$. Consider a finite partition $Q = \cup_{P \in \mathscr{P}} P$ of a subset of $\mathbb{R}^d$ and let ν be a measure on $\mathbb{R}^d$ with $\nu(Q) < +\infty$. Then*

$$\mathbb{E}L(\mathbf{1}_Q \cdot \nu) \leq \sum_{P \in \mathscr{P}} \mathbb{E}L(\mathbf{1}_P \cdot \nu) + 3C \operatorname{diam}(Q)^p \sum_{p \in \mathscr{P}} \sqrt{\nu(P)}.$$

Proof. Consider $\mathscr{X}, \mathscr{Y}$ two independent Poisson point processes with intensity ν. Note that $\mathscr{X} \cap P$ is a Poisson point process with intensity $\mathbf{1}_P \cdot \nu$, hence $\mathscr{X}(P)$ is a Poisson variable with parameter $\nu(P)$. We could apply the subadditivity property to $(\mathscr{X} \cap P)_{P \in \mathscr{P}}, (\mathscr{Y} \cap P)_{P \in \mathscr{P}}$, which yields

$$L(\mathscr{X} \cap Q, \mathscr{Y} \cap Q) \leq \sum_{P \in \mathscr{P}} L(\mathscr{X} \cap P, \mathscr{Y} \cap P) + C \operatorname{diam}(Q)^p \sum_{p \in \mathscr{P}} (1 + |\mathscr{X}(P) - \mathscr{Y}(P)|).$$

Nevertheless, doing this gives a contribution at least $C \operatorname{diam}(Q)^p$ to cells which do not intersect the multisets $\mathscr{X}, \mathscr{Y}$. To avoid this rough estimate, we consider the cells which meet at least one of the multisets:

$$\tilde{\mathscr{P}} := \{P \in \mathscr{P};\ \mathscr{X}(P) + \mathscr{Y}(P) \neq 0\}.$$

We get that

$$L(\mathscr{X} \cap Q, \mathscr{Y} \cap Q) \leq \sum_{P \in \tilde{\mathscr{P}}} L(\mathscr{X} \cap P, \mathscr{Y} \cap P)$$

$$+ C \operatorname{diam}(Q)^p \sum_{p \in \tilde{\mathscr{P}}} (1 + |\mathscr{X}(P) - \mathscr{Y}(P)|)$$

$$\leq \sum_{P \in \mathscr{P}} L(\mathscr{X} \cap P, \mathscr{Y} \cap P)$$

$$+ C \operatorname{diam}(Q)^p \sum_{p \in \mathscr{P}} \mathbf{1}_{\mathscr{X}(P) + \mathscr{Y}(P) \neq 0}(1 + |\mathscr{X}(P) - \mathscr{Y}(P)|)$$

$$\leq \sum_{P \in \mathscr{P}} L(\mathscr{X} \cap P, \mathscr{Y} \cap P)$$

$$+ C \operatorname{diam}(Q)^p \sum_{p \in \mathscr{P}} (\mathbf{1}_{\mathscr{X}(P) + \mathscr{Y}(P) \neq 0} + |\mathscr{X}(P) - \mathscr{Y}(P)|).$$

Since $\mathscr{X}(P)$ and $\mathscr{Y}(P)$ are independent Poisson variables with parameter $\nu(P)$,
$$\mathbb{P}\big(\mathscr{X}(P) + \mathscr{Y}(P) \neq 0\big) = 1 - e^{-2\nu(P)} \text{ and } \mathbb{E}\big|\mathscr{X}(P) - \mathscr{Y}(P)\big| \leq \sqrt{2\nu(P)}.$$

Hence, taking expectation and using the bound $1 - e^{-t} \leq \min(1, t) \leq \sqrt{t}$,
$$\mathbb{E}L(\mathbf{1}_Q \cdot \nu) \leq \sum_{P \in \mathscr{P}} \mathbb{E}L(\mathbf{1}_P \cdot \nu) + 2\sqrt{2}\, C \operatorname{diam}(Q)^p \sum_{p \in \mathscr{P}} \sqrt{\nu(P)}. \qquad \Box$$

The next statement deals with nested partitions, which are very useful in the study of combinatorial optimization problems, see e.g. [15, 19]. If $\mathscr{P}$ is a partition, we set $\operatorname{diam}(\mathscr{P}) = \max_{P \in \mathscr{P}} \operatorname{diam}(P)$ (the maximal diameter of its cells).

Corollary 3. *Assume that the functional L satisfies $(\mathscr{S}_p)$. Let $Q \subset \mathbb{R}^d$ and $\mathscr{Q}_1, \ldots, \mathscr{Q}_k$ be a sequence of nested finite partitions of Q. Let ν be a measure on $\mathbb{R}^d$ with $\nu(Q) < +\infty$. Then*
$$\mathbb{E}L(\mathbf{1}_Q \cdot \nu) \leq \sum_{q \in \mathscr{Q}_k} \mathbb{E}L(\mathbf{1}_q \cdot \nu) + 3C \sum_{i=1}^{k} \operatorname{diam}(\mathscr{Q}_{i-1})^p \sum_{q \in \mathscr{Q}_i} \sqrt{\nu(q)},$$
where by convention $\mathscr{Q}_0 = \{Q\}$ is the trivial partition.

Proof. We start with applying Proposition 6 to the partition $\mathscr{Q}_1$ of Q:
$$\mathbb{E}L(\mathbf{1}_Q \cdot \nu) \leq \sum_{q \in \mathscr{Q}_1} \mathbb{E}L(\mathbf{1}_q \cdot \nu) + 3C \operatorname{diam}(\mathscr{Q}_0)^p \sum_{q \in \mathscr{Q}_1} \sqrt{\nu(q)}.$$

Next for each $q \in \mathscr{Q}_1$ we apply the proposition again for the partition of q induced by $\mathscr{Q}_2$ and iterate the process $k - 2$ times. $\qquad \Box$

3 Uniform Cube Samples

We introduce a specific notation for $n \in (0, +\infty)$,
$$\bar{L}(n) := \mathbb{E}L\big(n\mathbf{1}_{[0,1]^d}\big).$$

In this section, we will prove that $\bar{L}(n)/n^{1-p/d}$ converges. This will be the basic ingredient in the proof of Theorem 3. We first point out the following easy consequence of the homogeneity properties of Poisson point processes.

Lemma 2. *If L satisfies the homogeneity property $(\mathscr{H}_p)$ then for all $a \in \mathbb{R}^d$, $\rho > 0$ and $n > 0$*

$$\mathbb{E} L\left(n\mathbf{1}_{a+[0,\rho]^d}\right) = \rho^p \bar{L}\left(n\rho^d\right).$$

The following theorem is obtained by adapting to our abstract setting the line of reasoning in the paper [3] which was devoted to the bipartite matching:

Theorem 4. *Let $d > 2p$ be an integer. Let L be a bipartite functional on $\mathbb{R}^d$ satisfying the properties $(\mathcal{H}_p)$, $(\mathcal{R}_p)$ and $(\mathcal{S}_p)$. Then there exists $\beta_L \geq 0$ such that*

$$\lim_{n \to \infty} \frac{\bar{L}(n)}{n^{1-\frac{p}{d}}} = \beta_L.$$

Proof. Let $m \geq 1$ be an integer. Let $K \in \mathbb{N}$ such that $2^K \leq m < 2^{K+1}$. Set $Q_0 = [0, a]^d$ where $a := 2^{K+1}/m > 1$. Let $\mathcal{Q}_0 = \{Q_0\}$. We consider a sequence of nested partitions $\mathcal{Q}_j$, $j \geq 1$ where $\mathcal{Q}_j$ is a partition of Q_0 into 2^{jd} cubes of size $a2^{-j}$ (throughout the paper, this means that the interior of the cells are open cubes of such size, while their closure is a closed cube of the same size. We do not describe precisely how the points in the boundaries of the cubes are partitioned, since it is not relevant for the argument). One often says that $\mathcal{Q}_j$, $j \geq 1$ is a sequence of dyadic partitions of Q_0.

A direct application of Corollary 3 for the partitions $\mathcal{Q}_1, \ldots, \mathcal{Q}_{K+1}$ and the measure $n\mathbf{1}_{[0,1]^d}(x)\,dx$ gives

$$\bar{L}(n) = \mathbb{E} L(n\mathbf{1}_{[0,1]^d}) \leq \sum_{q \in \mathcal{Q}_{K+1}} \mathbb{E} L(n\mathbf{1}_{q \cap [0,1]^d})$$

$$+ 3C \sum_{j=1}^{K+1} \operatorname{diam}(\mathcal{Q}_{j-1})^p \sum_{q \in \mathcal{Q}_j} \sqrt{n\operatorname{Vol}(q \cap [0,1]^d)}.$$

Note that $\mathcal{Q}_{K+1}$ is a partition into cubes of size $1/m$, so that its intersection with $[0, 1]^d$ induces an (essential) partition of the unit cube into m^d cubes of side-length $1/m$. Hence, in the first sum, there are m^d terms which are equal, thanks to translation invariance and Lemma 2 to $\mathbb{E} L(n\mathbf{1}_{[0,m^{-1}]^d}) = m^{-p}\bar{L}(nm^{-d})$. The remaining terms of the first sum vanish. In order to deal with the second sum of the above estimate, we simply use the fact that $\mathcal{Q}_j$ contains 2^{jd} cubical cells of size $a2^{-j} = 2^{K+1-j}/m \leq 2^{1-j}$. Hence their individual volumes are at most $2^{d(1-j)}$. These observations allow to rewrite the above estimate as

$$\bar{L}(n) \leq m^{d-p}\bar{L}(nm^{-d}) + 3C \sum_{j=1}^{K+1} \operatorname{diam}([0, 2^{2-j}]^d)^p 2^{jd} \sqrt{n\, 2^{d(1-j)}}$$

$$= m^{d-p}\bar{L}(nm^{-d}) + 3C \sqrt{n}\operatorname{diam}([0,1]^d)^p \sum_{j=1}^{K+1} 2^{p(2-j)+\frac{d}{2}(j+1)}.$$

Hence, there is a number D depending only on p, d and C such that

$$\bar{L}(n) \leq m^{d-p}\bar{L}(nm^{-d}) + D\sqrt{n}\, 2^{K(\frac{d}{2}-p)} \leq m^{d-p}\bar{L}(nm^{-d}) + D\sqrt{n}\, m^{\frac{d}{2}-p}.$$

Let $t > 0$. Setting, $n = m^d t^d$ and $f(u) = \bar{L}(u^d)/u^{d-p}$, the latter inequality reads as

$$f(mt) \leq f(t) + Dt^{p-\frac{d}{2}},$$

and is valid for all $t > 0$ and $m \in \mathbb{N}^*$. Since f is continuous (Proposition 2 shows that $u \mapsto \bar{L}(u)$ is Lipschitz) and $\lim_{t \to +\infty} t^{p-\frac{d}{2}} = 0$, it follows that $\lim_{t \to +\infty} f(t)$ exists (we refer to [3] for details). □

Remark 4. The above constant β_L is positive as soon as L satisfies the following natural condition: for all $x_1, \ldots, x_n, y_1, \ldots y_n$ in $\mathbb{R}^d$, $L(\{x_1, \ldots, x_n\}, \{y_1, \ldots, y_n\}) \geq c \sum_i \text{dist}(x_i, \{y_1, \ldots, y_n\})^p$. To see this, one combines Proposition 1 and the lower estimate given in [16].

4 Upper Bounds, Upper Limits

4.1 A General Upper Bound

Using nested partitions, it is possible to refine Corollary 1 to a sharp order of magnitude.

Lemma 3. *Let $d > 2p$ and let L be a bipartite functional satisfying $(\mathscr{S}_p)$, $(\mathscr{R}_p)$ and $L(\emptyset, \emptyset) = 0$. Then there exists a constant D such that, for all finite measures ν,*

$$\mathbb{E}L(\nu) \leq D \, \text{diam}(Q)^p \min\left(\nu(Q), \nu(Q)^{1-\frac{p}{d}}\right),$$

where Q contains the support of ν.

Proof. Thanks to Corollary 1, it is enough to deal with the case $\nu(Q) \geq 2^d$ (or any other positive number). First note that we may assume that Q is a cube (given a set of diameter Δ, one can find a cube containing it, with diameter no more than c times Δ where c only depends on the norm). We consider a sequence of dyadic partitions of Q, $(\mathscr{P}_\ell)_{\ell \geq 0}$, where for $\ell \in \mathbb{N}$, $\mathscr{P}_\ell$ divides Q into $2^{\ell d}$ cubes of side-length $2^{-\ell}$ times the one of Q. Let $k \in \mathbb{N}^*$ to be chosen later. By Corollary 3, we have the following estimate

$$\mathbb{E}L(\nu) \leq \sum_{P \in \mathscr{P}_k} \mathbb{E}L(1_P \cdot \nu) + 3C \sum_{\ell=1}^{k} \left(2^{-\ell+1}\text{diam}(Q)\right)^p \sum_{P \in \mathscr{P}_\ell} \sqrt{\nu(P)}. \quad (4)$$

Thanks to Corollary 1, the first term of the right-hand side of (4) is at most

$$\sum_{P \in \mathscr{P}_k} C \left(2^{-k} \operatorname{diam}(Q)\right)^p v(P) = C \, 2^{-kp} (\operatorname{diam}(Q))^p v(Q).$$

By the Cauchy–Schwarz inequality

$$\sum_{P \in \mathscr{P}_\ell} \sqrt{v(P)} \le (2^{\ell d})^{\frac{1}{2}} \left(\sum_{P \in \mathscr{P}_\ell} v(P) \right)^{\frac{1}{2}} = 2^{\frac{\ell d}{2}} \sqrt{v(Q)}.$$

Hence the second term of the right-hand side of (4) is at most

$$3C \left(2 \operatorname{diam}(Q)\right)^p \sum_{\ell=1}^{k} 2^{\ell \left(\frac{d}{2}-p\right)} \sqrt{v(Q)} \le C' 2^{k\left(\frac{d}{2}-p\right)} (\operatorname{diam}(Q))^p \sqrt{v(Q)}.$$

This leads to

$$\mathbb{E}L(v) \le (\operatorname{diam}(Q))^p \left(C 2^{-kp} v(Q) + C' 2^{k\left(\frac{d}{2}-p\right)} \sqrt{v(Q)} \right).$$

Choosing $k = \lfloor \frac{1}{d} \log_2 v(Q) \rfloor \ge 1$ completes the proof. $\square$

4.2 The Upper Limit for Densities

All ingredients have now been gathered in order to state our upper bound for measures μ which have a density.

Theorem 5. *Let $d > 2p$. Let L be a bipartite functional on $\mathbb{R}^d$ satisfying the properties $(\mathscr{H}_p)$, $(\mathscr{R}_p)$, $(\mathscr{S}_p)$. Let $f : \mathbb{R}^d \to \mathbb{R}^+$ be an integrable function with bounded support. Then*

$$\limsup_{n \to \infty} \frac{\mathbb{E}L(n\,f)}{n^{1-\frac{p}{d}}} \le \beta_L \int_{\mathbb{R}^d} f^{1-\frac{p}{d}},$$

where β_L is the constant appearing in Theorem 4.

Proof. By a scaling argument, we may assume that the support of f is included in $[0, 1]^d$ and $\int f = 1$ (the case $\int f = 0$ is trivial). We consider a sequence of dyadic partitions $(\mathscr{P}_\ell)_{\ell \in \mathbb{N}}$ of $[0, 1]^d$: for $\ell \in \mathbb{N}$, $\mathscr{P}_\ell$ divides $[0, 1]^d$ into $2^{\ell d}$ cubes of side-length $2^{-\ell}$. Let $k \in \mathbb{N}^*$ to be chosen later. Corollary 3 gives

$$\mathbb{E}L(n\,f) \leq \sum_{P \in \mathscr{P}_k} \mathbb{E}L(n\,f\mathbf{1}_P) + 3C \sum_{\ell=1}^{k} \left(2^{-\ell+1}\mathrm{diam}([0,1]^d)\right)^p \sum_{P \in \mathscr{P}_\ell} \sqrt{n \int_P f}. \tag{5}$$

By the Cauchy–Schwarz inequality

$$\sum_{P \in \mathscr{P}_\ell} \sqrt{\int_P f} \leq (2^{\ell d})^{\frac{1}{2}} \left(\sum_{P \in \mathscr{P}_\ell} \int_P f\right)^{\frac{1}{2}} = 2^{\frac{\ell d}{2}} \left(\int f\right)^{\frac{1}{2}} = 2^{\frac{\ell d}{2}}.$$

Hence the second term of the right-hand side of (5) is at most

$$3C \left(2\mathrm{diam}([0,1]^d)\right)^p \sqrt{n} \sum_{\ell=1}^{k} 2^{\ell\left(\frac{d}{2}-p\right)} \leq c_d n^{\frac{1}{2}} 2^{k\left(\frac{d}{2}-p\right)}.$$

Let α_P be the average of f on P, then applying Corollary 2 to the first terms of (5) leads to

$$\mathbb{E}L(n\,f) \leq \sum_{P \in \mathscr{P}_k} \left(\mathbb{E}L(n\,\alpha_P \mathbf{1}_P) + 2C\,n\,\mathrm{diam}(P)^p \int_P |f - \alpha_P|\right) + c_d n^{\frac{1}{2}} 2^{k\left(\frac{d}{2}-p\right)}.$$

Each P in the sum is a square of side length 2^{-k}, hence using homogeneity (see Lemma 2)

$$\mathbb{E}L(n\,f) \leq \sum_{P \in \mathscr{P}_k} \left(2^{-kp} M\left(n\,\alpha_P 2^{-kd}\right) + n\,c_d'\,2^{-kp} \int_P |f - \alpha_P|\right) + c_d n^{\frac{1}{2}} 2^{k\left(\frac{d}{2}-p\right)}. \tag{6}$$

Let us recast this inequality with more convenient notation. We set $g(t) = \bar{L}(t)/t^{1-p/d}$ and we define the piecewise constant function

$$f_k = \sum_{P \in \mathscr{P}_k} \alpha_P \mathbf{1}_P = \sum_{P \in \mathscr{P}_k} \frac{\int_P f(x)\,dx}{\mathrm{Vol}(P)} \mathbf{1}_P.$$

It is plain that $\int f_k = \int f < +\infty$. Moreover, by Lebesgue's theorem, $\lim_{k \to \infty} f_k = f$ holds for almost every point x. Inequality (6) amounts to

$$\frac{\mathbb{E}L(n\,f)}{n^{1-\frac{p}{d}}} \leq \sum_{P \in \mathscr{P}_k} \left(g(n\,\alpha_P 2^{-kd})\alpha_P^{1-\frac{p}{d}} 2^{-kd} + n^{\frac{p}{d}} c_d' 2^{-kp} \int_P |f - f_k|\right)$$

$$+ c_d n^{\frac{p}{d}-\frac{1}{2}} 2^{k\left(\frac{d}{2}-p\right)}$$

$$= \sum_{P \in \mathscr{P}_k} \left(\int_P g(n\,f_k 2^{-kd}) f_k^{1-\frac{p}{d}} + n^{\frac{p}{d}} c_d' 2^{-kp} \int_P |f - f_k|\right) + c_d n^{\frac{p}{d}-\frac{1}{2}} 2^{k\left(\frac{d}{2}-p\right)}$$

$$= \int g(n\,2^{-kd} f_k) f_k^{1-\frac{p}{d}} + c_d' n^{\frac{p}{d}} 2^{-kp} \int |f - f_k| + c_d \left(n^{\frac{1}{d}} 2^{-k}\right)^{p-\frac{d}{2}}.$$

If there exists k_0 such that $f = f_{k_0}$ then we easily get the claim by setting $k = k_0$ and letting n go to infinity (since g is bounded and converges to β_L at infinity, see Lemma 3 and Theorem 4). On the other hand, if f_k never coincides almost surely with f, we use a sequence of numbers $k(n) \in \mathbb{N}$ such that

$$\lim_n k(n) = +\infty, \quad \lim_n n^{\frac{1}{d}} 2^{-k(n)} = +\infty \quad \text{and} \quad \lim_n n^{\frac{1}{d}} 2^{-k(n)} \left(\int |f - f_{k(n)}| \right)^{\frac{1}{p}} = 0. \tag{7}$$

Assuming its existence, the claim follows easily: applying the inequality for $k = k(n)$ and taking upper limits gives

$$\limsup_n \frac{\mathbb{E} L(n\,f)}{n^{1-\frac{p}{d}}} \leq \limsup_n \int g\left(n\, 2^{-k(n)d}\, f_{k(n)}\right) f_{k(n)}^{1-\frac{p}{d}}.$$

Since $\lim f_{k(n)} = f$ a.e., it is easy to see that the limit of the latter integral is $\beta_L \int f^{1-\frac{p}{d}}$: first the integrand converges almost everywhere to $\beta_L f^{1-\frac{p}{d}}$ (if $f(x) = 0$ this follows from the boundedness of g; if $f(x) \neq 0$ then the argument of g is going to infinity). Secondly, the sequence of integrands is supported on the unit cube and is uniformly integrable since

$$\int \left(g\left(n\, 2^{-k(n)d}\, f_{k(n)}\right) f_k^{1-\frac{p}{d}} \right)^{\frac{d}{d-p}} \leq (\sup g)^{\frac{d}{d-p}} \int f_{k(n)} = (\sup g)^{\frac{d}{d-p}} \int f < +\infty.$$

It remains to establish the existence of a sequence of integers $(k(n))_n$ satisfying (7). Note that since $f_k \geq 0$, $\int f_k = \int f = 1$ and a.e. $\lim f_k = f$, it follows from Scheffé's lemma that $\lim_k \int |f - f_k| = 0$. Hence $\varphi(k) = (\sup_{j \geq k} \int |f - f_j|)^{-d/p}$ is nondecreasing with an infinite limit. We derive the existence of a sequence with the following stronger properties

$$\lim_n k(n) = +\infty, \quad \lim_n \frac{n}{(2^d)^{k(n)}} = +\infty \quad \text{and} \quad \lim_n \frac{n}{(2^d)^{k(n)} \varphi(k(n))} = 0 \tag{8}$$

as follows. Set $\gamma = 2^d$. Since $\gamma^k \sqrt{\varphi(k-1)}$ is increasing with infinite limit,

$$[\gamma \sqrt{\varphi(0)}, +\infty) = \cup_{k \geq 1} [\gamma^k \sqrt{\varphi(k-1)}, \gamma^{k+1} \sqrt{\varphi(k)}).$$

For $n \geq \gamma \sqrt{\varphi(0)}$, we define $k(n)$ as the integer such that

$$\gamma^{k(n)} \sqrt{\varphi(k(n) - 1)} \leq n < \gamma^{k(n)+1} \sqrt{\varphi(k(n))}.$$

This defines a nondecreasing sequence. It is clear from the above strict inequality that $\lim_n k(n) = +\infty$. Hence $n\gamma^{-k(n)} \geq \sqrt{\varphi(k(n)-1)}$ tends to infinity at infinity. Eventually $n/(\gamma^{k(n)} \varphi(k(n))) \leq \gamma / \sqrt{\varphi(k(n))}$ tends to zero as required. The proof is therefore complete. □

4.3 Purely Singular Measures

With Theorem 5 at hand, we should now understand what happens when μ has a singular part. Our next lemma states if μ is purely singular then $\mathbb{E}L(n\mu)$ is of order smaller than $n^{1-\frac{p}{d}}$.

Lemma 4. *Let $d > 2p$. Let L be a bipartite functional on $\mathbb{R}^d$ with properties $(\mathcal{R}_p)$ and $(\mathcal{S}_p)$. Let μ be a finite singular measure on $\mathbb{R}^d$ having a bounded support. Then*

$$\lim_{n\to\infty} \frac{\mathbb{E}L(n\mu)}{n^{1-\frac{p}{d}}} = 0.$$

Proof. Let Q be a cube which contains the support of μ. We consider a sequence of dyadic partitions of Q, $(\mathscr{P}_\ell)_{\ell\in\mathbb{N}}$. For $\ell \in \mathbb{N}$, $\mathscr{P}_\ell$ divides Q into $2^{\ell d}$ cubes of side length $2^{-\ell}$ times the one of Q. As in the proof of Lemma 3, a direct application of Corollary 3 gives for $k \in \mathbb{N}^*$:

$$\mathbb{E}L(n\mu) \leq \sum_{P\in\mathscr{P}_k} \mathbb{E}L(n\mathbf{1}_P \cdot \mu) + 3C \sum_{\ell=1}^{k} \left(2^{-\ell+1}\mathrm{diam}(Q)\right)^p \sum_{P\in\mathscr{P}_\ell} \sqrt{n\mu(P)}. \tag{9}$$

The terms of the first sum are estimated again thanks to the easy bound of Corollary 1: since each P in $\mathscr{P}_k$ is a cube of side length 2^{-k} times the one of Q, it holds

$$\sum_{P\in\mathscr{P}_k} \mathbb{E}L(n\mathbf{1}_P \cdot \mu) \leq \sum_{P\in\mathscr{P}_k} C\left(2^{-k}\mathrm{diam}(Q)\right)^p n\mu(P) = c_{p,Q} \, 2^{-kp} n|\mu|.$$

Here $|\mu|$ is the total mass of μ. We rewrite the second term in (9) in terms of the function

$$g_\ell = \sum_{P\in\mathscr{P}_\ell} \frac{\mu(P)}{\lambda(P)} \mathbf{1}_P,$$

where λ stands for Lebesgue's measure. Since $\lambda(P) = 2^{-\ell d}\lambda(Q)$, we get that

$$\mathbb{E}L(n\mu) \leq c_{p,Q} \, 2^{-kp} n|\mu|$$

$$+ 3C \left(2\mathrm{diam}(Q)\right)^p \sqrt{n} \sum_{\ell=1}^{k} 2^{-\ell p} \sum_{P\in\mathscr{P}_\ell} 2^{\frac{\ell d}{2}} \lambda(Q)^{-\frac{1}{2}} \lambda(P) \sqrt{\frac{\mu(P)}{\lambda(P)}}$$

$$= c_{p,Q} \, 2^{-kp} n|\mu| + 3C \left(2\mathrm{diam}(Q)\right)^p \lambda(Q)^{-\frac{1}{2}} \sqrt{n} \sum_{\ell=1}^{k} 2^{\ell\left(\frac{d}{2}-p\right)} \int \sqrt{g_\ell}.$$

By the differentiability theorem, for Lebesgue-almost every x, $g_\ell(x)$ tends to zero when ℓ tends to infinity (since μ is singular with respect to Lebesgue's measure). Moreover, g_ℓ is supported on the unit cube and $\int (\sqrt{g_\ell})^2 = \int g_\ell = |\mu| < +\infty$. Hence the sequence of functions $\sqrt{g_\ell}$ is uniformly integrable and we can conclude that $\lim_{\ell \to \infty} \int \sqrt{g_\ell} = 0$. By Cesaro's theorem, the sequence

$$\varepsilon_k = \frac{\sum_{\ell=1}^k 2^{\ell(\frac{d}{2}-p)} \int \sqrt{g_\ell}}{\sum_{\ell=1}^k 2^{\ell(\frac{d}{2}-p)}}$$

also converges to zero, using here that $d > 2p$. By an obvious upper bound of the latter denominator, we obtain that there exists a number c which does not depend on (k, n) (but depends on $C, p, d, Q, |\mu|$) such that for all $k \geq 1$

$$\mathbb{E} L(n\mu) \leq c \left(n 2^{-kp} + \sqrt{n}\, 2^{k(\frac{d}{2}-p)} \varepsilon_k \right),$$

where $\varepsilon_k \geq 0$ and $\lim_k \varepsilon_k = 0$. We may also assume that (ε_k) is nonincreasing (the inequality remains valid if one replaces ε_k by $\sup_{j \geq k} \varepsilon_j$). It remains to choose k in terms of n in a proper way. Define

$$\varphi(n) = \sqrt{\varepsilon_{\lfloor \frac{1}{d} \log_2 n \rfloor}}^{\frac{-1}{\frac{d}{2}-p}}.$$

Obviously $\lim_n \varphi(n) = +\infty$. For n large enough, define $k(n) \geq 1$ as the unique integer such that

$$2^{k(n)} \leq n^{\frac{1}{d}} \varphi(n) < 2^{k(n)+1}.$$

Setting $k = k(n)$, our estimate on the cost of the optimal matching yields

$$\frac{\mathbb{E} L(n\mu)}{n^{1-\frac{p}{d}}} \leq c(d) \left(\frac{2}{\varphi(n)^p} + \varepsilon_{k(n)} \varphi(n)^{\frac{d}{2}-p} \right).$$

It is easy to check that the right hand side tends to zero as n tends to infinity. Indeed, $\lim_n \varphi(n) = +\infty$, hence for n large enough

$$k(n) \geq \left\lfloor \log_2 \left(n^{\frac{1}{d}} \varphi(n)/2 \right) \right\rfloor \geq \left\lfloor \frac{1}{d} \log_2 n \right\rfloor.$$

Since the sequence (ε_k) is nonincreasing, it follows that

$$\varepsilon_{k(n)} \varphi(n)^{\frac{d}{2}-p} \leq \varepsilon_{\lfloor \frac{1}{d} \log_2 n \rfloor} \varphi(n)^{\frac{d}{2}-p} = \sqrt{\varepsilon_{\lfloor \frac{1}{d} \log_2 n \rfloor}}$$

tends to zero when $n \to \infty$. The proof is therefore complete. $\square$

4.4 General Upper Limits

We are now in position to conclude the proof the first statement of Theorem 3. It is a consequence of Propositions 1, 4, and the following result.

Theorem 6. *Let $d > 2p > 0$. Let L be a bipartite functional on $\mathbb{R}^d$ with the properties $(\mathcal{H}_p)$, $(\mathcal{R}_p)$, and $(\mathcal{S}_p)$. Consider a finite measure μ on $\mathbb{R}^d$ such that there exists $\alpha > \frac{2dp}{d-2p}$ with*

$$\int |x|^\alpha d\mu(x) < +\infty.$$

Let f be a density function for the absolutely continuous part of μ, then

$$\limsup_{n\to\infty} \frac{\mathbb{E}L(n\mu)}{n^{1-\frac{p}{d}}} \leq \beta_L \int f^{1-\frac{p}{d}}. \tag{10}$$

Remark 5. Observe that the hypotheses ensure the finiteness of $\int f^{1-\frac{p}{d}}$. Indeed Hölder's inequality gives

$$\int_{\mathbb{R}^d} f^{1-\frac{p}{d}} \leq \left(\int_{\mathbb{R}^d} (1+|x|^\alpha)f(x)dx\right)^{1-\frac{p}{d}} \left(\int_{\mathbb{R}^d} (1+|x|^\alpha)^{1-\frac{d}{p}}\right)^{\frac{p}{d}}$$

where the latter integral converges since $\alpha > \frac{2dp}{d-2p} > \frac{dp}{d-p}$.

Proof. Assume first that μ has a bounded support. Write $\mu = \mu_{ac} + \mu_s$ where μ_s is the singular part and $d\mu_{ac}(x) = f(x)\,dx$. Applying Proposition 5 to μ_{ac} and μ_s, dividing by $n^{1-p/d}$, passing to the limit and using Theorem 5 and Lemma 4 gives

$$\limsup_n \frac{\mathbb{E}L(n\mu)}{n^{1-\frac{p}{d}}} \leq \limsup_n \frac{\mathbb{E}L(n\mu_{ac})}{n^{1-\frac{p}{d}}} + \limsup_n \frac{\mathbb{E}L(n\mu_s)}{n^{1-\frac{p}{d}}} \leq \beta_L \int f^{1-\frac{p}{d}}.$$

Hence the theorem is established for measures with bounded supports.

Now, let us consider the general case. Let $B(t) = \{x \in \mathbb{R}^d : |x| \leq t\}$. Let $A_0 = B(2)$ and for integer $\ell \geq 1$, $A_\ell = B(2^{\ell+1})\setminus B(2^\ell)$. Now, let $\mathcal{X} = \{X_1,\cdots,X_{N_1}\}$, $\mathcal{Y} = \{Y_1,\cdots,Y_{N_2}\}$ be two independent Poisson process of intensity $n\mu$, and $T = \max\{|Z| : Z \in \mathcal{X} \cup \mathcal{Y}\}$. Applying the subadditivity property like in the proof of Proposition 6, we obtain

$$L(\mathcal{X},\mathcal{Y}) \leq \sum_{\ell \geq 0} L(\mathcal{X} \cap A_\ell, \mathcal{Y} \cap A_\ell) \tag{11}$$

$$+ CT^p \sum_{\ell \geq 0} \mathbf{1}_{\mathcal{X}(A_\ell)+\mathcal{Y}(A_\ell)\neq 0}\big(1 + |\mathcal{X}(A_\ell) - \mathcal{Y}(A_\ell)|\big).$$

Note that the above sums have only finitely many nonzero terms, since μ is finite. We first deal with the first sum in the above inequality. By Fubini's Theorem,

$$\mathbb{E}\sum_{\ell\geq 0}\frac{L(\mathcal{X}\cap A_\ell,\mathcal{Y}\cap A_\ell)}{n^{1-\frac{p}{d}}} = \sum_{\ell\geq 0}\mathbb{E}\frac{L(\mathcal{X}\cap A_\ell,\mathcal{Y}\cap A_\ell)}{n^{1-\frac{p}{d}}}.$$

Applying (10) to the compactly supported measure $\mu_{|A_\ell}$ for every integer ℓ gives

$$\limsup_n \mathbb{E}\frac{L(\mathcal{X}\cap A_\ell,\mathcal{Y}\cap A_\ell)}{n^{1-\frac{p}{d}}} \leq \beta_L \int_{A_l} f^{1-\frac{p}{d}}. \qquad (12)$$

By Lemma 3, for some constant c_d,

$$\mathbb{E}\frac{L(\mathcal{X}\cap A_\ell,\mathcal{Y}\cap A_\ell)}{n^{1-\frac{p}{d}}} \leq c_d 2^{\ell p}\mu(A_\ell)^{1-\frac{p}{d}}.$$

From Markov inequality, with $m_\alpha = \int |x|^\alpha d\mu(x)$,

$$\mu(A_\ell) \leq \mu(\mathbb{R}^d\setminus B(2^\ell)) \leq 2^{-\ell\alpha}m_\alpha.$$

Thus, since $\alpha > 2pd/(d-2p) > dp/(d-p)$, the series $\sum_\ell 2^{\ell p}\mu(A_\ell)^{1-\frac{p}{d}}$ is convergent. We may then apply the dominated convergence theorem, we get from (12) that

$$\limsup_n \mathbb{E}\sum_{\ell\geq 0}\frac{L(\mathcal{X}\cap A_\ell,\mathcal{Y}\cap A_\ell)}{n^{1-\frac{p}{d}}} \leq \beta_L \int f^{1-\frac{p}{d}}.$$

For the expectation of the second term on the right hand side of (11), we use Cauchy–Schwartz inequality,

$$\mathbb{E}\left[T^p \sum_{\ell\geq 0} \mathbf{1}_{\mathcal{X}(A_\ell)+\mathcal{Y}(A_\ell)\neq 0}(1+|\mathcal{X}(A_\ell)-\mathcal{Y}(A_\ell)|)\right]$$

$$\leq \sum_{\ell\geq 0}\sqrt{\mathbb{E}[T^{2p}]}\sqrt{\mathbb{E}(\mathbf{1}_{\mathcal{X}(A_\ell)+\mathcal{Y}(A_\ell)\neq 0}(1+|\mathcal{X}(A_\ell)-\mathcal{Y}(A_\ell)|)^2)}$$

$$\leq \sqrt{2}\sqrt{\mathbb{E}[T^{2p}]}\sum_{\ell\geq 0}\sqrt{\mathbb{P}(\mathcal{X}(A_\ell)+\mathcal{Y}(A_\ell)\neq 0)+\mathbb{E}[|\mathcal{X}(A_\ell)-\mathcal{Y}(A_\ell)|^2]}$$

$$= \sqrt{2}\sqrt{\mathbb{E}[T^{2p}]}\sum_{\ell\geq 0}\sqrt{1-e^{-2n\mu(A_\ell)}+2n\mu(A_\ell)}$$

$$\leq 2\sqrt{\mathbb{E}[T^{2p}]}\sqrt{n}\sum_{\ell\geq 0}\sqrt{\mu(A_\ell)},$$

where we have used $1 - e^{-u} \leq u$. As above, Markov inequality leads to

$$\sum_{\ell \geq 0} \sqrt{\mu(A_\ell)} \leq \sqrt{m_\alpha} \sum_{\ell \geq 0} 2^{-\ell \frac{\alpha}{2}} < +\infty.$$

Eventually we apply Lemma 1 with $\gamma := 2p < 2pd/(d-2) < \alpha$ to upper bound $\mathbb{E}[T^{2p}]$. We get that for some constant $c > 0$ and all $n > 0$,

$$n^{-1+\frac{p}{d}} \mathbb{E}\left[T^p \sum_{\ell \geq 0} \mathbf{1}_{\mathscr{X}(A_\ell) + \mathscr{Y}(A_\ell) \neq 0} \left(1 + |\mathscr{X}(A_\ell) - \mathscr{Y}(A_\ell)|\right)\right] \leq cn^{-\frac{1}{2}+\frac{p}{d}+\frac{p}{\alpha}}.$$

Since $\alpha > 2dp/(d-2p)$, the later and former terms tend to zero as n tends to infinity. The upper bound (10) is proved. □

5 Examples of Bipartite Functionals

The minimal bipartite matching is an instance of a bipartite Euclidean functional $M_1(\mathscr{X}, \mathscr{Y})$ over the multisets $\mathscr{X} = \{X_1, \ldots, X_n\}$ and $\mathscr{Y} = \{Y_1, \ldots, Y_n\}$. We may mention at least two other interesting examples: the bipartite traveling salesperson problem over $\mathscr{X}$ and $\mathscr{Y}$ is the shortest cycle on the multiset $\mathscr{X} \cup \mathscr{Y}$ such that the image of $\mathscr{X}$ is $\mathscr{Y}$. Similarly, the bipartite minimal spanning tree is the minimal edge-length spanning tree on $\mathscr{X} \cup \mathscr{Y}$ with no edge between two elements of $\mathscr{X}$ or two elements of $\mathscr{Y}$.

5.1 Minimal Bipartite Matching

Fix $p > 0$. Given two multisubsets of $\mathbb{R}^d$ with the same cardinality, $\mathscr{X} = \{X_1, \ldots, X_n\}$ and $\mathscr{Y} = \{Y_1, \ldots, Y_n\}$, the p-cost of the minimal bipartite matching of $\mathscr{X}$ and $\mathscr{Y}$ is defined as

$$M_p(\mathscr{X}, \mathscr{Y}) = \min_{\sigma \in \mathscr{S}_n} \sum_{i=1}^n |X_i - Y_{\sigma(i)}|^p,$$

where the minimum runs over all permutations of $\{1, \ldots, n\}$. It is useful to extend the definition to sets of different cardinalities, by matching as many points as possible: if $\mathscr{X} = \{X_1, \ldots, X_m\}$ and $\mathscr{Y} = \{Y_1, \ldots, Y_n\}$ and $m \leq n$ then

$$M_p(\mathscr{X}, \mathscr{Y}) = \min_{\sigma} \sum_{i=1}^m |X_i - Y_{\sigma(i)}|^p,$$

where the minimum runs over all injective maps from $\{1,\ldots,m\}$ to $\{1,\ldots,n\}$. When $n \leq m$ the symmetric definition is chosen $M_p(\mathscr{X},\mathscr{Y}) := M_p(\mathscr{Y},\mathscr{X})$.

The bipartite functional M_p is obviously homogeneous of degree p, i.e. it satisfies $(\mathscr{H}_p)$. The next lemma asserts that it also satisfies the subadditivity property $(\mathscr{S}_p)$. In the case $p = 1$, this is the starting point of the paper [3].

Lemma 5. *For any $p > 0$, the functional M_p satisfies property $(\mathscr{S}_p)$ with constant $C = 1/2$. More precisely, if $\mathscr{X}_1,\ldots,\mathscr{X}_k$ and $\mathscr{Y}_1,\ldots,\mathscr{Y}_k$ are multisets in a bounded subset $Q \subset \mathbb{R}^d$, then*

$$M_p\Big(\bigcup_{i=1}^k \mathscr{X}_i, \bigcup_{i=1}^k \mathscr{Y}_i\Big) \leq \sum_{i=1}^k M_p(\mathscr{X}_i,\mathscr{Y}_i) + \frac{\mathrm{diam}(Q)^p}{2} \sum_{i=1}^k |\mathrm{card}(\mathscr{X}_i) - \mathrm{card}(\mathscr{Y}_i)|.$$

Proof. It is enough to upper bound the cost of a particular matching of $\bigcup_{i=1}^k \mathscr{X}_i$ and $\bigcup_{i=1}^k \mathscr{Y}_i$. We build a matching of these multisets as follows. For each i we choose the optimal matching of $\mathscr{X}_i$ and $\mathscr{Y}_i$. The overall cost is $\sum_i M_p(\mathscr{X}_i,\mathscr{Y}_i)$, but we have left $\sum_i |\mathrm{card}(\mathscr{X}_i) - \mathrm{card}(\mathscr{Y}_i)|$ points unmatched (the number of excess points). Among these points, the less numerous species (there are two species: points from $\mathscr{X}_i$'s, and points from $\mathscr{Y}_i$'s) has cardinality at most $\frac{1}{2}\sum_i |\mathrm{card}(\mathscr{X}_i) - \mathrm{card}(\mathscr{Y}_i)|$. To complete the definition of the matching, we have to match all the points of this species in the minority. We do this in an arbitrary manner and simply upper bound the distance between matched points by the diameter of Q. □

The regularity property is established next.

Lemma 6. *For any $p > 0$, the functional M_p satisfies property $(\mathscr{R}_p)$ with constant $C = 1$.*

Proof. Let $\mathscr{X}, \mathscr{X}_1, \mathscr{X}_2, \mathscr{Y}, \mathscr{Y}_1, \mathscr{Y}_2$ be finite multisets contained in $Q = B(1/2)$. Denote by x, x_1, x_2, y, y_1, y_2 the cardinalities of the multisets and $a \wedge b$ for $\min(a,b)$. We start with an optimal matching for $M_p(\mathscr{X} \cap \mathscr{X}_2, \mathscr{Y} \cap \mathscr{Y}_2)$. It comprises $(x + x_2) \wedge (y + y_2)$ edges. We remove the ones which have a vertex in $\mathscr{X}_2$ or in $\mathscr{Y}_2$. There are at most $x_2 + y_2$ of them, so we are left with at least $((x + x_2) \wedge (y + y_2) - x_2 - y_2)_+$ edges connecting points of $\mathscr{X}$ to points of $\mathscr{Y}$. We want to use this partial matching in order to build a (suboptimal) matching of $\mathscr{X} \cap \mathscr{X}_1$ and $\mathscr{Y} \cap \mathscr{Y}_1$. This requires to have globally $(x + x_1) \wedge (y + y_1)$ edges. Hence we need to add at most

$$(x + x_1) \wedge (y + y_1) - \big((x + x_2) \wedge (y + y_2) - x_2 - y_2\big)_+$$

new edges. We do this in an arbitrary way, and simply upper bound their length by the diameter of Q. To prove the claim it is therefore sufficient to prove the following inequalities for nonnegative numbers:

$$(x+x_1)\wedge(y+y_1) - \big((x+x_2)\wedge(y+y_2) - x_2 - y_2\big)_+ \leq x_1 + x_2 + y_1 + y_2. \qquad (13)$$

This is obviously equivalent to

$$x + x_1 \le x_1 + x_2 + y_1 + y_2 + \big((x+x_2) \wedge (y+y_2) - x_2 - y_2\big)_+$$
$$\text{or } y + y_1 \le x_1 + x_2 + y_1 + y_2 + \big((x+x_2) \wedge (y+y_2) - x_2 - y_2\big)_+.$$

After simplification, and noting that $y_1 \ge 0$ appears only on the right-hand side of the first inequality (and the same for x_1 in the second one), it is enough to show that

$$x \wedge y \le x_2 + y_2 + \big((x+x_2) \wedge (y+y_2) - x_2 - y_2\big)_+.$$

This is obvious, as by definition of the positive part, $x \wedge y \le x_2 + y_2 + \big((x \wedge y) - x_2 - y_2\big)_+$. □

5.2 Bipartite Traveling Salesperson Tour

Fix $p > 0$. Given two multi-subsets of $\mathbb{R}^d$ with the same cardinality, $\mathscr{X} = \{X_1, \ldots, X_n\}$ and $\mathscr{Y} = \{Y_1, \ldots, Y_n\}$, the p-cost of the minimal bipartite traveling salesperson tour of $(\mathscr{X}, \mathscr{Y})$ is defined as

$$T_p(\mathscr{X}, \mathscr{Y}) = \min_{(\sigma,\sigma') \in S_n^2} \sum_{i=1}^n |X_{\sigma(i)} - Y_{\sigma'(i)}|^p + \sum_{i=1}^{n-1} |Y_{\sigma'(i)} - X_{\sigma(i+1)}|^p + |Y_{\sigma'(n)} - X_{\sigma(1)}|^p,$$

where the minimum runs over all pairs of permutations of $\{1, \ldots, n\}$. We extend the definition to sets of different cardinalities, by completing the longest possible bipartite tour: if $\mathscr{X} = \{X_1, \ldots, X_m\}$ and $\mathscr{Y} = \{Y_1, \ldots, Y_n\}$ and $m \le n$ then

$$T_p(\mathscr{X}, \mathscr{Y}) = \min_{(\sigma,\sigma')} \sum_{i=1}^m |X_{\sigma(i)} - Y_{\sigma'(i)}|^p + \sum_{i=1}^{m-1} |Y_{\sigma'(i)} - X_{\sigma(i+1)}|^p + |Y_{\sigma'(m)} - X_{\sigma(1)}|^p$$

where the minimum runs over all pairs (σ, σ'), with $\sigma \in S_m$ and σ' is an injective maps from $\{1, \ldots, m\}$ to $\{1, \ldots, n\}$. When $n \le m$ the symmetric definition is chosen $T_p(\mathscr{X}, \mathscr{Y}) := T_p(\mathscr{Y}, \mathscr{X})$. This traveling salesperson functional is an instance of a larger class of functionals that we now describe.

5.3 Euclidean Combinatorial Optimization Over Bipartite Graphs

For integers m, n, we define $[n] = \{1, \cdots n\}$ and $[n]_m = \{m+1, \cdots, m+n\}$. Let $\mathscr{B}_n$ be the set of bipartite graphs with common vertex set $([n], [n]_n)$: if $G \in \mathscr{B}_n$, the edge set of G is contained is the set of pairs $\{i, n+j\}$, with $i, j \in [n]$.

We should introduce some graph definitions. If $G_1 \in \mathscr{B}_n$ and $G_2 \in \mathscr{B}_m$ we define $G_1 + G_2$ as the graph in $\mathscr{B}_{n+m}$ obtained by the following rule : if $\{i, n + j\}$ is an edge of G_1 then $\{i, n + m + j\}$ is an edge of $G_1 + G_2$, and if $\{i, m + j\}$ is an edge of G_2 then $\{n + i, 2n + m + j\}$ is an edge of $G_1 + G_2$. Finally, if $G \in \mathscr{B}_{n+m}$, the restriction G' of G to $\mathscr{B}_n$ is the element of $\mathscr{B}_n$ defined by the following construction rule: if $\{i, n + m + j\}$ is an edge of G and $(i, j) \in [n]^2$ then add $\{i, n + j\}$ as an edge of G'.

We consider a collection of subsets $\mathscr{G}_n \subset \mathscr{B}_n$ with the following properties, there exist constants $\kappa_0, \kappa \geq 1$ such that for all integers n, m,

(A1) *(not empty)* If $n \geq \kappa_0$, $\mathscr{G}_n$ is not empty.
(A2) *(isomorphism)* If $G \in \mathscr{G}_n$ and $G' \in \mathscr{B}_n$ is isomorphic to G then $G' \in \mathscr{G}_n$.
(A3) *(bounded degree)* If $G \in \mathscr{G}_n$, the degree of any vertex is at most κ.
(A4) *(merging)* If $G \in \mathscr{G}_n$ and $G' \in \mathscr{G}_m$, there exists $G'' \in \mathscr{G}_{n+m}$ such that $G + G'$ and G'' have all but at most κ edges in common. For $1 \leq m < \kappa_0$, it also holds if G' is the empty graph of $\mathscr{B}_m$.
(A5) *(restriction)* Let $G \in \mathscr{G}_n$ and $\kappa_0 + 1 \leq n$ and G' be the restriction of G to $\mathscr{B}_{n-1}$. Then there exists $G'' \in \mathscr{G}_{n-1}$ such that G' and G'' have all but at most κ edges in common.

If $|\mathscr{X}| = |\mathscr{Y}| = n$, we define

$$L(\mathscr{X}, \mathscr{Y}) = \min_{G \in \mathscr{G}_n} \sum_{(i,j) \in [n]^2 : \{i,n+j\} \in G} |X_i - Y_j|^p.$$

With the convention that the minimum over an empty set is 0. Note that the isomorphism property implies that $L(\mathscr{X}, \mathscr{Y}) = L(\mathscr{Y}, \mathscr{X})$. If $m = |\mathscr{X}| \leq |\mathscr{Y}| = n$, we define

$$L(\mathscr{X}, \mathscr{Y}) = \min_{(G,\sigma)} \sum_{(i,j) \in [m]^2 : \{i,m+j\} \in G} |X_i - Y_{\sigma(j)}|^p, \qquad (14)$$

where the minimum runs over all pairs (G, σ), $G \in \mathscr{G}_m$ and σ is an injective maps from $\{1, \ldots, m\}$ to $\{1, \ldots, n\}$. When $n \leq m$ the symmetric definition is chosen $L(\mathscr{X}, \mathscr{Y}) := L(\mathscr{Y}, \mathscr{X})$.

The case of bipartite matchings is recovered by choosing $\mathscr{G}_n$ as the set of graphs in $\mathscr{B}_n$ where all vertices have degree 1. We then have $\kappa_0 = 1$ and $\mathscr{G}_n$ satisfies the merging property with $\kappa = 0$. It also satisfies the restriction property with $\kappa = 1$. The case of the traveling salesperson tour is obtained by choosing $\mathscr{G}_n$ as the set of connected graphs in $\mathscr{B}_n$ where all vertices have degree 2, this set is nonempty for $n \geq \kappa_0 = 2$. Also this set $\mathscr{G}_n$ satisfies the merging property with $\kappa = 4$ (as can be checked by edge switching). The restriction property follows by merging strings into a cycle.

For the minimal bipartite spanning tree, we choose $\mathscr{G}_n$ as the set of connected trees of $[2n]$ in $\mathscr{B}_n$. It satisfies the restriction property and the merging property with $\kappa = 1$. For this choice, however, the maximal degree is not bounded uniformly in n. We could impose artificially this condition by defining $\mathscr{G}_n$ as the set of connected graphs in $\mathscr{B}_n$ with maximal degree bounded by $\kappa \geq 2$. We would then get the minimal bipartite spanning tree with maximal degree bounded by κ. It is not hard to verify that the corresponding functional satisfies all the above properties.

Another interesting example is the following. Fix an integer $r \geq 2$. Recall that a graph is r-regular if the degree of all its vertices is equal to r. We may define $\mathscr{G}_n$ as the set of r-regular connected graphs in $\mathscr{B}_n$. This set is not empty for $n \geq \kappa_0 = r$. It satisfies the first part of the merging property (A4) with $\kappa = 4$. Indeed, consider two r-regular graphs G, G', and take any edge $e = \{x, y\} \in G$ and $e' = \{x', y'\} \in G'$. The merging property holds with G'', the graph obtained from $G + G'$ by switching (e, e') in $(\{x, y'\}, \{x', y\})$. Up to increasing the value of κ, the second part of the merging property is also satisfied. Indeed, if n is large enough, it is possible to find $rm < r\kappa_0 = r^2$ edges $e_1, \cdots, e_{rm}$ in G with nonadjacent vertices. Now, in G'', we add m points from each species, and replace the edge $e_{ri+q} = \{x, n+y\}, 1 \leq i \leq m, 0 \leq q < r$, by two edges: one between x and the i-th point of the second species, and one between y and the i-th point of the first species. G'' is then a connected r-regular graph in $\mathscr{B}_{n+m}$ with all but at most $2r^2$ edges in common with G. Hence, by taking κ large enough, the second part of the merging property holds.

Checking the restriction property (A5) for r-regular graphs requires a little more care. Let $r = \kappa_0 + 1 \leq n$ and consider the restriction G_1 of $G \in \mathscr{B}_n$ to $\mathscr{B}_{n-1}$. Our goal is to show that by modifying a small number of edges of G_1, one can obtain a connected r-regular bipartite graph on $\mathscr{B}_{n-1}$. We first explain how to turn G_1 into a possibly nonconnected r-regular graph. Let us observe that G_1 was obtained from G by deleting one vertex of each species and the edges to which these points belong. Hence G_1 has vertices of degree r, and vertices of degree $r - 1$ (r blue and r red vertices if the removed points did not share an edge, only $r - 1$ points of each species if the removed points shared an edge). In any case G_1 has at most $2r$ connected components and r vertices of each color with one edge missing. The simplest way to turn G_1 into a r regular graph is to connect each blue vertex missing an edge with a red vertex missing an edge. However this is not always possible as these vertices may already be neighbors in G_1 and we do not allow multiple edges. However given a red vertex v_R and a blue vertex v_B of degree $r - 1$ and provided $n - 1 > 2r^2$ there exists a vertex v in G_1 which is at graph distance at least 3 from v_B and v_R. Then open up an edge to which v belongs and connect its end-points to v_R and v_B while respecting the bipartite structure. In the new graph v_B and v_R have degree r. Repeating this operation no more than r times turns G_1 into a r regular graphs with at most as many connected components (and the initial and the final graph differ by at most $3r$ edges). Next we apply the merge operation at most $2r - 1$ times in order to glue together the connected components (this leads to modifying at most $4(2r-1)$ edges). As a conclusion, provided we choose $\kappa_0 > 2r^2$, the restriction property holds for $\kappa = 11r$.

We now come back to the general case. From the definition, it is clear that L satisfies the property ($\mathscr{H}_p$). We are going to check that it also satisfies properties ($\mathscr{S}_p$) and ($\mathscr{R}_p$).

Lemma 7. *Assume (A1-A4). For any $p > 0$, the functional L satisfies property ($\mathscr{S}_p$) with constant $C = (3 + \kappa_0)\kappa/2$.*

Proof. The proof of is an extension of the proof of Lemma 5. We can assume without loss of generality $k \geq 2$. Let $\mathscr{X}_1, \ldots, \mathscr{X}_k$ and $\mathscr{Y}_1, \ldots, \mathscr{Y}_k$ be multisets in $Q = B(1/2)$. For ease of notation, let $x_i = |\mathscr{X}_i|$, $y_i = |\mathscr{Y}_i|$ and $n = \sum_{i=1}^k x_i \wedge \sum_{i=1}^k y_i$. If $n < \kappa_0$, then from the bounded degree property (A3),

$$L\left(\bigcup_{i=1}^k \mathscr{X}_i, \bigcup_{i=1}^k \mathscr{Y}_i\right) \leq n\kappa \leq \kappa\kappa_0.$$

If $n \geq \kappa_0$, it is enough to upper bound the cost for an element G in $\mathscr{G}_n$. For each $1 \leq i \leq k$, if $n_i = x_i \wedge y_i \geq \kappa_0$, we consider the element G_i in $\mathscr{G}_{n_i}$ which reaches the minimum cost of $L(\mathscr{X}_i, \mathscr{Y}_i)$. From the merging property (A4), there exists G' in $\mathscr{G}_{\sum_i 1_{n_i \geq \kappa_0} n_i}$ whose total cost is at most

$$L' := \sum_i L(\mathscr{X}_i, \mathscr{Y}_i) + \kappa k.$$

It remains at most $\sum_i \kappa_0 + |x_i - y_i|$ vertices that have been left aside. The less numerous species has cardinal $m_0 \leq m = (\sum_i \kappa_0 + |x_i - y_i|)/2$. If $m_0 \geq \kappa_0$, from the nonempty property (A1), there exists a graph $G'' \in \mathscr{G}_{m_0}$ that minimizes the cost of the vertices that have been left aside. From the merging and bounded degree properties, we get

$$L\left(\bigcup_{i=1}^k \mathscr{X}_i, \bigcup_{i=1}^k \mathscr{Y}_i\right) \leq L' + \kappa + \kappa m \leq \sum_i L(\mathscr{X}_i, \mathscr{Y}_i) + \frac{\kappa}{2}\sum_i (3 + \kappa_0 + |x_i - y_i|).$$

If $m_0 < \kappa_0$, we apply to G' the merging property with the empty graph: there exists an element G in $\mathscr{G}_n$ whose total cost is at most

$$L\left(\bigcup_{i=1}^k \mathscr{X}_i, \bigcup_{i=1}^k \mathscr{Y}_i\right) \leq L' + \kappa \leq \sum_i L(\mathscr{X}_i, \mathscr{Y}_i) + (k+1)\kappa.$$

We have proved that property ($\mathscr{S}_p$) is satisfied for $C = (3 + \kappa_0)\kappa/2$. □

Lemma 8. *Assume (A1-A5). For any $p > 0$, the functional L satisfies property ($\mathscr{R}_p$) with constant $C = C(\kappa, \kappa_0)$.*

Proof. Let $\mathscr{X}, \mathscr{X}_1, \mathscr{X}_2, \mathscr{Y}, \mathscr{Y}_1, \mathscr{Y}_2$ be finite multisets contained in $B(1/2) = Q$. Denote by x, x_1, x_2, y, y_1, y_2 the cardinalities of the multisets. As a first step, let us prove that

$$L(\mathscr{X} \cup \mathscr{X}_1, \mathscr{Y} \cup \mathscr{Y}_1) \leq L(\mathscr{X}, \mathscr{Y}) + C(x_1 + y_1). \tag{15}$$

By induction, it is enough to deal with the cases $(x_1, y_1) = (1, 0)$ and $(x_1, y_1) = (0, 1)$. Because of our symmetry assumption, our task is to prove that

$$L(\mathcal{X} \cup \{a\}, \mathcal{Y}) \leq L(\mathcal{X}, \mathcal{Y}) + C. \tag{16}$$

If $\mathrm{card}(\mathcal{Y}) \leq \mathrm{card}(\mathcal{X})$, then the latter is obvious: choose an optimal graph for $L(\mathcal{X}, \mathcal{Y})$ and use it to upper estimate $L(\mathcal{X} \cup \{a\}, \mathcal{Y})$. Assume on the contrary that $\mathrm{card}(\mathcal{Y}) \geq \mathrm{card}(\mathcal{X}) + 1$. Then there exists $\mathcal{Y}' \subset \mathcal{Y}$ with $\mathrm{card}(\mathcal{Y}') = \mathrm{card}(\mathcal{X})$ and $L(\mathcal{X}, \mathcal{Y}')$. Let $b \in \mathcal{Y} \setminus \mathcal{Y}'$. In order to establish (16), it is enough to show that

$$L(\mathcal{X} \cup \{a\}, \mathcal{Y}' \cup \{b\}) \leq L(\mathcal{X}, \mathcal{Y}') + C,$$

but this is just an instance of the subadditivity property. Hence (15) is established.

In order to prove the regularity property, it remains to show that

$$L(\mathcal{X}, \mathcal{Y}) \leq L(\mathcal{X} \cup \mathcal{X}_2, \mathcal{Y} \cup \mathcal{Y}_2) + C(x_2 + y_2). \tag{17}$$

Again, using induction and symmetry, it is sufficient to establish

$$L(\mathcal{X}, \mathcal{Y}) \leq L(\mathcal{X} \cup \{a\}, \mathcal{Y}) + C. \tag{18}$$

If $\mathrm{card}(\mathcal{X}) \wedge \mathrm{card}(cY) < \kappa_0$, then by the bounded degree property $L(\mathcal{X}, \mathcal{Y}) \leq \kappa \kappa_0 \mathrm{diam}(Q)^p$ and we are done. Assume next that $\mathrm{card}(\mathcal{X}), \mathrm{card}(\mathcal{Y}) \geq \kappa_0$. Let us consider an optimal graph for $L(\mathcal{X} \cup \{a\}, \mathcal{Y})$. If a is not a vertex of this graph (which forces $\mathrm{card}(\mathcal{X}) \geq \mathrm{card}(\mathcal{Y})$) then one can use the same graph to upper estimate $L(\mathcal{X}, \mathcal{Y})$ and obtain (18). Assume on the contrary that a is a vertex of this optimal graph. Let us distinguish two cases: if $\mathrm{card}(\mathcal{X}) \geq \mathrm{card}(\mathcal{Y})$, then in the optimal graph for $L(\mathcal{X} \cup \{a\}, \mathcal{Y})$, at least a point $b \in \mathcal{X}$ is not used. Consider the isomorphic graph obtained by replacing a by b while the other points remain fixed (this leads to the deformation of the edges out of a. There are at most κ of them by the bounded degree assumption). This graph can be used to upper estimate $L(\mathcal{X}, \mathcal{Y})$, and gives

$$L(\mathcal{X}, \mathcal{Y}) \leq L(\mathcal{X} \cup \{a\}, \mathcal{Y}) + \kappa \, \mathrm{diam}(Q)^p.$$

The second case is when a is used but $\mathrm{card}(\mathcal{X}) + 1 \leq \mathrm{card}(\mathcal{Y})$. Actually, the optimal graph for $L(\mathcal{X} \cup \{a\}, \mathcal{Y})$ uses all the points of $\mathcal{X} \cup \{a\}$ and of a subset of same cardinality $\mathcal{Y}' \subset \mathcal{Y}$. Choose an element b in $\mathcal{Y}'$. Then $\mathcal{Y}'' = \mathcal{Y}' \setminus \{b\}$ has the same cardinality as $\mathcal{X}$. Obviously $L(\mathcal{X} \cup \{a\}, \mathcal{Y}) = L(\mathcal{X} \cup \{a\}, \mathcal{Y}'' \cup \{b\})$. Consider the corresponding optimal bipartite graph. By the restriction property, if we erase a and b and their edges, we obtain a bipartite graph on $(\mathcal{X}, \mathcal{Y}'')$ which differs from an admissible graph of our optimization problem by at most κ edges. Using this new graphs yields

$$L(\mathcal{X}, \mathcal{Y}) \leq \kappa \, \mathrm{diam}(Q)^p + L(\mathcal{X} \cup \{a\}, \mathcal{Y}'' \cup \{b\}) = \kappa \, \mathrm{diam}(Q)^p + L(\mathcal{X} \cup \{a\}, \mathcal{Y}).$$

This concludes the proof. □

6 Lower Bounds, Lower Limits

6.1 Uniform Distribution on a Set

In order to motivate the sequel, we start with the simple case where f is an indicator function. The lower bound is then a direct consequence of Theorems 4 and 5.

Theorem 7. *Let $d > 2p > 0$. Let L be a bipartite functional on $\mathbb{R}^d$ satisfying the properties $(\mathcal{H}_p)$, $(\mathcal{R}_p)$, $(\mathcal{S}_p)$. Let $\Omega \subset \mathbb{R}^d$ be a bounded set with positive Lebesgue measure. Then*

$$\lim_{n \to \infty} \frac{\mathbb{E}L(n\mathbf{1}_\Omega)}{n^{1-\frac{p}{d}}} = \beta_L \operatorname{Vol}(\Omega).$$

Proof. Theorem 5 gives directly $\limsup \mathbb{E}L(n\mathbf{1}_\Omega)/n^{1-\frac{p}{d}} \leq \beta_L \operatorname{Vol}(\Omega)$. By translation and dilation invariance, we may assume without loss of generality that $\Omega \subset [0,1]^d$. Let $\Omega_c := [0,1]^d \setminus \Omega$. Applying Proposition 6 for the partition $[0,1]^d = \Omega \cup \Omega_c$, gives after division by $n^{1-p/d}$

$$\frac{\mathbb{E}L(n\mathbf{1}_{[0,1]^d})}{n^{1-\frac{p}{d}}} - \frac{\mathbb{E}L(n\mathbf{1}_{\Omega_c})}{n^{1-\frac{p}{d}}}$$

$$\leq \frac{\mathbb{E}L(n\mathbf{1}_\Omega)}{n^{1-\frac{p}{d}}} + 3C \operatorname{diam}([0,1]^d) n^{\frac{p}{d}-\frac{1}{2}} \left(\operatorname{Vol}(\Omega)^{\frac{1}{2}} + \operatorname{Vol}(\Omega_c)^{\frac{1}{2}}\right).$$

Since $d > 2p$, letting n go to infinity gives

$$\liminf_n \frac{\mathbb{E}L(n\mathbf{1}_\Omega)}{n^{1-\frac{p}{d}}} \geq \lim_n \frac{\mathbb{E}L(n\mathbf{1}_{[0,1]^d})}{n^{1-\frac{p}{d}}} - \limsup_n \frac{\mathbb{E}L(n\mathbf{1}_{\Omega_c})}{n^{1-\frac{p}{d}}}$$

$$\geq \beta_L - \beta_L \operatorname{Vol}(\Omega_c) = \beta_L \operatorname{Vol}(\Omega),$$

where we have used Theorem 4 for the limit and Theorem 5 for the upper limit. □

The argument of the previous proof relies on the fact that the quantity $\lim n^{1-p/d} \mathbb{E}L(n\mathbf{1}_\Omega) = \beta_L \operatorname{Vol}(\Omega)$ is in a sense additive in Ω. This line of reasoning does not pass to functions since $f \mapsto \int f^{1-p/d}$ is additive only for functions with disjoint supports. The lower limit result requires more work for general densities.

6.2 Lower Limits for Matchings

In order to establish a tight estimate on the lower limit, it is natural to try and reverse the partition inequality given in Proposition 6. This is usually more difficult and there does not exist a general method to perform this lower bound. We shall first restrict our attention to the case of the matching functional M_p with $p > 0$, we define in this subsection

$$L = M_p.$$

6.2.1 Boundary Functional

Given a matching on the unit cube, one needs to infer from it matchings on the subcubes of a dyadic partition and to control the corresponding costs. The main difficulty comes from the points of a subcube that are matched to points of another subcube. In other words some links of the optimal matching cross the boundaries of the cells. As in the book by Yukich [19], a modified notion of the cost of a matching is used in order to control the effects of the boundary of the cells of a partition. Our argument is however more involved, since the good bound (2) used by Yukich is not available for the bipartite matching. We define

$$q = 2^{p-1} \wedge 1. \tag{19}$$

Let $S \subset \mathbb{R}^d$ and $\varepsilon \geq 0$. Given multisets $\mathscr{X} = \{X_1, \ldots, m\}$ and $\mathscr{Y} = \{Y_1, \ldots, Y_n\}$ included in S we define the penalized boundary-matching cost as follows

$$L_{\partial S, \varepsilon}(X_1, \ldots, X_m; Y_1, \ldots, Y_n) \tag{20}$$

$$= \min_{A,B,\sigma} \left\{ \sum_{i \in A} |X_i - Y_{\sigma(i)}|^p + \sum_{i \in A^c} q\left(d(X_i, \partial S)^p + \varepsilon^p\right) + \sum_{j \in B^c} q\left(d(Y_j, \partial S)^p + \varepsilon^p\right) \right\},$$

where the minimum runs over all choices of subsets $A \subset \{1, \ldots, m\}$, $B \subset \{1, \ldots, n\}$ with the same cardinality and all bijective maps $\sigma : A \to B$. When $\varepsilon = 0$ we simply write $L_{\partial S}$. Notice that in our definition, and contrary to the definition of optimal matching, all points are matched even if $m \neq n$. If $\mathscr{X}$ and $\mathscr{Y}$ are independent Poisson point processes with intensity ν supported in S and with finite total mass, we write $L_{\partial S, \varepsilon}(\nu)$ for the random variable $L_{\partial S, \varepsilon}(\mathscr{X}, \mathscr{Y})$.

The main interest of the notion of boundary matching is that it allows to bound from below the matching cost on a large set in terms of contributions on cells of a partition. The following Lemma establishes a superadditive property of $L_{\partial S}$ and it can be viewed as a counterpart to the upper bound provided by Proposition 6.

Lemma 9. *Assume $L = M_p$. Let ν be a finite measure on $\mathbb{R}^d$ and consider a partition $Q = \cup_{P \in \mathscr{P}} P$ of a subset of $\mathbb{R}^d$. Then*

$$\operatorname{diam}(Q)^p \sqrt{2\nu(\mathbb{R}^d)} + \mathbb{E}L(\nu) \geq \mathbb{E}L_{\partial Q}(\mathbf{1}_Q \cdot \nu) \geq \sum_{P \in \mathscr{P}} \mathbb{E}L_{\partial P}(\mathbf{1}_P \cdot \nu).$$

Proof. Let $\mathscr{X} = \{X_1, \ldots, X_m\}$, $\mathscr{Y} = \{Y_1, \ldots, Y_n\}$ be multisets included in Q and $\mathscr{X}' = \{X_{m+1}, \ldots, X_{m+m'}\}$, $\mathscr{Y}' = \{Y_{n+1}, \ldots, Y_{n+n'}\}$ be multisets included in Q^c. By considering an optimal matching of $\mathscr{X} \cup \mathscr{X}'$ and $\mathscr{Y} \cup \mathscr{Y}'$, we have the lower bound

$$\text{diam}(Q)^p |m + m' - n - n'| + L(\mathscr{X} \cup \mathscr{X}', \mathscr{Y} \cup \mathscr{Y}') \geq L_{\partial Q}(\mathscr{X}, \mathscr{Y}).$$

Indeed, if $1 \leq i \leq m$ and a pair (X_i, Y_{n+j}), is matched then $|X_i - Y_{n+j}| \geq d(X_i, \partial Q)$ and similarly for a pair (X_{m+i}, Y_j), with $1 \leq j \leq n$, $|X_{m+i} - Y_j| \geq d(Y_j, \partial Q)$. The term $\text{diam}(Q)^p |m + m' - n - n'|$ takes care of the points of $\mathscr{X} \cup \mathscr{Y}$ that are not matched in the optimal matching of $\mathscr{X} \cup \mathscr{X}'$ and $\mathscr{Y} \cup \mathscr{Y}'$. We apply the above inequality to $\mathscr{X}, \mathscr{Y}$ independent Poisson processes of intensity $\mathbf{I}_Q \cdot \nu$, and $\mathscr{X}', \mathscr{Y}'$, two independent Poisson processes of intensity $\mathbf{I}_{Q^c} \cdot \nu$, independent of $(\mathscr{X}, \mathscr{Y})$. Then $\mathscr{X} \cup \mathscr{X}', \mathscr{Y} \cup \mathscr{Y}'$ are independent Poisson processes of intensity ν. Taking expectation and bounding the average of the difference of cardinalities in the usual way, we obtain the first inequality.

Now, the second inequality will follow from the superadditive property of the boundary functional:

$$L_{\partial Q}(\mathscr{X}, \mathscr{Y}) \geq \sum_{P \in \mathscr{P}} L_{\partial P}(\mathscr{X} \cap P, \mathscr{Y} \cap P). \tag{21}$$

This is proved as follows. Let (A, B, σ) be an optimal triplet for $L_{\partial Q}(\mathscr{X}, \mathscr{Y})$:

$$L_{\partial Q}(\mathscr{X}, \mathscr{Y}) = \sum_{i \in A} |X_i - Y_{\sigma(i)}|^p + \sum_{i \in A^c} q d(X_i, \partial Q)^p + \sum_{j \in B^c} q d(Y_j, \partial Q)^p.$$

If $x \in Q$, we denote by $P(x)$ the unique $P \in \mathscr{P}$ that contains x. If $P(X_i) = P(Y_{\sigma(i)})$ we leave the term $|X_i - Y_{\sigma(i)}|$ unchanged. On the other hand if $P(X_i) \neq P(Y_{\sigma(i)})$, we find

$$|X_i - Y_{\sigma(i)}|^p \geq \left(d(X_i, \partial P(X_i)) + d(Y_{\sigma(i)}, \partial P(Y_{\sigma(i)}))\right)^p$$
$$\geq q\, d(X_i, \partial P(X_i))^p + q\, d(Y_{\sigma(i)}, \partial P(Y_{\sigma(i)}))^p,$$

where q was defined by (19) and, for $0 \leq p \leq 1$, we have used Jensen inequality $|x + y|^p \geq 2^{1-p}(|x|^p + |y|^p)$. Eventually, we apply the inequality

$$d(x, \partial Q) \geq d(x, \partial P(x))$$

in order to take care of the points in $A^c \cup B^c$. Combining these inequalities and grouping the terms according to the cell $P \in \mathscr{P}$ containing the points, we obtain that

$$L_{\partial Q}(\mathcal{X}, \mathcal{Y}) \geq \sum_{P \in \mathscr{P}} \left(\sum_{i \in A; \, X_i \in P, Y_{\sigma(i)} \in P} |X_i - Y_{\sigma(i)}|^p + \sum_{i \in A; \, X_i \in P, Y_{\sigma(i)} \notin P} q\, d(X_i, \partial P)^p \right.$$

$$+ \sum_{i \in A^c; \, X_i \in P} q\, d(X_i, \partial P)^p + \sum_{j \in B; \, Y_j \in P, \, j \notin \sigma(\{i; \, X_i \in P\})} q\, d(Y_j, \partial P)^p$$

$$\left. + \sum_{j \in B^c; \, Y_j \in P} q\, d(Y_j, \partial P)^p \right)$$

$$\geq \sum_{P \in \mathscr{P}} L_{\partial P}(\mathcal{X} \cap P, \mathcal{Y} \cap P),$$

and we have obtained the inequality (21). □

The next lemma on the regularity of $\mathbb{E}L_{\partial Q}(\mu)$ is the analog of Corollary 2. It will be used to reduce to uniform distributions on cubes.

Lemma 10. *Assume $L = M_p$. Let μ, μ' be two probability measures on $\mathbb{R}^d$ with supports in Q and $n > 0$. Then*

$$\mathbb{E}L_{\partial Q}(n\mu) \leq \mathbb{E}L_{\partial Q}(n\mu') + 4n \operatorname{diam}(Q)^p d_{\mathrm{TV}}(\mu, \mu').$$

Consequently, if f is a nonnegative locally integrable function on $\mathbb{R}^d$, setting $\alpha = \int_Q f / \operatorname{vol}(Q)$, it holds

$$\mathbb{E}L_{\partial Q}(nf\mathbf{1}_Q) \leq \mathbb{E}L_{\partial Q}(n\alpha\mathbf{1}_Q) + 2n \operatorname{diam}(Q)^p \int_Q |f(x) - \alpha|\, dx.$$

Proof. The functional $L_{\partial Q}$ satisfies a slight modification of property $(\mathscr{R}_p)$: for all multisets $\mathcal{X}, \mathcal{Y}, \mathcal{X}_1, \mathcal{Y}_1, \mathcal{X}_2, \mathcal{Y}_2$ in Q, it holds

$$L_{\partial Q}(\mathcal{X} \cup \mathcal{X}_1, \mathcal{Y} \cup \mathcal{Y}_1) \leq L_{\partial Q}(\mathcal{X} \cup \mathcal{X}_2, \mathcal{Y} \cup \mathcal{Y}_2)$$

$$+ \operatorname{diam}(Q)^p \big(\operatorname{card}(\mathcal{X}_1) + \operatorname{card}(\mathcal{X}_2) + \operatorname{card}(\mathcal{Y}_1) + \operatorname{card}(\mathcal{Y}_2) \big).$$

Indeed, we start from an optimal boundary matching of $L_{\partial Q}(\mathcal{X} \cup \mathcal{X}_2, \mathcal{Y} \cup \mathcal{Y}_2)$, we match to the boundary the points of $(\mathcal{X}, \mathcal{Y})$ that are matched to a point in $(\mathcal{X}_2, \mathcal{Y}_2)$. There are at most $\operatorname{card}(\mathcal{X}_2) + \operatorname{card}(\mathcal{Y}_2)$ such points. Finally we match all points of $(\mathcal{X}_1, \mathcal{Y}_1)$ to the boundary and we obtain a suboptimal boundary matching of $L_{\partial Q}(\mathcal{X} \cup \mathcal{X}_1, \mathcal{Y} \cup \mathcal{Y}_1)$. This establishes the above inequality. The statements follow then from the proofs of Proposition 3 and Corollary 2. □

We will need an analog of Lemma 2, i.e. an asymptotic for the boundary matching for the uniform distribution on the unit cube. Let $Q = [0,1]^d$ and denote

$$\bar{L}_{\partial Q}(n) = \mathbb{E}L_{\partial Q}(n\mathbf{1}_Q).$$

Lemma 11. *Assume $L = M_p$ and $0 < p < d/2$, then*

$$\lim_{n \to \infty} \frac{\bar{L}_{\partial Q}(n)}{n^{1-\frac{p}{d}}} = \beta'_L,$$

where $\beta'_L > 0$ is a constant depending on p and d.

Proof. Let $m \geq 1$ be an integer. We consider a dyadic partition $\mathscr{P}$ of Q into m^d cubes of size $1/m$. Then, Lemma 9 applied to the measure $n\mathbf{1}_{[0,1]^d}(x)\,dx$ gives

$$\bar{L}_{\partial Q}(n) \geq \sum_{q \in \mathscr{P}} \mathbb{E} L_{\partial q}(n\mathbf{1}_{q \cap [0,1]^d}).$$

However by scale and translation invariance, for any $q \in \mathscr{P}$ we have $\mathbb{E} L_{\partial q}(n\mathbf{1}_{q \cap [0,1]^d}) = m^{-p} \mathbb{E} L_{\partial Q}(nm^{-d}\mathbf{1}_Q)$. It follows that

$$\bar{L}_{\partial Q}(n) \geq m^{d-p} \bar{L}_{\partial Q}(nm^{-d}).$$

The proof is then done as in Theorem 4 where superadditivity here replaces subadditivity there. □

As already pointed, we conjecture that $\beta_L = \beta'_L$ where β_L is the constant appearing in Lemma 2 for $L = M_p$.

6.2.2 General Absolutely Continuous Measures

We are ready to state and prove

Theorem 8. *Assume $L = M_p$ and $0 < p < d/2$. Let $f : \mathbb{R}^d \to \mathbb{R}^+$ be an integrable function. Then*

$$\liminf_n \frac{\mathbb{E} L(nf)}{n^{1-\frac{p}{d}}} \geq \beta'_L \int_{\mathbb{R}^d} f^{1-\frac{p}{d}}.$$

Proof. Assume first that the support of f is bounded. By a scaling argument, we may assume that the support of f is included in $Q = [0,1]^d$. The proof is now similar to the one of Theorem 5. For $\ell \in \mathbb{N}$, we consider the partition $\mathscr{P}_\ell$ of $[0,1]^d$ into $2^{\ell d}$ cubes of side-length $2^{-\ell}$. Let $k \in \mathbb{N}^*$ to be chosen later. For $P \in \mathscr{P}_k$, α_P denotes the average of f over P. Applying Lemma 9, Lemma 10 and homogeneity, we obtain

$$2d^{\frac{p}{2}}\sqrt{n\int f} + \mathbb{E}L(nf) \geq \mathbb{E}L_{\partial Q}(nf)$$

$$\geq \sum_{P \in \mathscr{P}_k} \mathbb{E}L_{\partial P}(nf \mathbf{1}_P)$$

$$\geq \sum_{P \in \mathscr{P}_k} \left(\mathbb{E}L_{\partial P}(n\alpha_P \mathbf{1}_P) - 2nd^{\frac{p}{2}} 2^{-kp} \int_P |f - \alpha_P| \right)$$

$$= \sum_{P \in \mathscr{P}_k} \left(2^{-kp} \mathbb{E}L_{\partial Q}(n\alpha_P 2^{-kd} \mathbf{1}_Q) - 2nd^{\frac{p}{2}} 2^{-kp} \int_P |f - \alpha_P| \right).$$

Setting as before $f_k = \sum_{P \in \mathscr{P}_k} \alpha_P \mathbf{1}_P$ and $h(t) = \bar{L}_{\partial Q}(t)/t^{\frac{d-1}{d}}$ where $\bar{L}_{\partial Q}(t) = \mathbb{E}L_{\partial Q}(t\mathbf{1}_Q)$, the previous inequality reads as

$$2n^{\frac{p}{d}-\frac{1}{2}} d^{\frac{p}{2}} \sqrt{\int f} + \frac{\mathbb{E}L(nf)}{n^{1-\frac{p}{d}}} \geq \mathbb{E}L_{\partial Q}(nf)$$

$$\geq \int h(n2^{-kd} f_k) f_k^{1-\frac{p}{d}} - 2d^{\frac{p}{2}} n^{\frac{p}{d}} 2^{-kp} \int |f - f_k|.$$

As in the proof of Theorem 5 we may choose $k = k(n)$ depending on n in such a way that $\lim_n k(n) = +\infty$, $\lim_n n^{1/d} 2^{-k(n)} = +\infty$ and $\lim_n n^{\frac{1}{d}} 2^{-k(n)} \left(\int |f - f_{k(n)}| \right)^{\frac{1}{p}} = 0$. For such a choice, since $\liminf_{t \to +\infty} h(t) \geq \beta'_L$ by Lemma 11 and a.e. $\lim_k f_k = f$, Fatou's lemma ensures that

$$\liminf_n \int h(n2^{-k(n)d} f_{k(n)}) f_{k(n)}^{1-\frac{p}{d}} \geq \liminf_n \int_{\{f>0\}} h(n2^{-k(n)d} f_{k(n)}) f_{k(n)}^{1-\frac{p}{d}} \geq \beta'_L \int f^{1-\frac{p}{d}}.$$

Our statement easily follows.

Now, let us address the general case where the support is not bounded. Let $\ell \geq 1$ and $Q = [-\ell, \ell]^d$. By Lemma 9,

$$2\mathrm{diam}(Q)^p \sqrt{n\int f} + \mathbb{E}L(nf) \geq \mathbb{E}L_{\partial Q}(nf\mathbf{1}_Q).$$

Also, the above argument has shown that

$$\liminf_n \frac{\mathbb{E}L_{\partial Q}(nf\mathbf{1}_Q)}{n^{1-\frac{p}{d}}} \geq \beta'_L \int_Q f^{1-\frac{p}{d}}.$$

We deduce that for any $Q = [-\ell, \ell]^d$,

$$\liminf_n \frac{\mathbb{E} L(nf)}{n^{1-\frac{p}{d}}} \geq \beta'_L \int_Q f^{1-\frac{p}{d}}.$$

Taking ℓ arbitrary large we obtain the claimed lower bound. □

6.2.3 Dealing with the Singular Component

In this section we explain how to extend Theorem 8 from measures with densities to general measures. Given a measure μ, we consider its decomposition $\mu = \mu_{ac} + \mu_s$ into an absolutely continuous part and a singular part.

Our starting point is the following lemma, which can be viewed as an inverse subadditivity property.

Lemma 12. *Let $p \in (0, 1]$ and $L = M_p$. Let $\mathcal{X}_1, \mathcal{X}_2, \mathcal{Y}_1, \mathcal{Y}_2$ be four finite multisets included in a bounded set Q. Then*

$$L(\mathcal{X}_1, \mathcal{Y}_1) \leq L(\mathcal{X}_1 \cup \mathcal{X}_2, \mathcal{Y}_1 \cup \mathcal{Y}_2) + L(\mathcal{X}_2, \mathcal{Y}_2)$$
$$+ \mathrm{diam}(Q)^p \Big(|\mathcal{X}_1(Q) - \mathcal{Y}_1(Q)| + 2|\mathcal{X}_2(Q) - \mathcal{Y}_2(Q)| \Big).$$

Proof. Let us start with an optimal matching achieving $L(\mathcal{X}_1 \cup \mathcal{X}_2, \mathcal{Y}_1 \cup \mathcal{Y}_2)$ and an optimal matching achieving $L(\mathcal{X}_2, \mathcal{Y}_2)$. Let us view them as bipartite graphs $G_{1,2}$ and G_2 on the vertex sets $(\mathcal{X}_1 \cup \mathcal{X}_2, \mathcal{Y}_1 \cup \mathcal{Y}_2)$ and $(\mathcal{X}_2, \mathcal{Y}_2)$ respectively (note that if a point appears more than once, we consider its instances as different graph vertices). Our goal is to build a possibly suboptimal matching of $\mathcal{X}_1$ and $\mathcal{Y}_1$. Assume without loss of generality that $\mathcal{X}_1(Q) \leq \mathcal{Y}_1(Q)$. Hence we need to build an injection from $\sigma : \mathcal{X}_1 \to \mathcal{Y}_1$ and to upper bound its cost $\sum_{x \in \mathcal{X}_1} |x - \sigma(x)|^p$.

To do this, let us consider the graph G obtained as the union of $G_{1,2}$ and G_2 (allowing multiple edges when two points are neighbors in both graphs). It is clear that in G the points from $\mathcal{X}_1$ and $\mathcal{Y}_1$ have degree at most one, while the points from $\mathcal{X}_2$ and $\mathcal{Y}_2$ have degree at most 2. For each $x \in \mathcal{X}_1$, let us consider its connected component $C(x)$ in G. Because of the above degree considerations (and since no point is connected to itself in a bipartite graph) it is obvious that $C(x)$ is a path.

It could be that $C(x) = \{x\}$, in the case when x is a leftover point in the matching corresponding to $G_{1,2}$. This means that x is a point in excess and there are at most $|\mathcal{X}_1(Q) + \mathcal{X}_2(Q) - (\mathcal{Y}_1(Q) + \mathcal{Y}_2(Q))|$ of them.

Consider now the remaining case, when $C(x)$ is a nontrivial path. Its first edge belongs to $G_{1,2}$. If there is a second edge, it has to be from G_2 (since $G_{1,2}$ as degree at most one). Repeating the argument, we see that the edges of the path are alternately from $G_{1,2}$ and from G_2. Note also that the successive vertices are alternately from $\mathcal{X}_1 \cup \mathcal{X}_2$ and from $\mathcal{Y}_1 \cup \mathcal{Y}_2$ (see Fig. 1). There are three possibilities:

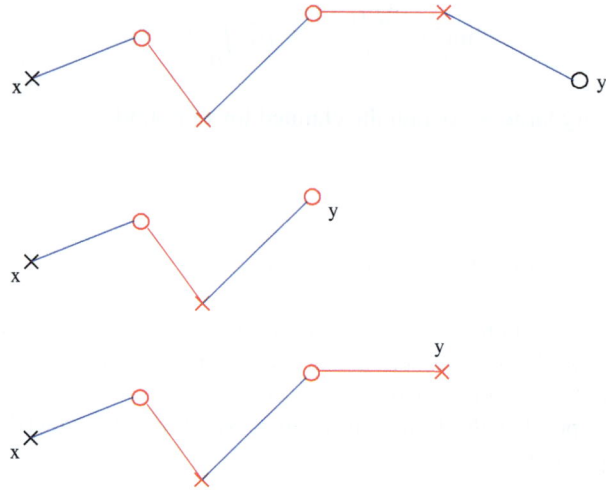

Fig. 1 The three possibilities for the path $C(x)$. In *blue*, $G_{1,2}$, in *red* G_2, the points in $\mathcal{X}_1 \cup \mathcal{X}_2$ are represented by a *cross*, points in $\mathcal{Y}_1 \cup \mathcal{Y}_2$ by a *circle*

- The other end of the path is a point $y \in \mathcal{Y}_1$. In this case we are done, we have associated a point $y \in \mathcal{Y}_1$ to our point $x \in \mathcal{X}_1$. By the triangle inequality and since $(a+b)^p \leq a^p + b^p$ due to the assumption $p \leq 1$, $|x-y|^p$ is upper bounded by the sum of the p-th powers of the length of the edges in $C(x)$.
- The other end of the path is a point $y \in \mathcal{Y}_2$. The last edge is from $G_{1,2}$. So necessarily, y has no neighbor in G_2. This means that it is not matched. There are at most $|\mathcal{X}_2(Q) - \mathcal{Y}_2(Q)|$ such points in the matching G_2.
- The other end of the path is a point $x' \in \mathcal{X}_2$. The last edge is from G_2. So necessarily, x' has no neighbor in $G_{1,2}$. This means that it is not matched in $G_{1,2}$. As already mentioned there are at most $|\mathcal{X}_1(Q) + \mathcal{X}_2(Q) - (\mathcal{Y}_1(Q) + \mathcal{Y}_2(Q))|$ such points.

Eventually we have found a way to match the points from $\mathcal{X}_1$, apart maybe $|\mathcal{X}_2(Q) - \mathcal{Y}_2(Q)| + |\mathcal{X}_1(Q) + \mathcal{X}_2(Q) - (\mathcal{Y}_1(Q) + \mathcal{Y}_2(Q))|$ of them. We match the latter points arbitrarily to (unused) points in $\mathcal{Y}_1$ and upper bound the distances between matched points by $\mathrm{diam}(Q)$. □

As a direct consequence, we obtain:

Lemma 13. *Let μ_1 and μ_2 be two finite measures supported in a bounded set Q. Let $p \in (0, 1]$ and $L = M_p$ be the bipartite matching functional. Then*

$$\mathbb{E}L(\mu_1) \leq \mathbb{E}L(\mu_1 + \mu_2) + \mathbb{E}L(\mu_2) + 3\,\mathrm{diam}(Q)^p\bigl(\sqrt{\mu_1(Q)} + \sqrt{\mu_2(Q)}\bigr).$$

Proof. Let $\mathcal{X}_1, \mathcal{X}_2, \mathcal{Y}_1, \mathcal{Y}_2$ be four independent Poisson point processes. Assume that for $i \in \{1,2\}$, $\mathcal{X}_i$ and $\mathcal{Y}_i$ have intensity measure μ_i. Consequently $\mathcal{X}_1 \cup \mathcal{X}_2$ and $\mathcal{Y}_1 \cup \mathcal{Y}_2$ are independent Poisson point processes with intensity $\mu_1 + \mu_2$. Applying the preceding Lemma 12 and taking expectations yields

$$\mathbb{E}L(\mu_1) \leq \mathbb{E}L(\mu_1 + \mu_2) + \mathbb{E}L(\mu_2)$$
$$+ 2\text{diam}(Q)^p \big(\mathbb{E}|\mathcal{X}_1(Q) - \mathcal{Y}_1(Q)| + \mathbb{E}|\mathcal{X}_2(Q) - \mathcal{Y}_2(Q)|\big).$$

As usual, we conclude using that

$$\mathbb{E}|\mathcal{X}_i(Q) - \mathcal{Y}_i(Q)| \leq \sqrt{\mathbb{E}\big((\mathcal{X}_i(Q) - \mathcal{Y}_i(Q))^2\big)} = \sqrt{2\text{var}(\mathcal{X}_i(Q))} = \sqrt{2\mu_i(Q)}.$$

□

Theorem 9. *Assume that $d \in \{1,2\}$ and $p \in (0, d/2)$, or that $d \geq 3$ and $p \in (0, 1]$. Let $L = M_p$ be the bipartite matching functional. Let μ be a finite measure on $\mathbb{R}^d$ with bounded support. Let f be the density of the absolutely continuous part of μ. Assume that there exists $\alpha > \frac{2dp}{d-2p}$ such that $\int |x|^\alpha d\mu(x) < +\infty$. Then*

$$\liminf_n \frac{\mathbb{E}L(n\mu)}{n^{1-\frac{p}{d}}} \geq \beta'_L \int_{\mathbb{R}^d} f^{1-\frac{p}{d}}.$$

Moreover if f is proportional to the indicator function of a bounded set with positive Lebesgue measure

$$\lim_n \frac{\mathbb{E}L(n\mu)}{n^{1-\frac{p}{d}}} = \beta_L \int_{\mathbb{R}^d} f^{1-\frac{p}{d}}.$$

Proof. Note that in any case, $p \leq 1$ is assumed. Let us write $\mu = \mu_{ac} + \mu_s$ where $d\mu_{ac}(x) = f(x)dx$ is the absolutely continuous part and μ_s is the singular part of μ.

The argument is very simple if μ has a bounded support: apply the previous lemma with $\mu_1 = n\mu_{ac}$ and $\mu_2 = n\mu_s$. When n tends to infinity, observing that $\sqrt{n}$ is negligible with respect to $n^{1-\frac{p}{d}}$, we obtain that

$$\liminf_n \frac{\mathbb{E}L(n\mu_{ac})}{n^{1-\frac{p}{d}}} \leq \liminf_n \frac{\mathbb{E}L(n\mu)}{n^{1-\frac{p}{d}}} + \limsup_n \frac{\mathbb{E}L(n\mu_s)}{n^{1-\frac{p}{d}}}.$$

Observe that the latter upper limit is equal to zero thanks to Theorem 6 applied to a purely singular measures. Eventually $\liminf_n \frac{\mathbb{E}L(n\mu_{ac})}{n^{1-\frac{p}{d}}} \geq \beta'_L \int_{\mathbb{R}^d} f^{1-\frac{p}{d}}$ by Theorem 8 about absolutely continuous measures.

If f is proportional to an indicator function, we simply use scale invariance and Theorem 7 in place of Theorem 8.

Let us consider the general case of unbounded support. Let $Q = [-\ell, \ell]^d$ where $\ell > 0$ is arbitrary. Let $\mathscr{X}_1, \mathscr{Y}_1, \mathscr{X}_2, \mathscr{Y}_2$ be four independent Poisson point processes, such that $\mathscr{X}_1$ and $\mathscr{Y}_1$ have intensity measure $n\mathbf{1}_Q \cdot \mu_{ac}$, and $\mathscr{X}_2$ and $\mathscr{Y}_2$ have intensity measure $n(\mu_s + \mathbf{1}_{Q^c} \cdot \mu_{ac})$. It follows that $\mathscr{X}_1 \cup \mathscr{X}_2$ and $\mathscr{Y}_1 \cup \mathscr{Y}_2$ are independent Poisson point processes with intensity $n\mu$. Set $T := \max\{|z|; z \in \mathscr{X}_1 \cup \mathscr{X}_2 \cup \mathscr{Y}_1 \cup \mathscr{Y}_2\}$. Applying Lemma 12 gives

$$L(\mathscr{X}_1, \mathscr{Y}_1) \leq L(\mathscr{X}_1 \cup \mathscr{X}_2, \mathscr{Y}_1 \cup \mathscr{Y}_2) + L(\mathscr{X}_2, \mathscr{Y}_2)$$
$$+ c_p T^p \big(|\mathrm{card}(\mathscr{X}_1) - \mathrm{card}(\mathscr{Y}_1)| - |\mathrm{card}(\mathscr{X}_2) - \mathrm{card}(\mathscr{Y}_2)|\big).$$

Taking expectations, applying the Cauchy–Schwarz inequality twice and Lemma 1 (note that $\alpha > 2p$) gives

$$\mathbb{E}L(nf\mathbf{1}_Q) \leq \mathbb{E}L(n\mu) + \mathbb{E}L\big(n(\mu_s + \mathbf{1}_{Q^c}\mu_{ac})\big)$$
$$+ c_p \sqrt{E[T^{2p}]} \Big(\sqrt{2n\mu_{ac}(Q)} + \sqrt{2n(\mu_s(\mathbb{R}^d) + \mu_{ac}(Q^c))}\Big)$$
$$\leq \mathbb{E}L(n\mu) + \mathbb{E}L\big(n(\mu_s + \mathbf{1}_{Q^c}\mu_{ac})\big) + c'_p n^{\frac{p}{\alpha} + \frac{1}{2}}.$$

Since $\alpha > \frac{2dp}{d-2p}$ we obtain

$$\liminf_n \frac{\mathbb{E}L(n\mu)}{n^{1-\frac{p}{d}}} \geq \liminf_n \frac{\mathbb{E}L(nf\mathbf{1}_Q)}{n^{1-\frac{p}{d}}} - \limsup_n \frac{\mathbb{E}L(n(\mu_s + \mathbf{1}_Q^c \cdot \mu_{ac}))}{n^{1-\frac{p}{d}}}$$
$$\geq \beta'_L \int_Q f^{1-\frac{p}{d}} - \beta_L \int_{Q^c} f^{1-\frac{p}{d}},$$

where we have used Theorem 8 for the lower limit for bounded absolutely continuous measures and Theorem 6 for the upper limit. Recall that $Q = [-\ell, \ell]^d$. It remains to let ℓ tend to infinity. □

Actually, using classical duality techniques (which are specific to the bipartite matching) we can derive the following improvement of Lemma 13, which can be seen as an average monotonicity property:

Lemma 14. *Let $p \in (0,1]$ and $L = M_p$. Let μ_1 and μ_2 be two finite measures supported on a bounded subset $Q \subset \mathbb{R}^d$. Then*

$$\mathbb{E}L(\mu_1) \leq \mathbb{E}L(\mu_1 + \mu_2) + 3\mathrm{diam}(Q)^p\big(\sqrt{\mu_1(Q)} + \sqrt{\mu_2(Q)}\big).$$

Proof. Since $p \in (0,1]$, the unit cost $c(x,y) := |x-y|^p$ is a distance on $\mathbb{R}^d$. The Kantorovich–Rubinstein dual representation of the minimal matching cost (or optimal transportation cost) is particularly simple in this case (see e.g. [11, 16, 18]): for $\{x_1, \ldots, x_n\}, \{y_1, \ldots, y_n\}$ two multisets in Q,

$$L(\{x_1,\ldots,x_n\},\{y_1,\ldots,y_n\}) = \sup_{f\in\mathrm{Lip}_{1,0}} \sum_i f(x_i) - f(y_i),$$

where $\mathrm{Lip}_{1,0}$ denotes the set of function $f: Q \to \mathbb{R}$ which are 1-Lipschitzian for the distance $c(x,y)$ (hence they are p-Hölderian for the Euclidean distance) and vanish at a prescribed point $x_0 \in Q$. Observe that any function in $\mathrm{Lip}_{1,0}$ is bounded by $\mathrm{diam}(Q)^p$ pointwise.

Let $\mathscr{X} = \{X_1,\ldots,X_{N_1}\}$ and $\mathscr{Y} = \{Y_1,\ldots,Y_{N_2}\}$ be independent Poisson point processes with intensity μ of finite mass and supported on a set Q of diameter $D < +\infty$. By definition, on the event $\{N_1 \leq N_2\}$,

$$L(\mathscr{X},\mathscr{Y}) = \inf_{A\subset\{1,\ldots,N_2\};\mathrm{card}(A)=N_1} L(\{X_i, 1\leq i\leq N_1\},\{Y_j, j\in A\})$$

$$= \inf_{A\subset\{1,\ldots,N_2\};\mathrm{card}(A)=N_1} \sup_{f\in\mathrm{Lip}_{1,0}} \left(\sum_{i\leq N_1} f(X_i) - \sum_{j\in A} f(Y_j)\right)$$

$$\geq \sup_{f\in\mathrm{Lip}_{1,0}} \left(\sum_{i\leq N_1} f(X_i) - \sum_{j\leq N_2} f(Y_j)\right) - D^p|N_1 - N_2|$$

where we have used Kantorovich–Rubinstein duality to express the optimal matching of two samples of the same size and used that every $f \in \mathrm{Lip}_{1,0}$ satisfies $|f| \leq D^p$ pointwise on Q. A similar lower bound is valid when $N_1 \geq N_2$. Hence, taking expectation and bounding $E|N_1 - N_2|$ in terms of the variance of the number of points in one process, one gets

$$\mathbb{E}L(\mu) \geq \mathbb{E}\sup_{f\in\mathrm{Lip}_{1,0}} \left(\sum_{i\leq N_1} f(X_i) - \sum_{j\leq N_2} f(Y_j)\right) - D^p\sqrt{2|\mu|}. \qquad (22)$$

A similar argument also gives the following upper bound

$$\mathbb{E}L(\mu) \leq \mathbb{E}\sup_{f\in\mathrm{Lip}_{1,0}} \left(\sum_{i\leq N_1} f(X_i) - \sum_{j\leq N_2} f(Y_j)\right) + D^p\sqrt{2|\mu|}. \qquad (23)$$

Let $\mathscr{X}_1, \mathscr{X}_2, \mathscr{Y}_1, \mathscr{Y}_2$ be four independent Poisson point processes. Assume that for $i \in \{1,2\}$, $\mathscr{X}_i$ and $\mathscr{Y}_i$ have intensity μ_i. As already mentioned, $\mathscr{X}_1 \cup \mathscr{X}_2$ and $\mathscr{Y}_1 \cup \mathscr{Y}_2$ are independent with common intensity $\mu_1 + \mu_2$. Given a compact set Q containing the supports of both measures, and $x_0 \in Q$ we define the set $\mathrm{Lip}_{1,0}$. Using (22),

$$\mathbb{E}L(\mu_1 + \mu_2) = \mathbb{E}L(\mathscr{X}_1 \cup \mathscr{X}_2, \mathscr{Y}_1 \cup \mathscr{Y}_2)$$

$$\geq \mathbb{E} \sup_{f \in \mathrm{Lip}_{1,0}} \left(\sum_{x_1 \in \mathscr{X}_1} f(x_1) - \sum_{y_1 \in \mathscr{Y}_1} f(y_1) + \sum_{x_2 \in \mathscr{X}_2} f(x_2) - \sum_{y_2 \in \mathscr{Y}_2} f(y_2) \right)$$
$$- D^p \sqrt{2|\mu_1 + \mu_2|}.$$

Now we use the easy inequality $\mathbb{E} \sup \geq \sup \mathbb{E}$ when $\mathbb{E}$ is the conditional expectation given $\mathscr{X}_1, \mathscr{Y}_1$. Since $(\mathscr{X}_2, \mathscr{Y}_2)$ are independent from $(\mathscr{X}_1, \mathscr{Y}_1)$, we obtain

$$\mathbb{E}L(\mu_1 + \mu_2) + D^p \sqrt{2|\mu_1 + \mu_2|}$$

$$\geq \mathbb{E} \sup_{f \in \mathrm{Lip}_{1,0}} \left(\sum_{x_1 \in \mathscr{X}_1} f(x_1) - \sum_{y_1 \in \mathscr{Y}_1} f(y_1) + \mathbb{E}\Big(\sum_{x_2 \in \mathscr{X}_2} f(x_2) - \sum_{y_2 \in \mathscr{Y}_2} f(y_2) \Big) \right)$$

$$= \mathbb{E} \sup_{f \in \mathrm{Lip}_{1,0}} \left(\sum_{x_1 \in \mathscr{X}_1} f(x_1) - \sum_{y_1 \in \mathscr{Y}_1} f(y_1) \right)$$

$$\geq \mathbb{E}L(\mu_1) - D^p \sqrt{2|\mu_1|},$$

where we have noted that the inner expectation vanishes and used (23). The claim easily follows. □

6.3 Euclidean Combinatorial Optimization

Our proof for the lower bound for matchings extends to some combinatorial optimization functionals L defined by (14). In this paragraph, we explain how to adapt the above argument at the cost of ad-hoc assumptions on the collection of graphs $(\mathscr{G}_n)_{n \in \mathbb{N}}$. As motivating example, we will treat completely the case of the bipartite traveling salesperson tour.

6.3.1 Boundary Functional

Let $S \subset \mathbb{R}^d$ and $\varepsilon, p \geq 0$. Set $q = 2^{p-1} \wedge 1$. In what follows, p is fixed and will be omitted in most places where it would appear as an index. Given multisets $\mathscr{X} = \{X_1, \ldots, X_n\}$ and $\mathscr{Y} = \{Y_1, \ldots, Y_n\}$ included in $\mathbb{R}^d$, we first set

$$L^0_{\partial S, \varepsilon}(\mathscr{X}, \mathscr{Y}) = \min_{G \in \mathscr{G}_n} \left\{ \sum_{(i,j) \in [n]^2 : \{i, n+j\} \in G} d_{S, \varepsilon, p}(X_i, Y_j) \right\},$$

where

$$d_{S,\varepsilon,p}(x,y) = \begin{cases} |x-y|^p & \text{if } x,y \in S, \\ 0 & \text{if } x,y \notin S, \\ q\big(\text{dist}(x,S^c)^p + \varepsilon^p\big) & \text{if } x \in S, y \notin S \\ q\big(\text{dist}(y,S^c)^p + \varepsilon^p\big) & \text{if } y \in S, x \notin S \end{cases} \quad (24)$$

Now, if $\mathscr{X}$ and $\mathscr{Y}$ are in S, we define the penalized boundary functional as

$$L_{\partial S,\varepsilon}(\mathscr{X},\mathscr{Y}) = \min_{A,B \subset S^c} L^0_{\partial S,\varepsilon}(\mathscr{X} \cup A, \mathscr{Y} \cup B), \quad (25)$$

where the minimum is over all multisets A and B in S^c such that $\text{card}(\mathscr{X} \cup A) = \text{card}(\mathscr{Y} \cup B) \geq \kappa_0$. When $\varepsilon = 0$ we simply write $L_{\partial S}$. The main idea of this definition is to consider all possible configurations outside the set S but not to count the distances outside of S (from a metric view point, all of S^c is identified to a point which is at distance ε from S).

The existence of the minimum in (25) is due to the fact that $L^0_{\partial S}(\mathscr{X} \cup A, \mathscr{Y} \cup B)$ can only take finitely many values less than any positive value (the quantities involved are just sums of distances between points of $\mathscr{X}, \mathscr{Y}$ and of their distances to S^c). Notice that definition (25) is consistent with the definition of the boundary functional for the matching functional M_p, given by (20). If $\mathscr{X}$ and $\mathscr{Y}$ are independent Poisson point processes with intensity ν supported in S and with finite total mass, we write $L_{\partial S,\varepsilon}(\nu)$ for the random variable $L_{\partial S,\varepsilon}(\mathscr{X},\mathscr{Y})$. Also note that $d_{S,0,p}(x,y) \leq |x-y|^p$. Consequently if $\text{card}(\mathscr{X}) = \text{card}(\mathscr{Y})$ then

$$L^0_{\partial S}(\mathscr{X},\mathscr{Y}) \leq L(\mathscr{X},\mathscr{Y}). \quad (26)$$

The next lemma will be used to reduce to uniform distributions on squares.

Lemma 15. *Assume (A1-A5). Let μ, μ' be two probability measures on $\mathbb{R}^d$ with supports in Q and $n > 0$. Then, for some constant c depending only on κ, κ_0,*

$$\mathbb{E}L_{\partial Q}(n\mu) \leq \mathbb{E}L_{\partial Q}(n\mu') + 2cn\,\text{diam}(Q)^p\,d_{\text{TV}}(\mu,\mu').$$

Consequently, if f is a nonnegative locally integrable function on $\mathbb{R}^d$, setting $\alpha = \int_Q f/\text{vol}(Q)$, it holds

$$\mathbb{E}L_{\partial Q}(nf\mathbf{1}_Q) \leq \mathbb{E}L_{\partial Q}(n\alpha\mathbf{1}_Q) + cn\,\text{diam}(Q)^p \int_Q |f(x) - \alpha|\,dx.$$

Proof. The functional $L_{\partial Q}$ satisfies a slight modification of property $(\mathscr{R}_p)$: for all multisets $\mathscr{X}, \mathscr{Y}, \mathscr{X}_1, \mathscr{Y}_1, \mathscr{X}_2, \mathscr{Y}_2$ in Q, it holds

$$L_{\partial Q}(\mathscr{X} \cup \mathscr{X}_1, \mathscr{Y} \cup \mathscr{Y}_1) \leq L_{\partial Q}(\mathscr{X} \cup \mathscr{X}_2, \mathscr{Y} \cup \mathscr{Y}_2)$$
$$+ C\,\text{diam}(Q)^p\big(\text{card}(\mathscr{X}_1) + \text{card}(\mathscr{X}_2) + \text{card}(\mathscr{Y}_1) + \text{card}(\mathscr{Y}_2)\big), \quad (27)$$

with $C = C(\kappa, \kappa_0)$. The above inequality is established as in the proof of Lemma 8. Indeed, by linearity and symmetry we should check (16) and (18) for $L_{\partial Q}$. To prove (16), we consider an optimal triplet (G, A, B) for $(\mathscr{X}, \mathscr{Y})$ and apply the merging property (A4) to G with the empty graph and $m = 1$: we obtain a graph G'' and get a triplet $(G'', A, B \cup \{b\})$ for $(\mathscr{X} \cup \{a\}, \mathscr{Y})$, where b is any point in ∂Q. To prove (18), we now consider an optimal triplet (G, A, B) for $(\mathscr{X} \cup \{a\}, \mathscr{Y})$ and move the point a to the a' in ∂Q in order to obtain a triplet $(G, A \cup \{a'\}, B)$ for $(\mathscr{X}, \mathscr{Y})$.

With (27) at hand, the statements follow from the proofs of Proposition 3 and Corollary 2. □

The next lemma gives a lower bound on L in terms of its boundary functional and states an important superadditive property of $L_{\partial S}$.

Lemma 16. *Assume (A1-A5). Let ν be a finite measure on $\mathbb{R}^d$ and consider a partition $Q = \cup_{P \in \mathscr{P}} P$ of a bounded subset of $\mathbb{R}^d$. Then, if $c = 4\kappa(1 + \kappa_0)$, we have*

$$c\sqrt{\nu(\mathbb{R}^d)} \operatorname{diam}(Q)^p + \mathbb{E}L(\nu) \geq \mathbb{E}L_{\partial Q}(\mathbf{1}_Q \cdot \nu) \geq \sum_{P \in \mathscr{P}} \mathbb{E}L_{\partial P}(\mathbf{1}_P \cdot \nu).$$

Proof. We start with the first inequality. Let $\mathscr{X} = \{X_1, \ldots, X_m\}$, $\mathscr{Y} = \{Y_1, \ldots, Y_n\}$ be multisets included in Q and $\mathscr{X}' = \{X_{m+1}, \ldots, X_{m+m'}\}$, $\mathscr{Y}' = \{Y_{n+1}, \ldots, Y_{n+n'}\}$ be multisets included in Q^c. First, let us show that

$$c_1|m + m' - n - n'|\operatorname{diam}(Q)^p + L(\mathscr{X} \cup \mathscr{X}', \mathscr{Y} \cup \mathscr{Y}') \geq L_{\partial Q}(\mathscr{X}, \mathscr{Y}), \quad (28)$$

with $c_1 = \kappa(1 + \kappa_0)$. To do so, let us consider an optimal graph G for $L(\mathscr{X} \cup \mathscr{X}', \mathscr{Y} \cup \mathscr{Y}')$. It uses all the points but $|m+m'-n-n'|$ points in excess. We consider the subsets $\mathscr{X}_0 \subset \mathscr{X}$ and $\mathscr{Y}_0 \subset \mathscr{Y}$ of points that are used in G and belong to Q. By definition there exist subsets $A, B \subset Q^c$ such that $\operatorname{card}(\mathscr{X}_0 \cup A) = \operatorname{card}(\mathscr{Y}_0 \cup B)$ and $L(\mathscr{X} \cup \mathscr{X}', \mathscr{Y} \cup \mathscr{Y}') = L(\mathscr{X}_0 \cup A, \mathscr{Y}_0 \cup B)$. By definition of the boundary functional and using (26),

$$L_{\partial Q}(\mathscr{X}_0, \mathscr{Y}_0) \leq L^0_{\partial Q}(\mathscr{X}_0 \cup A, \mathscr{Y}_0 \cup B) \leq L(\mathscr{X}_0 \cup A, \mathscr{Y}_0 \cup B) = L(\mathscr{X} \cup \mathscr{X}', \mathscr{Y} \cup \mathscr{Y}').$$

Finally, since there are at most $|n + n' - m - m'|$ points in $\mathscr{X} \cup \mathscr{Y}$ which are not in $\mathscr{X}_0 \cup \mathscr{Y}_0$ (i.e. points of Q not used for the optimal G), the modified ($\mathscr{R}_p$) property given by (27) yields (28). We apply the latter inequality to $\mathscr{X}$, $\mathscr{Y}$ independent Poisson processes of intensity $\mathbf{1}_Q \cdot \nu$, and $\mathscr{X}'$, $\mathscr{Y}'$, two independent Poisson processes of intensity $\mathbf{1}_{Q^c} \cdot \nu$, independent of $(\mathscr{X}, \mathscr{Y})$. Then $\mathscr{X} \cup \mathscr{X}'$, $\mathscr{Y} \cup \mathscr{Y}'$ are independent Poisson processes of intensity ν. Taking expectation, we obtain the first inequality, with $c = 4c_1$.

We now prove the second inequality. As above, let $\mathcal{X} = \{X_1, \ldots, X_m\}$, $\mathcal{Y} = \{Y_1, \ldots, Y_n\}$ be multisets included in Q. Let $G \in \mathcal{G}_k$ be an optimal graph for $L_{\partial Q}(\mathcal{X}, \mathcal{Y})$ and $A = \{X_{m+1}, \cdots, X_k\}$, $B = \{Y_{n+1}, \cdots, Y_k\}$ be optimal sets in Q^c. Given this graph G and a set S, we denote by E_S^0 the set of edges $\{i, k+j\}$ of G such that $X_i \in S$ and $Y_j \in S$, by E_S^1 the set of edges $\{i, k+j\}$ of G such that $X_i \in S$ and $Y_j \in S^c$, and by E_S^2 the set of edges $\{i, k+j\}$ of G such that $X_i \in S^c$ and $Y_j \in S$. Then by definition of the boundary functional

$$L_{\partial Q}(\mathcal{X}, \mathcal{Y}) = L_{\partial Q}^0(\mathcal{X} \cup A, \mathcal{Y} \cup B)$$
$$= \sum_{\{i,k+j\} \in E_Q^0} |X_i - Y_j|^p + \sum_{\{i,k+j\} \in E_Q^1} q \, d(X_i, Q^c)^p + \sum_{\{i,k+j\} \in E_Q^2} q \, d(Y_j, Q^c)^p.$$

Next, we bound these sums from below by considering the cells of the partition $\mathcal{P}$. If $x \in Q$, we denote by $P(x)$ the unique $P \in \mathcal{P}$ that contains x.

If an edge $e = \{i, k+j\} \in G$ is such that X_i, Y_j belong to the same cell P, we observe that $e \in E_P^0$ and we leave the quantity $|X_i - Y_j|^p$ unchanged.

If on the contrary, X_i and Y_j belong to different cells, from Hölder inequality,

$$|X_i - Y_j|^p \geq q \, d(X_i, P(X_i)^c)^p + q \, d(Y_j, P(Y_j)^c)^p.$$

Eventually, for any boundary edge in E_Q^1, we lower bound the contribution $d(X_i, Q^c)^p$ by $d(X_i, P(X_i)^c)^p$ and we do the same for E_Q^2. Combining these inequalities and grouping the terms according to the cell $P \in \mathcal{P}$ to which the points belong,

$$L_{\partial Q}(\mathcal{X}, \mathcal{Y})$$
$$\geq \sum_{P \in \mathcal{P}} \left(\sum_{\{i,k+j\} \in E_P^0} |X_i - Y_j|^p + \sum_{\{i,k+j\} \in E_P^1} q \, d(X_i, \partial P)^p + \sum_{\{i,k+j\} \in E_P^2} q \, d(Y_j, \partial P)^p \right).$$

For a given cell P, set $A' = (\mathcal{X} \cup A) \cap P^c$ and $B' = (\mathcal{Y} \cup B) \cap P^c$. We get

$$\sum_{\{i,k+j\} \in E_P^0} |X_i - Y_j|^p + \sum_{\{i,k+j\} \in E_P^1} q \, d(X_i, \partial P)^p + \sum_{\{i,k+j\} \in E_P^2} q \, d(Y_j, \partial P)^p$$
$$= L_{P^c}^0((\mathcal{X} \cap P) \cup A', (\mathcal{Y} \cap P) \cup B') \geq L_{\partial P}(\mathcal{X} \cap P, \mathcal{Y} \cap P).$$

So applying these inequalities to $\mathcal{X}$ and $\mathcal{Y}$ two independent Poisson point processes with intensity $\nu \mathbf{1}_Q$ and taking expectation, we obtain the claim. □

Let $Q = [0,1]^d$ and denote

$$\bar{L}_{\partial Q}(n) = \mathbb{E} L_{\partial Q}(n \mathbf{1}_Q).$$

Lemma 17. *Assume (A1-A5). Let $Q \subset \mathbb{R}^d$ be a cube of side-length 1. If $0 < 2p < d$, then*

$$\lim_{n \to \infty} \frac{\bar{L}_{\partial Q}(n)}{n^{1-\frac{p}{d}}} = \beta'_L,$$

where $\beta'_L > 0$ is a constant depending on L, p, and d.

Proof. The proof is the same than the proof of Lemma 11, with Lemma 16 replacing Lemma 9. □

6.3.2 General Absolutely Continuous Measures with Unbounded Support

Theorem 10. *Assume (A1-A5) and that $0 < 2p < d$. Let $f : \mathbb{R}^d \to \mathbb{R}^+$ be an integrable function. Then*

$$\liminf_n \frac{\mathbb{E}L(nf)}{n^{1-\frac{p}{d}}} \geq \beta'_L \int_{\mathbb{R}^d} f^{1-\frac{p}{d}}.$$

Proof. The proof is now formally the same than the proof of Theorem 8, invoking Lemmas 15, 16, and 17 in place of Lemmas 10, 9, and 11 respectively. □

Remark 6. Finding good lower bounds for a general bipartite functional L on $\mathbb{R}^d$ satisfying the properties $(\mathcal{H}_p)$, $(\mathcal{R}_p)$, $(\mathcal{S}_p)$ could be significantly more difficult. It is far from obvious to define a proper boundary functional $L_{\partial Q}$ at this level of generality. However if there exists a bipartite functional $L_{\partial Q}$ on $\mathbb{R}^d$ indexed on sets $Q \subset \mathbb{R}^d$ such that for any $t > 0$, $\mathbb{E}L_{\partial(tQ)}(n\mathbf{I}_{tQ}) = t^p \mathbb{E}L_{\partial Q}(nt^d \mathbf{I}_Q)$ and such that the statements of Lemmas 16, 15, and 17 hold, then the statement of Theorem 8 also holds for the functional L. Thus, the caveat of this kind of techniques lies in the good definition of a boundary functional $L_{\partial Q}$.

6.3.3 Dealing with the Singular Component: Example of the Traveling Salesperson Tour

Let $p \in (0, 1]$. We shall say that a bipartite functional L on $\mathbb{R}^d$ satisfies the inverse subadditivity property $(\mathcal{I}_p)$ if there is a constant C such that for all finite multisets $\mathcal{X}_1, \mathcal{Y}_1, \mathcal{X}_2, \mathcal{Y}_2$ included in a bounded set $Q \subset \mathbb{R}^d$,

$$L(\mathcal{X}_1, \mathcal{Y}_1) \leq L(\mathcal{X}_1 \cup \mathcal{X}_2, \mathcal{Y}_1 \cup \mathcal{Y}_2) + L(\mathcal{X}_2, \mathcal{Y}_2)$$
$$+ C \operatorname{diam}(Q)^p \left(1 + |\mathcal{X}_1(Q) - \mathcal{Y}_1(Q)| + |\mathcal{X}_2(Q) - \mathcal{Y}_2(Q)|\right).$$

Although it makes sense for all p, we have been able to check this property on examples only for $p \in (0, 1]$. Also we could have added a constant in front of $L(\mathcal{X}_2, \mathcal{Y}_2)$.

It is plain that the argument of Sect. 6.2.3 readily adapts to a functional satisfying $(\mathcal{I}_p)$, for which one already knows a general upper limit result and a limit result for absolutely continuous laws. It therefore provides a limit result for general laws. In the remainder of this section, we show that the traveling salesperson bipartite tour functional $L = T_p$, $p \in (0, 1]$ enjoys the inverse subadditivity property. This allows to prove the following result:

Theorem 11. *Assume that either $d \in \{1, 2\}$ and $0 < 2p < d$, or $d \geq 3$ and $p \in (0, 1]$. Let $L = T_p$ be the traveling salesperson bipartite tour functional. Let μ be a finite measure such that for some $\alpha > \frac{2dp}{d-2p}$, $\int |x|^\alpha d\mu < +\infty$. Then, if f is a density function for the absolutely continuous part of μ,*

$$\liminf_n \frac{\mathbb{E}L(n\mu)}{n^{1-\frac{p}{d}}} \geq \beta'_L \int_{\mathbb{R}^d} f^{1-\frac{p}{d}}.$$

Moreover if f is proportional to the indicator function of a bounded set with positive Lebesgue measure

$$\lim_n \frac{\mathbb{E}L(n\mu)}{n^{1-\frac{p}{d}}} = \beta_L \int_{\mathbb{R}^d} f^{1-\frac{p}{d}}.$$

All we have to do is to check property $(\mathcal{I}_p)$. More precisely:

Lemma 18. *Assume $p \in (0, 1]$ and $L = T_p$. For any set $\mathcal{X}_1, \mathcal{X}_2, \mathcal{Y}_1, \mathcal{Y}_2$ in a bounded set Q*

$$L(\mathcal{X}_1, \mathcal{Y}_1) \leq L(\mathcal{X}_1 \cup \mathcal{X}_2, \mathcal{Y}_1 \cup \mathcal{Y}_2) + L(\mathcal{X}_2, \mathcal{Y}_2)$$
$$+ 2 \operatorname{diam}(Q)^p \left(1 + |\operatorname{card}(\mathcal{X}_1) - \operatorname{card}(\mathcal{Y}_1)| + |\operatorname{card}(\mathcal{X}_2) - \operatorname{card}(\mathcal{Y}_2)|\right).$$

Proof. We may assume without loss of generality that $\operatorname{card}(\mathcal{X}_1) \wedge \operatorname{card}(\mathcal{Y}_1) \geq 2$, otherwise, $L(\mathcal{X}_1, \mathcal{Y}_1) = 0$ and there is nothing to prove. Consider an optimal cycle $G_{1,2}$ for $L(\mathcal{X}_1 \cup \mathcal{X}_2, \mathcal{Y}_1 \cup \mathcal{Y}_2)$. In $G_{1,2}$, $m = |\operatorname{card}(\mathcal{X}_1) + \operatorname{card}(\mathcal{Y}_1) - \operatorname{card}(\mathcal{X}_2) - \operatorname{card}(\mathcal{Y}_2)| \leq |\operatorname{card}(\mathcal{X}_1) - \operatorname{card}(\mathcal{Y}_1)| + |\operatorname{card}(\mathcal{X}_2) - \operatorname{card}(\mathcal{Y}_2)|$ points have been left aside. We shall build a bipartite tour G_1 on $(\mathcal{X}'_1, \mathcal{Y}'_1)$, the points of $(\mathcal{X}_1, \mathcal{Y}_1)$ that have not been left aside by $G_{1,2}$.

We consider an optimal cycle G_2 for $L(\mathcal{X}'_2, \mathcal{Y}'_2)$, where $(\mathcal{X}'_2, \mathcal{Y}'_2)$ are the points of $(\mathcal{X}_2, \mathcal{Y}_2)$ that have not been left aside by $G_{1,2}$. We define $(\mathcal{X}''_2, \mathcal{Y}''_2) \subset (\mathcal{X}'_2, \mathcal{Y}'_2)$ as the sets of points that are in G_2. Since $\operatorname{card}(\mathcal{X}'_1) + \operatorname{card}(\mathcal{X}'_2) = \operatorname{card}(\mathcal{Y}'_1) + \operatorname{card}(\mathcal{Y}'_2)$, we get $\operatorname{card}(\mathcal{X}'_1) - \operatorname{card}(\mathcal{Y}'_1) = -\operatorname{card}(\mathcal{X}'_2) + \operatorname{card}(\mathcal{Y}'_2)$. It implies that the same number of points from the opposite type need to be removed in $(\mathcal{X}'_1, \mathcal{Y}'_1)$ and $(\mathcal{X}'_2, \mathcal{Y}'_2)$ in order to build a bipartite tour. We fix an orientation on $G_{1,2}$. Assume for example that $\operatorname{card}(\mathcal{X}'_2) \geq \operatorname{card}(\mathcal{Y}'_2)$, if a point $x \in \mathcal{X}'_2 \setminus \mathcal{X}''_2$, we then remove the next point y on the oriented cycle $G_{1,2}$ of $\mathcal{Y}'_1$. Doing so, this defines a couple of sets $(\mathcal{X}'''_1, \mathcal{Y}'''_1) \subset (\mathcal{X}'_1, \mathcal{Y}'_1)$ of cardinality $\operatorname{card}(\mathcal{X}'_1) \wedge \operatorname{card}(\mathcal{Y}'_1)$ and

Fig. 2 In *blue*, the oriented cycle $G_{1,2}$, in *red* G_2, in *black* $G'_{1,2}$. The points in $\mathscr{X}_1 \cup \mathscr{X}_2$ are represented by a *cross*, points in $\mathscr{Y}_1 \cup \mathscr{Y}_2$ by a *circle*

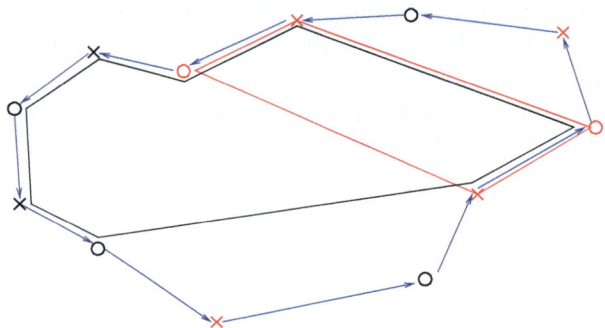

$$L(\mathscr{X}'_1, \mathscr{Y}'_1) \leq L(\mathscr{X}''_1, \mathscr{Y}''_1).$$

We define $G'_{1,2}$ as the cycle on $(\mathscr{X}''_1 \cup \mathscr{X}''_2, \mathscr{Y}''_1 \cup \mathscr{Y}''_2)$ obtained from $G_{1,2}$ by saying that the point after $x \in (\mathscr{X}''_1 \cup \mathscr{X}''_2, \mathscr{Y}''_1 \cup \mathscr{Y}''_2)$ in the oriented cycle $G'_{1,2}$ is the next point $y \in (\mathscr{X}''_1 \cup \mathscr{X}''_2, \mathscr{Y}''_1 \cup \mathscr{Y}''_2)$ in $G_{1,2}$. By construction, $G'_{1,2}$ is a bipartite cycle. Also, since $p \in (0, 1]$, we may use the triangle inequality: the distance between two successive points in the circuit $G'_{1,2}$ is bounded by the sum of the length of the intermediary edges in $G_{1,2}$. We get

$$L(\mathscr{X}''_1 \cup \mathscr{X}''_2, \mathscr{Y}''_1 \cup \mathscr{Y}''_2) \leq L(\mathscr{X}_1 \cup \mathscr{X}_2, \mathscr{Y}_1 \cup \mathscr{Y}_2).$$

Now consider the (multi) graph $G = G'_{1,2} \cup G_2$ obtained by adding all edges of $G'_{1,2}$ and G_2. This graph is bipartite, connected, and points in $(\mathscr{X}''_1, \mathscr{Y}''_1)$ have degree 2 while those in $(\mathscr{X}''_2, \mathscr{Y}''_2)$ have degree 4. Let k be the number of edges in G, we recall that an eulerian circuit in G is a sequence $E = (e_1, \cdots, e_k)$ of adjacent edges in G such that e_k is also adjacent to e_1 and all edges of G appears exactly once in the sequence E. By the Euler's circuit theorem, there exists an eulerian circuit in G. Moreover, this eulerian circuit can be chosen so that if $e_i = \{u_{i-1}, u_i\} \in G_2$ then $e_{i+1} = \{u_{i+1}, u_i\} \in G'_{1,2}$ with the convention that $e_{k+1} = e_1$.

This sequence E defines an oriented circuit of points. Now we define an oriented circuit on $(\mathscr{X}''_1, \mathscr{Y}''_1)$, by connecting a point x of $(\mathscr{X}''_1, \mathscr{Y}''_1)$ to the next point y in $(\mathscr{X}''_1, \mathscr{Y}''_1)$ visited by the oriented circuit E. Due to the property that $e_i \in G_2$ implies $e_{i+1} \in G$, if $x \in \mathscr{X}''_1$ then $y \in \mathscr{Y}''_1$ and conversely, if $x \in \mathscr{Y}''_1$ then $y \in \mathscr{X}''_1$. Hence, this oriented circuit defines a bipartite cycle G_1 in $(\mathscr{X}''_1, \mathscr{Y}''_1)$.

By the triangle inequality, the distance between two successive points in the circuit G_1 is bounded by the sum of the length of the intermediary edges in E. Since each edge of G appears exactly once in E, it follows that

$$L(\mathscr{X}'_1, \mathscr{Y}'_1) \leq L(\mathscr{X}_1 \cup \mathscr{X}_2, \mathscr{Y}_1 \cup \mathscr{Y}_2) + L(\mathscr{X}'_2, \mathscr{Y}'_2).$$

To conclude, we merge arbitrarily to the cycle G_1 the remaining points of $(\mathscr{X}_1, \mathscr{Y}_1)$, there are at most m of them (regularity $(\mathscr{R}_p)$ property). □

7 Variants and Final Comments

As a conclusion, we briefly discuss variants and possible extensions of Theorem 2. For $d > 2p$ and when μ is the uniform distribution on the cube $[0, 1]^d$, there exists a constant $\beta_p(d) > 0$ such that almost surely

$$\lim_{n \to \infty} n^{\frac{p}{d}-1} M_p(\{X_1, \ldots, X_n\}, \{Y_1, \ldots, Y_n\}) = \beta_p(d).$$

A natural question is to understand what happens below the critical line $d = 2p$, i.e. when $d \leq 2p$. For example for $d = 2$ and $p = 1$, a similar convergence is also expected in dimension 2 with scaling $\sqrt{n \ln n}$, but this is a difficult open problem. The main result in this direction goes back to Ajtai, Komlós, and Tusnády [1]. See also the improved upper bound of Talagrand and Yukich in [17]. In dimension 1, there is no such stabilization to a constant.

Recall that

$$\left(\frac{1}{n} M_p(\{X_i\}_{i=1}^n, \{Y_i\}_{i=1}^n)\right)^{\frac{1}{p}} = W_p\left(\frac{1}{n}\sum_{i=1}^n \delta_{X_i}, \frac{1}{n}\sum_{i=1}^n \delta_{Y_i}\right).$$

where W_p is the L_p-Wasserstein distance. A variant of Theorem 3 can be obtained along the same lines, concerning the convergence of

$$n^{\frac{1}{d}} W_p\left(\frac{1}{n}\sum_{i=1}^n \delta_{X_i}, \mu\right),$$

where μ is the common distribution of the X_i's. Such results are of fundamental importance in statistics. Also note that combining the triangle inequality and Jensen inequality, it is not hard to see that

$$\mathbb{E}W_1\left(\frac{1}{n}\sum_{i=1}^n \delta_{X_i}, \mu\right) \leq \mathbb{E}W_1\left(\frac{1}{n}\sum_{i=1}^n \delta_{X_i}, \frac{1}{n}\sum_{i=1}^n \delta_{Y_i}\right) \leq 2\mathbb{E}W_1\left(\frac{1}{n}\sum_{i=1}^n \delta_{X_i}, \mu\right),$$

(similar inequalities hold for $p \geq 1$). Hence it is clear that the behavior of this functional is quite close to the one of the two-sample optimal matching. However, the extension of Theorem 2 would require some care in the definition of the boundary functional.

Finally, it is worthy to note that the case of uniform distribution for $L = M_p$ has a connection with stationary matchings of two independent Poisson point processes of intensity 1, see Holroyd, Pemantle, Peres, and Schramm [6]. Indeed, consider mutually independent random variables $(X_i)_{i \geq 1}$ and $(Y_j)_{j \geq 1}$ having uniform distribution on $Q = [-1/2, 1/2]^d$. It is well known that for any x in the interior of Q, the pair of point processes

$$\left(\frac{1}{n} \sum_{i=1}^{n} \delta_{n^{\frac{1}{d}}(X_i - x)}, \frac{1}{n} \sum_{i=1}^{n} \delta_{n^{\frac{1}{d}}(Y_i - x)} \right)$$

converges weakly for the topology of vague convergence to (Ξ_1, Ξ_2), where Ξ_1 and Ξ_2 are two independent Poisson point processes of intensity 1. Also, we may write

$$n^{\frac{p}{d}-1} \mathbb{E} M_p(\{X_i\}_{i=1}^n, \{Y_i\}_{i=1}^n) = \frac{1}{n} \mathbb{E} \sum_{i=1}^{n} \left| n^{\frac{1}{d}}(X_i - x) - n^{\frac{1}{d}}(Y_{\sigma_n(i)} - x) \right|^p.$$

where σ_n is an optimal matching. Now, the fact that for $0 < p < 2d$, $\lim_n n^{\frac{p}{d}-1} \mathbb{E} M_p(\{X_i\}_{i=1}^n, \{Y_i\}_{i=1}^n) = \beta_p(d)$ implies the tightness of the sequence of matchings σ_n and it can be used to define a stationary matching σ on (Ξ_1, Ξ_2), see the proof of Theorem 1 (iii) in [6] for the details of such an argument. In particular, this matching σ will enjoy a local notion of minimality for the L_p-norm, as defined by Holroyd in [5] (for the L_1-norm). See also related work of Huesmann and Sturm [7].

Acknowledgements We are indebted to Martin Huesmann for pointing an error in the proof of a previous version of Theorem 2. This is also a pleasure to thank for its hospitality the Newton Institute where part of this work has been done (2011 Discrete Analysis program).

References

1. M. Ajtai, J. Komlós, G. Tusnády, On optimal matchings. Combinatorica **4**(4), 259–264 (1984)
2. J. Beardwood, J.H. Halton, J.M. Hammersley, The shortest path through many points. Proc. Camb. Phil. Soc. **55**, 299–327 (1959)
3. J. Boutet de Monvel, O. Martin, Almost sure convergence of the minimum bipartite matching functional in Euclidean space. Combinatorica **22**(4), 523–530 (2002)
4. V. Dobrić, J.E. Yukich, Asymptotics for transportation cost in high dimensions. J. Theor. Probab. **8**(1), 97–118 (1995)
5. A. Holroyd, Geometric properties of Poisson matchings. Probab. Theory Relat. Fields **150**(3–4), 511–527 (2011)
6. A. Holroyd, R. Pemantle, Y. Peres, O. Schramm, Poisson matching. Ann. Inst. Henri Poincaré Probab. Stat. **45**(1), 266–287 (2009)
7. M. Huesmann, K.T. Sturm, Optimal transport from Lebesgue to Poisson. arXiv preprint, arXiv:1012.3845 (2010). To appear in Ann. Probab.
8. J.F.C. Kingman, *Poisson Processes*. Oxford Studies in Probability, vol. 3. Oxford Science Publications (The Clarendon Press; Oxford University Press, New York, 1993)
9. C. Papadimitriou, The probabilistic analysis of matching heuristics, in *Proceedings of the 15th Allerton Conference on Communication, Control and Computing*, pp. 368–378, 1978
10. C. Papadimitriou, K. Steiglitz, *Combinatorial Optimization: Algorithms and Complexity* (Dover Publications, New York, 1998)
11. S.T. Rachev, L. Rüschendorf, *Mass Transportation Problems. Vol. I: Theory*. Probability and Its Applications (Springer, New York, 1998)
12. W.T. Rhee, A matching problem and subadditive Euclidean functionals. Ann. Appl. Probab. **3**, 794–801 (1993)

13. W.T. Rhee, On the stochastic Euclidean traveling salesperson problem for distributions with unbounded support. Math. Oper. Res. **18**(2), 292–299 (1993)
14. J.M. Steele, Subadditive Euclidean functionals and nonlinear growth in geometric probability. Ann. Probab. **9**(3), 365–376 (1981)
15. J.M. Steele, *Probability Theory and Combinatorial Optimization*. CBMS-NSF Regional Conference Series in Applied Mathematics, vol. 69 (Society for Industrial and Applied Mathematics (SIAM), Philadelphia, PA, 1997)
16. M. Talagrand, Matching random samples in many dimensions. Ann. Appl. Probab. **2**(4), 846–856 (1992)
17. M. Talagrand, J. Yukich, The integrability of the square exponential transport cost. Ann. Appl. Probab. **3**(4), 1100–1111 (1993)
18. C. Villani, *Topics in Optimal Transportation*. Graduate Studies in Mathematics, vol. 58 (American Mathematical Society, Providence, RI, 2003)
19. J. Yukich, *Probability Theory of Classical Euclidean Optimization Problems*. Lecture notes in mathematics, vol. 1675 (Springer, Berlin, 1998)

A Simple Proof of Duquesne's Theorem on Contour Processes of Conditioned Galton–Watson Trees

Igor Kortchemski

Abstract We give a simple new proof of a theorem of Duquesne, stating that the properly rescaled contour function of a critical aperiodic Galton–Watson tree, whose offspring distribution is in the domain of attraction of a stable law of index $\theta \in (1, 2]$, conditioned on having total progeny n, converges in the functional sense to the normalized excursion of the continuous-time height function of a strictly stable spectrally positive Lévy process of index θ. To this end, we generalize an idea of Le Gall which consists in using an absolute continuity relation between the conditional probability of having total progeny exactly n and the conditional probability of having total progeny at least n. This new method is robust and can be adapted to establish invariance theorems for Galton–Watson trees having n vertices whose degrees are prescribed to belong to a fixed subset of the positive integers.

Keywords Conditioned Galton–Watson tree • Stable continuous random tree • Scaling limit • Invariance principle

AMS 2000 subject classifications. Primary 60J80, 60F17, 60G52; secondary 05C05.

I. Kortchemski (✉)
Laboratoire de mathématiques, UMR 8628 CNRS, Université Paris-Sud, 91405 ORSAY Cedex, France
e-mail: igor.kortchemski@normalesup.org

Introduction

In this article, we are interested in the asymptotic behavior of critical Galton–Watson trees whose offspring distribution may have infinite variance. Aldous [1] studied the shape of large critical Galton–Watson trees whose offspring distribution has finite variance and proved that their properly rescaled contour functions converge in distribution in the functional sense to the Brownian excursion. This seminal result has motivated the study of the convergence of other rescaled paths obtained from Galton–Watson trees, such as the Lukasiewicz path (also known as the Harris walk) and the height function. In [20], under an additional exponential moment condition, Marckert and Mokkadem showed that the rescaled Lukasiewicz path, height function and contour function all converge in distribution to the same Brownian excursion. In parallel, unconditional versions of Aldous' result have been obtained in full generality. More precisely, when the offspring distribution is in the domain of attraction of a stable law of index $\theta \in (1, 2]$, Duquesne and Le Gall [8] showed that the concatenation of rescaled Lukasiewicz paths of a sequence of independent Galton–Watson trees converges in distribution to a strictly stable spectrally positive Lévy process X of index θ, and the concatenation of the associated rescaled height functions (or of the rescaled contour functions) converges in distribution to the so-called continuous-time height function associated with X. In the same monograph, Duquesne and Le Gall explained how to deduce a limit theorem for Galton–Watson trees conditioned on having at least n vertices from the unconditional limit theorem. Finally, still in the stable case, Duquesne [7] showed that the rescaled Lukasiewicz path of a Galton–Watson tree conditioned on having n vertices converges in distribution to the normalized excursion of the Lévy process X (thus extending Marckert and Mokkadem's result) and that the rescaled height and contour functions of a Galton–Watson tree conditioned on having n vertices converge in distribution to the normalized excursion of the continuous-time height function H^{exc} associated with X (thus extending Aldous' result).

In this work, we give an alternative proof of Duquesne's result, which is based on an idea that appeared in the recent papers [16, 18]. Let us explain our approach after introducing some notation. For every $x \in \mathbb{R}$, let $\lfloor x \rfloor$ denote the greatest integer smaller than or equal to x. If I is an interval, let $\mathcal{C}(I, \mathbb{R})$ be the space of all continuous functions $I \to \mathbb{R}$ equipped with the topology of uniform convergence on compact subsets of I. We also let $\mathbb{D}(I, \mathbb{R})$ be the space of all right-continuous with left limits (càdlàg) functions $I \to \mathbb{R}$, endowed with the Skorokhod J_1-topology (see [4, Chap. 3], [12, Chap. VI] for background concerning the Skorokhod topology). Denote by $\mathbb{P}_\mu$ the law of the Galton–Watson tree with offspring distribution μ. The total progeny of a tree τ will be denoted by $\zeta(\tau)$. Fix $\theta \in (1, 2]$ and let $(X_t)_{t \geq 0}$ be the spectrally positive Lévy process with Laplace exponent $\mathbb{E}[\exp(-\lambda X_t)] = \exp(t\lambda^\theta)$.

(0) We fix a critical offspring distribution μ in the domain of attraction of a stable law of index $\theta \in (1, 2]$. If $U_1, U_2, \ldots$ are i.i.d. random variables with

distribution μ, and $W_n = U_1 + \cdots + U_n - n$, there exist positive constants $(B_n)_{n\geq 0}$ such that W_n/B_n converges in distribution to X_1.

(i) Fix $a \in (0, 1)$. To simplify notation, set $\mathcal{W}^{a,(n)} = (\mathcal{W}_j^{a,(n)}, 0 \leq j \leq \lfloor na \rfloor)$ where $\mathcal{W}_j^{a,(n)} = \mathcal{W}_j(\tau)/B_n$ and $\mathcal{W}(\tau)$ is the Lukasiewicz path of τ (see Sect. 1.2 below for its definition). Then for every function $f_n : \mathbb{Z}^{\lfloor an \rfloor + 1} \to \mathbb{R}_+$, the following absolute continuity relation holds:

$$\mathbb{E}_\mu \left[f_n(\mathcal{W}^{a,(n)}) | \zeta(\tau) = n \right] = \mathbb{E}_\mu \left[f_n(\mathcal{W}^{a,(n)}) D_n^{(a)} \left(\mathcal{W}_{\lfloor an \rfloor}(\tau) \right) | \zeta(\tau) \geq n \right] \quad (1)$$

with a certain function $D_n^{(a)} : \{-1, 0, 1, 2, \ldots\} \to \mathbb{R}_+$.

(ii) We establish the existence of a measurable function $\Gamma_a : \mathbb{R}_+ \to \mathbb{R}_+$ such that the quantity $\left| D_n^{(a)}(j) - \Gamma_a(j/B_n) \right|$ goes to 0 as $n \to \infty$, uniformly in values of j such that j/B_n stays in a compact subset of $\mathbb{R}_+^*$. Furthermore, if H denotes the continuous-time height process associated with X and $\mathbf{N}$ stands for the Itô excursion measure of X above its infimum, we have for every bounded measurable function $F : \mathbb{D}([0, a], \mathbb{R}) \to \mathbb{R}_+$:

$$\mathbf{N}\left(F((H_t)_{0 \leq t \leq a}) \Gamma_a(X_a) | \zeta > 1 \right) = \mathbf{N}\left(F((H_t)_{0 \leq t \leq a}) | \zeta = 1 \right), \quad (2)$$

where ζ is the duration of the excursion under $\mathbf{N}$.

(iii) We show that under $\mathbb{P}_\mu[\cdot | \zeta(\tau) = n]$, the rescaled height function converges in distribution on $[0, a]$ for every $a \in (0, 1)$. To this end, we fix a bounded continuous function $F : \mathbb{D}([0, a], \mathbb{R}) \to \mathbb{R}_+$ and apply formula (1) with $f_n(\mathcal{W}^{a,(n)}) = F\left(\frac{B_n}{n} H_{\lfloor nt \rfloor}(\tau); 0 \leq t \leq a \right)$ where $H(\tau)$ is the height function of the tree τ. Using the previously mentioned result of Duquesne and Le Gall concerning Galton–Watson trees conditioned on having at least n vertices, we show that we can restrict ourselves to the case where $\mathcal{W}_{\lfloor an \rfloor}(\tau)/B_n$ stays in a compact subset of $\mathbb{R}_+^*$, so that we can apply (ii) and obtain that:

$$\lim_{n \to \infty} \mathbb{E}_\mu \left[F\left(\frac{B_n}{n} H_{\lfloor nt \rfloor}(\tau); 0 \leq t \leq a \right) \Big| \zeta(\tau) = n \right]$$

$$= \lim_{n \to \infty} \mathbb{E}_\mu \left[F\left(\frac{B_n}{n} H_{\lfloor nt \rfloor}(\tau); 0 \leq t \leq a \right) D_n^{(a)} \left(\mathcal{W}_{\lfloor an \rfloor}(\tau) \right) | \zeta(\tau) \geq n \right]$$

$$= \mathbf{N}\left(F(H_t; 0 \leq t \leq a) \Gamma_a(X_a) | \zeta > 1 \right)$$

$$= \mathbf{N}\left(F(H_t; 0 \leq t \leq a) | \zeta = 1 \right).$$

(iv) By using a relationship between the contour function and the height function which was noticed by Duquesne and Le Gall in [8], we get that, under $\mathbb{P}_\mu[\cdot | \zeta(\tau) = n]$, the scaled contour function converges in distribution on $[0, a]$.

(v) By using the time reversal invariance property of the contour function, we deduce that under $\mathbb{P}_\mu[\cdot|\zeta(\tau) = n]$, the scaled contour function converges in distribution on the whole segment $[0, 1]$.
(vi) Using once again the relationship between the contour function and the height function, we deduce that, under $\mathbb{P}_\mu[\cdot|\zeta(\tau) = n]$, the scaled height function converges in distribution on $[0, 1]$.

In the case where the variance of μ is finite, Le Gall gave an alternative proof of Aldous' theorem in [16, Theorem 6.1] using a similar approach based on a strong local limit theorem. There are additional difficulties in the infinite variance case, since no such theorem is known in this case.

Let us finally discuss the advantage of this new method. Firstly, the proof is simpler and less technical. Secondly, we believe that this approach is robust and can be adapted to other situations. For instance, using the same ideas, we have established invariance theorems for Galton–Watson trees having n vertices whose degrees are prescribed to belong to a fixed subset of the nonnegative integers [14].

The rest of this text is organized as follows. In Sect. 1, we present the discrete framework by defining Galton–Watson trees and their codings. We explain how the local limit theorem gives information on the asymptotic behavior of large Galton–Watson trees and present the discrete absolute continuity relation appearing in (1). In Sect. 2, we discuss the continuous framework: we introduce the strictly stable spectrally positive Lévy process, its Itô excursion measure $\mathbf{N}$ and the associated continuous-time height process. We also prove the absolute continuity relation (3). Finally, in Sect. 3 we give the new proof of Duquesne's theorem by carrying out steps (i–vi).

Notation and Main Assumptions. Throughout this work $\theta \in (1, 2]$ is a fixed parameter. We consider a probability distribution $(\mu(j))_{j \geq 0}$ on the nonnegative integers satisfying the following three conditions:

(i) μ is critical, meaning that $\sum_{k=0}^{\infty} k\mu(k) = 1$.
(ii) μ is in the domain of attraction of a stable law of index $\theta \in (1, 2]$. This means that either the variance of μ is positive and finite, or $\mu([j, \infty)) = j^{-\theta} L(j)$, where $L : \mathbb{R}_+ \to \mathbb{R}_+$ is a function such that $L(x) > 0$ for x large enough and $\lim_{x \to \infty} L(tx)/L(x) = 1$ for all $t > 0$ (such a function is called slowly varying). We refer to [5] or [9, Chap. 3.7] for details.
(iii) μ is aperiodic, which means that the additive subgroup of the integers $\mathbb{Z}$ spanned by $\{j; \mu(j) \neq 0\}$ is not a proper subgroup of $\mathbb{Z}$.

We introduce condition (iii) to avoid unnecessary complications, but our results can be extended to the periodic case.

In what follows, $(X_t)_{t \geq 0}$ will stand for the spectrally positive Lévy process with Laplace exponent $\mathbb{E}[\exp(-\lambda X_t)] = \exp(t\lambda^\theta)$ where $t, \lambda \geq 0$ and p_1 will denote the density of X_1. Finally, ν will stand for the probability measure on $\mathbb{Z}$ defined by $\nu(k) = \mu(k + 1)$ for $k \geq -1$. Note that ν has zero mean.

1 The Discrete Setting: Galton–Watson Trees

1.1 Galton–Watson Trees

Definition 1. Let $\mathbb{N} = \{0, 1, \ldots\}$ be the set of all nonnegative integers and $\mathbb{N}^* = \{1, \ldots\}$. Let also U be the set of all labels:

$$U = \bigcup_{n=0}^{\infty} (\mathbb{N}^*)^n,$$

where by convention $(\mathbb{N}^*)^0 = \{\emptyset\}$. An element of U is a sequence $u = u_1 \cdots u_j$ of positive integers, and we set $|u| = j$, which represents the "generation" of u. If $u = u_1 \cdots u_j$ and $v = v_1 \cdots v_k$ belong to U, we write $uv = u_1 \cdots u_j v_1 \cdots v_k$ for the concatenation of u and v. In particular, note that $u\emptyset = \emptyset u = u$. Finally, a *rooted ordered tree* τ is a finite subset of U such that:

1. $\emptyset \in \tau$.
2. if $v \in \tau$ and $v = uj$ for some $j \in \mathbb{N}^*$, then $u \in \tau$.
3. for every $u \in \tau$, there exists an integer $k_u(\tau) \geq 0$ such that, for every $j \in \mathbb{N}^*$, $uj \in \tau$ if and only if $1 \leq j \leq k_u(\tau)$.

In the following, by *tree* we will mean rooted ordered tree. The set of all trees is denoted by $\mathbb{T}$. We will often view each vertex of a tree τ as an individual of a population whose τ is the genealogical tree. The total progeny of τ will be denoted by $\zeta(\tau) = \text{Card}(\tau)$. Finally, if τ is a tree and $u \in \tau$, we set $T_u \tau = \{v \in U;\ uv \in \tau\}$, which is itself a tree.

Definition 2. Let ρ be a probability measure on $\mathbb{N}$ with mean less than or equal to 1 and such that $\rho(1) < 1$. The law of the Galton–Watson tree with offspring distribution ρ is the unique probability measure $\mathbb{P}_\rho$ on $\mathbb{T}$ such that:

1. $\mathbb{P}_\rho[k_\emptyset = j] = \rho(j)$ for $j \geq 0$.
2. for every $j \geq 1$ with $\rho(j) > 0$, conditionally on $\{k_\emptyset = j\}$, the shifted trees $T_1 \tau, \ldots, T_j \tau$ are i.i.d. with distribution $\mathbb{P}_\rho$.

A random tree whose distribution is $\mathbb{P}_\rho$ will be called a GW_ρ tree.

1.2 Coding Galton–Watson Trees

We now explain how trees can be coded by three different functions. These codings are crucial in the understanding of large Galton–Watson trees.

Definition 3. We write $u < v$ for the lexicographical order on the labels U (for example, $\emptyset < 1 < 21 < 22$). Consider a tree τ and order the individuals of τ in lexicographical order: $\emptyset = u(0) < u(1) < \cdots < u(\zeta(\tau) - 1)$. The height process

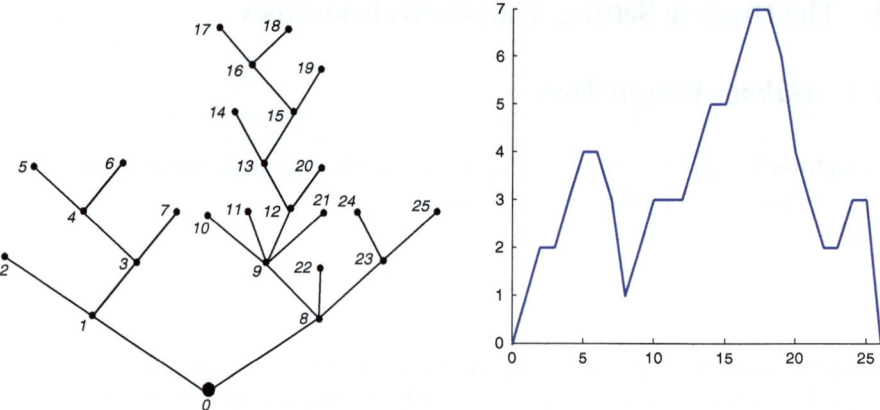

Fig. 1 A tree τ with its vertices indexed in lexicographical order and its contour function $(C_u(\tau); 0 \leq u \leq 2(\zeta(\tau) - 1)$. Here, $\zeta(\tau) = 26$

$H(\tau) = (H_n(\tau), 0 \leq n < \zeta(\tau))$ is defined, for $0 \leq n < \zeta(\tau)$, by $H_n(\tau) = |u(n)|$. For technical reasons, we set $H_k(\tau) = 0$ for $k \geq \zeta(\tau)$. We extend $H(\tau)$ to $\mathbb{R}_+$ by linear interpolation by setting $H_t(\tau) = (1 - \{t\})H_{\lfloor t \rfloor}(\tau) + \{t\}H_{\lfloor t \rfloor + 1}(\tau)$ for $0 \leq t \leq \zeta(\tau)$, where $\{t\} = t - \lfloor t \rfloor$.

Consider a particle that starts from the root and visits continuously all edges at unit speed (assuming that every edge has unit length), going backwards as little as possible and respecting the lexicographical order of vertices. For $0 \leq t \leq 2(\zeta(\tau) - 1)$, $C_t(\tau)$ is defined as the distance to the root of the position of the particle at time t. For technical reasons, we set $C_t(\tau) = 0$ for $t \in [2(\zeta(\tau) - 1), 2\zeta(\tau)]$. The function $C(\tau)$ is called the contour function of the tree τ. See Fig. 1 for an example, and [7, Sect. 2] for a rigorous definition.

Finally, the Lukasiewicz path $\mathcal{W}(\tau) = (\mathcal{W}_n(\tau), n \geq 0)$ of a tree τ is defined by $\mathcal{W}_0(\tau) = 0$, $\mathcal{W}_{n+1}(\tau) = \mathcal{W}_n(\tau) + k_{u(n)}(\tau) - 1$ for $0 \leq n \leq \zeta(\tau) - 1$, and $\mathcal{W}_k(\tau) = 0$ for $k > \zeta(\tau)$. For $u \geq 0$, we set $\mathcal{W}_u(\tau) = \mathcal{W}_{\lfloor u \rfloor}(\tau)$.

Note that necessarily $\mathcal{W}_{\zeta(\tau)}(\tau) = -1$. See Fig. 2 for an example.

Let $(W_n; n \geq 0)$ be a random walk which starts at 0 with jump distribution $\nu(k) = \mu(k+1)$ for $k \geq -1$. For $j \geq 1$, define $\zeta_j = \inf\{n \geq 0; W_n = -j\}$.

Proposition 1. $(W_0, W_1, \ldots, W_{\zeta_1})$ *has the same distribution as the Lukasiewicz path of a* GW_μ *tree. In particular, the total progeny of a* GW_μ *tree has the same law as* ζ_1.

Proof. See [17, Proposition 1.5]. □

We will also use the following well-known fact (see e.g. Lemma 6.1 in [21] and the discussion that follows).

Proposition 2. *For every integers* $1 \leq j \leq n$, *we have* $\mathbb{P}[\zeta_j = n] = \frac{j}{n}\mathbb{P}[W_n = -j]$.

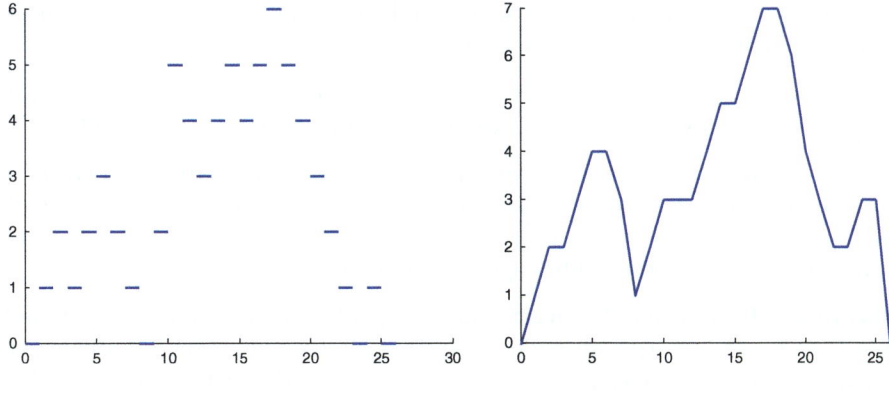

Fig. 2 The Lukasiewicz path $(\mathcal{W}_{\lfloor u \rfloor}(\tau); 0 \leq u < \zeta(\tau) + 1)$ and the height function $(H_u(\tau), 0 \leq u \leq \zeta(\tau))$ of the tree τ of Fig. 1

1.3 Slowly Varying Functions

Slowly varying functions appear in the study of domains of attractions of stable laws. Here we recall some properties of these functions in view of future use.

Recall that a positive measurable function $L : \mathbb{R}_+ \to \mathbb{R}_+$ is said to be slowly varying if $L(x) > 0$ for x large enough and, for all $t > 0$, $L(tx)/L(x) \to 1$ as $x \to \infty$. A useful result concerning these functions is the so-called Representation Theorem, which states that a function $L : \mathbb{R}_+ \to \mathbb{R}_+$ is slowly varying if and only if it can be written in the form:

$$L(x) = c(x) \exp\left(\int_1^x \frac{\epsilon(u)}{u} du\right), \qquad x \geq 0,$$

where c is a nonnegative measurable function having a finite positive limit at infinity and ϵ is a measurable function tending to 0 at infinity. See e.g. [5, Theorem 1.3.1] for a proof. The following result is then an easy consequence.

Proposition 3. *Fix $\epsilon > 0$ and let $L : \mathbb{R}_+ \to \mathbb{R}$ be a slowly varying function. There exist two constants $C > 1$ and $N > 0$ such that $\frac{1}{C}x^{-\epsilon} \leq L(nx)/L(n) \leq Cx^\epsilon$ for every integer $n \geq N$ and $x \geq 1$.*

1.4 The Local Limit Theorem

Definition 4. A subset $A \subset \mathbb{Z}$ is said to be lattice if there exist $b \in \mathbb{Z}$ and an integer $d \geq 2$ such that $A \subset b + d\mathbb{Z}$. The largest d for which this statement holds is called the span of A. A measure on $\mathbb{Z}$ is said to be lattice if its support is lattice, and a random variable is said to be lattice if its law is lattice.

Remark 1. Since μ is supposed to be critical and aperiodic, using the fact that $\mu(0) > 0$, it is an exercise to check that the probability measure ν is non-lattice.

Recall that p_1 is the density of X_1. It is well known that $p_1(0) > 0$, that p_1 is positive, bounded, and continuous, and that the absolute value of the derivative of p_1 is bounded over $\mathbb{R}$ (see e.g. [23, I. 4.]). The following theorem will allow us to find estimates for the probabilities appearing in Proposition 2.

Theorem 1 (Local Limit Theorem). *Let $(Y_n)_{n \geq 0}$ be a random walk on $\mathbb{Z}$ started from 0 such that its jump distribution is in the domain of attraction of a stable law of index $\theta \in (1, 2]$. Assume that Y_1 is non-lattice, that $\mathbb{E}[Y_1] = 0$ and that Y_1 takes values in $\mathbb{N} \cup \{-1\}$.*

(i) *There exists an increasing sequence of positive real numbers $(a_n)_{n \geq 1}$ such that Y_n/a_n converges in distribution to X_1.*

(ii) *We have* $\displaystyle\lim_{n \to \infty} \sup_{k \in \mathbb{Z}} \left| a_n \mathbb{P}[Y_n = k] - p_1\left(\frac{k}{a_n}\right) \right| = 0.$

(iii) *There exists a slowly varying function $l : \mathbb{R}_+ \to \mathbb{R}_+$ such that $a_n = n^{1/\theta} l(n)$.*

Proof. For (i), see [10, Sect. XVII.5, Theorem 3] and [5, Sect. 8.4]. The fact that (a_n) may be chosen to be increasing follows from [9, Formula 3.7.2]. For (ii), see [11, Theorem 4.2.1]. For (iii), it is shown in [11, p. 46] that a_{kn}/a_n converges to $k^{1/\theta}$ for every integer $k \geq 1$. Since (a_n) is increasing, a theorem of de Haan (see [5, Theorem 1.10.7]) implies that there exists a slowly varying function $l : \mathbb{R}_+ \to \mathbb{R}_+$ such that $a_n = l(n) n^{1/\theta}$ for every positive integer n. □

Let $(W_n)_{n \geq 0}$ be as in Proposition 1 a random walk started from 0 with jump distribution ν. Since μ is in the domain of attraction of a stable law of index θ, it follows that ν is in the same domain of attraction, and W_1 is not lattice by Remark 1. Since ν has zero mean, by the preceding theorem there exists an increasing sequence of positive integers $(B_n)_{n \geq 1}$ such that $B_n \to \infty$ and W_n/B_n converges in distribution towards X_1 as $n \to \infty$. In what follows, the sequence $(B_n)_{n \geq 1}$ will be fixed, and $h : \mathbb{R}_+ \to \mathbb{R}_+$ will stand for a slowly varying function such that $B_n = h(n) n^{1/\theta}$.

Lemma 1. *We have:*

(i) $\mathbb{P}_\mu[\zeta(\tau) = n] \underset{n \to \infty}{\sim} \dfrac{p_1(0)}{n^{1/\theta+1} h(n)},$ (ii) $\mathbb{P}_\mu[\zeta(\tau) \geq n] \underset{n \to \infty}{\sim} \dfrac{\theta p_1(0)}{n^{1/\theta} h(n)},$

where we write $a_n \sim b_n$ if $a_n/b_n \to 1$.

Proof. We keep the notation of Proposition 1. Proposition 2 gives that:

$$\mathbb{P}_\mu[\zeta(\tau) = n] = \frac{1}{n} \mathbb{P}[W_n = -1]. \tag{3}$$

For (i), it suffices to notice that the local limit theorem (Theorem 1) and the continuity of p_1 entail $\mathbb{P}[W_n = -1] \sim p_1(0)/(h(n)n^{1/\theta})$. For (ii), we use (i) to write:

$$\mathbb{P}_\mu [\zeta(\tau) \geq n] = \sum_{k=n}^{\infty} \left(\frac{1}{h(k)k^{1+1/\theta}} p_1(0) + \frac{1}{h(k)k^{1+1/\theta}} \delta(k) \right),$$

where $\delta(k)$ tends to 0 as $k \to \infty$. We can rewrite this in the form:

$$h(n)n^{1/\theta}\mathbb{P}_\mu [\zeta(\tau) \geq n] = \int_1^\infty du\, f_n(u), \tag{4}$$

where:

$$f_n(u) = \frac{h(n)n^{1/\theta+1}}{h(\lfloor nu \rfloor)\lfloor nu \rfloor^{1+1/\theta}} (p_1(0) + \delta(\lfloor nu \rfloor)).$$

For fixed $u \geq 1$, $f_n(u)$ tends to $\frac{p_1(0)}{u^{1/\theta+1}}$ as $n \to \infty$. Choose $\epsilon \in (0, 1/\theta)$. By Proposition 3, for every sufficiently large positive integer n we have $f_n(u) \leq C/u^{1+1/\theta-\epsilon}$ for every $u \geq 1$, where C is a positive constant. The dominated convergence theorem allows us to infer that:

$$\lim_{n \to \infty} \int_1^\infty du\, f_n(u) = \int_1^\infty du\, \frac{p_1(0)}{u^{1/\theta+1}} = \theta p_1(0),$$

and the desired result follows from (4). □

1.5 Discrete Absolute Continuity

The next lemma in another important ingredient of our approach.

Lemma 2 (Le Gall and Miermont). *Fix $a \in (0, 1)$. Then, with the notation of Proposition 2, for every $n \geq 1$ and for every bounded nonnegative function f_n on $\mathbb{Z}^{\lfloor an \rfloor+1}$:*

$$\mathbb{E}\left[f_n(W_0, \ldots, W_{\lfloor na \rfloor}) \mid \zeta_1 = n\right]$$
$$= \mathbb{E}\left[f_n(W_0, \ldots, W_{\lfloor na \rfloor}) \frac{\phi_{n-\lfloor an \rfloor}(W_{\lfloor an \rfloor} + 1)/\phi_n(1)}{\phi^*_{n-\lfloor an \rfloor}(W_{\lfloor an \rfloor} + 1)/\phi^*_n(1)} \middle| \zeta_1 \geq n \right], \tag{5}$$

where $\phi_p(j) = \mathbb{P}[\zeta_j = p]$ and $\phi^*_p(j) = \mathbb{P}[\zeta_j \geq p]$ for every integers $j \geq 1$ and $p \geq 1$.

Proof. This result follows from an application of the Markov property to the random walk W at time $\lfloor an \rfloor$. See [18, Lemma 10] for details in a slightly different setting. □

2 The Continuous Setting: Stable Lévy Processes

2.1 The Normalized Excursion of the Lévy Process and the Continuous-time Height Process

We follow the presentation of [7]. The underlying probability space will be denoted by $(\Omega, \mathcal{F}, \mathbb{P})$. Recall that X is a strictly stable spectrally positive Lévy process with index $\theta \in (1, 2]$ such that for $\lambda > 0$:

$$\mathbb{E}[\exp(-\lambda X_t)] = \exp(t\lambda^\theta). \tag{6}$$

We denote the canonical filtration generated by X and augmented with the $\mathbb{P}$-negligible sets by $(\mathcal{F}_t)_{t \geq 0}$. See [2] for the proofs of the general assertions of this subsection concerning Lévy processes. In particular, for $\theta = 2$ the process X is $\sqrt{2}$ times the standard Brownian motion on the line. Recall that X has the following scaling property: for $c > 0$, the process $(c^{-1/\theta} X_{ct}, t \geq 0)$ has the same law as X. In particular, the density p_t of the law of X_t enjoys the following scaling property:

$$p_t(x) = t^{-1/\theta} p_1(x t^{-1/\theta}) \tag{7}$$

for $x \in \mathbb{R}$, $t > 0$. The following notation will be useful: for $s < t$, we set $I_t^s = \inf_{[s,t]} X$ and $I_t = \inf_{[0,t]} X$. Notice that the process I is continuous since X has no negative jumps.

The process $X - I$ is a strong Markov process and 0 is regular for itself with respect to $X - I$. We may and will choose $-I$ as the local time of $X - I$ at level 0. Let $(g_i, d_i), i \in \mathcal{I}$ be the excursion intervals of $X - I$ above 0. For every $i \in \mathcal{I}$ and $s \geq 0$, set $\omega_s^i = X_{(g_i+s) \wedge d_i} - X_{g_i}$. We view ω^i as an element of the excursion space $\mathcal{E}$, which is defined by:

$$\mathcal{E} = \{\omega \in \mathbb{D}(\mathbb{R}_+, \mathbb{R}); \omega(0) = 0 \text{ and } \zeta(\omega) := \sup\{s > 0; \omega(s) > 0\} \in (0, \infty)\}.$$

From Itô's excursion theory, the point measure

$$\mathcal{N}(dt\,d\omega) = \sum_{i \in \mathcal{I}} \delta_{(-I_{g_i}, \omega^i)}$$

is a Poisson measure on $\mathbb{R}_+ \times \mathcal{E}$ with intensity $dt\mathbf{N}(d\omega)$, where $\mathbf{N}(d\omega)$ is a σ-finite measure on $\mathcal{E}$. By classical results, $\mathbf{N}(\zeta > t) = \Gamma(1 - 1/\theta)^{-1} t^{-1/\theta}$. Without risk of confusion, we will also use the notation X for the canonical process on the space $\mathbb{D}(\mathbb{R}_+, \mathbb{R})$.

We now define the normalized excursion of X. Let us first recall the Itô description of the excursion measure (see [6] or [2, Chap. VIII.4] for details). Define for $\lambda > 0$ the rescaling operator $S^{(\lambda)}$ on $\mathcal{E}$ by $S^{(\lambda)}(\omega) = \left(\lambda^{1/\theta}\omega(s/\lambda), s \geq 0\right)$. Then there exists a unique collection of probability measures $(\mathbf{N}_{(a)}, a > 0)$ on $\mathcal{E}$ such that the following properties hold.

(i) For every $a > 0$, $\mathbf{N}_{(a)}(\zeta = a) = 1$.
(ii) For every $\lambda > 0$ and $a > 0$, we have $S^{(\lambda)}(\mathbf{N}_{(a)}) = \mathbf{N}_{(\lambda a)}$.
(iii) For every measurable subset A of $\mathcal{E}$: $\mathbf{N}(A) = \int_0^\infty \mathbf{N}_{(a)}(A) \dfrac{da}{\theta \Gamma(1-1/\theta) a^{1/\theta+1}}$.

The probability distribution $\mathbf{N}_{(1)}$ on càdlàg paths with unit lifetime is called the law of the normalized excursion of X and will sometimes be denoted by $\mathbf{N}(\cdot \mid \zeta = 1)$. In particular, for $\theta = 2$, $\mathbf{N}_{(1)}$ is the law of $\sqrt{2}$ times the normalized excursion of linear Brownian motion. Informally, $\mathbf{N}(\cdot \mid \zeta = 1)$ is the law of an excursion conditioned to have unit lifetime.

We will also use the socalled continuous-time height process H associated with X which was introduced in [19]. If $\theta = 2$, H is set to be equal to $X - I$. If $\theta \in (1, 2)$, the process H is defined for every $t \geq 0$ by:

$$H_t := \lim_{\epsilon \to 0} \frac{1}{\epsilon} \int_0^t \mathbb{1}_{\{X_s < I_t^s + \epsilon\}} ds,$$

where the limit exists in $\mathbb{P}$-probability and in $\mathbf{N}$-measure on $\{t < \zeta\}$. The definition of H thus makes sense under $\mathbb{P}$ or under $\mathbf{N}$. The process H has a continuous modification both under $\mathbb{P}$ and under $\mathbf{N}$ (see [8, Chap. 1] for details), and from now on we consider only this modification. Using simple scale arguments one can also define H as a continuous random process under $\mathbf{N}(\cdot \mid \zeta = 1)$. For our purposes, we will need the fact that, for every $a \geq 0$, $(H_t)_{0 \leq t \leq a}$ is a measurable function of $(X_t)_{0 \leq t \leq a}$.

2.2 Absolute Continuity Property of the Itô Measure

We now present the continuous counterpart of the discrete absolute continuity property appearing in Lemma 2. We follow the presentation of [16] but generalize it to the stable case. The following proposition is classical (see e.g. the proof of Theorem 4.1 in [22, Chap. XII], which establishes the result for Brownian motion).

Proposition 4. *Fix $t > 0$. Under the conditional probability measure $\mathbf{N}(\cdot \mid \zeta > t)$, the process $(X_{t+s})_{s \geq 0}$ is Markovian with the transition kernels of a strictly stable spectrally positive Lévy process of index θ stopped upon hitting 0.*

We will also use the following result (see [3, Corollary 2.3] for a proof).

Proposition 5. Set $q_s(x) = \frac{x}{s} p_s(-x)$ for $x, s > 0$. For $x \geq 0$, let $T(x) = \inf\{t \geq 0; X_t < -x\}$ be the first passage time of $-X$ above x. Then $\mathbb{P}[T(x) \in dt] = q_t(x) dt$ for every $x > 0$.

Note that q_s is a positive continuous function on $(0, \infty)$, for every $s > 0$. It is also known that q_s is bounded by a constant which is uniform when s varies over $[\epsilon, \infty)$, $\epsilon > 0$ (this follows from e.g. [23, I. 4.]).

Proposition 6. For every $a \in (0, 1)$ and $x > 0$ define:

$$\Gamma_a(x) = \frac{\theta q_{1-a}(x)}{\int_{1-a}^\infty ds\, q_s(x)}.$$

Then for every measurable bounded function $G : \mathbb{D}([0, a], \mathbb{R}^2) \to \mathbb{R}_+$:

$$\mathbf{N}\left(G((X_t)_{0 \leq t \leq a}, (H_t)_{0 \leq t \leq a}) \Gamma_a(X_a) \mid \zeta > 1\right) = \mathbf{N}\left(G((X_t)_{0 \leq t \leq a}, (H_t)_{0 \leq t \leq a}) \mid \zeta = 1\right).$$

Proof. Since $(H_t)_{0 \leq t \leq a}$ is a measurable function of $(X_t)_{0 \leq t \leq a}$, it is sufficient to prove that for every bounded measurable function $F : \mathbb{D}([0, a], \mathbb{R}) \to \mathbb{R}_+$:

$$\mathbf{N}\left(F((X_t)_{0 \leq t \leq a}) \Gamma_a(X_a) \mid \zeta > 1\right) = \mathbf{N}\left(F((X_t)_{0 \leq t \leq a}) \mid \zeta = 1\right).$$

To this end, fix $r \in [0, a]$, let $f, g : \mathbb{R}_+ \to \mathbb{R}_+$ be two bounded continuous functions and let $h : \mathbb{R}_+^* \to \mathbb{R}_+$ be a continuous function. Using the notation of Proposition 5, we have:

$$\mathbf{N}\left(f(X_r) h(X_a) g(\zeta) \mathbb{1}_{\{\zeta > a\}}\right) = \mathbf{N}\left(f(X_r) \mathbb{1}_{\{\zeta > a\}} \mathbb{E}\left[h(x) g(a + T(x))\right]_{x = X_a}\right)$$

$$= \int_0^\infty ds\, g(a + s) \mathbf{N}\left(f(X_r) h(X_a) q_s(X_a) \mathbb{1}_{\{\zeta > a\}}\right)$$

$$= \int_a^\infty du\, g(u) \mathbf{N}\left(f(X_r) h(X_a) q_{u-a}(X_a) \mathbb{1}_{\{\zeta > a\}}\right), \quad (8)$$

where we have used Proposition 4 in the first equality and Proposition 5 in the second equality. Moreover, by property (iii) in Sect. 2.1:

$$\mathbf{N}\left(f(X_r) g(\zeta) \mathbb{1}_{\{\zeta > a\}}\right) = \int_a^\infty du \frac{g(u)}{\theta \Gamma(1 - 1/\theta) u^{1/\theta + 1}} \cdot \mathbf{N}_{(u)}(f(X_r)). \quad (9)$$

Now observe that (8) (with $h = 1$) and (9) hold for any bounded continuous function g. Since both functions $u \mapsto \mathbf{N}\left(f(X_r) q_{u-a}(X_a) \mathbb{1}_{\{\zeta > a\}}\right)$ and $u \mapsto \mathbf{N}_{(u)}(f(X_r))$ are easily seen to be continuous over (a, ∞), it follows that for every $u > a$:

$$\mathbf{N}\left(f(X_r) q_{u-a}(X_a) \mathbb{1}_{\{\zeta > a\}}\right) = \frac{1}{\theta \Gamma(1 - 1/\theta) u^{1/\theta + 1}} \mathbf{N}_{(u)}(f(X_r)).$$

In particular, for $u = 1$ we get:

$$\mathbf{N}\left(f(X_r)q_{1-a}(X_a)\mathbb{1}_{\{\zeta>a\}}\right) = \frac{1}{\theta\Gamma(1-1/\theta)}\mathbf{N}_{(1)}\left(f(X_r)\right). \quad (10)$$

On the other hand, applying (8) with $g(x) = \mathbb{1}_{\{x>1\}}$ and noting that $\mathbf{N}(\zeta > 1) = \frac{1}{\Gamma(1-1/\theta)}$, we get:

$$\mathbf{N}(f(X_r)h(X_a)\,|\,\zeta > 1) = \Gamma(1-1/\theta)\mathbf{N}\left(f(X_r)h(X_a)\mathbb{1}_{\{\zeta>a\}}\int_{1-a}^{\infty}ds\,q_s(X_a)\right). \quad (11)$$

By combining (11) and (10) we conclude that:

$$\mathbf{N}\left(f(X_r)\frac{\theta q_{1-a}(X_a)}{\int_{1-a}^{\infty}ds\,q_s(X_a)}\,\bigg|\,\zeta > 1\right) = \mathbf{N}_{(1)}(f(X_r)).$$

One similarly shows that for $0 \le r_1 < \cdots < r_n \le a$ and $f_1,\ldots,f_n : \mathbb{R}_+ \to \mathbb{R}_+$ continuous bounded functions:

$$\mathbf{N}\left(f_1(X_{r_1})\cdots f_n(X_{r_n})\frac{\theta q_{1-a}(X_a)}{\int_{1-a}^{\infty}ds\,q_s(X_a)}\,\bigg|\,\zeta > 1\right) = \mathbf{N}_{(1)}(f_1(X_{r_1})\cdots f_n(X_{r_n})).$$

The desired result follows since the Borel σ-field of $\mathbb{D}([0,a],\mathbb{R})$ is generated by the coordinate functions $X \mapsto X_r$ for $0 \le r \le a$ (see e.g. [4, Theorem 12.5 (iii)]). □

3 Convergence to the Stable Tree

3.1 An Invariance Theorem

We rely on the following theorem, which is similar in spirit to Donsker's invariance theorem (see the concluding remark of [8, Sect. 2.6] for a proof).

Theorem 2 (Duquesne and Le Gall). *Let $\mathbf{t}_n$ be a random tree distributed according to $\mathbb{P}_\mu[\,\cdot\,|\,\zeta(\tau) \ge n]$. We have:*

$$\left(\frac{1}{B_n}W_{\lfloor nt \rfloor}(\mathbf{t}_n), \frac{B_n}{n}H_{nt}(\mathbf{t}_n)\right)_{t\ge 0} \xrightarrow[n\to\infty]{(d)} (X_t, H_t)_{0\le t\le 1} \text{ under } \mathbf{N}(\cdot\,|\,\zeta > 1).$$

3.2 Convergence of the Scaled Contour and Height Functions

Recall the notation $\phi_n(j) = \mathbb{P}[\zeta_j = n]$ and $\phi_n^*(j) = \mathbb{P}[\zeta_j \geq n]$.

Lemma 3. *Fix $\alpha > 0$. We have:*

$$(i) \lim_{n \to \infty} \sup_{1 \leq k \leq \alpha B_n} \left| n\phi_n(k) - q_1\left(\frac{k}{B_n}\right) \right| = 0,$$

$$(ii) \lim_{n \to \infty} \sup_{1 \leq k \leq \alpha B_n} \left| \phi_n^*(k) - \int_1^\infty ds\, q_s\left(\frac{k}{B_n}\right) \right| = 0.$$

This has been proved by Le Gall in [16] when μ has finite variance. In full generality, the proof is technical and is postponed to Sect. 3.3.

Lemma 4. *Fix $a \in (0, 1)$. Let $\mathfrak{t}_n$ be a random tree distributed according to $\mathbb{P}_\mu[\cdot \mid \zeta(\tau) = n]$. Then the following convergence holds in distribution in the space $\mathcal{C}([0, a], \mathbb{R})$:*

$$\left(\frac{B_n}{n} H_{nt}(\mathfrak{t}_n); 0 \leq t \leq a\right) \xrightarrow[n \to \infty]{(d)} (H_t; 0 \leq t \leq a) \text{ under } \mathbf{N}(\cdot \mid \zeta = 1).$$

Proof. Recall the notation Γ_a introduced in Proposition 6. We start by verifying that, for $\alpha > 1$, we have:

$$\lim_{n \to \infty} \left(\sup_{\frac{1}{\alpha} B_n \leq k \leq \alpha B_n} \left| \frac{\phi_{n-\lfloor an \rfloor}(k+1)/\phi_n(1)}{\phi_{n-\lfloor an \rfloor}^*(k+1)/\phi_n^*(1)} - \Gamma_a\left(\frac{k}{B_n}\right) \right| \right) = 0. \quad (12)$$

To this end, we will use the existence of a constant $\delta > 0$ such that, for n sufficiently large,

$$\inf_{\frac{1}{\alpha} B_n \leq k \leq \alpha B_n} \int_1^\infty ds\, q_s\left(\frac{k+1}{B_{n-\lfloor an \rfloor}}\right) > \delta. \quad (13)$$

The existence of such δ follows from the fact that, for every $\beta > 1$, $\inf_{\frac{1}{\beta} \leq x \leq \beta} \int_1^\infty ds\, q_s(x) > 0$. We will also need the fact that for every $\beta > 1$ there exists a constant $C > 0$ such that:

$$\sup_{\frac{1}{\beta} \leq x \leq \beta} q_1(x) \leq C, \quad \sup_{\frac{1}{\beta} \leq x \leq \beta} \int_1^\infty ds\, q_s(x) \leq C. \quad (14)$$

This is a consequence of the fact that q_1 is bounded for the first inequality, and the second inequality follows from the scaling property (7) combined with the fact that p_1 is bounded (see e.g. [23, I. 4.]). To establish (12), we first show that

$$\lim_{n\to\infty}\left(\sup_{\frac{1}{\alpha}B_n\le k\le\alpha B_n}\left|\frac{\phi_{n-\lfloor an\rfloor}(k+1)/\phi_n(1)}{\phi^*_{n-\lfloor an\rfloor}(k+1)/\phi^*_n(1)}-\theta\,\frac{\frac{1}{1-a}q_1\left(\frac{k+1}{B_{n-\lfloor an\rfloor}}\right)}{\int_1^\infty ds\,q_s\left(\frac{k+1}{B_{n-\lfloor an\rfloor}}\right)}\right|\right)=0. \quad (15)$$

Since $B_{n-\lfloor an\rfloor}/B_n \to (1-a)^{1/\theta}$ as $n\to\infty$, Lemma 3 guaranties the existence of two sequences $(\varepsilon^{(1)}_{k,n},\varepsilon^{(2)}_{k,n})_{k,n\ge 1}$ such that

$$(n-\lfloor an\rfloor)\phi_{n-\lfloor an\rfloor}(k+1)=q_1\left(\frac{k+1}{B_{n-\lfloor an\rfloor}}\right)+\varepsilon^{(1)}_{k,n},$$

$$\phi^*_{n-\lfloor an\rfloor}(k+1)=\int_1^\infty ds\,q_s\left(\frac{k+1}{B_{n-\lfloor an\rfloor}}\right)+\varepsilon^{(2)}_{k,n}$$

and such that $\max(\varepsilon^{(1)}_{k,n},\varepsilon^{(2)}_{k,n})\to 0$ as $n\to\infty$, uniformly in $1/\alpha\cdot B_n\le k\le\alpha B_n$. To simplify notation set $m_n=n-\lfloor an\rfloor$. By (14) and the fact that $B_{m_n}/B_n\to(1-a)^{1/\theta}$, there exists $C>0$ such that for n sufficiently large and $1/\alpha\cdot B_n\le k\le\alpha B_n$:

$$\left|\frac{m_n\phi_{m_n}(k+1)}{\phi^*_{m_n}(k+1)}-\frac{q_1\left(\frac{k+1}{B_{m_n}}\right)}{\int_1^\infty ds\,q_s\left(\frac{k+1}{B_{m_n}}\right)}\right|=\left|\frac{\varepsilon^{(1)}_{k,n}\cdot\int_1^\infty ds\,q_s\left(\frac{k+1}{B_{m_n}}\right)-\varepsilon^{(2)}_{k,n}\cdot q_1\left(\frac{k+1}{B_{m_n}}\right)}{\int_1^\infty ds\,q_s\left(\frac{k+1}{B_{m_n}}\right)\cdot\left(\int_1^\infty ds\,q_s\left(\frac{k+1}{B_{m_n}}\right)+\varepsilon^{(2)}_{k,n}\right)}\right|$$

$$\le\frac{2C}{\delta^2}\cdot\sup_{\frac{1}{\alpha}B_n\le k\le\alpha B_n}\max(\varepsilon^{(1)}_{k,n},\varepsilon^{(2)}_{k,n}),$$

where we have used (13) for the last inequality. This, combined with the fact that $\phi^*_n(1)/(n\phi_n(1))\to\theta$ as $n\to\infty$ by Lemma 1, implies (15). Then our claim (12) follows the scaling property (7) and the continuity of Γ_a.

We shall now prove another useful result before introducing some notation. Fix $\alpha>1$. Let $g_n:\mathbb{R}^{\lfloor an\rfloor+1}\to\mathbb{R}_+$ be a bounded measurable function. To simplify notation, for $x_0,\ldots,x_{\lfloor an\rfloor}\in\mathbb{R}$, set

$$G_n(x_0,\ldots,x_{\lfloor an\rfloor})=g_n(x_0,\ldots,x_{\lfloor an\rfloor})\mathbb{1}_{x_{\lfloor an\rfloor}\in[\frac{1}{\alpha}B_n,\alpha B_n]}$$

and, for a tree τ, set

$$\tilde{G}_n(\tau)=g_n(W_0(\tau),W_1(\tau),\ldots,W_{\lfloor an\rfloor}(\tau))\mathbb{1}_{\{W_{\lfloor na\rfloor}(\tau)\in[\frac{1}{\alpha}B_n,\alpha B_n]\}}.$$

We claim that

$$\lim_{n\to\infty}\left|\mathbb{E}\left[\tilde{G}_n(\mathbf{t}_n)\right]-\mathbb{E}_\mu\left[\tilde{G}_n(\tau)\Gamma_a\left(\frac{W_{\lfloor an\rfloor}(\tau)}{B_n}\right)\Big|\zeta(\tau)\ge n\right]\right|=0. \quad (16)$$

Indeed, using successively Proposition 1 and (5), we have:

$$\mathbb{E}\left[\tilde{G}_n(t_n)\right] - \mathbb{E}_\mu\left[\left.\tilde{G}_n(\tau)\Gamma_a\left(\frac{\mathcal{W}_{\lfloor an\rfloor}(\tau)}{B_n}\right)\right|\zeta(\tau)\geq n\right]$$

$$= \mathbb{E}\left[G_n(W_0,\ldots,W_{\lfloor na\rfloor})|\,\zeta_1 = n\right] - \mathbb{E}\left[\left.G_n(W_0,\ldots,W_{\lfloor na\rfloor})\Gamma_a\left(\frac{W_{\lfloor an\rfloor}}{B_n}\right)\right|\zeta_1\geq n\right]$$

$$= \mathbb{E}\left[\left.G_n(W_0,\ldots,W_{\lfloor na\rfloor})\left(\frac{\phi_{n-\lfloor an\rfloor}(W_{\lfloor an\rfloor}+1)/\phi_n(1)}{\phi^*_{n-\lfloor an\rfloor}(W_{\lfloor an\rfloor}+1)/\phi^*_n(1)} - \Gamma_a\left(\frac{W_{\lfloor an\rfloor}}{B_n}\right)\right)\right|\zeta_1\geq n\right].$$

Our claim (16) then follows from (12).

We finally return to the proof of Lemma 4. Let $F : \mathbb{D}([0,a],\mathbb{R}) \to \mathbb{R}_+$ be a bounded continuous function. We also set $F_n(\tau) = F\left(\frac{B_n}{n}H_{\lfloor nt\rfloor}(\tau); 0\leq t\leq a\right)$. Since $(H_0(\tau),H_1(\tau),\ldots,H_{\lfloor an\rfloor}(\tau))$ is a measurable function of $(\mathcal{W}_0(\tau),\mathcal{W}_1(\tau),\ldots,\mathcal{W}_{\lfloor an\rfloor}(\tau))$ (see [17, Prop 1.2]), by (16) we get:

$$\lim_{n\to\infty}\left|\mathbb{E}\left[F_n(t_n)\mathbb{1}_{A^\alpha_n(t_n)}\right] - \mathbb{E}_\mu\left[\left.F_n(\tau)\mathbb{1}_{A^\alpha_n(\tau)}\Gamma_a\left(\frac{\mathcal{W}_{\lfloor an\rfloor}(\tau)}{B_n}\right)\right|\zeta(\tau)\geq n\right]\right| = 0.$$

By Theorem 2, the law of

$$\left(\left(\frac{B_n}{n}H_{\lfloor nt\rfloor}(\tau); 0\leq t\leq a\right), \frac{1}{B_n}\mathcal{W}_{\lfloor an\rfloor}(\tau)\right)$$

under $\mathbb{P}_\mu[\,\cdot\,|\,\zeta(\tau)\geq n]$ converges towards the law of $((H_t; 0\leq t\leq a), X_a)$ under $\mathbf{N}(\cdot\,|\,\zeta > 1)$ (for the convergence of the second component we have also used the fact that X is almost surely continuous at a). Thus:

$$\lim_{n\to\infty}\mathbb{E}\left[F_n(t_n)\mathbb{1}_{\{\mathcal{W}_{\lfloor na\rfloor}(t_n)\in[\frac{1}{\alpha}B_n,\alpha B_n]\}}\right]$$

$$= \mathbf{N}\left(F(H_t; 0\leq t\leq a)\Gamma_a(X_a)\mathbb{1}_{\{X_a\in[\frac{1}{\alpha},\alpha]\}}\,|\,\zeta > 1\right)$$

$$= \mathbf{N}\left(F(H_t; 0\leq t\leq a)\mathbb{1}_{\{X_a\in[\frac{1}{\alpha},\alpha]\}}\,|\,\zeta = 1\right), \tag{17}$$

where we have used Proposition 6 in the second equality.

By taking $F \equiv 1$, we obtain:

$$\lim_{n\to\infty}\mathbb{P}\left[\mathcal{W}_{\lfloor na\rfloor}(t_n) \in \left[\frac{1}{\alpha}B_n,\alpha B_n\right]\right] = \mathbf{N}\left(X_a\in\left[\frac{1}{\alpha},\alpha\right]\,\Big|\,\zeta = 1\right).$$

This last quantity tends to 1 as $\alpha \to \infty$. By choosing $\alpha > 1$ sufficiently large, we easily deduce from the convergence (17) that:

$$\lim_{n\to\infty}\mathbb{E}\left[F\left(\frac{B_n}{n}H_{\lfloor nt\rfloor}(t_n); 0\leq t\leq a\right)\right] = \mathbf{N}(F(H_t; 0\leq t\leq a)\,|\,\zeta = 1).$$

The path continuity of H under $\mathbf{N}(\,\cdot\mid \zeta = 1)$ then implies the claim of Lemma 4. □

Theorem 3. *Let $\mathbf{t}_n$ be a random tree distributed according to $\mathbb{P}_\mu[\,\cdot\mid \zeta(\tau) = n]$. Then:*

$$\left(\frac{B_n}{n} H_{nt}(\mathbf{t}_n), \frac{B_n}{n} C_{2nt}(\mathbf{t}_n)\right)_{0\le t\le 1} \xrightarrow[n\to\infty]{(d)} (H_t, H_t)_{0\le t\le 1} \text{ under } \mathbf{N}(\cdot\mid \zeta = 1).$$

Proof. The proof consists in showing that the scaled height process is close to the scaled contour process and then using a time-reversal argument in order to show that the convergence holds on the whole segment $[0, 1]$. To this end, we adapt [7, Remark 3.2] and [8, Sect. 2.4] to our context. For $0 \le p < n$ set $b_p = 2p - H_p(\mathbf{t}_n)$ so that b_p represents the time needed by the contour process to reach the $(p + 1)$-st individual of $\mathbf{t}_n$ (in the lexicographical order). Also set $b_n = 2(n - 1)$. Note that $C_{b_p} = H_p$. From this observation, we get:

$$\sup_{t\in[b_p,b_{p+1}]} |C_t(\mathbf{t}_n) - H_p(\mathbf{t}_n)| \le |H_{p+1}(\mathbf{t}_n) - H_p(\mathbf{t}_n)| + 1. \tag{18}$$

for $0 \le p < n$. Then define the random function $g_n : [0, 2n] \to \mathbb{N}$ by setting $g_n(t) = k$ if $t \in [b_k, b_{k+1})$ and $k < n$, and $g_n(t) = n$ if $t \in [2(n-1), 2n]$ so that for $t < 2(n-1)$, $g_n(t)$ is the index of the last individual which has been visited by the contour function up to time t if the individuals are indexed $0, 1, \ldots, n - 1$ in lexicographical order. Finally, set $\tilde{g}_n(t) = g_n(nt)/n$. Fix $a \in (0, 1)$. Then, by (18):

$$\sup_{t\le \frac{b_{\lfloor an\rfloor}}{n}} \left|\frac{B_n}{n} C_{nt}(\mathbf{t}_n) - \frac{B_n}{n} H_{n\tilde{g}_n(t)}(\mathbf{t}_n)\right| \le \frac{B_n}{n} + \frac{B_n}{n} \sup_{k\le \lfloor an\rfloor} |H_{k+1}(\mathbf{t}_n) - H_k(\mathbf{t}_n)|, \tag{19}$$

which converges in probability to 0 by Lemma 4 and the path continuity of (H_t). On the other hand it follows from the definition of b_n that:

$$\sup_{t\le \frac{b_{\lfloor an\rfloor}}{n}} \left|\tilde{g}_n(t) - \frac{t}{2}\right| \le \frac{1}{2B_n} \sup_{k\le an} \frac{B_n}{n} H_k(\mathbf{t}_n) + \frac{1}{n} \xrightarrow{(\mathbb{P})} 0$$

by Lemma 4. Finally, by the definition of b_n and using Lemma 4 we see that $\frac{b_{\lfloor an\rfloor}}{n}$ converges in probability towards $2a$ and that $\frac{B_n}{n} \sup_{t\le 2a} |H_{n\tilde{g}_n(t)}(\mathbf{t}_n) - H_{nt/2}(\mathbf{t}_n)|$ converges in probability towards 0 as $n \to \infty$. Using (19), we conclude that:

$$\frac{B_n}{n} \sup_{0\le t\le a} |C_{2nt}(\mathbf{t}_n) - H_{nt}(\mathbf{t}_n)| \xrightarrow{(\mathbb{P})} 0. \tag{20}$$

Together with Lemma 4, this implies:

$$\left(\frac{B_n}{n}C_{2nt}(\mathbf{t}_n); 0 \le t \le a\right) \xrightarrow{(d)} (H_t; 0 \le t \le a) \text{ under } \mathbf{N}(\cdot \mid \zeta = 1).$$

Since $(C_t(\mathbf{t}_n); 0 \le t \le 2n-2)$ and $(C_{2n-2-t}(\mathbf{t}_n); 0 \le t \le 2n-2)$ have the same distribution, it follows that:

$$\left(\frac{B_n}{n}C_{2nt}(\mathbf{t}_n); 0 \le t \le 1\right) \xrightarrow{(d)} (H_t; 0 \le t \le 1) \text{ under } \mathbf{N}(\cdot \mid \zeta = 1). \quad (21)$$

See the last paragraph of the proof of Theorem 6.1 in [16] for details.

Finally, we show that this convergence in turn entails the convergence of the rescaled height function of $\mathbf{t}_n$ on the whole segment $[0, 1]$. To this end, we verify that convergence (20) remains valid for $a = 1$. First note that:

$$\sup_{0 \le t \le 2}\left|\tilde{g}_n(t) - \frac{t}{2}\right| \le \frac{1}{2n}\sup_{k \le n}H_k(\mathbf{t}_n) + \frac{1}{n} = \frac{1}{2B_n}\sup_{k \le 2n}\frac{B_n}{n}C_k(\mathbf{t}_n) + \frac{1}{n} \xrightarrow{(\mathbb{P})} 0 \quad (22)$$

by (21). Secondly, it follows from (18) that:

$$\sup_{0 \le t \le 2}\left|\frac{B_n}{n}C_{nt}(\mathbf{t}_n) - \frac{B_n}{n}H_{n\tilde{g}_n(t)}\right| \le \frac{B_n}{n} + \frac{B_n}{n}\sup_{k \le n-1}|H_{k+1}(\mathbf{t}_n) - H_k(\mathbf{t}_n)|$$

$$= \frac{B_n}{n} + \frac{B_n}{n}\sup_{k \le n-1}\left|C_{\frac{b_{k+1}}{n}}(\mathbf{t}_n) - C_{\frac{b_k}{n}}(\mathbf{t}_n)\right|.$$

By (21), in order to prove that the latter quantity tends to 0 in probability, it is sufficient to verify that $\sup_{k \le n}\left|\frac{b_{k+1}}{n} - \frac{b_k}{n}\right|$ converges to 0 in probability. But by the definition of b_n:

$$\sup_{k \le n}\left|\frac{b_{k+1}}{n} - \frac{b_k}{n}\right| = \sup_{k \le n}\left|\frac{2 + H_k(\mathbf{t}_n) - H_{k+1}(\mathbf{t}_n)}{n}\right| \le \frac{2}{n} + 2\sup_{k \le n}\frac{H_k(\mathbf{t}_n)}{n}$$

which converges in probability to 0 as in (22). As a consequence:

$$\frac{B_n}{n}\sup_{0 \le t \le 1}|C_{2nt}(\mathbf{t}_n) - H_{n\tilde{g}_n(2t)}(\mathbf{t}_n)| \xrightarrow{(\mathbb{P})} 0.$$

By (21), we get that:

$$\left(\frac{B_n}{n}H_{n\tilde{g}_n(2t)}(\mathbf{t}_n)\right)_{0 \le t \le 1} \xrightarrow[n \to \infty]{(d)} (H_t)_{0 \le t \le 1} \text{ under } \mathbf{N}(\cdot \mid \zeta = 1).$$

Combining this with (22), we conclude that:

$$\left(\frac{B_n}{n}C_{2nt}(t_n), \frac{B_n}{n}H_{nt}(t_n)\right)_{0\leq t\leq 1} \xrightarrow[n\to\infty]{(d)} (H_t, H_t)_{0\leq t\leq 1} \text{ under } \mathbf{N}(\cdot\,|\,\zeta = 1).$$

This completes the proof. □

Remark 2. If we see the tree t_n as a finite metric space using its graph distance, this theorem implies that t_n, suitably rescaled, converges in distribution to the θ-stable tree, in the sense of the Gromov–Hausdorff distance on isometry classes of compact metric spaces (see e.g. [17, Sect. 2] for details).

Remark 3. When the mean value of μ is greater than one, it is possible to replace μ with a critical probability distribution belonging to the same exponential family as μ without changing the distribution of t_n (see [13]). Consequently, the theorem holds in the supercritical case as well. The case where μ is subcritical and $\mu(i) \sim L(i)/i^{1+\theta}$ as $i \to \infty$ has been treated in [15]. However, in full generality, the noncritical subcritical case remains open.

3.3 Proof of the Technical Lemma

In this section, we prove Lemma 3.

Proof (of Lemma 3). We first prove (i). By the local limit theorem (Theorem 1 (ii)), we have, for $k \geq 1$ and $j \in \mathbb{Z}$:

$$\left| B_n \mathbb{P}[W_n = j] - p_1\left(\frac{j}{B_n}\right) \right| \leq \epsilon(n),$$

where $\epsilon(n) \to 0$. By Proposition 2, we have $n\phi_n(j) = j\mathbb{P}[W_n = -j]$. Since $\frac{j}{B_n}p_1\left(-\frac{j}{B_n}\right) = q_1\left(\frac{j}{B_n}\right)$, we have for $1 \leq j \leq \alpha B_n$:

$$\left| n\phi_n(j) - q_1\left(\frac{j}{B_n}\right) \right| = \frac{j}{B_n}\left| B_n \mathbb{P}[W_n = -j] - p_1\left(\frac{j}{B_n}\right) \right| \leq \alpha\epsilon(n).$$

This completes the proof of (i).

For (ii), first note that by the definition of q_s and the scaling property (7):

$$\int_1^\infty ds\, q_s\left(\frac{j}{B_n}\right) = \int_1^\infty \frac{j/B_n}{s^{1/\theta+1}} p_1\left(-\frac{j/B_n}{s^{1/\theta}}\right) ds.$$

By Proposition 2 and the local limit theorem:

$$\left| \phi_n^*(j) - \sum_{k=n}^{\infty} \frac{j}{kB_k} p_1\left(-\frac{j}{B_k}\right) \right| = \left| \sum_{k=n}^{\infty} \left(\frac{j}{k}\mathbb{P}[W_k = -j] - \frac{j}{kB_k} p_1\left(-\frac{j}{B_k}\right) \right) \right|$$

$$\leq \sum_{k=n}^{\infty} \frac{j}{kB_k} \epsilon(k),$$

where $\epsilon(n) \to 0$. Then write:

$$\left| \sum_{k=n}^{\infty} \frac{j}{kB_k} p_1\left(-\frac{j}{B_k}\right) - \int_1^{\infty} ds \, \frac{j/B_n}{s^{1/\theta+1}} p_1\left(-\frac{j/B_n}{s^{1/\theta}}\right) \right|$$

$$\leq \int_1^{\infty} ds \left| \frac{jn}{B_{\lfloor ns \rfloor} \lfloor ns \rfloor} - \frac{j/B_n}{s^{1/\theta+1}} \right| p_1\left(-\frac{j}{B_{\lfloor ns \rfloor}}\right)$$

$$+ \int_1^{\infty} ds \, \frac{j/B_n}{s^{1/\theta+1}} \left| p_1\left(-\frac{j}{B_{\lfloor ns \rfloor}}\right) - p_1\left(-\frac{j/B_n}{s^{1/\theta}}\right) \right|.$$

Denote the first term of the right-hand side by $P(n, j)$ and the second term by $Q(n, j)$. Since p_1 is bounded by a constant which we will denote by M, we have for $1 \leq j \leq \alpha B_n$:

$$P(n, j) \leq \alpha M \int_1^{\infty} ds \, \frac{1}{s^{1/\theta+1}} \left| \frac{nB_n s^{1/\theta+1}}{B_{\lfloor ns \rfloor} \lfloor ns \rfloor} - 1 \right|.$$

For fixed $s \geq 1$, $\frac{1}{s^{1/\theta+1}} \left| \frac{nB_n s^{1/\theta+1}}{B_{\lfloor ns \rfloor} \lfloor ns \rfloor} - 1 \right|$ tends to 0 as $n \to \infty$, and using Proposition 3, the same quantity is bounded by an integrable function independent of n. The dominated convergence theorem thus shows that $P(n, j) \to 0$ uniformly in $1 \leq j \leq \alpha B_n$. Let us now bound $Q(n, j)$ for $1 \leq j \leq \alpha B_n$. Since the absolute value of the derivative of p_1 is bounded by a constant which we will denote by M', we have:

$$Q(n, j) \leq M' \int_1^{\infty} ds \, \frac{j/B_n}{s^{1/\theta+1}} \left| \frac{j}{B_{\lfloor ns \rfloor}} - \frac{j/B_n}{s^{1/\theta}} \right| \leq \alpha^2 M' \int_1^{\infty} ds \, \frac{1}{s^{2/\theta+1}} \left| \frac{B_n s^{1/\theta}}{B_{\lfloor ns \rfloor}} - 1 \right|.$$

The right-hand side tends to 0 by the same argument we used for $P(n, j)$. We have thus proved that:

$$\lim_{n \to \infty} \sup_{1 \leq j \leq \alpha B_n} \left| \sum_{k=n}^{\infty} \frac{j}{kB_k} p_1\left(-\frac{k}{B_k}\right) - \int_1^{\infty} ds \, q_s\left(\frac{j}{B_n}\right) \right| = 0.$$

One finally shows that $\sum_{k=n}^{\infty} \frac{j}{kB_k}\epsilon(k)$ tends to 0 as $n \to \infty$ uniformly in $1 \leq j \leq \alpha B_n$ by noticing that:

$$\sup_{n\geq 1} \sup_{1\leq j \leq \alpha B_n} \left(\sum_{k=n}^{\infty} \frac{j}{kB_k}\right) \leq \alpha \sup_{n\geq 1} \left(\sum_{k=n}^{\infty} \frac{B_n}{kB_k}\right) < \infty.$$

This completes the proof. □

Acknowledgements I am deeply indebted to Jean-François Le Gall for insightful discussions and for making many useful suggestions on first versions of this manuscript.

References

1. D. Aldous, The continuum random tree III. Ann. Probab. **21**, 248–289 (1993)
2. J. Bertoin, *Lévy Processes* (Cambridge University Press, Cambridge, 1996)
3. J. Bertoin, *Subordinators, Lévy Processes with No Negative Jumps and Branching Processes*. MaPhySto Lecture Notes Series No. 8, University of Aarhus (2000)
4. P. Billingsley, *Convergence of Probability Measures*, 2nd edn. Wiley Series in Probability and Statistics: Probability and Statistics (Wiley, New York, 1999)
5. N.H. Bingham, C.M. Goldie, J.L. Teugels, *Regular Variation*. Encyclopedia of Mathematics and Its Applications, vol. 27 (Cambridge University Press, Cambridge, 1987)
6. L. Chaumont, Excursion normalisée, méandre et pont pour les processus de Lévy stables. Bull. Sci. Math. **121**(5), 377–403 (1997)
7. T. Duquesne, A limit theorem for the contour process of conditioned Galton-Watson trees. Ann. Probab. **31**, 996–1027 (2003)
8. T. Duquesne, J.-F. Le Gall , Random trees, Lévy processes and spatial branching processes. Astérisque. **281** (2002)
9. R. Durrett, *Probability: Theory and Examples*, 4th edn. (Cambridge University Press, Cambridge, 2010)
10. W. Feller, *An Introduction to Probability Theory and Its Applications*, vol. 2, 2nd edn. (Wiley, New York, 1971)
11. I.A. Ibragimov, Y.V. Linnik, *Independent and Stationary Sequences of Independent Random Variables* (Wolters-Noordhoff, Groningen, 1971)
12. J. Jacod, A. Shiryaev, *Limit Theorems for Stochastic Processes*, 2nd edn. Grundlehren der mathematischen Wissenschaften, vol. 288 (Springer, Berlin, 2003)
13. D.P. Kennedy, The Galton-Watson process conditioned on the total progeny. J. Appl. Probab. **12**, 800–806 (1975)
14. I. Kortchemski, Invariance principles for Galton-Watson trees conditioned on the number of leaves. Stoch. Proc. Appl. **122**, 3126–3172 (2012)
15. I. Kortchemski, Limit theorems for conditioned non-generic Galton-Watson trees. arXiv preprint arXiv:1205.3145 (2012)
16. J.-F. Le Gall, Itô's excursion theory and random trees. Stoch. Proc. Appl. **120**(5), 721–749 (2010)
17. J.-F. Le Gall, Random trees and applications. Probab. Surv. **2**, 245–311 (2005)
18. J.-F. Le Gall, G. Miermont, Scaling limits of random planar maps with large faces. Ann. Probab. **39**(1), 1–69 (2011)
19. J.-F. Le Gall, Y. Le Jan, Branching processes in Lévy processes: the exploration process. Ann. Probab. **26**(1), 213–512 (1998)

20. J.-F. Marckert, A. Mokkadem, The depth first processes of Galton-Watson trees converge to the same Brownian excursion. Ann. Probab. **31**, 1655–1678 (2003)
21. J. Pitman, *Combinatorial Stochastic Processes*. Lecture Notes in Mathematics, vol. 1875 (Springer, Berlin, 2006)
22. D. Revuz, M. Yor, *Continuous Martingales and Brownian Motion*, 3rd edn. Grundlehren der Mathematischen Wissenschaften [Fundamental Principles of Mathematical Sciences], vol. 293 (Springer, Berlin, 1999)
23. V.M. Zolotarev, *One-Dimensional Stable Distributions*. Translations of Mathematical Monographs, vol. 65 (American Mathematical Society, Providence, 1986)

LECTURE NOTES IN MATHEMATICS  Springer

Edited by J.-M. Morel, B. Teissier; P.K. Maini

Editorial Policy (for Multi-Author Publications: Summer Schools / Intensive Courses)

1. Lecture Notes aim to report new developments in all areas of mathematics and their applications - quickly, informally and at a high level. Mathematical texts analysing new developments in modelling and numerical simulation are welcome. Manuscripts should be reasonably selfcontained and rounded off. Thus they may, and often will, present not only results of the author but also related work by other people. They should provide sufficient motivation, examples and applications. There should also be an introduction making the text comprehensible to a wider audience. This clearly distinguishes Lecture Notes from journal articles or technical reports which normally are very concise. Articles intended for a journal but too long to be accepted by most journals, usually do not have this "lecture notes" character.

2. In general SUMMER SCHOOLS and other similar INTENSIVE COURSES are held to present mathematical topics that are close to the frontiers of recent research to an audience at the beginning or intermediate graduate level, who may want to continue with this area of work, for a thesis or later. This makes demands on the didactic aspects of the presentation. Because the subjects of such schools are advanced, there often exists no textbook, and so ideally, the publication resulting from such a school could be a first approximation to such a textbook. Usually several authors are involved in the writing, so it is not always simple to obtain a unified approach to the presentation.

   For prospective publication in LNM, the resulting manuscript should not be just a collection of course notes, each of which has been developed by an individual author with little or no coordination with the others, and with little or no common concept. The subject matter should dictate the structure of the book, and the authorship of each part or chapter should take secondary importance. Of course the choice of authors is crucial to the quality of the material at the school and in the book, and the intention here is not to belittle their impact, but simply to say that the book should be planned to be written by these authors jointly, and not just assembled as a result of what these authors happen to submit.

   This represents considerable preparatory work (as it is imperative to ensure that the authors know these criteria before they invest work on a manuscript), and also considerable editing work afterwards, to get the book into final shape. Still it is the form that holds the most promise of a successful book that will be used by its intended audience, rather than yet another volume of proceedings for the library shelf.

3. Manuscripts should be submitted either online at www.editorialmanager.com/lnm/ to Springer's mathematics editorial, or to one of the series editors. Volume editors are expected to arrange for the refereeing, to the usual scientific standards, of the individual contributions. If the resulting reports can be forwarded to us (series editors or Springer) this is very helpful. If no reports are forwarded or if other questions remain unclear in respect of homogeneity etc, the series editors may wish to consult external referees for an overall evaluation of the volume. A final decision to publish can be made only on the basis of the complete manuscript; however a preliminary decision can be based on a pre-final or incomplete manuscript. The strict minimum amount of material that will be considered should include a detailed outline describing the planned contents of each chapter.

   Volume editors and authors should be aware that incomplete or insufficiently close to final manuscripts almost always result in longer evaluation times. They should also be aware that parallel submission of their manuscript to another publisher while under consideration for LNM will in general lead to immediate rejection.

4. Manuscripts should in general be submitted in English. Final manuscripts should contain at least 100 pages of mathematical text and should always include

   – a general table of contents;
   – an informative introduction, with adequate motivation and perhaps some historical remarks: it should be accessible to a reader not intimately familiar with the topic treated;
   – a global subject index: as a rule this is genuinely helpful for the reader.

   Lecture Notes volumes are, as a rule, printed digitally from the authors' files. We strongly recommend that all contributions in a volume be written in the same LaTeX version, preferably LaTeX2e. To ensure best results, authors are asked to use the LaTeX2e style files available from Springer's webserver at
   ftp://ftp.springer.de/pub/tex/latex/svmonot1/ (for monographs) and
   ftp://ftp.springer.de/pub/tex/latex/svmultt1/ (for summer schools/tutorials).
   Additional technical instructions, if necessary, are available on request from:
   lnm@springer.com.

5. Careful preparation of the manuscripts will help keep production time short besides ensuring satisfactory appearance of the finished book in print and online. After acceptance of the manuscript authors will be asked to prepare the final LaTeX source files and also the corresponding dvi-, pdf- or zipped ps-file. The LaTeX source files are essential for producing the full-text online version of the book. For the existing online volumes of LNM see:
   http://www.springerlink.com/openurl.asp?genre=journal&issn=0075-8434.
   The actual production of a Lecture Notes volume takes approximately 12 weeks.

6. Volume editors receive a total of 50 free copies of their volume to be shared with the authors, but no royalties. They and the authors are entitled to a discount of 33.3 % on the price of Springer books purchased for their personal use, if ordering directly from Springer.

7. Commitment to publish is made by letter of intent rather than by signing a formal contract. Springer-Verlag secures the copyright for each volume. Authors are free to reuse material contained in their LNM volumes in later publications: a brief written (or e-mail) request for formal permission is sufficient.

Addresses:
Professor J.-M. Morel, CMLA,
École Normale Supérieure de Cachan,
61 Avenue du Président Wilson, 94235 Cachan Cedex, France
E-mail: morel@cmla.ens-cachan.fr

Professor B. Teissier, Institut Mathématique de Jussieu,
UMR 7586 du CNRS, Équipe "Géométrie et Dynamique",
175 rue du Chevaleret,
75013 Paris, France
E-mail: teissier@math.jussieu.fr

For the "Mathematical Biosciences Subseries" of LNM:

Professor P. K. Maini, Center for Mathematical Biology,
Mathematical Institute, 24-29 St Giles,
Oxford OX1 3LP, UK
E-mail : maini@maths.ox.ac.uk

Springer, Mathematics Editorial I,
Tiergartenstr. 17,
69121 Heidelberg, Germany,
Tel.: +49 (6221) 4876-8259
Fax: +49 (6221) 4876-8259
E-mail: lnm@springer.com

MIX
Papier aus verantwortungsvollen Quellen
Paper from responsible sources
FSC® C105338

If you have any concerns about our products,
you can contact us on
ProductSafety@springernature.com

In case Publisher is established outside the EU,
the EU authorized representative is:
**Springer Nature Customer Service Center GmbH
Europaplatz 3, 69115 Heidelberg, Germany**

Printed by Libri Plureos GmbH
in Hamburg, Germany